SELECTED PAPERS OF DEMETRIOS G. MAGIROS

DEMETRIOS G. MAGIROS

Selected Papers of Demetrios G. Magiros

Applied Mathematics, Nonlinear Mechanics, and Dynamical Systems Analysis

Edited by

S. G. Tzafestas

Electrical Engineering Department,
National Technical University, Athens, Greece

D. REIDEL PUBLISHING COMPANY

A MEMBER OF THE KLUWER  ACADEMIC PUBLISHERS GROUP

DORDRECHT / BOSTON / LANCASTER

Published on behalf of the Greek Mathematical Society

Library of Congress Cataloging in Publication Data

CIP

Magiros, Demetrios G.
Selected papers of Demetrios G. Magiros.

"Published on behalf of the Greek Mathematical Society."
Bibliography: p.
1. Mathematics–Collected works. 2. Nonlinear mechanics–Collected works. 3. System analysis–Collected works. I. Tzafestas, S. G., 1939- . II. Hellénikè Mathématikè Hetaireia (Greece)
QA3.M277 1985 510 85-2204
ISBN 90-277-2003-7

Published by D. Reidel Publishing Company,
P.O. Box 17, 3300 AA Dordrecht, Holland.

Sold and distributed in the U.S.A. and Canada
by Kluwer Academic Publishers,
190 Old Derby Street, Hingham, MA 02043, U.S.A.

In all other countries, sold and distributed
by Kluwer Academic Publishers Group,
P.O. Box 322, 3300 AH Dordrecht, Holland.

Printed in The Netherlands

FOREWORD

The theory of nonlinear oscillations and stability of motion is a fundamental part of the study of numerous real world phenomena. These phenomena, particularly auto-oscillations of the first and second kind, capture, parametric, subharmonic and ultraharmonic resonance, asymptotic behavior and orbits' stability, constitute the core of problems treated in "Nonlinear Mechanics", and their study is connected with the names of H. Poincaré, A.M. Lyapunov, N.M. Krylov and N.N. Bogolyubov.

Professor Demetrios Magiros, a widely known scientist in the theories of oscillations and nonlinear differential equations, has devoted his numerous works to this significant part of modern physical science. His scientific results can be classified in the following way:

1) creation of methods of analysis of subharmonic resonances under the nonlinear effect,
2) determination and analysis of the main modes of nonlinear oscillations on the basis of infinite determinants,
3) analysis of problems of celestial mechanics,
4) classification of stability of solutions of dynamic systems concepts,
5) mathematical analogs of physical and social systems.

He has developed new methods and solutions for a great number of difficult problems of nonlinear mechanics making a significant contribution to the theory and applications of the field.

Urgency, depth of perception of the considered phenomena, and practical directness are characteristics of his work.

D.G. Magiros' scientific findings have been utilized in the research of many scientists and can be applied to a variety of branches of mechanics, physics and engineering, in which nonlinear oscillating processes and problems of stability are involved.

Dr Magiros participated in several International Conferences, among which the International Conferences on Nonlinear Oscillations held in Kiev (1969, 1981).

We are sure that the publication of Professor D.G. Magiros' Selected Papers will be accepted by numerous specialists, working on the theory and application of nonlinear oscillations, with great interest.

Yu. A. Mitropolsky, Academician
A.A. Martynyuk
Ukrainian Academy of Sciences
Kiev, USSR

IN MEMORY OF DIMITRIS MAGIROS

The Greek Mathematical Society is pleased to sponsor the publication of the "Selected Papers" of a prominent member Dimitris G. Magiros, honoring his memory and his work.

The distinguished mathematician graduated from the National Capodistrian University of Athens and continued his studies and work in the USA, where he dedicated his time principally to research in applied mathematics and mechanics.

Our colleague Dimitris, in spite of the fact that he had spent most of his scientific life abroad, was an untiring collaborator of our Society and has particularly shared our efforts for the advancement of mathematics in Greece.

We believe that this volume, where most of his publications -scattered in various scientific sources- are made collectively accessible, will facilitate the propagation of the thoughts of an inspiring researcher for the benefit of the international mathematical and systems science community.

The Governing Board of the
Greek Mathematical Society

PREFACE

It is a distinct honor and privilege to be the editor of the Selected Papers of Dr Demetrios G. Magiros.

Magiros' scientific work continued for about forty years bearing important contributions of theoretical and practical significance. His enormous interest in the applications of mathematics, and the versatility of problems which he has solved in the fields of his applications are reflected in these papers.

Magiros' thoughts have left an impact on a variety of fields such as nonlinear differential equations, subharmonic oscillations, principal modes of nonlinear systems, celestial and orbital mechanics, stability analysis of nonlinear dynamical systems, precessional phenomena, and other.

In his investigation, Magiros starts from a philosophical basis and follows an axiomatic principle aiming always at the mathematical description and interpretation of the reality with emphasis on the applications.

Dr Magiros was an enthusiastic teacher with the talent to transmit both the depth and width of his knowledge. I personally remember his excellent lectures at NRC "Demokritos" in Athens (1973) on mathematical modeling and stability of dynamical systems. His impressive set of lecture notes, listed at the end of this book, covers a variety of topics from special functions and transforms to numerical methods for solution of nonlinear differential equations and optimal control problems.

We have decided with Mrs Helen Magiros to publish this volume hoping that it would serve both as an important reference in the field of nonlinear mechanics and dynamical systems, and as a source of inspiration to all young mathematicians and systems scientists who are challenged to solve real problems of our space age.

Many thanks are expressed to the Greek Mathematical Society for encouraging and supporting this publication.

Spyros G. Tzafestas

ACKNOWLEDGEMENTS

The publishers, the Greek Mathematical Society, the Editor, and Mrs Helen Magiros wish to thank the following publishers for granting permission to reprint the papers which appear in this volume (the identification of papers refers to the List of Contents):

Athens Academy of Sciences (Practika-Proceedings of Athens Academy of Sciences) for papers 2, 3, 4, 5, 8, 9, 10, 12, 13, 14, 15, 17, 21, 22, 23, 24, 29, 33, 34, 39 and 40.

Ukrainian Academy of Sciences (Ukrainian Mathematical Journal and Matematicheskaya Fizika) for papers 41, 42 and 43.

John Wiley and Sons, Inc. (Communications of Pure and Applied Mathematics) for paper 7.

Academic Press, Inc. (Information and Control) for papers 16 and 31.

The American Institute of Physics (Journal of Mathematical Physics) for paper 20.

The Franklin Institute (Journal of the Franklin Institute) for papers 25, 26, 27 and 32.

It is also acknowledged that papers 1, 18, 19 and 30 are reprinted from Proceedings of the National Academy of Sciences (U.S.A.), papers 28, 37 and 38 from the Proceedings of International Astronautical Congresses held in Madrid (1966), Buenos Aires (1969) and Brussels (1971), respectively, and paper 36 from the Proceedings of the VIIIth Symposium on Space Technology and Science (Tokyo, 1969).

The work described in the papers of the volume was done by D.G. Magiros while he was with General Electric Co. (ReEntry Systems Operation, Space Systems Division, Philadelphia, PA., U.S.A.).

Finally, the publishers, The Greek Mathematical Society, the Editor and Mrs Magiros would like to thank A.J. Dennison, Joseph B. Keller and George Reehl, co-authors of D.G. Magiros, for granting permission to reprint the respective papers (7, 28, 29, 32, 37).

CONTENTS

PART I
APPLIED MATHEMATICS AND MODELLING

* *Numbers in brackets refer to the Complete Chronological List of Magiros' publications.*

PART II
NONLINEAR MECHANICS

APPENDIX: PAPERS IN RUSSIAN

PART I

APPLIED MATHEMATICS AND MODELLING

In this Part, papers dealing with differential equations (DE), mathematical modelling, and linearization of nonlinear models are presented.

In paper 1 the reduction of a class of nonautonomous DE to another class of autonomous DE is studied, via an exact method which uses an appropriate transformation of the independent variable. A necessary and sufficient condition for this reduction is derived, and the equivalence of the above classes of DE as well as the concept of "irreducibility strips" of the reducible DE are discussed. In paper 2 (and paper 41 published in Russian) the existence and construction of several general solutions of nonlinear ordinary DE is investigated. To this end, appropriate restrictions of quantities of the DE, or the factorization of the DE, are employed. In paper 3, the way in which the arbitrary constants enter the general solutions of nonlinear DE is studied. It is found that the arbitrary constants appear either "linearly" in a few cases, or "rationally linearly" in some other cases, or in a "complicated nonlinear manner" in the majority of the situations. The singularities, particularly the essential singularities, make the general solutions as well as the introduction of the arbitrary constants in them, very complicated.

Paper 4 presents a "classical procedure" for determining "actual" mathematical solutions of well-posed (real) systems, either physical or social. Three basic steps in the procedure can be distinguished, namely: contruction of models of the system, selection of a well-posed model, and solution of this model. Paper 5 provides a selection of applications from various fields in an attempt to illustrate the procedure outlined in paper 4. Paper 6 gives some more material along the lines of the two previous papers.

In paper 7, the diffraction of a cylindrical wave by a perfectly conducting, semi-infinite thin screen with a cylindrical tip is analyzed by three different methods. These methods are: (i) the geometrical theory of diffraction, (ii) expansion in radial eigenfunctions, and (iii) the Watson transformation of the angular eigenfunction expression. Using the third method it is shown that the geometrical theory of diffraction is correct for the problem at hand, giving the asymptotic form of the exact solution. The result determines how the field in the shadow depends upon the wavelength and the curvature of the shadow forming object.

In papers 8 and 9 linearization methods for nonlinear models of physical phenomena are presented. Paper 8 refers to linearization by exact methods (exact transformations of variables, restriction of variables and parameters) which lead to general solutions of the nonlinear systems in closed form. Paper 9 refers to linearization by approximate methods and conditions are imposed for the linearization to be accepted. In paper 10, various properties which characterize linear and nonlinear physical systems are discussed.

In the final paper of this part (Paper 11) the creation of mathematical models of physical, technological and real-life systems is discussed. A "classical method" of creating mathematical models is proposed (see Papers 4 and 5) which is clarified by a variety of remarks and illustrated applications (of "complete" and "incomplete" cycle) from many fields.

A CLASS OF NONAUTONOMOUS DIFFERENTIAL EQUATIONS REDUCIBLE TO AUTONOMOUS ONES BY AN EXACT METHOD

By Demetrios G. Magiros

GENERAL ELECTRIC COMPANY, VALLEY FORGE SPACE TECHNOLOGY CENTER, PHILADELPHIA

Communicated by Leon Brillouin, June 23, 1967

1. *Introduction.*—This paper is concerned with the reduction of a class of nonautonomous differential equations to another class of autonomous ones by using an exact method based on an appropriate transformation of the independent variable. A condition necessary and sufficient for the reducibility is found and proved.

By autonomous differential equations we mean differential equations where there is no term containing explicitly the independent variable, and by nonautonomous ones those where at least one term contains explicitly the independent variable.

The reduction of nonautonomous differential equations into autonomous ones is important in theory and very desirable in practice. Some nonautonomous equations may be reduced into autonomous ones by either analytical exact methods or approximate ones, but the first are complicated and the second are based on assumptions rarely found in reality. The ideal methods for this reducibility are those which are short and exact and of practical usefulness, if such methods exist.

In this paper we discuss the conditions under which a nonautonomous equation of the form

$$\ddot{x} + p_1(t)\dot{x} + p_2(t)x = 0 \tag{1}$$

can be reduced to an autonomous one of the form

$$x'' + a_1x' + a_2x = 0 \tag{2}$$

by using an exact method. The coefficients p_1, p_2 in equation (1) are time-dependent, and the coefficients a_1 and a_2 in equation (2) are constants. The "dots" and "primes" in these equations denote differentiation with respect to the original time t and to the transformed time τ, respectively; p_1 and p_2 are single-valued, real-valued, and differentiable functions of t in an interval I: $t_0 < t < t_1$, where equation (1) is valid; t_0 may be zero and t_1 infinite.

The following theorem gives a condition under which equation (1) is reducible to equation (2).

2. Theorem. *A necessary and sufficient condition for the equation* (1) *to be reducible to equation* (2) *is that the coefficients p_1 and p_2 of equation* (1) *must satisfy the identity*

$$\frac{1}{p_2}\left\{2p_1 + \frac{\dot{p}_2}{p_2}\right\}^2 \equiv \gamma^2, \tag{3}$$

where γ^2 is a nonnegative fixed constant, p_1 and p_2 have the properties indicated above, and in addition, p_2 is positive.

Proof: In equation (1), we change the independent variable t into another independent variable τ according to a given transformation:

$$\tau = \varphi(t). \tag{4}$$

Reprinted from the *Proceedings of the National Academy of Sciences* **58,** No. 2 (1967), 412–419.

By using this transformation, the derivatives $\dot{x}$ and $\ddot{x}$ of equation (1) are given by

$$\dot{x} = x'\dot{\varphi}, \qquad \ddot{x} = x'\ddot{\varphi} + x''(\dot{\varphi})^2 \tag{5}$$

when, inserting (5) into (1), we can get the equation

$$x'' + \frac{p_1\dot{\varphi} + \ddot{\varphi}}{(\dot{\varphi})^2}x' + \frac{p_2}{(\dot{\varphi})^2}x = 0, \tag{6}$$

where the independent variable is τ, its coefficients are dependent on p_1, p_2, φ, and then they are functions of t, and φ is nonconstant.

If equation (1) is reducible to equation (2), then the functions p_1, p_2, φ must be restricted in such a way that the coefficients of equation (6) are constants, that is:

$$\frac{p_1\dot{\varphi} + \ddot{\varphi}}{(\dot{\varphi})^2} = a_1, \qquad \frac{p_2}{(\dot{\varphi})^2} = a_2. \tag{7i), (7ii}$$

From (7*ii*), we have

$$\dot{p}_2 = 2a_2\dot{\varphi}\ddot{\varphi}, \tag{8}$$

and the elimination of $\dot{\varphi}$ and $\ddot{\varphi}$ between (7) and (8) gives the condition of the reducibility of equation (1) into equation (2).

From (7*ii*), we have

$$(\dot{\varphi})^2 = \frac{p_2}{a_2}, \qquad \dot{\varphi} = \pm\sqrt{\frac{p_2}{a_2}}, \tag{9}$$

and inserting (9) into (8) we have

$$\ddot{\varphi} = \pm\frac{\dot{p}_2}{2\sqrt{a_2p_2}}. \tag{10}$$

Finally, inserting (9) and (10) into (7*i*) we have

$$\frac{1}{p_2}\left\{2p_1 + \frac{\dot{p}_2}{p_2}\right\}^2 \equiv \gamma^2, \tag{11}$$

where

$$\gamma^2 = \frac{4a_1^2}{a_2}. \tag{11a}$$

The quantity of the left-hand-member of (11) is the "reducibility quantity," the nonnegative constant γ^2 is the "reducibility constant," and the condition (11) the "reducibility condition" of equation (1).

By the above we prove that if equation (1) is reducible to equation (2), then p_1 and p_2 must satisfy condition (11), which then is a necessary condition for the reducibility. Now we can see that the converse is true; that is, if p_1 and p_2 satisfy the condition (11), then equation (1) is reducible to equation (2).

We select the transformation φ in such a way that $(\dot{\varphi})^2$ is a constant multiple of p_2, as formula (7*ii*) shows, $(\ddot{\varphi})^2 = p^2/a^2$, then we will have

$$\dot{\varphi} = \pm\frac{1}{\sqrt{a_2}}\sqrt{p_2}, \qquad \ddot{\varphi} = \pm\frac{1}{2\sqrt{a_2}}\frac{1}{\sqrt{p_2}}, \qquad \varphi = \pm\frac{2}{2\sqrt{a_2}}p_2^{3/2} + K, \tag{12}$$

K is an arbitrary constant. Inserting the first two of (12) into formula (7*i*), we can have the coefficient a_1 in terms of the functions p_1, p_2, the transformation φ, and the arbitrary coefficient a_2, when equation (2), corresponding to equation (1), is constructed.

If ρ_1 and ρ_2 are the eigenvalues of equation (2), its general solution is given by

$$x(\tau) = K_1 e^{\rho_1 \tau} + K_2 e^{\rho_2 \tau}. \tag{13}$$

where K_1, K_2 are arbitrary constants, and inserting $\tau = \varphi(t)$ in (13), we have the general solution of equation (1) in the form

$$x(t) = K_1 e^{\rho_1 \varphi(t)} + K_2 e^{\rho_2 \varphi(t)}. \tag{14}$$

Remark 1: The reducibility condition (11) of equation (1) into equation (2) is independent of the transformation φ used.

Remark 2: The above procedure shows that p_2 and a_2 must be positive.

Remark 3: If the reducibility constant γ^2 exists, then, by using (11*a*), we can determine either a_1 in terms of a_2 and γ^2, or a_2 in terms of a_1 and γ^2, without using the transformation function φ. Then the constant coefficients of equation (2) will be either a_1, $a_2 = 4a_1^2/\gamma^2$, or $a_1 = \pm(\gamma/2)\sqrt{a_2}$, a_2, if $\gamma \pm 0$; and $a_1 = 0$. $a_2 =$ arbitrary, if $\gamma = 0$; or $a_1 =$ arbitrary, $a_2 = 0$, if $\gamma = \infty$.

3. COROLLARY. *From the reducibility condition* (11) *we have*

$$p_1 = \frac{1}{2}\left\{\pm \gamma\sqrt{p_2} - \frac{\dot{p}_2}{p_2}\right\} \tag{15i}$$

$$\dot{p}_2 + 2p_1p_2 = \pm\gamma p_2^{3/2} \tag{15ii}$$

Then, if the coefficient p_2 of equation (1) *is given, we can calculate the other coefficient p_1 of equation* (1) *in such a way that equation* (1) *is reducible to equation* (2). *The coefficient p_1 is given by* (15*i*). *γ^2 is the nonnegative fixed reducibility constant. If the coefficient p_1 of equation* (1) *is given, we can calculate the coefficient p_2 of equation* (1) *for its reducibility. The coefficient p_2 is a solution of the specific Bernoulli equation* (15*ii*).

Remark 4: Formula (15*i*) implies that if p_2 is a positive constant, then, for the reducibility, p_1 must be a real constant; consequently, in case p_1 is variable and p_2 a constant, equation (1) is irreducible to equation (2). The Hermite equation: $\ddot{x} - 2t\dot{x} + cx = 0$, is an example of the last case. But, if p_1 is a constant, then there are variable coefficients p_2 of equation (1) for its reducibility, and these coefficients are solutions of the Bernoulli equation (15*ii*).

4. *Illustrated Problems.*—The above discussion is clarified by the following problems.

PROBLEM 1. *Given the equation*

$$t^2\ddot{x} + 2t\dot{x} + \mu x = 0, \qquad t \neq 0, \tag{16}$$

where the coefficients are variable and μ is a fixed positive constant:

(i) *check the possibility for its reduction to an equation* (2) *with constant coefficients;*

(ii) *find the appropriate transformation of the independent variables of these equations for the reducibility of equation* (16) *to equation* (2);

(iii) *determine the constants a_1 and a_2 of the reduced equation* (2), *and*

(iv) *give the form of the general solution of equation* (16), *using the general solution of the reduced equation* (2).

Solution: (*i*) Equation (16) can be written in the form

$$\ddot{x} + \frac{2}{t}\dot{x} + \frac{\mu}{t^2}x = 0, \tag{16a}$$

and we have

$$p_1 = \frac{2}{t}, \qquad p_2 = \frac{\mu}{t^2}$$

when condition (11) gives $\gamma^2 = (4/\mu)$, and then equation (16) is reducible to equation (2).

(*ii*) The transformation $\tau = \varphi(t)$, which reduces equation (16) to equation (2), can be found by using (7*ii*) when

$$(\dot{\varphi})^2 = \frac{\mu}{a_2}\frac{1}{t^2}, \qquad \dot{\varphi} = \pm\sqrt{\frac{\mu}{a^2}}\cdot\frac{1}{t}, \qquad \ddot{\varphi} = \mp\sqrt{\frac{\mu}{a_2}}\cdot\frac{1}{t^2} \tag{16b}$$

$$\tau = \varphi(t) = \pm\sqrt{\frac{\mu}{a_2}}\log t + K. \tag{16c}$$

(*iii*) Inserting (16*b*) into (7*i*), we can find

$$a_1 = \pm\sqrt{\frac{a_2}{\mu}}, \qquad a_2 = \text{arbitrary constant}, \tag{16d}$$

then equation (2) is one of the following equations:

$$\left.\begin{aligned} x'' \pm \sqrt{\frac{a_2}{\mu}}x' + a_2x &= 0 \\ x'' + a_1x' + \mu a_1^2x &= 0 \end{aligned}\right\}, \tag{16e}$$

a_2 and a_1 are arbitrary constants in the first and second of equations (16*e*), respectively.

(*iv*) The characteristic roots of, say, the second of (16*e*) are

$$\rho_{1,2} = a_1\lambda_{1,2}, \qquad \lambda_{1,2} = \frac{1}{2}\left\{-1 \pm \sqrt{1-4\mu}\right\}, \tag{16f}$$

when its general solution is

$$x(\tau) = K_1e^{a_1\lambda_1\tau} + K_2e^{a_1\lambda_2\tau}. \tag{16g}$$

By using transformation (16*c*), function (16*g*) becomes

$$x(t) = \bar{K}_1e^{\lambda_1\log t} + \bar{K}_2e^{\lambda_2\log t}, \tag{16h}$$

which is the general solution of equation (16).

Remark 5: Equation (16) is of Euler's type and, as we know, this class of equations accepts solutions of the form $y = t^\lambda$, where λ is a root of the characteristic equation, which in the case of equation (16) is $\lambda = \frac{1}{2}\{-1 \pm \sqrt{1 - 4\mu}\}$, and then the general solution of equation (16), by using this method, which characterizes the Euler's equation, is

$$x(t) = c_1 t^{\lambda_1} + c_2 t^{\lambda_2}, \tag{16i}$$

where c_1 and c_2 are arbitrary constants. Since $t^\lambda = e^{\lambda \log t}$, solutions (16*h*) and (16*i*) are the same.

Remark 6: We can check that the reducibility quantity of the following equations is time-dependent, then these equations are irreducible to equations with constant coefficients:

(*i*) $t^2\ddot{x} + t\dot{x} + (t^2 - n^2)x = 0, \quad t \neq 0$ (Bessel).

(*ii*) $t\ddot{x} + (1 - t)\dot{x} + tx = 0, \quad t \neq 0$ (Laguerre).

(*iii*) $(1 - t^2)\ddot{x} - 2t\dot{x} + n(n - 1)x = 0, \quad t \neq \pm 1$ (Legendre).

(*iv*) $(1 - t^2)\ddot{x} - t\dot{x} + n^2 x = 0, \quad t \neq \pm 1$ (Tschebycheff).

PROBLEM 2. *In the equation*

$$\ddot{x} + \frac{2}{t}\dot{x} + p_2(t)x = 0, \qquad t \neq 0, \tag{17}$$

calculate the coefficient p_2 in such a way that equation (17) *is reducible to an equation* (2) *with constant coefficients and find the general solution of equation* (17).

Solution: The Bernoulli equation (15*ii*), of which the solution p_2 gives the coefficient of equation (17) for its reducibility to equation (2), becomes

$$\dot{p}_2 + \frac{4}{t}p_2 = \pm\gamma p_2^{3/2}, \tag{17a}$$

where γ^2 is the fixed reducibility constant.

Equation (17*a*), by using the transformation

$$Z = \frac{1}{\sqrt{p_2}} \tag{17b}$$

can be reduced to the linear equation:

$$\dot{Z} - \frac{2}{t}Z = \pm\frac{1}{2}\gamma \tag{17c}$$

of which the general solution can be found to be

$$Z = t(c_3 t \pm \frac{1}{2}\gamma), \tag{17d}$$

where c_3 is an arbitrary constant.

By using transformation (17*b*), function (17*d*) becomes

$$p_2 = t^{-2}(c_3 t \pm {}^1/_2\gamma)^{-2}, \tag{17e}$$

which is the coefficient of equation (17) for its reducibility.

For $c_3 = 0$, $\frac{1}{2}\gamma = (1/\sqrt{\mu})$, (17*e*) gives $p_2 = \mu t^{-2}$, the special case of Problem 1.

To calculate the constant coefficients of the reduced equation (2), we determine the appropriate transformation $\varphi(t)$.

Combining (7*ii*) and (17*e*), we have

$$(\dot{\varphi})^2 = \frac{1}{a_2\{t^2(c_3t \pm \frac{1}{2}\gamma)^2\}} = \frac{1}{t^2\{c_3\sqrt{a_2}\,t \pm \frac{1}{2}\gamma\sqrt{a_2}\}^2},$$

where a_2, c_3 are arbitrary constants.

If we take the arbitrary constants a_2 and c_3 such that $c_3 = (1/\sqrt{a_2})$, and put $c = \pm\ \frac{1}{2}\gamma\sqrt{a_2}$, we have:

$$(\dot{\varphi})^2 = \frac{1}{t^2(t+c)^2}, \qquad \dot{\varphi} = \pm\frac{1}{t(t+c)}, \qquad \ddot{\varphi} = \mp\frac{2t+c}{t^2(t+c)^2}. \tag{17f}$$

$$\varphi(t) = \pm\int\frac{dt}{t(t+c)} = \pm\frac{1}{c}\log\frac{t}{t+c} + K. \tag{17g}$$

We can check that $a_1 = \pm c = \pm\ \frac{1}{2}\gamma\sqrt{a_2}$, then the corresponding differential equation with constant coefficients is either:

$$x'' \pm \tfrac{1}{2}\gamma\sqrt{a_2}\,x' + a_2x = 0, \qquad \text{or:}\quad x'' + a_1x' + \frac{4a_1^2}{\gamma^2}x = 0. \tag{17h}$$

The characteristic roots of (17*h*) are

$$\rho_{1,2} = \frac{a_1}{2}\left\{-1 \pm \sqrt{1 - \left(\frac{4}{\gamma}\right)^2}\right\}, \tag{17i}$$

then

$$x(\tau) = K_1e^{\rho_1\tau} + K_2e^{\rho_2\tau} \tag{17j}$$

is the general solution of (17*h*), when

$$x(t) = \bar{K}_1e^{\rho_1\varphi(t)} + \bar{K}_2e^{\rho_2\varphi(t)} \tag{17k}$$

is the general solution of equation (17), with $\varphi(t)$ given by (17*g*).

PROBLEM 3. *Calculate the coefficient p_2 in the equation*

$$\ddot{x} + p_2(t)x = 0 \tag{18}$$

in such a way that this equation is reducible to an equation (2) *with constant coefficients and find the general solution of equation* (18).

Solution: The reducibility condition (11) in the case of this problem becomes $\dot{p}_2 = \pm\gamma p_2^{3/2}$ where γ^2 is the reducibility fixed constant, then the function $p_2(t)$ for the reducibility of equation (18) is a positive nonoscillatory function of t given by the formula

$$p_2(t) = (\pm\ \tfrac{1}{2}\gamma t + K_1)^{-2} \tag{18a}$$

where K_1 is an arbitrary constant. The appropriate reducibility transformation $\tau = \varphi(t)$ can be found from

$$(\dot{\varphi})^2 = \frac{1}{a_2}(\pm\ \tfrac{1}{2}\gamma t + K_1)^{-2}, \qquad \text{or}\quad \dot{\varphi} = \pm\frac{1}{\sqrt{a_2}}(\pm\ \tfrac{1}{2}\gamma t + K_1)^{-1},$$

then:

$$\tau = \varphi(t) = \frac{2}{\gamma\sqrt{a_2}} \log\left(\pm \tfrac{1}{2}\gamma t + K_1\right). \tag{18b}$$

Calculating a_1, we find $a_1 = \pm \frac{1}{2}\gamma\sqrt{a_2}$, then $a_2 = (4a_1^2/\gamma^2)$, when the corresponding equation with constant coefficients is

$$x'' \pm \tfrac{1}{2}\gamma\sqrt{a_2}\, x' + a_2 x = 0, \qquad \text{or} \quad x'' + a_1 x' + \frac{4a_1^2}{\gamma^2} x = 0. \tag{18c}$$

The characteristic roots of (18c) are

$$\rho_{1,2} = \tfrac{1}{2}a_1 \left\{ -1 \pm \sqrt{1 - \left(\frac{4}{\gamma}\right)^2} \right\}, \tag{18d}$$

the general solution of (18) is given by

$$x(t) = K_1 e^{\rho_1\varphi(t)} + K_2 e^{\rho_2\varphi(t)}, \tag{18e}$$

where K_1 and K_2 are arbitrary constants, ρ_1 and ρ_2 are given by (18*d*), and $\varphi(t)$ by (18*b*).

Remark 7: By the above, it is found that if the function $p_2(t)$ of equation (18) is given by (18*a*), the equation (18) is reducible to equation (18*c*), when its general solution is given by (18*e*). If $p_2(t)$ of equation (18) has a form different than (18*a*), equation (18) is irreducible to an equation with constant coefficients, and in case of irreducibility of equation (18), its general solution, if it is possible to be found, cannot have the form (18*e*).

This happens when, say, $p_2(t)$ is oscillatory or periodic in t, as, e.g., in the case of the Mathieu equation, when $p_2 = k_1 + k_2 \cos 2t$. The solution of a Mathieu equation, as we know, is given by

$$x(t) = K_1(t)e^{r_1 t} + K_2(t)e^{r_2 t}, \tag{18f}$$

where $K_1(t)$ and $K_2(t)$ are purely periodic functions of t of period π, and the calculation of the constants r_1 and r_2 necessitates no simple methods. The solution (18*f*) is not of the form (18*e*).

5. *Equivalence of the Classes.—Irreducibility strips of the reducible equation* (1): (*a*) The solutions of equation (1) with variable coefficients lead in general to special functions expressed in terms of infinite series or continued fractions or integrals, as, e.g., the solutions of the special differential equations referred to in *Remark* 6. The reducible class of equation (1) has the important property that it can be fully and explicitly treated as the class of equation (2) with constant coefficients, then its solution can be expressed in terms of elementary functions.

Stability properties of the solutions of these two classes of differential equations are preserved when we go from one class to the other through the suitable transformation, and then these classes have a "stability equivalence."

(*b*) We supposed in the preceding that the coefficients $p_1(t)$ and $p_2(t)$ of equation (1) are of ordinary behavior throughout the interval I of the time t where equation (1) is valid, and then all values of this interval are "ordinary points" of the reducible equation (1).

There may be values of the interval I where at least one of these coefficients has a nonordinary behavior, and these values are "singular points" of the equation (1), and equation (1) is irreducible at these points.

If at least one of the coefficients have the form of a ratio, and at a point $t = \bar{t}$ the denominator of the ratio has a zero value but the numerator has a nonzero value, this point $\bar{t}$ is a singular point of the coefficient, then it is a singular point of the solution $x(t)$ of equation (1), either "regular" or "irregular" singular point.

A small neighborhood around any singular point of the solution $x(t)$ corresponds to a narrow strip in the "t,x-plane," which is an "irreducibility strip" of equation (1). The solution of a reducible equation (1), outside the irreducibility strips, can be expressed by elementary functions according to formula (14).

Inside these strips, the solution may be expressed by a special series convergent or not, [1, 3(a), 4(a)] or by formulas similar to "Kramer's connection formulas," known from quantum mechanics (cf. ref. 4(b)).

The subject of the present paper was inspired by (b) of reference 3, and was especially prepared to cover practical needs by giving short criteria useful in many fields. It must be supplemented by an extension to some other classes of equations (cf. refs. 2, 5).

[1] Dennery, P., and A. Krzywicki, *Mathematics for Physicists* (New York: Harper and Row, 1967), pp. 298–301.

[2] Erugin, N., *Linear Systems of Ordinary Differential Equations* (New York: Academic Press, 1966), chap. 19.

[3] Ince, E., *Ordinary Differential Equations* (New York: Dover Publications, 1944): (a) pp. 160–161, 168–169; (b) p. 131.

[4] Kemble, E., *The Fundamental Principles of Quantum Mechanics* (New York: McGraw-Hill Book Co., 1937): (a) pp. 140–142; (b) pp. 93–103.

[5] Roseau, M., *Vibrations Nonlinéaires et Théorie de la stabilit é* (New York: Springer-Verlag, 1966), pp. 27–34.

ΕΦΗΡΜΟΣΜΕΝΑ ΜΑΘΗΜΑΤΙΚΑ.— **Nonlinear differential equations with several general solutions,** *by Demetrios G. Magiros* *.

Ἀνεκοινώθη ὑπὸ τοῦ Ἀκαδημαϊκοῦ κ. Ἰω. Ξανθάκη.

I. INTRODUCTION

Nonlinear differential equations constitute today a field of scientific knowledge basic for the investigation of the majority of physical phenomena, technological systems, social problems.

The general solutions of these DE are the most desirable, especially in applications, but only for some nonlinear DE the general solutions are known to exist, and only in a few classes of these DE the general solutions can be determined in a closed form.

While any linear ordinary DE has one general solution, a nonlinear ODE may have either one or several general solutions.

It is necessary here to make clear some concepts related to that of the general solution. We consider a nonlinear ODE (F) valid in a region R of the space of its variables, and a function (Φ) of the variables of (F) containing a number of arbitrary constants independent to each other, which can take the values in a region C of the space of these arbitrary constants. A function coming from (Φ) by a specification of all its arbitrary constants, if it satisfies the DE (F), is called a «particular solution» of (F). The function (Φ), a totality of all the particular solutions of (F) coming from (Φ), is called a «general solution» of (F).

Any function, coming from the general solution (Φ) of (F) by a specification of some only arbitrary constants of (Φ), which, then, contains the unspecified constants of (Φ) as arbitrary parameters, is called a «part of the general solution» of (F).

If there are functions (Φ_i), $i = 1, 2, \ldots, n$, which have properties similar to that of (Φ) in connection to the DE (F), and, in addition, particular solutions of one of the (Φ_i) are not identical to particular solutions of any of the others (Φ_i), these functions must be considered as «distinct general solutions» of (F), and, in this case, we say that the DE

* Δ. Γ. ΜΑΓΕΙΡΟΥ, **Μὴ γραμμικαὶ διαφορικαὶ ἐξισώσεις μὲ πολλὰς γενικὰς λύσεις.**

Reprinted from the *Proceedings of the Athens Academy of Sciences* **52** (1977), 221–229.

(F) has «several general solutions». By this definition, is not excluded that a «common part» among some of (Φ_i) may exist.

In this paper remarks are given concerning the existence and determination of several general solutions of some nonlinear ODE.

The «restriction of the solutions of the DE», and the «factorization of the DE» can be used as methods for investigation. These methods are explained and illustrated by examples from which we draw certain conclusions.

II. THE METHOD OF RESTRICTING THE SOLUTIONS

By this method, if it is applicable, one can determine several general solutions of a given nonlinear DE in a closed form. We give an example.

Example 1: $y'''(1+y'^2) - 3y'y''^2 = 0$ (1.1)

One can have two appropriate restrictions of the unknown solution of (1.1), and to each restriction a general solution of (1.1) corresponds.

(a): The solution y of (1.1) is restricted by $y'' = 0$, which implies $y''' = 0$. In this case, (1.1) is satisfied simultaneously by:

$$y'' = 0, \quad y''' = 0 \tag{1.2}$$

when the function

$$y = \alpha_1 x + \alpha_2 \tag{1.3}$$

which satisfies (1.2) is a general solution of (1.1). The α_1 and α_2 are independent to each other and the two-parameter family (1.3) represents «all straight lines» in the x, y - plane.

(b): The solution y of (1.1) is restricted by $y'' \neq 0$, and, in this case, one can determine a general solution of (1.1) which is different than (1.3). The DE (1.1) in this case can be written in the form: [1]

$$\frac{y'''}{y''} = \frac{3y'y''}{1+y'^2} = \frac{3}{2} \cdot \frac{d(1+y'^2)}{1+y'^2}. \tag{1.4}$$

An integration of this equation gives:

$$\frac{y''}{(1+y'^2)^{3/2}} = c_1.$$

A new integration leads to:

$$y' = \frac{c_1x + c_2}{\sqrt{1 - (c_1x + c_2)^2}} \tag{1.5}$$

and a third integration to:

$$c_1y + c_3 = -\sqrt{1 - (c_1x + c_2)^2} \tag{1.6}$$

which can get the form:

$$x^2 + y^2 + C_1x + C_2y + C_3 = 0. \tag{1.7}$$

The C_1, C_2, C_3, functions of c_1, c_2, c_3, are arbitrary constants independent to each other, and the three-parameter family (1.7), which represents «all circles» in x, y-plane, is a general solution of (1.1). There is no specification of the arbitrary constants of (1.3) and (1.7) by which a member of (1.3) becomes identical with a member of (1.7), then (1.3) and (1.7) are two distinct general solutions of (1.1).

III. THE METHOD OF FACTORIZING THE DE

A possible application of this method can give several general solutions of a DE. We give examples.

Example 2: $x^3y''y''' + x^2y''^2 - 2xy'y'' + 2yy'' = 0.$ (2.1)

The factorization gives:

$$y''(x^3y''' + x^2y'' - 2xy' + 2y) = 0 \tag{2.2}$$

then:

$$\left.\begin{aligned} &(a): \quad y'' = 0 \\ &(b): \quad x^3y''' + x^2y'' - 2xy' + 2y = 0, \quad x \neq 0 \end{aligned}\right\} \tag{2.3}$$

the general solutions of which are:

$$\left.\begin{aligned} &(a): \quad y = a_1x + a_2 \\ &(b): \quad y = c_1x + c_2x^2 + c_3x^{-1} \end{aligned}\right\} \tag{2.4}$$

and these functions are the general solutions of (2.1).

The family of straight lines through the origin is a «common part» of the general solutions (2.4) of the DE (2.1).

Example 3: $y'^2 + y(x^2y - xy) - x^3y^2 = 0$ (3.1)

By factorizing one has:

$$(y' - xy) \cdot (y' + x^2y) = 0 \tag{3.2}$$

then:

$$\left.\begin{array}{ll}(a): & y' = xy \\ (b): & y' = -x^2y\end{array}\right\} \tag{3.3}$$

from which:

$$\left.\begin{array}{ll}(a): & y = c_1 e^{x^2/2} \\ (b): & y = c_2 e^{-x^3/3}\end{array}\right\} \tag{3.4}$$

that is the general solutions of the DE (3.1):

We remark that from each point of the x, y-plane two, in general, integral curves of (3.1) pass with different slopes, one belonging to the family (3.4.a) corresponding to a specific value of c_1, and one belonging to the family (3.4.b) for a specific value of c_2, and these values of c_1 and c_2 are different.

The equation $(xy = -x^2y)$ gives the singular lines: $x = 0$, $x = -1$, $y = 0$, at the points of which the integral curves have just one slope, the slope $-y$ at $x = -1$, and the zero-slope at the coordinate axes, which are the «singular solutions» of the DE (3.1).

Example 4:

«The general nonlinear DE of the first order:

$$F(x, y, y') = 0 \tag{4.1}$$

in case F is a polynomial for y' of degree m".

In this case, (4.1) can have the form:

$$F(x, y, y') \equiv y'^m + P_1 y'^{m-1} + \ldots + P_{m-1} y' + P_m = 0 \tag{4.2}$$

where $P_1, \ldots, P_m$, functions of x and y, are continuously differentiable in the region of validity of (4.2).

$$F_{y'}(x, y, y') \neq 0.$$

The DE (4.2) if $F_{y'}(x, y, y') \neq 0$, solved for y', gives m simple roots: $y'_1, \ldots, y'_m$, when one can write:

$$F(x, y, y') \equiv (y'_1 - p_1) \cdot (y'_2 - p_2) \ldots (y'_m - p_m) = c \tag{4.3}$$

where $p_1, \ldots, p_m$ are functions of x and y.

cases. Nevertheless, in the majority of the cases, the arbitrary constants enter the general solutions in a «complicated nonlinear way», which characterizes the nature of the general solutions and their corresponding DE.

The singularities, especially the essential singularities, appearing in the general solutions, make these solutions, as well as the manner in which the arbitrary constants enter them, very complicated.

I. GENERAL SOLUTIONS AS LINEAR FUNCTIONS OF THEIR ARBITRARY CONSTANTS

In a linear DE, the general solution is a linear combination of the fundamental set of solutions, and this general solution contains the arbitrary constants linearly. Such a general solution interprets the «principle of superposition», which holds only in linear DE.

As an example, we take the «Bessel DE»:

$$x^2 y'' + xy' + (x^2 - n^2)\,y = 0$$

of which the general solution is of the form:

$$y = c_1 J_n(x) + c_2 Y_n(x)$$

where c_1, c_2 are the two arbitrary constants, and $J_n(x)$, $Y_n(x)$ the «Bessel functions» of the first and second kind, respectively, which are «special functions» particular solutions of the DE.

We remark that the principle of superposition is sometimes difficult to apply, as, e.g. the general solution of the Mathieu equation:

$$y'' + (a + b\cos 2x)\,y = 0$$

is not known containing two arbitrary constants in the above way **[2]**.

In the following we find classes of nonlinear DE of which the general solutions are linear functions of their arbitrary constants.

Some special methods are applied, illustrated by proper examples.

I. 1. The Factorization Method.

By applying the factorization method to a nonlinear DE, if this method is applicable, one can have several general solutions of the DE, which contain the arbitrary constants linearly.

A general example is the first order DE: $F(x, y, y') = 0$ in case it is a polynomial for y' of order m In this case, one can write:

$$F(x, y, y') \equiv y'^m + P_1(x, y) \cdot y'^{m-1} + \ldots + P_{m-1}(x, y) \cdot y' + P_m(x, y) = 0. \quad (a.1)$$

This DE can be solved under the restriction $F_{y'}(x, y, y') \neq 0$, and if $y_1', \ldots, y_m'$ are the m simple roots, one can get F in the product form:

$$F \equiv [y_1' - p_1(x, y)] \cdot [y_2' - p_2(x, y)] \ldots [y_m' - p_m(x, y)] = 0, \quad (a.2)$$

which is equivalent to the m DE:

$$y_1' = p_1(x, y), \quad y_2' = p_2(x, y), \quad \ldots, \quad y_m' = p_m(x, y). \quad (a.3)$$

Integrating, the m general solutions of (a. 1) are:

$$\varphi_1(x, y, c_1) = 0, \quad \varphi_2(x, y, c_2) = 0, \quad \ldots, \quad \varphi_m(x, y, c_m) = 0. \quad (a.4)$$

There are forms of the functions p_i in (a. 3) which lead to functions φ_i as linear functions of the arbitrary constants; for instance, in the cases where DE (a. 3) are with separable variables, or they are exact DE, or p_i are homogeneous functions of, say, degree n, etc.

Example 1: $$y'^2 + y'(x^2y - xy) - x^3y^2 = 0. \quad (1.1)$$

The factorization gives: $(y' - xy) \cdot (y' + x^2y) = 0$, when the two distinct general solutions of (1. 1), containing the arbitrary constants linearly, are:

$$y = c_1 e^{x^2/2}, \quad y = c_2 e^{-x^3/3}. \quad (1.2)$$

Example 2: $$x^3y''y''' + x^2y''^2 - 2xy'y'' + 2yy'' = 0. \quad (2.1)$$

The factorization gives: $y''(x^3y''' + x^2y'' - 2xy' + 2y) = 0$, when the two general solutions of (2. 1) are:

$$\text{(a)} \quad y = a_1 x + a_2, \qquad \text{(b)} \quad y = c_1 x + c_2 x^2 + c_3 x^{-1}. \quad (2.2)$$

These two general solutions contain the family of straight lines through the origin ($a_2 = c_2 = c_3 = 0$, $a_1 = c_1$) as a «common part».

Example 3: $$y'^2 y''' + y^2 y''' + y''' = 0, \quad (3.1)$$

y''' is a common factor of the terms, then $y'''(y'^2 + y^2 + 1) = 0$, and since $y'^2 + y^2 + 1 \neq 0$, the DE (3.1) is equivalent to $y''' = 0$, of which the general solution is:

$$y = c_1 x^2 + c_2 x + c_3, \tag{3.2}$$

which consists of two different families of curves, the parabolas ($c_1 \neq 0$) and the straight lines ($c_1 = 0$).

Example 4: $$y'^3 - \frac{y^2 - x^2}{2xy} y'^2 - y' + \frac{y^2 - x^2}{2xy} = 0. \tag{4.1}$$

Factorizing, one can get:

$$(y'^2 - 1) \cdot \left(y' - \frac{y^2 - x^2}{2xy}\right) = 0 \tag{4.2}$$

which is equivalent to:

$$\text{(a)} \quad y' = \pm 1, \qquad \text{(b)} \quad y' = \frac{y^2 - x^2}{2xy} \tag{4.3}$$

The first of (4.3) gives two general solutions of (4.1):

$$\text{(a)} \quad y = x + c_1, \qquad \text{(b)} \quad y = -x + c_2. \tag{4.4}$$

The second DE of (4.3) has its right hand member as a homogeneous function of second degree, and can be written as:

$$y' = \frac{(y/x)^2 - 1}{2(y/x)}. \tag{4.5}$$

By using the transformation $y = xu$, the DE (4.5) can be solved and its general solution can be found to be:

$$x^2 + y^2 - cx = 0. \tag{4.6}$$

The three functions (4.4, a, b) and (4.6) are the three general solutions of the DE (4.1), containing the arbitrary constants linearly.

I. 2. The Method of Restricting Quantities of the DE.

By restricting quantities of the DE one may have several general solutions of the DE containing the arbitrary constants linearly. We give two examples.

Example 5: $$y'(1 + y'^2) - 2y'y'' = 0. \tag{5.1}$$

We restrict the unknown function y of (5. 1) as follows:

(a): If y is such that $y''=0$, which implies $y'''=0$, then the general solution of (5. 1) is:

$$y = a_1 x + a_2. \tag{5. 2}$$

(b): If y is such that $y'' \neq 0$, then one can integrate (5. 1) exactly, and its new general solution is:

$$x^2 + y^2 + c_1 x + c_2 y + c_3 = 0. \tag{5. 3}$$

Both functions (5 2) and (5. 3) are distinct general solutions of (5. 1) and contain the arbitrary constants linearly.

Example 6:

$$y'' + \frac{2}{x} y' + y^n = 0. \tag{6. 1}$$

This is the famous Emden equation, coming from his investigation on basic problems of astrophysics. The solutions of this DE in a closed form are **[1]**:

$$\left.\begin{aligned} &(a) \quad n = 0: \quad y = a_1 + \frac{a_2}{x} - \frac{x^2}{6} \\ &(b) \quad n = 1: \quad y = c_1 \frac{\sin x}{x} + c_2 \frac{\cos x}{x} \\ &(c) \quad n = 5: \quad y = \left(\frac{3a}{x^2 + 3a^2}\right)^{1/2}. \end{aligned}\right\} \tag{6. 2}$$

a, a_1, a_2, c_1, c_2 are arbitrary constants. The first two functions are the only known general solutions of the corresponding linear DE of (6. 1) containing the arbitrary constants linearly. The third function of (6. 2), containing one arbitrary constant nonlinearly, is a «part» of the unknown general solution of (6. 1) in case $n = 5$.

On the occasion of this DE, we remark that any attempt to find the general solutions of (6. 1) for other values of n will be governed rather by a theoretical curiosity than by its usefulness, since, even if we know it, it has, according to Emden, no physical meaning in the Emden problems of astrophysics. Emden and his followers in astrophysics found a solution of (6. 1) in the form of a Taylor series, which interprets the reality very adequately.

I. 3. **A class of nonlinear DE with general solutions containing the arbitrary constants linearly.**

We can find a class of nonlinear DE of which the general solution contains the arbitrary constants linearly.

Starting from the simple primitive:

$$y^2 = 4x\,; \quad y \neq 0, \quad x > 0$$

by differentiation one gets the DE: $yy' - 2 = 0$ of which the general solution is $y^2 = 4x + c_1$. Another differentiation gives: $yy'' + y'^2 = 0$ with general solution: $y^2 = 4x + c_1 x + c_2$. Continuing the differentiation up, say, to the order n, a nonlinear DE of order n results of which the general solution is:

$$y^2 = 4x + c_1 x^{n-1} + \ldots + c_{n-1} x + c_n\,,$$

where the arbitrary constants $c_1, \ldots, c_n$ enter linearly.

Generalizing the above, one can see that a nonlinear DE of order n:

$$F(y, y', \ldots, y^{(n)}) = \Phi(x), \tag{7.1}$$

where:

$$F = \frac{d^n}{dx^n} f(y), \qquad \Phi(x) = \frac{d^n}{dx^n} \varphi(x) \tag{7.2}$$

has as general solution the function:

$$f(y) = \varphi(x) + c_1 x^{n-1} + \ldots + c_{n-1} x + c_n\,. \tag{7.3}$$

We remark that the «principle of superposition» may be applicable to some nonlinear DE of which the general solutions are linear functions of the arbitrary constants, and this is an important problem.

II. GENERAL SOLUTIONS AS RATIONALLY LINEAR FUNCTIONS OF THE ARBITRARY CONSTANTS

We distinguish a class of nonlinear DE of which the general solutions are rationally linear functions of the arbitrary constants, that is the general solutions are ratios of functions, where the arbitrary constants enter linearly in the numerators and denominators.

This class of nonlinear DE is the class of «Riccati equations of any order».

II. 1. Let us start from the simple case:

$$y = \frac{a_1 + cb_1}{a_2 + cb_2} \tag{8.1}$$

where c is the arbitrary constant, and a_1, a_2, b_1, b_2 functions of x.

Eliminating c between (8.1) and its derivative, the resulting DE has (8.1) as a general solution. The differentiation gives:

$$y' = \frac{a_1' + cb_1' - y(a_2' + cb_2')}{a_2 + cb_2}. \tag{8.2}$$

From (8.1) and (8.2) one can get, respectively:

$$c = -\frac{a_1 - ya_2}{b_1 - yb_2}, \qquad c = -\frac{a_1' - ya_2' - y'a_2}{b_1 - yb_2' - y'b_2} \qquad (8.2)$$

and equating these values of c, one has:

$$(b_1 a_2 - a_1 b_2) y' + (a_1' b_2 - a_1 b_2' + b_1 a_2' - b_1' a_2) y + \\ + (a_2 b_2' - a_2' b_2) y^2 = a_1' b_1 - a_1 b_1',$$

which is of the form:

$$y' = A_0(x) + A_1(x) y + A_2(x) y^2, \tag{8.3}$$

where:

$$\left.\begin{array}{ll} A_0 = (a_1' b_1 - a_1 b_1') / D, & A_1 = (a_1 b_2' - a_1' b_2 + b_1' a_2 - b_1 a_2') / D \\ A_2 = (a_2' b_2 - a_2 b_2') / D, & D = b_1 a_2 - a_1 b_2 \neq 0. \end{array}\right\} \tag{8.4}$$

The DE (8.3) is the «Riccati DE of order first», and the primitive (8.1) is its general solution, which contains the arbitrary constant c rationally linearly. We remark that the transformation:

$$y = \frac{u'}{A_3 u}, \tag{8.5}$$

applied to (8.3), leads to a linear DE of order two in u, when the DE (8.3) can be regarded as a solved DE.

As a simple example of the above is the DE: $y' = -2xy + xy^2$, which is of the form (8. 3), and which has as a general solution the function: $y = 2/(1 + ce^{x^2})$ of the form (8. 1).

II. 2. A natural generalization of (8. 1) is the function:

$$y = \frac{c_1 v_1 + \ldots + c_n v_n}{c_1 w_1 + \ldots + c_n w_2} \tag{8. 6}$$

where c_i are the arbitrary constants, and v_i and w_i arbitrary functions of x. This generalization was introduced by E. Vessiot (1895) and G. Wallenberg (1899) **[1]**. The elimination of c_i gives a nonlinear DE of order n, called a «Riccati DE of order n», of which (8. 6) is its general solution.

By a proper transformation, the solution (8. 6) of the Riccati DE of order n can be expressed in terms of the solutions of a linear DE of order $(n + 1)$, which corresponds to this Riccati DE.

The function (8. 6) in case all of w_i are zero, except one of them, say $w_n \neq 0$, becomes:

$$y = \bar{c}_1 \frac{v_1}{w_n} + \ldots + \bar{c}_{n-1} \frac{v_{n-1}}{w_n} + \frac{v_n}{w_n}, \tag{8. 7}$$

which contains the arbitrary constants $\bar{c}_i = (c_i/c_n)$, $i = 1, \ldots, (n-1)$ linearly, and it is the general solution of a linear DE of order $(n - 1)$.

The polynomial DE:

$$y' = A_0(x) + A_1(x)y + \ldots + A_n(x)y^n, \tag{8. 8}$$

which is a natural generalization of the Riccati DE (8. 3), in case all of A's are constants, can be integrated exactly and its general solution contains the arbitrary constant linearly.

III. THE GENERAL CASE OF THE GENERAL SOLUTIONS

In the general case of nonlinear DE the arbitrary constants enter into their general solutions in a «complicated nonlinear way», which characterizes the nature of the general solutions and their corresponding DE.

The singularities appearing in the general solutions make these solutions, as well as the manner in which the arbitrary constants enter them, very complicated.

An investigation on this line of problems is of theoretical and practical interest.

The writer has arrived at some results on these problems, but he considers these results as not yet sufficiently decisive to be communicated.

ΠΕΡΙΛΗΨΙΣ

Εἰς τὴν παροῦσαν ἐργασίαν ἐρευνᾶται ὁ τρόπος μὲ τὸν ὁποῖον αἱ αὐθαίρετοι σταθεραὶ εἰσέρχονται εἰς τὰς γενικὰς λύσεις τῶν μὴ γραμμικῶν διαφορικῶν ἐξισώσεων.

Αἱ μὴ γραμμικαὶ ΔΕ δύνανται νὰ ὑπαχθοῦν εἰς τρεῖς κατηγορίας.

Ἡ πρώτη ἀποτελεῖται ἀπὸ τὰς κλάσεις τῶν ΔΕ τῶν ὁποίων αἱ γενικαὶ λύσεις περιέχουν τὰς αὐθαιρέτους σταθερὰς «γραμμικῶς».

Ἡ δευτέρα περιέχει τὰς ΔΕ εἰς τῶν ὁποίων τὰς γενικὰς λύσεις αἱ αὐθαίρετοι σταθεραὶ ὑπεισέρχονται «ρητῶς - γραμμικῶς».

Ἡ τρίτη περιέχει ὅλας τὰς ἄλλας ΔΕ εἰς τῶν ὁποίων τὰς γενικὰς λύσεις αἱ αὐθαίρετοι σταθεραὶ ὑπεισέρχονται κατὰ «πολύπλοκον μὴ γραμμικὸν τρόπον».

Ἡ διερεύνησις τῶν συνθηκῶν, ὑπὸ τὰς ὁποίας ἡ «ἀρχὴ τῆς ἐπιπροσθέσεως» (principle of superposition) τῶν λύσεων τῶν μὴ γραμμικῶν ΔΕ δύναται νὰ ἐφαρμοσθῆ, ἕνα σημαντικὸν πρόβλημα, ὑπάγεται εἰς τὴν πρώτην κατηγορίαν.

Ἡ δευτέρα κατηγορία τῶν ΔΕ ἀναφέρεται εἰς τὰς γενικὰς λύσεις τῶν ἐξισώσεων τοῦ Riccati οἱασδήποτε τάξεως.

REFERENCES

1. H. Davis, «Introduction to the nonlinear differential and integral equations», Dover Publications, Inc. New York (1962), pp. 76.
2. E. Halle, «Lectures on ordinary differential equations», Addison - Wesley Publications Co, Reading, Mass, U.S.A., (1969), pp. 358 - 370.

ΠΡΑΚΤΙΚΑ ΤΗΣ ΑΚΑΔΗΜΙΑΣ ΑΘΗΝΩΝ

ΣΥΝΕΔΡΙΑ ΤΗΣ 26ΗΣ ΝΟΕΜΒΡΙΟΥ 1970

ΠΡΟΕΔΡΙΑ ΛΕΩΝ. Θ. ΖΕΡΒΑ

ΑΝΑΚΟΙΝΩΣΙΣ ΜΗ ΜΕΛΟΥΣ

ΜΑΘΗΜΑΤΙΚΑ.— **Actual Mathematical Solutions of Problems Posed by Reality, I. (A classical Procedure)** *, *by D. G. Magiros* **.
Ἀνεκοινώθη ὑπὸ τοῦ Ἀκαδημαϊκοῦ κ. Ἰω. Ξανθάκη.

INTRODUCTION

We discuss here phenomena or situations posed by reality which are changes of variable quantities that influence and interact to each other in an organized behavior. We call such a situation a «real system». A real system, either physical or social, corresponds to a «physical or social problem», mathematically expressed.

To describe quantitatively and explain the entire spectrum of the functional behavior of a real system, a «theory» of the system is needed, that is a set of statements concerning the behavior of the objects of the system. This theory, in general, implies a mathematical expression of the relationships between certain quantities of the system, that is a «mathematical model» of the system, associated with the «data» of the system.

The main requirement that the solution of a model is expected to satisfy is: «to interpret the real system in an adequate way». Such a

* Δ. Γ. ΜΑΓΕΙΡΟΥ, **Δεκταὶ μαθηματικαὶ λύσεις φυσικῶν προβλημάτων, I.**

** Consulting scientist, General Electric Company (RESD), Philadelphia, Pa., U.S.A.

Reprinted from the *Proceedings of the Athens Academy of Sciences* **45** (1970), 179–187.

solution of the model we call an «actual solution» of the system. There are non-actual solutions, called «formal solutions», which only satisfy the equations of the model and the data.

The above requirement would be fulfilled by the actual solution if this solution satisfies certain mathematical restrictions, called «Hadamard's restrictions». If this is so, the correspondent model must have an appropriate mathematical structure. This structure of the model shows the existence of an actual solution, although this solution is unknown.

Models with actual solutions are called: «well-posed», and with formal solutions: «non-well-posed».

In this paper we discuss a procedure by which «well-posed-models» may be constructed, and «actual solutions» be determined. Three important phases in the procedure can be distinguished: (a) the formulation of a theory concerning the system, which will lead to the construction of a model of the system; (b) the selection of a «well-posed-model», by applying the «Hadamard's restrictions»; and (c) the application of mathematical methods to find the solution of the «well-posed-model». The discussion, interpretation and evaluation of the results will complete the cycle of the research.

THE CLASSICAL PROCEDURE

1. Theory and models of real systems. (3), (4), (9).

The construction of a mathematical model is, in many systems, the most important step of the study of the systems. It presupposes a «theory» of the system, that is, a set of statements some of which are verified by the experimental observations, while some others may be postulated. The objects and the formulation of the statements are the two constituents of the theory.

For a real system, the investigator needs a detailed knowledge of the observational and experimental facts, the pertinent laws, a penetrating insight, a mature judgment.

The data of the system, that is the results of observation, experiments and measurements related to the system, must be completely stated and known approximately within an accepted error, thus called «admissible data».

The domain on which the system operates must be known. This is related to the selection of the «major» variables, among the variables of the system, which define the process of the motion of the system, that is the main subject of the theory. The other variables of the system, the «minor» variables, are either «parameters» of the system, thus defining the «environment» of the system, or a «noise». A good theory of the system depends on the appropriate selection of the major and minor variables of the system.

All the above considerations, and many others of special interest, constitute the theory of a real system, and a satisfactory theory of the system makes the system «correctly stated».

Correctly stated systems can be represented in a mathematical form, called «model of the system», which gives the functional behavior of the system in a quantitative way.

The «strong interactions» between the variables of the system can be expressed explicitly in a mathematical form and give the basis of a completely deterministic (non-statistical) model. The «weak interactions» are not represented explicitly in the equations, but they are an essential part of the system related to statistical mechanics.

Arbitrary assumptions, decisions and choices in developing a theory and the model of the system, have as a result the construction of various models to the same real system.

The model summarizes the data of the system, and if one repeats an experiment or gets numbers, by using the model, and these results agree with the data assumed in constructing the model, the model is acceptable.

The applicability of a model depends on the possibility of estimating their parameters from the data.

Models, parts of which cannot explain given experimental facts of the system, are not acceptable. But if these parts of the model are cancelled, then every feature of it is related in some way to the experimental data and the model becomes acceptable.

Usually, the models are «boundary and/or initial value problems».

2. Hadamard's restrictions, well-posed-models.

Among the possible models of a real system correctly stated, one can distinguish the «well-posed» ones of which the solutions are «actual solution» of the system (**7**).

As we know, if the equations of the system and its data are such that:

a. the model has a solution corresponding to the data,

b. the solution is unique, and

c. the solution depends continuously on the data,

we say that the solution satisfies these three «Hadamard's restrictions», and that the equations of the model and the data give a «well-posed» or «reasonable» problem, and we have a definition of «correctness» of the problem in the sense of Hadamard (**2a**).

If the solution fails to satisfy even one of the «Hadamard's restrictions», it is a formal solution, and the problem a «non-well-posed» or «non-reasonable», or «improperly posed» problem.

Much attention has been given to such problems in recent years (**6**), (**8**).

Appropriate continuity and differentiability properties of the mathematical expression of the «well-posed-problem», by means of known existence, uniqueness and continuity theorems, assure the existence of a solution which satisfies the Hadamard's restrictions.

It remains now to see that the solutions which satisfy the Hadamard's restrictions are actual solutions of the real problem.

We remark that (**3**), (**5**):

First: The well-posed-problem must necessarily possess a solution.

The existence of a solution and its determination are different concepts, and we try to determine a solution only if we know it exists.

Second: The solution must be just one.

The existence and uniqueness properties of the solution express our belief «in causality» or «in determinism», a principle according to which one can repeat experiments with the expectation to get consistent results.

Third: The continuous dependence of the solution on the data has as a result that small changes in the data imply small changes in the solution.

The data, as results of observation, experiments and measurements, are given with small errors, when the solution has an uncertainty and its estimation an error (**2b**). If the above continuity property of the solution holds, then «the smaller the error in the data, the smaller the error in the estimation of the solution», and «admissible solutions» correspond to «admissible data» (**1**). The above continuity property holds for finite time. If it holds for any time, this property becomes a «stability property» of the solution in the sense of Liapunov in a «parameter-space», which is here the «data-space» (**5**).

The above remarks make clear that solutions which satisfy the Hadamard's restrictions are actual solutions of real systems.

3. The determination of the actual solution.

Having now found the «well-posed-problem» corresponding to the real system, we try, next, to determine its solution. Heuristic scientific reasoning toward the ultimate solution can be frequently used. The «principle of approximation» can be introduced to achieve the solution needed. The continuity property of the Hadamard's restrictions helps to apply this principle.

To the «well-posed-problem» M we try to find an appropriate «approximate problem» M_n, of which the solution A_n, containing the index n, will be determined. The limit A of the solution A_n as $n \to \infty$, $A = \lim_{n \to \infty} A_n$, is the solution of the well-posed-problem, that is the actual solution of the real system (**2a**).

4. Summary and Conclusion.

Summarizing the preceding we see that the research for finding

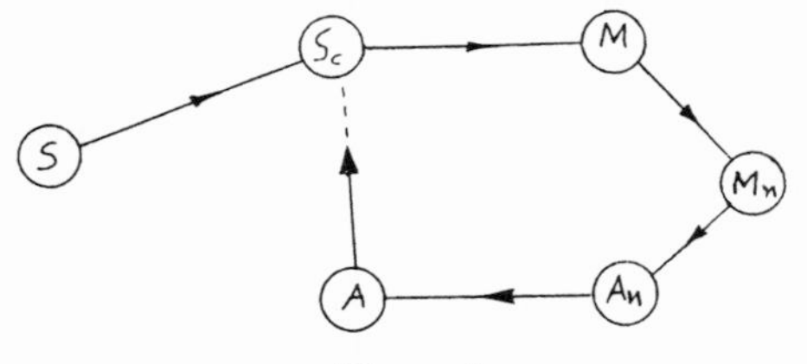

Figure 1.

S : real system
S_c : system correctly stated
M : well-posed-model
M_n : approximate problem to M
A_n : solution of M_n
$A = \lim_{n \to \infty} A_n$: solution of M

actual solutions of real systems is constituted by a set of steps, which makes a complete cycle, shown schematically in Figure 1.

(i) Step : $S \rightarrow S_c$; We need the theory of the system, which will make the system correctly stated and the modeling possible.
(ii) Step : $S_c \rightarrow M$; We use the Hadamard's restrictions, which will help to get the well-posed-model of the system.
(iii) Step : $M \rightarrow M_n$; We select the appropriate approximation to « w - p - m ».
(iv) Step : $M_n \rightarrow A_n$; We find the solution A_n the problem M_n.
(v) Step : $A_n \rightarrow A$; We find the solution required of the system, by using the limiting process, as $n \rightarrow \infty$.

The step (i) is a decisive step. It requires ingenuity, experience and knowledge of the field to which the real system belongs;

The step (ii) helps to get the reasonable model on which the whole investigation is based;

The steps (iii) and (iv) need a constructive imagination and a complete knowledge of mathematics to be applied;

The step (v) is an application of the principle of approximation, which can be succeeded by applying the continuity property of the Hadamard's restrictions.

5. Remarks on the classical procedure.

The classical procedure for finding actual solutions of real systems shows a general orientation of thought on the problem. It is based on the Hadamard's definition of «correctness», of «well-poseness».

It was found that the Hadamard's definition of «correctness» rules out as «non-well-posed-problems» important real problems, as, e.g., problems of geophysics, and some authors at the present time present different notions of «correctness» and various approaches to the formulation and investigation of the «non-well-posed-problems» (**6**).

There are difficulties in the application of the classical procedure in some scientific fields. Problems of modern physics especially present these difficulties (**2b**).

The exact laws governing the behavior of the system under consideration may not be known, and, in this case, a complete theoretical description of the system is impossible.

Problems of fundamental particle structure and reaction present the difficulties in an acute form.

More often the situation is analogous to that of problems of atomic and molecular structure, where the interactions are known, but where the structure in a particular problem may be complicated.

In problems of classical dynamics of compressible fluids, the differential equations supplemented by boundary conditions are not always a sufficiently complete framework for an adequate description of physical reality.

The above remarks must be considered, when one tries to present a mathematical description of real systems.

Selected applications from different scientific fields should show the importance of the above procedure and remarks.

REFERENCES

1. Brillouin, L.: «Scientific Uncertainty and Information», Academic Press, New York, 1964.
2. Courant, R.: (a) «Methods of Applied Mathematics», in the book: «Recent Advances in Science»; Editors: M. H. Shammos and G. M. Murphy, New York University Press, New York, 1956.
   (b) «Boundary Value Problems in Modern Fluid Dynamics», Proc., Intern. Congress of Mathematicians, Harvard University, Volume II, 1956.
3. «Education in Applied Mathematics»: Proc. of Conference, May 24-27, 1966. Siam Review, Volume 9, No. 2, April 1967.
4. Goodwin, B. C.: «Temporal Organization in Cells», Academic Press. New York, 1963.
5. Hale, J. K.: «Ordinary Differential Equations», Wiley-Interscience, New York, 1963.
6. Lavrentiev, M.: «Some Improperly Posed Problems of Mathematical Physics», Springer-Verlag, New York, 1967.
7. Magiros, D. G.: «Physical problems discussed mathematically», Bulletin, Greek Mathematical Society, New Series, Volume 6, II, No. 1, 143-156, 1965.
8. Miranker, W.: «A Well-Posed-Problem for a Backward Heat Equation», Proc., American Mathem. Society, April, 1961.
9. Mesarovič, M. D., Editor: «System Theory and Biology», Proc., III Systems Symposium, Springer-Verlag, New York, 1968.

ΠΕΡΙΛΗΨΙΣ

Εἰς τὴν παροῦσαν ἐργασίαν περιγράφεται μέθοδος ἐρεύνης ἡ ὁποία ἀφορᾷ εἰς τὴν εὕρεσιν «φυσικῶς δεκτῶν» μαθηματικῶν λύσεων εἰς φυσικὰ καὶ κοινωνικὰ προβλήματα. Αἱ κύριαι φάσεις τῆς μεθόδου εἶναι : (α) Ἡ ἔρευνα ὅπως καθιερωθῇ μία «θεωρία τοῦ συστήματος», ἡ ὁποία νὰ καθιστᾷ τὸ σύστημα ἀφ' ἑνὸς μὲν καλῶς ἐκπεφρασμένον, ἀφ' ἑτέρου δὲ νὰ ὑποβοηθῇ εἰς τὴν εὕρεσιν μαθηματικῶν προτύπων τοῦ συστήματος· (β) Ἡ ἐκλογὴ ἑνὸς «καλῶς τεθειμένου προτύπου» τοῦ συστήματος, τὸ ὁποῖον νὰ ὁδηγῇ εἰς ἓν «καλῶς τεθειμένον μαθηματικὸν πρόβλημα», τοῦ ὁποίου νὰ προσδιορισθῇ ἡ λύσις· καὶ (γ) Ἡ εὕρεσις τῆς λύσεως τοῦ ἀρχικοῦ συστήματος, βάσει τῆς εὑρεθείσης λύσεως τοῦ καλῶς τεθέντος μαθηματικοῦ προβλήματος.

Ὑπάρχουν πεδία ἐρεύνης, ὅπου ἡ μέθοδος εἶναι ἢ δύσκολον ἢ ἀδύνατον νὰ ἐφαρμοσθῇ, ἀλλ' ὅμως ὅπου αὕτη δύναται νὰ ἐφαρμοσθῇ τὰ ἀποτελέσματα δύνανται νὰ ἑρμηνεύσουν τὴν φυσικὴν πραγματικότητα κατὰ πολὺ ἱκανοποιητικὸν τρόπον.

Εἰς ἑπομένην ἐργασίαν θὰ δοθοῦν ἐφαρμογαὶ — ἀπὸ διάφορα πεδία ἐρεύνης — διὰ τῶν ὁποίων θὰ διασαφηνίζεται ἡ πορεία τῆς μεθόδου καὶ θὰ καταδεικνύεται ἡ σημασία της.

*

Ὁ Ἀκαδημαϊκὸς κ. **Ἰω. Ξανθάκης** κατὰ τὴν ἀνακοίνωσιν τῆς ἐργασίας τοῦ κ. Δ. Μαγείρου εἶπε τὰ ἑξῆς :

Ἔχω τὴν τιμὴν νὰ παρουσιάσω εἰς τὴν Ἀκαδημίαν ἐργασίαν τοῦ κ. Δημ. Μαγείρου ὑπὸ τὸν τίτλον : «Δεκταὶ Μαθηματικαὶ λύσεις φυσικῶν προβλημάτων». Εἰς τὴν ἐργασίαν ταύτην ὁ συγγραφεὺς ἐκθέτει μίαν μέθοδον ἐρεύνης ἀφορῶσαν εἰς τὴν εὕρεσιν μαθηματικῶν λύσεων φυσικῶν προβλημάτων ἀποδεκτῶν ἀπὸ φυσικῆς ἀπόψεως. Αἱ κύριαι φάσεις τῆς μεθόδου εἶναι αἱ ἑξῆς :

α) Ἡ διατύπωσις μιᾶς «θεωρίας τοῦ συστήματος» ἡ ὁποία νὰ καθιστᾷ ἀφ' ἑνὸς μὲν καλῶς ἐκπεφρασμένον τὸ σύστημα, ἀφ' ἑτέρου δὲ νὰ ὑποβοηθῇ εἰς τὴν εὕρεσιν μαθηματικῶν προτύπων τοῦ συστήματος.

β) Ἡ ἐκλογὴ ἑνὸς καταλλήλου προτύπου τοῦ συστήματος, τὸ ὁποῖον νὰ ὁδηγῇ εἰς ἓν καλῶς τεθειμένον μαθηματικὸν πρόβλημα καὶ

γ) Ἡ εὕρεσις τῆς λύσεως τοῦ ἀρχικοῦ συστήματος βάσει τῆς εὑρεθείσης λύσεως τοῦ καλῶς τεθέντος μαθηματικοῦ προβλήματος.

Ὁ συγγραφεὺς ἀναφέρει ὅτι ὑπάρχουν πεδία ἐρεύνης, ὅπου ἡ μέθοδος αὕτη εἶναι εἴτε δύσκολον εἴτε ἀδύνατον νὰ ἐφαρμοσθῇ. Ὅπου ὅμως αὕτη δύναται νὰ ἐφαρμοσθῇ τὰ ἀποτελέσματα, κατὰ τὸν συγγραφέα, δύνανται νὰ ἑρμηνεύσουν τὴν φυσικὴν πραγματικότητα κατὰ τρόπον λίαν ἱκανοποιητικόν.

Τέλος ὁ κ. Δ. Μάγειρος προτίθεται εἰς προσεχῆ ἀνακοίνωσίν του νὰ παρουσιάσῃ ἐφαρμογὰς ἐπὶ διαφόρων πεδίων ἐρεύνης διὰ τῶν ὁποίων θὰ διαγράφεται ἡ πορεία τῆς μεθόδου καὶ θὰ καταδεικνύεται ἡ σημασία της.

ΜΑΘΗΜΑΤΙΚΑ.— **Actual Mathematical Solutions of Problems Posed by Reality, II. (Applications)***, *by D. G. Magiros***. Ἀνεκοινώθη ὑπὸ τοῦ Ἀκαδημαϊκοῦ κ. Ἰω. Ξανθάκη.

INTRODUCTION

In the previous paper **3(a)**, we discussed a classical procedure for finding actual mathematical solutions of real systems in many physical or social fields. The main phases of the procedure were:

A. The creation of a theory of the system, which helps its modeling;
B. The selection of a «well-posed-model» of the system, which gives a well-posed mathematical problem, and
C. The construction of the solution of this problem, which is the «actual solution» of the system.

In the present paper we give some applications of the above classical method, by which we can see the difficulties of its application and its advantages in case this method can be applied. We select the applications from thermodynamics, astrodynamics, non-linear mechanics, biology, etc.

1st Application **Problem of Thermodynamics: (4)**

Forward and Backward Heat Flow Problem.

The Step: «$S_c \rightarrow M$» of the classical method characterizes the whole study of the problem. The problem is: *«To study the heat flow in a given medium»*. To make this physical problem correctly stated, one accepts for the medium to be homogenious and isotropic with respect to the heat flow, and that the heat flow is towards the decreasing temperature. Based on these hypotheses, the mathematical idealization, the model, is the partial differential equation:

---

* Δ. Γ. ΜΑΓΕΙΡΟΥ, **Δεκταὶ μαθηματικαὶ λύσεις φυσικῶν προβλημάτων, II. (Ἐφαρμογαί)**.

** Consulting scientist, General Electric Company (RESD), Philadelphia, Pa., U.S.A.

Reprinted from the *Proceedings of the Athens Academy of Sciences* **46** (1971), 21–31.

$$U_{xx} + U_{yy} + U_{zz} = U_t \tag{1}$$

where : $u = u(x, y, z, t)$ is the temperature in the x, y, z -space and t-time. In the equation (1) there is a coefficient depending on density, specific heat, and thermal conductivity and this coefficient is here taken equal to unity.

In case of a «one-dimensional medium», if the «data-initial condition» is :

$$u(x, 0) = n \,.\, \sin nx, \qquad n = \text{integer} \tag{2}$$

one can check that the solution of equation (1), satisfied by (2), is :

$$u(x, t) = n \,.\, e^{-n^2 t} \,.\, \sin nx \tag{3}$$

and it is unique, when the first two Hadamard's restrictions are satisfied. We distinguish here two cases :

a. It $t>0$, when one has the «forward heat problem», the solution (3) $\rightarrow 0$ and the condition (2) $\rightarrow \infty$, as $n \rightarrow \infty$, then the solution (3) satisfies also the third Hadamard's restriction, when the function (3) is accepted as an «actual solution» of the «forward heat problem», which is a «well-posed-problem».

b. If $t<0$, when one has the «backward heat problem», the solution (3) and the condition (2) $\rightarrow \infty$, as $n \rightarrow \infty$, then the solution (3) violates the third Hadamard's restriction, when the function (3) is a «formal solution» of the «backward heat problem», which is a «non-well-posed-problem».

2nd Application **Problem of Orbital Mechanics: 3(b)**

An artificial celestial body is moving under the influence of a central force obeying the inverse square Newton's law toward the attractive center. A general force is applied, acts for an interval of time, then it is removed. Find the motion of the body during the action of the general force.

A model of this problem is:

$$\left.\begin{aligned} &\ddot{\underline{r}} = -\frac{\mu}{r^3(\tau)}\,\underline{r}(\tau) + \underline{T}(\tau) \\ &\underline{r}(0) = \underline{r}_0\,, \quad \dot{\underline{r}}(0) = \dot{\underline{r}}_0 + \underline{I}_0 \\ &D_1 : |\underline{r}(\tau)| < M_1\,, \quad |\dot{\underline{r}}| < M_2 \\ &D: \; 0 \leqslant \tau \leqslant \tau' \end{aligned}\right\} \tag{4}$$

where $\underline{T}$ the general force, $\underline{r}$ the radial vector from the attractive center to the center of mass of the body, $\underline{I}_0$ the impulse, which is given by:

$$\underline{I}_0 = \int_0^{t_0} T(t)\,dt\,, \qquad \tau = t - t_0\,. \tag{4. 1}$$

If we take the function:

$$\underline{r}(\tau) = a_1(\tau)\,\underline{r}_0^* + a_2(\tau)\,\underline{s}_0^* + a_3(\tau)\,\underline{T}_0^* \tag{5}$$

as a «trial solution», where $\underline{r}_0^*$, $\underline{s}_0^*$, $\underline{T}_0^*$ are special unit vectors, the coefficients a_1, a_2, a_3 must satisfy the following conditions in order that the function (5) is a «formal solution» of (4):

$$\left.\begin{aligned} &\ddot{a}_1 + \frac{\mu}{r^3}a_1 = T_1\,; \quad a_1(0) = r_0\,, \quad \dot{a}_1(0) = 0 \\ &\ddot{a}_2 + \frac{\mu}{r^3}a_2 = T_2\,; \quad a_2(0) = 0\,, \quad \dot{a}_2(0) = s_0 \\ &\ddot{a}_3 + \frac{\mu}{r^3}a_3 = T_3\,; \quad a_3(0) = 0\,, \quad \dot{a}_3(0) = 0 \end{aligned}\right\} \tag{6}$$

If T_1, T_2, T_3 are differentiable, $\dot{T}_1$, $\dot{T}_2$, $\dot{T}_3$ continuous, $r \neq 0$; a_1, a_2, a_3 twice differentiable, and $\ddot{a}_1$, $\ddot{a}_2$, $\ddot{a}_3$ continuous, we see that equations (6) satisfy the Hadamard's restrictions, when the functions: $a_1(\tau)$, $a_2(\tau)$, $a_3(\tau)$ can be uniquely determined from equations (6), and are continuous functions of the initial conditions of (6). Therefore, the solution (5) of equation (4), after the above restrictions of the force T and its derivative, is unique and depends continuously on the initial conditions of (4), then it can be accepted as an actual solution of the equation (4).

3rd Application **Problem of Non-Linear Mechanics: 3 (c)**

The Problem of Principal Modes of Non-Linear Systems.

The concept of «principal modes» of linear systems plays a predominant role in the analysis of the oscillatory systems of many fields.

The principal modes in linear systems are, by definition, the fundamental set of solutions of which a linear combination gives the general solution of the linear system; or, physically speaking, they are the special modes of oscillations of the linear system in terms of which we can discuss any kind of oscillations of the system.

Since the «principle of superposition» does not hold in non-linear systems, the concept of principal modes, as given above, is meaningless in non-linear systems, and the following problem may arise: *«Has the problem of principal modes of non-linear systems a physical meaning?»*; or *«How one can make the problem of principal modes of non-linear systems a well-posed problem?»*

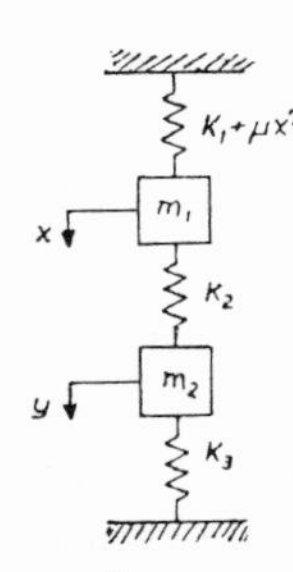

Figure 1.

The writer has publisched some papers in connection with this important problem, and transfers here some appropriate thoughts, techniques and results in order to give this problem as an example of the classical approach of the preceding paper.

We can find a new definition of the concept of principal modes for both the linear and non-linear systems, and such that the known definition in linear systems comes as a result from the new definition. The writer gave two new definitions which, under some conditions, are equivalent.

After that we try to make the physical problem correctly stated and the mathematical idealization well-posed.

We take a trial solution and make it formal, first, and then actual.

If we restrict ourselves to a «two-degrees-of-freedom» mechanical non-linear system, as shown in Figure 1, the equations of motion of the «two-masses-three springs» non-linear system are:

$$\left.\begin{aligned} \ddot{x} + \omega_1^2 x - \lambda_2 y + \lambda_1 x^3 = 0 \\ \ddot{y} + \omega_2^2 y - \lambda_3 x = 0 \end{aligned}\right\} \qquad (7)$$

where :

$$\omega_1^2 = \frac{K_1 + K_2}{m_1}, \quad \omega_2^2 = \frac{K_2 + K_3}{m_2}, \quad \lambda_1 = \frac{\mu}{m_1}, \quad \lambda_2 = \frac{K_2}{m_1}, \quad \lambda_3 = \frac{K_2}{m_2} \tag{7a}$$

and μ characterizes the non-linearity of one anchor spring.

By using the transformation :

$$x = x_1, \quad \dot{x} = x_2, \quad y = x_3, \quad y = x_4 \tag{8}$$

the system (7) can be reduced to its normal form :

$$\left.\begin{array}{l} x_i = f_i\,(x_1, x_2, x_3, x_4), \quad i = 1, 2, 3, 4 \\ f_1 = x_2, \quad f_2 = -\,\omega_1^2 x_1 + \lambda_2 x_3 - \lambda_1 x_1^3, \quad f_3 = x_4, \quad f_4 = \lambda_3 x_1 - \omega_2^2 x_3 \end{array}\right\} \tag{9}$$

valid in a region R :

$$R: \ |x_i| < h, \quad i = 1, 2, 3, 4 \tag{9a}$$

The appropriate initial conditions for «principal modes» are in R :

$$x_1(0) = x_{10}, \quad x_2(0) = 0, \quad x_3(0) = x_{30}, \quad x_4(0) = 0 \tag{9b}$$

where x_{10} and x_{30} are appropriately related to each other.

Now we remark that the nature of the functions f_i of (9) are such that all Hadamard's restrictions are satisfied. These functions f_i are continuous in R, then bounded; they have continuous partial derivatives $\partial f_i / \partial x_k$ in R, when they satisfy Lipschitz conditions with respect to x_i in R for a Lipschitz constant $s = \mathrm{l.u.b}\, /\partial f_i / \partial x_k/$. The above properties assure the unique existence of the solution of (9) and (9b) in a region $R' c R$. As the initial point x_{10}, $i = 1, 2, 3, 4$ varies in R', the solution satisfies the three Hadamard's restrictions, and the problem is «well-posed».

4th Application **A Problem of Underwater Warfare:**

The Problem of Domes. 1

The problem of domes arose in the winter of 1942 - 1943 in connection with «underwater warfare». As is known, underwater sound ranging depends on sending out a sound beam in water and, attached to a fast-moving ship, the water steaming around the plate causes serious disturbances. For elimination of these disturbances, the projector is closed in a so-called «dome», Figure 2, which is a convex shell of metal or other material filled with water. Such domes interfere only slightly with the formation of a concentrated sound beam. During 1942 - 1943, a large number of small submarines chases were built and equipped with sound gear similar to, but smaller than, the gear used before. While the manufacture of domes to fit this smaller gear was underway, it was discovered that these smaller domes led to an intolerable diffusion on the sound beam. At that time, a quick remedy was imperative, and a mathematical analysis of the problem was needed to support and speed-up experimented work.

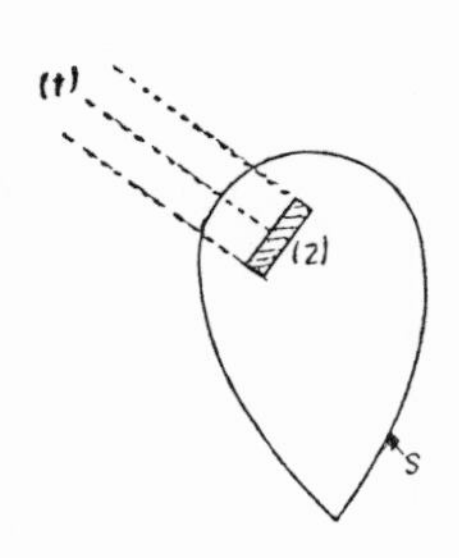

Figure 2.
(1) Axis of beam sound
(2) Projector
S Surface of Dome

The mathematical problem, related to the above physical problem, was to solve the differential equation:

$$\left.\begin{aligned} &\nabla^2 P + K^2 P = 0 \\ &\nabla^2 = \frac{\partial^2}{\partial x^2} + \frac{\partial^2}{\partial y^2} + \frac{\partial^2}{\partial z^2} \end{aligned}\right\} \qquad (10)$$

in which $K = \omega/c$, ω the frequency, c the sound velocity and K has for our problem, unfortunately, different values within the shell of the dome and outside.

This mathematical idealization was not a suitable one for the problem.

They found the suitable mathematical idealization by the following process. The actual dome of small finite thickness was replaced by an extremely thin surface, then the influence of the dome was simply replaced by conditions for jump discontinuities of the disturbance q of the beam across the surface.

These conditions are:

$$\left.\begin{aligned} [q] &= \frac{p_1}{p_0^{-1}} \cdot \frac{\partial p}{\partial n} \\ \left[\frac{\partial q}{\partial n}\right] &= \frac{p_0}{p}(K_0^2 - K_1^2)\, p - \left(1 - \frac{p_0}{p}\right)\left(\frac{\partial^2 p}{\partial n^2} + 2H\frac{\partial p}{\partial n}\right) \end{aligned}\right\} \qquad (11)$$

where the symbol $[\cdot]$ means jump of the quantity of the symbol across the surface, q is the disturbance of the acoustic pressure p caused by the dome and the normal derivatives $\frac{\partial}{\partial n}$ are to be evaluated on the surface S of the dome. The quantity H is the mean curvature of S, i. e. the average of the curvature of any two normal plane sections at right angles to each other. In addition to conditions (11) to be satisfied by q on S, q should be a solution of the equation :

$$\nabla^2 q + K_0^2 q = 0 \tag{12}$$

same behavior as P at ∞. This problem possess the unique solution :

$$q = -\frac{1}{4\pi}\iint_S \left[\frac{\partial q}{\partial n}\right] \frac{e^{ik_0 r'}}{r'} ds + \frac{1}{4\pi}\iint_S [q] \frac{\partial}{\partial n}\left(\frac{e^{ik_0 r'}}{r'}\right) dS \tag{13}$$

The quantities in the brackets are given by conditions (11), r' is the distance from a fixed point (x, y, z) at which q (x, y, z) is to be determined to the point of integration on S. This formula yields the disturbance as the effect due to a layer of point sources and a layer of dipoles disturbed on S with intensities which are known as soon as the original pressure p is known, since the quantitites in brackets are fixed in value due to conditions (11). The relative directional disturbance :

$$\left|\frac{p_1}{p}\right| R.c.h\left(\frac{q_1}{p_1} - \frac{q_0}{p_0}\right)$$

would, finally, be obtained from (13). The solution (13) is valid for a shell of constant thickness, but it could be extended without essential error to cases in which the dome shell is made up of a not too large number of pieces, each of which is of constant thickness. All that would be necessary would be to insert a numerical factor d in the integrands on the right-hand side of expression (13), which would be precise constant on S. By this formula, one can analyze the contribution to the distortion of various factors, such as the curvature of the dome and the density and sould velocity within it.

The above kind of mathematical idealization, even without detailed numerical computation, proved helpful to the designing engineer.

5th Application **Biology, Ecology, Economics : 2, 5**

The Problem of Mixed Populations: Two Species Competing for a Common Food Supply.

For the study of the growth of two mixed populations of species in mutual interdependence of any kind, e. g. in competing for a common food supply, several models have been proposed. One of these models, of which the formulation is based on determinizing the time-rate of change of quantities as a function of the quantities and some parameters, is :

$$\left.\begin{aligned} \dot{x} &= a\,[b - x - f_1(y)]\,x \\ \dot{y} &= c\,[d - y - f_2(x)]\,y \end{aligned}\right\} \tag{14}$$

x and y are the numbers (or masses) of individuals of the species present at any time, and a, b, c, d parameters of which the domain of possible change define the environment of the model.

The model (14) is either «well-posed» or «non-well-posed», depending on properties of the functions $f_1(y)$ and $f_2(x)$.

Physically, the quantities x and y are non-negative, when the region D of the validity of the model (14) is the first quadrant of the x, y - plane. The initial conditions x_0, y_0 of (14) is the starting point of the solution, if this solution exists, and this point lies in the region D.

If the functions $f_1(y)$ and $f_2(x)$ are defined, single-valued and continuous in the region D, then the right-hand numbers of (14) are continuous functions of all their arguments, when a solution of (14) necessarily exists through the point (x_0, y_0), and the first Hadamard's restriction is satisfied.

If, in addition, the functions $f_1(y)$ and $f_2(x)$ have continuous derivatives in y and x, respectively, then the right-hand numbers of (14) have continuous partial derivatives with respect to all their arguments, and the solution through (x_0, y_0) is unique and depends continuously on the x_0, y_0, when the second and third Hadamard's restrictions are satisfied, and the model (14) is a «well-posed» one.

We remark that : the solution of the model (14), which starts from the inital point (x_0, y_0), tends, as t increases, to a point $\bar{x}$, $\bar{y}$, and we may have three cases. First, the point $(\bar{x}, \bar{y})$ may be a point inside the region D, both $\bar{x}$ and $\bar{y}$ positive, when one can speak about the «co-existence of the species». Second, the point $(\bar{x}, \bar{y})$ may be identical with the origin, when one can speak about the «extinction of the species». Third, one of $\bar{x}$ and $\bar{y}$ may be zero and the other positive, and this case corresponds to the «principle of competitive exclusion», a principle much used in ecology, but which has been much criticized.

We remark that if the variables x and y of the model (14) are numbers of individuals of the populations they are restricted to be (positive) integers, when they are «step functions» of time, and the functions f_1 and f_2 of (14) are restricted to assume values according to permitted values of x and y. The functions f_1 and f_2 in this case have no properties, as mentioned above, which make the model (14) a «well-posed» one. In this case, the model (14) is not a «continuous system», but a «discrete system». If we assume that x and y in the model (14) are the masses of the populations, we can remove the above restriction of x and y and the function f_1 and f_2 regain the properties needed in order for the model to be a «well-posed» one.

All the above remarks and results can be applied to different social problems, if the competitive species and the limiting resources are appropriately specified.

To apply the above in the field of economics, the variables x and y must denote the size or extent of two commercial enterprises competing for common sources and for a common market.

6th Application Modern Physics, Dynamic Meterology : 1

The classical procedure, discussed in the preceding, and especially the step to find the «well-posed-model», combined with numerical analysis and the use of high-speed computers, gave and may give much success in the investigation of problems of great contemporary interest.

The «Synchrotron» and the «weather prediction» can be used as examples.

a. *Synchrotron.* The recently discovered «strong-focusing-principle» is the basis for the study of the multibillion-volt proton accelerators. This principle is related to the stability of solutions of ordinary linear differential equations of second order with periodic coefficients. The actual orbits, because of unavoidable imperfections of magnets and other causes, follow, approximately, linear periodic differential equations, and a modified nonlinear model is not possible. Experimental studies, under various assumptions, the use of computing and mathematical analysis, give encouragement to the designers for success.

b. *Weather Prediction.* According especially to Bjerkness, one may formulate the laws of atmospheric phenomena by models which are partial differential equations. Based on the today's data and using the Bjerkness model as a «well-posed» one, the prediction of tomorrow's weather would require qualified computer men with desk computing machines for much time.

REFERENCES

1. COURANT, R.: «Methods of Applied Mathematics» in the book: «Recent Advances of Science»; Editors: M. H. Shammos and G. M. Murphy, New York University Press, New York, 1956.
2. CUNNINGHAM, W.: «Simultaneous Non-Linear Equations of Growth», Bulletin, Mathematical Biophysics, Volume 17, 101 - 110, 1955.
3. MAGIROS, D. G.: (a) «Actual Mathematical Solutions of Problems Posed by Reality, I», Proc., Athens Academy of Sciences, Volume 45, 179-187, 1970.
(b) «The Motion of an Artificial Celestial Body under the Influence of a Newtonian Center and a General Force», Proc., XV Intern. Astronautical Congress, Warsaw, Poland, 1964.
(c) «Methods for Finding Principal Modes of Non-Linear Systems Utilizing Infinite Determinants», Journal of Mathematical Physics, 2, No. 6, 869 - 875, 1961.
4. MIRANKER, W.: «A Well-Posed Problem for a Backward Heat Equation», Proc., American Mathematical Society, April, 1961.
5. VOLTERRA, V.: «Leçons sur la Théorie Mathématique de la lutte pour la vie», Gauthier-Villars et Cie, Paris, 1931.

ΠΕΡΙΛΗΨΙΣ

Εἰς προηγουμένην ἐργασίαν **3(α)** ἀνεπτύχθη μέθοδος ἐρεύνης «φυσικῶς δεκτῶν» μαθηματικῶν λύσεων εἰς φυσικὰ καὶ κοινωνικὰ συστήματα καὶ ὑπεδείχθησαν δυσκολίαι ἐφαρμογῆς τῆς μεθόδου, ὅπως καὶ τὰ πλεονεκτήματά της. Εἰς τὴν παροῦσαν ἐργασίαν δίδονται ἐφαρμογαὶ τῆς μεθόδου εἰς διάφορα πεδία ἐρεύνης, ὡς, λ. χ., εἰς τὴν θερμοδυναμικήν, Ἀστροδυναμικήν, μή-γραμμικὴν μηχανικήν, βιολογίαν, κ.λ.π.

★

Ὁ Ἀκαδημαϊκὸς κ. **Ἰω. Ξανθάκης** κατὰ τὴν ἀνακοίνωσιν τῆς ἀνωτέρω ἐργασίας εἶπε τὰ κάτωθι :

Ἔχω τὴν τιμὴν νὰ παρουσιάσω εἰς τὴν Ἀκαδημίαν Ἀθηνῶν τὸ δεύτερον μέρος τῆς ἐργασίας τοῦ κ. Δημητρίου Μαγείρου, ὑπὸ τὸν τίτλον :

«Δεκταὶ Μαθηματικαὶ Λύσεις Φυσικῶν Προβλημάτων».

Εἰς τὴν προηγηθεῖσαν ἀνακοίνωσίν του ὁ κ. Μάγειρος παρουσίασε μίαν μέθοδον ἐρεύνης μαθηματικῶν λύσεων φυσικῶν καὶ κοινωνικῶν συστημάτων «φυσικῶς ἀποδεκτῶν». Ὑπέδειξε δὲ τὰς δυσκολίας ἐφαρμογῆς τῆς μεθόδου ταύτης, ὅπως καὶ τὰ πλεονεκτήματά της, εἰς τὰς περιπτώσεις καθ' ἃς δύναται νὰ ἐφαρμοσθῇ.

Εἰς τὴν παροῦσαν ἀνακοίνωσιν παρέχονται αἱ ἐφαρμογαὶ τῆς ἐν λόγῳ μεθόδου εἰς διάφορα πεδία ἐρεύνης, ὅπως λ. χ. εἰς τὴν θερμοδυναμικήν, εἰς τὴν ἀστροδυναμικήν, ἐπὶ τοῦ προβλήματος τῆς κινήσεως τεχνητοῦ δορυφόρου ὑπὸ τὴν ἐπίδρασιν μιᾶς κεντρικῆς δυνάμεως πληρούσης τὸν νόμον τοῦ Νεύτωνος καὶ μιᾶς ὠστικῆς τοιαύτης ἐπενεργούσης ἐπί τι χρονικὸν διάστημα, καθὼς καὶ ἐπὶ προβλημάτων μή-γραμμικῶν Μηχανικῆς καὶ Βιολογίας.

PHYSICAL PROBLEMS DISCUSSED MATHEMATICALLY

By

DEMETRIOS G. MAGIROS (in Philadelphia U.S.A.)

INTRODUCTION

Applied mathematics within the last few years has been developed into a distinguished branch of science with completely established purpose, principles, and approach. The human need created the effort of man for a penetration of mathematical methods for study of physical and human phenomena, that is to use mathematics in real life, and this gave rise to this branch of science.

The purpose of this paper is to provide a brief discussion of what the writer conciders of special interest as an introduction to Applied Mathematics.

In the first part of the report a general approach is indicated for reasonable mathematical solutions of physical problems, while in the second part examples illustrated show the approach in application.

PART A

CLASSICAL APPROACH

I. Pure and Applied Mathematics.

It is a scientific fact today that «*applied mathematics*» is distinct from «*pure mathematics*», the distinction being a matter of attitude and motivation and not a subject matter.

Pure mathematics is directed towards logical crystallization, abstraction, generalization. Applied mathematics is an interconnection of mathematical methods for interpretation of phycical phenomena.

Only great scientists were able in the past, as e.g. Newton, Maxwell, Gauss, Riemann, Poincaré, etc, to work in pure and at the same time in applied mathematics, although at their

Reprinted from the *Bulletin of the Greek Mathematical Society, New Series*, V. 6II, No. 1 (1965), 143–156.

time *«the principles of applied Mathematics»* were not completely established.

The mathematical needs during World Way II helped to distinguish applied mathematics from the other scientific pursuits of man and to get their principles clearly stated.

Applied mathematics is based on two principles; *«the principle of idealization»* and *«the principle of approximation»*. The principle of idealization is fundamental even for the formulation of the basic concepts and laws of nature The definition of *«the density of a fluid»* at a point p can be considered as an example of this principle. The total fluid mass in a sphere of small radius ε about **P**, divided by the volume of the sphere, by forcing ε to tend to zero, is, by definition, the density of the fluid at the point P. The limiting process associated to this definition is an idealization of unrealistic type, since in a very small volume the fluid molecules are irregularly distributed. Idealized concepts of the above type are necessary in physical sciences. The physical laws deal with idealized concepts and they by themselves are idealizations valid «in a limited field of application» and correct «within certain possible errors.»

II. Statement of A Physical Problem. Correctly stated Problems.

Any physical problem posed by reality presupposes a physical situation with given data, and questions to be answered by using the data.

For the physical problem one must know the pertinent physical laws, the appropriate coordinate system, especially for astronomical problems, and the data, that is the initial and/or boundary conditions.

The initial conditions prescribe the state of a system at a particular time, while the boundary conditions prescribe the physical behavior of a system at the frontier of a region independently of time. The data must be admissible, i.e. completely stated and known approximately within an accepted error.

Any incorrect assumption in the problem will be preserved as incorrect in the procedure for the solution.

If all the above requirements for the problem are satisfied, one has a *«physical problem correctly stated»*.

Only for such a physical problem one tries to get a mathematical solution.

Problems of this type are the subject matter of applied mathematics

III. Mathematical Idealization of Physical Problems Correstly Stated. Formal Solution.

A physical problem correctly stated may be capable of being idealized in different mathematical ways, and it is important to distinguish the reasonable idealization. Usually one has as a result of the idealization differential equations with initial and/or boundary conditions.

For physical problems related to motion of a system, one has «initial value problems», while for equilibrium problems one has «bounbary value problems.» In the «mixed problems» the data are initial and boundary conditions.

The differential equations of a problem are local restrictions of a function - physical quantity, local in space and local in time. It is the task of mathematicians to deduce from these local restictions a picture of the phenomenon in the large. Infinitely many functions satisfy a differential equation. The particular solution desired, which represents a particular physical phenomenon, will be selected from the manifold of all solutions, the selection depending on the data.

Incorrect mathematical formulation of a correctly stated physical problem will lead to a non - accepted mathematical answer, so that one must get a *«well formulated mathematical problem»* corresponding to the physical problem.

By purely mathematical operations in the well formulated mathematical problem certain mathematical results and a solution will be obtained, that is the *«Formal solution.»*

By having the proper mathemartical idealization A, one first finds an appropriate approximation A_n of A, and if S_n is the solution of A_n and by using the limiting process one has $S_n \to S_f$ as $n \to \infty$, then S_f is the solution of A, the formal solution.

The discussion of convergense of the formal solution in the form of series aud the region of the convergence of the solution give a deep insight to the problem.

Not only convergent solutions but some non-convergent ones are sometimes accepted, as e.g., «asymptotic expansions».

In the formal process we do not ask whether there exists a counterpart in nature, or an interpretation into the physical order of things. Such an isolation of the proccess of mathematics is called *«pure mathematics.»*

It is the endeavor which keeps in closer touch with the inherent order of the physical nature which distinguishes the field of *«applied mathematics»*.

IV. **Physical Interpretation of the Mathematical Results.**

The mathematical idealization selected for the physical problem must be *«reasonable idealization»*, and the requirements for the reasonability are: (a) *«the existence of the solution»*, (b) *«the uniqueness of the solution»*, and (c) *«the continuous dependence on the data»*.

We discuss these criteria in detail.

(a) *«the existence»*, The well formulated mathematical problem must possess a solution. We remark that the existence of a solution does not necessarily imply the possibility of the solution. Example: *«the three-body problem»*. The existence of the solution of this problem has been proved by Sundman, but the problem has not been solved yet in a form which permits one to deal with questions of stability, in spite of all the tremendous efforts of mathematicians during past centuries. The question of stability in Celestial mechanics is closely related to the form of the solution and to the divergence of the series employed as its solution.

(b) *«The uniqueness»*. The solution of the mathematical problem must be unique.

The existence of a unique solution expresses the belief of causality or determinism without which experiments could not be repeated with the expectation of consistent results.

(c) *«The continuous dependence on the data»*: The solution must be such that small changes in the data must produce small changes in the solution.

The data, as results of experiments or measurements, contain certain small errors and then one always has an uncertainty in the solution, and error in its estimation, and it is required that «the smaller the error in the data the smaller the error in the estimation of the solution».

Also, the solution S_n of the approximating problem A_n for large n must be a good approximation of the «true» solution. These requirements are satisfied if «the solution is continuous function of the data».

This postulate corresponds to the Liapunov stability definition in parameter space, where the parameter space is the *«initial conditions space»*.

The above requirements give a perfect mathematical description of the reality and are called the *«Hadamard's postulates»*.

The solution which satisfies these postulates is an *«actual solution»* of the problem, which is called *«well posed* or *properly posed* or *reasonable problem»*.

Violation of any postulate of the above makes the solution *«nonactual»*, when the problem is *«unreasonable»* or *«improperly posed»*.

In recent years much attention has been given to these problems, especially when the violated postulate is the last one.

The above *classical approach* for finding actual solutions of physical problems shows a general orientation of thought. We remark that there are difficulties in the application of this approach for a mathematical description of a phenomenon in some scientific fields. Problem of modern physics show these difficulties.

The exact physical laws governing the behavior of a system may not be known making it impossible to arrive at a complete theoretical description of the system.

Problems of fundamental particle structure and reaction present the difficulties in an acute form. More often the situation is analogous to that of problems of atomic and molecular structure, where the interactions are known, but where the structure in a particular problem may be so complicated.

In problems of classical dynamics of compressible fluids the differential equation supplemented by boundary conditions are not always a sufficiently complete framework for an adequate description of physical reality.

The above remarks must be always remembered when one tries to present a mathematical description of physical systems of this kind or to solve physical problems.

V. **Conclusion**.

The above classical approach for actual solutions of physical problems is shown schematically in Fig. 1.

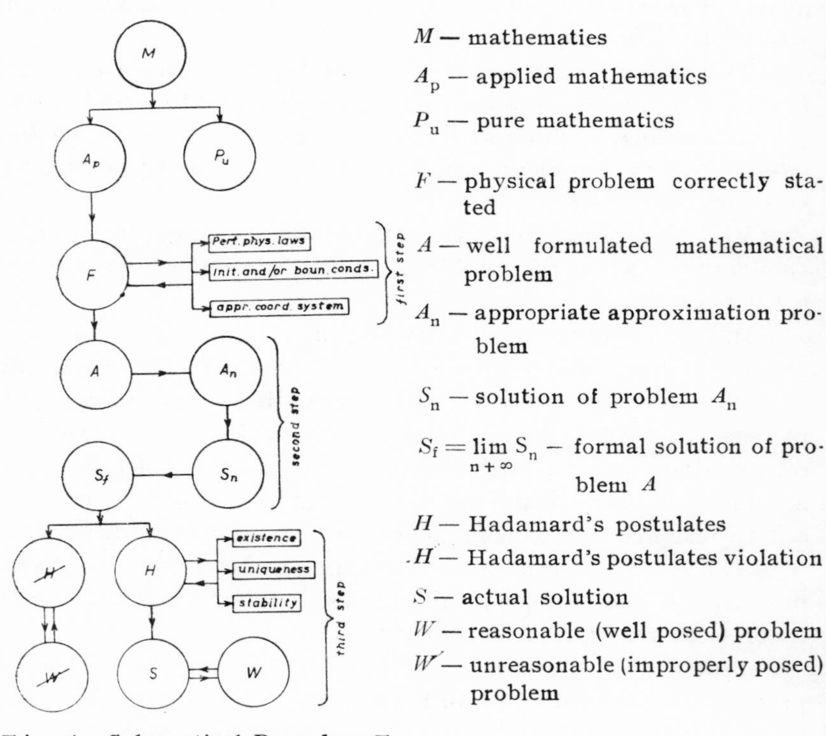

Fig. 1. Schematical Procedure For An Actual Solution Of A Physical Problem.

One may have three steps in the approach.

The *«first step»* is to get the physical problem correctly stated.

The *«second step»* is to get the well formulated mathematical problem and to determine the formal solution.

The *«third step»* to check whether the formal solution is an actual solution.

The first step is a decisive step and requires experience, intuitive understanding of the realities of physics, engineering and other fields.

The second step leaves open many possibilities for mathematical success.

The third step characterizes the procedure of applied mathematics.

PART B

ILLUSTRATED EXAMPLES

The classical approach of the preceding section for finding reasonable mathematical solutions of physical problems is successfully applied to many fields of science and technology. We give here some examples.

Example 1. *Forward and backward heat flow problem* Ref. (5).

This classical problem of Thermodynamics clarifies especially the third step of the approach.

The problem is *«to study the heat flow in a given medium»*.

To make this physical problem correctly stated, one accepts for the medium to be «homogenious» and «isotropic» with respect to the heat flow, and that the heat flow is towards the decreasing temperature.

Based on these hypotheses, the mathematical idealization is the partial differential equation:

$$u_{xx} + u_{yy} + u_{zz} = u_t \tag{1}$$

where $u = u(x, y, z, t)$ is the temperature in the x, y, z—space and t—time.

In the equation (1) there is a coefficient, taken equal to unity, depending on density, specific heat thermal conductivity of the medium.

If, for one—dimensional medium, one takes an initial condition

$$u(x, 0) = x \sin nx, \quad n = \text{integer} \tag{2}$$

one can check that the system (1) and (2) accepts as a unique solution the function:

$$u(x, t) = n \cdot e^{-n^2 t} \cdot \sin nx \tag{3}$$

then the first two Hadamard's postulates are satisfied by this function.

a) For $t>0$, when one has the «*forward heat problem*», the solution (3) satisfies also the third Hadamard's postulates, and the function (3) in this case is accepted as an «*actual solution*» of the problem, which is a «*reasonable problem*».

b) For $t<0$, when one has the «*backward heat problem*», the solution (3) violates the third Hadamard's postulate, then the function (3) in this case is a «*formal solution*», and not an «*actual solution*», of the Backward heat problem, which is «*unreasonable problem*».

Example 2. *Principal modes of nonlinear systems* Ref. (4a).

The concept of «*principal modes*» of linear systems plays a predominant role in the analysis of the oscillatory systems of many fields.

The principal modes in linear systems are, by definition, the fundamental set of solutions of which a linear combination gives the general solution of the linear system; or, physically speaking, they are that special modes of oscillation of the linear system in terms of which we can discuss any kind of oscillation of the system.

Since the «*principle of superposition*» does not hold in nonlinear systems, the concept of principal modes, as given above, is meaningless in nonlinear systems, and the following problem may arise: «*Has the problem of principal modes of nonlinear systems a physical meaning?*»; or «*How one can make the problem of principal modes of nonlinear systems a reasonable problem?*».

The writer has published some papers in connection with this important problem, and transfers here some appropriate thoughts, techniques and results example of the classical approach of the previous section.

We can find a new definition of the concept of principal modes for both the linear and nonlinear systems, and such that the known definition in linear systems is a property resulting from the new definition. The writer gave two new definitions which, under some conditions, are equivalent.

Then we can make the physical problem correctly stated and the mathematical idealization reasonable.

We take a trial solution and make it formal first and then actual.

If we restrict ourselves to a «*two-degrees-of-freedom*» mechanical nonlinear system, as shown in Fig. 2, the equation of motion of the «*two-masses-three springs*» nonlinear system are:

$$\left.\begin{aligned} &\ddot{x} + \omega_1^2 x - \lambda_2 y + \lambda_1 x^3 = 0 \\ &\ddot{y} + \omega_2^2 y - \lambda_3 x = 0 \end{aligned}\right\} \tag{1}$$

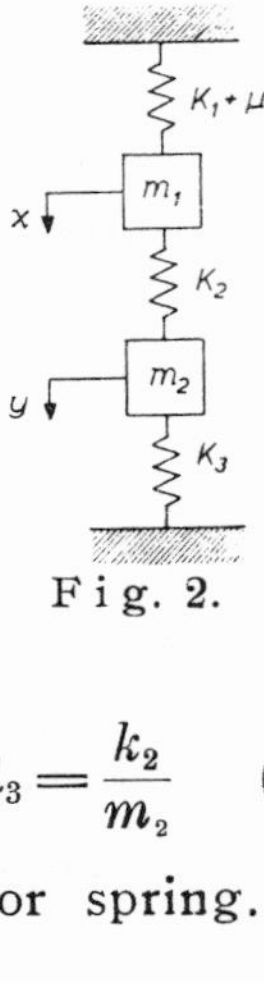

Fig. 2.

where:

$$\omega_1^2 = \frac{k_1 + k_2}{m_1}, \quad \omega_2^2 = \frac{k_2 + k_3}{m_2}, \quad \lambda_1 = \frac{\mu}{m_1}, \quad \lambda_2 = \frac{k_2}{m_1}, \quad \lambda_3 = \frac{k_2}{m_2} \tag{1a}$$

and μ characterizes the nonlinearity of one anchor spring. By using the transformation:

$$x = x_1, \quad x = x_2, \quad y = x_3, \quad \dot{y} = x_4 \tag{2}$$

the system (1) can be reduced to its normal form:

$$\left.\begin{aligned} &x_i = f_i(x_1, x_2, x_3, x_4), \; i = 1, 2, 3, 4 \\ &f_1 = x_2, \; f_2 = -\omega_1^2 x_1 + \lambda_2 x_3 - \lambda_1 x_1^3, \; f_3 = x_4, \; f_4 = \lambda_3 x_1 - \omega_2^2 x_3 \end{aligned}\right\} \tag{3}$$

valid in a region R:

$$R: |x_i| < h, \; i = 1, 2, 3, 4. \tag{3a}$$

The appropriate initial conditions for «principal modes» are in R:

$$x_1(0) = x_{10}, \quad x_2(0) = 0, \quad x_3(0) = x_{30}, \quad x_4(0) = 0 \tag{3b}$$

where x_{10} and x_{30} are appropriately related to each other.

Now we remark that the nature of the functions f_i of (3) are such that all Hadamard's postulates are satisfied. These functions f_i are continuous in R, then bounded; they have continuous partial derivatives $\partial f_i / \partial x_k$ in R, when they satisfy Lipschitz conditions with respect to x_i in R for a Lipschitz constant $s = l.u.b \,|\, \partial f_i / \partial x_k \,|$. The above properties quarantee the uniquely existing solution of the system (3) and (3b) in a region $R' \subset R$. As the initial point x_{10}, $i = 1, 2, 3, 4$ varies in R', the solution satisfies the three Hadamard's postulates, and the problem is *«well posed»*.

Example 3. *The motion of an artificial celestial body under the Influence of a Newtonian Center and a general force* (Ref. 4).

We take this problem from orbital mechanics. It is a special two-body problem treated by the writer in his own way and presented to the «*15th International Astronautical Congress*» Warsaw, Poland, Sept. 1964.

The problem is: «*An artificial celestial body is moving under the influence of a central force obeying the inverse square Newton's law toward an attractive center. A general force is applied, acts for an interval of time, then it is removed. Find the motion of the body during the action of the general force*».

The mathematical formulation of the problem is:

$$\left.\begin{array}{l} \ddot{\underline{r}}(\tau) = - \dfrac{\mu}{r^3(\tau)} \underline{r}(\tau) + \underline{T}(\tau) \\ \underline{r}(0) = \underline{r}_0, \ \dot{\underline{r}}(0) = \dot{r}_0 + \underline{I}_0 \\ D: 0 \leqslant \tau \leqslant \tau' \\ D_1: |\underline{r}(\tau)| < M_1, \quad |\dot{\underline{r}}(\tau)| < M_2 \end{array}\right\} \tag{1}$$

T is the general force, $\underline{I}_0$ the impulse, is given by

$$\left.\begin{array}{l} \underline{I}_0 = \int_0^{t_0} T(t) dt \\ \tau = t - t_0 \end{array}\right\} \tag{1a}$$

If we take the function:

$$\underline{r}(\tau)=\alpha_1(\tau)\underline{r}_0^*+\alpha_2(\tau)\underline{s}_0^*+\alpha_3(\tau)\underline{T}_0^* \tag{2}$$

as a trial solution, where $\underline{r}_0^*$, $\underline{s}_0^*$, $\underline{T}_0^*$ are special unit vectors, the coefficients α_1, α_2, α_3 must satisfy the following conditions in order the function (2) is a formal solution of (1):

$$\left.\begin{array}{lll} \ddot{\alpha}_1+\frac{\mu}{r^3}\alpha_1=T_1; & \alpha_1(0)=r_0, & \dot{\alpha}_1(0)=0 \\ \ddot{\alpha}_2+\frac{\mu}{r^3}\alpha_2=T_2; & \alpha_2(0)=0, & \dot{\alpha}_2(0)=s_0 \\ \ddot{\alpha}_3+\frac{\mu}{r^3}\alpha_3=T_3; & \alpha_3(0)=0, & \dot{\alpha}_3(0)=0 \end{array}\right\} \tag{3}$$

The system of equations (3) satisfy the Hadamard's postulates, as easily can be proved, when the functions $\alpha_1(\tau)$, $\alpha_2(\tau)$, $\alpha_3(\tau)$ can be uniquely determined and are continuous functions of the initial conditions of (3).

Therefore the solution (2) of (1) is a unique solution and continuous function of the initial conditions of (1).

Example 4. *The problem of Domes of Underwater Warfare in 1942–1943* Ref. (2a).

By this problem one can illustrate the diffilculties during the transition from the physical problem to a reasonable mathematical idealization. The problem arose in the winter *1942–1943* in connection with «underwater warfare». As is known, underwater sound ranging depends on sending out a sound beam in water and attached to a fast-moving ship, the water steaming around the plate causes serious disturbances. For elimination of these disturbances the projector is closed in a so called «*dome*», Fig. 3, which is a convex shell of metal or other material filled with water. Such domes interfere only slightly with the formation of a concentrated sound beam.

During *1942–1943* a large number of small submarines chases were built and equipped with sound gear similar to but smaller than the gear used before. While the manufacture of domes to fit this smaller gear was under way, it was discovered that these smaller domes led to an intolerable diffusion of the sound beam. At that time a quick remedy was imperative, and a mathematical analysis of the problem was needed to support and speed - up experimented work.

The mathematical problem, related to the above physical problem, was to solve the differential equation:

$$\left.\begin{aligned} &\nabla^2 P + k^2 P = 0 \\ &\nabla^2 = \frac{\partial^2}{\partial x^2} + \frac{\partial^2}{\partial y^2} + \frac{\partial^2}{\partial z^2} \end{aligned}\right\} \tag{1}$$

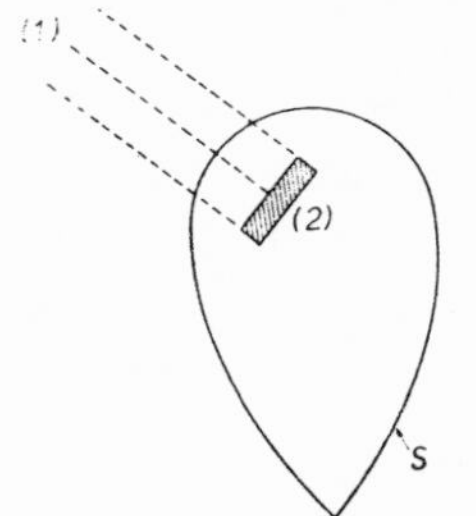

(1) axis of beam sound
(2) projector
S surface of dome
Fig. 3. Dome.

in which $k = \omega / c$, ω the frequency, c the sound velocity, and k has for our problem unfortunately different values within the shell of the dome and outside.

This mathematical idealization was not a suitable one for the problem.

They found the suitable mathematical idealization by the following process.

The actual dome of small finite thickness was replaced by an extremally thin surface, then the influence of the dome was simply replaced by conditions for jump discontinuities of the disturbance q of the beam across the surface.

These conditions are:

$$\left.\begin{aligned} &[q] = \frac{\varrho_1}{\varrho_0 - 1} \cdot \frac{\partial p}{\partial n} \\ &\left[\frac{\partial q}{\partial n}\right] = \frac{\varrho_0}{\varrho_1} (k_0^2 - k_1^2) p - \left(1 - \frac{\varrho_0}{\varrho_1}\right)\left(\frac{\partial^2 p}{\partial n^2} + 2H \frac{\partial p}{\partial n}\right) \end{aligned}\right\} \tag{2}$$

where the symbol $[\ ]$ means jump of the quantity of the symbol across the surface, q is the disturbance of the acoustic pressure p caused by the dome and the normal derivatives $\frac{\partial}{\partial n}$ are to be evaluated on the surface S of the dome. The quantity H is the mean curvature of S, i.e. the average of the curvature of any two normal plane sections at right angles to each other. In addition to conditions (2) to be satisfied by q on S, q should be a solution of the equation:

$$\nabla^2 q + k_0^2 q = 0. \tag{3}$$

which is regular everywhere except on S and which has the same behavior as P at ∞. This problem possesses the unique solution :

$$q = -\frac{1}{4\pi}\iint_S \left[\frac{\partial q}{\partial n}\right]\frac{e^{ik_0 r'}}{r'}\,ds + \frac{1}{4\pi}\iint_S [q]\frac{\partial}{\partial n}\left(\frac{e^{ik_0 r'}}{r'}\right)dS. \quad (4)$$

The quantities in the brackets are given by condition (2), r' is the distance from a fixed point (x, y, z) at which $q\ (x, y, z)$ is to be determined to the point of integration on S. This formula yields the disturbance as the effect due to a layer of point sources and a layer of dipoles disturbed on S with intensities which are known as soon as the original pressure p is known, since the quantities in brackets are fixed in value due to conditions (2). The relative directional disturbance :

$$\left|\frac{p_1}{p}\right| Rch\left(\frac{q_1}{p_1} - \frac{q_0}{p_0}\right)$$

would, finally, be obtained from (4). The solution (4) is valid for a shell of constant thickness, but it could be extended without essential error to cases in which the dome shell is made up of a not too large number of pieces, each of which is of constant thickness. All that would be necessary would be to insert a numerical factor d in the integrands on the right-hand side of expression (4), which would be precise constant on S. By this formula one can analyse the contribution to the distortion of various factors, such as the curvature of the dome and the density and sound velocity within it.

The above kind of mathematical idealization, even without detailed numerical computation, proved helpful to the designing engineer.

REFERENCES

[1] *L. Brillouin*: «Scientific Uncertainty and Information», (1964).

[2] *R. Courant*: (a) «Methods in Applied Mathematics», Rec. Adv. in Science, 1—14 (1956).

(b) «Boundary value probleme in Modern Fluid Dynamics», Rroc. Intern Congress of Math., Vol. II (1956).

156 DEMETRIOS G. MAGIROS

[3] *R. Courant* and *C. Friedrichs*: «Supersonic Flow and Shock Waves», 367 ft, (1948).

[4] *D. Magiros*: (a) «Method for Defining Principal Modes of Nonlinear Systems Utilizing Infinite Determinants», J. Math. Phys. 2, No 6, 869—875 (1961).

(b) «The Motion of an Artificial Celestial Body under the Influence of a Newtonian Center and a General Force». Proc. XVth International Astr. Congress, Warsaw, Poland (1964).

[5] *W. Miranker*: «A Well Posed Problem for a Backward Heat Equation» Proc. Am. Math. Soc. (Apr. 1961).

[6] *F. Weyl* «Applied Mathematic in U. S.», Monographs of the Soc. for Ind. and Appl. Math., No 1 (1956).

Mathematical Consultant of General Electric Comp.
M.S.D. Re-entry Systems Department, Philadelphia, PA.

K. O. Friedrichs anniversary issue

Diffraction by a Semi-Infinite Screen With a Rounded End*

JOSEPH B. KELLER and DEMETRIOS G. MAGIROS

1. Introduction

Suppose a cylindrical wave is incident upon a semi-infinite thin screen with a rounded end, such as that shown in Figure 1. The rounded end is obtained by placing a cylindrical tip of radius a on the end of the screen.

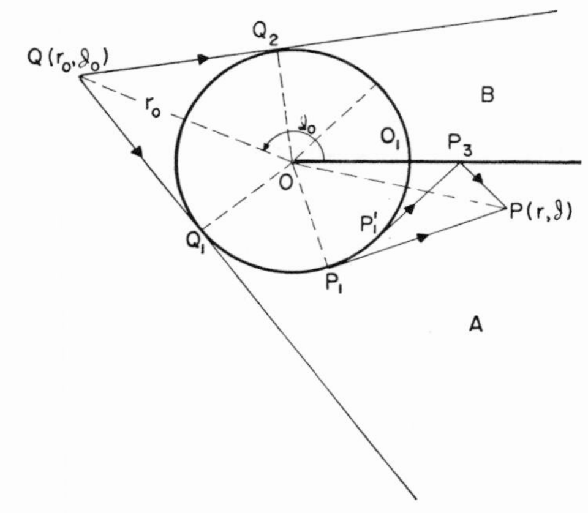

Figure 1
A semi-infinite thin screen, represented by the heavy line with its end at 0, is attached to a circular cylinder of radius a, represented by the circle. A line source parallel to the cylinder is at Q, which has coordinates r_0 and θ_0. Two rays from Q are tangent to the cylinder at Q_1 and Q_2. Their extensions bound the two parts of the shadow, A below the screen and B above it. The tangent ray at Q_1 produces a diffracted surface ray which proceeds along the cylinder in the counterclockwise direction. The two diffracted rays which it sheds at P_1 and P_1' are shown passing through P, the second ray having been reflected from the screen at P_3.

If a is large compared to the wavelength λ of the incident radiation, then the resulting field can be determined by means of the geometrical theory of

*The research reported in this paper has been sponsered by the Electronics Research Directorate of the Air Force Cambridge Research Laboratories, Air Research and Development command, under Contract No. AF 19(604)5238. Reproduction in whole or in part permitted for any purpose of the United States Government.

Reprinted from the *Communications on Pure and Applied Mathematics* **XIV** (1961), 457–471.

diffraction [2, 3]. This theory provides a simple geometrical procedure for constructing the field. The construction is carried out in Section 2. In making this construction, it is assumed that the field u is a scalar which satisfies the reduced wave equation and either it or its normal derivative vanishes on the screen. Thus u may represent either the component of electric or magnetic field parallel to the generators of the cylinder, and the screen may represent a perfect conductor. In the second case, u may also represent acoustic pressure and the screen may represent a rigid body.

Another method for constructing the field is that of expansion in radial eigenfunctions. This method is employed in Section 3. It is then shown that for large values of ka, where $k = 2\pi/\lambda$, the asymptotic form of the field so constructed coincides exactly with the diffracted field given by the geometrical theory of diffraction. This agreement confirms the geometrical theory, which is based on several hypotheses about wave propagation. However, it is not a conclusive proof of the correctness of that theory because the eigenfunctions are not complete.

In Section 4, we prove that the radial eigenfunction expansion does represent the field u. We do this by first expanding u in angular eigenfunctions, which are known to be complete. This expansion was first given by S. N. Karp [4]. Then we convert this series into the radial eigenfunction series by applying the Watson transformation. In this way we prove that the latter series does represent u. Therefore, the result of the geometrical theory of diffraction, which is asymptotic to it, is indeed the asymptotic form of the exact solution.

The diffracted field in the shadow has been computed [1] from the geometrical solution and also from the exact solution as a function of ka. The two results agree well for $ka \geqq 2$ for both boundary conditions. In the case in which $u = 0$ on the boundary, the field in the shadow is found to decrease as ka increases, being greatest when $ka = 0$, i.e. when the cylindrical tip is absent. In the other case, in which $\partial u/\partial n = 0$ on the boundary, the field in the shadow is found to oscillate as ka increases, but ultimately, for sufficiently large values of ka, it also decreases monotonically. These results are interpreted in terms of the rays of the geometrical theory of diffraction.

It is to be noted that the problem under consideration is two dimensional, so that we may confine our attention to a single plane.

2. Geometrical Solution

According to the geometrical theory of diffraction, the field u can be represented as the sum of a geometrical optics field u_g and a diffracted field u_d. The field $u_g(P)$ at a point P is the sum of the fields on the incident and reflected rays through P. The field $u_i(P)$ on an incident ray through P is given by

$$(1) \qquad u_i(P) = (8\pi kR)^{-1/2}\exp\left\{ikR+\frac{i\pi}{4}\right\}.$$

Here R denotes the distance from the source Q to the point P. The constants in (1) have been chosen to correspond to a source of unit strength for which the exact field would be given by $(i/4)H_0^{(1)}(kR)$. The field on a reflected ray can be constructed by a well known procedure, so we need not consider it further. Instead, we turn to the diffracted field $u_d(P)$ which is the sum of the fields on all the diffracted rays through P.

The diffracted rays are produced by the two rays from Q which are tangent to the tip of the screen. The two points of tangency are denoted by Q_1 and Q_2 (see Figure 1). The surface ray produced at Q_1 will shed a diffracted ray at P_1 which will travel along the tangent from P_1 to P. It will also shed a diffracted ray at P_1' which will be reflected at P_3 and then reach P. The surface ray itself will be reflected from the screen at O_1, encircle the tip and be reflected at O_1 again, and repeat this process infinitely many times. Each time it passes through P_1 and P_1' in the clockwise direction, the rays it sheds will pass through P. Thus infinitely many diffracted rays will pass through P, all originally produced by the ray incident at Q_1. In a similar manner, the surface ray produced at Q_2 will be reflected at O_1 and then shed rays at P_1 and P_1' which will pass through P. It will also give rise to infinitely many rays through P as it continues to travel back and forth along the surface of the tip.

As a consequence of the preceding considerations, we see that there are four types of diffracted rays which pass through a point P in the shadow. We may describe them as follows:

I. Rays produced at Q_1 and diffracted at P_1. The ray QQ_1P_1P is the simplest ray of this type. For each non-negative value of the integer n, there is one of these rays which makes n trips around the circle. This ray is reflected $2n$ times at O_1. Now the reflection coefficient R has the value -1 for the case in which $u = 0$ on the screen and $+1$ for the case in which $\partial u/\partial n = 0$ on the screen. Thus in either case $R^{2n} = 1$. The path length of such a ray along the circle is $4\pi an+t$, where $t = Q_1P_1$.

II. Rays produced at Q_1 and diffracted at P_1'. The ray $QQ_1P_1P_1'P_3P$ is the simplest ray of this type. For each $n \geqq 0$, the ray which makes n round trips is reflected $2n+1$ times, including the reflection at P_3. Thus $R^{2n+1} = -1$ if $u = 0$ on the screen and $R^{2n+1} = 1$ if $\partial u/\partial n = 0$. The path length of such a ray along the circle is $4\pi an+t+t_1$, where $t_1 = P_1P_1'$.

III. Rays produced at Q_2 and diffracted at P_1. The ray $QQ_2O_1Q_2Q_1P_1P$

is the simplest ray of this type. For each $n \geqq 0$, the ray which makes n round trips is reflected $2n+1$ times. Its path length on the circle is $4\pi an+t_2$, where $t_2 = Q_2O_1+O_1Q_2Q_1P_1$.

IV. Rays produced at Q_2 and diffracted at P_1'. The ray $QQ_2O_1Q_2Q_1P_1P_1'P_3P$ is the simplest ray of this type. The ray which makes $n \geqq 0$ round trips is reflected $2n+2$ times, including the reflection at P_3. Its path length along the circle is $4\pi an+t_2+t_1$.

According to equation (10a) of [3], the field on a diffracted ray of type I is given by

$$u_i(Q_1)\frac{e^{iks}}{s^{1/2}}\sum_m D_m^2 \exp\{(ik-\alpha_m)T_n\}. \tag{2}$$

In (2), s denotes distance along the diffracted ray from the point where it leaves the surface, $T_n = 4\pi an+t$ denotes its arclength along the surface, and α_m and D_m are certain constants. The field on the rays of other types is also given by (2) with the appropriate values of s and T_n in each case, and with an appropriate power of the reflection coefficient R.

Let the coordinates of Q be (r_0, θ_0) and those of P be (r, θ) in a polar coordinate system with its origin at the center of the tip and the polar axis along the screen. Then $R = (r_0^2-a^2)^{1/2}$ and $s = (r^2-a^2)^{1/2}$. We insert these values into (2) using (1) for $u_i(Q_1)$, and we sum over n. This summation merely involves a geometric series, which can be summed explicitly. Then we add together the corresponding sums for the four types of rays and obtain the diffracted field at a point P in the shadow

$$\begin{aligned} u_d(P) &= \frac{i(8\pi k)^{1/2}\exp\left\{ik(r_0^2-a^2)^{1/2}+ik(r^2-a^2)^{1/2}-\dfrac{i\pi}{4}\right\}}{(r_0^2-a^2)^{1/4}(r^2-a^2)^{1/4}} \\ &\times \sum_m \frac{D_m^2}{1-\exp\{4\pi a(ik-\alpha_m)\}}\,[\exp\{(ik-\alpha_m)t\}\mp\exp\{(ik-\alpha_m)(t+t_1)\} \\ &\qquad \mp\exp\{(ik-\alpha_m)t_2\}+\exp\{(ik-\alpha_m)(t_2+t_1)\}]. \end{aligned} \tag{3}$$

In (3) the upper signs apply in case $u = 0$ on the screen while the lower signs apply if $\partial u/\partial n = 0$ on the screen.

In terms of the coordinates of Q and P, the distances t, t_1 and t_2 are given by

$$\begin{aligned} t &= a\left[\theta-\theta_0-\cos^{-1}\frac{a}{r_0}-\cos^{-1}\frac{a}{r}\right], \\ t_1 &= 2a(2\pi-\theta), \\ t_2 &= a\left[\theta+\theta_0-\cos^{-1}\frac{a}{r_0}-\cos^{-1}\frac{a}{r}\right]. \end{aligned} \tag{4}$$

The diffraction coefficients D_m and the decay constants α_m are given by [3, 5]

$$\alpha_m = e^{-\pi i/6}\left(\frac{k}{6a^2}\right)^{1/3} q_m, \tag{5}$$

$$D_m^2 = \left[\frac{\pi^9}{2^5\,3^8}\frac{a^2}{k}\right]^{1/6} e^{\pi i/12}[A'(q_m)]^{-2}, \tag{6a}$$

$$D_m^2 = \left[\frac{\pi^9}{2^5\,3^2}\frac{a^2}{k}\right]^{1/6} e^{\pi i/12} q_m^{-1}[A(q_m)]^{-2}. \tag{6b}$$

In these equations $A(q)$ denotes the Airy function. For the boundary condition $u = 0$, q_m is the m-th positive zero of $A(q)$ while for the boundary condition $\partial u/\partial n = 0$, q_m is the m-th positive zero of $A'(q)$.

We now insert the preceding results into (3) and obtain in the cases $u = 0$ and $\partial u/\partial n = 0$ the respective results for a point P in the shadow

$$\begin{aligned} u_d(P) = {} & \frac{\pi}{6}\left(\frac{a}{6k^2}\right)^{1/3} \frac{\exp\left\{ik(r_0^2-a^2)^{1/2}+ik(r^2-a^2)^{1/2}+\dfrac{5\pi i}{6}\right\}}{(r_0^2-a^2)^{1/4}(r^2-a^2)^{1/4}} \\ & \times \sum_m \frac{\sin[(ka+i\alpha_m a)(2\pi-\theta)]\sin[(ka+i\alpha_m a)\theta_0]}{[A'(q_m)]^2 \sin[(ka+i\alpha_m a)2\pi]} \\ & \times \exp\left\{-i(ka+i\alpha_m a)\left[\cos^{-1}\frac{a}{r_0}+\cos^{-1}\frac{a}{r}\right]\right\}, \end{aligned} \tag{7a}$$

$$\begin{aligned} u_d(P) = {} & \frac{\pi}{18}\left(\frac{a}{6k^2}\right)^{1/3} \frac{\exp\left\{ik(r_0^2-a^2)^{1/2}+ik(r^2-a^2)^{1/2}+\dfrac{5\pi i}{6}\right\}}{(r_0^2-a^2)^{1/4}(r^2-a^2)^{1/4}} \\ & \times \sum_m \frac{\cos[(ka+i\alpha_m a)(2\pi-\theta)]\cos[(ka+i\alpha_m a)\theta_0]}{q_m A^2(q_m)\sin[(ka+i\alpha_m a)2\pi]} \\ & \times \exp\left\{-i(ka+i\alpha_m a)\left[\cos^{-1}\frac{a}{r_0}+\cos^{-1}\frac{a}{r}\right]\right\}. \end{aligned} \tag{7b}$$

These results hold when $\theta > \theta_0$. When $\theta < \theta_0$ we obtain the same results with θ and θ_0 interchanged. Slightly different results can be obtained in the same way for a point P in the illuminated region.

3. Radial Eigenfunction Expansion

We shall now formulate a boundary value problem for the total field

u and solve it by the method of expansion in radial eigenfunctions. The conditions to be satisfied by $u(r, \theta)$ are

$$(\Delta+k^2)u = r^{-1}\delta(r-r_0)\delta(\theta-\theta_0), \tag{1}$$

$$\lim_{r\to\infty} r^{1/2}(u_r-iku) = 0, \tag{2}$$

$$u(a, \theta) = u(r, 0) = u(r, 2\pi) = 0, \tag{3a}$$

$$u_r(a, \theta) = u_\theta(r, 0) = u_\theta(r, 2\pi) = 0. \tag{3b}$$

In order to construct u we introduce the radial eigenfunctions $H^{(1)}_{\nu_m}(kr)$. These Hankel functions are the radial factors of product solutions of the homogeneous equation corresponding to (1). They also satisfy the radiation condition (2). The eigenvalues ν_m are determined by requiring these functions to satisfy the first of the boundary conditions (3a) or (3b) which become, respectively,

$$H^{(1)}_{\nu_m}(ka) = 0, \tag{4a}$$

$$H^{(1)\prime}_{\nu_m}(ka) = 0. \tag{4b}$$

For large values of ka, and fixed m, the roots of (4a) and of (4b) are given by

$$\nu_m = ka+q_m(\tfrac{1}{6}ka)^{1/3}e^{i\pi/3}+O((ka)^{-1/3}). \tag{5}$$

In the case of (4a), q_m is the m-th positive zero of the Airy function $A(q)$ while, in the case of (4b), q_m is the m-th positive zero of $A'(q)$. We now assume that u can be represented by the series

$$u = \sum_{m=1}^{\infty} f_m(\theta)H^{(1)}_{\nu_m}(kr). \tag{6}$$

To determine the coefficient functions $f_m(\theta)$ we insert (6) into (1) and make use of Bessel's equation to obtain

$$\sum_{m=1}^{\infty} r^{-2}H^{(1)}_{\nu_m}(kr)[f''_m(\theta)+\nu_m^2 f_m(\theta)] = r^{-1}\delta(r-r_0)\delta(\theta-\theta_0). \tag{7}$$

It is easy to prove that the radial eigenfunctions $H^{(1)}_{\nu_m}(kr)$ are mutually orthogonal, which may be expressed in the form

$$\int_a^\infty r^{-1}H^{(1)}_{\nu_m}(kr)H^{(1)}_{\nu_n}(kr)dr = N_n\delta_{mn}. \tag{8}$$

An analysis similar to that of Sommerfeld [6] yields for the constant N_n the following results in the respective cases $u = 0$ and $\partial u/\partial n = 0$ on the screen:

(9a) $$N_n = \left[\frac{2i}{\pi\nu}\frac{\frac{\partial}{\partial\nu}H_\nu^{(1)}(ka)}{H_\nu^{(2)}(ka)}\right]_{\nu=\nu_n},$$

(9b) $$N_n = \left[\frac{2i}{\pi\nu}\frac{\frac{\partial}{\partial\nu}H_\nu^{(1)\prime}(ka)}{H_\nu^{(2)\prime}(ka)}\right]_{\nu=\nu_n}.$$

We now multiply (7) by $r^{-1}H_{\nu_n}^{(1)}(kr)$ and integrate the resulting equation from a to ∞ with respect to r. Then by using (8) we obtain

(10) $$f_n''(\theta)+\nu_n^2 f_n(\theta) = N_n^{-1}H_{\nu_n}^{(1)}(kr_0)\delta(\theta-\theta_0).$$

In order to determine $f_n(\theta)$ from (10), we must impose the second and third of the boundary conditions (3a) or (3b) on u, which is given by (6). By again utilizing the orthogonality condition (8) these boundary conditions yield, in the respective cases,

(11a) $$f_n(0) = f_n(2\pi) = 0,$$

(11b) $$f_n'(0) = f_n'(2\pi) = 0.$$

The solutions of (10) which satisfy (11a) or (11b), respectively, are

(12a) $$f_n(\theta) = -\nu_n^{-1}\csc 2\pi\nu_n N_n^{-1}H_{\nu_n}^{(1)}(kr_0)\sin\nu_n\theta_<\sin\nu_n(2\pi-\theta_>),$$

(12b) $$f_n(\theta) = +\nu_n^{-1}\csc 2\pi\nu_n N_n^{-1}H_{\nu_n}^{(1)}(kr_0)\cos\nu_n\theta_<\cos\nu_n(2\pi-\theta_>).$$

In (12a) and (12b), $\theta_<$ and $\theta_>$ denote, respectively, the smaller and larger of the two angles θ and θ_0, both of which lie between 0 and 2π.

We now insert (12a) into (6) and use (9a) for N_n. Thus we obtain for the solution u in the case of the boundary condition $u = 0$, the result

(13a) $$u = +\frac{i\pi}{2}\sum_{n=1}^{\infty}H_{\nu_n}^{(1)}(kr_0)H_{\nu_n}^{(1)}(kr)\frac{\sin\nu_n\theta_<}{\sin 2\pi\nu_n}\sin\nu_n(2\pi-\theta_>)\frac{H_{\nu_n}^{(2)}(ka)}{\frac{\partial}{\partial\nu}H_{\nu_n}^{(1)}(ka)}.$$

In the case of the boundary condition $\partial u/\partial n = 0$ we obtain instead, from (12b), (9b) and (6), the result

(13b) $$u = -\frac{i\pi}{2}\sum_{n=1}^{\infty}H_{\nu_n}^{(1)}(kr_0)H_{\nu_n}^{(1)}(kr)\frac{\cos\nu_n(2\pi-\theta_>)}{\sin 2\pi\nu_n}\cos\nu_n\theta_<\frac{H_{\nu_n}^{(2)\prime}(ka)}{\frac{\partial}{\partial\nu}H_{\nu_n}^{(1)\prime}(ka)}.$$

Let us now expand these expressions asymptotically for large values of ka. If $r > a$ and $r_0 > a$, we may use the Debye asymptotic expansions of the Hankel functions with arguments kr and kr_0. However, for the Hankel functions with argument ka, which is nearly equal to the order ν_n, we must use the expansion in terms of Airy functions. When we insert these expansions into (13a) and (13b), we obtain the results

$$u(r,\theta) \sim \frac{\pi}{6^{4/3}}(ka)^{1/3}\sum_{n=1}^{\infty}\frac{\exp\left\{i(k^2r^2-\nu_n^2)^{1/2}+i(k^2r_0^2-\nu_n^2)^{1/2}+\dfrac{5\pi i}{6}\right\}}{(k^2r^2-\nu_n^2)^{1/4}(k^2r_0^2-\nu_n^2)^{1/4}} \times \frac{\sin\nu_n\theta_< \sin\nu_n(2\pi-\theta_>)}{[A'(q_n)]^2\sin 2\pi\nu_n}\exp\left\{-i\nu_n\left(\cos^{-1}\frac{\nu_n}{kr}+\cos^{-1}\frac{\nu_n}{kr_0}\right)\right\}, \tag{14a}$$

$$u(r,\theta) \sim \frac{\pi}{3\cdot 6^{4/3}}(ka)^{1/3}\sum_{n=1}^{\infty}\frac{\exp\left\{i(k^2r^2-\nu_n^2)^{1/2}+i(k^2r_0^2-\nu_n^2)^{1/2}+\dfrac{5\pi i}{6}\right\}}{(k^2r^2-\nu_n^2)^{1/4}(k^2r_0^2-\nu_n^2)^{1/4}} \times \frac{\cos\nu_n\theta_< \cos\nu_n(2\pi-\theta_>)}{q_nA^2(q_n)\sin 2\pi\nu_n}\exp\left\{-i\nu_n\left(\cos^{-1}\frac{\nu_n}{kr}+\cos^{-1}\frac{\nu_n}{kr_0}\right)\right\}. \tag{14b}$$

By using (5) for ν_n in these equations, we can simplify them to the form

$$u(r,\theta) \sim \frac{\pi}{6^{4/3}}\frac{(k^{-2}a)^{1/3}e^{5\pi i/6}}{(r^2-a^2)^{1/2}(r_0^2-a^2)^{1/2}}\exp\{ik[(r^2-a^2)^{1/2}+(r_0^2-a^2)^{1/2}]\} \times \sum_{n=1}^{\infty}\frac{\sin\nu_n\theta_< \sin\nu_n(2\pi-\theta_>)}{[A'(q_n)]^2\sin 2\pi\nu_n}\exp\left\{-i\nu_n\left(\cos^{-1}\frac{a}{r}+\cos^{-1}\frac{a}{r_0}\right)\right\}, \tag{15a}$$

$$u(r,\theta) \sim \frac{\pi}{3\cdot 6^{4/3}}\frac{(k^{-2}a)^{1/3}e^{5\pi i/6}}{(r^2-a^2)^{1/2}(r_0^2-a^2)^{1/2}}\exp\{ik[(r^2-a^2)^{1/2}+(r_0^2-a^2)^{1/2}]\} \times \sum_{n=1}^{\infty}\frac{\cos\nu_n\theta_< \cos\nu_n(2\pi-\theta_>)}{q_nA^2(q_n)\sin 2\pi\nu_n}\exp\left\{-i\nu_n\left(\cos^{-1}\frac{a}{r}+\cos^{-1}\frac{a}{r_0}\right)\right\}. \tag{15b}$$

The results (15) coincide exactly with the results (7) of Section 2 which were obtained by the geometrical theory of diffraction for a point in the shadow. This verifies that the geometrical theory is correct in this case since (13), from which (15) was obtained, does indeed represent the solution as we shall show in the next section. However, the asymptotic evaluation of (13), which led to (15), is correct only for a finite number of terms in the series. This is so because in the later terms ν_n becomes large compared to ka, kr and kr_0 and this was not taken into account in the asymptotic evaluation. Furthermore, for n large ν_n is not given by (5). But as we shall see, (15) converges very rapidly for points in the shadow so that the error it contains in the later terms is asymptotically negligible. For points in the illuminated region, however, (15) does not converge although (13) does.

Therefore a different asymptotic evaluation must be made in this region. This will be done in the next section.

4. Angular Eigenfunction Expansion

The eigenvalue problem for the radial eigenfunctions involves the radiation condition and, therefore, it is not of Sturm-Liouville type. Consequently, we do not know whether these eigenfunctions are complete. Therefore, we cannot be sure that the solution u can be represented by the series (6) of Section 3. We shall now prove that it can be so represented and that it is given by (13a) or (13b).

We begin with a representation of u in terms of the angular eigenfunctions which are complete, since they are solutions of a Sturm-Liouville problem. This representation was given by S. N. Karp [4]. By modifying his result to correspond to our problem (1)—(3) of Section 3, or by direct expansion of u in angular eigenfunctions, we find, in the two cases,

$$\text{(1a)} \quad u = \frac{i}{2} \sum_{n=1}^{\infty} H_{n/2}^{(1)}(kr_>) \left[J_{n/2}(kr_<) - \frac{J_{n/2}(ka)}{H_{n/2}^{(1)}(ka)} H_{n/2}^{(1)}(kr_<) \right] \sin\frac{n}{2}\theta_< \sin\frac{n}{2}\theta_> ,$$

$$\text{(1b)} \quad u = -\frac{i}{2} \sum_{n=1}^{\infty} H_{n/2}^{(1)}(kr_>) \left[J_{n/2}(kr_<) - \frac{J'_{n/2}(ka)}{H_{n/2}^{(1)\prime}(ka)} H_{n/2}^{(1)}(kr_<) \right] \sin\frac{n}{2}\theta_< \sin\frac{n}{2}\theta_> .$$

These series are useful for small values of ka but not for large values.

To obtain representations of u useful for large ka we apply the Watson transformation to the series. First we use in (1a) and (1b) the identities

$$\sin\frac{n}{2}\theta_> = \frac{-\sin\frac{n}{2}(2\pi-\theta_>)}{\cos n\pi},$$

$$\cos\frac{n}{2}\theta_> = \frac{\cos\frac{n}{2}(2\pi-\theta_>)}{\cos n\pi}.$$

Then we represent each series as a contour integral. This is accomplished by first choosing a contour L in the μ-plane which encircles the non-negative integers $\mu = n$, $n = 0, 1, 2, \cdots$, and then an integrand which has as its only singularities within L poles at these integers. In addition for each n, the residue at the n-th pole must be the n-th term in the series. In this way we arrive at the following integral representations:

$$u = \frac{1}{4}\int_L \frac{1}{\sin\mu\pi}\,[J_{\mu/2}(kr_<)H^{(1)}_{\mu/2}(ka) - J_{\mu/2}(ka)H^{(1)}_{\mu/2}(kr_<)] \times \frac{H^{(1)}_{\mu/2}(kr_>)}{H^{(1)}_{\mu/2}(ka)} \sin\frac{\mu\theta_<}{2}\sin\frac{\mu}{2}(2\pi-\theta_>)d\mu, \tag{2a}$$

$$u = \frac{1}{4}\int_L \frac{1}{\sin\mu\pi}\,[J_{\mu/2}(kr_<)H^{(1)\prime}_{\mu/2}(ka) - J'_{\mu/2}(ka)H^{(1)}_{\mu/2}(kr_<)] \times \frac{H^{(1)}_{\mu/2}(kr_>)}{H^{(1)\prime}_{\mu/2}(ka)} \cos\frac{\mu\theta_<}{2}\cos\frac{\mu}{2}(2\pi-\theta_>)d\mu. \tag{2b}$$

The contour L encircles the positive real axis of the μ-plane in the clockwise direction as is shown in Figure 2.

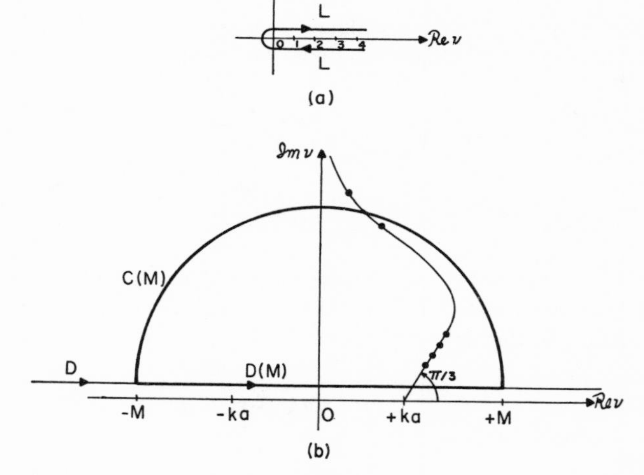

Figure 2

a. The contour L in the ν-plane encircles the positive real axis in the clockwise direction. b. The contour D is shown just above the real axis of the ν-plane. The segment $D(M)$ of it extends from $\mathcal{R}e\,\nu = -M$ to $\mathcal{R}e\,\nu = M$. The semicircle $C(M)$ into which $D(M)$ is deformed, is also shown. The poles of the integrand in the half plane $\mathcal{I}m\,\nu > 0$ lie on the curved line shown in the first quadrant.

Let us replace μ by $\nu = \mu/2$ in (2). Let us also change ν to $-\nu$ in that part of the integral for which $\mathcal{I}m\,\nu < 0$. Then that part of the contour is transformed into a path from $-\infty$ to 0 lying just above the negative real axis. The integrand is unchanged as we see from the relations

$$
(3) \qquad \begin{aligned} H_{-\nu}^{(1)} &= e^{i\pi\nu} H_{\nu}^{(1)}, \\ H_{-\nu}^{(2)} &= e^{i\pi\nu} H_{\nu}^{(2)}. \end{aligned}
$$

Thus the contour L can be replaced by the contour D which is a line from $-\infty$ to ∞ lying just above the real axis of the ν-plane and (2) becomes

$$
(4a) \qquad u = \tfrac{1}{2}\int_D \frac{1}{\sin 2\pi\nu} [J_\nu(kr_<)H_\nu^{(1)}(ka) - J_\nu(ka)H_\nu^{(1)}(kr_<)] \times \frac{H_\nu^{(1)}(kr_>)}{H_\nu^{(1)}(ka)} \sin \nu\theta_< \sin \nu(2\pi-\theta_>)d\nu,
$$

$$
(4b) \qquad u = \tfrac{1}{2}\int_D \frac{1}{\sin 2\pi\nu} [J_\nu(kr_<)H_\nu^{(1)\prime}(ka) - J_\nu'(ka)H_\nu^{(1)}(kr_<)] \times \frac{H_\nu^{(1)}(kr_>)}{H_\nu^{(1)\prime}(ka)} \cos \nu\theta_< \cos \nu(2\pi-\theta_>)d\nu.
$$

Let $D(M)$ denote that part of the contour from $-M$ to $+M$ just above the real axis (see Figure 2). Let $C(M)$ denote a semi-circle in the upper half plane with radius M and center at the origin. We wish to deform the partial contour $D(M)$ into $C(M)$. This is permissible since the integrands in (4) are analytic functions of ν. They are regular in the upper half of the ν-plane except for poles at the zeroes of their denominators. These zeroes are the values of ν for which either

$$
(5a) \qquad H_\nu^{(1)}(ka) = 0,
$$

or

$$
(5b) \qquad H_\nu^{(1)\prime}(ka) = 0.
$$

In each case we shall denote these values of ν by ν_n, $n = 1, 2, \cdots$. Let $N(M)$ denote the number of these values whose distance from the origin is less than M and let R_n denote the residue of the integrand at ν_n. Then we have in each case, omitting the integrands, and choosing M so that no zero lies on $C(M)$,

$$
(6) \qquad \int_{D(M)} = \sum_{n=1}^{N(M)} R_n + \int_{C(M)} .
$$

From (4) and (6) we have in each case

$$
(7) \qquad u = \lim_{M\to\infty} \int_{D(M)} = \sum_{n=1}^{\infty} R_n + \lim_{M\to\infty} \int_{C(M)} .
$$

We shall now assume that the limit of the last integral in (7) is zero. Then (7) yields for u the following series, in which the residues R_n have been written explicitly:

$$(8a)\quad u(r,\theta)=\frac{i\pi}{2}\sum_{n=1}^{\infty}H_{\nu_n}^{(1)}(kr_<)H_{\nu_n}^{(1)}(kr_>)\frac{\sin\nu_n\theta_<}{\sin 2\pi\nu_m}\sin\nu_n(2\pi-\theta_>)\frac{H_{\nu_n}^{(2)}(ka)}{\frac{\partial}{\partial\nu}H_{\nu_n}^{(1)}(ka)},$$

$$(8b)\quad u(r,\theta)=-\frac{i\pi}{2}\sum_{n=1}^{\infty}H_{\nu_n}^{(1)}(kr_<)H_{\nu_n}^{(1)}(kr_>)\frac{\cos\nu_n\theta_<}{\sin 2\pi\nu_n}\cos\nu_n(2\pi-\theta_>)\frac{H_{\nu_n}^{(2)\prime}(ka)}{\frac{\partial}{\partial\nu}H_{\nu_n}^{(2)\prime}(ka)}.$$

These series are identical with the radial eigenfunction expansions of u given by (13) of Section 3. Thus the validity of those expansions is proved.

Another representation of u can be obtained by using in (4) the identities

$$(9a)\quad \sin\nu(2\pi-\theta_>)=-e^{i2\pi\nu}\sin\nu\theta_>+e^{i\nu\theta_>}\sin 2\pi\nu,$$

$$(9b)\quad \cos\nu(2\pi-\theta_>)=e^{i2\pi\nu}\cos\nu\theta_>-ie^{i\nu\theta_>}\sin 2\pi\nu.$$

The resulting integrands contain one term proportional to $e^{i\nu\theta_>}$ and another proportional to $\sin\nu\theta_>$ or $\cos\nu\theta_>$. The integral of the former term is kept as an integral over D, while the integral of the latter term is converted into a residue series as before. In this way we obtain the alternative representations

$$(10a)\quad \begin{aligned} u=&\frac{i\pi}{2}\sum_{n=1}^{\infty}H_{\nu_n}^{(1)}(kr_<)H_{\nu_n}^{(1)}(kr_>)\frac{\sin\nu_n\theta_<\sin\nu_n\theta_>}{\sin 2\pi\nu_n}e^{i2\pi\nu_n}\frac{H_{\nu_n}^{(2)}(ka)}{\frac{\partial}{\partial\nu}H_{\nu_n}^{(1)}(ka)}\\ &-\frac{1}{4}\int_D H_{\nu}^{(1)}(kr_<)H_{\nu}^{(1)}(kr_>)\sin\nu\theta_<e^{i\nu\theta_>}\frac{H_{\nu}^{(2)}(ka)}{H_{\nu}^{(1)}(ka)}\,d\nu,\end{aligned}$$

$$(10b)\quad \begin{aligned} u=&\frac{-i\pi}{2}\sum_{n=1}^{\infty}H_{\nu_n}^{(1)}(kr_<)H_{\nu_n}^{(1)}(kr_>)\frac{\cos\nu_n\theta_<\cos\nu_n\theta_>}{\sin 2\pi\nu_n}e^{i2\pi\nu_n}\frac{H_{\nu_n}^{(2)\prime}(ka)}{\frac{\partial}{\partial\nu}H_{\nu_n}^{(1)\prime}(ka)}\\ &+\frac{i}{4}\int_D H_{\nu}^{(1)}(kr_<)H_{\nu}^{(1)}(kr_>)\cos\nu\theta_<e^{i\nu\theta_>}\frac{H_{\nu}^{(2)\prime}(ka)}{H_{\nu}^{(1)\prime}(ka)}\,d\nu.\end{aligned}$$

We shall see in the next section that this representation is more useful in the illuminated region, while (8) is more useful in the shadow region.

5. Convergence of the Series

To investigate the convergence of the various series, we follow the procedure of W. Franz [7]. The Hankel function $H_{\nu}^{(1)}(x)$ of large complex order ν and fixed real argument x has the asymptotic form [7]

$$H_\nu^{(1)}(x) \sim -2\sqrt{\frac{2}{\pi\nu}} \sinh\left[\nu \log\frac{2\nu}{x} - \nu + i\frac{\pi}{4}\right]. \tag{1}$$

Consequently the values of ν at which $H_\nu^{(1)}(x)$ vanishes are asymptotically determined by

$$\nu\left(\log\frac{2\nu}{x} - 1\right) = (n - \tfrac{1}{4})i\pi. \tag{2}$$

In (2), n denotes a large integer. To solve (2) we set $\nu = \rho e^{i\phi}$. Then since the real part of the left side of (2) must be zero, we find that for large ρ ϕ and ρ must be related by

$$\phi \sim \frac{\pi}{2}\left(1 - \frac{1}{\log 2\rho x^{-1}}\right). \tag{3}$$

From the imaginary parts in (2) we then find, denoting ρ by ρ_n,

$$\rho_n \sim \frac{(n-\frac{1}{4})\pi}{\log 2(n-\frac{1}{4})\pi x^{-1}}. \tag{4}$$

Thus for large positive n the zeros ν_n of $H_\nu^{(1)}(x)$ are

$$\nu_n \sim \frac{(n-\frac{1}{4})\pi}{\log 2(n-\frac{1}{4})\pi x^{-1}} \exp\left\{i\frac{\pi}{2}\left(1 - \frac{1}{\log\left[2(n-\frac{1}{4})\pi/x \log\left(2(n-\frac{1}{4})\pi x^{-1}\right)\right]}\right)\right\}. \tag{5}$$

From (1) we have

$$H_\nu^{(1)\prime}(x) \sim \frac{2}{x}\sqrt{\frac{2\nu}{\pi}} \cosh\left[\nu \log\frac{2\nu}{x} - \nu + \frac{i\pi}{4}\right]. \tag{6}$$

Then (2)—(5) hold for the zeroes of $H_\nu^{(1)\prime}(x)$ if n is replaced by $n+\frac{1}{2}$.

By using (1) and (5) with $x = kr_>$, we find

$$|H_{\nu_n}^{(1)}(kr_>)| \sim \sqrt{\frac{2}{\pi\nu_n}} \exp\left\{\rho_n \frac{\pi}{2} \frac{\log\left(\frac{a}{r_>}\right)}{\log\left(\frac{2\rho_n}{ka}\right)}\right\}. \tag{7}$$

To evaluate $H_{\nu_n}^{(2)}(ka)$ we use the relation $H_\nu^{(2)}(x) = \overline{H_{\bar{\nu}}^{(1)}(x)}$ valid for x real. Then we see that (1) holds for $H_\nu^{(2)}(x)$ with $i\pi/4$ replaced by $-i\pi/4$. Therefore we have

$$H_{\nu_n}^{(2)}(ka) \sim 2\sqrt{\frac{2}{\pi\nu_n}}(-1)^n. \tag{8}$$

From (1) we also obtain

$$\frac{\partial}{\partial \nu} H^{(1)}_{\nu_n}(ka) \sim -2\sqrt{\frac{2}{\pi \nu_n}} \log\left(\frac{2\nu_n}{ka}\right) \cdot (-1)^n. \tag{9}$$

Upon combining (7)—(9) we obtain for the asymptotic form of the absolute value of the n-th term in (8a) of Section 4 the expression

$$\frac{1}{2\rho_n \log \frac{\rho_n}{ka}} \exp\left\{-\rho_n\left[\theta_> - \theta_< - \frac{\pi}{2}\frac{\log \frac{a^2}{r_< r_>}}{\log \frac{2\rho_n}{ka}}\right]\right\}. \tag{10}$$

This expression diminishes sufficiently rapidly as n increases to guarantee that the series (8a) of Section 4 converges absolutely for $\theta_> > \theta_<$, i.e. for $\theta \neq \theta_0$. The same result applies to the series (8b) of Section 4.

We shall now examine the leading terms in these series in order to show how rapidly they diminish. For any fixed value of n and ka large, ν_n is asymptotically given by (5) of Section 3. Therefore, for fixed n the order and argument of the Hankel functions in the series (8) of Section 4 are both large. Thus for large ka we may evaluate the Hankel functions with argument $kr_<$ or $kr_>$ by using the Debye asymptotic form which is

$$H^{(1)}_\nu(x) \sim \sqrt{\frac{2}{\pi}}\,(x^2-\nu^2)^{-1/4} \exp\left\{i(x^2-\nu^2)^{1/2} - i\nu \cos^{-1}\frac{\nu}{x} - \frac{i\pi}{4}\right\}. \tag{11}$$

For the Hankel functions of argument ka we must use instead the Airy function representation

$$H^{(1)}_\nu(x) \sim \frac{2}{\pi} e^{-i\pi/3}\left(\frac{6}{x}\right)^{1/3} A\left[\left(\frac{6}{x}\right)^{1/3} e^{-i\pi/3}(\nu - x)\right]. \tag{12}$$

By using these expansions in (8) of Section 4 we obtain the series (15) of Section 3. If we further retain only the leading terms in the trigonometric factors in (15a) the summand becomes

$$\frac{1}{2i[A'(q_n)]^2} \exp\left\{i\nu_n\left[\theta_> - \theta_< - \cos^{-1}\frac{a}{r} - \cos^{-1}\frac{a}{r_0}\right]\right\}. \tag{13}$$

The absolute value of (13) is asymptotically

$$\frac{1}{2[A'(q_n)]^2} \exp\left\{-q_n\left(\frac{ka}{6}\right)^{1/3} \sin\frac{\pi}{3}\left[\theta_> - \theta_< - \cos^{-1}\frac{a}{r} - \cos^{-1}\frac{a}{r_0}\right]\right\}. \tag{14}$$

The q_n are positive and increase with n like $n^{2/3}$. Therefore, the absolute value (14) decreases with increasing n provided that

$$\theta_> - \theta_< - \cos^{-1}\frac{a}{r} - \cos^{-1}\frac{a}{r_0} > 0. \tag{15}$$

For a point in the shadow region B shown in Figure 2, $\theta_> = \theta_0, \theta_< = \theta$ and

$$\theta_0 > \theta + \cos^{-1}\frac{a}{r} + \cos^{-1}\frac{a}{r_0}. \tag{16}$$

Therefore, (15) is satisfied for points in B. Similarly for points in the shadow region A of Figure 2, $\theta_> = \theta, \theta_< = \theta$ and

$$\theta > \theta_0 + \cos^{-1}\frac{a}{r} + \cos^{-1}\frac{a}{r_0}. \tag{17}$$

Thus (15) is also satisfied for points in A. We conclude that (15) is satisfied for points in the shadow region but not for points in the illuminated region. The same results apply to the series (15b). Consequently, for points in the shadow the terms in the radial eigenfunction series (13) of Section 3 or (8) of Section 4 decrease rapidly like $\exp\{-n^{2/3}(ka)^{1/3}c\}$, where c is a positive constant. Hence the radial eigenfunction series converges rapidly only in the shadow although it also converges in the illuminated region. The series (7) of Section 2, which results from the geometrical theory of diffraction for points in the shadow, also converges rapidly. In the shadow it coincides with the asymptotic expansion (15) of the radial eigenfunction series. Outside the shadow the series (15) diverges. The representation (10) of Section 5 is useful in the illuminated region since the series in it converge rapidly there, as can be shown by an analysis like that we have just given. In fact the asymptotic expansion of the series in (10) yields exactly the same result as that given by the geometrical theory of diffraction for points in the illuminated region. The asymptotic evaluation of the integrals in (10) yields the incident and reflected fields. However, we shall not carry out these calculations.

Bibliography

[1] Keller, J. B., *How dark is the shadow of a round-ended screen?* J. Appl. Phys., Vol. 30, 1959, pp. 1452–1454.

[2] Keller, J. B., *A geometrical theory of diffraction, Calculus of Variations and its Applications*, Proc. Symposia Appl. Math., VIII, McGraw-Hill, New York, 1958, pp. 27–52.

[3] Keller, J. B., *Diffraction by a convex cylinder*, I.R.E. Trans. Antennas and Prop., AP-4, 1956, pp. 312–321.

[4] Karp, S. N., *Diffraction by a tipped wedge*, New York Univ., Math. Res. Group, Wash. Square Coll., Res. Rep. No. EM-52, 1953.

[5] Levy, B. R., and Keller, J. B., *Diffraction by a smooth object*, Comm. Pure Appl. Math., Vol. 12, 1959, pp. 159–209.

[6] Sommerfeld, A., *Partial Differential Equations*, Academic Press, New York, 1949.

[7] Franz, W., *Über die Greenschen Funktionen des Zylinders und der Kugel* Z. Naturforschung, Vol. 9a, 1954, pp. 705–716.

Received February, 1961.

ΕΦΗΡΜΟΣΜΕΝΑ ΜΑΘΗΜΑΤΙΚΑ.— **On the linearization of nonlinear models of the phenomena. First part: Linearization by exact methods,** *by Demetrios G. Magiros* *. Ἀνεκοινώθη ὑπὸ τοῦ Ἀκαδημαϊκοῦ κ. Ἰω. Ξανθάκη.

A B S T R A C T

This paper deals with remarks on the linearization methods of nonlinear mathematical models of the phenomena.

We can classify the linear methods into two distinct types, the «exact methods» and the «approximate methods».

The advantages and disadvantages of these methods are clarified by appropriately selected examples.

1. INTRODUCTION

The phenomena, either physical or natural or social, are interrelated variations of certain variable quantities and their rate of change, and their nature is usually discussed by mathematical models, where dominant concepts are the linearities and nonlinearities of the models, which usually are differential equations nonlinear in their variables (NLDE).

By «linearization» of a NLDE we mean a reduction to a linear differential equation (LDE), which is either «equivalent» or «almost equivalent» to the NLDE, that is the solution of the LDE may give the solution of the NLDE either «exactly» or «approximately» by an error of small order.

The linearization is a tool for a simplified and easy discussion of the nonlinear phenomena. This tool is helpful in a few cases, but in many cases, especially of practical importance, the linearized models leave out essential features of the NLDE, or they contain properties which are not properties of the NLDE, when the linearization is not acceptable for an adequate description of the real phenomena.

---

* ΔΗΜΗΤΡΙΟΥ Γ. ΜΑΓΕΙΡΟΥ, **Γραμμικοποίησις μὴ γραμμικῶν μαθηματικῶν μοντέλων τῶν φαινομένων. Μέρος Ι: Γραμμικοποίησις μὲ ἀκριβεῖς μεθόδους.**

Reprinted from the *Proceedings of the Athens Academy of Sciences* **51** (1976), 659–668.

The aim of this paper is to provide a sketch of ideas and techniques associated with the notion of linearization of NLDE and to exhibit advandages and disadvantages of the linearization methods by using examples of practical significance.

We distinguish two classes of linearization methods : the «exact methods», and the «approximate methods».

The exact methods give exact solutions, essentially general solutions in closed form, and the approximate methods give approximate solutions within an accepted error.

In this part of the paper we deal with «exact methods».

2. LINEARIZATION BY EXACT METHODS

NLDE may be reduced to LDE by «exact transformations of the variables». By this linearization we may obtain general solutions of the NLDE in a «closed form», by using the solutions of the LDE and the transformation formulae of the variables.

Also by exact linearizations, corresponding to some restrictions of the variables or of the parameters, it is possible to find general solutions of special cases of the original NLDE.

The following examples may be sufficient to show the essence and the applicability of the exact linearization methods and their important results.

In some of the examples use is made of appropriate transformations of the variables, and in some other examples appropriate restrictions of the variables or of the parameters.

Example 1. The Bernoulli equation :

$$y' + ay = by^n \tag{1}$$

where a and b are functions of x, and $b \neq 0$, $n \neq 0$, $\neq 1$, is a monumental example of exact linearization.

We change the variable y into a new variable z, keeping the same independent variable x, by the transformation formula :

$$z = y^{1-n} \tag{1.1}$$

when the linear equation in z results :

$$z' + (1 - n)\alpha z = (1 - n)b \tag{1.2}$$

By solving (1.2), then using (1.1), we can get the solution of (1) in an exact and closed form.

In the specific example :

$$y' - \frac{2}{x}y = 4x^2y^{1/2}, \quad x \neq 0$$

which is of Bernoulli type, the transformation $z = y^{1/2}$ leads to the LDE:

$$z' - \frac{1}{x}z = zx^2$$

of which, by using $\mu = \frac{1}{x}$ as an integrating factor, we find the general solution : $z = (cx + x^3)$, when the general solution of the original NL equation is : $y = (cx + x^3)^2$.

As we see, the single constant of integration enters the general solution of (1) in a nonlinear way.

Example 2. The Ricatti equation :

$$y' = \alpha y^2 + by + c \tag{2}$$

where α, b, c are functions of x, and $\alpha \neq 0$.

Liouville proved (1841) that this simple NLDE in its general form can not be solved by elementary exact methods.

(a): This equation can be linearized to a LDE of second order.

Transforming (2) by

$$y(x) = \frac{v(x)}{\alpha(x)} \tag{2.1}$$

we get :

$$v' = v^2 + \left(\frac{\alpha'}{\alpha} + b\right)v + \alpha c \tag{2.2}$$

which is NL in the new vrriable v. Now, transforming (2.2) by:

$$v(x) = -\frac{u'(x)}{u(x)} \tag{2.3}$$

we get the LDE of second order in u(x) with variable coefficients:

$$u'' - \left(\frac{\alpha'}{\alpha} + b\right)u' + \alpha cu = 0 \tag{2.4}$$

of which the solution can not, in general, be expressed in exact form in term of a finite number of elementary functions.

.(b): We can linearize the Ricatti equation (2) by specific «restrictions», e.g. when we know a particular solution $y_1(x)$.

By the transformation :

$$y(x) = y_1(x) + z(x) \tag{2.5}$$

the equation (2) leads to the Bernoulli equation in z :

$$z' - (b + 2\alpha y_1) z = \alpha z^2 \tag{2.6}$$

which can be solved by linearization, and from its solution and the transformation (2.5) we can get the general solution of (2).

Example 3. The NLDE :

$$y = \frac{X + yZ}{Y + xZ} \tag{3}$$

where X, Y, Z are homogeneous functions of x, y, the X, Y of degree μ of homogeneity, and Z of degree v, can be solved by an exact linearization. Transforming by :

$$y = xz \tag{3.1}$$

and using the homogeneity property of X, Y, Z, equation (3) becomes :

$$\frac{dx}{dz} + A(z)\, x = B(z)\, x^{v-\mu+2} \tag{3.2}$$

which is of Bernoulli type. The functions A(z) and B(z) are known functions, calculated in the process to find (3.2).

Example 4. The Lagrange equation :

$$y = x\,\varphi(y') + \psi(y') \tag{4}$$

which is linear in x and y, but nonlinear in y', can be solved by an exact linearization.

Differentiating (4) with respect to x, putting $y' = p$, then considering x as dependent variable and p as independent variable, we get:

$$(\varphi(p) - p)\frac{dx}{dp} + x\,\varphi'(p) + \psi'(p) = 0 \tag{4.1}$$

which is linear in x. If $x = x(p)$ is the solution of (4.I), inserting $x = x(p)$ into (4), we have the function:

$$y = x(p)\,\varphi(p) + \psi(p) \tag{4.2}$$

which is the general solution of (4) in an exact closed form with p as a parameter.

Examples 5. The NLDE: (a): $f(y') = 0$, (b): $f(y'') = 0, \ldots$ can be solved by an exact linearization.

.(a): If in (a) we put $y' = k$, when $y = kx + c$, that is $k = \frac{y-c}{x}$ then the general solution of (a) is: $f\left(\frac{y-c}{x}\right) = 0$, $c =$ parameter. E.g. the equation: $(y')^4 - 1 = 0$ has $\left(\frac{y-c}{x}\right)^4 - 1 = 0$ as general solution. We have:

$$\left(\frac{y-c}{x}\right)^4 - 1 = \left\{\left(\frac{y-c}{x}\right)^2 + 1\right\} \cdot \left(\frac{y-c}{x} - 1\right)\left(\frac{y-c}{x} + 1\right).$$

The first factor is not zero, the other factors may be zero, then the general solution of this specific example is the couple of the two families of lines: $y = x + c$, and $y = -x + c$ that is all lines in the x, y —plane parallel to the first and second bisector.

.(b): If in (b) we put $y'' = k$, when $y = \frac{k}{2}x^2 + c_1 x + c_2$, or $k = \frac{2}{x^2}(y - c_1 x - c_2)$ then the general solution of (b) is:

$$f\left(\frac{2}{x^2}(y - c_1 x - c_2)\right) = 0, \quad c_1 \text{ and } c_2 \text{ parametrs.}$$

Example 6. We may have a linearization of a complicated NLDE by factorization, e.g., if the NLDE is factorized as:

$$(y^2 + y'^2 + 1) \cdot y'' \cdot (x^3 y''' + x^2 y'' - 2xy' + 2y) = 0 \tag{6}$$

This equation is equivalent to the system:

$$\text{(a)}: \quad y'' = 0, \qquad \text{(b)}: \quad x^3 y''' + x^2 y'' - 2xy' + 2y = 0 \tag{6.1}$$

The first factor can not be zero. The second equation of (6.1) is of Euler type. The solutions of (6.1) are:

$$(a): \quad y = c_1 x + c_2, \qquad (b): \quad y = c_3 x + c_4 x^{-1} + c_5 x^2 \qquad (6.2)$$

where $c_1, \ldots, c_5$ are parameters. Therefore the general solution of (6) is given by both equations (6.2), and from each point of the x, y -plane two solution curves of (6.2) pass, one from the family (6.2.a) and one from the family (6.2.b).

We may have other cases of exact linearization of NLDE [1], [2].

In the following two examples the exact linearization is of different nature:

Example 7. The problem of mechanics of the free rotation of a rigid body with any «mass distribution» is governed by the Euler system of NLDE:

$$\left.\begin{aligned} I_1 \dot{\omega}_1 + (I_3 - I_2)\,\omega_3\,\omega_2 &= 0 \\ I_2 \dot{\omega}_2 + (I_1 - I_3)\,\omega_1\,\omega_3 &= 0 \\ I_3 \dot{\omega}_3 + (I_2 - I_1)\,\omega_2\,\omega_1 &= 0 \end{aligned}\right\} \qquad (7)$$

where $\omega_1, \omega_2, \omega_3$ are the unknown angular velocity components, and the parameters I_1, I_2, I_3 are the moments of inertia of the given rigid body about the body coordinate system with origin the mass center of the body. We can linearize this nonlinear system, if we accept a «special mass distribution», which corresponds to a «special restriction of the parameters» I_1, I_2, I_3. If the mass distribution of the rotating body is such that the body has an axis of symmetry, say the ω_1 -axis, then $I_2 = I_3 = I$, and (7) leads to the linear coupled system:

$$\dot{\omega}_2 = c_1 \omega_3, \qquad \dot{\omega}_3 = -c_1 \omega_2 \qquad (7.1)$$

where: $\quad c_1 = \bar{\omega}_1 (I - I_1)/I = \text{constant}.$

The above linearization, occurring by a restriction of the parameters, corresponds to a physical problem, which is a special case of the initial problem expressed by (7), and the solution of the linearized system (7.1) is a special case of (7).

Example 8. The «pure Keplerian motion» of a body of mass m around a central body of mass M is due to the attractive Newtonian force of the bodies, in the absence of perturbing forces. This motion is governed by the NLDE in vector form:

$$\ddot{x} + \frac{k^2}{r^3} x = 0 \tag{8}$$

where $k = K(M+m) =$ constant, $x = (x_1, x_2, x_3)$, $r^2 = x_1^2 + {}_2^2 + x_3^2$.

In the following we see two types of exact linearizations of (8), one by «regularization», and the other by «restriction of the variables» [3].

(a): Regularization of (8). We use an auxiliary equation, and change the independent variable.

The equation (8) is «singular» with the origin as the singularity.

When the motion of m is close to M, it is a «near collision» motion, when large gravitational forces appear and sharp bends of the orbit.

Such a phenomenon occurs when, e.g., an artificial space vehicle is at its start or at its destination.

By appropriate transformation of (8), it is possible to get a regular equation free ot singularities. This is called: «regularization». In the one-dimension case, (8) becomes:

$$\ddot{x} + \frac{k^2}{x^2} = 0 \tag{8.1}$$

and the energy function of the pure Keplerian motion is:

$$h_k = \frac{k}{x} - \frac{1}{2}\dot{x}^2 \tag{8.2}$$

where the energy h_k is a negative constant.

If, in these equations, instead of the «natural time» t, the «artificial time» τ is taken according to:

$$dt = x\,d\tau \tag{8.3}$$

the equations (8.1) and (8.2) become:

$$xx'' - x'^2 + k^2 x = 0 \tag{8.4}$$

$$x'^2 = 2(k^2 x + h_k x^2) \tag{8.5}$$

where τ is the independent variable.

Inserting now (8.5) into (8.4) we have the LDE

$$x'' + 2 h_k x = k^2 \tag{8.6}$$

which is the «regularization» of (8).

(b): Restriction of the Variables in (8). By restricting the variables x_1, x_2, x_3 according to:

$$r^2 = x_1^2 + x_2^2 + x_3^2 = \text{constant} \tag{8.7}$$

we linearize (8), when it reduces to the LDE

$$\ddot{x} + vx = 0 \tag{8.8}$$

with $v = kr^{-3/2}$. The solution of (8.8) is:

$$x = a \cos vt + b \sin vt \tag{8.9}$$

with a and b are constant vectors.

The restriction (8.7) specializes the motion of m to be a motion on the surface of the sphere (8.7), and, since the motion of m is only under the influence of a central force, the motion of m is circular on a plane through the origin with period of revolution: $T = \frac{2\pi}{v} = \frac{2\pi}{k} r^{3/2}$ and velocity

$$\dot{x} = v(-c_1 \sin vt + c_2 \cos vt) \tag{8.10}$$

of magnitude:

$$U = vr = \frac{k}{\sqrt{r}} = \text{constant}. \tag{8.11}$$

By considering r of (8.7) as a parameter, the above linearization gives various circular motions of m around M inside a sphere of radius the maximum of the parameter r.

SUMMARY

The mathematical models of the phenomena are usually NLDE, which is very difficult to be solved. Some classes of these equations can be treated by special methods, among which are the linearization methods, either exact or approximate.

In the exact linearization methods one may use appropriate transformation of the variables, when one may have general solutions of the NLDE in closed form, which is an ideal case.

Also, one may use restrictions of the variables or of the parameters of the NLDE, when one may have some special subclasses of the general solutions of the NLDE.

For a single NLDE one may have more than one formula for its general solution, and a NLDE may have, in addition, singular solutions.

ΠΕΡΙΛΗΨΙΣ

Τὰ μαθηματικὰ μοντέλα τῶν διαφόρων φαινομένων εἶναι, συνήθως, μὴ γραμμικαὶ διαφορικαὶ ἐξισώσεις, τῶν ὁποίων ἡ λύσις εἶναι, ἐν γένει, εἴτε ἀδύνατον εἴτε πολὺ δύσκολον νὰ εὑρεθῇ. Διὰ μερικὰς κατηγορίας τῶν μὴ γραμμικῶν ἐξισώσεων δύνανται νὰ ἐφαρμοσθοῦν εἰδικαὶ μέθοδοι, μεταξὺ τῶν ὁποίων εἶναι καὶ ἡ «γραμμικοποίησις», εἴτε ἀκριβὴς (exact) εἴτε κατὰ προσέγγισιν (approximate). Παλαιότερα, ἡ σπουδὴ τῶν μὴ γραμμικῶν ἐξισώσεων διὰ γραμμικοποιήσεως (διαγραφῆς τῶν μὴ γραμμικῶν ὅρων τῶν ἐξισώσεων) ἦτο ἡ δεσπόζουσα μέθοδος, ἰδίως εἰς προβλήματα ἐφαρμογῶν. Πρὸ μερικῶν δεκαετηρίδων ὅμως ἀπεδείχθη ὅτι ἡ γραμμικοποίησις ἐξαλείφει βασικὰς ἰδιότητας τῶν λύσεων τῶν μὴ γραμμικῶν ἐξισώσεων καὶ ὅτι οἱ μὴ γραμμικοὶ ὅροι τῶν ἐξισώσεων παίζουν δεσπόζοντα ρόλον εἰς τὴν ἔρευναν τῶν φαινομένων.

Διὰ τῆς παρούσης ἐργασίας, ὅπως καὶ διὰ μιᾶς ἄλλης ποὺ θὰ ἐπακολουθήσῃ, δίδονται παρατηρήσεις ἐπὶ τῶν μεθόδων γραμμικοποιήσεως, κυρίως ἐπὶ τῆς καταλληλότητος ἢ μὴ τῆς γραμμικοποιήσεως ὡς μεθόδου ἐρεύνης τῶν φαινομένων. Εἰδικῶς, εἰς τὴν παροῦσαν ἐργασίαν ἐξετάζονται ἀκριβεῖς μέθοδοι γραμμικοποιήσεως. Εἰς αὐτὰς τὰς μεθόδους γίνεται χρῆσις καταλλήλων μετασχηματισμῶν τῶν μεταβλητῶν, ὁπότε εἶναι δυνατὸν νὰ ἐπιτευχθοῦν «γενικαὶ λύσεις» τῶν μὴ γραμμικῶν ἐξισώσεων ὑπὸ «κλειστὴν μορφὴν» (closed form), τὸ ὁποῖον εἶναι ἰδεῶδες ἐπίτευγμα.

Ἐπίσης, μὲ κατάλληλον περιορισμὸν τῶν μεταβλητῶν ἢ τῶν συντελεστῶν τῶν μὴ γραμμικῶν ἐξισώσεων, εἶναι δυνατὸν νὰ ἐπιτευχθοῦν γενικαὶ λύσεις, αἱ ὁποῖαι εἶναι εἰδικαὶ περιπτώσεις τῆς γενικῆς λύσεως τῶν μὴ γραμμικῶν ἐξισώσεων.

Μία μὴ γραμμικὴ ἐξίσωσις ἐνδέχεται νὰ ἔχῃ περισσοτέρας τῆς μιᾶς γενικὰς λύσεις, ὅπως ἐπίσης ἐνδέχεται νὰ ἔχῃ, ἐπὶ πλέον, καὶ ἀνωμάλους (singular) λύσεις.

REFERENCES

1. E. Kamke, «Differentialgleichungen reeler Functionen», 2nd edition, Leipzig (1952).
2. D. Magiros, «Exact solutions of nonlinear differential equations», General Electric Co., 73A5240, 11/26/73.
3. E. Stiefel and G. Scheifele, «Linear and regular celestial mechanics», Springer-Verlag, New York (1971).

*

Ὁ Ἀκαδημαϊκὸς κ. **Ἰω Ξανθάκης,** παρουσιάζων τὴν ἀνωτέρω ἀνακοίνωσιν, εἶπε τὰ ἑξῆς :

Ἔχω τὴν τιμὴν νὰ ἀνακοινώσω εἰς τὴν Ἀκαδημίαν ἐργασίαν τοῦ κ. Δημητρίου Μαγείρου διὰ τὴν γραμμικοποίησιν μὴ γραμμικῶν Μαθηματικῶν Προτύπων τῶν Φαινομένων.

Ὡς γνωστὸν τὰ μαθηματικὰ πρότυπα (μοντέλα) διαφόρων φυσικῶν φαινομένων μᾶς ὁδηγοῦν συνήθως εἰς μὴ γραμμικὰς διαφορικὰς ἐξισώσεις, τῶν ὁποίων αἱ λύσεις εἶναι, κατὰ γενικὸν κανόνα, εἴτε ἀδύνατοι εἴτε πολὺ δύσκολοι. Λόγῳ τῆς δυσχερείας ταύτης, μέχρι πρό τινων ἐτῶν, διεγράφοντο οἱ μὴ γραμμικοὶ ὅροι τῶν ἀντιστοίχων διαφορικῶν ἐξισώσεων ἰδίως εἰς προβλήματα ποὺ ἀφεώρων ἐφαρμογάς· ἀλλὰ ἡ γραμμικοποίησις τῶν διαφορικῶν ἐξισώσεων διὰ τῆς μεθόδου ταύτης, δηλαδὴ τῆς διαγραφῆς τῶν μὴ γραμμικῶν ὅρων, ἐξαλείφει βασικὰς ἰδιότητας τῶν λύσεων τῶν μὴ γραμμικῶν διαφορικῶν ἐξισώσεων, διότι διαγραφόμενοι μὴ γραμμικοὶ ὅροι διαδραματίζουν ὡς ἐπὶ τὸ πλεῖστον δεσπόζοντα ρόλον εἰς τὴν ἔρευναν τῶν φαινομένων. Εἰς ὡρισμένας κατηγορίας μὴ γραμμικῶν διαφορικῶν ἐξισώσεων δύνανται νὰ ἐφαρμοσθοῦν εἰδικαὶ μέθοδοι, ποὺ στηρίζονται εἰς γραμμικοποίησιν κατὰ τρόπον ἀκριβῆ ἢ κατὰ προσέγγισιν.

Εἰς τὴν παροῦσαν ἀνακοίνωσιν ὁ κ. Μάγειρος ἀσχολεῖται μὲ ἀκριβεῖς μεθόδους γραμμικοποιήσεως, τὰς ὁποίας ἐφαρμόζει εἰς τὰς ἐξισώσεις Bernouilli, Ricatti, καθὼς καὶ εἰς διαφόρους ἄλλας μορφὰς διαφορικῶν ἐξισώσεων.

ΠΡΑΚΤΙΚΑ ΤΗΣ ΑΚΑΔΗΜΙΑΣ ΑΘΗΝΩΝ

ΣΥΝΕΔΡΙΑ ΤΗΣ 10ΗΣ ΙΟΥΝΙΟΥ 1976

ΠΡΟΕΔΡΙΑ ΝΙΚ. Κ. ΛΟΥΡΟΥ

ΕΦΗΡΜΟΣΜΕΝΑ ΜΑΘΗΜΑΤΙΚΑ.— **On the linearization of nonlinear models of phenomena. Second part: Linearization by approximate methods,** *by Demetrios G. Magiros* *. Ἀνεκοινώθη ὑπὸ τοῦ Ἀκαδημαϊκοῦ κ. Ἰ. Ξανθάκη.

INTRODUCTION

In a previous paper, contained in this volume, we discussed exact linearization techniques for solving nonlinear differential equations, when we get exact general solutions of NLDE in a closed form.

These ideal methods can be applied to only a few cases, when approximate linearization methods are suggested for approximate particular solutions of NLDE.

By a variety of examples, we point out different approximate linearization methods, cases where the linearization is not permitted, and cases where the linearization gives useful results.

In many fields of research, we see advantages and disadvantages of the linearization methods, as well as the importance and the influence of the nonlinearities.

---

* Δ. Γ. ΜΑΓΕΙΡΟΥ, **Γραμμικοποίησις μὴ γραμμικῶν μαθηματικῶν μοντέλων τῶν φαινομένων. Μέρος II: Γραμμικοποίησις μὲ κατὰ προσέγγισιν μεθόδους.**— Scientific Consultant, General Electric Company, (RESD) Philadelphia, PA., U. S. A.

Reprinted from the *Proceedings of the Athens Academy of Sciences* **51** (1976), 669–683.

LINEARIZATION BY APPROXIMATE METHODS

Whenever exact methods are not applicable to NLDE, and this case is the usual one in applications, approximate methods are suggested, by which one may get approximations of particular solutions.

The approximate methods are either geometrical, or analytical, or numerical, and the concept of linearization may enter to any of these methods.

The linearization by approximate methods leads to results, which are acceptable in some cases, but not acceptable in some other cases, this depending mainly on the nature of the problem which is associated with the linearization. In some linearization approximate methods the concepts of the «error» and the «norm», which measures the error, play essential role and characterize the methods, but in some other methods the concepts of the «error» and «norm» are not necessary. By the following examples and appropriate remarks we try to clarify the above statements.

Example 1. Let us pose the following problem:

«Find a curve such that the product of the distances from two different points E and E′ to anyone of its tangents is a non-zero constant, b^2». We take the line EE′ as x-axis, the middle point of EE′ as the origin, and $EE' = 2c$. The property of the curve of the problem leads to DE:

$$(y - cy')^2 - c^2y'^2 = b^2(1 + y'^2)$$

then we have:

$$y = xy' \pm (b^2 + \alpha^2 y'^2)^{1/2}, \qquad \alpha^2 = b^2 + c^2 \tag{1}$$

The equation (1) is of Clairaut type and by exact methods (not exact linearization) its solution can be found to be:

General Solution: $$y = cx \pm (b^2 + \alpha^2 c^2)^{1/2} \tag{1.1}$$

Singular Solution: $$\frac{x^2}{\alpha^2} + \frac{y^2}{b^2} = 1 \tag{1.2}$$

The true solution of the problem is the ellipse (1.2), and not any particular solution coming from the general solution (1.1).

If we try to solve the NLDE (1) by linearization, the only way is to omit the nonlinear term in (1), when the corresponding linear equation is $y = xy' + b$ of which the solution is $y = cx/b$, that is the family of straight lines in the x, y - plane, but not the solution (1.2) (ellipse).

By this example we see that the linearization by omitting the nonlinearities leads in general to unaccepted results.

Example 2. The NLDE.

$$y' = (x^2 + \varepsilon^2 y^2)^{1/2}, \quad \varepsilon = \text{parameter} \tag{2}$$

can not be solved by exact methods, but by a linearization coming from an «appropriate restriction of the variables», which leads to an approximate geometrical method, called «method of isoclines».

The linearization of (2) by omitting the nonlinearity leads to the linearized equation: $y' = \pm x$, of which the general solution is the family of parabolas $y = c \pm \frac{1}{2} x^2$, c = parameter. Such a linearization is not accepted.

The linearization of (2) by restricting the variables x and y according to the restriction:

$$(x^2 + \varepsilon^2 y^2)^{1/2} = k, \quad k = \text{constant} \tag{2.1}$$

gives the linearized equation:

$$y' = k \tag{2.2}$$

with general solution:

$$y = kx + c \tag{2.3}$$

The restriction (2.1) of the variables of (2) is written in the form:

$$\frac{x^2}{k^2} + \frac{y^2}{(k/\varepsilon)^2} = 1 \tag{2.4}$$

The above linearization of (2) is equivalent to taking curves in the x, y - plane through any point of which the unknown solution of (2) has the same slope, that is the same inclination angle with x-axis.

This idea gives a «graphical method» for construction of the solution of a NLDE, called «method of isoclines».

The «isocline curves» of (2) are the ellipses (2.4), by which one can get approximately the solution of (2) in the form of a «directed field».

Example 3. The study of the multivibrator, a basic electronic circuit, with nonlinear resistance and in the absence of external forces, leads to the famus Van der Pol equation :

$$\ddot{x} - \varepsilon (1 - x^2) \dot{x} + x = 0 \tag{3}$$

which is equivalent to the normal system :

$$\dot{x}_1 = x_2, \qquad \dot{x}_2 = \varepsilon (1 - x_1^2) x_2 - x_1 \tag{3.1}$$

The time t is the independent variable.
This equation can not be solved by exact methods.
Van der Pol found graphically that this equation has one isolated periodic solution, its «limit cycle».

We can linearize the system (3.1) by considering x^2 as an infinitesimal stronger than x, that is by taking the «nonlinearity condition» :

$$x^2 = o(x) \tag{3.2}$$

when the linearized system is :

$$\dot{x}_1 = x_2, \qquad \dot{x}_2 = -x_1 + \varepsilon x_2 \tag{3.3}$$

The origin of (3.3) is an «unstable equilibrium» for $\varepsilon > 0$, and it is a «globally asymptotically stable» for $\varepsilon < 0$.

According to a theorem of Liapunov, the stability situation of the origin of (3.1) is «topologically» the same with the stability situation of the origin of the linearized system (3.3), and, then, if the point of interest in our investigation is the stability situation of the origin of the NLS (3.1), the above linearization by the condition (3.2), is permitted.

But, the NLS (3.1) has one «limit cycle», while the linearized system (3.3) has no such solution. In addition, if the origin of (3.3) is «asymptotically stable», it will be «globally asymptotically stable», while the origin of (3.1) will be «asymptotically stable» in a finite region, the «region of attraction», of which the determination of the boundary needs the knowledge of the nonlinearity of (3.1). Therefore, the linearization, from the above point of interest, is not permitted, and the nonlinearity is necessary to be taken into account.

Linearization and the Stability of Equilibrum Points of Dynamical Systems.

The results of the previous example (the Van der Pol equation) can be generalized and supplemented by considering the general system of NLDE in its normal form :

$$\dot{x}_i = f_i(t, x_1, \ldots, x_n); \quad i = 1, \ldots, n \tag{a}$$

where the functions f_i have the properties that guarantee the existence and uniqueness of the solution of (a) through any point x_o in the region of the validity of (a) :

$$D: \quad 0 \leqslant t, \quad -\infty < x_1, \ldots, x_n < \infty \tag{a. 1}$$

In a large number of cases, the system (a) can be written in the form :

$$\dot{x} = Ax + X \tag{a. 2}$$

where x, $\dot{x}$, X are n-column matrices, A is a $(n \times n)$ - matrix either constant or time-dependent ; X does not contain linear terms in the variables $x_1, \ldots, x_n$ then X represents the set of the nonlinearities of (a), and $X(t, 0) \equiv 0$. The system :

$$\dot{x} = Ax \tag{a. 3}$$

is the linear part of (a. 2), that is its «first approximation».

For the applications, especially of engineering kind, the deduction of the NLDE (a. 2) to the LDE (a. 3), that is the linearization of (a. 2), has been universally prevailed in the past. This lasted until some decades ago, when stringent requirements and serious demands of new phenomena, as, e. g., phenomena of vacuum tubes, made clear the inadequacy of the linearization and the importance of the nonlinearities [**1**].

To study properties of the solutions of the NLDE (a. 2), (e. g., their continuity, periodicity, boundedness, stability, oscillation) the linearization of (a. 2) helps only in a few cases, and its acceptance is permitted only «under restrictions of the nature of the matrix A and of the nature and smallness of the nonlinearities X».

The study of the stability of the trivial solution of (a. 2) gives a typical area of research for a verifi cation of the above statements and remarks. Let us first state some results and apply them to appropriate examples in order to show possibilities for a linearization and the influence of the nonlinearities.

The systems (a. 2) and (a. 3) may be either «critical» or «noncritical». In the «noncritical systems» the real parts of the eigenvalues of the matrix A are all nonzero; and in the «critical systems» there are eigenvalues of A with zero real part.

a. Criteria of the First Approximation.

The following classical «Liapunov criteria of the first approximation» are based on the nature of the eigenvalues of the system.

Criterion a. «If in a noncritical linear system (a. 3) all eigenvalues have negative real part, then this system is «asymptotically stable» at the origin. If, in addition, the nonlinearity X of the system (a. 2) satisfies the condition:

$$\lim_{|x| \to 0} \frac{|X|}{|x|} = 0 \qquad \text{(a. 4)}$$

then, (a. 2) is also «asymptotically stable» at the origin».

Criterion b: «If in a noncritical linear system (a. 3) there are eigenvalues, at least one, with positive real part, then (a. 3) and (a. 2) are «unstable» at the origin, even if the nonlinearity condition (a. 4) holds».

Criterion c: «In a critical system one can not decide about the stability or instability on the system at the origin without taking into account the nonlinearities of the system».

As we can see, the linearization of the system in the first two cases gives results accepted for the nonlinear system, but in the third case, the «undecided case», the linearization is not permitted.

The above criteria are referred to «autonomous systems». In «nonautonomous systems», the linearization in the case of the first criterion necessitates nonlinearity conditions different than (a. 4).

E.g., if the matrix A is either constant or periodic function of time t, and $X = X(t, x)$, the «nonlinearity condition» will be:

$$\lim_{|x| \to 0} \frac{|X(t, x)|}{|x|} \leqslant m\, e^{\alpha t} |x|^{b} \quad (a.5)$$

where α, b, m are positive constants and $t \geqslant 0$.

In implicit equations, e.g., if $X = X(x, \dot{x})$, the «nonlinearity condition» will be:

$$\lim_{\substack{|x| \to 0 \\ |\dot{x}| \to 0}} \frac{|X(x_1 \dot{x})|}{|x| + |\dot{x}|} = 0 \quad (a.6)$$

There are «nonlinearity conditions» of integral type.

b. Criteria by Using a Liapunov Function.

The use of Liapunov functions, which gives the «second Liapunov method» in stability theory, answers the question of the stability situation of the origin in critical systems.

A function $V = V(t, x_1, \ldots, x_n)$, of which the time derivative is calculated by using the NLDE (a):

$$\dot{V} = \frac{\partial v}{\partial t} + \sum_{i=1}^{n} \frac{\partial v}{\partial x_i} \frac{\partial x_i}{\partial t} = \frac{\partial v}{\partial t} + \sum_{i=1}^{n} \frac{\partial v}{\partial x_1} f_i \quad (a.7)$$

is called a «Liapunov function».

To find a «Liapunov fuction» for a nonlinear system is a very important problem. The following criteria by using Liapunov functions consist of the classical «second group of Liapunov criteria».

Criterion d: «If in a region D, where the origin of the system is included, there is a Liapunov function V such that V is positive and its time derivative $\dot{V}$ is negative, except at the origin where both V and $\dot{V}$ are zero, then the origin (as an equilibrium point of the system) is «asymptotically stable».

Criterion e: «If in a region D, which includes the origin, there exists a Liapunov V such that V is positive in D, except at the origin where it is zero, and $\dot{V}$ is either zero everywhere in D or negative in D with the exception of some points of D (the origin included) where it is zero, then the origin is «stable» in D».

Criterion f: «If V and $\dot{V}$ are both positive in D, but both zero at the origin, then the origin is «unstable» in D».

Of all the above criteria there are modifications and extensions, but these criteria are sufficient to show possibilities for a successful linearization of nonlinear systems, the influence of the nonlinearities for the stability of the origin of the systems, and indicate an approximate calculation of the «stability regions» of the origin.

By the following examples we analyze and accomplish all these [**3**].

Example 4. The Duffing equation:

$$\ddot{x} + \alpha_1 \dot{x} + \alpha_2 x + b x^3 = 0 \tag{4}$$

which in a normal form, can be written as:

$$x = x_1, \quad \dot{x}_1 = x_2, \quad \dot{x}_2 = -\alpha_2 x_1 - \alpha_1 x_2 - b x_1^3 \tag{4.1}$$

or in a matrix form:

$$\begin{pmatrix} \dot{x}_1 \\ \dot{x}_2 \end{pmatrix} = \begin{pmatrix} 0 & 1 \\ -\alpha_2 & -\alpha_1 \end{pmatrix} \begin{pmatrix} x_1 \\ x_2 \end{pmatrix} = \begin{pmatrix} 0 \\ -b x_1^3 \end{pmatrix} \tag{4.2}$$

has three equilibrium points, among which is the origin of which we ask for the stability situation.

The nonlinearity condition (a. 4) is satisfied, for

$$\lim_{|x| \to 0} \frac{|X|}{|x|} = \lim_{|x_1| \to 0} \frac{|b|\,|x_1^3|}{|x_1|} = \lim_{|x_1| \to 0} |b|\, x_1^2 = 0 \tag{4.3}$$

The eigenvalues are given by:

$$\lambda = \frac{1}{2}\left(-\alpha_1 \pm \sqrt{\alpha_1^2 - 4\alpha_2}\right) \tag{4.4}$$

then the origin of (4), that is its state $(x, \dot{x})$ at the origin, is

(a): «asymptotically stable», if $\alpha_1 > 0$, $\alpha_2 > 0$

(b): in the «undecided case», if $\alpha_1 = \alpha_2 = 0$, or $\alpha_1 = 0$, $\alpha_2 > 0$, or $\alpha_1 \neq 0$, $\alpha_2 = 0$

(c): «unstable» in the other cases of α_1 and α_2.

E x a m p l e 5. The equation:

$$\ddot{x} + \dot{x} + x - \sqrt{x} = 0, \qquad x > 0 \tag{5}$$

can be written in the form:

$$x = x_1, \quad \dot{x}_1 = x_2, \quad \dot{x}_2 = -x_1 - x_2 + \sqrt{x_1} \tag{5.1}$$

or:

$$\begin{pmatrix} \dot{x}_1 \\ \dot{x}_2 \end{pmatrix} = \begin{pmatrix} 0 & 1 \\ -1 & -1 \end{pmatrix} \begin{pmatrix} x_1 \\ x_2 \end{pmatrix} = \begin{pmatrix} 0 \\ \sqrt{x_1} \end{pmatrix} \tag{5.2}$$

The equilibrium points of this system are the origin and $(x = 1, x_2 = 0)$, and we are interested in the stability of the origin.

The eigenvalues are:

$$\lambda = \frac{1}{2}(-1 \pm \sqrt{3}\, i), \quad i = \sqrt{-1} \tag{5.3}$$

The nonlinearity condition (a. 4) is not satisfied, for:

$$\lim_{|x| \to 0} \frac{|X|}{|x|} = \lim_{|x_1| \to 0} \frac{\sqrt{x_1}}{|x_1|} = \lim_{|x_1| \to 0} \frac{1}{\sqrt{x_1}} = \infty \tag{5.4}$$

Then, the state $(x = 0, \dot{x} = 0)$ is «asymptotically stable» for the linear system corresponding to (5), but this state is «unstable» for the system (5) because of its nonlinearity.

E x a m p l e 6. The linear equation:

$$\ddot{x} + k\dot{x} + (\omega^2 + \varepsilon \cos t)\, x = 0 \tag{6}$$

with k positive constant, and ε small parameter, is equivalent to:

$$\dot{x} = y, \quad \dot{y} = -(\omega^2 + \varepsilon \cos t)\, x - ky \tag{6.1}$$

Its eigenvalues are:

$$\lambda = \frac{1}{2}\left(-k \pm \sqrt{k^2 - 4(\omega^2 + \varepsilon \cos t)}\right) \tag{6.2}$$

and the origin of (6) is «asymptotically stable».

If we add to (6) a nonlinearity $X = X(x)$ such that:

$$\lim_{|x| \to 0} (|X| / |x|) = 0$$

then the state $x = 0$, $x = 0$ will be «asymptotically stable».

Also, if we add a nonlinearity $X = X(t, x)$ such that:

$$\lim_{|x| \to 0} (|X(t, x)| / |x|) = 0$$

holds uniformly in t for $t \geqslant 0$, then the state $(x, \dot{x})$ is «asymptotically stable» at the origin.

Example 7.

$$\ddot{x} + \alpha x - e^t x^2 = 0 \tag{7}$$

The nonlinearity $X = -e^t x^2$ satisfies the condition (a. 5) for $m = \alpha = b = 1$, for:

$$\lim_{|x| \to 0} \frac{-e^t x^2}{|x|} = \lim_{|x| \to 0} e^t |x| = 0 \tag{7.1}$$

The eigenvalues are $\lambda = \pm \sqrt{-\alpha}$, then we have two cases:

(a): for $\alpha > 0$, $\lambda = \pm \sqrt{\alpha}\, i$, (b): for $\alpha < 0$, $\lambda = \pm \sqrt{\alpha}$

Therefore, although the nonlinearity is favorable for stability of the origin, if $\alpha > 0$ we have the «undecided case», and if $\alpha < 0$ the origin is «unstable».

Example 8. We consider the nonlinear singular equation:

$$\ddot{x} + \left(\frac{1}{1+x}\right)\dot{x} + \left(\frac{1}{1+x}\right)x = 0 \tag{8}$$

For the discussion of this equation by linearization, we need to separate in it the linear and nonlinear parts. We can write (8) as:

$$\ddot{x} + \left(1 - \frac{x}{1+x}\right)\dot{x} + \left(1 - \frac{x}{1+x}\right)x = 0$$

when:

$$\ddot{x} + \dot{x} + x - \left[\frac{x}{1+x}(x + \dot{x})\right] = 0 \tag{8.1}$$

The nonlinearity: $X = -\left[\frac{x}{1+x}(x + \dot{x})\right]$ satisfies the condition needed for stability of the origin, for:

$$\lim_{\substack{|x| \to 0 \\ |\dot{x}| \to 0}} \frac{\frac{x}{1+x}(|x| + |\dot{x}|)}{|x| + |\dot{x}|} = \lim_{|x| \to 0} \frac{|x|}{|1+x|} = 0 \tag{8.2}$$

The eigenvalues of the equation (8.1) are:

$$\lambda = \frac{1}{2}(-1 \pm \sqrt{3}\, i) \tag{8.3}$$

then the state $(x, \dot{x})$ is «asymptotically stable» at the origin.

The previous examples are treated by using the criteria of the «first approximation», that is, by using the nature of the eigenvalues. In the following two examples we use the criteria of the «Liapunov second method», which are related to Liapunov functions.

Example 9.

$$\left.\begin{aligned} \dot{x} &= -y + \alpha x^3 \\ \dot{y} &= x + \alpha y^3 \end{aligned}\right\} \tag{9}$$

The constant α characterizes the nonlinearity of the system. The origin is an equilibrium point, and the eigenvalues are $\lambda = \pm i$, then the stability of the origin can not be decided by linearization. We use a Liapunov function and apply the «criteria of the second group».

The function:

$$V = x^2 + y^2 \tag{9.1}$$

can be taken as a Liapunov function.

(a): For the linear system $(\alpha = 0)$,

$$\dot{x} = -y, \qquad \dot{y} = x \tag{9.2}$$

the derivative of V becomes:

$$\dot{V} = 2(x\dot{x} + y\dot{y}) = 2(-xy + xy) = 0 \tag{9.3}$$

Both V and $\dot{V}$ are zero at the origin, V is positive in the x, y - plane, and $\dot{V}$ is everywhere zero, then the origin is «stable» (but not «asymptotically stable») for the linear system (9.2).

(b): The function $\dot{V}$ for the nonlinear system (9) becomes:

$$\dot{V} = 2\{x(-y+\alpha x^3) + y(x+\alpha y^3)\} = 2\alpha(x^4+y^4) \tag{9.4}$$

Both V and $\dot{V}$ are zero at the origin, $V > 0$; and $\dot{V} > 0$ for $\alpha > 0$, $\dot{V} < 0$ for $\alpha < 0$.

Then, for $a > 0$ the origin is «unstable» for the system (9), and for $\alpha < 0$ it is «asymptotically stable».

This example shows the great influence of the nonlinearity, even if it is very small in magnitude. The stability of the origin of a linear system is completely changed if we add to the linear system a small nonlinearity, and in the nonlinear system which results, the nature of the stability depends not only on the magnitude of the nonlinearity but also on the sign of the nonlinearity.

Example 10. The linearization is unable to help the calculation of the «region of asymptotic stability» of the origin of a nonlinear system. The nonlinearities play a decisive role in this problem.

The Liapunov functions help to find an approximation to the region. The Van der Pol equation can be used as an example.

The Van der Pol equation, written in a normal form, is:

$$\dot{x}_1 = x_2, \qquad \dot{x}_2 = -x_1 + \varepsilon(1-x_1^2)x_2 \tag{10}$$

and the origin of this system is, for $\varepsilon < 0$, «asymptotically stable». An appropriate Liapunov fuction for this system is:

$$V = \frac{1}{3}(x_1^2 + x_2^2) \tag{10.1}$$

when, by using (10), its derivative is:

$$\dot{V} = \varepsilon x_2^2(1-x_1^2) \tag{10.2}$$

The function V is zero at the origin and positive everywhere in the x, y - plane. The function $\dot{V}$ is zero at the origin; and since $\varepsilon < 0$, it is $\dot{V} < 0$ for $1 - x_1^2 > 0$, that is inside the unit circle:

$$x_1^2 + v_2^2 = 1 \tag{10.3}$$

This circle is a part of the unknown «region of attraction» of the origin of the Van der Pol equation, that is this circle is an approximation of this region.

Example 11. Linearization of Almost Linear Systems.

In the previous examples, we saw different types of approximate linearization methods, which are not explicitely related to the concept of the «error of the approximation». In this example of «linearization of almost linear systems» we have a case of linearization which is characterized by the «error of approximation».

The nonlinear systems of the special form:

$$m\ddot{x} + kx = \varepsilon f(t, x, \dot{x}) \tag{11}$$

where m and k are positive constants, ε a small parameter, and f a nonlinear function, is called «almost linear» or «quasi-linear» system.

It appears in many fields and its treatment is very difficult.

By the «method of Krylor-Bogaliubov» one can replace this system by an «equivalent linear system» [2].

The linearization by this method is succeeded by the use of a combination of restrictions and transformations of variables and coefficients, and introducing some integrator operators.

This method can be successfully applied to modern control systems, where certain linear loops are interlinked with nonlinear ones.

The starting point of the method are formulae known from the «first approximation» of the system.

Let us see the essentials of the method.

By a series of mathematical operations during the process of linearization two parameters λ_1 and λ_2 are introduced, and, by using them, the equations (11) can be reduced to the linear system:

$$m\ddot{x} + \lambda_1\dot{x} + \lambda_2 x = 0(\varepsilon^2) \tag{11.1}$$

The solution of (11.1) gives the solution of (11) with accuracy of ε^2, and (11.1) and (11) are «almost equivalent» with error ε^2.

λ_1 and λ_2 are given in integral form by:

$$\left.\begin{aligned} \lambda_1 &= \frac{\varepsilon}{\pi\alpha\omega}\int_0^{2\pi} f(\alpha\cos\psi,\ -\alpha\omega\sin\psi)\sin\psi\, d\psi \\ \lambda_2 &= k - \frac{\varepsilon}{\pi\alpha}\int_0^{2\pi} f(\alpha\cos\psi,\ -\alpha\omega\sin\psi)\cos\psi\, d\psi \end{aligned}\right\} \tag{11.2}$$

where $\alpha = \alpha(t)$ and $\psi = \psi(t)$ are the α and ψ of $x = \alpha \cos \psi$, which is the solution of the first approximation of (11), that is when $\varepsilon = 0$ and $\omega = \sqrt{k/m}$ is the linear frequency.

λ_1 is the equivalent coefficient of damping, and λ_2 the equivalent coefficient of restoring force. λ_1 and λ_2 depend on the amplitude α.

The integral representation (11.2) of λ_1 and λ_2 is found formally, but one can give a justification of these formulae by using the «principle of energy balance», and the «principle of harmonic balance».

The calculation of λ_1 and λ_2 is the main subject of the above linearization «method of Krylov - Bogaliubov».

ΠΕΡΙΛΗΨΙΣ

Εἰς προηγουμένην ἐργασίαν, ἡ ὁποία περιέχεται εἰς τὸν παρόντα τόμον, ἔχομεν ἐξετάσει «ἀκριβεῖς μεθόδους γραμμικοποιήσεως» μὴ γραμμικῶν μοντέλων τῶν φαινομένων, αἱ ὁποῖαι δίδουν «γενικὰς λύσεις» τῶν μοντέλων.

Εἰς τὴν παροῦσαν ἐργασίαν ἐξετάζομεν «κατὰ προσέγγισιν μεθόδους γραμμικοποιήσεως», ὁπότε λαμβάνομεν προσεγγίσεις «μερικῶν λύσεων» τῶν μοντέλων. Ἡ σπουδὴ τῶν ἰδιοτήτων τῶν λύσεων μὴ γραμμικῶν διαφορικῶν ἐξισώσεων μόνον εἰς ὀλίγας περιπτώσεις ὑποβοηθεῖται μὲ τὴν γραμμικοποίησιν. Εἰς τὰς περισσοτέρας περιπτώσεις ἡ γραμμικοποίησις ὁδηγεῖ εἰς συμπεράσματα ἀσύμφωνα μὲ τὴν πραγματικότητα. Διὰ μίαν γραμμικοποίησιν, ἡ ὁποία ἐνδέχεται νὰ εἶναι ἐπωφελής, εἶναι ἀναγκαῖον νὰ λαμβάνεται ὑπ' ὄψιν ἡ φύσις τοῦ γραμμικοῦ μέρους τῶν μοντέλων, καθὼς καὶ κατάλληλοι περιορισμοὶ τῶν μὴ γραμμικοτήτων τῶν μοντέλων.

Τὸ πρόβλημα τῆς εὐσταθείας καταστάσεων ἰσορροπίας δυναμικῶν συστημάτων, αἱ ὁποῖαι δύνανται νὰ θεωροῦνται ὡς ἡ ἀρχὴ τῶν συστημάτων ἀναφορᾶς τῶν μοντέλων, δίδει ἓν τυπικὸν παράδειγμα, ποὺ ὑποδεικνύει περιπτώσεις δεκτῆς γραμμικοποιήσεως, καθὼς καὶ τὴν ἐπίδρασιν τῶν μὴ γραμμικοτήτων εἰς τὴν ἐξέτασιν τῶν φαινομένων.

Ἀκόμη καὶ εἰς μὴ γραμμικὰ συστήματα, τὰ ὁποῖα εἶναι «σχεδὸν γραμμικά», ἡ γραμμικοποίησις ἐμφανίζει ἀφαντάστους δυσκολίας.

Ἡ ἀλήθεια τῶν ἀνωτέρω παρατηρήσεων δεικνύεται μὲ κατάλληλα παραδείγματα.

REFERENCES

1. S. Lefschetz, «Geometric Differential Equations: Recent Past and Proximate Future», Differential Equations and Dynamical Systems, Academic Press, New York, pp. 1-14, 1967.
2. N. Minorsky, «Nonlinear Oscillations», D. Van Nostrum and Co., New York, pp. 349-355, 1962.
3. R. Struble, «Nonlinear Differential Equations», McGraw-Hill Book Co., New York, Chapter 5, 1962.

★

Ὁ Ἀκαδημαϊκὸς κ. **Ἰω. Ξανθάκης,** παρουσιάζων τὴν ἀνωτέρω ἐργασίαν, εἶπε τὰ ἑξῆς:

Ἔχω τὴν τιμὴν νὰ παρουσιάσω ἐργασίαν τοῦ κ. Δημητρίου Μαγείρου ὑπὸ τὸν τίτλον: «Κατὰ προσέγγισιν Μέθοδοι γραμμικοποιήσεως μὴ γραμμικῶν διαφορικῶν ἐξισώσεων».

Εἰς προγενεστέραν ἀνακοίνωσιν ὁ κ. Μάγειρος ἀναφέρεται εἰς ἀκριβεῖς μεθόδους γραμμικοποιήσεως μὴ γραμμικῶν διαφορικῶν ἐξισώσεων προτύπων τῶν φαινομένων, αἱ ὁποῖαι μᾶς ὁδηγοῦν εἰς γενικὰς λύσεις τῶν ἐν λόγῳ προτύπων.

Εἰς τὴν παροῦσαν ἐργασίαν ἐξετάζονται «Κατὰ προσέγγισιν μέθοδοι αἱ ὁποῖαι μᾶς ὁδηγοῦν εἰς μερικὰς λύσεις τῶν ἐν λόγῳ διαφορικῶν ἐξισώσεων». Ἡ σπουδὴ τῶν ἰδιοτήτων τῶν λύσεων μὴ γραμμικῶν διαφορικῶν ἐξισώσεων εἰς ὀλίγας μόνον περιπτώσεις ὑποβοηθεῖται μὲ τὴν γραμμικοποίησιν. Πράγματι εἰς τὰς περισσοτέρας τῶν περιπτώσεων ἡ γραμμικοποίησις μᾶς ὁδηγεῖ εἰς συμπεράσματα ποὺ δὲν συμφωνοῦν μὲ τὴν πραγματικότητα. Εἴς τινας ὅμως περιπτώσεις αἱ κατὰ προσέγγισιν μέθοδοι γραμμικοποιήσεως μᾶς παρέχουν συμπεράσματα ἀποδεκτά.

Ὁ συγγραφεὺς παρέχει μίαν ποικιλίαν ἐφαρμογῶν, ὅπου αἱ κατὰ προσέγγισιν μέθοδοι γραμμικοποιήσεως δὲν εἶναι ἐπιτρεπταί, καθὼς καὶ περιπτώσεις ὅπου αἱ ἐν λόγῳ μέθοδοι μᾶς ὁδηγοῦν εἰς χρήσιμα συμπεράσματα.

ΜΑΘΗΜΑΤΙΚΑ.— **Characteristic Properties of Linear and Nonlinear Systems,** *by Demetrios G. Magiros* *. Ἀνεκοινώθη ὑπὸ τοῦ Ἀκαδημαϊκοῦ κ. Ἰω. Ξανθάκη.

INTRODUCTION

In two previous papers (published in: Practica of Athens Academy, June 1976; and as GE Reports: (a) 76SDRO26, 6/28/76; and b) 76SDRO27, 7/1/76, we examined exact and approximate methods of linearization of nonlinear systems.

In the present paper the main characteristic properties of linear and nonlinear systems will be discussed. Some of these properties characterize only the linear systems, and some others only the nonlinear systems. Some properties of NLS disappear by its reduction to a LS, and, therefore, the nature of the problem associated with the NLS, accompanied by the knowledge of the properties of the systems, decide whether the linearization of the NLS is permitted or not.

Nonlinear systems, that is phenomena whose behavior can be described by models which are nonlinear differential equations (NLDE), become increasingly important in many fields, as in astronomy, space flight, automatic control, biology, economics.

Since by linearization of nonlinear systems important features of the phenomena are neglected, it is necessary to know general features, basic striking properties and characteristic peculiarities of linear and nonlinear systems.

We discuss this subject here, and the discussion is illustrated by simple examples.

I. THE PRINCIPLE OF SUPERPOSITION

This principle, first stated by D. Bernoulli (1775) and used by Fourier (1822) in his theorem, holds in linear systems (LS) and characterizes them, but in general, does not hold in nonlinear systems (NLS).

---

* Δ. Γ. ΜΑΓΕΙΡΟΥ, **Χαρακτηριστικαὶ Ἰδιότητες Γραμμικῶν καὶ μὴ Γραμμικῶν Συστημάτων.**

Reprinted from the *Proceedings of the Athens Academy of Sciences* **51** (1976), 907–935.

This principle consists of two properties :

(a): The sum of any number of linearly independent particular solutions of a DE is also a solution of the DE ; and

(b): Any constant multiple of a solution is also a solution.

In homogeneous LS the above principle holds and it characterizes completely these systems. By using this principle, one can obtain the general solution of these systems as a linear combination of some easy to get special solutions, the fundamental set of solutions.

In nonhomogeneous LS :

$$\dot{x}_i = Ax_i + h_i(t) \tag{1}$$

where A is a $(n \times n)$ matrix and $h_i(t)$ a «forcing function» or «input», the above principle means that: if x_1 is a solution of system (1) with input h_1, and x_2 another solution with input h_2, then $c_1 x_1 + c_2 x_2$ is solution of (1) with input: $c_1 h_1 + c_2 h_2$. The c_1 and c_2 are arbitrary constants.

• In NLS the principle of superposition does not hold, and, then, its general solution, if it exists, can not be formulated in the simple way as in L. S.

There are NLDE where only the property (b) of the above holds, but not the property (a). For example, for the equation :

$$y''^2 + yy' + y'^2 = 0 \tag{1.1}$$

ot which the terms have the same degree (two) in y, y', y'' we see that:

(i) : If y_1 and y_2 are «linearly independent» solutions, then $y=y_1+y_2$ is not a solution, but $y = c_1 y_1$ and $y = c_2 y_2$ are solutions ;

(ii): If y_1 and y_2 are «Linearly dependent» solutions, that is $y_2 = cy_1$, then this equation has $y = y_1 + y_2$ as a solution.

II. THE GLOBAL PROPERTY

The global (or predictability, or provincial) property characterizes the LS, but not the NLS.

• In LS the local behavior of the solutions implies their global behavior. That is, the global behavior can be predicted from the local behavior, then the LS, which may be defined for all values of time, are by nature «provincial».

• In NLS this property does not hold, the global behavior of NLS can not be implied from their local behavior, that is, the «unpredictability» characterizes the NLS. In NLS it may not be possible to extend the solutions beyond a certain time, or these solutions need not be defined for all values of time.

The linearization of a NLS may help to get local properties of NLS. By the following simple example the above are clarified. [4]

Let us take a LS and a NLS :

$$(a): \ \dot{x} = -x, \quad x(0) = x_0\,; \qquad (b): \ \dot{x} = -x + \varepsilon x^2, \quad x(0) = x_0 \tag{2}$$

where ε is a parameter. Their solutions are, respectively:

$$(a): \ x(t) = x_0 e^{-t}\,; \qquad (b): \ x(t) = \frac{x_0}{\varepsilon x_0 - (\varepsilon x_0 - 1)\, e^{t}} \tag{2.1}$$

graphically shown in Figure 1.

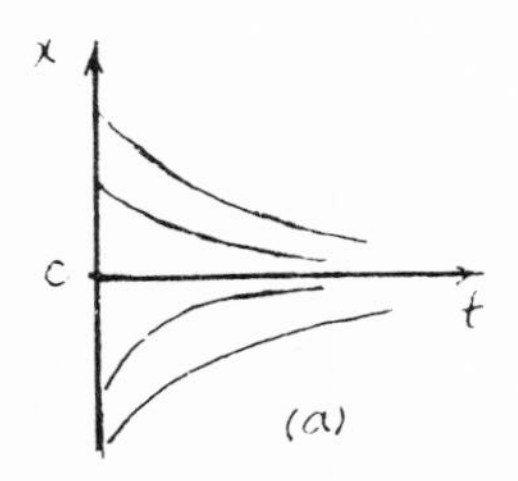

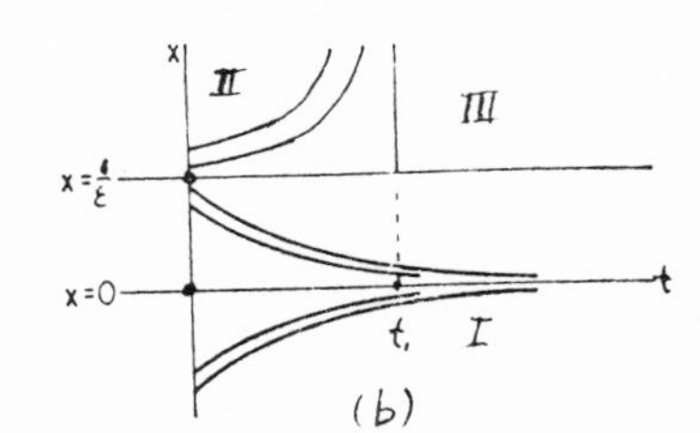

Fig. 1.

The equilibrium point of (2. a) is $x = 0$, and of (2. b) $x = 0$, $x = \frac{1}{\varepsilon}$.

For both systems (2), $x = 0$ is «asymptotically stable», but $x = \frac{1}{\varepsilon}$ is «unstable».

The solution (2. 1b) for $x > \frac{1}{\varepsilon}$ becomes infinite for the finite time:

$$t_1 = \log \frac{\varepsilon x_0}{\varepsilon x_0 - 1}\,. \tag{2.2}$$

In Figure (1. b) we distinguish three regions: I, II, III.

In region I $\left(0 \leqslant t,\ x < \frac{1}{\varepsilon}\right)$ all solutions have the t-axis as an asymptote;

In region II $\left(0 \leqslant t \leqslant t_1,\ x > \frac{1}{\varepsilon}\right)$ the solutions become unbounded for the finite time t_1; and

In region III $\left(t_1 \leqslant t,\ x > \frac{1}{\varepsilon}\right)$ there are no solutions.

We see that:

(i): In region I and close to origin, the solutions (2.1.b) have the same topological behavior as in the linear case (2.1.a), which means that the linearization of (2.b) in the region I gives useful information, and the local behavior of the solutions of both system (2) in I implies their global behavior.

(ii): In region II, the nonlinear term εx^2 causes the existence of a new phenomenon, and the linearization makes this phenomenon disappear.

From the local behavior of the solutions close to $x = \frac{1}{\varepsilon}$ one can not predict the future behavior.

As $\varepsilon \to 0$, the line $x = \frac{1}{\varepsilon}$ and the regions II, III tend to disappear at infinity, and the NLS tends to become a LS.

By the above example it becomes clear that in NLS the linearization has limitations, and, in general, the future behavior cannot be predicted.

III. LIMIT CYCLES

Limit cycles may be a phenomenon of NLS, but never a phenomenon of LS. Periodic phenomena of LS or NLS correspond to closed trajectories, called «cycles» with a period or frequency a finite number.

The cycles may constitute a «continuum spectrum», but may be isolated cycles, called «limit cycles», when in a neighborhood of them no other cycles exist.

The limit cycles must not be confused with some nonperiodic closed trajectories, which are members of a special class of solution-curves of systems, the «separatrices».

If, in a LS, a periodic solution y exists, then, due to the principle of superposition, which holds in LS, cy must be also a periodic solution, and since c is an arbitrary constant, no limit cycles exist in LS.

In some NLS, due to the special nature of their nonlinearities, limit cycles may exist, and for each NLDE the problems of existence of one or several limit cycles, their uniqueness and stability, as well as the construction of their boundaries and the calculation of their periods are important and, in general, difficult problems.

Stable limit cycles correspond to important physical phenomena, as, for example, to self-excited oscillations.

Example 1. The system: [8]

$$\left.\begin{aligned} \dot{x} &= \quad y + \frac{x}{\sqrt{x^2+y^2}}\{1-(x^2+y^2)\} \\ \dot{y} &= -x + \frac{y}{\sqrt{x^2+y^2}}\{1-(x^2+y^2)\} \end{aligned}\right\} \tag{3}$$

has the unit circumference $x^2+y^2=1$ as a «stable limit cycle». Indeed transforming (3) into polar coordinates and using: $x\dot{x}+y\dot{y}=r\dot{r}$, we have:

$$\dot{r} = 1-r^2, \quad \dot{\theta} = 1 \tag{3.1}$$

of which the solution is:

$$r = \frac{ce^{2t}-1}{ce^{2t}+1}, \quad \theta = t + c_1 \tag{3.2}$$

where $c = \dfrac{1+r_0}{1-r_0}$ and c_1 are the integration constants, r_0 the initial condition. If $r_0 < 1$, then, as $t \to \infty$, $r \to 1$ from inside. If $r_0 > 1$, then, as $t \to \infty$, also $r \to 1$ but from outside, when $x^2+y^2=1$ is a «stable limit cycle» of (3).

Example 2. The system:

$$\left.\begin{aligned} \dot{x} &= -y + x(x^2+y^2-1) \\ \dot{y} &= \quad x + y(x^2+y^2-1) \end{aligned}\right\} \tag{3.3}$$

has the unit circumference $x^2+y^2=1$ as an «unstable limit cycle».

This system, in polar coordinates, is written in the form:

$$\dot{r} = r(r^2-1), \quad \dot{\theta} = -1 \tag{3.4}$$

with a solution :

$$r = \frac{1}{\sqrt{1 - ce^{2t}}}, \quad \theta = t + c_1 \tag{3.5}$$

where $c = \frac{t_0^2 - 1}{r_0^2}$ and c_1 are the integration constants, and t_0, r_0 the initial conditions. We can check that $x^2 + y^2 = 1$ is an «unstable limit cycle» of (3.3).

Example 3. The NLDE ;

$$\ddot{x} + 3x - 4x^3 + x^5 = 0 \tag{3.6}$$

has infinitely many cycles, and some closed trajectories of special type (separatrices), but no limit cycles. We explain this statement.

The points : $x = 0$, $x = \pm 1$, $x = \pm \sqrt{3}$ on the x-axis are singular points of (3.6) ; the points : $x = 0$, $x = \pm \sqrt{3}$ are centers, and $x = \pm 1$ saddle points. **[7. b]**

The phase portrait of (3.6) is shown in Figure 2. There are special solution curves from a saddle point to another one, the «separatrices»

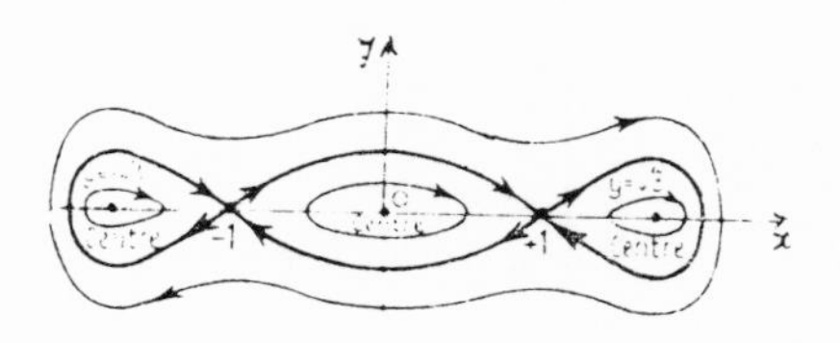

Fig. 2.

of (3.6), which separate the whole x, y-plane into four distinct regions in each of which there is a continuum spectrum of cycles. To go from one saddle point to the other, following a separatrice, theoretically infinite time is needed, and (3.6) has no limit cycles.

IV. SELF-EXCITED OSCILLATIONS

Self-excited (or self-sustained) oscillations are special periodic phenomena corresponding to stable limit cycles. They do not exist in LS, but may exist in NLS. They can be produced in NLS, where the nonlinearities appear in the damping forces, without the influence of external forces, that is in NLS of the form: $\ddot{x} + \varepsilon\varphi(x, \dot{x}) + kx = 0$.

In particular, the form:

$$\ddot{x} + \varepsilon\varphi(\dot{x}) + kx = 0 \tag{4}$$

is very useful. ε and k are constants.

To the nonlinearity of this equation we can give another useful form. We differentiate (4) with respect to the independent variable t then put $\dot{x} = y$, when (4) reduces to:

$$\ddot{y} + \varepsilon\psi(y)\dot{y} + \varkappa y = 0 \tag{4. 1}$$

where $\psi(y)$ is the derivative of $\varphi(\dot{x})$ with respect to $\dot{x} = y$.

The above NLE can be transformed into another form where $k = 1$, by changing the independent variable t into a new one τ, according to $\tau = kt$, when, e.g., (4) can be written in the form:

$$x'' + \varepsilon\varphi_1(x') + x = 0 \tag{4. 2}$$

where $\varphi_1(x') = \frac{1}{k^2}\varphi(kx')$, and the derivatives are taken with respect to τ.

Electrical systems involving vacuum tubes, mechanical systems of action of solid friction, the Froude's pendulum, and other systems, which can be formulated as special cases of the above NLDE, can execute selfexcited oscillations.

Rayleigh (1883) first studied this kind of oscillations in connection with acoustical phenomena, then Van der Pol (1927) in connection with electrical phenomena.

The Rayleigh equation is:

$$\ddot{x} + (-\alpha + b\dot{x}^2)\dot{x} + kx = 0 \tag{4. 3}$$

which is of the form (4). The Van der Pol equation is:

$$\ddot{y} - \varepsilon(1 - y^2)\dot{y} + y = 0 \tag{4. 4}$$

wich is of the form (4. 1).

The Rayleigh equation (4. 3) can be reduced to the Van der Pol equation (4. 4), by changing the variables t and x in (4. 3) into new variables τ and y, respectively, according to formulae:

$$\tau = \sqrt{k}\,t, \qquad y = \sqrt{\frac{3bk}{\alpha}}\,\dot{x} \tag{4. 5}$$

and getting $\varepsilon = \frac{\alpha}{\sqrt{k}}$.

- We give some special examples.

Example 1. We take a special case of the Rayleigh equation (4.3) with $\alpha = 1$, $b = \frac{1}{3}$, $k = 1$, that is the NLDE:

$$\ddot{x} - \dot{x} + x + \frac{1}{3}\dot{x}^3 = 0. \tag{4.6}$$

The linear part of (4.6):

$$\ddot{x} - \dot{x} + x = 0 \tag{4.7}$$

has $\lambda = \frac{1}{2} \pm i\frac{\sqrt{3}}{2}$ as eigenvalues, then its origin is unstable and the solutions around the origin are spirals that wind away from the origin.

If we add to (4.7) the nonlinearity $\frac{1}{3}\dot{x}^3$, we have (4.6) which in the phase plane can be written in the form:

$$\frac{dy}{dx} = \frac{-x + y - \frac{1}{3}y^3}{y}. \tag{4.8}$$

Applying the Liénard's graphical method, Figure (3a), we find a single stable limit cycle as a closed solution, Figure (3b), corresponding to self-excited oscillations.

-10-

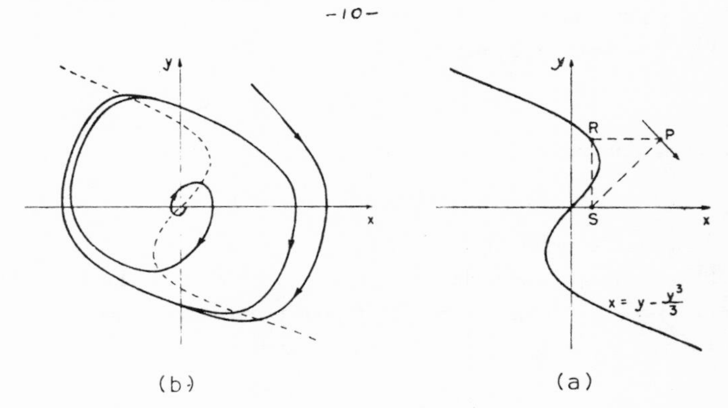

Fig. 3.

Example 2. The Van der Pol equation:

$$\ddot{x} - \varepsilon(1 - x^2)\dot{x} + x = 0 \tag{4.9}$$

is equivalent to the equation:

$$\frac{dy}{dx} = \frac{-x + \varepsilon(1 - x^2)y}{y}. \tag{4.10}$$

For $\varepsilon = 0$, the general solution of (4.10) is the family of concentric circles with center the origin. For $\varepsilon = 1$, application of isocline method shows the limit cycle as in Figure 4, corresponding to self-excited oscillations.

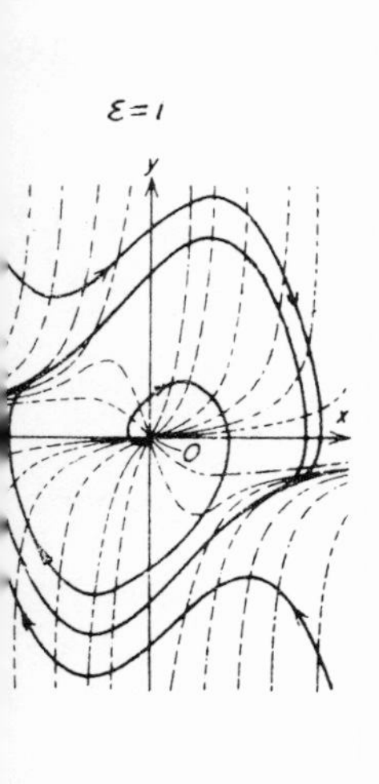

Fig. 4.

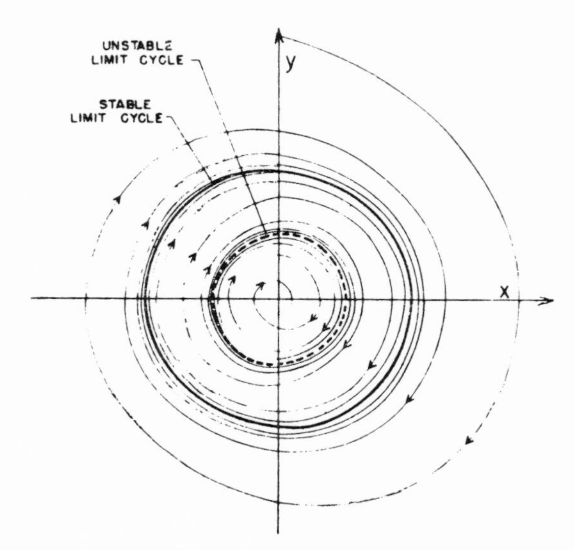

Fig. 5.

Example 3. The equation: [2]

$$\ddot{x} + \varepsilon(1 - \alpha\dot{x}^2 + b\dot{x}^4)\dot{x} + x = 0 \tag{4.11}$$

for $\varepsilon = 1$ is equivalent to:

$$\frac{dy}{dx} = \frac{-x - (1 - \alpha y^2 + by^4)y}{y} \tag{4.12}$$

of which the graphical solution, as shown in Figure 5, has two limit cycles, one unstable and the other stable, corresponding to self-excited oscillations.

Example 4. The equation:

$$\ddot{x} + x = \alpha\dot{x} + bx\dot{x} + c\dot{x}^2 + dx^2 \qquad (4.13)$$

can be produced in the theory of a common cathode generator, taking into account the anode reaction, if the value characteristic is represented by a quadratic polynomial. [11]

This equation in case: $\alpha = 0.2$, $b = 1$, $c = -1$, $d = c$ becomes:

$$\ddot{x} = -x + (0.2 + x - \dot{x})\dot{x} \qquad (4.14)$$

which, in the phase plane, is equivalent to

$$\frac{dy}{dx} = \frac{-x + (0.2 + x - y)y}{y}. \qquad (4.15)$$

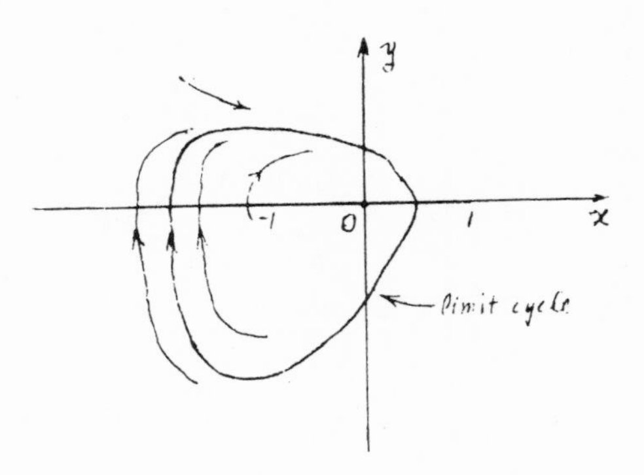

Fig. 6.

Equation (4.15) has one equilibrium point, the origin, which is unstable, and the method of isoclines, applied to (4.15), leads to Figure 6, where we see that a stable limit cycle exists, corresponding to self-excited oscillations. By using a graphical method, we can check that the shape of the limit cycle is distorted with the increase of the parameter α, becoming more and more non-sinusoidal.

V. THE PHENOMENA OF SUBHARMONIC RESPONSE

Subharmonic phenomena occur in some NLS but, in general, not in LS. They appear when the systems are subjected to external periodic forces, say sinusoidal.

If the frequency of the external force is ω, the system, under the influence of such a force, may exhibit periodic motions with frequency $\frac{\omega}{n}$, where $n = 2, 3, \ldots$, and such motions are, by definition, «subharmonic oscillations» or simply «subharmonics» of order $\frac{1}{n}$, $n = 2, 3, \ldots$

The existence of subharmonics can be found theoretically and checked experimentally.

Mechanical, electrical, acoustical, aerodynamical phenomena, and so on, exhibit subharmonic response. We refer some physical examples.

- The «loudspeaker» can be considered as a physical example of subharmonics of order $\frac{1}{2}$. A sinusoidal current in the coil causes the loudspeaker diaphragm to vibrate axially about a central position. These vibrations may, under certain circumstances, be with frequency half of that of the driving current.
- An aerodynamical model of subharmonics could be based on the fact that cartain parts of an airplane can be excited to violent oscillations by an engine running with a frequency much larger than the natural frequency of the oscillating parts.
- An electrical model of subharmonics might be an electrical oscillatory circuit in which the nonlinear oscillation take place because of a saturablecore inductance under the impression of an alternating electromotive force of sinusoidal type.

In connection with subharmonics of a nonlinear forced system, important problems of current interest are: to find conditions for the existence of one or several subharmonics and their appropriate order, to calculate their amplitudes, to discuss their stability and determine their region of stability. These problems lead to restrictions of the coefficients of the system and of its nonlinearities, and to restrictions of the amplitude and frequency of the external forces.

By the following examples we clarify some of the above statements.

Example 1. The Linear Forced System of One Degree of Freedom.

We consider the linear system of one degree of freedom under the influence of an external sinusoidal force in two cases, one with constant coefficients and the other with variable coefficients.

Case A: Linear System With Constant Coefficients. In this case we have:

$$\ddot{x} + 2r\dot{x} + p^2x = B\cos(\omega t + \varphi) \tag{5}$$

where r, p, B, ω, φ are constants; B, ω and φ are the amplitude, the frequency and the phase of the external force.

We investigate the possibility of the existence of subharmonics of this system.

The general solution of (5) is of the form:

$$x = A_1 e^{\lambda_1 t} + A_2 e^{\lambda_2 t} + B_1 \cos(\omega t + \varphi - \delta) \tag{5.1}$$

where δ is the phase shift, A_1 and A_2 are arbitrary constants, and the amplitude B_1 of the last term is due to the external force.

The eigenvalues λ_1 and λ_2 are given by:

$$\lambda_{1,2} = -r \pm \sqrt{r^2 - p^2} = -r \pm iq, \qquad q = \sqrt{p^2 - r^2}. \tag{5.2}$$

The subharmonics should come from the first two terms of (5.1), which must be periodic, then we exclude $q = 0$, and $q =$ imaginary, and we accept the case of a real q, that is $p^2 > r^2$, then, by calculating B_1, the solution (5.1) will have the form:

$$x = e^{-rt}(c_1 \cos qt + c_2 \sin qt) + \frac{B\cos(\omega t + \varphi - \delta)}{\sqrt{(p^2 - \omega^2)^2 + 4r^2\omega^2}} \tag{5.3}$$

with c_1, c_2 arbitrary constants.

The first part of (5.3) is the «free oscillations», and the second part the «forced oscillations» of the system (5).

The amplitudes of these two oscillations must be bounded, then we exclude for the first term of (5.3) $r < 0$, and for the second term $p^2 = \omega^2$, $r = 0$. Also, we exclude the case $r > 0$, because, in this case, the free oscillations of (5.3) are damped out and only the forced oscillations can be observed. Therefore, it remains the only case, the undamped case, $r = 0$, when (5.3) assumes the form:

$$x = c_1 \cos pt + c_2 \sin pt + \frac{B \cos(\omega t + \varphi - \delta)}{|p^2 - \omega^2|} \tag{5.4}$$

with the restriction $p \neq \omega$. δ in this formula is either zero (if $\omega < p$), or 2π (if $\omega > p$), when the «free oscillations» are either «in phase», or «180^0 out of phase» with the «forced oscillations». We can select ω and p such that $\omega = np$, n = integer, when the free oscillations of (5.4) should be «subharmonics of order $\frac{1}{n}$» of the undamped system (5).

But in the actual cases no system is undamped, and we can tell that we have «unstable subharmonics», which are not acceptable in practice. As a result, the system (5) with constant coefficients has no subharmonic oscillations.

Case B. Linear Systems With Variable Coefficients. In this case, subharmonics may exist even in the presence of viscous damping, but they rest on hypotheses which are not always met exactly by reality.

The systems with coefficients varying periodically in time are of great interest, and the external forces may depend not only on time, but on displacement and velocity as well.

We give two physical examples.

B.1: The Problem of Transverse Vibrations of a Rod Under the Action of a Longitudinal Periodic Force. [1]

This problem, after some hypotheses and transformations, leads to the «Mathieu equation»

$$\ddot{x} + \omega^2 (1 - h \cos vt)\, x = 0 \tag{5.5}$$

where

$$\omega^2 = \frac{g\pi^4 EI}{\gamma A l^4}, \qquad h = \frac{Pl^2}{\Pi EI} \ll 1, \tag{5.6}$$

l is the length of the rod, A its cross-section, γ its density, EI its rigidity, ω the free frequency of (5. 5), and the external force is $F(t) = P\cos\nu t$. We can calculate a solution of the equation (5. 5) which is a subharmonic of order $\frac{1}{2}$, of the form :

$$x = \alpha\cos\left(\frac{\nu}{2}t + \theta\right) \tag{5. 7}$$

by calculating α and θ as appropriate functions of time.

The calculation shows that oscillations of the form (5. 7) will be automatically excited, if the parameters ω, h of (5. 5) and the frequency ν of the external force are restricted according to :

$$2\omega\left(1 - \frac{h}{4}\right) < \nu < 2\omega\left(1 + \frac{h}{4}\right) \tag{5. 8}$$

which represents a zone for the existence of such a subharmonic.

The solution (5. 7) is in its «first approximation».

The solution of (5. 5) in its «second approximation» is :

$$x = \alpha\cos\left(\frac{\nu}{2}t + \theta\right) - \frac{\alpha h\omega}{8\left(\omega + \frac{\nu}{2}\right)}\cos\left(\frac{3}{2}\nu t + \theta\right) \tag{5. 9}$$

under the restriction :

$$2\omega\left(1 - \frac{h}{4} - \frac{h^2}{64}\right) < \nu < 2\omega\left(1 + \frac{h}{4} + \frac{h^2}{64}\right) \tag{5. 10}$$

We remark that ω and h are taken as parameters of (5. 5), and the subharmonic of (5. 5) may be called «parametric subharmonic».

B. 2. A Conical Loudspeaker Diaphragm. [7a]

The mechanism, shown in Figure 7, of which a simplified version may be a device for a conical loudspeaker diaphragm, can execute a parametric subharmonic oscillation of order $\frac{1}{2}$ of the driving force.

The mass m slides over a frictionless horizontal plane. The links and the spring are massless. The pin-joints at O, A, B are frictionless.

OA and AB are long enough for motion parallel to BD to be negligible in comparison with that along the axis of the spring. The driving force $F_0 = lf_0 \cos 2vt$ is applied to the cross-head B. It may be resolved into

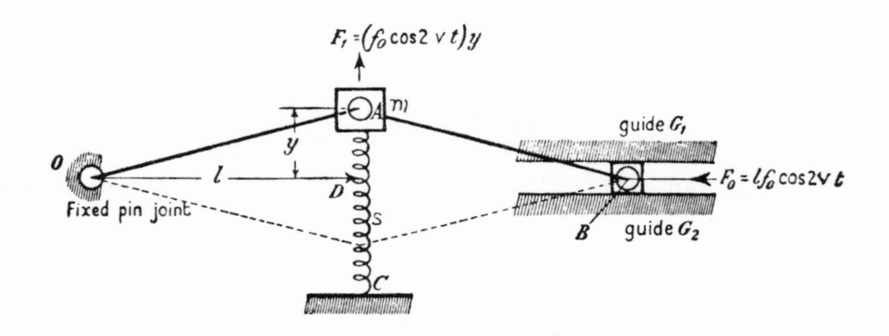

Fig. 7.

two components, one along AB, the other along DA. The latter is nearly $(f_0 \cos 2vt)y$ and it causes m to slide along the line CDA. There are three forces associated with m: one the inertia force $m\ddot{y}$, one the constraint sy due to the spring, and the driving force $(f_0 \cos 2nt)y$. The equation of the motion of m is:

$$m\ddot{y} + sy = (f_0 \cos 2v)\, y \tag{5.11}$$

which, if $\omega^2 = \frac{s}{m}$, $h = \frac{f_0}{s}$, becomes

$$\ddot{y} + \omega(1 - h \cos 2vt)\, y = 0 \tag{5.12}$$

a «Mathieu equation». As we can check, this equation enhibits a parametric subharmonic oscillation of order $\frac{1}{2}$. Figure 7 can be considered as a schematic plan of mechanism illustrating Mathieu equations.

We now refer to some examples of subharmonics of NLS.

E x a m p l e 2. Existence of Subharmonics of NL Nondissipative Systems.

The motion of a material point along a line is under the influence of a nonlinear restoring force, g(x), and an external time-dependent force, f(t), in the absence of resistance, when the equation of its motion is:

$$\ddot{x} + g(x) = f(t) \tag{5.13}$$

f(t) is periodic, $f(t) = f(t+T)$, and g(x) satisfies a Lipschitz condition. From the theorems of existence of subharmonic of (5.13) we refer only to the following two: [9]

I: «If f(t) is an even function, $f(t) = f(-t)$, n a natural number, and $\dot{x}(0) = \dot{x}\left(\frac{n}{2}T\right) = 0$, then (5.13) has a subharmonic of order $\frac{1}{n}$».

II: «If f(t) and g(x) are odd functions, $f(-t) = -f(t)$, $g(-x) = -g(x)$, and $x(0) = x\left(\frac{n}{2}T\right) = 0$, then (5.13) has a subharmonic of order $\frac{1}{n}$».

Example 3. The NLDE:

$$\ddot{y} + \omega_0^2 y - 2\varepsilon(1 - by^2)\dot{y} = -\frac{4\varepsilon\omega}{\sqrt{b}}\cos 3\omega t \tag{5.14}$$

with NL damping has a subharmonic of order $\frac{1}{3}$.

Indeed, by using the trigonometric identity:

$$4\cos^3\omega t = 3\cos\omega t + \cos 3\omega t$$

we can check that the function

$$y = \frac{2}{\sqrt{b}}\sin\omega t$$

satisfies (5.14).

Example 4. A generalization of the equation (5.14) is:

$$\ddot{y} + \omega_0^2 y + \varepsilon f(y)\dot{y} = A\cos(n\omega t + \varphi) \tag{5.15}$$

with $\varepsilon \ll 1$. By applying the «Poincaré method» and restricting A, f(y) and the integer n appropriately, we can find stable subharmonics of this equation. [3]

Example 5. We consider the NLS [6a, b]

$$\ddot{Q} + \bar{k}\dot{Q} + \bar{c}_1 Q + \bar{c}_2 Q^2 + \bar{c}_3 Q^3 = A \sin 2t \tag{5.16}$$

coming from electrical problems. By changing the coefficients according to:

$$\bar{k} = \varepsilon k, \quad 1 - \bar{c}_1 = \varepsilon c_1, \quad \bar{c}_2 = \varepsilon c_2, \quad \bar{c}_3 = \varepsilon c_3 \tag{5.17}$$

the system can be written in the useful form:

$$\ddot{Q} + Q = \varepsilon f(Q, \dot{Q}) + A \sin 2t \tag{5.18}$$

where:

$$f(Q, \dot{Q}) = -k\dot{Q} + c_1 Q - c_2 Q^2 - c_3 Q^3 \tag{5.19}$$

Case A: For $\varepsilon = 0$, the solution of (5.18) is:

$$Q = \bar{x}_1 \sin t + \bar{x}_2 \cos t - \frac{A}{3} \sin 2t \tag{5.20}$$

where $\bar{x}_1$ and $\bar{x}_2$ are arbitrary constants. The first two terms of (5.20) give the «subharmonic component» of order $\frac{1}{2}$ of the solution (5.20) with amplitude: $\bar{r} = \sqrt{\bar{x}_1^2 + \bar{x}_2^2}$.

This subharmonic is, as we know, without practical importance.

Case B: For $\varepsilon \neq 0$, we try to establish a solution of (5.18) of the form (5.20), where, instead of the constants $\bar{x}_1$ and $\bar{x}_2$, we calculate appropriate functions $x_1 = x_1(\varepsilon, t)$ and $x_2 = x_2(\varepsilon, t)$, such that their limits, as $\varepsilon \to 0$, are the constants $\bar{x}_1$ and $\bar{x}_2$, respectively.

The calculation of the amplitude $r = \sqrt{x_1^2 + x_2^2}$ of the subharmonic of order $\frac{1}{2}$ in the nonlinear case leads to appropriate restrictions for the existence of a real and stable amplitude r, and for the existence of two subharmonics of order $\frac{1}{2}$ with two amplitudes r_1 and r_2, which can be calculated.

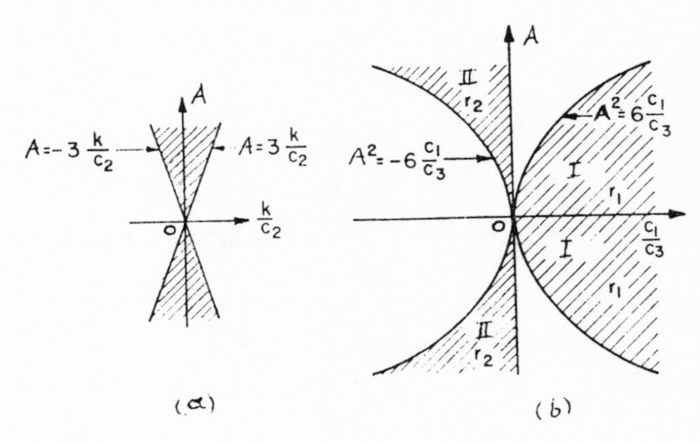

Fig. 8.

In Figure 8(a), the shaded regions in the $\left(\frac{k}{c_2}, A\right)$ -plane corresponds to a real amplitude r of subharmonics.

In Figure 8(b), the shaded regions I and II in the $\left(\frac{c_1}{c_3}, A\right)$ -plane correspond to real amplitudes r_1 and r_2, respectively, of the subharmonics.

Example 6. The forced system which is linearly damping and has a nonlinear restoring force expressed by : [5]

$$\ddot{y} + y = \varepsilon\{-k\dot{y} - f(y) + A\sin(\omega t + \varphi) \qquad (5.21)$$

where $\varepsilon \ll 1$, $f(y)/y \geqslant 0$, and ε, A, k positive, can be studied by applying the «Cartwright-Littlewood method», when important results in connection with its subharmonics can be found. Some of these results are :

(a) : Subharmonics of a given order exist in certain frequency bands.

(b) : If, at a given frequency, several states of subharmonics are possible, the amplitude is largest and the frequency band smallest for the lower order of subharmonics.

(c) : For a given form of the non-linearity f(y), the highest order obtainable is a function of the ratio A / k, which also determines the width of the frequency bands.

We remark that the subharmonic oscillations of a NLS belong to a general class of periodic motions characterized by the property that the ratio of the frequencies ω_0 (free frequency) and ω (frequency of the external force) is a rational number, that is : $\frac{\omega_0}{\omega} = \frac{n_0}{n}$, where n_0 and n are mutually prime integers.

We may have the following cases :

(a) If $n_0 = n = 1$, then: $\omega_0 = \omega$: harmonic oscillation of the system

(b) If $n = 1$, then: $\omega_0 = n_0 \omega$: harmonic oscillation of order n_0

(c) If $n_0 = 1$, then: $\omega_0 = \frac{\omega}{n}$: subharmonic oscillation of order $\frac{1}{n}$

(d) If $1 < n_0 < n$, then: $\omega_0 = n_0 \left(\frac{\omega}{n}\right)$: n_0 multiple of subharmonic of order $\frac{1}{n}$.

VI. AMPLITUDE AND FREQUENCY OF PERIODIC SOLUTIONS OF FREE LINEAR AND NON-LINEAR SYSTEMS

The periodic solutions of free (unforced) LS have amplitude independent of the frequency, and the frequency is the same for all trajectories. On the contrary, in the periodic solutions of free NLS the amplitude depends on the frequency, and the frequency changes from trajectory to trajectory. We see examples.

Example 1. The free LS :

$$\ddot{x} + 2r\dot{x} + p^2 x = 0 \tag{6}$$

has as general oscillatory solution the function :

$$x = e^{-rt}(c_1 \cos \beta t + c_2 \sin \beta t), \quad \beta = \sqrt{p^2 - r^2} \tag{6.1}$$

where c_1 and c_2 arbitrary constants.

(a): In the damped case, $r \neq 0$, the factor e^{-rt} of (6.1), which characterizes the amplitude of the oscillatory motion (6.1), either tends to infinity (for $r < 0$), or tends to zero (for $r > 0$) as $r \to \infty$, and the motion is not periodic.

(b): In the undamped case, $r = 0$, (6.1) becomes:

$$x = c_1 \cos pt + c_2 \sin pt \tag{6.2}$$

which is a family of periodic motions with the same frequency p, but amplitude $c = \sqrt{c_1^2 + c_2^2}$ calculated from the initial conditions. These amplitudes are independent of the frequency.

Example 2. The free NLS: [10 a]

$$\ddot{x} + p^2 x + b x^3 = 0 \tag{6.3}$$

can be solved exactly, and the period of the closed trajectories can be calculated. (6.3) is equivalent to:

$$\dot{x} = y, \qquad \dot{y} = -(p^2 x + b x^3)$$

then to:

$$\frac{dy}{dx} = -\frac{p^2 x + b x^3}{y} \tag{6.4}$$

of which the solution is the family of the closed trajectories:

$$y^2 + p^2 x^2 + \frac{b}{2} x^4 = c \tag{6.5}$$

c is the integration constant.

For the period T of the trajectories (6.5), we insert $y = \dot{x}$ and we take into account the symmetry of these trajectories, when:

$$T = 4 \int_0^{\alpha} \frac{dx}{\sqrt{c - \left(p^2 x^2 + \frac{b}{2} x^4\right)}} . \tag{6.6}$$

Transforming according to:

$$x = A \sin \theta \tag{6.7}$$

we can find:

$$T = 4\sqrt{2} \int_0^{\pi/2} \frac{dx}{\sqrt{2p^2 + bA^2 + bA^2 \sin^2 \theta}} . \tag{6.8}$$

As a result, the period T, or the frequency (T^{-1}), corresponding to the periodic motion on a trajectory is dependent on the amplitude of this motion, and the period, or the frequency, changes from trajectory to trajectory. In the special case $b = 0$ the system (6.3) becomes linear, and its period, which comes from (6.8) for $b = 0$, is a constant for all trajectories.

VII. THE RESONANCE PHENOMENA

The resonance phenomena occur in forced LS and NLS, in case the free frequency of the system is equal or very close to the frequency of the external force.

The damping in a LS and the nonlinearities in a NLS play a very important role in the resonance phenomena.

By the following examples is shown the influence of the damping and of the nonlinearities to the resonance phenomena.

Example 1. The general oscillatory solution of a damped forced LS:

$$\ddot{x} + 2r\dot{x} + p^2x = \alpha \cos \omega t \tag{7}$$

is the function:

$$x = e^{-rt}(c_1 \cos qt + c_2 \sin qt) + A \cos(\omega t + \varphi) \tag{7.1}$$

where:

$$q = \sqrt{p^2 - r^2}, \qquad A = \frac{\alpha}{\sqrt{(p^2 - \omega^2)^2 + 4r^2\omega^2}} \tag{7.2}$$

c_1 and c_2 in (7.1) are arbitrary constants.

The system (7) is in «resonance» with the external force, if $\omega = p$, and «near resonance» is $\omega - p \ll 1$.

We distinguish two cases:

(a): If (7) is «undamped», $r = 0$, then the solution (7.1) becomes:

$$x = c_1 \cos pt + c_2 \sin pt + \frac{\alpha}{|p^2 - \omega^2|} \cos(\omega t + \varphi). \tag{7.3}$$

If, in (7.3), ω is equal or close to p, the amplitude of the term of forced oscillations of (7.3) becomes infinite or very large, when the undamped system (7) is in «resonance» or «near resonance» with the external force.

(b): If (7) is positively damped, $r > 0$, the free oscillations of (7.1) are oscillatory motions but not periodic, since for $t \to \infty$, are damped out, when in the case of «resonance», $p = \omega$, the solution (7.1) becomes:

$$x = \frac{\alpha}{2\omega r} \cos \omega t. \tag{7.4}$$

This is a periodic motion with finite amplitude, which becomes very large if the damping coefficient r becomes very small.

By the above example is shown that the damping in LS can prevent resonance, and that a weak damping force can be capable of sustaining oscillations of large amplitude.

Example 2. Consider the LS and NLS:

$$(a): \ \ddot{x} + x = 0, \qquad (b): \ \ddot{x} + x + \frac{1}{6} x^3 = 0 \tag{7.5}$$

of which the general solutions are:

$$(a): \ x^2 + y^2 = c^2, \qquad (b): \ x^2 + y^2 + \frac{1}{12} x^4 = c^2 \tag{7.6}$$

All solutions of the LS are periodic with the same period.

But the period of the solutions of the NLS changes from trajectory to trajectory, and the period varies with amplitude. A periodic disturbance in the NLS will become out of phase with the free motion, and the forcing function should be an obstacle to increasing amplitude. The period varies with amplitude and non periodic solutions are possible.

As a result, the nonlinearity can prevent resonance, even in the absence of damping.

Due to the nonlinearity, the frequency will be changed, then resonance will be stopped.

The nonlinear terms exert, in general, a stabilizing influence until the motion has passed.

«Resonance phenomena» are in many cases unavoidable. They are dangerous, but sometimes controllable, and, although uncomfortable, they are not in all cases undesirable.

We refer to some physical examples related to resonance.

- If an elastic machine part vibrates in resonance with a sinusoidal force, it may become the source of vibrations with large amplitudes, which, in turn, may produce excessive stresses and lead to possible failure. It is, then, vital to design machine parts, or other engineering structures, in such a way as to avoid resonance with periodic forces.

• Tacoma Narrows bridge offers an example of a big failure in engineering history, due to resonance. This suspension bridge, just after its opening, started to exhibit a marked flexibility and a series of torsicnal oscillations, the amplitude of which steadily increased until the convolutions tore several suspenders loose, and the span of the bridge broke up (November 7, 1940) four months after its building. The wind created aerodynamical forces, which, at the time, were insufficiently understood.

• When a group of soldiers marches in step over a suspension bridge, the feet of the group exert a periodic force on the road bed. If the period of marching is equal to the natural period of the bridge resonance occurs and the sustained bridge oscillations may become dangerous.

• Resonance is sometimes not undesirable. One can in fact utilize it to produce large vibrations by means of small forces. E. g., the vibrations of a string can be sustained by means of an electro-magnet which is activated from a weak alternating current.

VIII. JUMP, OR HYSTERESIS, PHENOMENA

Jump discontinuities are phenomena in damped forced NLS and not of LS. They are found and explained mathematically and checked experimentally, especially in electrical and mechanical systems.

There are frequency regions where the amplitude of the oscillations jumps discontinuously and, in these regions, the oscillations have a kind of instability.

Let us take, as an example, the system : **[10b]**

$$\ddot{x} + c\dot{x} + p^2x + bx^3 = F\cos(\omega t + \varphi) \qquad (8)$$

The investigation of a periodic solution of this equation of the form : $x = A\cos\omega t$ loads to the formula :

$$\left\{(p^2 - \omega^2)A + \frac{3}{4}bA^3\right\}^2 + cA^2\omega^2 = F^2 \qquad (8.1)$$

where we consider A versus ω, p and c constants, and F as a parameter.

In the undamped case, $c = 0$, Figure 9 shows the amplitude curves which are curves without closed branches.

In the damped case, $c > 0$, Figure 10 shows these curves having a single branch for each value of F.

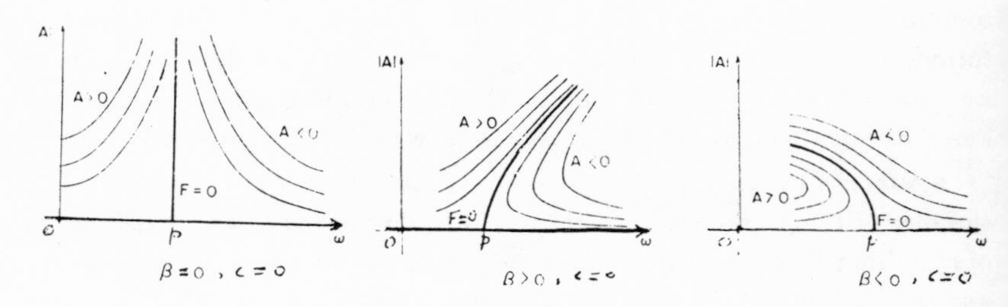

Fig. 9.

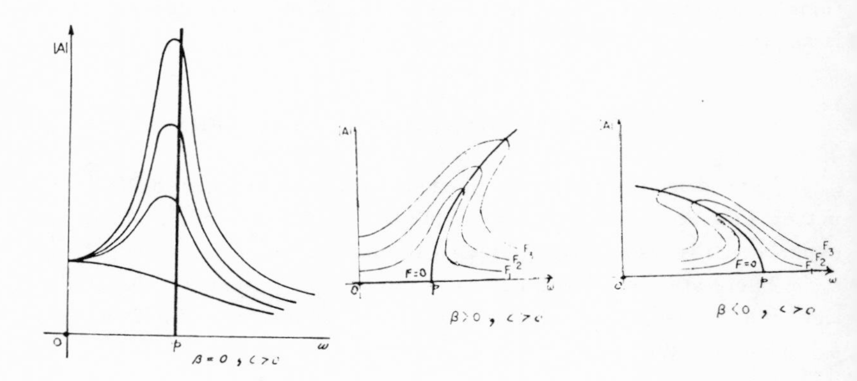

Fig. 10.

Suppose we keep $F =$ constant and vary ω from large values to smaller values, starting, say, at point 1, Figure 11 (a). As ω decreases, A increases slowly until point 3 (tangent point to the curve), when a further decrease of ω causes a jump up to the amplitude from point 3 to point 5 of the curve, and after that the amplitude decreases with ω.

If we start increasing ω from a value corresponding to point 6, the amplitude follows the portion $6 \rightarrow 5 \rightarrow 4$ of the curve, when, if 4 is the tangent point to the curve, the amplitude jumps down to point 2, and after that decreases slowly with increasing ω.

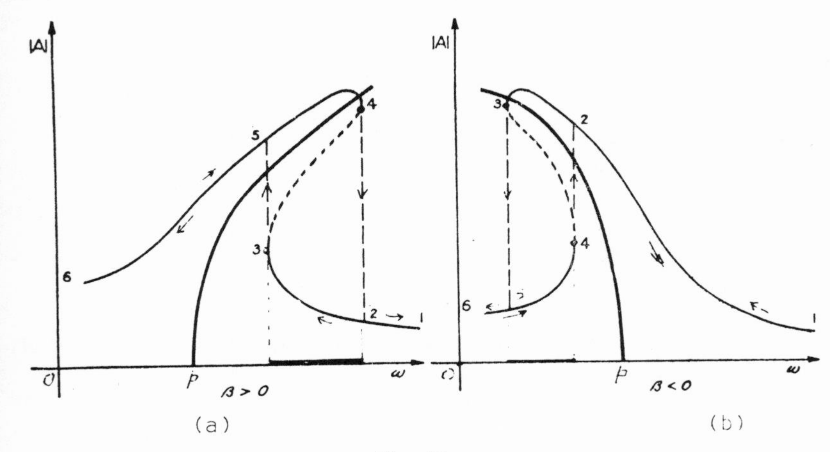

Fig. 11.

The same phenomenon occurs in Figure 11(b), but in the reverse direction. The above is the jump, a hysteresis, phenomenon, corresponding to an interval of ω, in which the oscillations are unstable.

The portion $3 \rightarrow 4$ of the curve is a «dead portion».

IX. COMBINATION FREQUENCIES

Helmholtz in his acoustical studies, and Poincaré in his studies of NLDE established that, in addition to certain fundamental frequencies ω_1 and ω_2 in a NLS, there exist solutions of the same DE with the frequencies: $\omega = m\omega_1 + n\omega_2$, where m and n are integers.

These are called «combination frequencies», or «combination tones» of the system, and the phenomena of combination frequencies are phenomena of NLS.

E x a m p l e 1. The oscillations of the LS:

$$\ddot{x} + c\dot{x} + x = F_1 \cos \omega_1 t + F_2 \cos \omega_2 t \equiv H(t) \tag{9}$$

where $\omega_1 \neq \omega_2 \neq 1$, are superposition of a damped free oscillation and a forced oscillation, which, in turn, is a superposition of the fundamental oscillations due to each of the two separate tones of the excitation H(t) individually. That is essentially the oscillations of the LS (9) are simply a superposition of the two harmonies with frequencies ω_1 and ω_2.

Example 2. Consider now the NLS: **[10c]**

$$\ddot{x} + c\dot{x} + x - bx^3 = H(t). \tag{9.1}$$

We have two cases:

Case a: $\frac{\omega_1}{\omega_2}$ = rational number. In this case the excitation H(t) is a periodic function of time, and (9.1) has periodic solutions of various kinds.

Case b: $\frac{\omega_1}{\omega_2}$ = irrational number. In this case H(t) is an almost periodic function of time, and since $\frac{\omega_1}{\omega_2} \neq \frac{p}{q}$, where p and q are mutually prime integers, we will have solutions with frequency: $\omega = q\omega_1 - p\omega_2 \neq 0$.

In this case, by using approximate methods for calculation of approximations of the solutions of (9.1), we find that these approximations of the solutions contain terms with denominators powers of $(\pm m\omega_1 \pm n\omega_2)$, where m and n are integers.

But, by virtue of the «Kronecker theorem» it is known that the expressions $(\pm m\omega_1 \pm n\omega_2)$ are arbitrarily close to zero for infinitely many different integers m and n, when the approximations of the solutions of (9.1) are divergent, because of the «difficulty of small divisors», which was first pointed out by Poincaré in discussing perturbations methods with problems in Celestial Mechanics.

The «difficulty of small divisors» can be circumvented in some cases if we use viscous damping in the system.

In case the excitation of the NLS has a single frequency, which is an irrational multiple of the free frequency of the system, we have the same situation as above.

RESULTS AND REMARKS

• The linear and nonlinear systems have striking properties which characterize them. Some properties are properties of LS and some others only of NLS.

• The principle of superposition of the solutions, the global property, the independence of amplitude and frequency of an oscillation, characterize LS.

The existence of limit cycles, the change of frequency from trajectory to trajectory, the jump discontinuities of the amplitude of oscillations, the development of stable self-excited oscillations, the phenomenon of combination tones, characterize NLS.

The resonance phenomena are related to LS as well as to NLS. The damping in LS can prevent resonance, and the nonlinearities in NLS can stop resonance, and by a proper selection of nonlinearities the system can become stabilized.

• The reduction of a NLS to a LS implies that some properties of NLS will be lost, and, therefore, if a problem in connection with the NLS is related to a property which will be lost by linearization, then the linearization as a method to solve the problem is not applicable.

ΠΕΡΙΛΗΨΙΣ

Τὰ γραμμικὰ (ΓΣ) καὶ μὴ - γραμμικὰ (ΜΓΣ) συστήματα ἔχουν ἰδιότητας αἱ ὁποῖαι ἄλλαι μὲν χαρακτηρίζουν τὰ ΓΣ ἄλλαι τὰ ΜΓΣ.

• Εἰς τὰ ΓΣ ἡ γενικὴ λύσις δύναται νὰ δοθῇ ὡς γραμμικὸς συνδυασμὸς μερικῶν ἁπλῶν λύσεων, καὶ αἱ σταθεραὶ ὁλοκληρώσεως εἰσέρχονται εἰς τὰς γενικὰς λύσεις γραμμικῶς. Εἰς τὰ ΜΓΣ ἢ δὲν ὑπάρχουν γενικαὶ λύσεις, ἤ, ἐὰν ὑπάρχουν, ἔχουν πολύπλοκον μορφήν, ὅπου αἱ σταθεραὶ εἰσέρχονται ἐν γένει μὴ - γραμμικῶς.

• Εἰς τὰ ΓΣ αἱ λύσεις ἔχουν ὁλικὸν χαρακτῆρα, ἐνῷ εἰς τὰ ΜΓΣ ἔχουν τοπικὸν χαρακτῆρα.

• Εἰς τὸ ΓΣ τὸ πλάτος περιοδικῶν κινήσεων εἶναι ἀνεξάρτητον τῆς συχνότητος, καὶ εἰς ἓν συνεχὲς πεδίον περιοδικῶν κινήσεων αἱ τροχιαὶ ἀντιστοιχοῦν εἰς τὴν αὐτὴν συχνότητα. Τοὐναντίον, εἰς τὰ ΜΓΣ τὸ πλάτος ἐξαρτᾶται ἀπὸ τὴν συχνότητα, καὶ ἡ συχνότης ἀλλάζει ἀπὸ τροχιᾶς εἰς τροχιάν.

● Ἡ ὕπαρξις ἀνωμαλιῶν εἰς τὰ πλάτη περιοδικῶν κινήσεων συναρτήσει τῆς συχνότητος, εἶναι φαινόμενον τῶν ΜΓΣ, καὶ τοιοῦτον φαινόμενον δὲν ὑπάρχει εἰς τὰ ΓΣ.

● Ἡ ὕπαρξις μεμονωμένων περιοδικῶν κινήσεων (limit cycles), ὅπως καὶ «εὐσταθῶν ταλαντώσεων αὐτοαναπτυσσομένων» χαρακτηρίζει τὰ ΜΓΣ, καὶ τοιαῦτα φαινόμενα δὲν ὑφίστανται εἰς τὰ ΓΣ.

● Τὸ φαινόμενον τοῦ «συνδυασμοῦ συχνοτήτων» (combination frequencies) εἶναι φαινόμενον τῶν ΜΓΣ καὶ ὄχι τῶν ΓΣ.

● Αἱ δυνάμεις «damping» εἰς τὰ ΓΣ δύνανται νὰ ἐμποδίσουν φαινόμενα «resonance», ἐνῷ εἰς τὰ ΜΓΣ αἱ μὴ-γραμμικότητες δύνανται νὰ χρησιμοποιηθοῦν δι' εὐστάθειαν ἔναντι «resonance».

REFERENCES

1. N. Bogoliubov and Y. Mitropolsky, «Asymptotic method in the theory of nonlinear oscillations», Hindustan Publishing Corp., India, Delhi-6 (1961), pp. 267-280.
2. F. Clauser, «The behavior of nonlinear systems», J. of the Aeronautical Sciences (May 1956), pp. 411-434.
3. H. Cohen, «On subharmonic synchronization of nearly linear systems», Quart. Appl. Math., Vol. XIII (April 1955), pp. 102-105.
4. J. Hale and J. La Salle, «Differential equations: linearity vs. nonlinearity», SIAM Review, Vol. 5 (July 1963), pp. 249-273.
5. S. Lundqwist, «Subharmonic oscillations in a nonlinear system with positive damping», Quart. Appl. Math., Vol. XIII (Oct. 1955), pp. 305-310.
6. D. Magiros, (a) «Subharmonics of any order of nonlinear systems of one degree of freedom: Application to subharmonics of order 1/3», J. Information and Control, Vol. 1, No. 3 (September 1958), pp. 198-227.
   (b) «On a problem of nonlinear mechanics», J. Information and Control, Vol. 2, No. 3 (September 1959), pp. 297-309.
7. N. McLachlan, «Ordinary nonlinear differential equations», 2nd edition, Oxford (1955), (a) pp. 130-131, (b) p. 196.
8. N. Minorsky, «Nonlinear oscillations», D. Van Nostrand Co., New York (1962), pp. 72-73.
9. V. Pliss, «Nonlocal problems of the theory of oscillations», Acad. Press, New York (1966), pp. 219-227.

10. J. Stoker, «Nonlinear vibrations», Interscience Publishers, Inc., New York (1950), (a) pp. 20 - 23, (b) pp. 94 - 96, (c) pp. 112 - 114.
11. N. Bautin, «On a Certain Differential Equation Having a Limit Cycle», Gorkij Physico-Technical Institute of the University (1937), pp. 229-243.

★

Ὁ Ἀκαδημαϊκὸς κ. **Ἰωάννης Ξανθάκης,** παρουσιάζων τὴν ἀνωτέρω ἀνακοίνωσιν, εἶπε τὰ ἑξῆς:

Εἰς τὴν παροῦσαν ἀνακοίνωσιν ὁ κ. Μάγειρος μελετᾷ τὰς χαρακτηριστικὰς ἰδιότητας τῶν γραμμικῶν καὶ μὴ γραμμικῶν συστημάτων. Ἡ διερεύνησις αὕτη τὸν ὡδήγησεν εἰς τὰ ἑξῆς γενικὰ συμπεράσματα:

1. Εἰς τὰ γραμμικὰ συστήματα ἡ γενικὴ λύσις δύναται νὰ εἶναι ἕνας γραμμικὸς συνδυασμὸς μερικῶν ἁπλῶν λύσεων, ὅπου αἱ σταθεραὶ ὁλοκληρώσεως εἰσέρχονται εἰς τὰς γενικὰς λύσεις γραμμικῶς. Ἀντιθέτως εἰς τὰ μὴ γραμμικὰ συστήματα ἢ δὲν ὑπάρχουν γενικαὶ λύσεις ἢ ἐὰν ὑπάρχουν ἔχουν πολύπλοκον μορφήν, αἱ δὲ σταθεραὶ εἰσέρχονται μὴ γραμμικῶς.

2. Εἰς τὰ γραμμικὰ συστήματα τὸ πλάτος τῶν περιοδικῶν κινήσεων εἶναι ἀνεξάρτητον τῆς συχνότητος, αἱ τροχιαὶ δέ, εἰς ἓν συνεχὲς πεδίον περιοδικῶν κινήσεων, ἀντιστοιχοῦν εἰς τὴν αὐτὴν συχνότητα. Ἀντιθέτως εἰς τὰ μὴ γραμμικὰ συστήματα τὸ πλάτος ἐξαρτᾶται ἀπὸ τὴν συχνότητα, ἡ ὁποία ἀλλάζει ἀπὸ τροχιᾶς εἰς τροχιάν.

3. Ἡ ὕπαρξις ἀνωμαλιῶν εἰς τὰ πλάτη περιοδικῶν κινήσεων συναρτήσει τῆς συχνότητος παρατηρεῖται μόνον εἰς τὰ μὴ γραμμικὰ συστήματα, οὐδόλως δὲ εἰς τὰ γραμμικὰ τοιαῦτα.

4. Ἡ ὕπαρξις μεμονωμένων περιοδικῶν κινήσεων καθὼς καὶ εὐσταθῶν ταλαντώσεων αὐτοαναπτυσσομένων εἶναι χαρακτηριστικὸν τῶν μὴ γραμμικῶν συστημάτων, δὲν παρατηρεῖται δὲ τὸ φαινόμενον τοῦτο εἰς τὰ γραμμικὰ συστήματα.

Τέλος τὸ φαινόμενον τοῦ «Συνδυασμοῦ συχνοτήτων» παρατηρεῖται μόνον εἰς τὰ μὴ γραμμικὰ συστήματα.

MATHEMATICAL MODELS OF

PHYSICAL AND SOCIAL SYSTEMS

By: Demetrios G. Magiros*

ABSTRACT

In this paper a "classical procedure" is discussed for building mathematical models of physical, technological and real-life phenomena. The method is accompanied by appropriate remarks and applications.

In Chapter 1 the procedure is discussed for the model building, which is characterized by some steps of the investigation.

In Chapter 2 remarks are given related to concepts involved, to applicability of the method, to reasonability of the problem, etc. By these remarks the investigator faces views for examination and problems for solution.

In Chapter 3 applications are illustrated, by which one can see how the method for model building can be applied.

The applications are distinguished into applications of "complete cycle" and of "incomplete cycle".

The paper is dedicated to Leon Brillouin, the outstanding scientist and human, whose memory will always be a source for inspiration in my work, and for whom my admiration, respect and gratitude are unlimited.

---

*Scientific Consultant
General Electric Co., RSD, Philadelphia, PA.
September, 1980

Reprinted from a *Technical Report, Genl. Electric Co.,* RSD, Philadelphia (Sept. 1980), 1–56.

TABLE OF CONTENTS

TABLE OF CONTENTS

TABLE OF CONTENTS

-1-

INTRODUCTION

"Model building" is a significant part of applied mathematics and of great importance in science and technology.

Mathematical methods and reasoning, properly applied to the investigation of a phenomenon, may present theoretical conclusions, which, by an empirical verification, may lead to new insights of the phenomenon and to important contributions to human knowledge.

Research in an empirical field may reach a state of "axiomatization."

This state of the field is of such a perfection that it is possible to formalize in the field a set of propositions, which satisfy certain conditions free of contradictions. In such a situation of the field mathematics is indispensable for the investigation.

For a successful study of a phenomenon, one may follow a set of steps of research. The description of the phenomenon, its understanding and explanation, its prediction and control, are "steps of the modeling process."

Facts and data, hypotheses and axioms, basic laws and theories, tables and diagrams, all these related to the phenomenon, serve as "tools for modeling" the phenomenon.

Many phenomena of different nature, phenomena with a variety of style, purpose, effect, have factors in common, and lead to the same mathematical structure, so the modeling may be used as a "unification method" for the treatment of groups of phenomena.

In modeling, heuristic and non-rigorous reasoning are often employed, and, therefore, the results might be incomplete and of doubt. But, realistic problems often require techniques that cannot, at the moment, be rigorously justified. A rigorous reasoning, which is able to penetrate the essentials of the subject matter in question, is always needed, but, for the sake of progress, the mathematical and experimental analysis must proceed, even if, at the moment, the logical structure is incomplete.

In this paper the building of mathematical models of phenomena or systems will be discussed. An exposition of the modeling process shall be given first, and, then, through remarks and applications, attempts will be made to clarify the steps of the process and the concepts entering into the process.

CHAPTER ONE

A CLASSICAL PROCEDURE FOR THE CREATION OF MATHEMATICAL MODELS OF SYSTEMS

[Ref. 7(a), 18(a, d), 20]

The building of mathematical models is an important method of studying the behavior of complicated systems and forecasting their future. In developing the procedure to find mathematical models one may have the following steps.

STEP I: MAKING THE SYSTEM WELL EXPRESSED PHYSICALLY.

The starting point for the investigation of models of systems is the knowledge of the original data of the system, that is the knowledge of correct observations, measurements and experimental facts related to the system, which must be completely stated and known approximately within an accepted error, thus being: "admissible data". It is also necessary to know the "pertinent laws" to the system and all the information needed in order to make the system physically clear, when we call it "well expressed".

Abstractions and generalizations, which violate the essence of reality, must be avoided in this step.

STEP II: MAKING THE SYSTEM CORRECTLY STATED MATHEMATICALLY

This step is the most decisive for the investigation because it helps to go from the reality to mathematics. It often requires ingenuity, experience, intuitive understanding of the scientific field to which the system belongs, and ability of the investigator to apply his mathematical knowledge.

The selection of the main variables, among all the variables of the system, and the creation of the theory of the system characterize this step.

2.1 Major and Minor Variables of the System.

For a system which is clearly stated physically, one must select an appropriate number of appropriate variables of the system suitable for the description of the process of the system. The selected variables are the "major variables", or "generalized coordinates" of the system, also called "state variables", because they are used as components for the description of the "state" of the system.
The other variables of the system, the "minor variables", called "parameters", define the "environment" of the system.
Restrictions on the magnitude of the state variables and of the parameters give the "regions" of the state variables and the parameters, and the system is considered operating in these regions.

2.2 Theory and Model of the System.

After the selection of the state variables of the system, one must try to seek out a set of statements in connection with the behavior of these

variables of the system, based on appropriate hypotheses. If the investigator is successful to get relationships between the variables of the system, based on the statements and the hypotheses, he has by all these created a "theory" of the system, and a theory leads, in general, to mathematical expressions called "mathematical model" of the system, which gives a functional behavoir of the system in a quantitative way.
Well-expressed systems, for which one was successful to create a theory and a correspondent model, we call "correctly stated" systems.
Correctly stated models are, in general, differential equations, because, in general, rates of change of the variables enter the models.
These differential equations are associated with appropriate admissible data, and these data are the initial and/or boundary conditions.

STEP III: GETTING A REASONABLE MODEL.

The hypotheses, decisions and choices in developing a theory and the corresponding model of a real system referred in the preceding step, were arbitrary, then, one may build various models for the same real system, and these models do not necessarily give actual solutions, that is, solutions which interpret the reality in an adequate way.
One needs to find a way to distinguish, among the correctly stated models of the system, a model which ought to be the most preferable from the point of view of interpreting the reality, that is a model with actual solution, called "reasonable model", or "properly posed", or "well posed".

Model reasonable in the Sense of Hadamard. A way to check whether a correctly stated model is a "reasonable" one, is to examine whether the solution of the constructed model satisfies the conditions:

- 5 -

(i) the solution of the model exists,

(ii) the solution is unique, and

(iii) the solution depends continuously on the data.

These conditions for the solution of the model are called "Hadamard restrictions", and the model with solution satisfying the Hadamard restrictions is called: "reasonable in the sense of Hadamard".

If the solution of a model fails to satisfy even one of the Hadamard restrictions, it is a formal solution of a model called "non-reasonable", or "non-properly posed", or "non-well posed" in the sense of Hadamard.

In a "reasonable model", the "existence" and "uniqueness" of its solution express the belief in "determinism", a principle according to which one can repeat experiments with the expectation to get consistent results.

For the "reasonability" of the model the third Hadamard restriction is necessary. The data, as results of observations, experiments,and measurements, are given with small errors, when the solution has an uncertainty and its estimation an error. The acceptance of the "continuity property" of the solution with the data implies that "the smaller the error in the data, the smaller the error in the estimation of the solution", then "admissible data" will correspond to "admissible solutions", that is solutions "physically accepted".

We remark that: in the case where a model is expressed by a system of ordinary differential equations in its normal form, it is possible to check the reasonability of the model, without knowing its solution, by considering the form of the functions of the right-hand members of the equations. If these functions and their first partial derivatives with respect to the unknown variables are continuous, the solution satisfies the three Hadamard restrictions, and the model is "reasonable".

- 6 -

STEP IV: DETERMINING THE ACTUAL SOLUTION.

The determination of the solution of a "reasonable model" is a mathematical step leading in many cases to possibilities for constructive imagination. Sometimes it is possible to solve the model equations explicitely and to get the solution in a closed form in terms of simple functions, when the behavior of the system is given in a general form. But more often it happens not to be able to do it, and in such circumstances one uses approximate methods for solutions. In many cases the wide availability of higher speed computers eliminates tne need for a solution to be in closed form.

Heuristic reasoning toward the ultimate solution in many cases may be used.

The principle of approximation to achieve the solution is in some cases needed, and the continuity property of the solution, that is the third Hadamard restriction, helps to apply this principle.

In some cases, the reasonable model M is reduced to an auxiliary "approximate model" M_n of which the solution A_n can be found more easily, when the required solution is the $\lim_{n \to \infty} A_n = A$.

STEP V: COMPARING THE RESULTS COMING FROM THE MODEL WITH THE EMPIRICAL RESULTS.

The theoretical results coming from the reasonable model must agree with the empirical results, and this agreement guarantees the establishment of the model.

-7-

SUMMARY

Summarizing the preceding, one can see that the procedure to get a physically accepted mathematical solution of a real system must follow some steps, shown in Figure 1. These steps are:

I. The real system (S_r) must become physically complete, that is "well-expressed" (S_w), by taking into account all the physical information needed.

II. The well-expressed system (S_w) must be done "correctly stated" mathematically (S_c), by selection of the appropriate state variables, and creation of the theory and the corresponding model of the system.

III. The most "reasonable model" (M), among the correctly stated models (S_c) of the system, must be selected, by imposing the Hadamard restrictions to the solution of the model.

IV. The (actual) solution (A) of the reasonable model (M) must be determined, and

V. The theoretical results coming from the reasonable model must be compared with empirical results, when their agreement establishes reasonable model (M_e).

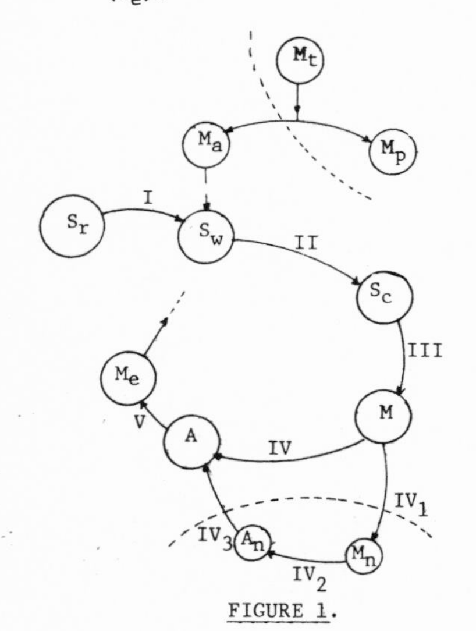

FIGURE 1.

M_t : mathematics
M_p : pure mathematics
M_a : applied mathematics
S_r : real system
S_w : well-expressed system (physically)
S_c : correctly stated model (mathematically)
M : reasonable model
M_n : auxiliary model
A_n : solution of M_n
A : solution of M
M_e : model empirically checked

CHAPTER TWO

REMARKS ON THE CREATION OF MATHEMATICAL MODELS

[Ref. 1(a), 2, 4, 7(b), 8, 12, 13, 15, 16, 18(d), 19, 20, 23, 24, 25]

In the preceding chapter we discussed an organized procedure of a method for building mathematical models of real systems.
But, many points of the method need a clarification and the concepts involved in the method, as well as the results given by it, require a thorough analysis.
By the remarks of this chapter we try to accomplish these necessities.
In this chapter remarks are given on the variables of the system and its data, on controllable systems, on the theory and the corresponding model of the systems and on the hypotheses on which theory and model are based, on deterministic and stochastic systems and models, on the applicability of the method, on the reasonability of problems, etc.
These remarks may put to the investigator views for examination and problems for solution.

1. Remarks on the nature of a system and its isolation

The definitions of systems and models, given in the preceding, help the investigator's attempts for modeling. The nature of a system is characterized by the nature of its elements and parts. So, a watch is a mechanical system, a radio an electrical system, a university an education system, a family a social system.

- 9 -

The system under investigation was, in the preceding, silently accepted as isolated, that is, under no external influence.
But this is rarely an actual situation. The coupling of the system under investigation with other systems, which are not of an immediate interest, may be unsuspected and uncontrollable.
Such a situation exists, for example, in biological and economic systems. Biological components may have different performance characteristics in isolation than they do when they are connected within the system of interest. In economic systems, where the data are poor, unforeseen changes: Political, social, technological, psychological, climatic, make the prediction of the economic situation not possible.

2. On the steps of modeling.

By the step I one makes the system "well-expressed," that is "physically complete." Data, inputs, assumptions and conjectures of qualitative nature, are constituents of this step. By this step, one obtains, in general, to get the system as a specified order set of interconnected operations performed by its elements, which participate in the process of transformation. Historical realization and the concept of regularization are included in the above situation. The well-expressed system is a formalization of the physical knowledge, and it expresses the so-called "structure" of the system, which will lead to the quantitative analysis of the next steps.

By the steps II and III, one has a mathematical idealization of completely given physical systems, namely, one has "correctly stated" and "reasonable" systems and models. The model describes the special manner of transformation of inputs to outputs, that is, shows the "mode of functioning" of the system, that is its "activity", which involves a manner of transformation of inputs to outputs and the limits of this process.

It is the nature of this activity which defines distinct systems. The invariance in the structure of a system implies invariant laws of modes of the transformation.

By the step IV, the solution of a "reasonable" model is given and this solution is the time path of the "response" of the system. From this response many conclusions can be obtained, especially a prediction is available.

3. On controlled systems

If a system A, in its native form, does not operate in a desirable form, that is, its outputs x deviate from the demands, it is possible sometimes to modify or "control" the system A by introducing appropriate inputs u, that is, special "control functions", which are either "forcing functions" (time dependent) or "feedback controls" (functions of state variables). We may use a "controller" B, Figure 2, of which the "operating law" produces the appropriate inputs u of the system A. The "driving action" x*, input on the controller B, realizes the purpose of the control, that is, represents the instruction of what must be the product x of the system A. z is the noise.

- 11 -

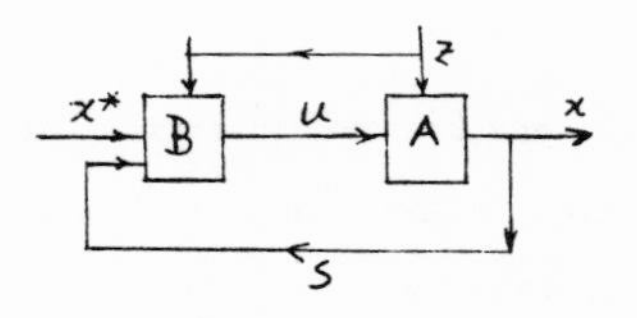

FIGURE 2

If the controller B does not get information about the actual state x of the controlled system A, the system A is an "open-loop" system; and if there exists a "feedback path" S by which the controller B receives the information x of A, the system is a "closed-loop" or "feedback" system.

The study of control systems, although of engineering discipline, is a mathematically oriented science. Probability theory, calculus of variations, dynamic programming, maximum principle and other mathematical branches, give mathematical methods applicable to control problems and especially to problems of "optimization of controls", the most interesting problems.

By "optimization of controls" we mean a mathematical description of control systems in such a way that the application of rigorous procedures of this representation results in the best performance of the systems within the framework of the relevant limitations and variables.

The solution of optimal problems permits a maximum effectiveness in manufacturing, it increases the productivity, it provides the quantity of production, the economy of electric energy, and so on.

The "completely controllable" and "completely observable" controlled systems are the most important systems.
In a "completely controllable" system it is possible to move any state of the system to any other state during any finite time interval by a suitably chosen input;and in a "completely observable" system, the knowledge of the input and output on a finite time interval is sufficient to determine uniquely the state of the system.

4. On the hypotheses

The hypotheses play a basic role in the investigation and they are an important weapon in the investigator's hands.
But, the useful hypotheses must be always under empirical control, that is, an agreement is needed between the results coming from the hypotheses and the observed facts.
It is an important method of research to form hypotheses and then to compare the results with the reality.
Hypotheses on a subject having empirical support are of active interest in the subject. In many cases they lead to discoveries, to theories, to laws.
Hypotheses which are not under experimental control cannot be accepted for scientific purposes.
It is by experiments that one hopes to get confirmation of hypotheses and new ideas for better theories, and such interplay constitutes a most desirable development.

Hypotheses are a mental technique, a product of imagination and they are productive, if it is possible to be a basis for discovery of the truth and lead to observations and experiments.

Fruitful hypotheses lead, in general, to a generalization, and a generalization leads to a mathematical model able for prediction.

5. On the variables of a system

The major variables of a system are significant varieties of the system and omission of an important major variable from their selection will result a wrong notation of the relationships, then a wrong model of the system.

In systems of physical sciences, the selection of the major variables is not a difficult subject, for the quantities in these sciences are expressible in terms of well-defined and standard units which are time-invariant.

But, in systems of, say, biology, economics, sociology, one meets a difficult situation for the selection of the major variables, since the variables in these fields are, in general, time-dependent, when the selection of major variables is an elusive subject. For the above case, one may have an indication for the selection of the major variables if one has information about the dependence of the increments of the variables with time.

If a system, after a small disturbance, can get its original situation, this will be in a finite time called a "relaxation time" of the system. A system with small (considerable) relaxation time is called "fast" (slow) system.
The relaxation time of the variables of a system is not the same for each variable.
Variables with fast (slow) increment can be taken as major variables (parameters).
Two selections of the major variables for the same phenomenon will lead to two different models with different relaxation time,when the selection of the major variables is more successful in the faster system. In other words, a successful selection of the major variables makes the system as faster as possible.

The specification of the number of major variables is also very important. The smaller the number of the major variables, the simpler the model of the system. If the number of the major variables is inappropriately small, the corresponding model will be oversimplified and, as such, the model loses the essentials of the system and it is unrealistic. To know where and when to simplify a complicated system, this is a very important subject for an investigator.

6. <u>On the data, and deterministic and stochastic systems</u>

The original empirical facts of a system, that is the data, were considered in the preceding as completely given and accurately observed

and measured in order to state the problem at the beginning and to give a preliminary study of the system.
By such data and same hypotheses on the parameters of the system, the procedure leads to models with the possibility of exact prediction and the models are termed "deterministic".

But in the actual scientific work, the above situation is unrealistic. In scientific fields, as e.g., in biology, economics, social sciences, unusual uncertainties occur and variabilities involve, when the deterministic concepts and techniques are, in many cases, inappropriate.
There appear inevitable errors and difficulties. Errors in the experiments and measurements; errors and difficulties in getting good quantifiable data, or non-quantifiable ones, whenever needed; errors in the determination of the parameters and difficulties in fitting them to the data. Also, in some cases, the actual values of the variables emerge from the chaos of predestination, when the variables are considered as "random variables".
Due to the above, the "determinism" must be replaced by the "probabilistic causality", and the investigation will, in general, be concerned with probabilistic theories, stochastic processes, statistical methods, methods of statistical mechanics, decision analysis.

7. On the theories of a system

The theory of a system is a set of statements and a collection of propositions about the behavior of the system, based on the data and some hypotheses. Some of the statements are verified by experimental observations, but some others are postulated or based on hypotheses.

The theory is a human creation much of which is a work of imagination,and nothing human is permanent. As a product of human brain, the theory is open to discussion and possible revision or modification,and sometimes rejection,if new empirical facts are contradictory to the theory. Experiments and theory must be intimately related and none can make significant progress without the other.

A theory on a system may be considered acceptable,if

(i) the theory is able to produce rules for modeling;

(ii) the region of the applicability of the theory,which is useful for predictions, is known;

(iii) the limits of the errors in this region are known.

The bigger the region of applicability of a theory, the more opportunities for practical uses of the theory.

8. On the models of the systems

The construction of a model deals with finding, in mathematical terms, the interrelationships between the variables or parts of the system and its inputs and outputs,and this construction presupposes a theory of the system to which the model corresponds.

Without a conjunction of systematic quantitative data and a general framework of the theoretical orientation and theoretical ideas about description and processes, the development of models should be impossible. The model usually explains the anatomy of the system.

The procedure of modeling is sometimes to start building a simple and descriptive model, then penetrating more and more into the causal relationship, and, then, modifying the starting model according to new hypotheses and new empirical facts. One selects such a construction of the model that will facilitate deductive investigation of the properties of interest. One builds models for specific purposes, and one may have for the same system different models for different purposes.

Due to the complexity of the real systems, the building of their models must be necessarily restricted only to important features of the system and some specific details will have to be ignored for a manageable description.

Among the different types of models of systems, one can distinguish the "time-delay" systems, and the "deterministic" and "stochastic" systems.

Systems of which the process at a time depends not only on the state at that time but also on the past history of the systems, that is the process is characterized by "after-effects", are called "time-delay" or "time-lag" systems, when we will have "time-delay" or "time-lag" models. For example, in problems of long-range planning in economics one has "after-effects" situations.

This kind of systems is expressed by "differential equations with deviating arguments", of which important classes are the "difference-differential equations" and the "retarded differential equations" with "delay periods" or "retardations".

In deterministic systems the effect of any change in the system can be predicted with certainty.

In practice, especially in social sciences, this is not the case, and, either because the system is not fully specified, or because of the unpredictable character of much human behavior, there are usually elements of uncertainty in any prediction. These uncertainties can be accommodated if we introduce probability distributions into the model in place of the variables, or, more precisely, if the equations of the model include "random variables". Such a model is called "stochastic".
"Stochastic processes" occur widely in nature, say in economics, in psychology, in social sciences, and they are developed in time according to probabilistic laws.
By the stochastic processes one cannot predict the future behavior of the system with constancy, but only attach probabilities to various possible future states.
By probabilistic methods one can specify the state of an individual only in terms of its probabilities of choices, and go no further, thus probabilistic methods form no basis for prediction in cases where two responses contradict.
Any model describing human behavior must be formulated in stochastic terms.
In case one needs the solution of the models, it may be advisable to use deterministic approximations.
The stochastic models give a way of testing the fit of the model to data.
The deterministic models are very simple compared to stochastic ones, and the stochastic equations are usually cumbersome.

There are relationships between deterministic and probabilistic points of view.

The deterministic form is the same as the mean probability of distribution. We can pass on to more precise stochastic formulation by giving special attention to special features suggested by deterministic models.

The models are mixed, deterministic and stochastic, that is, sometimes we have stochastic behavior out of deterministic models, and sometimes the models are partly stochastic.

If in a deterministic model, one admits random variables, the model becomes a stochastic one, when it raises problems of identifiability, that is we may write down a model when certain constants cannot be estimated, however many observations we have, unless we introduce extraneous information.

Deterministic treatments may be regarded as approximately valid in certain circumstances and worth examining in any given situation. The deterministic models are preferred in case they give the expected values of the distribution.

In chemical kinetics there exists a set of mathematical theories or models to account for reaction rates. Problems in physical chemistry deal with models, and, since physical chemists, dealing with aggregates of billions of molecules, do not need to make their model a stochastic one, they will treat the phenomena as deterministic.

Finally, we remark that the model is not the reality but it is an instrument, a method, to investigate the reality.

A model is for the reality what a geographical map is for the part of the surface of the earth that it maps. Both, models and maps, give methods for the investigation of the reality,in order to satisfy the eagerness and attempts of the investigator.

9. On the applicability of the classical method

The classical method for finding models of systems, wherever this method is applicable, gives good results.

But, there are scientific fields where the method either cannot be applicable, or it is difficult to be applied.

If the exact laws, which govern the system under investigation, are not completely known, then a theoretical description of the system is not possible and the classical method is inapplicable. This occurs e.g., in problems of atomic and molecular structure, where the interactions are known, but where the structure may be complicated.

In problems of classical dynamics of compressible fluids, the differential equations, even if supplemented by boundary conditions, are not always a complete framework for an adequate description of the physical reality.

10. On the reasonability of models

The classical method for finding "reasonable" models and "actual" solutions of real systems, shows a general orientation of thought on the problem and it is not always literally binding. It is based on the "Hadamard restrictions" which are necessary and sufficient for deterministic models.

The concept of reasonability of a problem plays an important role in the solution of a system. This concept was introduced by Hadamard at the beginning of this century and a variety of very interesting problems, reasonable in the sense of Hadamard, can be successfully treated.
But, it was found that the Hadamard concept of reasonability rules out, as unreasonable problems, important real problems.
Various developments in physics point up the need for restrictions of the validity of Hadamard definition of reasonability, and this indicates an important consequence for natural philosophy as well as for practical applicability.
There are available today different notions of reasonability, and various approaches to the formulation of an investigation of problems which are unreasonable in the sense of Hadamard.
The Russians Tychonov and Lavrentiev have a success in treating such problems. Probability methods are used in their study.

11. On the Concept of Probability

In case probabilistic models are used, the role of probability definition is basic for the investigation.

To a real stochastic event α , one can assign a probability $p(\alpha)$, for the concept of which one may have four essentially definitions

(a) The "classical probability definition", that is the ratio of a favorable to total number of alternations,

(b) The "probability definition as a measure of belief", which is of the form of inductive reasoning,

- 22 -

(c) The "axiomatic probability definition", for which measure theory is needed, and which, according to Kolmogoroff, is a superior definition, and

(d) The "relative frequency probability definition", which is very popular and most used among physicists and engineers.

If an experiment is repeated n times, and the event α is occurring n_α times, the relative frequency of occurrence is $\frac{n_\alpha}{n}$ and it is directly observable.

To make statements about relative frequency, one uses probability and defines the probability of the event α as the limit of the relative frequency as $n \to \infty$, $p(\alpha) = \lim_{n \to \infty} \frac{n_\alpha}{n}$.

Such a limit is not an experimental quantity.

The existence of this limit presupposes the possibility of repeating experiments infinitely many times, which is unrealistic.

Therefore, the relative frequency definition of probability presupposes that the "principle of approximation", a basic principle of applied mathematics, and the "law of large numbers", a law of probability theory, are workable.

CHAPTER THREE

APPLICATIONS

The creation of models is, in general, very difficult work, and the establishment of a model needs in many cases investigation of many years. The applications of the classical method, which was developed in Chapter I, can be classified into two classes: Applications of "complete cycle" and of "incomplete cycle".

For applications of "complete cycle" it is necessary:

(i) to confront the physical problem, and construct its mathematical model,

(ii) to compare the results coming from the model with experimental results, and these results must agree.

We can consider as belonging to this class models of great historical interest, developed after years of effort by brilliant men.

In applications of "incomplete cycle" either the model is not yet completely established, or the interests of the investigator are only on some steps of the procedure.

A. APPLICATIONS OF COMPLETE CYCLE. (Ref. 3, 27)

We give some applications of models of complete cycle of deterministic type.

1. The Bernoulli Model of a Vibrating Beam.

Daniel Bernoulli (1700-1782), working with the problem of a vibrating beam:

(i) derived the model of the vibrating beam in the form of differential equations,

(ii) solved the resulting equation in the case of vibrating cantilever beam,

(iii) calculated the frequencies,

(iv) sketched out the shapes of the model,

(v) made experiments with vibrating splines,

(vi) traced their shapes, and finally.

(vii) compared the experimental results with the results predicted by the model.

Bernoulli did all these approximately two centuries ago, and, until today, the theoretical results, coming from his model, are in a complete agreement with the experimental results, and the Bernoulli model is a perfect example of complete cycle model. We notice that the work of Bernoulli plays an important role in the "theory of music" and the "theory of musical instruments".

We now mention some models of complete cycle, which are of high originality, created by talented persons.

2. The Newton Mathematical Model of the Laws of Gravitation, which gives a deep explanation of the Kepler empirical laws of the motion of planets, and it is the basis for astronomical and astronautical investigations. During the seventeenth and eighteenth centuries the Newton model, according to the observations and measurements of this time, has been established as the only exact and true model. But, during the nineteenth century, new perfections and modifications in the technique of observations and measurements made possible the creation of the Einstein theory of relativity, which completes the Newton model.

3. The Einstein Mathematical Model of the Relativity Theory, which completes our knowledge on time and space, and it is the basis for so many contemporary investigations.

- 25 -

4. The Maxwell mathematical formulation of electromagnetic theory, which is the basis for the study of electromagnetic phenomena, and explains the basic principles of electric power transmission and radio technology. Mathematical reasoning led Maxwell (1801-1865) to produce the existence of electro-magnetic waves including the many familiar types since discovered radio, radar, cosmic rays, etc.

5. The Yucawa mathematical theory of nuclear forces (Nobel Prize, 1949), by which Yucawa predicted the existence of mesons, and which later was confirmed by experiments.

6. The Frish and Tinbergen mathematical model in econometrics (Nobel Prize 1969), by which these economists described and predicted economical phenomena.

7. The mathematical development of electronic computer, in which Von Neuman was a deep investigator.
The electronic computer changed the functioning of today's society and it will influence significantly future life.
Many contemporary problems of great importance are expected to be solved by the creation of special models by using high-speed electronic computers.
But, for this scientists are needed with the ability to apply mathematical thinking and methods to the computer science, as exactly this was the case in physics, where scientists like Newton, Laplace, Maxwell, Einstein, a.o, by using mathematics in the physical reality, created today's progress in physics.
The weather prediction for some years is a very important problem for the solution of which the use of high-speed computers will play, as it is believed, a decisive role.

The laws of atmospheric phenomena can be expressed, according especially to Bjerkness, by mathematical models which are partial differential equations. Based on meteorological data and using these models, if they become reasonable, the prediction of tomorrow's weather for some years will become, as it is believed, a reality, by using computer machines in the hands of qualified computer men.

8. The mathematical work for the discovery of Neptune by Adams and Leverrier. Adams in England and Leverrier in France, working, independently to each other, on the problem of the "irregularities in the motion of the planet Uranus," discovered (1846) Neptune by an appropriate mathematical model.

We remark that each of the above models, which are considered as models of complete cycle, is associated with a history, by which can be shown the power of mathematics as a research method, and let us discuss the history of one of them, say the discovery of Neptune.
The discovery in 1846 of the planet Neptune was a very impressive and spectacular achievement of mathematics applied to astronomy. The existence of this new member of the solar system with specified properties and specified location in the heavens was demonstrated with pencil and paper, by finding the appropriate mathematical model to the problem. There was left to observers only the routine task of pointing their telescopes at the spot the mathematicians marked.

As early as about 1820 the astronomers noticed irregularities in the motion of the planet Uranus, that is, deviations of its observed orbit from its calculated positions.

- 27 -

Adams in England, and Leverrier in France, working independently to each other, solved the above problem by the discovery of the planet Neptune, which was the reason for the irregularities in the motion of the planet Uranus. The Adams memoire on the above problem was a masterpiece of remarkable mathematical maturity. But Airy, the Royal astronomer in England, didn't give too much attention to Adams' work, or the observers in England didn't have appropriate astronomical maps, or didn't find favorable conditions to observe the new star at the place and in time indicated by Adams.
Leverrier, several months later of the Adams' success, finished his calculations and asked the astronomers Galle and D'Arrest in Berlin, who had good astronomical maps, to see the new star at a place and in time indicated by his calculations. On September 23, 1846, these German astronomers saw first the new star.

B. APPLICATIONS OF INCOMPLETE CYCLE

In each of the following applications our interest is to emphasize appropriately some steps of the classical method for modeling and to use remarks and hypotheses for modifications of models.

1. APPLICATION 1 (MECHANICS). THE SIMPLE PENDULUM.

We discuss the solution of the problem of the motion of a simple pendulum in the way to see how, by changing the hypotheses we modify the model in order that the modified models satisfy practical needs.
We follow the Newton theory of "motion and force", and not the Lagrange theory of "equilibrium and work". These theories, although give different models, have the same results.

A mass m suspended from a fixed point O by means of a thread of length ℓ is a simple pendulum. The mass is displaced from its equilibrium position with an angle ϑ_0 with respect to the vertical, and then is free to move. We examine the motion of such a simple system by imposing the concrete problem: "how the motion of the pendulum depends on m, ℓ, ϑ_0."
To solve this problem, we accept special conditions and assumptions, and we try, based on them, to make the system "well-expressed" and "correctly stated".
We assume

(i) The mass m is a material point,

(ii) The mass m is free to move, without initial velocity, when it is in the place with angle ϑ_0 ,

(iii) The thread is absolutely solid with constant length and negligible weight, and fixed in its end O ,

(iv) The resistance of the environment of the pendulum, as well as the friction at O , are negligible,

- 29 -

(v) The motion of m is only under the influence of gravity forces,

(vi) The motion of m is on a vertical plane through O .

Under the above assumptions, the motion of m is on an arc of the circumference with center O and radius ℓ , and the place of m , at any time t , is known, if the corresponding angle ϑ is a known function of time t , that is $\vartheta=\vartheta(t)$.

The data are the initial conditions: $\vartheta(0)=\vartheta_0$, $\dot{\vartheta}(0)=0$.

So, for the problem, the time t is selected as "independent variable", the angle ϑ as the "main variable", and as "parameters" the length ℓ , the mass m , and the acceleration of gravity g .

The problem becomes a "static problem" by including "inertial forces" and using the appropriate "Newton law".

Under all the above the problem is "well expressed" and "correctly stated", and, as we know, the corresponding mathematical model, consistent with the above hypotheses, is

$$\ddot{\vartheta}+(g/\ell)\sin\vartheta=0, \quad \vartheta(0)=\vartheta_0, \quad \dot{\vartheta}(0)=0. \tag{1.1}$$

The model (1.1) is based on the above hypotheses, but not all these hypotheses are accepted in practice, then the results coming from this model can not be satisfied by experiments and measurements. The model (1.1) needs appropriate modifications, in order to interpret the reality adequately. The guide for the modifications is the special interests of the investigator and his physical intuition, the appropriate change of the hypotheses, and the mathematical needs. We see below some modifications.

(a) We suppose that the resistance of the environment of the pendulum is considerable. In case this resistance is considered as a force analogous to the velocity of the moving mass, the term $(\frac{K}{m}\dot{\vartheta})$ must be added to the equation (1.1) and the modified model is:

$$\ddot{\vartheta} + \frac{\kappa}{m}\dot{\vartheta} + \frac{g}{\ell}\sin\vartheta = 0 \tag{1.2}$$

κ characterizes the magnitude of the resistance of the environment.

(b) By supposing that the end O of the thread is moving during the motion of m, we will have a new model coming from (1.1). In case O moves horizontally and periodically, the modified model becomes an equation of Mathieu type.

(c) Another modification of (1.1) comes, if the length ℓ of the thread is changed during the motion of m. In case ℓ is taken analogous to the time and the velocity, and if we have small oscillation angles of the pendulum, then, under appropriate transformations of the corresponding equation, we lead to an equation of Bessel type as the corresponding modified model.

(d) The linearization of the nonlinear equation (1.1) gives another modified model. That is, under the hypothesis of small oscillation angles of the pendulum, ϑ and $\sin\vartheta$ are equivalent infinitesimals, and the modified model is:

$$\ddot{\vartheta} + \frac{g}{\ell}\vartheta = 0. \tag{1.3}$$

In this case we must remember that the linearization of nonlinear models eliminates, in general, important characteristics of the solution of the nonlinear model.

2. APPLICATION 2 (THERMODYNAMICS)

THE HEAT FLOW PROBLEM. (Ref. 21)

The concept of reasonability and unreasonability in the sense of Hadamard is, by this application, clarified.

To make the problem of heat flow in a given medium correctly stated in the sense of Hadamard, one accepts that the medium is homogeneous and isotropic with respect to the heat flow, and that the heat flow is towards the decreasing temperature. Based on these assumptions, the model of the problem is expressed by the elliptic partial differential equation:

$$u_{xx} + u_{yy} + u_{zz} = u_t \tag{2.1}$$

where $u = u(x, y, z, t)$ is the temperature. In the equation (2.1) there is a factor-coefficient depending on the density, specific heat and thermal conductivity, and this coefficient is here taken equal to unity. If the data are:

$$u(x, 0) = n \sin nx \, , \quad n = \text{integer} \tag{2.2}$$

the equation (2.1), in a one-dimension medium, has as a solution the function:

$$u(x, t) = n \cdot e^{-n^2 t} \cdot \sin nx \tag{2.3}$$

which is a unique solution, then the two Hadamard restrictions are satisfied. As for the third Hadamard restriction, one can distinguish two cases

(i) For $t > 0$, when one has the "forward heat problem", the solution (2.3), for any fixed n and $t \to \infty$, tends to zero, and the condition (2.2) is bounded; then the third Hadamard restriction is satisfied, the problem is "reasonable" in the sense of Hadamard, and the solution (2.3) is "physically accepted".

(ii) For $t < 0$, when one has the "backward heat problem", the solution (2.3) for any fixed n and $t \to \infty$, tends to infinity, and the condition (2.2) is bounded; then the third Hadamard restriction is violated, the problem is "unreasonable" in the sense of Hadamard, and the solution (2.3) is a "formal solution" of the heat problem, that is "not physically accepted".

3. APPLICATION 3 (NONLINEAR MECHANICS)

THE PROBLEM OF PRINCIPAL MODES OF NONLINEAR SYSTEMS. [Ref. 18(b)]

The concept of "principal modes" of nonlinear systems plays a predominant role in the analysis of oscillatory systems in many fields.

The principal modes of linear systems are, by definition, the fundamental set of solutions of which a linear combination gives the general solution of the linear system; or, physically speaking, they are the special modes of oscillations of the linear system in terms of which one can discuss any kind of oscillations of the system.

Since the "principle of superposition" does not hold in nonlinear systems, the concept of principal modes, as is given above, is meaningless in nonlinear systems, and the following question arises:
"Has the problem of principal modes of nonlinear systems a physical meaning"?
or: "How one can make the problem of principal modes of nonlinear systems a reasonable problem"?
For the solution of this problem, one must find a new definition of the concept of principal modes valid for both the linear and nonlinear systems, and such that the known definition in linear systems to be a special case. We can find two such new definitions of principal modes of nonlinear systems which, under some special conditions, are equivalent.
After we fix the subject of the definition of principal modes, we must make the problem "well expressed", "correctly stated" and "reasonable".

Let us restrict ourselves to a "two degrees of freedom" mechanical system, Figure 3. The motion of this system is governed by the model:

- 33 -

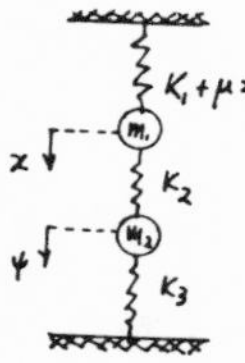

$K_1+\mu x^2$, m_1, K_2, m_2, K_3, x, ψ

FIGURE 3.

$$\left.\begin{aligned} \ddot{x} + \omega_1^2 x - \lambda_2 \psi + \lambda_1 x^3 &= 0 \\ \ddot{\psi} + \omega_2^2 \psi - \lambda_3 x &= 0 \end{aligned}\right\} \quad (3.1)$$

where:

$$\left.\begin{aligned} \omega_1^2 = \frac{K_1+K_2}{m_1}, \quad \omega_2^2 = \frac{K_2+K_3}{m_2} \\ \lambda_1 = \frac{\mu}{m_1}, \quad \lambda_2 = \frac{K_2}{m_1}, \quad \lambda_3 = \frac{K_2}{m_2} \end{aligned}\right\} \quad (3.2)$$

and m_1 , m_2 are the oscillating masses, μ characterizes the nonlinearity of the first anchor spring.

By using the transformation:

$$x = x_1, \quad \dot{x} = x_2, \quad \psi = x_3, \quad \dot{\psi} = x_4 \quad (3.3)$$

the system (3.1) takes the normal form:

$$\left.\begin{aligned} &(a): \quad \dot{x}_i = f_i(x_1, x_2, x_3, x_4), \quad i = 1,2,3,4 \\ &(b): \quad f_1 = x_2, \quad f_2 = -\omega_1^2 x_1 + \lambda_2 x_3 - \lambda_1 x_1^3, \quad f_3 = x_4, \quad f_4 = \lambda_3 x_1 - \omega_2^2 x_3 \end{aligned}\right\} \quad (3.4)$$

valid in a region R

$$R: \quad |x_i| < \ell, \quad i = 1, 2, 3, 4. \quad (3.5)$$

The appropriate initial conditions for existence of principal modes in R are:

$$x_1(0) = x_{10}, \quad x_2(0) = 0, \quad x_3(0) = x_{30}, \quad x_4(0) = 0 \quad (3.6)$$

where the non-zero x_{10} and x_{30} are appropriately related to each other.

We now remark that the functions f_i , given by (3.4, b), are such that the Hadamard restrictions are satisfied. These functions are continuous in R , then bounded, they have continuous partial derivatives $\frac{\partial f_i}{\partial x_k}$ in R , when they satisfy Lipschitz conditions with respect to x_i in R for a Lipschitz constant: $s = \ell.u.b.\left|\frac{\partial f_i}{\partial x_k}\right|$.

The above properties of f_i assure the unique existence of the solution of the model (3.4) associated with the conditions (3.6) in a region $R' \subset R$. As the initial point x_{i0}, $i=1,2,3,4$ varies in R' , the solution, which is yet unknown, satisfies the three Hadamard restrictions, and the model (3.4), subject to initial conditions (3.6), is "reasonable" in Hadamard sense.

APPLICATION 4 (ASTRODYNAMICS)

We discuss the following problem

"An artificial celectial body is moving under the influence of a central force obeying the inverse square Newton law toward the attractive center. A general perturbing force is applied, acts for an interval of time, then it is removed. Find the motion of the body during the action of the perturbing force". [Ref 18(c)]

According to known reasoning, this problem can be made "correctly stated", when the model is:

$$\left.\begin{aligned} &\ddot{\underline{r}}(\tau) = -\frac{\mu}{r^3(\tau)}\,\underline{r}(\tau) + \underline{T}(\tau) \\ &\underline{r}(0) = \underline{r}_0\,, \quad \dot{\underline{r}}(0) = \dot{\underline{r}}_0 + \underline{I}_0 \\ &D_1 : \quad |\underline{r}(\tau)| < M_1 \\ &D_2 : \quad |\dot{r}| < M_1 \\ &D_3 : \quad 0 \leq \tau \leq \tau' \end{aligned}\right\} \tag{4.1}$$

$\underline{T}$ is the perturbing force, $\underline{r}$ the radial vector from the attractive center to the center of the mass of the body, and $\underline{I}_0$, the impulse, is given by:

$$\underline{I}_0 = \int_0^{t_0} \underline{T}(t)\,dt\,, \quad \tau = t - t_0 \tag{4.2}$$

Forces $\underline{T}$ can be produced, e.g., by drag, by asphericity of the central body, etc.

The force $\underline{T}$ may depend on the position and velocity of the moving body, and on the time explicitly. We accept here that the force $\underline{T}$ is "considerable" with respect to central attraction, and that it is only a "function of time".

We select appropriate coordinate axes on which the unit vectors are: $\underline{r}_0^*$, $\underline{s}_0^*$, $\underline{T}_0^*$, shown in Figure 4.

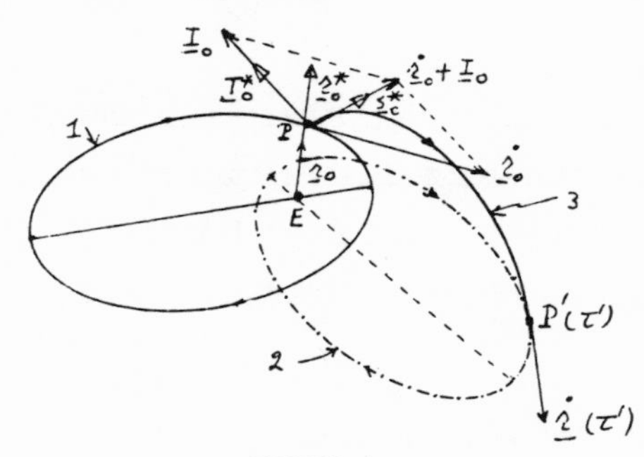

FIGURE 4.

Referred to this coordinate system, we take the function:

$$\underline{r}(\tau) = \alpha_1(\tau)\, \underline{r}_0^* + \alpha_2(\tau)\, \underline{s}_0^* + \alpha_3(\tau)\, \underline{T}_0^* \tag{4.3}$$

as a "trial solution" of the system (4.1), and by an appropriate calculation of the coefficients: $\alpha_1(\tau)$, $\alpha_2(\tau)$, $\alpha_3(\tau)$, we make this function first a "formal solution", and then an "actual solution" of (4.1).

The investigation shows that the function (4.3) becomes a "formal solu tion" of the system (4.1), if α_1 , α_2 , α_3 of (4.3) satisfy the equations

$$\left.\begin{aligned} \ddot{\alpha}_1 + \frac{\mu}{r^3}\alpha_1 &= T_1 \;; \quad \alpha_1(0) = r_0 \,, \quad \dot{\alpha}_1(0) = 0 \\ \ddot{\alpha}_2 + \frac{\mu}{r^3}\alpha_2 &= T_2 \;; \quad \alpha_2(0) = 0 \,, \quad \dot{\alpha}_2(0) = s_0 \\ \ddot{\alpha}_3 + \frac{\mu}{r^3}\alpha_3 &= T_3 \;; \quad \alpha_3(0) = 0 \,, \quad \dot{\alpha}_3(0) = 0 \end{aligned}\right\} \tag{4.4}$$

The solutions $\alpha_1(\tau)$, $\alpha_2(\tau)$, $\alpha_3(\tau)$ of (4.4) exist and are unique if the components T_1, T_2, T_3 of $\underline{T}$ are continuous functions of time and differentiable, also, if $\dot{T}_1$, $\dot{T}_2$, $\dot{T}_3$ are continuous functions of time, and also $\ell \neq 0$.

Under the above restrictions of T_1, T_2, T_3, the functions α_1, α_2, α_3 satisfy the Hadamard restrictions, and the function (4.3) is an "actual solution" of the model (4.1), then "physically accepted".

5. APPLICATION 5 (PROBLEM OF UNDERWATER WARFARE)

THE PROBLEM OF DOMES [Ref. 7(a)]

The problem of domes arose in the winter of 1942-43 in connection with "underwater warfare", and it gives a good example of the difficulties to go from reality to an acceptable idealized mathematical model.

As is known, underwater sound ranging depends on sending out a sound beam in water and, attached to a fast-moving ship, the water steaming around the plate causes serious disturbances. For elimination of these disturbances, the projector is enclosed in a "dome", Figure 5, which is a convex shell of metal, or other material, filled with water. Such domes interfere only slightly with the formation of a concentrated sound beam.

During 1942-43, a large number of small submarine chasers were built and equipped with sound gear similar to, but smaller than, the gear used before. While the manufacture of domes to fit this smaller gear was under way, it was discovered that these smaller domes led to an intolerable diffusion on the sound beam. At that time, a quick remedy was imperative, and a mathematical analysis of the problem was needed to support and speed-up experimental work. The mathematical problem, related to the above physical problem, was to solve the differential equation:

-37-

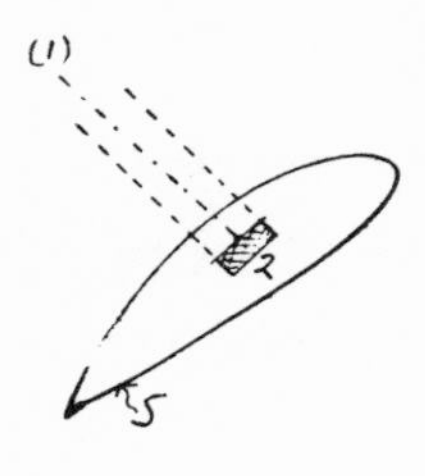

FIGURE 5.

(1) axis of beam sound

(2) projector

S surface of dome

$$\left.\begin{aligned} &\nabla^2 P + K^2 P = 0 \\ &\nabla^2 = \frac{\partial^2}{\partial x^2} + \frac{\partial^2}{\partial y^2} + \frac{\partial^2}{\partial z^2} \end{aligned}\right\} \quad (5.1)$$

in which $K = \omega / c$, ω the frequency c the sound velocity.

K has for our problem different values within the shell of the dome and outside.

This model was not a suitable one for the problem. A suitable model was found by the following process.

The actual dome of small finite thickness was replaced by an extremely thin surface, then the influence of the dome was simply replaced by conditions for jump discontinuities of the disturbance q of the beam across the surface. These conditions are

$$\left.\begin{aligned} [q] &= \frac{\rho_1}{\rho_0 - 1} \frac{\partial p}{\partial n} \\ \frac{\partial q}{\partial n} &= \frac{\rho_0}{\rho_1}(K_0^2 - K_1^2)p - \left(1 - \frac{\rho_0}{\rho_1}\right)\left(\frac{\partial^2 p}{\partial n^2} - 2H\frac{\partial p}{\partial n}\right) \end{aligned}\right\} \quad (5.2)$$

where the symbol $[\cdot]$ means jump of the quantity of the symbol across the surface, q is the disturbance of the acoustic pressure p caused by the dome, and the normal derivatives $\frac{\partial}{\partial n}$ are to be evaluated on the surface S of the dome. The quantity H is the mean curvature of S , i.e. the average of the curvature of any two normal plane sections at right angles to each other. In addition to conditions (5.2) to be satisfied by q on S , q should be a

- 38 -

solution of the equation:

$$\nabla^2 q + K_o^2 q = 0 \tag{5.3}$$

which is regular everywhere except on S and which has the same behavior as P at ∞ . This problem possesses the unique solution:

$$q = -\frac{1}{4\pi}\iint_S \left[\frac{\partial q}{\partial n}\right] \frac{e^{iK_o r'}}{r'} ds + \frac{1}{4\pi}\iint_S [q]\frac{\partial}{\partial n}\left(\frac{e^{iK_o r'}}{r'}\right) ds \tag{5.4}$$

The quantities in the brackets are given by conditions (5.2), r' is the distance from a fixed point (x, y, z) , at which $q(x, y, z)$ is to be determined, to the point of integration on S . This formula yields the disturbance as the effect due to a layer of point sources and a layer of dipoles disturbed on S with intensities which are known as soon as the original pressure p is known, since the quantities in brackets are fixed in value due to conditions (5.2). The relative directional disturbance

$$\left|\frac{p_1}{p}\right| Re. k\left(\frac{q_1}{A} - \frac{q_o}{p_o}\right)$$

would, finally, be obtained from (5.4). The solution (5.4) is valid for a shell of constant thickness, but it could be extended without essential error to cases in which the dome shell is made up of a not too large number of pieces, each of which is of constant thickness. All that would be necessary would be to insert a numerical factor α in the integrands on the right-hand side of expression (5.4), which would be piecewise constant on S . By this formula, one can analyze the contribution to the distortion of various factors, such as the curvature of the dome and the density of sound velocity within it. The above kind of mathematical idealization, even without detailed numerical computation, proved very helpful to the designing engineer.

6. APPLICATION 6. (BIOLOGY, ECOLOGY, ECONOMICS, ETC.)

(Ref. 1, 5, 9, 10, 11, 14, 17, 26)

There are many assemblies around us of which the elements influence each other through competition and cooperation. Some examples are: population of various biological species; components of the nervous system; coupled reacting chemical components in the atmosphere, in bodies of water, in organisms as a whole or in part; business; political parties; countries.

With the currently developing interest in the investigation of biological, social and economic mechanisms, it is of considerable importance to find mathematical models, which might be amenable to a detailed investigation.

Let us discuss

THE PROBLEM OF MODELING OF POPULATION GROWTH,

as a biological problem, but, by appropriate changes of the discussion and appropriate changes in the meaning of the variables involved, the discussion can be used for modeling in other fields.

Lofka and Volterra are pioneers in the problem and their models are the basis for modifications or generalizations of the models.

We first discuss the modeling in the case of a single population, then the case of two or more populations in coexistence.

6.1 GROWTH OF A SINGLE POPULATION.

Let $N(t)$ be the number of individuals in the population, or population density per unit area, at a certain time t. $N(t)$ is a positive function of time, a "step function" of t, and it is constant in an interval of time, when the population does not change, and it is discontinuous, when a birth or death changes the population.

To create a model for the growth of a single population, one may start from two assumptions

(i) $N(t)$ is a continuous function of time t,

(ii) The instantaneous rate of growth of $N(t)$ depends on $N(t)$ itself, that is

$$\dot{N} = f(N). \tag{6}$$

The function $f(N)$ must be specified according to new assumptions and observation facts.

We distinguish the following two cases.

(a) Single Nonisolated Population.

By nonisolated population we mean population with unlimited food supply and space.

If n is the birth coefficient and m the death coefficient, then, in a small interval of time Δt, by assuming that the increment of the population ΔN is analogous of Δt and N itself, one will have in the limiting case

$$\dot{N} = (n-m)\ N \tag{6.1}$$

that is we assume as function $f(N)$ in (6) the function $f(N) = (n-m)N$. The coefficient $(n-m)$ characterizes the nature of the population. If $N(0) = N_0$ is the initial population, then the solution of (6.1) is:

$$N = N_0\, e^{(n-m)t}. \tag{6.2}$$

In case $n < m$, the population tends to disappear. But in case $n > m$ the population increases exponentially to infinity under the hypothesis of unlimited food supply and space for the population.

The above conclusion does not agree with observations and experimental facts, then the model (6.1) must be changed.

(b) Single Isolated Population

In any population with a large number of individuals, the nourishment is bad, the food supply limited, the space also limited, the population is

- 41 -

"isolated", and the growth of the population meets a resistance. Its mortality increases and the population can not pass an upper limit but tends to approach a maximum N_M . The model (6.1), which is good for N close to N_0 , needs a modification in order to accomplish the observations.

For a modification of the model (6.1) one needs to insert a coefficient for the resistance effect. This comes by changing the law of mortality, that is by changing the mortality coefficient m into

$$m' = m + bN \tag{6.3}$$

when the new model is

$$\dot{N} = (n-m)N - bN^2 \tag{6.4}$$

when the function $f(N)$ of (6) is now $f(N) = (n-m)N - bN^2$.

The equation (6.4) is an equation of Bernoulli's type, is called a "Verhalst equation", and its general solution, called "logistic curve", is

$$N = \frac{n-m}{b + e^{-c} \cdot e^{-(n-m)t}} \tag{6.5}$$

where c is a constant.

This solution can be modified in order to contain the initial N_0 and the maximum N_M populations.

The solution (6.5) for $t=0$ gives $e^{-c} = (n-m-bN_0)/N_0$, when (6.5) becomes

$$N = \frac{(n-m)N_0}{bN_0 + (n-m-bN_0)\, e^{-(n-m)t}} \tag{6.6}$$

which includes N_0 .

For $n>m$, as $t \to \infty$, N becomes $N_M = (n-m)/b$, when (6.6) becomes

$$N = \frac{N_0 N_M}{N_0 + (N_M - N_0)\, e^{-(n-m)t}} \tag{6.7}$$

which includes N_0 and N_M .

- 42 -

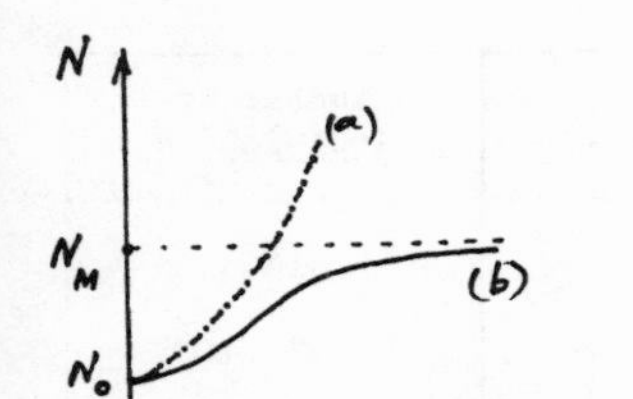

FIGURE 6. Curve (a) is the graph of (6.2), and (b) that of (6.7)

In Figure 6, the solid curve (b) is the graph of the solution (6.7), and the dotted curve (a) the graph of (6.2).

According to the formula (6.7), the population N increases when $n > m$, and finally reaches its maximum N_M, which is the "equilibrium of the population", when $m' = m + bN_M = n$.

If $n = m$, then $N = N_0$ for any t; and if $n < m$, then $N \to 0$ for $t \to \infty$.

For N_0 small, compared to N_M, and N close to N_0, then the term bN^2 of (6.3) is negligible compared to the other term, when the model (6.4) near N_0 is identical to the model (6.1).

If N increases, the term bN^2 becomes considerable, the model (6.1) stops to be a good model, and the model (6.4) is the appropriate model.

The model (6.4) is an appropriate one for a variety of single population growth problems, and its solution (6.7), that is the curve (b) in Figure 6, fits quite well the data on population growth of many species.

For this, we mention two examples:

(i) The formula (6.7) has been applied to population growth of U.S.A. and gave accurate predictions, although the creation of the model (6.4) is based on very simplified assumptions.

(ii) The growth of yeast cells in a laboratory culture, for which the data of the following table are given, corresponds to the curve of the Figure 7, which agrees with the curve (b) of Figure 6.

Time In Hours	Number of Individuals	Time In Hours	Number of Individuals
0	9.6	10	513.3
2	29.0	12	594.4
4	71.1	14	640.8
6	174.6	16	655.9
8	350.7	18	661.8

Table of Data

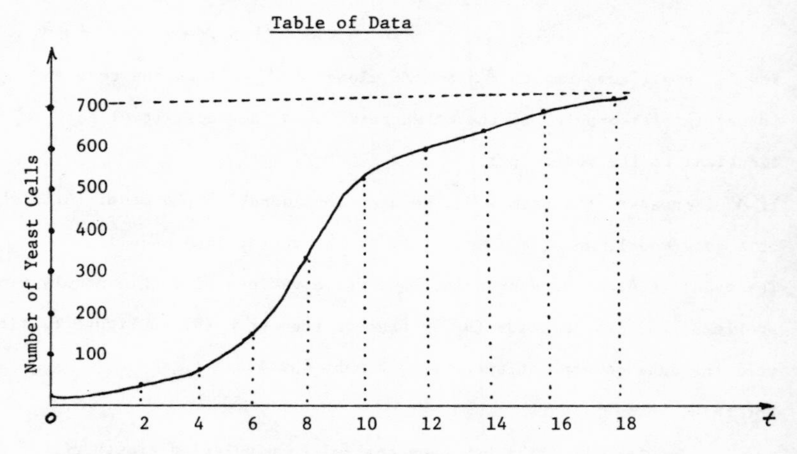

FIGURE 7. The Growth Curve of Yeast Cells in Laboratory

We remark that different assumptions for specification of the function $f(N)$ of the general original model (6) will give different models, which may be proved to be in many cases better than, the above models. E.g., if, instead of N^2 in the model (6.4), one takes $N \cdot \log N$, one has another useful model.

6.2 THE CASE OF TWO SPECIES

We first clarify some notions usually appearing in coexisting populations.

In any model of two species with population densities N_1 and N_2 in community, its solution starts from an initial point (N_{10}, N_{20}) of the first quadrant of the N_1, N_2 -plane and terminates to a point $(\bar{N}_1, \bar{N}_2)$ in finite or infinite time, and we may have the following three cases

(i) The point ($\bar{N}_1$, $\bar{N}_2$) may coincide with the origin, when one speaks about "extinction of the species";

(ii) One of $\bar{N}_1$, $\bar{N}_2$ may be zero and the other nonzero, that is the point ($\bar{N}_1$, $\bar{N}_2$) may be a point of one of the axes, and this case corresponds to the "principle of competitive exclusion", a principle much used in ecology, but which has been much critisized.

(iii) The point ($\bar{N}_1$, $\bar{N}_2$) may be a point with nonzero coordinates, then may be inside a region of the first quadrant of N_1, N_2 -plane, that is all points (N_1 , N_2) may be in the region and N_1 , N_2 may be changed periodically in the region. In this case one speaks about "coexistence of two species", "symbiosis", "parasitism".

Let us discuss now the modeling of two species of populations inhabiting the same environment and competing for limiting food supply and space.

Lotka investigated the problem in case of living species, the "host", in the presence of "parasites" capable of resisting the development of the host. Volterra discussed a similar problem, that of the evaluation in a closed sea of two kinds of fish, the first "soles" and the second "sharks", the second eating the first. Experimental and statistical facts in the Adriatic sea

inspired Volterra in the mathematical discussion of the problem.
The Lotka and Volterra problems are equivalent from a theoretical point of view. The soles are the hosts and the sharks their parasites.

Let n_1, m_1, b_1 be the birth, death and competition coefficients, respectively, of the population N_1, the host (prey). The increase of N_1 in time dt, due to the births, is $n_1 N_1 dt$. This is an hypothesis not always true, but we consider it as true temporarily. Let N_2 be the population of parasites (predators). The number of meetings will be proportional to $N_1 N_2$, say, $a N_1 N_2$, when the increment of N_1 is

$$\dot{N}_1 = (n_1 - m_1) N_1 - b_1 N_1^2 - a N_1 N_2 \tag{6.8}$$

which is a DE corresponding to the host population. This formula differs from (6.4) by the term $-a N_1 N_2$ of deaths after the meeting of host and parasite individuals.

Now, the equation corresponding to the parasite population. The births will be proportional to N_1 and to N_2, then proportional to their product $N_1 N_2$. If m_2 is the coefficient of natural deaths, and b_2 the competition coefficient, then the evolution of parasites is given by the equation:

$$\dot{N}_2 = k a N_1 N_2 - m_2 N_2 - b_2 N_1^2. \tag{6.9}$$

The two simultaneous DE (6.8) and (6.9) is the model of the above "prey-predator system". This system has been generalized to the set of n-species. If the effect of the species j on the growth $\dot{N}_i$ of the species i is proportional to the number of meetings between them, which we assume to be the product $N_i N_j$ of their population sizes, then the model of this system is

$$\dot{N}_i = K_i N_i + \sum_{j=1}^{n} a_{ij} N_i N_j \quad , \quad i = 1, \ldots, n. \tag{6.10}$$

For different sets of the values of the coefficients of (6.10) and different signs of these values, one may have different situations of the population growth in coexisting species, that is different physical problems.

The above models, called "Lotka-Volterra models", are "reasonable" in the sense of Hadamard.

6.3 REVISIONS, MODIFICATIONS AND GENERALIZATIONS OF THE LOTKA-VOLTERRA MODELS OF POPULATION GROWTH.

(a) Some Remarks.

The Lotka-Volterra models for the population of interacting species are created based mainly on the assumptions

(i) All the members of a species have the same age,

(ii) Any member of the species reacts instantaneously to any change in the environment, and

(iii) The prey-predator interactions affect instantaneously the population of both the prey and the predator.

Such assumptions may be realized in laboratory situations, but, in general, are not realistic. Observations and experiments contradict some results coming from the Lotka-Volterra models.

The growth of a population may be affected by temperature, humidity, age, distribution of the species, time of the year, time lags in population, and other ecological factors, and the biological interactions are competition, parasitism, symbiosis, complex interactions between the organisms and its physical environment.

The attempts to manage the population growth will profit from the use of mathematical models, even when these models were found with very simplified assumptions, as this happens with Lotka-Volterra models.

We must not expect to find a model of population growth to satisfy all the above ecological factors, and to include all the important features about the population. We must restrict ourselves to modified models, which satisfy a specific program of research and to be realistic.

There are no clear-cut guidelines for choosing one interaction over the other in population dynamics, and it is the experiment which would help the decision.

Depending on the problem and on the purpose of the investigation, one may use techniques of statistical mechanics, or one may take into account statistical aspects of the population growth of individual species and the correlation between population of different species.

Usually, one can not have information about detailed interactions between various species, then their statistical mechanics treatment is desirable. The statistical treatment provides an empirical method for calculating the interactions between two species and the stability of the population.

In many problems the interaction coefficients of the models is difficult to be evaluated, when the prediction is imprecise. One must assume in these cases that the coefficients of the models are random variables, when the populations are characterized by a probability distribution, which can be used to calculate the average population.

Each modification or generalization of existing models must have as a purpose to cover questions of specific research and to be more realistic. Models exactly solvable are preferable, because their exact solutions provide a testing ground for general principles used in the study of population dynamics.

We now give some modified and generalized models.

(b) Exactly Solvable Models.

The system of equations (6.8) and (6.9) can be solved only by using approximate methods. We see here that, by specific modifications and transformations, these equations will lead to equations exactly solvable. We have two cases.

(b_1) Case Without Mutual Interactions of the Two Species.

In the equations (6.8) and (6.9), in case of no mutual interactions, $b_1 = b_2 = 0$, when, if instead of the product $N_1 N_2$ we take $N_1 \log N_2$ in the first equation and $N_2 \log N_1$ in the second, the model reduces to an exactly solvable one.

In this case, if, for simplicity, we take a_1, a_2, λ_1, λ_2 instead of $n_1 - m_1$, m_2, α, $k\alpha$, respectively, the new model becomes

$$\left.\begin{aligned} \dot{N}_1 &= a_1 N_1 - \lambda_1 N_1 \log N_2 \\ \dot{N}_2 &= a_2 N_2 + \lambda_2 N_2 \log N_1 \end{aligned}\right\} \tag{6.12}$$

with initial values N_{10} and N_{20}.

By the transformation

$$X = \log N_1 - \frac{a_2}{\lambda_2}, \quad X_2 = \log N_2 - \frac{a_1}{\lambda_1} \tag{6.13}$$

the model (6.12) becomes

$$\dot{X}_1 = -\lambda_1 X_2, \quad \dot{X}_2 = \lambda_2 X_1 \tag{6.14}$$

of which the initial conditions are:

$$X_{10} = \log N_{10} - \frac{a_2}{\lambda_2}, \quad X_{20} = \log N_{20} - \frac{a_1}{\lambda_1} \tag{6.15}$$

The solution of (6.14), subject to conditions (6.15) is

$$\left.\begin{aligned} X_1 &= X_{10} \cos\omega t - \frac{\lambda_1}{\omega} \sin\omega t \\ X_2 &= X_{20} \cos\omega t + \frac{\lambda_2}{\omega} \sin\omega t \end{aligned}\right\} \tag{6.16}$$

where $\omega = \sqrt{\lambda_1 \lambda_2}$.

Eliminating t in (6.16), one gets the family of orbits

$$\left.\begin{aligned} &\text{(a)} \quad \lambda_2 X_1^2 + \lambda_1 X_2^2 = c \\ &\text{(b)} \quad c = \lambda_2 X_{10}^2 + \lambda_1 X_{20}^2 \end{aligned}\right\} \tag{6.17}$$

which are ellipses in the X_1, X_2-plane.

- 49 -

By using (6.13) and (6.16), the solution of (6.12) can be written in the form

$$\left.\begin{aligned} N_1 &= \exp\left\{X_{10}\cos\omega t - \frac{\lambda_1}{\omega} X_{20}\sin\omega t + \frac{a_2}{\lambda_2}\right. \\ N_2 &= \exp\left\{X_{20}\cos\omega t + \frac{\lambda_2}{\omega} X_{10}\sin\omega t + \frac{a_1}{\lambda_1}\right. \end{aligned}\right\} \tag{6.18}$$

Then the corresponding orbits are members of the family of the complicated curves in the N_1, N_2-plane

$$\lambda_2\left(\log N_1 - \frac{a_2}{\lambda_2}\right)^2 + \lambda_1\left(\log N_2 - \frac{a_1}{\lambda_1}\right)^2 = 2c \tag{6.19}$$

where c is given by (6.17.b).

Through the transformation (6.13), the equilibrium $\left(N_1 = e^{a_2/\lambda_2}, N_2 = e^{a_1/\lambda_1}\right)$ of (6.12) is transformed into the center of the ellipses (6.17), which is the origin in the X_1, X_2-plane.

By using the above exact solution of the model (6.12) and restricting suitably the values of the constants, one may find conditions for "extinction" of the species, for "competitive exclusion", or for their "stable coexistence". Some results of the investigation by using (6.12) are different from results by using the corresponding Lotka-Volterra model, and experiments decide on the choice of the model (6.12).

(b_2) Case with Mutual Interactions of the Two Species.

The model of (6.8) and (6.9) with mutual-interactions and with self-interactions, if modified according to

$$\left.\begin{aligned} \dot{N}_1 &= a_1 N_1 - \lambda_{11} N_1 \log N_1 - \lambda_{12} N_1 \log N_2 \\ \dot{N}_2 &= a_2 N_2 - \lambda_{22} N_2 \log N_2 - \lambda_{21} N_2 \log N_1 \end{aligned}\right\} \tag{6.20}$$

Such a model can be solved exactly.

We remark that both models (6.12) and 6.20) are reasonable in the sense of Hadamard.

- 50 -

(c) Some Modified and Generalized Models.

An important generalization, which includes many classes of problems of population growth, is the model

$$\dot{N}_i = f_i(N_1, \ldots, N_n, P_1, \ldots, P_m) \, , \quad i = 1, \ldots, n \tag{6.21}$$

of n species in community with population densities N_i , $i = 1, \ldots, n$. The parameters P_j , $j = 1, \ldots, m$ define the environment where the evolution of N_i takes place.

One can incorporate the fact " $f_i \equiv 0$ for $N_i = 0$ " in the system (6.21), if one writes this system as

$$\dot{N}_i = N_i \, g_i(N_1, \ldots, N_n, P_1, \ldots, P_m), \; i = 1, \ldots, n . \tag{6.22}$$

The parameters P_j in (6.22) influence the populations, their rate of reproduction, the duration of their life. They are related to topological and geographical features of the species, or to climate conditions, etc. These parameters may depend on each other, when they may exist relationships between the parameters themselves, when the number of the parameters in (6.22) may be decreased.

In the above general model, the populations N_i may in general be also subject to certain relationships, which enable one to decrease the number of N_i. We assume that the model (6.22) is taken after using the restraint relationships, when the number of species and the number of parameters are the smallest possible to determine the phenomenon.

The model (6.22) must be a "reasonable" one, when the functions must be appropriately restricted. The g_i 's and their first partial derivatives with respect to N_i and P_j must be defined, single-valued and continuous functions of N_i and P_j and of the initial values N_{0i} and P_{0j} in the state space and the parameter space, when there exists only one solution of the model (6.22), starting from the initial state N_{0i} for any initial parameter point P_{0j} , and this solution is physically accepted.

We might mention a special class of models created by relaxing the known assumptions on which the Lotka-Volteva models are based, the class of models with "time-lags" populations.

In the special case of a "single species" of this class, the following equations with "time-lag" where proposed as models

$$\dot{N}(t) = K\left\{1-(1/\theta)\,N(t-\tau)\right\}N(t) \tag{6.23}$$

and

$$\dot{N}(t) = K\left\{1-(1/\theta)\,N(t-\tau)\right\}N(t-\tau)-K'N(t) \tag{6.24}$$

of which there are some generalizations.

In these models it is assumed that the birth rate coefficient is diminished by a quantity proportional to the population of the preceding generation, the positive τ being the generation time, that is time required in going from an egg-stage to the adult-stage.

The solution of the above models shows a variety of properties depending on the parameters included. Of these properties some are more of mathematical interest than of practical usefulness.

The "birth-death process", applied to population growth, to social and physical contagion, as well as to other phenomena, illustrates well some of the advantages and disadvantages of a deterministic or probabilistic approach.

For example, let us see the deterministic and stochastic models in the case of a single population.

For a single population of size N at time t, if n and m are the birth and death coefficients, respectively, the deterministic model is

$$\dot{N} = (n-m)\,N \tag{6.25}$$

which is a very simple model clearly used, as we know, in some circumstances.

For the probabilistic version of this particular process, that is for the corresponding probabilistic model, we express the rate of change $\dot{p}_i$ in the probability p_i of being in state i in terms of the three conditional probabilities: $n(i-1)$, $(m+n)i$, $m(i+1)$, since the probabilities of birth or death are proportional to the existing sizes: $(i-1)$, i, $(i+1)$, respectively, for the three states, when we have the stochastic model of the above process

$$\dot{p}_i = n(i-1)\,p_{i-1} - (m+n)\,i\,p_i + m(i+1)\,p_{i+1} \tag{6.26}$$

This stochastic model is, by neglecting some factors, an accurate one, but it is difficult to be treated analytically.

The stochastic model (6.26) gives the probability that a population will die out due to change fluctuation, an information much important in some applications, but which cannot be given by the deterministic model (6.25).

The deterministic model (6.25) is the "expected mean value" of the probability distribution, and this model is preferred in cases one only needs the expected value of the probability distribution.

In general, in spite of the obvious stochastic nature of processes, much of the existing theory is deterministic, because the mathematics of a full stochastic version of many of the models is much intractable. One must prefer the deterministic version of the model in order to make progress by taking an approximation of the model, which first is stochastically formulated.

We, finally, remark that the previous discussion, if suitably modified, that is by appropriate changes to the problem and appropriate specification of the competitive species and the limiting resources, might be useful for an investigation of a problem of nature different than the above. E.g., one can have a problem in the field of economics, if the variables denote the size or extent of commerical enterprises competing for a common source and for a common market.

- 54 -

BIBLIOGRAPHY

1. Ayala, F.: (a) "Biological evaluation: Natural selection or random walk"? American Scientist, (Nov-Dec. 1974), pp. 672-701, U.S.A.
(b) "Competition between species", American Scientist (1972), pp. 348-357, U.S.A.

2. Bartholomew, D.: "Stochastic models for social processes", 2nd Ed., John Wiley and Sons, New York (1973), Chapter I, U.S.A.

3. Birckoff, Garrett: "Mathematics and computer science", American Scientist,(Jan-Feb. 1975) pp. 83-91, U.S.A.

4. Brillouin, L.: (a) "Scientific uncertainty and information", Academic Press, New York (1946), U.S.A.
(b) "Relativity re-examined", Academic Press, New York (1970), U.S.A.

5. Coleman, S.: "Introduction to mathematical sociology", The Free Press of Glencoe, Collier-Macmillan Lim., London (1964), pp. 526-528, England.

6. Cortes, F., Przeworski, A. and Sprague, J.: "Systems analysis for social scientists", John Wiley and Sons, New York (1974), U.S.A.

7. Courant, R.: (a) "Methods of applied mathematics", in the book: "Recent advances in science" of Shammos, M. and Murphy, G. Science Editions, Inc., New York (1961), U.S.A.
(b) "Boundary value problems in modern dynamics", Proc., Intern. Congress of Math, Vol. II (1950)

8. Education in Applied Mathematics: SIAM Review, Vol. 9, No. w (April 1967), U.S.A.

9. Gandolfo, G.: "Mathematical methods and models in economic dynamics," North Holland Publ. Co., Amsterdám (1971).

10. Goel, N., Maitra, S. and Montroll, E.: "On the Volterra and other nonlinear models on interacting populations", Review of Modern Physics, Vol. 43, No. 2, Part 1 (April 1971) pp. 231-276, U.S.A.

- 55 -

11. Gomatan, J.: "A new model for interacting populations",
I: "Two species system", pp. 347, 353
II: "Principle of competitive exclusion", pp. 355-364;
Bulletin of Mathematical Biology, Vol. 36 (1974), U.S.A.

12. Goodwin, B.: "Temporal organization in cells",
Academic Press, New York (1963), U.S.A.

13. Keisler, H.: "Model theory", Proc., Intern. Congress of Math, Vol. I (1970)

14. Kostitzin, V.: "Biologie mathématique," Librarie Armand Colin, Paris (1937), France.

15. Lavrentiev, M.: "Some improperly-posed problems of mathematical physics",
Springer-Verlag, New York (1967), U.S.A.

16. Lin, C. and Segel, L.: "Mathematics applied to deterministic problems in
the Natural Sciences", Macmillan Publ. Co., New York (1974), U.S.A.

17. Lotka, A.: "Elements of physical biology", Baltimore, Williams (1974), U.S.A.

18. Magiros, D.: (a) "Physical problems discussed mathematically", Bulletin
Greek Math. Soc., Vol. 6, II, No. 1 (1956), Athens, Greece.
(b) "Methods for defining principal modes of nonlinear
systems utilizing infinite determinants", Journal of Math.
Physics, Vol. 2, No. 66 (Nov-Dec., 1960), Americal Institute
of Physics, U.S.A.
(c) "Motion in a Newtonian field modified by a general force",
Journal of Franklin Institute (Sept. 1964), U.S.A.
(d) "The role of mathematics in the investigation of physical
and social phenomena," Greek Math. Soc. "The Euclide",
(May-June, 1973), Athens, Greece.

19. Marčhac, G.: "Methods and problems of computational mathematics",
Proc., Intern. Congress of Math, Vol. I (1970).

20. Mesaronič, M.: "Systems theory and biology", Proc., III Systems Symposium
Spring-Verlag, New York (1968), U.S.A.

21. Miranker, N.: "A well-posed problem for a backward heat equation",
Proc. Amer Math. Soc. (April 1961), U.S.A.

22. Neumann, John Von: "The mathematician", Collected Works, Vol. I, pp. 1-9, Oxford, New York (1961).

23. Rapoport, A.: "Mathematics outside the physician sciences", SIAM Review (Apr. 1973), pp. 481-502, U.S.A.

24. Papoulis, A.: "Probability random variables and stochastic processes", McGraw-Hill Book Co., New York (1965), Chapter 1, U.S.A.

25. Poincare, H.: (a) "Science and hypothesis", Dover (1952), New York, U.S.A.
(b) "The value of science", Dover, (1958), New York, U.S.A.

26. Volterra, V.: "Leçons sur la théorie mathematique de la lutte pour la vie ," Gauthier Villars et Cie, Paris (1931), France

27. Will, G.: "Gravitation theory", Scientific American (Nov. 1974), U.S.A.

PART II

NONLINEAR MECHANICS

This part contains papers dealing with subharmonic oscillations, principal modes, and celestial and orbital mechanics.

Papers 12 through 17 treat various problems of subharmonic oscillations of any order, especially of order 2 and 3. In these problems, the nonlinearity appears in the elastic forces. A very important fact is that the coefficients of the equations are not assumed to be necessarily small. Steady-state and transient type responses are examined. The conditions under which subharmonics exist are established, and the stability of these subharmonics in the steady-state is investigated. Various problems and applications related to subharmonics are discussed. Magiros' studies on subharmonics of order 1/3 formed the theoretical basis of the IBM research on the "Atomic clock".

Papers 18 through 21 are devoted to the study of principal modes which play the predominant role in the analysis of the oscillating systems in all fields. In these papers a method for calculating principal modes of linear and nonlinear systems is established. The author uses two definitions of principal modes in nonlinear systems which are shown to be equivalent under certain restrictions. These definitions filled-in a gap in linear and nonlinear system theory and helped in the understanding of the general behavior of solutions of nonlinear systems. The method yields the possibility of getting the principal modes in the form of a convergent series, all coefficients of which can be calculated. The problem of principal modes in Magiros' sense is a well-posed problem. The singularities of the system, i.e. the internal resonant frequencies of the components of principal modes solution, are also discussed.

Finally Papers 22 through 29 are devoted to celestial and orbital mechanics (two-body) problems which are of great interest both in theory and practice. Magiros' contribution in this field is especially significant. In papers 22, 23, 24 and 26 the motion of man-made celestial body under the influence of an attractive center, according to the inverse square of Newton's law and a "general force", is indicated. A solution of special form, referred to a specific inertial coordinate system is established. Taylor series in time are employed. It is shown that the solution found satisfies Hadamard's postulates. The region of convergence of the solution is determined and the cases of sudden or gradual application of the force are examined. In paper 25 the impulse (or impulsive force), required to effectuate a new Keplerian orbit around an attractive center at a given point in space is calculated. The solution of the problem is based on an auxiliary problem (solved geometrically), on a projection property of the "admissible impulse" and on trigonometric calculations. Finally, papers 27, 28 and 29 discuss the solution derived in the previous papers, when there are singularities. By an appropriate restriction of the new forces, the finite region of convergence of the solution is calculated. The case of collision is also examined in some instances. Furthermore, the re-entry problem is discussed as an application of the general procedure developed in paper 26.

ΠΡΑΚΤΙΚΑ
ΤΗΣ ΑΚΑΔΗΜΙΑΣ ΑΘΗΝΩΝ

ΕΤΟΣ 1957: ΤΟΜΟΣ 32ΟΣ

ΑΝΑΤΥΠΟΝ
ΣΕΛ. 77 - 85

Subharmonics of any order in case of non linear restoring force. Part I.,

by ***Dem. G. Magiros****.

Ἀνεκοινώθη ὑπό τοῦ κ. Βασ. Αἰγινήτου.

Introduction.

We discuss here the subharmonics of any order in the case of linear damping, sinusoidal external force, and cubic type restoring force, with coefficients not necessarily small. By using ideas of Van der Pol[1], Mandels-

* ΔΗΜ. Γ. ΜΑΓΕΙΡΟΥ, Περὶ τῶν ὑποαρμονικῶν ταλαντώσεων οἱασδήποτε τάξεως. Μέρος Α΄.

[1] VAN DER POL, *Phil. Mag.*, **3**, 1927, 65.

Reprinted from the *Proceedings of the Athens Academy of Sciences* **32** (1957), 77–85.

tam - Papalexi[1], Andronow - Witt[2], we get a proper transformation of the equation, and for its «steady state» and «transient» solutions use of the Poincaré's method for periodic solutions is made.

The conditions for the existence of the subharmonics and their stability in a steady state are discussed by considering that of the singularities of the corresponding equation.

The formulae given here can be used for investigation of the subharmonics of any order.

§ 1. *The problem.*

Many problems in Physics lead to the differential equation of the form:

$$\ddot{Q} + \bar{k}\dot{Q} + \bar{c}_1 Q + \bar{c}_2 Q^2 + \bar{c}_3 Q^3 = B \sin n\tau, \qquad n = 2, 3, \ldots, \tag{1}$$

where the coefficients $\bar{k}$, $\bar{c}_1$, $\bar{c}_2$, $\bar{c}_3$, B are not necessarely small. The solution of (1) is know when the coefficients are small, but it is unknown in case of not necessarily small coefficients.

We intend to find the solution in this last case.

§ 2. *Proper transformation.*

We transform (1) by taking a parameter ε such that:

$$\bar{k} = \varepsilon k, \quad 1 - \bar{c}_1 = \varepsilon c_1, \quad \bar{c}_2 = \varepsilon c_2, \quad \bar{c}_3 = \varepsilon c_3 \tag{2}$$

The result of this transformation is:

$$\ddot{Q} + Q = \varepsilon f(Q, \dot{Q}) + B \sin n\tau \tag{3}$$

$$f(Q, \dot{Q}) = -k\dot{Q} + c_1 Q - c_2 Q^2 - c_3 Q^3 \tag{3α}$$

The coefficients and ε in (2) are finite.

a) In case ε=0, the solution of (1) is:

$$Q = x \sin\tau - y \cos\tau + \frac{B}{1 - n^2} \sin n\tau, \tag{4}$$

with period 2π; $n \neq 1$. The arbitrary constants x and y can be determined by using the initial conditions of (1). x and y are the components of the subharmonic of order $\frac{1}{n}$, of which $r = (x^2 + y^2)^{1/2}$ is the amplitude. The third part in (4) is the harmonic part of the solution due to the forcing term of (1).

[1] L. Mandelstam - N. Papalexi, *Techn. Phys.*, U. S. S. R., 1935, 415.

[2] A. Andronow - A. Witt, Arch. für Electrotechn., 1930.

b) In case $\varepsilon \neq 0$, we attempt to determine a periodic solution of (3) of the form (4), in which x and y are functions of ε and τ and such that the limits:

$$\lim_{\varepsilon \to 0} x, \quad \lim_{\varepsilon \to 0} y \tag{5}$$

are the constants of the «generating solution» in case $\varepsilon = 0$.

§ 3. *Reduction to a first order system.*

We take a new variable q according to:

$$q = Q - \frac{B}{1-n^2} \sin n\tau. \tag{6}$$

Substituting (6) into (3) we get:

$$\ddot{q} + q = \varepsilon f(q, \dot{q}) \tag{7}$$

where:

$$\begin{aligned} f(q, \dot{q}) = \Big[& -k\dot{q} + c_1 q - c_2 q^2 - c_3 q^3 - \frac{kn}{1-n^2} B \cos n\tau \\ & + c_1 \frac{B}{1-n^2} \sin n\tau - c_2 \frac{B^2}{(1-n^2)^2} \sin^2 n\tau - c_3 \frac{B^3}{(1-n^2)^3} \sin^3 n\tau \\ & - 2c_2 \frac{B}{1-n^2} q \sin n\tau - 3c_3 \frac{B}{1-n^2} q^2 \sin n\tau \\ & - 3c_3 \frac{B^2}{(1-n^2)^2} q \sin^2 n\tau \Big]. \end{aligned} \tag{7α}$$

Introduce into (7) new variables u_1 und u_2 defined by:

$$\begin{cases} u_1 = \dot{q} \cos \tau + q \sin \tau \\ u_2 = \dot{q} \sin \tau - q \cos \tau, \end{cases} \tag{8}$$

from which we get:

$$\begin{cases} q = u_1 \sin \tau - u_2 \cos \tau \\ \dot{q} = u_1 \cos \tau + u_2 \sin \tau, \end{cases} \tag{9}$$

and

$$\begin{cases} \dot{u}_1 = (\ddot{q} + q) \cos \tau \\ \dot{u}_2 = (\ddot{q} + q) \sin \tau, \end{cases} \tag{10}$$

when according to (7), we have:

$$\begin{cases} \dot{u}_1 = \varepsilon f_1(u_1, u_2, \tau) \\ \dot{u}_2 = \varepsilon f_2(u_1, u_2, \tau), \end{cases} \tag{11}$$

where:

$$(11\alpha) \qquad \begin{cases} f_1(u_1, u_2, \tau) = f(u_1, u_2, \tau) \cdot \cos\tau \\ f_2(u_1, u_2, \tau) = f(u_1. u_2, \tau) \cdot \sin\tau ; \end{cases}$$

The function $f(u_1, u_2, \tau)$ is given from (7α) by using (9).

The equation (3) is replaced by the first order system (11), which gives advantages in the analysis.

From (6) and the first of (9) we obtain:

$$(12) \qquad \dot{Q} = u_1 \sin\tau - u_2 \cos\tau + \frac{B}{1-n^2} \sin n\tau .$$

The expressions (4) and (12) are of the same form, then we ask for a determination of the limits:

$$(5\alpha) \qquad \lim_{\varepsilon \to 0} u_1 , \quad \lim_{\varepsilon \to 0} u_2$$

§ 4. *Discussion of a general system.*

We briefly discuss, as it is needed for our purpose here, the solution of the general system:

$$(N) \qquad \begin{cases} \dot{u}_1 = \varepsilon \bar{f}_1(u_1, u_2, \tau) \\ \dot{u}_2 = \varepsilon \bar{f}_2(u_1, u_2, \tau) \\ u_1(\tau_0) = u_{1\tau_0}, \ u_2(\tau_0) = u_{2\tau_0} , \end{cases}$$

where the functions $\bar{f}_1$ and $\bar{f}_2$ are analytic in u_1 and u_2, and periodic of period 2π and continuous in τ, hence continuous in u_1, u_2, τ, and therefore $|\bar{f}_1|$ and $|\bar{f}_2|$ have upper bounds M_1 and M_2 respectively in the domain:

$$(D) \qquad |u_1 - u_{1\tau_0}| < r_1 , \ |u_2 - u_{2\tau_0}| < r_2 , \qquad \tau_0 \leq \tau \leq T ,$$

and in this domain $\bar{f}_1$ and $\bar{f}_2$ are expansible as power series in $(u_1 - u_{1\tau_0})$ and $(u_2 - u_{2\tau_0})$ and convergent.

The number ε is real. The system (11) of our problem is a special case of the system (N).

α) *The formal solution.*

We want to find a solution u_1 and u_2 of the system (N) such that, if $\varepsilon \to 0$, u_1 and u_2 tend to the constants x and y. For $\varepsilon \neq 0$, u_1 aud u_2 depend on ε and τ, and assume that for $\tau = \tau_0$ they differ a little from x and y,

$$(13) \qquad u_{1\tau_0} = x + \xi , \quad u_{2\tau_0} = y + \eta$$

ξ and η are very small.

Take as formal solutions u_1 and u_2 of (N) the expressions:

$$(14)\qquad \begin{cases} u_1 = X_1^{(0)} + \varepsilon X_1^{(1)} + \varepsilon^2 X_1^{(2)} + \dots \\ u_2 = X_2^{(0)} + \varepsilon X_2^{(1)} + \varepsilon^2 X_2^{(2)} + \dots \end{cases}$$

where:

$$(14\alpha)\qquad X_1^{(0)} = x + \xi(\varepsilon, \tau_0) \qquad X_2^{(0)} = y + \eta(\varepsilon, \tau_0).$$

The coefficients X of (14) are regular functions of x, y, ξ, η, τ if $|x|$, $|y|$, $|\xi|$, $|\eta|$, $|\tau-\tau_0|$ are less than certain constants.

β) *The coefficients of the formal solution.*

By presupposing the series (14) as convergent, we can find that the coefficients X of (14) are given by:

$$X_1^{(1)} = \int_{\tau_0}^{\tau} [\bar{f}_1] d\tau,$$

$$X_2^{(1)} = \int_{\tau_0}^{\tau} [\bar{f}_2] d\tau,$$

$$(15)\qquad X_1^{(2)} = \int_{\tau_0}^{\tau} \left\{ X_1^{(1)} [\bar{f}_1]_{u_1} + X_2^{(1)} [\bar{f}_1]_{u_2} \right\} d\tau,$$

$$X_2^{(2)} = \int_{\tau_0}^{\tau} \left\{ X_1^{(1)} [\bar{f}_2]_{u_1} + X_2^{(1)} [\bar{f}_2]_{u_2} \right\} d\tau,$$

$$X_i^{(3)} = \int_{\tau_0}^{\tau} \sum_{j=1}^{2} [\bar{f}_i]_{u_j} X_j^{(2)} + \frac{1}{2} \int_{\tau_0}^{\tau} \sum_{j=1}^{2} \sum_{l=1}^{2} [\bar{f}_i]_{u_j u_l} X_j^{(1)} X_l^{(1)},$$

. . . .

$$j, l, i = 1, 2.$$

The brackets indicate that the corresponding functions and their derivatives are taken at $u_1 = u_{1\tau_0} = x + \xi$, $u_2 = u_{2\tau_0} = y + \eta$. By carrying out the integrations in the first two of (15) we find the functions $X_1^{(1)}$, $X_2^{(1)}$. Upon substituting the $X_1^{(1)} X_2^{(1)}$ into the second two of (15), the integrands become known continuous functions of τ, then $X_1^{(1)}$, $X_2^{(2)}$ are defined by quadratures. With the same procedure we can find $X_1^{(3)}$, $X_2^{(3)}$, ...

γ) *The convergence of the formal sotution.*

To prove the convergence and to find the domain of the validity of the series (14), we use the «method of dominants». We can prove that the condition for the convergence is:

$$(16)\qquad |\varepsilon| < \frac{r}{4M(\tau - \tau_0)},$$

$$(16\alpha)\qquad M \geq M_i, \quad r \leq r_i, \qquad i = 1, 2.$$

In many cases the domain of convergence of (14) is much larger than the condition (16) shows.

The inequality (16) can be satisfied by imposing restrictions upon both ε and τ, or by taking τ arbitrarily and restricting ε or by taking ε arbitrarily and then restricting τ.

§ 5. *Use of the periodicity.*

Take now the solution u_1 and u_2 of (N) as periodic one in τ with period 2π, that is:

$$u_1(\tau_o + 2\pi) - u_1(\tau_o) = 0, \quad u_2(\tau_o + 2\pi) - u_2(\tau_o) = 0. \tag{17}$$

Applying this periodicity condition to the series (14) we have:

$$\begin{aligned} &\{X_1^{(1)}(\tau_o + 2\pi) - X_1^{(1)}(\tau_o)\} + \varepsilon\{X_1^{(2)}(\tau_o + 2\pi) - X_1^{(2)}(\tau_o)\} + \ldots = 0, \\ &\{X_2^{(1)}(\tau_o + 2\pi) - X_2^{(1)}(\tau_o)\} + \varepsilon\{X_2^{(2)}(\tau_o + 2\pi) - X_2^{(2)}(\tau_o)\} + \ldots = 0. \end{aligned} \tag{18}$$

If the Fourier series developments in τ of $\bar{f}_1$ and $\bar{f}_2$ are:

$$\begin{aligned} \bar{f}_1 &= A_o + A_1 \cos\tau + B_1 \sin\tau + \ldots + A_m \cos m\tau + B_m \sin m\tau + \ldots \\ \bar{f}_2 &= C_o + C_1 \cos\tau + D_1 \sin\tau + \ldots + C_m \cos m\tau + D_m \sin m\tau + \ldots, \end{aligned} \tag{19}$$

where the coefficients A, B, C, D are functions of: $u_{1\tau_o} = x + \xi$, $u_{2\tau_o} = y + \eta$, and such that the developments of (19) are convergent. By taking into account the expression of X given by (15), the conditions (18) give:

$$\begin{aligned} &A_o(x + \xi, y + \eta) + \varepsilon\varphi_1(x + \xi, y + \eta, \tau_o) + \ldots = 0, \\ &C_o(x + \xi, y + \eta) + \varepsilon\varphi_2(x + \xi, y + \eta, \tau_o) + \ldots = 0. \end{aligned} \tag{20}$$

Provided that the jakobian of (20) is not zero, i. e.

$$\begin{vmatrix} \dfrac{\partial A_o}{\partial \xi} & \dfrac{\partial C_o}{\partial \xi} \\ \dfrac{\partial A_o}{\partial n} & \dfrac{\partial C_o}{\partial n} \end{vmatrix} \neq 0 \tag{21}$$

we can solve the system (20) in ξ and η interms of ε and τ_o:

$$\xi = \xi(\varepsilon, \tau_o), \quad \eta = \eta(\varepsilon, \tau_o), \tag{22}$$

with the conditions:

$$\xi = \xi(o, \tau_o) = 0, \quad \eta = \eta(o, \tau_o) = 0,$$

If the Jakobian is zero, we consider in (20) terms of 1, 2,... degree in ε, that is the functions $\varphi_1, \varphi_2, \ldots$[1]

From (22) and (13) we get:

---

[1] P. Fatou, *Bulletin Société Math. de France*, 1928 - 30, pp. 112 - 115.

(23) $$u_{1\tau_0} = x + \xi(\varepsilon, \tau_0), \quad u_{2\tau_0} = y + \eta(\varepsilon, \tau_0);$$

ξ and η are given in series in ε, τ_0 in the interval $[0,2\pi]$. ε is under the condition (16). Then is the series (14) the only unknown are the constants x and y.

§ 6. *The components of the amplitude of the subharmonics. Conditions of existence of the subharmonics.*

The equations (20) must be satisfied in case: $\varepsilon = \xi = \eta = 0$. In this case the equations (20) give:

(24) $$A_0(x, y) = 0, \quad C_0(x, y) = 0.$$

The solutions (x, y) of the system (24) give the limits x and y, when the steady state solutions of the original equation are known in the form (4).

The conditions for real intersections of the curves (24) in the x, y-plane give the conditions of the existence of the subharmonics of our equation (3).

§ 7. *The stability of the steady state subharmonics.*

For the stability of the steady state subharmonics we study the stability of the singularities of the equation:

(25) $$\frac{du_2}{du_1} = \frac{\bar{f}_2}{\bar{f}_1},$$

which comes from the system (N). The difficulty is that $\bar{f}_1$ and $\bar{f}_2$ depend on time τ.

But by taking into account the developements of $\bar{f}_1, \bar{f}_2$ given by (19), and that *their mean values with respect to time* τ *over the period* 2π *are* A_0 *and* C_0 *respectively,* the singularities of (25) are that of the equation:

(25α) $$\frac{dy}{dx} = \frac{C_0(x, y)}{A_0(x, y)},$$

then the singularities are given by the solutions (x, y) of the system (24).

According to the corresponding theory of Poincaré[1] and Bendixson[2] the distinction between the different kinds of the singularities depends on two numbers ϱ_1 and ϱ_2, the roots of the characteristic equation:

(26) $$\begin{vmatrix} \alpha_1 - \varrho & b_1 \\ \alpha_2 & b_1 - \varrho \end{vmatrix} = 0,$$

[1] H. Poincaré, Sur les courbes définies par une équation différentielle. *Œuvres*, Gauthier-Villars, Paris, Vol. 1892.

[2] I. Bendixson, *Acta Math.*, **24**, 1901.

where:

$$(26\alpha) \qquad \alpha_1 = \frac{\partial A_o}{\partial x}, \quad \alpha_2 = \frac{\partial A_o}{\partial y}, \quad b_1 = \frac{\partial C_o}{\partial x}, \quad b_2 = \frac{\partial C_o}{\partial y}$$

For non zero roots ϱ_1 and ϱ_2 of (26) the «simple singularities» are classified in the following classes:

I: «*nodal points*», when ϱ_1, ϱ_2, are real and of the same sign,

II: «*saddle points*», when ϱ_1, ϱ_2, are real but of opposite sign,

III: «*spiral points*», when ϱ_1, ϱ_2, are complex conjugates, aud

IV: «*spiral points* or *centers*», when ϱ_1, ϱ_2, are pure imaginaries.

The condition of the roots being pure imaginaries, which is a necessary condition for being a center, it is not a sufficient condition. There is the Poincaré's criterion[1] for distinguishing spiral from center in this case.

We define the above singularities as «*stable*» or «*unstable*», when any point on any integral curve moves into the said singularity or not with increasing time τ, i.e. according as the real part of the roots is negative or positive respectively.

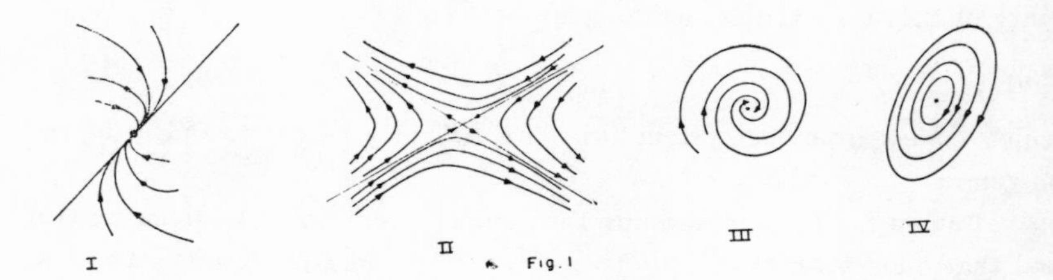

Fig. 1

The singularities are shown in Fig. 1, where I is a «nodal stable», III a «stable spiral», IV a (neutral) center, and II a «saddle point intrinsically unstable».

§ 8. *Application to the system* (N).

Let us apply the previous theory in the case of the equation $\bar{f}_1$ and $\bar{f}_2$ given by (11α).

The important thing here is to give to functions f_1 und f_2 a proper form from which we can get the development in Fourier series. If we replace the

[1] J. Hadamard, *Rice Institute Pamphlet*, **20**, 1, 1933, 9-28.

powers and products of sines and **cosines** of multiple angles in constructing the functions f_1 and f_2 from the function f, we get the following results, if **we restrict** ourselves to the coefficients which are useful for the construction of the functions A_0 and C_0, which depend on the number **n** characterizing the order of the subharmonics:

$$
\begin{aligned}
f_1(u_1, u_2, \tau) = & \left\{ -\frac{1}{2} k u_1 - \frac{1}{2} c_1 u_2 + \frac{3}{8} c_3 u_2^3 + \frac{3}{8} c_3 u_1^2 u_2 + \frac{3}{4} c_3 \frac{B}{(1-n^2)^2} u_2 \right\} + \\
& + \dots \\
& + \left\{ -\frac{1}{2} c_2 \frac{B}{1-n^2} u_1 \right\} \cos(n-2)\tau + \\
& + \left\{ \frac{3}{4} c_3 \frac{B}{1-n^3} u_1 u_2 \right\} \cos(n-3)\tau + \dots
\end{aligned} \tag{27}
$$

$$
\begin{aligned}
f_2(u_1, u_2, \tau) = & \left\{ \frac{1}{2} c_1 u_1 - \frac{1}{2} k u_2 - \frac{3}{8} c_3 u_1^3 - \frac{3}{8} c_3 u_1 u_2^2 - \frac{3}{4} c_3 \frac{B^2}{(1-n^3)^2} u_1 \right\} + \\
& + \dots \\
& + \left\{ \frac{1}{2} c_2 \frac{B}{1-n^2} u_2 \right\} \cos(n-2)\tau + \dots \\
& + \left\{ -\frac{3}{8} c_3 \frac{B}{1-n^2} (-u_1^2 + u_2^2) \right\} \cos(n-3)\tau + \dots
\end{aligned} \tag{28}
$$

All these are **referred** to any order $\frac{1}{n}$ of subharmonics. In a next paper we shall apply the above theory for the subharmonics of order one third.

ΠΕΡΙΛΗΨΙΣ

Εἰς τὴν ἐργασίαν ταύτην μελετῶνται αἱ ὑποαρμονικαὶ ταλαντώσεις οἱασδήποτε τάξεως εἰς τὴν περίπτωσιν μὴ γραμμικῆς ἐλαστικῆς δυνάμεως. Ἡ ἀντίστοιχος διαφορικὴ ἐξίσωσις μετασχηματίζεται καταλλήλως διὰ χρησιμοποιήσεως ἰδεῶν τῶν Van der Pol, Mandelstam - Papalexi καὶ Andronow - Witt. Σπουδάζονται λύσεις «εὐσταθεῖς» καὶ «μεταβατικαὶ» διὰ χρησιμοποιήσεως τῆς μεθόδου περιοδικῶν λύσεων τοῦ Poincaré. Ἡ σπουδὴ τῶν συνθηκῶν διὰ τὴν ὕπαρξιν τῶν ὑποαρμονικῶν ταλαντώσεων καὶ τὴν εὐστάθειάν των ἀνάγεται εἰς τὴν σπουδὴν τῶν ἀνωμάλων σημείων τῆς ἀντιστοίχου ἐξισώσεως. Εἰς ἑπομένην ἀνακοίνωσιν θὰ ἐκτεθῇ ἡ ἔρευνα παρ' ἐμοῦ τῶν ὑποαρμονικῶν ταλαντώσεων τάξεως ἑνὸς πρὸς τρία, ὡς ἐφαρμογὴ τῶν γενικῶν σκέψεων τῆς παρούσης ἀνακοινώσεως.

ΠΡΑΚΤΙΚΑ
ΤΗΣ ΑΚΑΔΗΜΙΑΣ ΑΘΗΝΩΝ

ΕΤΟΣ 1957: ΤΟΜΟΣ 32ος

ΑΝΑΤΥΠΟΝ
ΣΕΛ. 101 - 108

Subharmonics of order one third in the case of cubic restoring force. Part II,

by ***Dem. G. Magiros****.

Ἀνεκοινώθη ὑπὸ τοῦ κ. Βασιλ. Αἰγινήτου.

Introduction

In this paper we discuss briefly the **subharmonics** of order one third in the case of a cubic restoring force.

The properly transformed equations, that give the components of the amplitude of the subharmonics, contain, if the amplitude of the external

---

*** ΔΗΜ. ΜΑΓΕΙΡΟΥ, Περὶ τῶν ὑποαρμονικῶν ταλαντώσεων τάξεως ἑνὸς πρὸς τρία.**

Reprinted from the *Proceedings of the Athens Academy of Sciences* **32** (1957), 101–108.

force takes the values of an interval given in the paper, the ratios of the coefficients of the damping and the restoring force, and these ratios, under certain condition, can have any values, the coefficients themselves need not necessarily be small, a case very important in many engineering problems. We solve here the problem of subharmonics, their existence and stability, in case of given coefficients of the damping and the cubic restoring force, and the amplitude of the sinusoidal external force. An example is illustrated and the corresponding sketch of the singularities and the integral curves in the whole plane, which is separated into proper regions, is given.

§ I. *The equations for the components of the amplitude of subharmonics.*

The equation to be solved is:

$$\ddot{Q} + \bar{k}\dot{Q} + \bar{c}_1 Q + \bar{c}_2 Q^2 + \bar{c}_3 Q^3 = B \sin 3\tau. \tag{1}$$

By using:

$$\bar{k} = \varepsilon k, \quad 1 - \bar{c}_1 = \varepsilon c_1, \quad \bar{c}_2 = \varepsilon c_2, \quad \bar{c}_3 = \varepsilon c_3, \tag{2}$$

the equation is transformed into:

$$\ddot{Q} + Q = \varepsilon f(Q, \dot{Q}) + B \sin 3\tau, \tag{3}$$

$$f(Q, \dot{Q}) = -k\dot{Q} + c_1 Q - c_2 Q^2 - c_3 Q^3. \tag{3α}$$

In case $\varepsilon = 0$, the solution of (3) is given by:

$$Q = x \sin\tau - y \cos\tau - \frac{B}{8} \sin 3\tau, \tag{4}$$

where x and y are coustants, known for given initial conditions.

In case $\varepsilon \neq 0$ we try to determine the steady state solutions of the equation (3), i. e. the constant limits: $\lim_{\varepsilon \to 0} x(\varepsilon, \tau)$, $\lim_{\varepsilon \to 0} y(\varepsilon, \tau)$, according to the previous paper, part I.[1]

For this we have to find the functions $A_0(x, y)$ and $C_0(x, y)$, of the paper [I]. These functions come from the equations (27) aud (28) of the paper [I], if we put $n = 3$ and x and y instead of u_1 and u_2 respectively, when the result is:

---

[1] D. G. Magiros, Subharmonics of any order in case of nonlinear restoring forces. *Praktika of Athens Academy* **32**, 1957, pp. 77.

$$(5)\quad \begin{aligned} A_o(x, y) &= \frac{1}{2}\left\{-kx - c_1y + \frac{3}{4}c_3y\left(x^2+y^2+\frac{B^2}{32}\right) - \frac{3}{16}c_3B\,xy\right\}, \\ C_o(x, y) &= \frac{1}{2}\left\{c_1x - ky - \frac{3}{4}c_3x\left(x^2+y^2+\frac{B^2}{32}\right) + \frac{3}{32}c_3B(-x^2+y^2)\right\}, \end{aligned}$$

and the equations which give the unknown x and y are:

$$(6)\quad \begin{aligned} &kx + c_1y - \frac{3}{4}c_3y\left(x^2+y^2+\frac{B^2}{32}\right) + \frac{3}{16}Bxy = 0, \\ &c_1x - ky - \frac{3}{4}c_3x\left(x^2+y^2+\frac{B^2}{32}\right) + \frac{3}{32}c_3B(-x^2+y^2) = 0. \end{aligned}$$

In the case $c_3 \neq 0$, (6) can be written as:

$$(7)\quad \begin{aligned} &\mu x + \lambda y - y\left(x^2+y^2+\frac{B^2}{32}\right) + \frac{1}{4}Bxy = 0, \\ &\lambda x - \mu y - x\left(x^2+y^2+\frac{B^2}{32}\right) + \frac{1}{8}B(-x^2+y^2) = 0, \end{aligned}$$

with:

$$(7\alpha)\quad \lambda = \frac{4}{3}\frac{c_1}{c_3}, \quad \mu = \frac{4}{3}\frac{k}{c_3}.$$

We ask for «real solutions» (x, y) of the system (7).

Remarks. From the prescribed initial conditions of the equation (3), say Q_o and $\dot{Q}_o$ at $\tau=0$, we have, according to (4),

$$(8\alpha)\quad Q_o = -y, \quad \dot{Q}_o = x - \frac{3}{8}B,$$

when the given initial conditions correspond to the point:

$$(8\beta)\quad x = \dot{Q}_o + \frac{3}{8}B, \quad y = -Q_o,$$

in the x, y−plane, which is the «starting point».

Starting from the «starting point» and following the corresponding integral curve with the lapse of time we can terminate to a «final point», which corresponds to the proper steady solution. The coordinates of the «final point» are solutions of the system (7). Given the initial conditions and the amplitude B of the external force, the «starting point» in the x, y−plane is defined; conversly, any point of x, y−plane can be taken as «starting point» by properly choosing the initial conditions and the amplitude B.

If the «starting point» is selected in coincidence with a «stable final point», no «transient phenomena» may exist.

§ II. *Restrictions to the coefficients of the equation* (1).

If:

$$(8)\quad A = r^2 + \frac{B^2}{32}, \quad r^2 = x^2 + y^2,$$

the system (7) is written as:

$$\mu x + \lambda y - Ay = -\frac{1}{8} B(2xy) \tag{9}$$
$$\lambda x - \mu y - Ax = -\frac{1}{8} B(-x^2 + y^2).$$

Squaring and adding (9) we find:

$$\lambda^2 + \mu^2 + A^2 - 2A\lambda = \frac{B^2}{64} r^2; \tag{10}$$

eliminating A between (10) and (8) we have.

$$r^4 + \left(\frac{3}{64} B^2 - 2\lambda\right) r^2 + \left(\lambda^2 + \mu^2 + \frac{B^4}{32^2} - \lambda \frac{B^2}{16}\right) = 0, \tag{11}$$

the roots of which are:

$$r^2 = \frac{1}{2}\left\{2\lambda - \frac{3}{64} B^2 \pm \sqrt{\left(2\lambda - \frac{3}{64} B^2\right)^2 - 4\left(\lambda^2 + \mu^2 + \frac{B^4}{32^2} - \lambda \frac{B^2}{16}\right)}\right\}. \tag{11α}$$

The reality of r^2 requires:

$$I \equiv 7B^4 - 2^8 \lambda B^2 + 2^{14} \mu^2 \leq 0. \tag{12}$$

The roots of: $I = 0$ are:

$$B^2 = \frac{2^7}{7}\left(\lambda \pm \sqrt{\lambda^2 - 7\mu^2}\right), \tag{13}$$

then the condition (12) requires the following conditions to be fulfilled:

$$(14) \begin{cases} \alpha) & \lambda^2 - 7\mu^2 > 0, \\ \beta) & \frac{2^7}{7}\left(\lambda - \sqrt{\lambda^2 - 7\mu^2}\right) \leq B^2 \leq \frac{2^7}{7}\left(\lambda + \sqrt{\lambda^2 - 7\mu^2}\right). \end{cases}$$

By using (7α) and (2) we find the following restrictions for $\bar{k}, \bar{c}_1, \bar{c}_3$, B:

$$(15) \begin{cases} \alpha) & \left(\frac{1-\bar{c}_1}{\bar{k}}\right)^2 > 7, \\ \beta) & \frac{2^9}{3 \cdot 7}\left(\frac{1-\bar{c}_1}{\bar{c}_3} - \sqrt{\left(\frac{1-\bar{c}_1}{\bar{c}_3}\right)^2 - 7\left(\frac{\bar{k}}{\bar{c}_3}\right)^2}\right) \leq B^2 \leq \frac{2^9}{3 \cdot 7}\left(\frac{1-\bar{c}_1}{\bar{c}_3} + \sqrt{\left(\frac{1-\bar{c}_1}{\bar{c}_3}\right)^2 - 7\left(\frac{\bar{k}}{\bar{c}_3}\right)^2}\right). \end{cases}$$

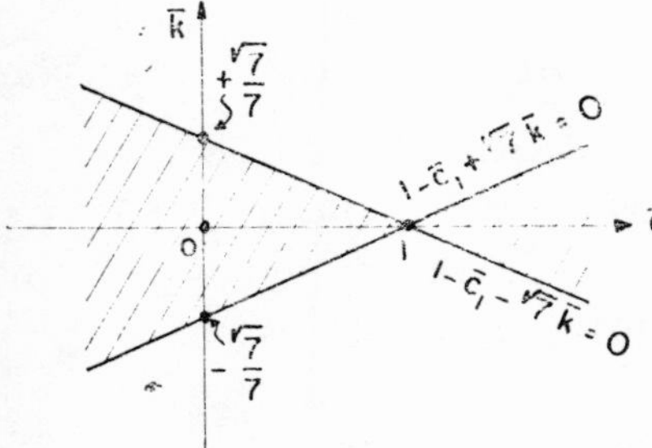

Fig. 1

The inequality (15α) can be written as:

$$\left(1 - \bar{c}_1 - \sqrt{7}\,\bar{k}\right)\left(1 - \bar{c}_1 + \sqrt{7}\,\bar{k}\right) > 0,$$

then only the shaded region in Fig. 1 is valid in the $\bar{c}_1, \bar{k}$ – plane.

From (11) or (11α) we can draw r^2, versus B^2, and by using (15β) we can have the arcs of the diagram which are valid in our problem.

§ III. *The solutions of the system (9).*

The system (9) can be written as:

$$(16)\qquad \begin{aligned} &\left(\mu + \frac{1}{4} By\right) x + (\lambda - A)y = 0 , \\ &\frac{1}{8} Bx^2 - (\lambda - A)x + \left(\mu - \frac{1}{8} By\right)y = 0 \cdot \end{aligned}$$

The vanishing of, say, the Sylvester's eliminant, which is the condition for common roots of (16), leads, for non-zero roots, to the cubic:

$$(17)\qquad y^3 - 3\left(\frac{4}{B}\right)^2 \left\{(\lambda - A)^2 + \mu^2\right\} y - 2\left(\frac{4\mu}{B}\right)^3 - 2\left(\frac{4}{B}\right)^3 \mu (\lambda - A)^2 = 0$$

By knowing the coefficients of (1), we know, from (7α) and (2), λ and μ, then we know the amplitude r from (11α) in its two values. On the circumference with radius r there are one or three singularities, of which the ordinates y are the real roots of (17), when for their abscissas we apply the Pythagoras theorem. The singularities (x, y) are therefore at most seven, included the origin, which, in every case, is a singularity.

§ IV. *Example :*

Given: « $B = 4$, $\bar{k} = \frac{3}{16}$, $\bar{c}_1 = 1\frac{3}{4}$, $\bar{c}_3 = -\frac{1}{2}$ ».

We can find is this special case:

$$(18)\qquad \begin{cases} \frac{1 - \bar{c}_1}{\bar{c}_3} = \frac{c_1}{c_3} = \frac{3}{2} , \quad \frac{\bar{k}}{\bar{c}_3} = \frac{k}{c_3} = -\frac{3}{8} , \quad \lambda = 2 , \quad \mu = -\frac{1}{2} , \\ r_1^2 \simeq 2 , \quad r_2^2 \simeq 1{,}25 , \quad r_1 \simeq 1{,}414 , \quad r_2 \simeq 1{,}118 , \\ A_1 \simeq 2{,}5 , \quad A_1^2 \simeq 6{,}25 , \quad A_2 \simeq 1{,}75 , \quad A_2^2 \simeq 3{,}06 , \end{cases}$$

then:

$$(19)\qquad \begin{aligned} &\alpha) \quad y_1^3 - \frac{3}{2} y_1 + \frac{1}{2} = 0 , \\ &\beta) \quad y_2^3 - \frac{15}{16} y_2 + \frac{5}{16} = 0 . \end{aligned}$$

Each of these cubic equations has three real unequal roots:

$$(20\alpha)\qquad y_{11} \simeq 0{,}9996 , \quad y_{12} \simeq -1{,}3645 , \quad y_{13} \simeq 1{,}3645 ,$$

the first, and

$$(21\alpha)\qquad y_{21} \simeq 0{,}7032 , \quad y_{22} \simeq -1{,}1034 , \quad y_{23} \simeq 1{,}0431 ,$$

the second, when the corresponding abscissas are:

$$(20\beta)\qquad x_{11} \simeq 0{,}9996 , \quad x_{12} \simeq 0{,}3648 , \quad x_{13} \simeq -0{,}3648 ,$$

$$(21\beta)\qquad x_{21} \simeq 0{,}8672 , \quad x_{22} \simeq 0{,}1755 , \quad x_{23} \simeq -0{,}3991 .$$

§ V. *The stability of the solutions.*

For the study of the stability of the solutions the number ε enters. The number ε must be such that:

$$|\varepsilon| < \frac{r}{4M(\tau - \tau_0)}, \tag{22}$$

according to paper [1], § IV, γ.

If in (22) the initial time $\tau_0 = 0$, by taking arbitrarily $\varepsilon = 1$, this means that the max $\tau = T$ is taken according to:

$$T < \frac{r}{4M}. \tag{23}$$

Take now the partial derivatives with respect to x and y of the functions $A_0(x,y)$ and $C_0(x,y$ given by (5). By establishing the restriction (23), which corresponds to $\varepsilon = 1$, these partial derivatives can be written as follows:

$$\begin{aligned}
\frac{\partial A_0}{\partial x} &\equiv a_1 = \frac{1}{2}\bar{c}_3 \left\{ -\frac{\bar{k}}{\bar{c}_3} + \frac{3}{2}xy - \frac{3}{16}By \right\}, \\
\frac{\partial A_0}{\partial y} &\equiv a_2 = \frac{1}{2}\bar{c}_3 \left\{ -\frac{1-\bar{c}_1}{\bar{c}_3} + \frac{3}{4}\left(x^2 + 3y^2 + \frac{B^2}{32}\right) - \frac{3}{16}Bx \right\}, \\
\frac{\partial C_0}{\partial x} &\equiv b_1 = \frac{1}{2}\bar{c}_3 \left\{ \frac{1-\bar{c}_1}{\bar{c}_3} - \frac{3}{4}\left(3x^2 + y^2 + \frac{B^2}{32}\right) - \frac{3}{16}Bx \right\}, \\
\frac{\partial C_0}{\partial y} &\equiv b_2 = \frac{1}{2}\bar{c}_3 \left\{ -\frac{\bar{k}}{\bar{c}_3} - \frac{3}{2}xy + \frac{3}{16}By \right\}.
\end{aligned} \tag{24}$$

The characteristic roots, which help to find the type of the singularity, according to § VII of paper [1], is:

$$p_{1,2} = \frac{1}{2}\left\{ a_1 + b_2 \pm \sqrt{(a_1 - b_2)^2 + 4a_2b_1} \right\}. \tag{25}$$

The computation for the singularities of our example, the coordinates of which are given by (20α,β) and (21α,β), gives:

(26)

0:	The origin	$x = 0$	$y = 0$	:	«*stable spiral*»
I:	The point	$x_{11} = 0{,}9996$	$y_{11} = 0{,}9996$	:	» »
II:	» »	$x_{12} = 0{,}3648$	$y_{12} = -1{,}3645$	:	» »
III:	» »	$x_{13} = -0{,}3648$	$y_{13} = 1{,}3645$	:	«*saddle point*»
IV:	» »	$x_{21} = 0{,}8675$	$y_{21} = 0{,}7032$	:	» »
V:	» »	$x_{22} = 0{,}1755$	$y_{22} = -1{,}1034$	:	» »
VI:	» »	$x_{23} = -0{,}3991$	$y_{23} = 1{,}0431$	:	» »

The origin corresponds to «harmonic solution», which, as stable, is acceptable. The points I, II, III are on the circumference with radius $r_1 = 1{,}414$. The ponts I, II correspond to acceptable stable subharmonic solutions. The points IV, V, VI, which are on the circumference with radius $r_2 = 1{,}118$, are «intrinçically unstable».

§ VI. *Non-existence of limiting cycles.*

From (24) we have:

$$\frac{\partial A_o}{\partial x} + \frac{\partial C_o}{\partial y} = -\bar{k} \tag{27}$$

valid in the whole x,y-plane, and according to Bendixson's[1] criterion no limit cycles can exist in the whole x,y-plane.

For $\bar{k} = 0$, some of the singularities may be «centers», then we may have «closed integral curves».

§ VII. *Sketch corresponding to the above example.*

Applying the «method of isoclines» to the differential equation:

$$\frac{dy}{dx} = \frac{C_o(x,y)}{A_o(x,y)}, \tag{28}$$

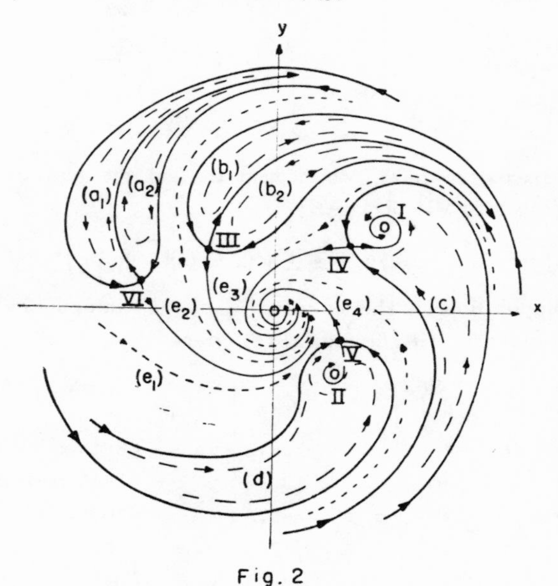

Fig. 2

[1] I. Bendixson, *Acta Math.* **24** (1901), 1-88.

where the functions $A_o(x,y)$ and $C_o(x,y)$ are given by (5), we can have a figure showing the singularities, the integral curves and the separation into regions corresponding to the above example. In Fig. 2 a sketch of these things is given.

The solid lines are the boundaries from the saddle points, the dotted lines are the integral curves. The regions: (α_1), (α_2), (β_1) and (β_2) correspond to no solutions of our example. The regions: (e_1), (e_2), (e_3) and (e_4) correspond to «stable harmonic solution»: $Q = -\frac{1}{2}\sin 3\tau$. The region (c) correspond to «stable solution»: $Q = 0{,}9996 \sin\tau - 0{,}9996 \cos\tau - \frac{1}{2} \sin 3\tau$; and the region (d) correspond to «stable solution»: $Q = 0{,}3648 \sin\tau + 1{,}3645 \cos\tau - \frac{1}{2} \sin 3\tau$. The amplitude of the subharmonic term in the last two stable solutions is the same: $r_1 = 1{,}414$.

ΠΕΡΙΛΗΨΙΣ

Εἰς τὴν ἐργασίαν ταύτην συζητεῖται ἐν συντομίᾳ τὸ ζήτημα τῶν ὑποαρμονικῶν ταλαντώσεων τάξεως ἑνὸς πρὸς τρία εἰς τὴν περίπτωσιν κυβικῆς συναρτήσεως τῆς ἐλαστικῆς δυνάμεως. Οἱ συντελεσταὶ τῆς διαφορικῆς ἐξισώσεως εἶναι ὄχι κατ' ἀνάγκην μικρῶν τιμῶν.

Αἱ ἐξισώσεις αἱ δίδουσαι τὰς συνιστώσας τῶν ὑποαρμονικῶν περιέχουν τὰ πηλίκα τῶν συντελεστῶν τῆς ἐλαστικῆς δυνάμεως καὶ τῆς ἀντιστάσεως (damping) καὶ τὰ πηλίκα αὐτὰ ὑπὸ δεδομένας συνθήκας δύνανται νὰ ἔχουν οἱασδήποτε τιμάς, χωρὶς νὰ εἶναι ἀναγκαῖον νὰ δεχθῶμεν μικρὰς τιμὰς διὰ τοὺς συντελεστάς, περίπτωσις πολὺ σημαντικὴ εἰς πολλὰ προβλήματα τῶν μὴ γραμμικῶν ταλαντώσεων. Εὑρίσκονται ἐνταῦθα αἱ συνιστῶσαι τῶν ὑποαρμονικῶν, ἐρευνᾶται τὸ ζήτημα τῆς ὑπάρξεως καὶ εὐσταθείας των εἰς τὴν περίπτωσιν ἐξωτερικῆς δυνάμεως ἡμιτονοειδοῦς τύπου ὑπὸ πλάτος μεταβλητὸν εἰς δεδομένον διάστημα, διδομένων τῶν συντελεστῶν τῆς ἐλαστικῆς δυνάμεως καὶ τῆς ἀντιστάσεως. Δίδεται παράδειγμα ἀριθμητικὸν ὡς ἐφαρμογὴ τῆς θεωρίας, καθὼς καὶ σχεδιάγραμμα ἀντιστοιχοῦν εἰς τὸ παράδειγμα αὐτό.

Remarks on a problem of subharmonics*,

by ***Dem. G. Magiros.***

Ἀνεκοινώθη ὑπὸ τοῦ κ. Βασιλ. Αἰγινήτου.

Introduction.

This paper is a supplement of the author's previous paper under the title: «*Subharmonics of order one third in the case of cubic restoring force*», contained in this volume as the author's second work on subharmonics and called in the following «paper B», and its previous «paper A». In the first chapter we discuss the conditions under which the basic equation (1) of the «paper B» accepts the harmonic solution $\left(-\frac{B}{8}\sin 3\tau\right)$ as a stable one, the subharmonic term of the solution being zero. In the second chapter the «inverse problem» of that of the «paper B» is discussed. This «inverse problem», so simple from a mathematical point of view, according to the equations found, seems to be of importance from an engineering point of view.

I. *The singularity of the origin in the general case.*

The basic equation is:

$$\ddot{Q} + \bar{k}\dot{Q} + \bar{c}_1 Q + \bar{c}_2 Q^2 + \bar{c}_3 Q^3 = B\sin 3\tau, \tag{1}$$

and it can be written in the form:

$$\ddot{Q} + Q = \varepsilon f(Q, \dot{Q}) + B\sin 3\tau, \tag{2}$$

with:

$$f(Q, \dot{Q}) = -k\dot{Q} + c_1 Q - c_2 Q^2 - c_3 Q^3, \tag{2α}$$

if:

$$\bar{k} = \varepsilon k, \quad 1 - \bar{c}_1 = \varepsilon c_1, \quad \bar{c}_2 = \varepsilon c_2, \quad \bar{c}_3 = \varepsilon c_3. \tag{3}$$

The steady state solution is:

$$Q = x\sin\tau - y\cos\tau - \frac{B}{8}\sin 3\tau, \tag{4}$$

and the components x and y of the amplitude r of the subharmonic of order one third are given, according to «paper B», by the two equations:

$$\begin{cases} \mu x + \lambda y - y\left(x^2 + y^2 + \frac{B^2}{32}\right) + \frac{1}{4}Bxy = 0, \\ \lambda x - \mu y - x\left(x^2 + y^2 + \frac{B^2}{32}\right) + \frac{1}{8}B(-x^2 + y^2) = 0, \end{cases} \tag{5}$$

* Παρατηρήσεις ἐπὶ προβλήματος τῶν ὑποαρμονικῶν.

Reprinted from the *Proceedings of the Athens Academy of Sciences* **32** (1957), 143–146.

where:

$$\lambda = \frac{4}{3}\frac{c_1}{c_3} = \frac{4}{3}\frac{1-\bar{c}_1}{\bar{c}_3}, \qquad \mu = \frac{4}{3}\frac{k}{c_3} = \frac{4}{3}\frac{\bar{k}}{\bar{c}_3}. \tag{5α}$$

From (5) we see that the origin is a singularity of our equation; in other wordes, the function: $Q = -\frac{B}{8}\sin 3\tau$ is always a solution of the equation, the harmonic solution.

For the stability of this harmonic solution we take the derivatives given by (24) in the «paper B», established under the conditions:

$$\varepsilon = 1, \qquad T = \max\tau < \frac{r}{4M}. \tag{6}$$

From (24) of the «paper B», in the case $x=0$, $y=0$, we get:

$$\begin{aligned} a_1 &= -\frac{1}{2}k, \qquad a_2 = -\frac{1}{2}c_1 + 3c_3\left(\frac{B}{16}\right)^2, \\ b_1 &= \frac{1}{2}c_1 - 3c_3\left(\frac{B}{16}\right)^2, \qquad b_2 = -\frac{1}{2}k, \end{aligned} \tag{7}$$

when, from (26) of the «paper A», the characteristic roots are given by:

$$p_{1,2} = -\frac{1}{2}k \pm i\left\{\frac{1}{2}c_1 - 3c_3\left(\frac{B}{16}\right)^2\right\}. \tag{8}$$

We, therefore, have, according to the definitions on the singularities of the «paper A» the following:

A. If the imaginary part of the characteristic roots is not zero, that is if:

$$\frac{c_1}{c_3} \neq 6\left(\frac{B}{16}\right)^2, \tag{9}$$

then:

α) for $k>0$, the origin is a «stable spiral point»;

β) for $k<0$, the origin is an «unstable spiral point»;

γ) for $k=0$, the origin is either a «center» or a «spiral point».

B. If the imaginary part is zero, that is if:

$$\frac{c_1}{c_3} = 6\left(\frac{B}{16}\right)^2, \tag{10}$$

then:

α) for $k>0$, the origin is a «nodal stable point»;

β) for $k<0$, the origin is an «unstable nodal point»;

γ) for $k=0$, the origin is not a simple singularity of the kind we know from «paper A» and «paper B», since the characteristic roots are zero.

The result from the above is that: the origin is a stable singularity when $k>0$, of the spiral type under the condition (9), and of the nodal type under the condition (10). In other words, the function $(Q = -\frac{B}{8}\sin 3\tau)$ is the harmonic stable solution of the equation (1) if $k>0$.

We plot in the: $\frac{1-\bar{c}_1}{\bar{c}_3}$, $\bar{k}$ — plane the above results. The left half of this plane corresponds to the instability of the zero-subharmonic, that is to instability to the solution $Q = -\frac{B}{8}\sin 3\tau$; the right half to the stability,

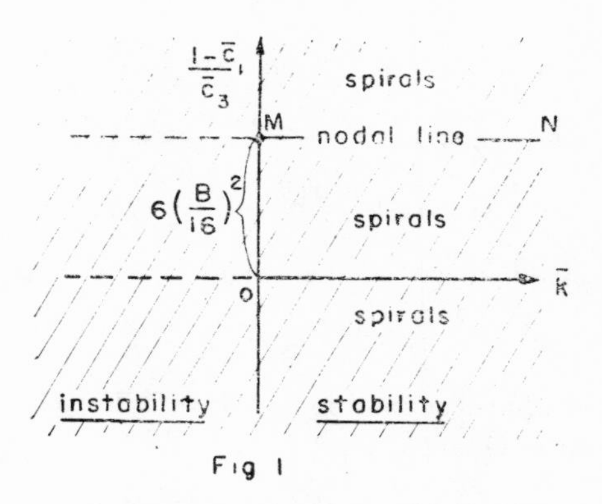

Fig 1

with all «spirals», except the points of the line $\frac{1-\bar{c}_1}{\bar{c}_3} = 6\left(\frac{B}{16}\right)^2$ which are «nodals».

The points of the: $\frac{1-\bar{c}_1}{\bar{c}_3}$ — axis may be either «centers» or «spirals», except the point M which is not a simple singularity but of an advanced order.

The distance of the «nodal line» MN from the $\bar{k}$ — axis has a maximum and a minimum, due to the restrictions of B, given in the «paper B», and in the case of free vibrations $(B=0)$ the «nodal line» is the $\bar{k}$ — axis itself.

2. *The inverse problem.*

The inverse problem of that of «paper B» is the following: «*Given the amplitude r of the subharmonic vibrations, find the coefficients of the differential equation, and study the stability of the solutions obtained*».

The solution of this problem corresponds to determine the numbers λ and μ in terms of r and B, by using the additional equation:

$$x^2 + y^2 - r^2 = 0. \tag{11}$$

This determination is impossible due to the form of the equations (5).

From (5) and (11) we can have λ and μ in terms of x, y, r, B; then the determination of λ and μ needs to know B and two of x, y, r. If we know B, x, y, then the numbers λ and μ are known by solving the system (5) in λ and μ, (11) being a restriction between x, y, r. By knowing λ and μ, we know the ratios $\frac{1-\bar{c}_1}{\bar{c}_3}$, $\frac{\bar{k}}{\bar{c}_3}$, then any two coefficients from $\bar{c}_1$, $\bar{c}_3$, $\bar{k}$ can be determined in terms of their third one, which can have arbitrary values, and the «inverse problen» is solved, since the subject of the stability can be treated as is shown in «paper B».

ΠΕΡΙΛΗΨΙΣ

Ἡ ἐργασία αὕτη ἀποτελεῖ συμπλήρωμα τῆς δευτέρας ἐργασίας ἡμῶν ἐπὶ τῶν ὑποαρμονικῶν ταλαντώσεων (βλ. σελ. 77 κέξ. τοῦ παρόντος τόμου).

Εἰς ταύτην α΄) δίδονται αἱ συνθῆκαι ὑπὸ τὰς ὁποίας ἡ βασικὴ διαφορικὴ ἐξίσωσις δέχεται λύσιν συνισταμένην ἀπὸ μόνον τὸ ἁρμονικόν της μέρος (ἄνευ ὑποαρμονικοῦ) καὶ δὴ εὐσταθὲς καὶ ἑπομένως φυσικῶς δεκτὴν λύσιν· β΄) ἐξετάζεται τὸ «ἀντίστροφον πρόβλημα» τῆς δευτέρας, ἀνωτέρω μνημονευθείσης, ἐργασίας. Τὸ πρόβλημα τοῦτο ἐμφανίζεται ἐδῶ, βάσει τῆς σειρᾶς συλλογισμῶν τῶν προηγουμένων καὶ τῆς παρούσης ἐργασίας, ὡς ἁπλούστατον μαθηματικῶς, ὅμως ἀποτελεῖ πρόβλημα πολλῆς σπουδαιότητος ἀπὸ τεχνικῆς ἀπόψεως.

ΑΝΑΤΥΠΟΝ
ΣΕΛ. 448 - 451

On the singularities of a system of differential equations, where the time figures explicitly,

by ***Dem. G. Magiros****.

᾿Ανεκοινώθη ὑπὸ τοῦ κ. Βασ. Αἰγινήτου.

1. In my first paper of this volume is referred, without any explanation, that the singular points of the system:

$$\frac{du_1}{dt} = \varepsilon F_1(u_1, u_2, t), \qquad \frac{du_2}{dt} = \varepsilon F_2(u_1, u_2, t), \tag{1}$$

fulfill the equations:

$$A_o(u_1, u_2) = 0, \qquad C_o(u_1, u_2) = 0, \tag{2}$$

where A_o and C_o are the first terms of the Fourier series expansions of the functions F_1 and F_2 respectively.

In the following we discuss the above subject. We restrict ourselves to the system (1), although the theory can be applied to more general systems.

A *constant solution* $\{u_1, u_2\}$, of the system (1), determines a point in the u_1, u_2 - plane independent of the time t, and this point is, by definition, a *singular point* of the system (1).

In the following we try to find how to determine approximately the singular points of the system (1).

2. Suppose we are given that the functions F_1 and F_2 fulfill the *expansibility conditions* into Fourier series in t[1], according to which we have:

$$\left.\begin{aligned} \frac{du_1}{dt} &= \varepsilon\{A_o + A_1 \cos t + B_1 \sin t + \ldots + A_m \cos mt + B_m \sin mt + \ldots\} \\ \frac{du_2}{dt} &= \varepsilon\{C_o + C_1 \cos t + D_1 \sin t + \ldots + C_m \cos mt + D_m \sin mt + \ldots\} \end{aligned}\right\} \tag{3}$$

where the coefficients A, B, C, D are functions of u_1, and u_2.

By the above we mean that F_1 and F_2 fulfill the conditions of the *Fourier's theorem*[1], then:

a) F_1, F_2 are periodic in t of period, say, 2π,

b) F_1, F_2 are integrable, say Riemann - integrable, in $[t_o, t_o + 2\pi]$,

c) F_1, F_2 have limited total fluctuations in $[t_o, t_o + 2\pi]$, and

d) the coefficients in (3) can be found according to the standard manner.

* ΔΗΜ. ΜΑΓΕΙΡΟΥ, ᾿Επὶ τῶν ἀνωμάλων σημείων διαφορικοῦ συστήματος, ὅπου ὁ χρόνος εἰσέρχεται ἐκπεφρασμένως.

Reprinted from the *Proceedings of the Athens Academy of Sciences* **32** (1957), 448–451.

Let us take the system:

$$\frac{du_1}{dt} = \varepsilon A_o(u_1, u_2), \qquad \frac{du_2}{dt} = \varepsilon C_o(u_1, u_2), \tag{4}$$

where the time t does not figure explicitly.

Each of the above systems accepts a unique solution $\{u_1, u_2\}$, which assumes given values: $\{u_{1o}, u_{2o}\}$ at $t = t_o$, provided that the functions in their right-hand sides fulfill a Lipschitz condition, when their arguments are restricted to be in the domain:

$$D: \qquad |u_1 - u_{1o}| \leqq k_1, \qquad |u_2 - u_{2o}| \leqq k_2, \qquad |t - t_o| < T.$$

Apply Picard's method of succesive approximations for calculation of the solution $\{\bar{u}_1, \bar{u}_2\}$ of the system (4), by taking as zeroth approximation arbitrary conditions $\{\bar{u}_{1o}, \bar{u}_{2o}\}$. The successive approximations, which converge to the solution $\{\bar{u}_1, \bar{u}_2\}$ of (4), are:

$$\left.\begin{aligned}
\overset{(1)}{\bar{u}_1} &= \bar{u}_{1o} + \varepsilon \int_{t_o}^{t} A_o(\bar{u}_{1o}, \bar{u}_{2o})\, dt, \qquad \overset{(1)}{\bar{u}_2} = \bar{u}_{2o} + \varepsilon \int_{t_o}^{t} C_o(\bar{u}_{1o}, \bar{u}_{2o})\, dt, \\
&\cdots\cdots\cdots\cdots\cdots\cdots\cdots\cdots \\
\overset{(n)}{\bar{u}_1} &= \bar{u}_{1o} + \varepsilon \int_{t_o}^{t} A_o(\overset{(n-1)}{\bar{u}_1}, \overset{(n-1)}{\bar{u}_2})\, dt, \qquad \overset{(n)}{\bar{u}_2} = \bar{u}_{2o} + \varepsilon \int_{t_o}^{t} C_o(\overset{(n-1)}{\bar{u}_1}, \overset{(n-1)}{\bar{u}_2})\, dt
\end{aligned}\right\} \tag{5}$$

Apply also the above method for calculation of the solution $\{u_1, u_2\}$ of (3) by taking as zeroth approximarion arbitrary initial conditions $\{u_{1o}, u_{2o}\}$ at $t = t_o$. The successive approximations are:

$$\left.\begin{aligned}
\overset{(1)}{u_1} &= u_{1o} + \varepsilon \int_{t_o}^{t} A_o(u_{1o}, u_{2o})\, dt + \varepsilon \int_{t_o}^{t} A_1(u_{1o}, u_{2o}) \cos t\, dt + \varepsilon \int_{t_o}^{t} B_1(u_{1o}, u_{2o}) \sin t\, dt + \ldots \\
\overset{(1)}{u_2} &= u_{2o} + \varepsilon \int_{t_o}^{t} C_o(u_{1o}, u_{2o})\, dt + \varepsilon \int_{t_o}^{t} C_1(u_{1o}, u_{2o}) \cos t\, dt + \varepsilon \int_{t_o}^{t} D_1(u_{1o}, u_{2o}) \sin t\, dt + \ldots \\
&\cdots\cdots\cdots\cdots\cdots\cdots\cdots\cdots \\
\overset{(n)}{u_1} &= u_{1o} + \varepsilon \int_{t_o}^{t} A_o(\overset{(n-1)}{u_1}, \overset{(n-1)}{u_2})\, dt + \varepsilon \int_{t_o}^{t} A_1(\overset{(n-1)}{u_1}, \overset{(n-1)}{u_2}) \cos t\, dt + \varepsilon \int_{t_o}^{t} B_1(\overset{(n-1)}{u_1}, \overset{(n-1)}{u_2}) \sin t\, dt + \ldots \\
\overset{(n)}{u_2} &= u_{2o} + \varepsilon \int_{t_o}^{t} C_o(\overset{(n-1)}{u_1}, \overset{(n-1)}{u_2})\, dt + \varepsilon \int_{t_o}^{t} C_1(\overset{(n-1)}{u_1}, \overset{(n-1)}{u_2}) \cos t\, dt + \varepsilon \int_{t_o}^{t} D_1(\overset{(n-1)}{u_1}, \overset{(n-1)}{u_2}) \sin t\, dt + \ldots
\end{aligned}\right\} \tag{6}$$

which couverge to the unique solution $\{u_1, u_2\}$ of the system (3).

Let us take the same initial conditions:

$$u_{1o} = \bar{u}_{1o}, \quad u_{2o} = \bar{u}_{2o}, \tag{7}$$

in the approximations (5) and (6), and subtract properly; the result for the nth approximation is:

$$\left.\begin{aligned} \overset{(n)}{u}_1-\overset{(n)}{\bar{u}}_1 &= \varepsilon\int_{t_o}^{t}\{A_o(\overset{(n-1)}{u_1},\overset{(n-1)}{u_2})-A_o(\overset{(n-1)}{\bar{u}_1},\overset{(n-1)}{\bar{u}_2})\}dt+\varepsilon\int_{t_o}^{t}A_1(\overset{(n-1)}{u_1},\overset{(n-1)}{u_2})\cos t\,dt+\\ &\qquad+\varepsilon\int_{t_o}^{t}B_1(u_1,u_2)\sin t\,dt+\ldots\\ \overset{(n)}{u}_2-\overset{(n)}{\bar{u}}_2 &= \varepsilon\int_{t_o}^{t}\{C_o(\overset{(n-1)}{u_1},\overset{(n-1)}{u_2})-C_o(\overset{(n-1)}{\bar{u}_1},\overset{(n-1)}{\bar{u}_2})\}dt+\varepsilon\int_{t_o}^{t}C_1(\overset{(n-1)}{u_1},\overset{(n-1)}{u_2})\cos t\,dt+\\ &\qquad+\varepsilon\int_{t_o}^{t}D_1(\overset{(n-1)}{u_1},\overset{(n-1)}{u_2})\sin t\,dt+\ldots \end{aligned}\right\}\quad(8)$$

The integrals in (8) are bounded and the right-hand sides contain ε as a common factor, then the approximations, and consequently their limits, are for small ε of order ε, that is:

$$|u_1-\bar{u}_1|=0(\varepsilon)\,,\qquad |u_2-\bar{u}_2|=0(\varepsilon)\,. \tag{9}$$

In (9) the $\{u_1, u_2\}$ and $\{\bar{u}_1, \bar{u}_2\}$ are solutions of (3) and (4) respectively, then any solution of (4) can be considered as an approximation of the solution of (3) of the first order in ε.

3. The constant solutions of (3) come when, in (1), εF_1 and εF_2 tend to zero then, since, the time t figures explicitly in F_1 and F_2, when $\varepsilon \rightarrow 0$. But in (4) the time t does not figure explicitly, then we can get constant solutions of (4), if ε is not necessarily zero, by taking proper initial conditions in the approximations (5), namely the initial conditions $\{\bar{u}_{1o}, \bar{u}_{2o}\}$ which fulfill the conditions:

$$A_o(\bar{u}_{1o}, \bar{u}_{2o})=0\,,\qquad C_o(\bar{u}_{1o}, \bar{u}_{2o})=0\,, \tag{10}$$

when the integrals in (5) are zero, and the solution $\{\bar{u}_1, \bar{u}_2\}$ of (4) is the constant $\{\bar{u}_{1o}, \bar{u}_{2o}\}$ for any ε, included $\varepsilon = 0$.

A constant solution $\{\bar{u}_1, \bar{u}_2\}$ of the approximate system (4), which fulfills (10), in considered as an approximate solution of the exact system (3) of first order in ε.

4. The above technique of replacing the system (3) where the time t figures explicity, by the approximate system (4), where the time t does not figure explicitly, consists essentially of substituting a function by its «*mean value*» over an interval[2], which is called «*moving average*» or «*sliding *mean»* [4]. Since the *«moving average»* is, in general, smother than the original function, the above technique, which is known as the *«averaging principle»* [3], offers advantages in the study of the original system, and it is realized in practice with good results, say in economics, or in electrical problems, e. g. in the photoelectric reproduction of sound, in television images, etc. [4].

Acknowledgments.

Most of the work of the papers Part I (above p. 77 - 85) and part II (above p. 101 - 108) was carried outwhen the author was on a project, sponsord by I. B. M. Watson Laboratory of Columbia University. The author is deeply indebted to Prof. L. H. THOMAS *of Columbia University for his interest and many invaluable discussions concerning the problems of these papers, and to Prof.* J. B. KELLER *of New-York University for his helpful critisism.*

ΠΕΡΙΛΗΨΙΣ

Ἡ ἐργασία αὕτη ἀναφέρεται ἐπὶ τῶν ἀνωμάλων σημείων τοῦ συστήματος (1), τῶν ὁποίων ἡ σπουδὴ γίνεται διὰ τῆς σπουδῆς τῶν ἀνωμάλων σημείων τοῦ συστήματος (4), τὰ ὁποῖα πληροῦν τὰς συνθήκας (2). Διὰ τῆς χρήσεως τῆς ἀνωτέρω μεθόδου, ἡ ὁποία εἶναι γνωστὴ ὡς *«ἀρχὴ τοῦ μέσου ὅρου»*, παρακάμπτονται μεγάλαι μαθηματικαὶ δυσκολίαι, αἱ δὲ λαμβανόμεναι κατὰ προσέγγισιν λύσεις εἶναι εἰς τὴν πρᾶξιν λίαν ἱκανοποιητικαί.

REFERENCE

[1] E. WITTAKER - S. WATSON, A Treatise of Modern Analysis, 4th ed. (1952), 164.
[2] P. FATOU, *Société Math. de France,* Bulletin (1929).
[3] N. KRYLOFF & N. BONGOLIUBOFF, Introduction of Noulinear Mechanics, p. 12, *Ann. of Math. Studies,* 11.
[4] B. VANDER POL & H. BREMMER, Operational Calculus, chapt. XIV, § 2 (1955).

Subharmonics of Any Order in Nonlinear Systems of One Degree of Freedom: Application to Subharmonics of Order $^1/_3$*

DEMETRIOS G. MAGIROS†

New York University, New York

This paper consists of two parts. In the first one we discuss the subharmonics of any order in the case where the nonlinearity enters in the elastic forces. The basic differential equation is with coefficients not necessarily small. The "steady state" and "transient" solutions of the differential equation, and the conditions for the existence of the subharmonics and their stability in a steady state are examined. In the second part of the paper an application is given, namely, the investigation of the subharmonics of order ⅓ according to the theory of the first part. An illustrated example is given, the "inverse problem" is examined, and the conditions for the stability and instability of the "harmonic solution" are found.

INTRODUCTION

The behavior of an oscillatory system with one degree of freedom is governed by an equation of the form

$$\ddot{Q} = \phi(Q, \dot{Q}, t), \tag{a}$$

ϕ being a function nonlinear in the variables Q and $\dot{Q}$, and periodic or almost periodic in t. In some cases Eq. (a) leads to the equation

$$\ddot{Q} + Q = \epsilon f(Q, \dot{Q}) + kg(t), \tag{b}$$

from which we may get the first order equation

$$\frac{du_2}{du_1} = \frac{f_2(u_1, u_2, t)}{f_1(u_1, u_2, t)}, \tag{c}$$

where u_1 and u_2 are related to Q and $\dot{Q}$ by proper relations.

* Most of the work of the paper was carried out when the author was employed on a project sponsored by I.B.M. Watson Laboratory at Columbia University. The author is deeply indebted to Prof. L. H. Thomas of Columbia University for his interest and the many invaluable discussions concerning the problems of the paper, and to Prof. J. B. Keller of New York University for his helpful criticism.

† Present address: Hofstra College, Hempstead, New York.

Reprinted from *Information and Control* 1 (1958), 198–227.

In this paper we treat a special case of the Eq. (a), namely, the problem of subharmonics of any order when the system is governed by an equation of the form

$$\ddot{Q} + \bar{k}\dot{Q} + \bar{c}_1 Q + \bar{c}_2 Q^2 + \bar{c}_3 Q^3 = B \sin nt, \tag{d}$$

where the coefficients are not necessarily very small. This case is very important from a mathematical point of view and very useful in nonlinear engineering problems.

A mechanical model of a system, governed by Eq. (d), might be a mass under the action of a viscous damping force linear in the velocity, of an elastic force which is a cubic function of the deflection, and of a simple harmonic forcing function of a given frequency and not necessarily small amplitude. An electrical model might be an electrical oscillatory circuit in which the nonlinear oscillations take place because of a saturable-core inductance under the impression of an alternating electromotive force of sinusoidal type (Hayachi, 1953). An aerodynamical model could be based on the fact that certain parts of an airplane can be excited to violent oscillations by an engine running with a number of revolutions much larger than the natural frequency of the oscillating parts (Von Kármán, 1940).

By using ideas of Van der Pol (1927), Mandelstam and Papalexi (1935), and Andronow and Witt (1930), we transform Eq. (d) to the forms (b) and (c). We deal with the steady state and transient solutions, by using Poincaré's method for periodic solutions (Friedrichs, 1953; Poincaré, 1890; Tsien, 1956). The condition for the existence of the subharmonics and their stability in a steady state are discussed by considering the stability of the singularities of the corresponding equation of the form (c). In the appendix (Magiros, 1957d) we discuss the fact that the singularities of an ordinary system, where the time enters explicitly, are the same as those of an "approximate system," where the time does not enter explicitly. As an application of the theory we discuss the subharmonics of order $\frac{1}{3}$. In this case, the equations that give the components of the amplitude of the subharmonics in steady state are properly transformed. The components of the amplitude are given in terms of the amplitude of the external force, and the ratios of the coefficients of the linear damping and the cubic elastic force. Under certain conditions given here, these ratios can have any values. The coefficients themselves need not necessarily be small—a case important in theoretical and practical problems. Some bounds are given for the amplitude of the external force, and some restrictions for the coefficients of the differential equation. An example is illustrated and the corresponding sketch of the integral curves in the whole plane, which is separated into appropriate regions, is given. The "converse problem" is also discussed. Finally, we discuss the conditions under which we have the stability or instability of the "harmonic part" of the solution, the subharmonic part being zero.

PART A: THE GENERAL CASE

1. The Problem (Magiros, 1957a)

We discuss subharmonics of any order $1/n(n = 2, 3, \cdots)$ of the differential equation

$$\ddot{Q} + \bar{k}\dot{Q} + \bar{c}_1 Q + \bar{c}_2 Q^2 + \bar{c}_3 Q^3 = B \sin nt, \tag{1}$$

in their steady and transient states, in the case when the coefficients of (1) are not necessarily small—a case very important in engineering problems. Dots in (1) denote derivatives with respect to time t. By taking a parameter ϵ such that

$$\bar{k} = \epsilon k, \qquad 1 - \bar{c}_1 = \epsilon c_1, \qquad \bar{c}_2 = \epsilon c_2, \qquad \bar{c}_3 = \epsilon c_3, \tag{2}$$

Eq. (1) can be written as

$$\ddot{Q} + Q = \epsilon f(Q, \dot{Q}) + B \sin nt, \tag{3}$$

$$f(Q, \dot{Q}) = -k\dot{Q} + c_1 Q - c_2 Q^2 - c_3 Q^3. \tag{3a}$$

The parameter ϵ is a constant with respect to the variables Q, $\dot{Q}$. All the coefficients in (2) must be finite as well as the parameter ϵ. The case $\epsilon = 0$ corresponds to that in which all the coefficients in left-hand members of (2) are zero, except $\bar{c}_1 = 1$, and the coefficients in right-hand members can take any finite value. When $\epsilon \neq 0$, even if ϵ is small, the coefficients of left-hand members of (2) are not necessarily small, if we take appropriate values for the coefficients of the right-hand members. If we know the values of Q and $\dot{Q}$ at $t = t_0$, say $Q_{t=t_0} = Q_t$, $\dot{Q}_{t=t_0} = \dot{Q}_0$; and if the function $f(Q, \dot{Q})$, given by (3a), is continuous and has continuous derivatives with respect to Q and $\dot{Q}$ in a domain D,

$$D: \quad Q_0 - \bar{\zeta}_1 \leqq Q \leqq Q_0 + \bar{\zeta}_1, \quad \dot{Q}_0 - \bar{\zeta}_2 \leqq \dot{Q} \leqq \dot{Q}_0 + \bar{\zeta}_2, \quad t_0 \leqq t \leqq T;$$

then $| f(Q, \dot{Q}) |$ has an upper bound in the domain of the variables, which depends on the domain.

In case $\epsilon = 0$, Eq. (3) reduces to

$$\ddot{Q} + Q = B \sin nt, \tag{4}$$

of which the solution of period 2π is

$$Q = x \sin t - y \cos t + (B/1 - n^2) \sin nt, \tag{5}$$

where $n \neq \pm 1$ and x, y are arbitrary constants which can be determined by using the initial conditions.

If $\epsilon \neq 0$, we attempt to determine a periodic solution of Eq. (3) of the form (5), in which x and y are functions of ϵ and t, say $x = u_1(\epsilon, t)$, $y = u_2(\epsilon, t)$, and such that when $\epsilon \to 0$ the limits

$$\lim_{\epsilon \to 0} u_1, \qquad \lim_{\epsilon \to 0} u_2 \tag{6}$$

are the constants of the "generating solution" in case $\epsilon = 0$. Our type of periodic solution, given by (5), consists of two parts. One part, the term $(B/1 - n^2) \sin nt$, with the same frequency as that of the external force, is called the harmonic part of the solution. The other part, $x \sin t - y \cos t$, with frequency a fraction $1/n$ of that of the external force, is called the "subharmonic part," and its amplitude is $\zeta = (x^2 + y^2)^{1/2}$. Given the amplitude and the frequency of the external force, the amplitude of the harmonic part of the solution is constant. But this does not happen with the components x and y of the amplitude of the subharmonic part of the solution.

When $\epsilon = 0$, and for given initial conditions, x and y are constants, and in this case we speak about "subharmonics in steady states of oscillation," either stable or unstable. If $\epsilon \neq 0$, the components x and y are functions of time t, we have a "transient phenomenon," and we speak about "subharmonics in transient states." The steady states of oscillation, i.e., the equilibrium positions of the system, are correlated with the singularities of the differential equation; the transient states with their integral curves, which do or do not terminate on the singularities, depending on the stability or instability of the singularities. The solution in the transient state yields, with the lapse of time, ultimately to steady state solution (stable) (Hayachi, 1953; Stoker, 1955).

2. Reduction of Eq. (3) to an Equivalent Normal System

Take a new variable q according to the relation

$$q = Q - (B/1 - n^2) \sin nt. \tag{7}$$

Substituting (7) into (3) we obtain

$$\ddot{q} + q = \epsilon f(q, \dot{q}), \tag{8}$$

with

$$\begin{aligned} f(q, \dot{q}) = & -k\dot{q} + c_1 q - c_2 q^2 \\ & - \frac{kn}{1-n^2} B \cos nt + c_1 \frac{B}{1-n^2} \sin nt \\ & - c_2 \frac{B^2}{(1-n^2)^2} \sin^2 nt - c_3 \frac{B^3}{(1-n^2)^3} \sin^3 nt \\ & - 2c_2 \frac{B}{1-n^2} q \sin nt - 3c_3 q^2 \sin nt - 3c_3 \frac{B^2}{(1-n^2)^2} q \sin^2 nt. \end{aligned} \tag{8a}$$

Introduce into (8) new variables, u_1 and u_2, defined by

$$u_1 = \dot{q} \cos t + q \sin t, \qquad u_2 = \dot{q} \sin t - q \cos t. \tag{9}$$

From (9) we get

$$q = u_1 \sin t - u_2 \cos t, \qquad \dot{q} = u_1 \cos t + u_2 \sin t, \tag{10}$$

$$\dot{u}_1 = (\ddot{q} + q) \cos t, \qquad \dot{u}_2 = (\ddot{q} + q) \sin t, \tag{11}$$

when, according to (8), we have

$$\dot{u}_1 = \epsilon f_1(u_1, u_2, t), \qquad \dot{u}_2 = \epsilon f_2(u_1, u_2, t), \tag{12}$$

where

$$\begin{aligned} f_1(u_1, u_2, t) &= f(u_1, u_2, t) \cdot \cos t, \\ f_2(u_1, u_2, t) &= f(u_1, u_2, t) \cdot \sin t, \end{aligned} \tag{12a}$$

with

$$\begin{aligned} f(u_1, u_2, t) = \Big[& -k(u_1 \cos t + u_2 \sin t) + c_1(u_1 \sin t - u_2 \cos t) \\ & - u_2(u_1^2 \sin^2 t + u_2^2 \cos^2 t - 2u_1 u_2 \sin t \cos t) \\ & - c_3(u_1^3 \sin^3 t - u_2^3 \cos^3 t + 3u_1 u_2^2 \sin t \cos^2 t \\ & - 3u_1^2 u_2 \sin^2 t \cos t) - k \frac{nk}{1-n^2} \cos nt + c_1 \frac{B}{1-n^2} \sin nt \end{aligned}$$

$$- c_2 \frac{B^2}{(1 - u^2)^2} \sin^2 nt - c_3 \frac{B^3}{(1 - u^2)^3} \sin^3 nt \tag{12b}$$

$$- 2c_2 \frac{B}{1 - n^2} (u_1 \sin t - u_2 \cos t) \sin nt$$

$$- 3c_3 \frac{B}{1 - n^2} (u_1^2 \sin^2 t + u_2^2 \cos^2 t - 2u_1u_2 \sin t \cos t) \sin nt$$

$$\left. - 3c_3 \frac{B^2}{(1 - n^2)^2} (u_1 \sin t - u_2 \cos t) \sin^2 nt \right].$$

The function $f(u_1, u_2, t)$ is found by inserting (10) into (8a). The system (12), which takes the place of Eq. (3), gives advantages in the analysis. From (7) and the first of (10) we obtain

$$Q = u_1 \sin t - u_2 \cos t + (B/1 - n^2) \sin nt. \tag{13}$$

The expressions (5) and (13) are of the same form, then a solution $\{u_1, u_2\}$ of the system (12) gives the components of the subharmonics in the transient states, and the limits $\lim_{\epsilon \to 0} u_1$, $\lim_{\epsilon \to 0} u_2$ give the components of the subharmonics in the steady states.

3. The General System

In this section we shall discuss the system

$$\begin{aligned} \dot{u}_1 &= \epsilon F_1(u_1, u_2, t), \qquad & \dot{u}_2 &= \epsilon F_2(u_1, u_2, t), \\ u_1(t_0) &= u_{10}, \qquad & u_2(t_0) &= u_{20}, \end{aligned} \tag{14}$$

in a general way, as it is needed for our purposes. The functions F_1 and F_2 are taken with the following properties: They are analytic in u_1 and u_2, continuous and periodic with period 2π in t, and hence continuous in u_1, u_2, t; therefore $|F_1|$, $|F_2|$ have least upper bounds M_1, M_2 respectively, in a domain

$$D_1: \quad |u_1 - u_{10}| < \zeta_1, \qquad |u_2 - u_{20}| < \zeta_2, \qquad t_0 < t < T;$$

also F_1, F_2 are expansible as power series in $(u_1 - u_{10})$, $(u_2 - u_{20})$ and convergent in the domain D_1. The number ϵ is real.

(a) Try to find a solution $\{u_1, u_2\}$ of the system (14) such that if $\epsilon \to 0$, u_1 and u_2 tend to constants x and y. When $\epsilon \neq 0$, u_1 and u_2 depend on ϵ and t, and we assume that for $t = t_0$ they differ a little from x and y; that is,

$$u_{10} = x + \xi, \qquad u_{20} = y + \eta, \tag{15}$$

where ξ, η are very small. These conditions can be considered as initial conditions for the system (14).

Take as a formal solution $\{u_1, u_2\}$ of (14) the following expressions:

$$\begin{aligned} u_1 &= X_1^0 + \epsilon X_1^1 + \epsilon^2 X_1^2 + \cdots \\ u_2 &= X_2^0 + \epsilon X_2^1 + \epsilon^2 X_2^2 + \cdots \end{aligned} \tag{16}$$

where

$$X_1^0 = x + \xi, \qquad X_2^0 = y + \eta, \tag{16a}$$

and the other coefficients of (16) are regular functions of x, y, ξ, η, t when $|x|$, $|y|$, $|\xi|$, $|\eta|$, $|t - t_0|$ are less than certain constants. We determine the coefficients X of (16) by presupposing the convergence of the series (16).

(b) For the determination of the coefficients X, substitute (16) into the system (14) after the right members of (14) have been developed as power series in $(u_1 - u_{10})$, $(u_2 - u_{20})$, ϵ. Then rearrange the terms according to powers of ϵ, and equate coefficients of corresponding powers of ϵ. We obtain an infinite series of systems of differential equations, from which, by integrating, we have

$$\begin{aligned} X_1^1 &= \int_{t_0}^t [F_1]\,dt, \qquad X_2^1 = \int_{t_0}^t [F_2]\,dt, \\ X_1^2 &= \int_{t_0}^t \{X_1^1[F_1]_{u_1} + X_2^1[F_2]_{u_2}\}\,dt, \\ X_2^2 &= \int_{t_0}^t \{X_1^1[F_2]_{u_1} + X_2^1[F_2]_{u_2}\}\,dt, \\ X_i^3 &= \int_{t_0}^t \sum_{j=1}^2 X_i^2[F_i]_{u_i}\,dt + \frac{1}{2}\int_{t_0}^t \sum_{j=1}^2 \sum_{l=1}^2 X_j^1 X_l^1 [F_i]_{u_j u_l}\,dt, \\ X_i^k &= \int_{t_0}^t P_i^k(X_i^1, \cdots X_i^{k-1})\,dt, \\ & j, l, i = 1, 2 \qquad k = 1, 2, \cdots \end{aligned} \tag{17}$$

The brackets indicate that the corresponding functions and their derivatives are taken at: $u_1 = u_{10} = x + \xi$, $u_2 = u_{20} = y + \eta$. The P_i^k have the following properties. (i) They are polynomials in $X_1^1, X_2^1, \cdots, X_1^{k-1}, \cdots$, not in $(t - t_0)$. (ii) Their coefficients are linear functions of the coefficients of the expansion of F_1, F_2 given in power series of $(u_1 - u_{10})$, $(u_2 - u_{20})$. (iii) The numerical multipliers of f_1 and f_2 are positive numbers, namely, the numbers arising in Taylor's expansion and in forming products of various power series. All X, according to (17), are zero at $t = t_0$.

By carrying out the integrations in the first two of (17), we find X_1^1, X_2^1. Upon substituting the values of X_1^1, X_2^1 into the second two of (17), the integrands become known continuous functions of t, when X_1^2, X_2^2 are defined by quadratures. With the same procedure we can find $X_1^3, X_2^3, \cdots$.

(c) The convergence of the formal solution (16) in our domain was presupposed for the determination of the coefficients X, given by (17). This convergence and the domain of its validity come from using the "method of dominants," as follows. There always exists (Petrovsky, 1954) a function which dominates F_1, F_2 of (14); that is, there exists a function F such that the nonnegative coefficients of its expansion in power series, in the neighborhood of any point of the domain D_1, are not smaller than the absolute values of the corresponding coefficients of the power series expansion of F_1, F_2 in the neighborhood of the point. Such a dominating function of F_1, F_2 in D_1 is the function

$$F = \frac{M}{1 - [(x_1 - u_{10}) + (x_2 - u_{20})]\zeta^{-1}}, \tag{18}$$

where

$$M \geqq M_i, \qquad \zeta \leqq \zeta_i, \qquad i = 1, 2. \tag{18a}$$

This gives a power series expansion that converges when

$$| x_i - u_{i0} | < \tfrac{1}{2}\zeta, \qquad i = 1, 2.$$

Consider the auxiliary system

$$\dot{x}_i = \epsilon \frac{M}{1 - \frac{1}{2}[(x_1 - u_{10}) + (x_2 - u_{20})]} \tag{19}$$

$$x_i(t_0) = u_{i0}, \qquad i = 1, 2.$$

The system (19) gives $\dot{x}_1(t_0) = \dot{x}_2(t_0)$. On making use of the initial conditions, we have

$$x_1 - u_{10} = x_2 - u_{20} = \bar{x}, \tag{20}$$

and $\bar{x}$ satisfies the differential equation

$$\dot{\bar{x}} = \epsilon \frac{M}{1 - (2\bar{x}/\zeta)} \qquad \bar{x}(t_0) = 0. \tag{21}$$

On separating the variables in (21) and performing the quadratures, we find that

$$\bar{x} - \bar{x}^2/\zeta = \epsilon M(t - t_0).$$

Then

$$\bar{x} = \frac{1}{2}\zeta\left\{1 - \left[1 - \frac{4\epsilon M(t - t_0)}{\zeta}\right]^{1/2}\right\}, \tag{22}$$

by selecting the negative sign before the radical, in order to satisfy the condition $\bar{x}(t_0) = 0$.

According to (20), the solution of (19) is

$$x_1 = u_{10} + \bar{x}, \qquad x_2 = u_{20} + \bar{x}, \tag{23}$$

$\bar{x}$ being given by (22).

The expansion of the right-hand member of (22) as power series in ϵ has the singularity: $\epsilon = \zeta/4M(t - t_0)$. Therefore the expansion converges for all values of ϵ such that

$$|\epsilon| < \frac{\zeta}{4M(t - t_0)}. \tag{24}$$

The expansions of x_1, x_2 as power series in ϵ converge under the same condition (24). The second members of (19) have all properties of the second members of (14), so that the system (19) can be solved directly by using power series in ϵ of the form (16). This solution is unique, and so it is identical with that given by (22) and (23), if we take the expansion of $\bar{x}$ in powers of ϵ. Then the power series solution of (19) in ϵ is valid under the condition (24).

If the series solutions of (19) and (21) are

$$\begin{aligned} \bar{x} &= \bar{x}^0 + \epsilon\bar{x}^1 + \epsilon^2\bar{x}^2 + \cdots \\ \bar{x}^0 &= 0 \end{aligned} \tag{25a}$$

$$\begin{gathered} x_i = x_i^0 + \epsilon x_i^1 + \epsilon^2 x_i^2 + \cdots \\ x_1^0 = u_{10}, \qquad x_2^0 = u_{20}, \qquad x_1^1 = x_2^1 = \bar{x}^1, \qquad x_1^2 = x_2^2 = \bar{x}^2, \cdots \end{gathered} \tag{25b}$$

and we consider the formulas (17) that give the coefficients X_1^1, X_2^1, and the formulas that give the coefficients x_1^1, x_2^1, then the integrand of the former is dominated by the integrand of the latter in these definite integrals. Therefore

$$x_i^1 \geqq |X_i^1|, \qquad t_0 \leqq t \leqq T, \qquad i = 1, 2. \tag{26a}$$

The corresponding result is true successively for terms with higher indices. Thus

$$x_i^k \geqq |X_i^k|, \qquad t_0 \leqq t \leqq T, \qquad i = 1, 2, \tag{26b}$$

and the investigation of the convergence of (16) is completed.

(d) In many special cases the domain of convergence of (16) is much larger than the condition (24) shows. This is due to additional properties that the differential equations possess, which give to their solution a wider domain of convergence. The inequality (24) can be satisfied (i) by imposing restrictions upon both ϵ and t, or (ii) by taking t arbitrarily and then, for the convergence, restricting ϵ, or (iii) by taking ϵ arbitrarily and then restricting t.

For nonnegative ϵ, (24) gives

$$t - t_0 \leqq \frac{\zeta}{4\epsilon M} = T - t_0 .$$

Then

$$T = \max t = t_0 + \frac{\zeta}{4\epsilon M}, \tag{27a}$$

and if t_0 varies in $0 \leqq t_0 \leqq 2\pi$, we have

$$\frac{\zeta}{4\epsilon M} \leqq T \leqq 2\pi + \frac{\zeta}{4\epsilon M}. \tag{27b}$$

4. THE USE OF THE PERIODICITY OF THE SOLUTION $\{u_1, u_2\}$

In the theory of the previous section no use of the periodicity of the solution $\{u_1, u_2\}$ of the system (14) has been made. Let us assume now that u_1, u_2 are periodic in t with period 2π; that is,

$$u_1(t_0 + 2\pi) - u_1(t_0) = 0, \qquad u_2(t_0 + 2\pi) - u_2(t_0) = 0. \tag{28}$$

Applying (28) to (16) we have

$$\begin{aligned} &[X_1^1(t_0 + 2\pi) - X_1^1(t_0)] + \epsilon[X_1^2(t_0 + 2\pi) - X_1^2(t_0)] + \cdots = 0, \\ &[X_2^1(t_0 + 2\pi) - X_2^1(t_0)] + \epsilon[X_2^2(t_0 + 2\pi) - X_2^2(t_0)] + \cdots = 0. \end{aligned} \tag{29}$$

Now take the Fourier series development in t of the function F_1 and F_2 :

$$\begin{aligned} F_1 &= A_0 + A_1 \cos t + B_1 \sin t + \cdots A_m \cos mt + B_m \sin mt + \cdots \\ F_2 &= C_0 + C_1 \cos t + D_1 \sin t + \cdots C_m \cos mt + D_m \sin mt + \cdots, \end{aligned} \tag{30}$$

where the coefficients A, B, C, D are functions of u_1, u_2 and such that the series

$$\begin{aligned} &| A_0 | + | A_1 | + \cdots + | A_m | + | B_m | + \cdots, \\ &| C_0 | + | C_1 | + \cdots + | C_m | + | D_m | + \cdots, \end{aligned} \tag{30a}$$

are uniformly convergent with least upper bound $\bar{M}$; also the series of partial derivatives with respect to u_1, u_2 :

$$\begin{aligned} &| \partial A_0/\partial u_1 | + | \partial A_1/\partial u_1 | + \cdots, \\ &| \partial C_0/\partial u_2 | + | \partial C_1/\partial u_2 | + \cdots, \end{aligned} \tag{30b}$$

are uniformly convergent with least upper bound $\bar{H}$.

By taking into account the expressions of X given by (17) and the series (30), the conditions (29) give

$$\begin{aligned} A_0(x + \xi, y + \eta) + \epsilon\phi_1(x + \xi, y + \eta, t_0) + \cdots = 0, \\ C_0(x + \xi, y + \eta) + \epsilon\phi_2(x + \xi, y + \eta, t_0) + \cdots = 0, \end{aligned} \tag{31}$$

where we suppose

$$A_0(0, 0) = 0, \qquad C_0(0, 0) = 0. \tag{31a}$$

Provided that the Jacobian of the system (31) is not zero,

$$\begin{vmatrix} \partial A_0/\partial \xi & \partial C_0/\partial \xi \\ \partial A_0/\partial \eta & \partial C_0/\partial \eta \end{vmatrix} \neq 0, \tag{32}$$

we can solve the system (31) in ξ and η in terms of ϵ and t;

$$\xi = \xi(\epsilon, t_0), \qquad \eta = \eta(\epsilon, t_0), \quad \text{say,} \tag{33}$$

with the conditions

$$\xi = \xi(0, t_0) = 0, \qquad \eta = \eta(0, t_0) = 0. \tag{33a}$$

If the Jacobian is zero, it is necessary to consider in (31) terms of degree 1, 2, $\cdots$ in ϵ, that is, the function $\phi_1, \phi_2, \cdots$, (Faton, 1929). ξ, η in (33) are given in series in ϵ, where t_0 can take any value of the interval $0 \leqq t_0 \leqq 2\pi$ and ϵ is in accordance with the condition (24). From (15), (16a), and (33) we get

$$X_1^0 = u_{10} = x + \xi(\epsilon, t_0), \qquad X_2^0 = u_{20} = y + \eta(\epsilon, t_0), \tag{34}$$

where x and y are unknown. If we find the constants x and y, the transient solution (16) of the system (14) is completely determined.

5. Determination of the Constants x and y: Conditions of Existence of Steady States Subharmonics

Equations (31) must be satisfied when $\epsilon = \xi = \eta = 0$. In that case (31) give

$$A_0(x, y) = 0, \qquad C_0(x, y) = 0. \tag{35}$$

The solution (x, y) of the system (35) gives the constants x and y, which characterize the steady state subharmonics. The conditions for real intersections of the curves (35) in the x, y-plane give the conditions of the existence of the steady state subharmonics.

6. The Stability of the Steady State Subharmonics

The singular points of the general system (14), where the time enters explicitly, are given, according to the appendix of the paper, by Eq. (35). The stability of these singularities gives the stability of the steady state solution of the system (14). The following definitions of the singular points shall be used in the second part of this paper.

According to Poincaré (1892) and Bendixson (1901), the distinction between the different kinds of singularities depends on two numbers ρ_1, ρ_2, the roots of the characteristic equation

$$\begin{vmatrix} a_1 - \rho & b_1 \\ a_2 & b_2 - \rho \end{vmatrix} = 0; \tag{36}$$

that is, on the numbers

$$\rho_{1,2} = \tfrac{1}{2}\{a_1 + b_2 \pm [(a_1 - b_2)^2 + 4a_2b_1]^{1/2}\}, \tag{36a}$$

where:

$$a_1 = \partial A_0/\partial x, \qquad a_2 = \partial A_0/\partial y, \qquad b_1 = \partial C_0/\partial x, \qquad b_2 = \partial C_0/\partial y. \tag{36b}$$

For nonzero roots ρ_1, ρ_2 the "simple singularities" are classified in four classes as follows:

I. "Nodal points," when ρ_1, ρ_2 are real and of same sign:

$$(a_1 - b_2)^2 + 4a_2b_1 \geqq 0, \qquad a_1b_2 - a_2b_1 > 0. \tag{37a}$$

II. "Saddle (pass) points," when ρ_1, ρ_2 are real but of opposite sign:

$$(a_1 - b_2)^2 + 4a_2b_1 \geqq 0, \qquad a_1b_2 - a_2b_1 < 0. \tag{37b}$$

III. "Spiral (focus) points," when ρ_1, ρ_2 are complex conjugates:

$$(a_1 - b_2)^2 + 4a_2b_1 < 0, \qquad a_1b_2 - a_2b_1 \neq 0. \tag{37c}$$

IV. "Spiral points" or "centers," when ρ_1, ρ_2 are pure imaginaries:

$$(a_1 - b_2)^2 + 4a_2b_1 < 0, \qquad a_1b_2 - a_2b_1 = 0. \tag{37d}$$

Under the conditions (37d) the singularity "may be a center." These conditions are necessary, but not sufficient. There is "Poincaré's criterion" in this case for distinguishing a spiral point from a center (Hadamard, 1933).

Define the above singularities as "stable" or "unstable" when, with increasing time t, any point on any integral curve does or does not move into the said singularity, that is, according as the real part of the roots is negative or positive, respectively. In Fig. 1, I shows a "stable node," II a saddle point which is "intrinsically unstable," III a "stable spiral point," IV a (neutral) center. By "intrinsically unstable" we mean that

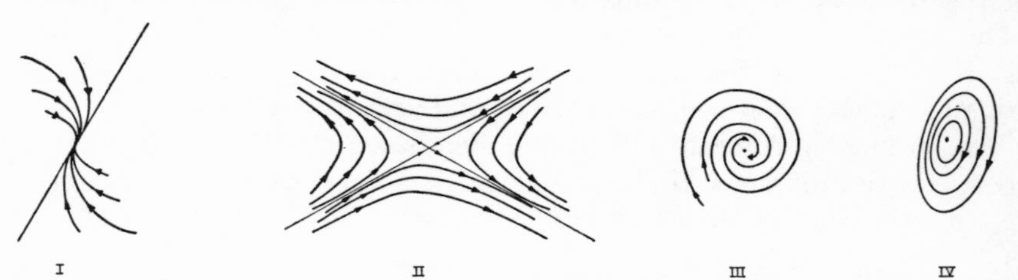

FIG. 1. The four kinds of simple singularities. I is the stable node, II the saddle, III the stable spiral, IV the center.

the saddle point is terminated by four trajectories forming two distinct integral curves. Two of these trajectories approach the saddle point with increasing time t, and correspond to the stable curves, while the others move away from it with increasing time and correspond to the unstable curves. There exist four regions containing continua of hyperbolically shaped integral curves which do not approach the saddle point (Hayachi, 1953).

7. Application to the System (14)

Now, coming back to our original problem, we see that the system (12) is a special case of the system (14). To apply the theory of the previous sections concerning the system (14) to our special case of the system (12), we give to the functions f_1 and f_2 of (12a) and (12b) an appropriate form from which we can get the development into Fourier series for any order of the subharmonics. By replacing the powers and products of sines and cosines of multiple angles in constructing f_1 and f_2, according to (12a) and (12b), we can get the following results, if we restrict ourselves to coefficients which are useful for the construction of the corresponding functions A_0 and C_0, which depend on the number n characterizing the order of the subharmonics:

$$f_1(u_1, u_2, t) = \left[-\frac{1}{2} ku_1 - \frac{1}{2} c_1 u_2 + \frac{3}{8} c_3 u_2^3 + \frac{3}{8} c_3 u_1^2 u_2 \right.$$
$$\left. + \frac{3}{4} c_3 \frac{B^2}{(1-n^2)^2} u_2 \right] + \left[-\frac{1}{2} c_2 \frac{B}{1-n^2} u_1 \right] \cos (n-2)t + \cdots \tag{38a}$$
$$+ \left[\frac{3}{4} c_3 \frac{B}{1-n^2} u_1 u_2 \right] \cos (n-3)t + \cdots$$

$$f_2(u_1, u_2, t) = \left[\frac{1}{2} c_1 u_1 - \frac{1}{2} ku_2 - \frac{3}{8} c_3 u_1^3 - \frac{3}{8} c_3 u_1 u_2^2 \right.$$
$$\left. - \frac{3}{4} c_3 \frac{B^2}{(1-n^2)^2} u_1 \right] + \left[\frac{1}{2} c_2 \frac{B}{1-n^2} u_2 \right] \cos (n-2)t + \cdots \tag{38b}$$
$$+ \left[-\frac{3}{8} c_3 \frac{B^2}{1-n^2} (-u_1^2 + u_2^2) \right] \cos (u-3)t + \cdots$$

For any given n, that is, for any order of the subharmonics, the corresponding functions $A_0(x, y)$, $C_0(x, y)$ are the independent-of-time terms of the right-hand members of (38a) and (38b), respectively, if we put x and y instead of u_1 and u_2, respectively. If we know f_1 and f_2, the equations $A_0 = 0$ and $C_0 = 0$ are known, and their solutions give the steady states subharmonics of the initial equations. The stability of the steady state solutions comes from the study of the singularities (x, y), according to what has been quoted above on the singularities. The terms of the functions $A_0(x, y)$ and $C_0(x, y)$ are, as we can see from the formulas (38a) and (38b), such that we can have x and y, from $A_0 = 0$ and $C_0 = 0$, in terms of B and the ratios k/c_3, c_1/c_3, c_2/c_3 for $c_3 \neq 0$.

PART B. APPLICATION: SUBHARMONICS OF ORDER 1/3 (MAGIROS, 1957b)

8. THE EQUATIONS FOR THE COMPONENTS OF THE AMPLITUDE OF THE SUBHARMONICS

From Eq. (1) and its solution (5) for $n = 3$ we deduce

$$\ddot{Q} + \bar{k}\dot{Q} + \bar{c}_1 Q + \bar{c}_2 Q^2 + \bar{c}_3 Q^3 = B \sin 3t, \tag{39}$$

$$Q = x \sin t - y \cos t - \tfrac{1}{8} B \sin 3t. \tag{40}$$

For $\epsilon \neq 0$ the components x and y of the subharmonics depend on time t, and this transient state yields, with the lapse of the time, to the steady state, either stable or unstable. Since the coefficients of (39) can have arbitrary values, Eq. (39) may contain various types of periodic solutions for different initial values. For the constant values of x and y, which characterize the subharmonics in the "steady state," we must find the functions $A_0(x, y)$ and $C_0(x, y)$. If x and y are substituted for u_1 and u_2, Eqs. (38a) and (38b) give for $n = 3$:

$$\begin{aligned} A_0(x, y) &= \tfrac{1}{2}[-kx + c_1 y + \tfrac{3}{4} c_3 y(x^2 + y^2 + \tfrac{1}{32}B^2) - \tfrac{3}{16} c_3 Bxy], \\ C_0(x, y) &= \tfrac{1}{2}[c_1 x - ky - \tfrac{3}{4} c_3 x(x^2 + y^2 + \tfrac{1}{32}B^2) \\ &\qquad + \tfrac{3}{32} c_3 B(-x^2 + y^2)]. \end{aligned} \tag{41}$$

Setting the right-hand members of (41) equal to zero, we have the following equations for the determination of x and y:

$$\begin{aligned} \mu x + \lambda y - y(x^2 + y^2 + \tfrac{1}{32}B^2) + \tfrac{1}{4}Bxy &= 0, \\ \lambda x - \mu y + x(x^2 + y^2 + \tfrac{1}{32}B^2) + \tfrac{1}{8}B(-x^2 + y^2) &= 0, \end{aligned} \tag{42}$$

where

$$\lambda = \frac{4}{3}\frac{c_1}{c_3} = \frac{4}{3}\frac{1 - \bar{c}_1}{\bar{c}_3}, \qquad \mu = \frac{4}{3}\frac{k}{c_3} = \frac{4}{3}\frac{\bar{k}}{\bar{c}_3}. \tag{42a}$$

To any real solution of the system (42) corresponds a solution of (39) in the form of (40).

Remarks

For prescribed initial conditions of (39), say Q_0 and $\dot{Q}_0$ at $t = 0$, we have, according to (40),

$$Q_0 = -y, \qquad \dot{Q}_0 = x - 3/8 B, \tag{43}$$

and the given initial conditions correspond to the point

$$x = \dot{Q}_0 + 3/8 B, \qquad y = -Q_0 \tag{43a}$$

in the x, y-plane. Starting from the point (43a), the "starting point," and following the corresponding integral curve with the lapse of time, we may terminate to a "final point," which corresponds to the proper steady solution. The coordinates of this final point are solutions of the system (42) and the stability of this point can be investigated according to Section 6.

Given the initial condition and the amplitude B of the external force, the "starting point" in the x, y-plane is defined; conversely, any point of the x, y-plane can be taken as a starting point by properly choosing the initial conditions and the amplitude B. If the "starting point" is selected in coincidence with a "stable final point," no "transient phenomena" must exist. This last remark may be of importance in engineer ing problems.

9. Restrictions of the Coefficients of Eq. (39)

The system (42) can be written as

$$\begin{aligned} \mu x + \lambda y - Ay &= -1/8 B(2xy) \\ \lambda x - \mu y - Ax &= -1/8 B(-x^2 + y^2), \end{aligned} \tag{44}$$

with

$$A = \zeta^2 + 1/32 B^2, \qquad \zeta^2 = x^2 + y^2. \tag{44a}$$

Squaring, and adding (44) we find

$$\lambda^2 + \mu^2 + A^2 - 2\lambda A = 1/64 B^2\zeta^2,$$

which, by using (44a), can give

$$\zeta^4 + \left(\frac{3}{64} B^2 - 2\lambda\right)\zeta^2 + \left(\lambda^2 + \mu^2 + \frac{B^4}{32^2} - \lambda\frac{B^2}{16}\right) = 0, \tag{45}$$

with roots

$$\zeta^2 = \frac{1}{2}\left\{2\lambda - \frac{3}{64}B^2 \pm \left[\left(2\lambda - \frac{3}{64}B^2\right)^2 - 4\left(\lambda^2 + \mu^2 + \frac{B^4}{32^2} - \lambda\frac{B^2}{16}\right)\right]^{1/2}\right\}. \tag{46}$$

The reality of ζ^2 requires

$$I \equiv 7B^4 - 2^8\lambda B^2 + 2^{14}\mu^2 \leqq 0. \tag{47}$$

The roots of $I = 0$ are

$$B^2 = \frac{2^7}{7}[\lambda \pm (\lambda^2 - 7\mu^2)^{1/2}] \tag{48}$$

Then the condition (47) is fulfilled only when the following inequalities are fulfilled:

$$\lambda^2 - 7\mu^2 > 0, \tag{49a}$$

$$\frac{2^7}{7}\{\lambda - (\lambda^2 - 7\mu^2)^{1/2}\} \leqq B^2 \leqq \frac{2^7}{7}\{\lambda + (\lambda^2 - 7\mu^2)^{1/2}\}. \tag{49b}$$

The inequalities (49), by using (42a), can be written as

$$\left(\frac{1 - \bar{c}_1}{\bar{k}}\right)^2 > 7, \tag{50a}$$

$$\frac{2^9}{21}\left\{\frac{1 - \bar{c}_1}{\bar{c}_3} - \left[\left(\frac{1 - \bar{c}_1}{\bar{c}_3}\right)^2 - 7\left(\frac{\bar{k}}{\bar{c}_3}\right)^2\right]^{1/2}\right\} \leqq B^2 \leqq \frac{2^9}{21} \cdot \left\{\frac{1 - \bar{c}_1}{\bar{c}_3} + \left[\left(\frac{1 - \bar{c}_1}{\bar{c}_3}\right)^2 - 7\left(\frac{\bar{k}}{\bar{c}_3}\right)^2\right]^{1/2}\right\}. \tag{50b}$$

The inequalities (50) are restrictions of the values of the coefficients of the differential equation (39) for real amplitude of the subharmonics of order $1/3$.

From (46), if λ and μ are given, we can draw the diagram of ζ^2 versus B^2, taking into account the arc of this diagram which is valid for B^2, according to the restriction (50b).

As for the condition (50a), it can be written in the form:

$$(1 - \bar{c}_1 - \sqrt{7}\bar{k}) \cdot (1 - \bar{c}_1 + \sqrt{7}\bar{k}) > 0, \tag{51}$$

which corresponds to the shaded region of the $\bar{c}_1$, $\bar{k}$-plane (Fig. 2).

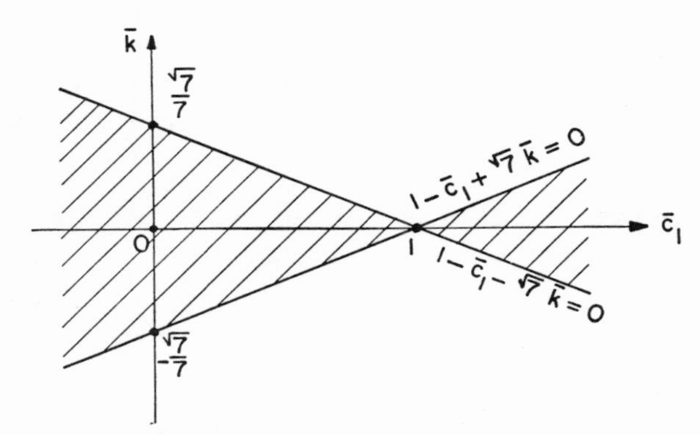

FIG. 2. Graphic representation of the restriction (50a) for stability. Accepted values of $\bar{c}_1$ and $\bar{k}$ must represent points in the shaded region of $\bar{c}_1$, $\bar{k}$-plane.

10. Solution of the System (42)

The real roots of the system (42) may be at most nine. By using the number A, given by (44a), we reduce to the system (44), of which the real roots are acceptable. By regarding the equations (44) as equations in x, whose coefficients are functions of y, we get the vanishing of the eliminant of this system, say the Sylvester's eliminant, which is the condition for common roots of the system. The system (42) can be written as

$$\begin{aligned} (\mu + \tfrac{1}{4}By)x + (\lambda - A)y &= 0, \\ \tfrac{1}{8}Bx^2 - (\lambda - A)x + (\mu - \tfrac{1}{8}By)y &= 0, \end{aligned} \tag{52}$$

and its Sylvester's eliminant is

$$\begin{vmatrix} \mu + \tfrac{1}{4}By & (\lambda - A)y & 0 \\ 0 & \mu + \tfrac{1}{4}By & (\lambda - A)y \\ \tfrac{1}{8}B & -(\lambda - A) & (\mu - \tfrac{1}{8}By)y \end{vmatrix} = 0, \tag{53}$$

which, for nonzero roots y, leads to the cubic:

$$y^3 - 3(4/B)^2[(\lambda - A)^2 + \mu^2]y - 2(4\mu/B)^3 - 2(4/B)^3\mu(\lambda - A)^2 = 0. \tag{54}$$

If we know the coefficients of the differential equation (39), we find the values λ and μ according to (42a). We can then obtain from (46) the amplitude of the subharmonics, which are at most two. On the circumference with radius ζ there are one or three singular points, of which the ordinates y are the real roots of the cubic (54), when for their abscissas x we apply the Pythagorean theorem. Therefore, the singular points (x, y) are at most seven, including the origin, which in every case is a singular point, as we shall see in Section 16.

11. Requirements for Stability of the Solution (x, y)

For the stability of (x, y), according to Section 6, we need the partial derivatives of the functions $A_0(x, y)$ and $C_0(x, y)$, given by (41), with respect to x and y. These derivatives are

$$\begin{aligned}
\partial A_0/\partial x &= a_1 = \tfrac{1}{2}[-k + \tfrac{3}{2}c_3xy - \tfrac{3}{16}c_3By],\\
\partial A_0/\partial y &= a_2 = \tfrac{1}{2}[-c_1 + \tfrac{3}{4}c_3(x^2 + 3y^2 + \tfrac{1}{32}B^2) - \tfrac{3}{16}c_3Bx],\\
\partial C_0/\partial x &= b_1 = \tfrac{1}{2}[c_1 - \tfrac{3}{4}c_3(3x^2 + y^2 + \tfrac{1}{32}B^2) - \tfrac{3}{16}c_3Bx],\\
\partial C_0/\partial y &= b_2 = \tfrac{1}{2}[-k - \tfrac{3}{2}c_3xy + \tfrac{3}{16}c_3By].
\end{aligned} \tag{55}$$

If in the inequality (24), which is the condition for convergence of the solution (5) and then of (40), we take $t_0 = 0$, $\epsilon = 1$, we thus establish the restriction for the $T = \max t$:

$$T < \frac{\zeta}{4M}. \tag{56}$$

Under this restriction, and by the use of (2), the partial derivatives (55) can be written as

$$\begin{aligned}
a_1 &= \frac{1}{2}\bar{c}_3\left\{-\frac{\bar{k}}{\bar{c}_3} + \frac{3}{2}xy - \frac{3}{16}By\right\},\\
a_2 &= \frac{1}{2}\bar{c}_3\left\{-\frac{1-\bar{c}_1}{\bar{c}_3} + \frac{3}{4}\left(x^2 + 3y^2 + \frac{B^2}{32}\right) - \frac{3}{16}Bx\right\},\\
b_1 &= \frac{1}{2}\bar{c}_3\left\{\frac{1-\bar{c}_1}{\bar{c}_3} - \frac{3}{4}\left(3x^2 + y^2 + \frac{B^2}{32}\right) - \frac{3}{16}Bx\right\},\\
b_2 &= \frac{1}{2}\bar{c}_3\left\{-\frac{\bar{k}}{\bar{c}_3} - \frac{3}{2}xy + \frac{3}{16}By\right\}.
\end{aligned} \tag{57}$$

12. Illustrated Example

With the above we have everything we need for treating a numerical example. For the components x and y of the amplitude of the subharmonics of order $\tfrac{1}{3}$ of the differential equation (39), we do not need, as we can see from (42) and (42a), the coefficient $\bar{c}_2$ of the elastic force, since $\bar{c}_3 \neq 0$. Such a thing happens for the components of any order of the subharmonics, except in the case of subharmonics of order $1/2$, as we can see from the equations (38a) and (38b).

Let us take

$$B = 4, \qquad \bar{k} = 3/16, \qquad \bar{c}_1 = 1\tfrac{3}{4}, \qquad \bar{c}_3 = -1/2.$$

For this example we have

$$\frac{1-\bar{c}_1}{\bar{c}_3} = \frac{3}{2}, \qquad \frac{\bar{k}}{\bar{c}_3} = -\frac{3}{8}, \qquad \lambda = 2, \qquad \mu = -\frac{1}{2},$$

and these values, and $B = 4$, are in accordance with the restrictions (50a, b).

We have

$$\zeta_1^2 \cong 2, \qquad \zeta_1 \cong 1.414, \qquad \zeta_2^2 \cong 1.25, \qquad \zeta_2 \cong 1.118,$$

$$A_1 = \zeta_1^2 + 1/2 \cong 2.5, \qquad A_2 = \zeta_2^2 + 1/2 \cong 1.75,$$

$$A_1^2 \cong 6.25, \qquad A_2^2 \cong 3.06.$$

The cubic (54) gives

$$\begin{aligned} y_1^3 - 3/2\,y_1 + 1/2 &= 0, \\ y_2^3 - 15/16\,y_2 + 5/16 &= 0. \end{aligned} \tag{58}$$

These cubics have three real unique roots. The roots of the first are

$$y_{11} = 0.9999, \qquad y_{12} = -1.3649, \qquad y_{13} = 1.3645; \tag{59a}$$

those of the second

$$y_{21} = 0.7032, \qquad y_{22} = -1.1034, \qquad y_{23} = 1.0431. \tag{60a}$$

The corresponding abscissas are

$$x_{11} = 0.9999, \qquad x_{12} = 0.3648, \qquad x_{13} = -0.3648 \tag{59b}$$

$$x_{21} = 0.8672, \qquad x_{22} = 0.1755, \qquad x_{23} = -0.3991 \tag{60b}$$

Thus, the origin and the above points (59a, b) and (60a, b) are the singular points in our numerical example.

The kind of singularity of these points can be computed from the corresponding characteristic roots, according to (36a) and (57), by using the coordinates of the above points. The results of the computation are

TABLE I

THE SINGULAR POINTS AND THEIR CLASSES FOR THE NUMERICAL EXAMPLE OF SECTION 12.

	Point		Class
0	$x = 0$	$y = 0$	Stable spiral point
I	$x_{11} = 0.9999$	$y_{11} = 0.9999$	Stable spiral point
II	$x_{12} = 0.3648$	$y_{12} = -1.3645$	Stable spiral point
III	$x_{13} = -0.3648$	$y_{13} = 1.3645$	Saddle point
IV	$x_{21} = 0.8675$	$y_{21} = 0.7032$	Saddle point
V	$x_{22} = 0.1755$	$y_{22} = -1.1034$	Saddle point
VI	$x_{23} = -0.3991$	$y_{23} = 1.0431$	Saddle point

given in Table I. The singularities I, II, III are on the circumference with radius 1.414; the others, IV, V, VI, on the circumference with radius 1.118.

13. Nonexistence of Poincaré's Cycles

In order to give a sketch of the behavior of the integral curves and the singularities in the x, y-plane of the above example, it is useful to know the existence or nonexistence of Poincaré's cycles. "Bendixson's criterion" can be applied here. Bendixson's criterion states that: "If the expression $\partial A_0/\partial x + \partial C_0/\partial y$ does not change its sign within a domain in the x, y-plane, no Poincaré cycle can exist in the domain." Application of this criterion gives, by using (57),

$$\partial A_0/\partial x + \partial C_0/\partial y = -\bar{k},$$

which is valid in the whole x, y-plane. Then no Poincaré cycles can exist in the whole x, y-plane. Should $\bar{k}$ not equal zero none of the singularities can be a "center." However, we may have centers if $\bar{k} = 0$, when "closed integral curves" appear, which have nothing to do with the Poincaré cycles.

14. Sketch Corresponding to Our Example

By applying the method of isoclives to the equation

$$\frac{dy}{dx} = \frac{C_0(x, y)}{A_0(x, y)}, \tag{61}$$

we can sketch the regions and integral curves of our example of Section 12, as is shown in Fig. 3. The solid lines are the boundaries from the

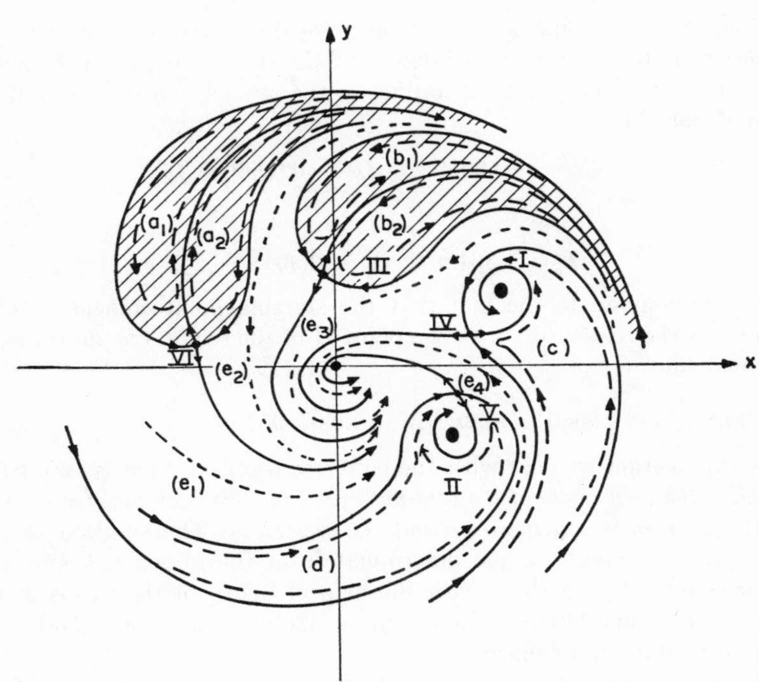

FIG. 3. The regions and the singularities in the x, y-plane. The solid lines through the saddle points III, IV, V, VI are the boundaries of the regions. The shaded regions (a_1), (a_2) and (b_1), (b_2) do not correspond to any solution. The points I, II and the origin are stable spirals and correspond to the regions (c), (d) and (e_1, e_2, e_3, e_4), respectively. The dotted lines are integral curves.

saddle points, the dotted lines the integral curves. The regions (a_1), (a_2), (b_1), (b_2) correspond to no solution of our example. The regions (e_1), (e_2), (e_3), (e_4) correspond to the stable harmonic solution

$$Q = -\tfrac{1}{2} \sin 3t.$$

The region (c) corresponds to the stable solution

$$Q = 0.9999 \sin t - 0.9999 \cos t - \tfrac{1}{2} \sin 3t,$$

and the region (d) to the stable solution

$$Q = 0.3648 \sin t - 1.3645 \cos t - \tfrac{1}{2} \sin 3t.$$

By "corresponds" we mean that if the starting point is a point of a region, the final point of this region gives the stable solution, which is appropriate to the starting point, and all the starting points of the region give the same final point. According to the remarks of Section 8, if the initial conditions are

$$Q_0 = -0.9999, \qquad \dot{Q}_0 = 0.9999 - 3/2,$$

or

$$Q_0 = 1.3645, \qquad \dot{Q}_0 = 0.3648 - 3/2,$$

which correspond to the fact that the starting point coincides to the point I of the region (*c*), or to the point II of the region (*d*), no transient phenomena can exist.

15. The Converse Problem (Magiros, 1957c)

In this section we deal with the converse problem; that is, *We try to establish the coefficients of the original equation (39) such that for the subharmonic of order* $1/3$ *the amplitude is prescribed.* The solution of this problem corresponds to the determination of the numbers λ and μ in the system (42) and then, with the help of (42a), of the ratios of the coefficients. This determination is to be made in terms of B and ζ, by using the additional equation

$$x^2 + y^2 - \zeta^2 = 0. \tag{62}$$

The determination of λ and μ in this way is impossible because of the form of Eqs. (42). From Eqs. (42) and (62) we can obtain λ and μ in terms of x, y, ζ, and B. Then the determination of λ and μ requires us to know B and two of x, y, ζ. But, if we know B and x, y, the numbers λ and μ are known by solving the system (42) in λ and μ, (62) being a restriction between x, y, ζ. Now, by knowing λ and μ, we know the ratios $(1 - \bar{c}_1)/\bar{c}_3$, $\bar{k}/\bar{c}_3$. Thus any two coefficients from $\bar{c}_1$, $\bar{c}_3$, $\bar{k}$ can be determined in terms of a third one, which can get arbitrary values. Since the stability can be treated as in the previous example, the converse problem is finished.

To illustrate the above, let us take two cases with the same amplitude: (a) $x = 1, y = 1$ and (b) $x = 0.3648, y = -1.3645$.

(a) $x = 1, y = 1, B = 4$. According to (40) the corresponding solution is

$$Q = \sin t - \cos t - 1/2 \sin 3t. \tag{63}$$

From (42) and (42a) we find that

$$\frac{1-\bar{c}_1}{\bar{c}_3} = -\frac{3}{8}, \qquad \frac{\bar{k}}{\bar{c}_3} = \frac{3}{2}. \tag{63a}$$

The test for the stability of the singularity (1, 1) shows that this point is a saddle point. Thus the solution (63) of Eq. (39), when the coefficients satisfy the relations (63a), is not acceptable.

(b) $x = 0.3648$, $y = -1.3645$, $B = 4$. The corresponding ratios of the coefficients are

$$\frac{1-\bar{c}_1}{c_3} = \frac{3}{2}, \qquad \frac{\bar{k}}{\bar{c}_3} = -\frac{3}{8}, \tag{64a}$$

and the solution is

$$Q = 0.3648 \sin t + 1.3645 \cos t - \tfrac{1}{2} \sin 3t. \tag{64}$$

This is acceptable, since the singularity $x = 0.3648$, $y = -1.3645$ is a stable spiral point.

16. The Singularity at the Origin in the General Case (Magiros, 1957c)

From the system (42) we see that the origin $x = 0$, $y = 0$ is, for any case of the coefficients of (39), a singularity of (39). In other words, the function $-\frac{1}{8}B \sin 3t$ is always a solution of (39), the harmonic solution, either stable or unstable. For the stability of this harmonic solution we have, first, from (55):

$$\begin{aligned} a_1 &= -\frac{1}{2}\bar{k} \\ a_2 &= -\frac{1}{2}(1-\bar{c}_1) + \frac{3}{16^2}\bar{c}_3 B^2, \\ b_1 &= \frac{1}{2}(1-\bar{c}_1) - \frac{3}{16^2}\bar{c}_3 B^2 \\ b_2 &= -\tfrac{1}{2}\bar{k}, \end{aligned} \tag{65}$$

when, from (36a) and (65), we obtain

$$\rho_{1,2} = -\frac{1}{2}\bar{k} \pm i\left[\frac{1}{2}(1-\bar{c}_1) - 3\bar{c}_3\frac{B^2}{16^2}\right]. \tag{66}$$

By using the definitions of the singularities of Section 6 we get the following:

(a) If the imaginary part of (66) is not zero, that is, if

$$\frac{1 - \bar{c}_1}{\bar{c}_3} \neq 6 \cdot \frac{B^2}{16^2}, \tag{67a}$$

then (i) for $\bar{k} > 0$, the origin is a stable spiral point; (ii) for $\bar{k} < 0$, the origin is an unstable spiral point; (iii) for $\bar{k} = 0$, the origin is either a center or a spiral.

(b) If the imaginary part of (66) is zero, that is, if

$$\frac{1 - \bar{c}_1}{\bar{c}_3} = 6 \cdot \frac{B^2}{16^2}, \tag{67b}$$

then (i) for $\bar{k} > 0$, the origin is a nodal stable point; (ii) for $\bar{k} < 0$, the origin is an unstable nodal point; (iii) for $\bar{k} = 0$, the origin is not a simple singularity of the kind we know in Section 6, since the characteristic roots are zero.

The result of this section is that *the origin is a stable singularity when* $\bar{k} > 0$, *and of spiral type under the condition* (67*a*), *or of nodal type under the condition* (67b). This is shown in Fig. 4. In Fig. 4, the left-hand of

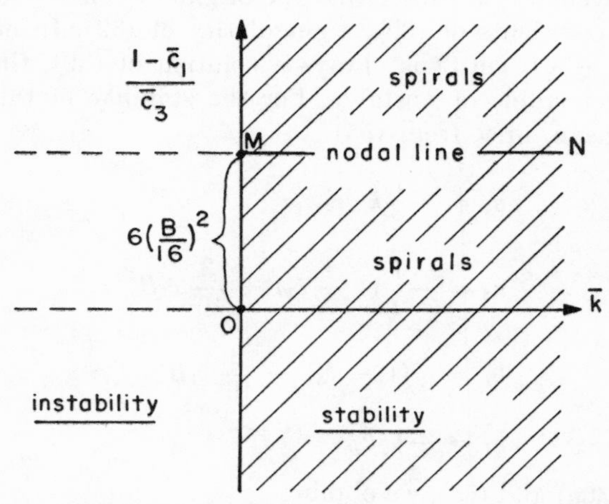

FIG. 4. The kinds of singularity of the origin of Eq. (39). For $\bar{k} > 0$ the origin is stable and either of spiral or of nodal kind. For $\bar{k} < 0$ the origin is an unstable singularity.

the $(1 - \bar{c}_1)/\bar{c}_3$, $\bar{k}$-plane corresponds to instability of the harmonic solution. The right-hand of this plane corresponds to stability, with all spiral points, except the points of the nodal line MN, of which all points are nodal points, except the point M which is a singularity but not of simple type. The points of the $[(1 - \bar{c}_1)/\bar{c}_3]$-axis may be either centers or spirals, except M. The distance OM has a maximum and a minimum due to the restriction of B, given by (50b). For $B = 0$, the $\bar{k}$-axis is the nodal line.

APPENDIX

The following are concerned with the singular points, that is, the constant solutions, of the system (14). The functions F_1 and F_2 of (14) satisfy the conditions of Fourier's theorem (17), so:

$$\begin{aligned} \dot{u}_1 &= \epsilon(A_0 + A_1 \cos t + B_1 \sin t + \cdots \\ &\qquad + A_m \cos mt + B_m \sin mt + \cdots) \\ \dot{u}_2 &= \epsilon(C_0 + C_1 \cos t + D_1 \sin t + \cdots \\ &\qquad + C_m \cos mt + D_m \sin mt + \cdots) \end{aligned} \tag{68}$$

where the coefficients A, B, C, D depend on u_1 and u_2.

Take the system:

$$\begin{aligned} \dot{u}_1 &= \epsilon A_0(u_1, u_2), \\ \dot{u}_2 &= \epsilon C_0(u_1, u_2), \end{aligned} \tag{68a}$$

where the time t does not enter explicitly. Each of the above systems accepts a unique solution $\{u_1, u_2\}$ which assumes given values

$$\{u_1(t_0) = u_{10}, u_2(t_0) = u_{20}\}$$

at $t = t_0$, provided that the functions in their right-hand sides satisfy a Lipschitz condition, when their arguments are restricted to lie in the domain

$$\Delta: \quad |u_1 - u_{10}| \leqq k_1, \quad |u_2 - u_{20}| \leqq k_2, \qquad |t - t_0| < T. \tag{69}$$

Apply Picard's method of successive approximations for the calculation of the solution $\{\bar{u}_1, \bar{u}_2\}$ of the system (68a), by taking as zeroth approximation arbitrary initial conditions $\{\bar{u}_{10}, \bar{u}_{20}\}$. The successive approximations, which converge to the unique solution $\{\bar{u}_1, \bar{u}_2\}$ of (68a), are

$$\bar{u}_1^{(1)} = \bar{u}_{10} + \epsilon \int_{t_0}^{t} A_0(\bar{u}_{10}, \bar{u}_{20})\, dt, \qquad \bar{u}_2^{(1)} = \bar{u}_{20} + \epsilon \int_{t_0}^{t} C_0(\bar{u}_{10}, \bar{u}_{20})\, dt,$$

$$\cdots\cdots\cdots\cdots\cdots\cdots\cdots\cdots \qquad (70a)$$

$$\bar{u}_1^{(n)} = \bar{u}_{10} + \epsilon \int_{t_0}^{t} A_0(\bar{u}_1^{(n-1)}, \bar{u}_2^{(n-1)})\, dt, \qquad \bar{u}_2^{(n)} = \bar{u}_{20} + \epsilon \int_{t_0}^{t} C_0(\bar{u}_1^{(n-1)}, \bar{u}_2^{(n-1)})\, dt.$$

Apply also the above method for the calculation of the solution $\{u_1, u_2\}$ of the system (68), by taking as zeroth solution arbitrary initial conditions $\{u_1(t_0) = u_{10}, u_2(t_0) = u_{20}\}$. The successive approximations are

$$\begin{aligned}
u_1^{(1)} &= u_{10} + \epsilon \int_{t_0}^{t} A_0(u_{10}, u_{20})\, dt \\
&\qquad + \epsilon \int_{t_0}^{t} A_1(u_{10}, u_{20}) \cos t\, dt + \cdots \\
u_2^{(1)} &= u_{20} + \epsilon \int_{t_0}^{t} C_0(u_{10}, u_{20})\, dt \\
&\qquad + \epsilon \int_{t_0}^{t} C_1(u_{10}, u_{20}) \cos t\, dt + \cdots \\
u_1^{(n)} &= u_{10} + \epsilon \int_{t_0}^{t} A_0(u_1^{(n-1)}, u_2^{(n-1)})\, dt \\
&\qquad + \epsilon \int_{t_0}^{t} A_1(u_1^{(n-1)}, u_2^{(n-1)}) \cos t\, dt + \cdots \\
u_2^{(n)} &= u_{20} + \epsilon \int_{t_0}^{t} C_0(u_1^{(n-1)}, u_2^{(n-1)})\, dt \\
&\qquad + \epsilon \int_{t_0}^{t} C_1(u_1^{(n-1)}, u_2^{(n-1)}) \cos dt + \cdots
\end{aligned} \qquad (70)$$

which converge to the unique solution $\{u_1, u_2\}$ of the system (68).

Let us take the same initial conditions

$$u_{10} = \bar{u}_{10}, \qquad u_{20} = \bar{u}_{20}, \tag{71}$$

of the above approximation (70) and (70a), and subtract. We get

$$\begin{aligned} u_1^{(n)} - \bar{u}_1^{(n)} &= \epsilon \int_{t_0}^{t} \{A_0(u_1^{(n-1)}, u_2^{(n-1)}) - A_0(\bar{u}_1^{(n-1)}, \bar{u}_2^{(n-1)})\}\, dt \\ &\quad + \epsilon \int_{t_0}^{t} A_1(u^{(n-1)}, u_2^{(n-1)}) \cos t\, dt \\ &\quad + \epsilon \int_{t_0}^{t} B_1(u_1^{(n-1)}, u_2^{(n-1)}) \sin t\, dt + \cdots \\ u_2^{(n)} - \bar{u}_2^{(n)} &= \epsilon \int_{t_0}^{t} \{C_0(u_1^{(n-1)}, u_2^{(n-1)}) - C_0(\bar{u}_1^{(n-1)}, \bar{u}_2^{(n-1)})\}\, dt \\ &\quad + \epsilon \int_{t_0}^{t} C_1(u_1^{(n-1)}, u_2^{(n-1)}) \cos t\, dt \\ &\quad + \epsilon \int_{t_0}^{t} D_1(u_1^{(n-1)}, u_2^{(n-1)}) \cos t\, dt + \cdots \end{aligned} \tag{72}$$

The integrals in (72) are bounded and the right-hand sides contain ϵ or a common factor, so the approximations, and consequently their limits, are, for small ϵ, of order ϵ; that is,

$$| u_1 - \bar{u}_1 | = O(\epsilon), \qquad | u_2 - \bar{u}_2 | = O(\epsilon). \tag{73}$$

In (73) $\{u_1, u_2\}$ and $\{\bar{u}_1, \bar{u}_2\}$ are solutions of (68) and (68a), respectively. Thus any solution of the system (68a) can be considered as an approximation of the solution of the system (68) of the first order ϵ.

The solutions of the system (14), i.e., of the system (68), are constant if ϵF_1 and ϵF_2 tend to zero, or, since the time t enters explicitly in F_1 and F_2, if $\epsilon \to 0$. But in the system (68a) the time t does not enter explicitly, so we can get constant solutions of (68a), if ϵ is not necessarily zero, by taking appropriate initial conditions in the approximations (70a), namely, the initial conditions $\{\bar{u}_{10}, \bar{u}_{20}\}$ which satisfy the conditions

$$A_0(\bar{u}_{10}, \bar{u}_{20}) = 0, \qquad C_0(\bar{u}_{10}, \bar{u}_{20}) = 0, \tag{74}$$

when the integrals in (70a) are zero and the solution $\{\bar{u}_1, \bar{u}_2\}$ of (68a) is the constant $\{\bar{u}_{10}, \bar{u}_{20}\}$ for any ϵ, including $\epsilon = 0$. A constant solution

$\{\bar{u}_1, \bar{u}_2\}$ of the approximate systems (68a), which satisfies the condition (74), is considered as an approximate constant solution $\{u_1, u_2\}$ of the exact system (68), or (14), of the first order in ϵ.

The technique of substituting the system (68), where the time t enters explicitly, by the approximate system (68a), where the time does not enter explicitly, consists essentially of replacing a function by its "mean value" over an interval (Fatou, 1929), which is called "moving average" or "sliding mean" (Van der Pol and Bremmer, 1955). Since the moving average is generally smoother than the original function, the above technique, which is known as the "averaging principle" (Kryloff and Bogoliuboff, 1947), offers advantages in the study of the original system. In practice, this technique brings good results, for example, in economics or in electrical problems such as the photoelectric reproduction of sound, television images (Van der Pol and Bremmer, 1955).

RECEIVED: November 27, 1957.

REFERENCES

ANDRONOW, A., AND WITT, A. (1930). Zur Theorie des Mitnehmens von B. van der Pol. *Arch. Elektrotechn.* **24,** 99.

BENDIXSON, I. (1901). Sur les courbes définies par les équations différentielles. *Acta Math.* **24,** 1–88.

FATOU, P. (1929). Sur le mouvement d'un système soumis à des forces à courte période. *Bull. Soc. math. France* pp. 98–139.

FRIEDRICHS, C. (1953). Fundamentals of Poincaré's theory. "Proceedings of the Symposium on Nonlinear Circuit Analysis," Vol. II, pp. 56–67. Polytechnic Institute of Brooklyn, New York.

HADAMARD, J. (1933). The latest work of Poincaré. *Rice Inst. Pamphlet* **20**(1), 9–28.

HAYACHI, C. (1953). "Forced Oscillations in Nonlinear Systems," p. 23. Nippon Printing and Publishing Company, Osaka, Japan.

KRYLOFF, N. AND BOGOLIUBOFF, N. (1947). Introduction of nonlinear mechanics. *Ann. Math. Studies* **11,** 12.

MAGIROS, D. (1957a). "Subharmonics of any order in case of nonlinear restoring force." *Prakt. Athens Acad. Sci.* **32,** 77–85.

MAGIROS, D. (1957b). "Subharmonics of order one third in case of cubic restoring force." *Prakt. Athens Acad. Sci.* **32,** 101–108.

MAGIROS, D. (1957c). "Remarks on a problem of subharmonics; its inverse problem." *Prakt. Athens Acad. Sci.* **32,** 143–146.

MAGIROS, D. (1957d). "On the singularities of a system of differential equations where the time enters explicitly." *Prakt. Athens Acad. Sci.* **32,** 448–451.

MANDELSTAM, L. AND PAPALEXI, N. (1935). Über einige nichstationäre schwingungsvorgänge." *Tech. Phys. U.S.S.R.* **1,** 415–428.

PETROVSKY, I. (1954). Partial Differential Equations, p. 21. Interscience, New York.

POINCARÉ, H. (1890). Sur le problème des trois corps et les équations différentielles. *Acta Math.* **13,** 88.

POINCARÉ, H. (1892). Sur les courbes définies par une équation différentielle, Vol. 1 of "Oeuvres." Gauthier-Villars, Paris,

STOKER, J. J. (1955). "On the stability on mechanical systems," *Communs. Pure and Appl. Math.* **7,** 132–142.

TSIEN, H. (1956). The "Poincaré-Lighthill-Kuo" method. *Advances in Appl. Mech.* **4,** 281–349.

VAN DER POL, B. (1927). Forced oscillations in a circuit with non-linear resistance. *Phil. Mag.* [7] **3,** 65–80.

VAN DER POL, B. AND BREMMER, H. (1955). "Operational Calculus," 2nd ed., Chapter XIV, §2. Cambridge Univ. Press, London and New York.

VON KÁRMÁN, T. (1940). The engineer grapples with nonlinear problems. *Bull. Am. Math. Soc.* **46,** 615–683.

WHITTAKER, E. T., AND WATSON, G. N. (1952). "A course of Modern Analysis," 4th ed., p. 164. MacMillan, New York.

On a Problem of Nonlinear Mechanics

DEMETRIOS G. MAGIROS

Republic Aviation Corporation, Farmingdale, New York

The behavior of a forced oscillatory system which is linearly damped but nonlinear in the restoring force is investigated according to the author's previous papers (Magiros, 1957, 1958). It is shown under which conditions the system may contain subharmonics of order $\frac{1}{2}$. The amplitudes of the subharmonics and their components, and the bounds for the amplitude of the external force, are given in terms of the coefficients of the differential equation of the system, which are not necessarily very small, as well as the regions in the $(c_1/c_3, I)$-plane, where we have subharmonics with two, one, or neither amplitudes. Also discussed are the stability of the subharmonics, the free vibrations of the system, and the case when one of the coefficients of the nonlinear terms is zero.

INTRODUCTION

The investigation of the oscillations of a nonlinear system with frequency half of that of the external force—subharmonics of order $\frac{1}{2}$—is of current interest (Mandelstam, 1935; Melikjan, 1935; Reuter, 1949). A physical example, to mention just one, is the loudspeaker. A sinusoidal current in the coil causes the loudspeaker diaphragm to vibrate axially about a central position. These vibrations may, under certain circumstances, be with frequency half of that of the driving current.

In this paper we study the subharmonics of order $\frac{1}{2}$ when the system is linearly damped and the nonlinear restoring force of cubic type, that is when the system is governed by a differential equation of the form (1), where the coefficients are not necessarily very small. The amplitudes of the subharmonics and their components are found in terms of the coefficients of the differential equation. The regions in the $(c_1/c_3, I)$-plane, where we have subharmonics with two, one, or neither amplitudes, are also discussed. I is a polynomial in the coefficients of the differential equation. Bounds for the amplitude of the external force are given. The stability of the subharmonics, the free vibrations of the system, and the

Reprinted from *Information and Control* **2** (1959), 297–309.

case where one of the coefficients of the nonlinear terms is zero are also treated.

I. THE AMPLITUDE OF THE SUBHARMONICS

We ask for oscillations with frequency half of that of the excitation of the differential equation

$$\ddot{Q} + \bar{k}\dot{Q} + \bar{c}_1 Q + \bar{c}_2 Q^2 + \bar{c}_3 Q^3 = B \sin 2t. \tag{1}$$

By using

$$\bar{k} = \epsilon k, \qquad 1 - \bar{c}_1 = \epsilon c_1, \qquad \bar{c}_2 = \epsilon c_2, \qquad \bar{c}_3 = \epsilon c_3, \qquad \epsilon > 0, \tag{2}$$

Eq. (1) can be written as

$$\ddot{Q} + Q = \epsilon f(Q,\dot{Q}) + B \sin 2t, \tag{3}$$

the solution of which, in case $\epsilon = 0$, is given by

$$Q = x \sin t - y \cos t - \frac{B}{3} \sin 2t, \tag{4}$$

where x and y are arbitrary constants which depend on the initial conditions of the differential equation. x and y are the components of the amplitude r of the subharmonics. If $\epsilon \neq 0$, (4) also can give the solution of (3) or (1), but x and y are appropriate functions of t, say $u_1(t)$ and $u_2(t)$ respectively, which, as $\epsilon \to 0$, must have as limits x and y.

By applying Eqs. (38a) and (38b) of Magiros (1958), we can find that x and y satisfy the equations

$$\begin{aligned} A_0(x,y) &\equiv \tfrac{1}{2}[\tfrac{1}{3}c_2B - k]x + \tfrac{1}{2}[\tfrac{1}{6}c_3B^2 - c_1]y + \tfrac{3}{8}c_3 y(x^2 + y^2) = 0, \\ C_0(x,y) &= \tfrac{1}{2}[c_1 - \tfrac{1}{6}c_3B^2]x - \tfrac{1}{2}[k + \tfrac{1}{3}c_2B]y - \tfrac{3}{8}c_3 x(x^2 + y^2) = 0. \end{aligned} \tag{5}$$

For $c_3 \neq 0$, Eqs. (5) can be written as

$$(\mu - \tfrac{1}{3}\nu B)x + (\lambda - A)y = 0, \qquad (\lambda - A)x - (\mu + \tfrac{1}{3}\nu B)y = 0 \tag{6}$$

with

$$\begin{aligned} \lambda &= \frac{4}{3}\frac{c_1}{c_3} = \frac{4}{3}\frac{1 - \bar{c}_1}{\bar{c}_3}, \qquad \mu = \frac{4}{3}\frac{k}{c_3} = \frac{4}{3}\frac{\bar{k}}{\bar{c}_3}, \\ \nu &= \frac{4}{3}\frac{c_2}{c_3} = \frac{4}{3}\frac{\bar{c}_2}{\bar{c}_3}, \qquad A = r^2 + \frac{2}{9}B^2, r^2 = x^2 + y^2. \end{aligned} \tag{6a}$$

The condition for nonzero amplitude r of the subharmonics, that is for nonzero roots x and y of the system (6), is the vanishing of the determinant of the coefficients of the system (6), which gives

$$r_{1,2}^2 = \lambda - \tfrac{2}{9}B^2 \mp \sqrt{\tfrac{1}{9}\nu^2 B^2 - \mu^2}. \tag{7}$$

II. THE REALITY OF THE AMPLITUDE r

The reality of r, as given by (7), implies the following conditions to be satisfied: (a) the expression under the radical must be nonnegative and, (b) the right-hand side of (7) must be positive.

The condition (a) gives the restrictions

$$| B | \geqq 3 | \mu/\nu | = 3 | k/c_2 |, \tag{8a}$$

$$(B - 3k/c_2)(B + 3k/c_2) \geqq 0. \tag{8b}$$

The restriction (8a) gives bounds for B, and the shaded regions in Fig. 1 are the permissible ones, according to (8b).

The condition (b) gives the following restrictions:

(A) The existence of nonzero r_1 implies that

$$\sqrt{\tfrac{1}{9}\nu^2 B^2 - \mu^2} < \lambda - \tfrac{2}{9}B^2, \tag{9}$$

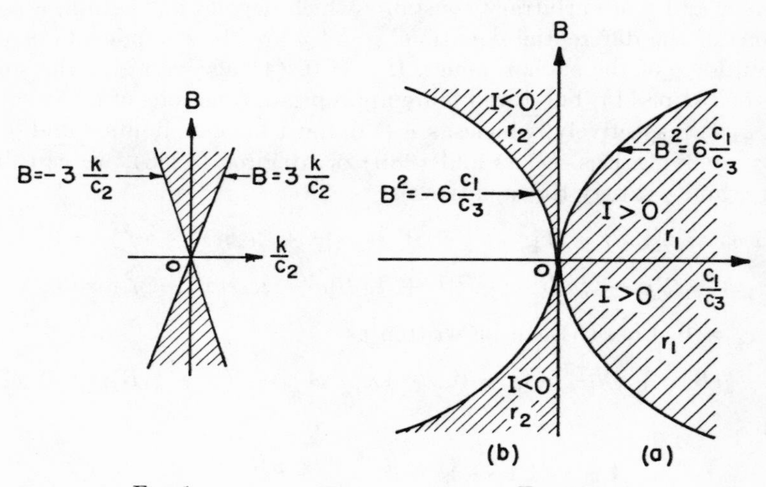

FIG. 1. FIG. 2.

FIG. 1. The shaded regions in the $(k/c_2, B)$-plane are the appropriate ones for the reality of the amplitude r of the subharmonics.

FIG. 2. The shaded region (a) in the $(c_1/c_3, B)$-plane, for $I > 0$, corresponds to the existence of the amplitude r_1 of the subharmonics; the shaded regions (b), for $I < 0$, to the existence of the amplitude r_2.

then

$$\lambda - \tfrac{2}{9}B^2 > 0 \quad \text{and} \quad \lambda > 0,$$

or

$$B^2 < \frac{6c_1}{c_3}, \qquad \frac{c_1}{c_3} > 0 \tag{10}$$

Squaring (9) we can get $I > 0$ if

$$I = \tfrac{4}{81}B^4 - \tfrac{1}{9}(\nu^2 + 4\lambda)B^2 + \lambda^2 + \mu^2, \tag{11}$$

and r_1 exists only for points of the shaded region Fig. 2(a) with $I > 0$.

(B) The existence of nonzero r_2 implies that

$$\sqrt{\tfrac{1}{9}\nu^2B^2 - \mu^2} > \tfrac{2}{9}B^2 - \lambda \tag{12}$$

and we have three cases for the existence of r_2 :

(i) For $\lambda < 0$, we have $\frac{2}{9}B^2 - \lambda > 0$, and squaring (12) we get $I < 0$. Then r_2 always exists for points of the shaded region in Fig. 2(b) with $I < 0$.

(ii) For $\lambda > 0$, if $| \frac{2}{9}B^2 - \lambda |^2 < \frac{1}{9}\nu^2B^2 - \mu^2$, we have $I < 0$.

(iii) For $\lambda > 0$, if $| \frac{2}{9}B^2 - \lambda |^2 > \frac{1}{9}\nu^2B^2 - \mu^2$, we have $I > 0$.

The above results can be summarized in Fig. 3.

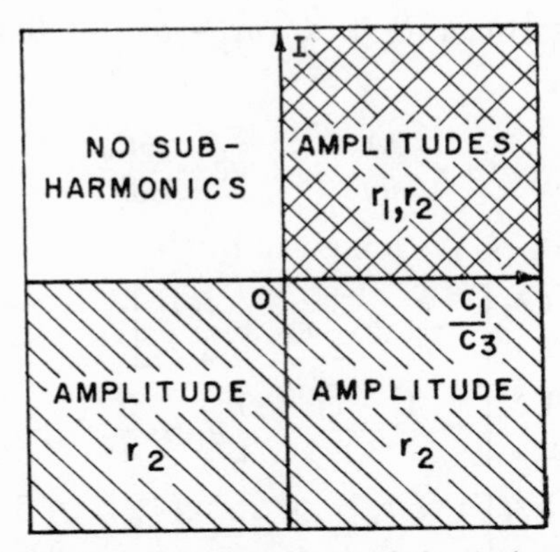

FIG. 3. Only points of the first quadrant of the $(c_1/c_3, I)$-plane can give subharmonics with two different nonzero amplitudes r_1, r_2. Points of the third and fourth quadrant give subharmonics with one amplitude r_2, and points of the second quadrant correspond to *no* subharmonics.

III. THE COMPONENTS OF THE AMPLITUDES r

The system (6) gives

$$\frac{y}{x} = \frac{\mu - \frac{1}{3}\nu B}{A - \lambda} = \frac{\lambda - A}{\mu + \frac{1}{3}\nu B} = \theta \tag{13}$$

when the components x and y are given as solutions of the system

$$y = \theta x, \qquad x^2 + y^2 = r^2 \tag{14}$$

Then

$$\begin{aligned} x_1 &= \frac{r}{\sqrt{1+\theta^2}}, & y_1 &= \frac{r\theta}{\sqrt{1+\theta^2}}, \\ x_2 &= -\frac{r}{\sqrt{1+\theta^2}}, & y_2 &= -\frac{r\theta}{\sqrt{1+\theta^2}}. \end{aligned} \tag{15}$$

These solutions correspond to two symmetric points with respect to the origin in the x,y-plane. Since we have two, in general, values of r, and θ depends on r, (15) give four points, then the singularities of Eq. (1) are, in general, five, the origin included.

IV. THE STABILITY OF THE SUBHARMONICS

The coordinates (x,y) of the five singular points of Eq. (1) are given by the solutions of the system: $A_0(x,y) = 0$, $C_0(x,y) = 0$; then the origin and the points (15) are the singular points. Their kind depends on the numbers

$$\rho_{1,2} = \tfrac{1}{2}\{a_1 + b_2 \mp \sqrt{(a_1 - b_2)^2 + 4a_2b_1}\}, \tag{16}$$

where

$$a_1 = \frac{\partial A_0}{\partial x}, \qquad a_2 = \frac{\partial A_0}{\partial y}, \qquad b_1 = \frac{\partial C_0}{\partial x}, \qquad b_2 = \frac{\partial C_0}{\partial y}.$$

For $c_3 \neq 0$, the a_1, a_2, b_1, b_2 are given by

$$\begin{aligned} a_1 &= \frac{1}{2}\frac{\bar{c}_3}{\epsilon}\left[-\frac{\bar{k}}{\bar{c}_3} + \frac{3}{2}xy + \frac{B}{3}\frac{\bar{c}_2}{\bar{c}_3}\right], \\ a_2 &= \frac{1}{2}\frac{\bar{c}_3}{\epsilon}\left[-\frac{1-\bar{c}_1}{\bar{c}_3} + \frac{9}{4}y^2 + \frac{3}{4}x^2 + \frac{B^2}{6}\right], \\ b_1 &= \frac{1}{2}\frac{\bar{c}_3}{\epsilon}\left[\frac{1-\bar{c}_1}{\bar{c}_3} - \frac{9}{4}x^2 - \frac{3}{4}y^2 - \frac{B^2}{6}\right], \\ b_2 &= \frac{1}{2}\frac{\bar{c}_3}{\epsilon}\left[-\frac{\bar{k}}{\bar{c}_3} - \frac{3}{2}xy - \frac{B}{3}\frac{\bar{c}_2}{\bar{c}_3}\right]. \end{aligned} \tag{17}$$

The tangential direction of the integral curves at the neighborhood of a nodal or a saddle singular point is characterized by the numbers

$$\mu_{1,2} = \frac{1}{2a_2}\left[b_2 - a_1 \mp \sqrt{(a_1 - b_2)^2 + 4a_2 b_1}\right]. \tag{18}$$

For a "stable singular point," a point on the integral curve near the singular point, tends, with the increase of time, to that singular point; otherwise the singular point is "unstable." For a "stable nodal point," the values of μ_1, μ_2 must be real and of different signs. For a "spiral point," when $\rho_{1,2}$, $\mu_{1,2}$ are complex numbers, the negative real part of ρ gives information about its "stability." The integral curves around a "stable spiral point" are of the "clockwise direction," if $a_2 > 0$ (Hayashi, 1953). Let us illustrate the above by the following example.

V. NUMERICAL EXAMPLE

Take $B = 1$, $\bar{k} = 0.1$, $\bar{c}_1 = 1\frac{3}{4}$, $\bar{c}_2 = \frac{3}{8}$, $\bar{c}_3 = -\frac{1}{2}$. Notice first that $c_1/c_3 = (1 - \bar{c}_1)/\bar{c}_3 = \frac{3}{2}$, $I = 3.1205$. Then, according to Fig. 3, we expect two subharmonics with different amplitudes. The result of the computation is shown in Table II below.

VI. FREE NONLINEAR VIBRATIONS

The previous forced vibrations are dependent on, or governed by, the characteristics of the system and the external force. For the free nonlinear vibrations we must put in (1) $B = 0$. In this case the amplitude of the free vibrations of the system is given by:

$$r_{1,2}^2 = \lambda \mp \sqrt{-\mu^2} \tag{19}$$

The reality of r implies $\mu = 0$, $\lambda > 0$, that is $\bar{k} = 0$, $c_1/c_3 = (1 - \bar{c}_1)/\bar{c}_3 > 0$; then the free vibrations of (1) occur in the system when we have either

$$\begin{array}{llll} \text{(i)} & k = 0, & c_1 > 0, & c_3 > 0, \quad \text{or} \\ \text{(ii)} & k = 0, & c_1 < 0, & c_3 < 0. \end{array} \tag{20}$$

TABLE I

$\lambda = 2$	$\mu = -\frac{4}{15}$	$\nu = -1$	
$r_1^2 = 1.57778$	$r_2^2 = 1.97778$	$r_1 = 1.25610$	$r_2 = 1.40634$
$\theta_1 = 0.3333$	$\theta_1^2 = 0.1111$	$\sqrt{1 + \theta_1^2} = 1.05409$	
$\theta_2 = 0.3333$	$\theta_2^2 = 0.1111$	$\sqrt{1 + \theta_2^2} = 1.05409$	

TABLE II
THE SINGULAR POINTS AND THEIR CLASSES FOR THE EXAMPLE IN SECTION V

Point			
0*	$x_0 = 0$	$y_0 = 0$	$\rho_{1,2} = \frac{1}{4\epsilon}(0.1 + i\,0.65484)$
$I^\dagger$	$x_1 = 1.19164$	$y_1 = -0.39721$	$\rho_{1,2} = \frac{1}{4\epsilon}(-0.1 + i\,1.82791)$
$II^\dagger$	$x_2 = -1.19164$	$y_2 = 0.39721$	$\rho_{1,2} = \frac{1}{4\epsilon}(-0.1 + i\,1.82791)$
$III^\dagger$	$x_3 = 1.33417$	$y_3 = 0.44471$	$\rho_{1,2} = \frac{1}{4\epsilon}(-0.1 + i\,2.40058)$
$IV^\dagger$	$x_4 = -1.33417$	$y_4 = -0.44472$	$\rho_{1,2} = \frac{1}{4\epsilon}(-0.1 - i\,2.40058)$

* Unstable spiral with clockwise direction.
† Stable spiral with counterclockwise direction.

or, in terms of the coefficients of (1),

$$\begin{aligned} &\text{(i)} && \bar{k} = 0, \quad 1 > \bar{c}_1, \quad \bar{c}_3 > 0. \\ &\text{(ii)} && \bar{k} = 0, \quad 1 < \bar{c}_1, \quad \bar{c}_3 < 0. \end{aligned} \tag{21}$$

In Fig. 4 we have the inequalities (21) graphically. The amplitude r of the free vibrations becomes

$$r = 2\sqrt{(1 - \bar{c}_1)/3\bar{c}_3}\,. \tag{22}$$

From (15) we see that θ is of undetermined form; then the components x and y of the amplitude r are also undetermined, but such that

$$x^2 + y^2 = \frac{4}{3}\,\frac{1 - \bar{c}_1}{\bar{c}_3}\,. \tag{23}$$

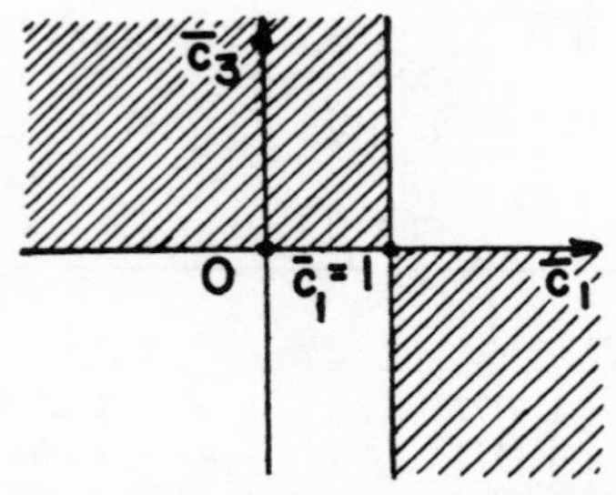

FIG. 4. The shaded regions in $\bar{c}_1$, $\bar{c}_2$-plane correspond to free undamped nonlinear vibrations.

The solution (4) in this case is

$$Q = x \sin t - y \cos t. \tag{24}$$

If the initial conditions are, for $t = 0$, Q_0 and $\dot{Q}_0$, then

$$Q_0 = -y, \qquad \dot{Q}_0 = x, \tag{25}$$

the solution becomes

$$Q = \dot{Q}_0 \sin t + Q_0 \cos t, \tag{26}$$

the restriction of the initial conditions Q_0, $\dot{Q}_0$ in the $(Q_0, \dot{Q}_0)$-plane being of the cyclic type:

$$Q_0^2 + \dot{Q}_0^2 = \frac{4}{3} \frac{1 - \bar{c}_1}{\bar{c}_3}. \tag{27}$$

The results from the above are the following:

(a) the necessary condition for the existence of free oscillations is the system to be undamped, $\bar{k} = 0$.

(b) the free oscillations do not depend on the coefficient $\bar{c}_2$;

(c) the amplitude r of the free vibrations is given by (22), where the coefficients $\bar{c}_1$ and $\bar{c}_3$ must be such that r is positive,

(d) the components x and y of the amplitude r may be arbitrary but they must obey the restriction (23).

(e) the initial conditions Q, $\dot{Q}_0$ must be of the cyclic restriction (27); then if one of them is prescribed the other must have a value given by (27).

VII. DUFFING'S EQUATION

In case $\bar{c}_2 = 0$, we have the Duffing's equation (Stoker, 1950). In this case the formula (7) gives

$$r^2 = \frac{4}{3} \frac{1 - \bar{c}_1}{\bar{c}_3} - \frac{2}{9} B^2 \mp \sqrt{-(4\bar{k}/3\bar{c}_3)^2} \tag{28}$$

For the existence of the subharmonics we must have

$$\bar{k} = 0, B^2 < 6 \frac{1 - \bar{c}_1}{\bar{c}_3} \tag{29}$$

which, for positive values of $(1 - \bar{c}_1)/\bar{c}_3$, that is for values of $\bar{c}_1$, $\bar{c}_2$ in the shaded region of Fig. 4, gives bounds for B:

$$|B| < \left(6 \frac{1 - \bar{c}_1}{\bar{c}_3}\right)^{1/2}. \tag{30}$$

The subharmonics have the single amplitude

$$r = \left(\frac{4}{3}\frac{1-\bar{c}_1}{\bar{c}_3} - \frac{2}{9}B^2\right)^{1/2}. \tag{31}$$

VIII. THE CASE $\bar{c}_2 \neq O$

For the nonlinearity of the system either one or both the coefficients $\bar{c}_2$ and $\bar{c}_3$ must be nonzero. In the foregoing we took $\bar{c}_3 \neq 0$, when $\bar{c}_2$ might be negligible or zero. Let us consider now the case $\bar{c}_2 \neq 0$ when $\bar{c}_3$ might be negligible or zero.

If

$$\begin{gathered} \frac{c_1}{c_2} = \frac{1-\bar{c}_1}{\bar{c}_2} = l, \qquad \frac{k}{c_2} = \frac{\bar{k}}{\bar{c}_2} = m, \qquad \frac{c_3}{c_2} = \frac{\bar{c}_3}{\bar{c}_2} = n, \\ A = r^2 + \frac{2}{9}B^2, \qquad r^2 = x^2 + y^2 \end{gathered} \tag{32}$$

the system (5) can be written as:

$$\begin{aligned} (\tfrac{1}{3}B - m)x + (-l + \tfrac{3}{4}nA)y &= 0, \\ (l - \tfrac{3}{4}nA)x - (\tfrac{1}{3}B + m)y &= 0. \end{aligned} \tag{33}$$

The vanishing of the determinant of the coefficients of the system (33), which is the requirement for the existence of nonzero roots, gives

$$A^2 - \frac{8}{3}\frac{l}{n}A + \frac{16}{9n^2}\left(l^2 - m^2 + \frac{B^2}{9}\right) = 0, \tag{34}$$

We can obtain from this, by solving for A, the amplitudes r of the subharmonics:

$$r_{1,2}^2 = -\frac{2}{9}B^2 + \frac{4}{3n}\left(l \mp \sqrt{(B/3)^2 - m^2}\right). \tag{35}$$

The requirement for the reality of r gives the same restriction as that established in Section 3. Here, instead of the expression (11) for I, we have

$$I = \frac{1}{36}B^4 - \frac{1}{3}\left(\frac{l}{n} + \frac{1}{3n^2}\right)B^2 + \left(\frac{l}{n}\right)^2 + \left(\frac{m}{n}\right)^2. \tag{36}$$

The components of the amplitude r of (35) are given by (15), where the quantity θ, calculated from (33), is the same as that given by (13).

As a special case of the above let us treat the case when $n = \bar{c}_3/\bar{c}_2$ is negligible and especially when $n \to 0$ either from positive or from negative values. We write (35) in the form

$$r_{1,2}^2 = \frac{4/3[l \mp \sqrt{(B/3)^2 - m^2}] - \frac{2}{9}B^2 n}{n}. \tag{37}$$

We have two cases:

(A) If n goes to zero from positive values, the reality of r implies that l and m vary in such a way that $[l \mp \sqrt{(B/3)^2 - m^2}]$, being positive and larger than $B^2 n$, goes to zero.

$$l \mp \sqrt{(B/3)^2 - m^2} > 0,$$

or

$$\frac{1 - \bar{c}_1}{\bar{c}_2} > \pm \sqrt{(B/3)^2 - (\bar{k}/\bar{c}_2)^2};$$

and

(i) for $(1 - \bar{c}_1)/\bar{c}_2 \geqq 0$, that is, for $1 \geqq \bar{c}_1$ and $\bar{c}_2 < 0$, or $1 \leqq \bar{c}_1$ and $\bar{c}_2 > 0$, we get

$$\left(\frac{1 - \bar{c}_1}{\bar{c}_2}\right)^2 + \left(\frac{\bar{k}}{\bar{c}_2}\right)^2 > \left(\frac{B}{3}\right)^2; \tag{38a}$$

(ii) for $(1 - \bar{c}_1)/\bar{c}_2 \leqq 0$, that is, for $1 \geqq \bar{c}_1$ and $\bar{c}_2 < 0$, or $1 \leqq \bar{c}_1$ and $\bar{c}_2 > 0$ we have

$$\left(\frac{1 - \bar{c}_1}{\bar{c}_2}\right)^2 + \left(\frac{\bar{k}}{\bar{c}_2}\right)^2 < \left(\frac{B}{3}\right)^2. \tag{38b}$$

Then as $\bar{c}_3 \to 0$, subharmonics may exist in the region (a_1), Fig. 5a, if $\bar{c}_2 > 0$, $1 \geqq \bar{c}_1$, or $\bar{c}_2 < 0$, $1 \leqq \bar{c}_1$; as well as in **the region (b_1) Fig. 5b**, if $\bar{c}_2 > 0$, $1 \leqq \bar{c}_1$, or $\bar{c}_2 < 0$, $1 \geqq \bar{c}_1$.

(B) If n goes to zero from negative values, the reality of r implies that $l \mp \sqrt{(B/3)^2 - m^2} < 0$, then, in the same manner, we can find the region (a_2) and (b_2) of Fig. 6a and Fig. 6b where we may have subharmonics. In the undamped case, $\bar{k} = 0$, the subharmonics correspond to the appropriate segment of $[(1 - \bar{c}_1)/\bar{c}_2]$-axis in the shaded regions.

In the case $\bar{c}_3 = 0$, by combining appropriately the above figures, we have that subharmonics may exist outside the circle if $(1 - \bar{c}_1)/\bar{c}_2 > 0$, and inside if $(1 - \bar{c}_1)/\bar{c}_2 < 0$, Fig. 7(a), 7(b).

Let us call "resonance" the situation where the amplitude r of the

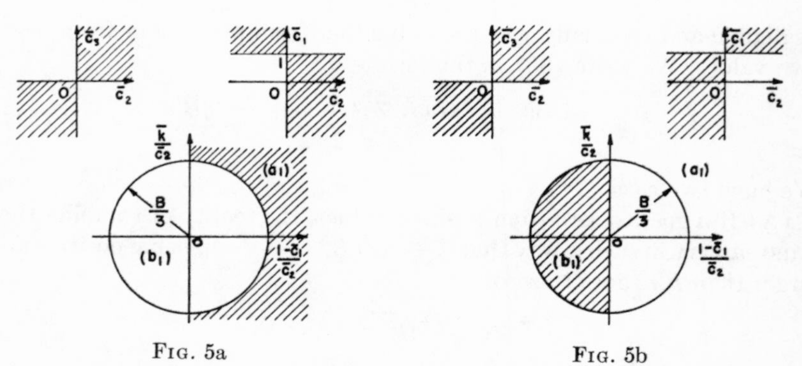

FIG. 5a FIG. 5b

FIG. 5. The shaded regions (a_1) and (b_1) outside and inside the circle respectively correspond to $\bar{c}_3/\bar{c}_2 > 0$. For the region (a_1) we have $(1 - \bar{c}_1)/\bar{c}_2 > 0$, and for region (b_1): $(1 - \bar{c}_1)/\bar{c}_2 < 0$.

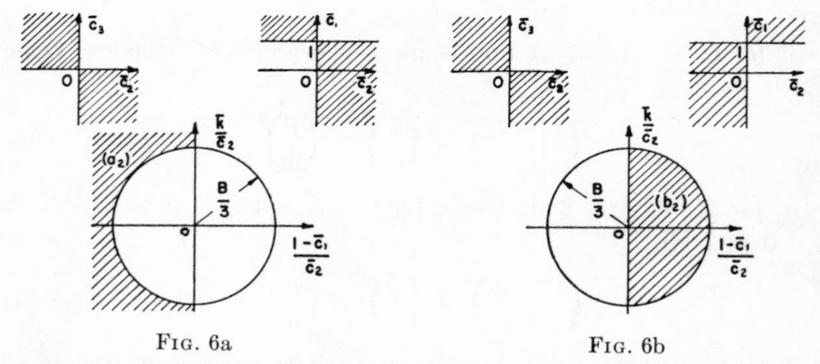

FIG. 6a FIG. 6b

FIG. 6. The shaded regions (a_2) and (b_2) outside and inside the circle respectively correspond to $\bar{c}_3/\bar{c}_2 < 0$. For the region (a_2) we have $(1 - \bar{c}_1)/\bar{c}_2 > 0$, and for the region (b_2): $(1 - \bar{c}_1)/\bar{c}_2 < 0$.

subharmonics is not finite. The resonance here is characterized by the ratio

$$\frac{l \pm \sqrt{(B/3)^2 - m^2}}{n}.$$

This ratio is not finite, then resonance appears, if $l \neq 0$, $|m| \to B/3$, $n = 0$; that is, if, for $\bar{c}_2 \neq 0$, $\bar{c}_1 \neq 1$, $|\bar{k}/\bar{c}_2| \neq B/3$, $\bar{c}_3 = 0$. Also we have resonance if $l \to 0$, $|m| \to B/3$, $n \to 0$ in such a way that n goes to zero

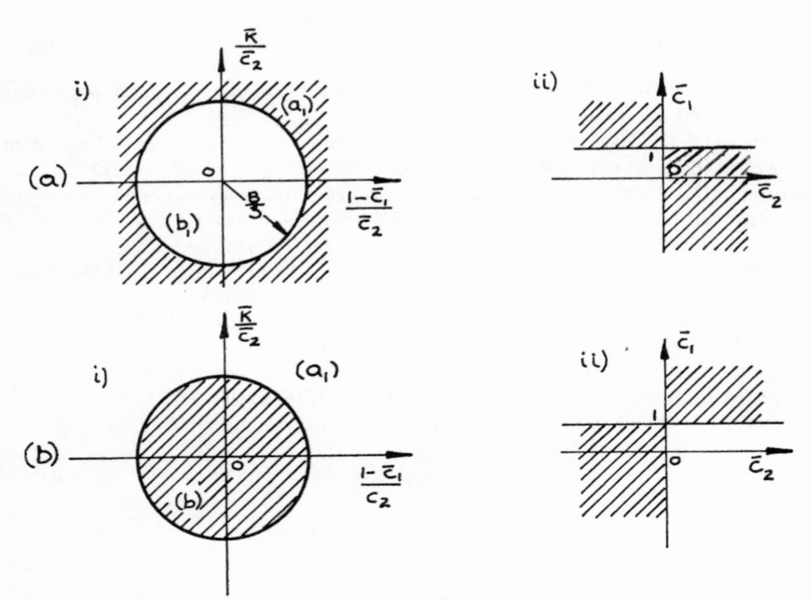

FIG. 7. The shaded region outside the circle, under the condition of the shaded region (a, ii), and the shaded region inside the circle, under the condition of the shaded region (b, ii), correspond to subharmonics.

faster than the expression $l \pm \sqrt{(B/3)^2 - m^2}$ does. If the above ratio is finite and bigger than the number $B^2/6$, subharmonics exist.

G. Reuter (1949) studied the subharmonics of order $\frac{1}{2}$ in a special case, the case when the damping coefficient $\bar{k}$ is small and positive, the nonlinear coefficient $\bar{c}_2$ small, the linear coefficient $\bar{c}_1$ near to 1, and the amplitude B of the external force not too large. His results about the the existence of the subharmonics in this special case agree with the above results, and this fact gives a direct verification and a natural check of the validity of the results given above in our general case.

RECEIVED: May, 1959.

REFERENCES

HAYASHI, C. (1953). "Forced Oscillations in Nonlinear Systems," pp. 84–88, 143–147. Nippon Printing and Publishing Company, Osaka, Japan.

MAGIROS, D. (1957a). Subharmonics of any order in case of nonlinear restoring force. Prakt. Athens Acad. Sci. **32,** 77–85.

MAGIROS, D. (1957b). Subharmonics of order one third in case of cubic restoring force. *Prakt. Athens Acad. Sci.* **32,** 101–108.

MAGIROS, D. (1957c). Remarks on a problem of subharmonics; its inverse problem. *Prakt. Athens Acad. Sci.* **32,** 143–146.

MAGIROS, D. (1957d). On the singularities of a system of differential equations where the time enters explicity. *Prakt. Athens Acad. Sci.* **32,** 448–451.

MAGIROS, D. (1958). Subharmonics of any order in nonlinear systems of one degree of freedom: Application to subharmonics of order $\frac{1}{3}$. *Inform. and Control* **1,** 198–227.

MANDELSTAM,. L, AND PAPALEXI, N. (1935). "Über einige nichtstationäre Schwingungsvorgänge". *Tech. Phys. U.S.S.R.* **1,** 415–428.

MELIKJAN, A. (1935). Über das anwachsen der Amplitude bei Resonanzerscheinungen 2-ter Art," *Tech. Phys. U.S.S.R.* 429–448.

REUTER, G. (1949). "Subharmonics in a nonlinear system with unsymmetrical restoring force." *Quart. J. Mech. and Appl. Math.* **2,** 198–207.

STOKER, J. J. (1950). "Nonlinear Vibrations." Interscience, New York.

A METHOD FOR DEFINING PRINCIPAL MODES OF NONLINEAR SYSTEMS UTILIZING INFINITE DETERMINANTS

BY DEMETRIOS G. MAGIROS

REPUBLIC AVIATION CORPORATION, FARMINGDALE, N. Y.

Communicated by L. Brillouin, October 3, 1960

1. The concept of "principal modes" plays the predominant role in the analysis of oscillatory systems, no matter what field the systems occur in. For the calculations of principal modes of systems, either linear or nonlinear, two appropriate definitions may be used, namely, the "proportionality definition" of principal modes and their definition as solutions of special "initial value problems."

It is the purpose of this note to provide a procedure for the calculation of principal

Reprinted from the *Proceedings of the National Academy of Sciences* **46** (1960), 1608–1611.

modes of nonlinear systems utilizing "infinite determinants." For a "two-mass-three-spring" free nonlinear system, R. Rosenberg[1] carried out the calculation of principal modes by using the "potential function" of the corresponding conservative system. We restrict ourselves, without loss of generality, to the above "dual-mode" nonlinear system for reasons of comparison with the results obtained by using other methods. The motion of this system is governed by two second-order ordinary differential equations one of which is nonlinear. When the solutions of these equations are assumed to have the form of complex exponential series, the calculation of the frequency and the coefficients leads to recursion formulae, which give rise to "infinite determinants."

2. In this note, the following "proportionality definition" of "principal modes" of an oscillatory nonlinear system is used. By using the terminology of mechanical systems, the "principal modes" are defined as those oscillations for which the amplitudes of the fundamental and the corresponding harmonics of the displacements of any two of the oscillating masses have, separately, a constant ratio at all times. For such motions, the masses all oscillate about their equilibrium positions with the same frequency, the "principal frequency" of the system. If x_i, $i = 1, 2, \ldots s$ are the displacements of the masses m_i, $i = 1, 2, \ldots s$ from their equilibrium positions and x_{i_n}, $x_{i'_n}$ the corresponding amplitudes of the nth harmonic of the displacements x_i, x_i' of the two masses m_i, $m_{i'}$, and if there exist constants c_n such that the conditions,

$$x_{i_n}t\ /x_{i'_n}(t) \equiv c_n,\ i \neq i',\ i, i' = 1, 2, \ldots s\ , c_n \neq 0, \infty \tag{1}$$

are satisfied, then these motions are, by definition, the "principal modes" of the system.

3. If m_1 and m_2 are the oscillating masses, K_2 and K_3 the spring constants of the coupling and the second anchor springs, $K_1 = K_1 + \mu x^2$ that of the first anchor spring, μ a constant which characterizes the nonlinearity, and x and y the displacements of m_1 and m_2 from their equilibrium positions, the equations of motion of this system are

$$\begin{aligned} \ddot{x} + \omega_1^2 x + \lambda_1 x^3 - \lambda_2 y &= 0 \\ \ddot{y} + \omega_2^2 y - \lambda_3 x &= 0 \end{aligned} \tag{2}$$

where

$$\omega_1^2 = \frac{K_1 + K_2}{m_1},\ \omega_2^2 = \frac{K_2 + K_3}{m_2},\ \lambda_1 = \frac{\mu}{m_1},\ \lambda_2 = \frac{K_2}{m_1},\ \lambda_3 = \frac{K_2}{m_2}. \tag{2a}$$

Assume solution of (2) to be represented by the exponential series,

$$x(t) = \sum_{n=-\infty}^{\infty} \alpha_n e^{in\omega t},\quad y(t) = \sum_{n=-\infty}^{\infty} A_n e^{in\omega t}. \tag{3}$$

For the reality of the solution, one takes $\bar{\alpha}_n = \alpha_{-n}$, $\bar{A}_n = A_{-n}$, where $\bar{\alpha}_n$ and $\bar{A}_n$ are conjugates of α_n and A_n. Also assume that

$$\sum_{n=-\infty}^{\infty} |\alpha_n| < \infty,\ \sum_{n=-\infty}^{\infty} |A_n| < \infty,\ \sum_{u=-\infty}^{\infty} n^2|\alpha_n| < \infty,\ \sum_{n=-\infty}^{\infty} n^2|A_n| < \infty.$$

The first two of these inequalities guarantee the convergence of the series (3), while the second ones the existence of the second derivatives $\ddot{x}$, $\ddot{y}$. Inserting (3) into (2), if one puts x^3 into the form,

$$x^3 = \sum_n \sum_{\rho_1} \sum_{\rho_2} \alpha_{\rho_1}\alpha_{\rho_2}\alpha_{n-\rho_1-\rho_2}e^{in\omega t},$$

one can get the system,

$$\left(\omega_1^2 - n^2\omega^2 - \frac{\lambda_2\lambda_3}{\omega_2^2 - n^2\omega^2}\right)\alpha_n + \lambda_1 \sum_{\rho_1}\sum_{\rho_2} \alpha_{\rho_1}\alpha_{\rho_2}\alpha_{n-\rho_1-\rho_2} = 0 \tag{4i}$$

$$A_n/\alpha_n = \frac{\lambda_3}{\omega_2^2 - n^2\omega^2} \tag{4ii}$$

where n, ρ_1, ρ_2 are integers. This system consists of infinitely many nonlinear ($\lambda_1 \neq 0$) homogeneous equations for the infinitely many unknown coefficients α_n and A_n.

If α_1 is the dominant coefficient of the first of the series (3), the approximate value of the double series in (4i) is

$$c_1\alpha_1^2\alpha_{n-2} + c_2|\alpha_1|^2\alpha_n + c_3\bar{\alpha}_1^2\alpha_{n+2}$$

with an error $o(\alpha_1^2)$. c_1, c_2, c_3 are appropriately calculated integers. Then one can get from (4i) the following recursion formula:

$$p_n\alpha_{n-2} + \alpha_n + q_n\alpha_{n+2} = 0 \tag{5}$$

where

$$p_n = \frac{c_1\lambda_1\alpha_1^2}{K_n},\ q_n = \frac{c_3\lambda_1\bar{\alpha}_1^2}{K_n},\ K_n = \omega_1^2 + c_2\lambda_1|\alpha_1|^2 - n^2\omega^2 - \frac{\lambda_2\lambda_3}{\omega_2^2 - n^2\omega^2}. \tag{5a}$$

For non-zero α_n's, the following necessary and sufficient condition in infinite determinant form must be satisfied:

$$\Delta \equiv \begin{Vmatrix} \cdots & \cdots & \cdots & \cdots & \cdots & \cdots & \cdots & \cdots & \cdots \\ \cdots & 1 & 0 & q_{-3} & 0 & 0 & 0 & 0 & \cdots \\ \cdots & 0 & 1 & 0 & q_{-2} & 0 & 0 & 0 & \cdots \\ \cdots & p_{-1} & 0 & 1 & 0 & q_{-1} & 0 & 0 & \cdots \\ \cdots & 0 & p_0 & 0 & 1 & 0 & q_0 & 0 & \cdots \\ \cdots & 0 & 0 & p_1 & 0 & 1 & 0 & q_1 & \cdots \\ \cdots & 0 & 0 & 0 & p_2 & 0 & 1 & 0 & \cdots \\ \cdots & 0 & 0 & 0 & 0 & p_3 & 0 & 1 & \cdots \\ \cdots & \cdots & \cdots & \cdots & \cdots & \cdots & \cdots & \cdots & \cdots \end{Vmatrix} = 0 \tag{6}$$

The "principal frequency" ω can be determined as a root of the "frequency equation" (6).

To calculate the coefficients α_n, in the case α_1 dominant, we see that by using the notation,

$$u_n = \frac{\alpha_n}{\alpha_{n-2}}, \tag{7}$$

the recursion formula (5) gives

$$u_n = -\frac{p_n}{1 + q_n u_{n+2}}$$

which leads to the infinite continued fraction,

$$\alpha_n/\alpha_{n-2} = -\cfrac{p_n}{1 - \cfrac{p_{n+2}q_n}{1 - \cfrac{p_{n+4}q_{n+2}}{1 - \dots}}} \qquad (8)$$

The infinite continued fraction (8) can be expressed by means of infinite determinants, and the coefficients α_n can be calculated in terms of the arbitrary α_0 and α_1

Having now calculated the principal frequency ω and the coefficients α_n, the solution (3), by using (4ii), reads:

$$x(t) = \sum_{n=-\infty}^{\infty} \alpha_n e^{in\omega t}, \quad y(t) = \sum_{n=-\infty}^{\infty} \frac{\lambda_3}{\omega_2^2 - n^2\omega^2} \alpha_n e^{in\omega t} \qquad (9)$$

The principal frequency ω is the same for both oscillators, and since the coefficients of (9) satisfy the condition (1), then (9) gives the "principal modes" of our nonlinear system.

The method followed is based on the use of exponential series with complex coefficients leading to infinite determinants when one calculates either the frequency or the coefficients of the series. It is similar to the discussion given by L. Brillouin for Hill and Mathieu functions.[2] Next will be discussed the infinite determinants involved in the present note and the solution in its final form given. It may be mentioned that Hill, in his "Lunar Theory," brought into notice the infinite determinants, and Poincaré first gave conditions of their convergence.

The author is indebted to Dr. G. Nomicos for a comment on a point of the note.

[1] Rosenberg, R., ASME, Appl. Mech. Div. paper No. 59-A-93.

[2] Brillouin, L., *Wave Propagation in Periodic Structures*, 2nd ed. (New York: Dover Publications, Inc., 1953).

A METHOD FOR DEFINING PRINCIPAL MODES OF NONLINEAR SYSTEMS UTILIZING INFINITE DETERMINANTS, II

By Demetrios G. Magiros

MISSILE AND SPACE VEHICLE DEPARTMENT, GENERAL ELECTRIC COMPANY, PHILADELPHIA

Communicated by Leon Brillouin, April 17, 1961

1. (*a*) In a recent paper,[1] the writer has discussed the "principal modes" of a "dual-mode" nonlinear system utilizing "infinite determinants." However, this paper did not give the calculation of the infinite determinants involved and the solution in its final form; also, it did not give the "principal modes" defined as solutions of special "initial-value problems." This is the object of the present paper.

(*b*) We can state for reference that in a "dual-mode" nonlinear system, examined in the previous paper, the "principal modes" are given by

$$\chi(t) = \sum_{n=-\infty}^{\infty} \alpha_n e^{in\,\omega t}, \qquad \psi(t) = \sum_{n=-\infty}^{\infty} \frac{\lambda_3}{\omega_2^2 - n^2\omega^2}\,\alpha_n e^{in\,\omega t}, \tag{1}$$

and the "recursion formula," for the calculation of the "principal frequency" ω and the coefficients α_n, is

$$p_n\alpha_{n-2} + \alpha_n + q_n\alpha_{n+2} = 0 \tag{2}$$

where

$$p_n = \frac{3\lambda_1\alpha_1^2}{k_n + 6\lambda_1|\alpha_1|^2}, \quad q_n = \frac{3\lambda_1\bar{\alpha}_1^2}{k_n + 6\lambda_1|\alpha_1|^2}, \quad k_n = \omega_1^2 - n^2\omega^2 - \frac{\lambda_2\lambda_3}{\omega_2^2 - n^2\omega^2}. \tag{2a}$$

The factors 3, 6, 3 of the expressions (2*a*) are the coefficients of the trinomial (A_1) of the Appendix, which gives the dominant sum of the double series of the formula:

$$k_n\alpha_n + \lambda_1 \sum_{\rho_1}\sum_{\rho_2} \alpha_{\rho_1}\alpha_{\rho_2}\alpha_{n-\rho_1-\rho_2} = 0. \tag{3}$$

For the convergence of the series (1), it is necessary that $(n^2\omega^2 - \omega_2^2)$ be neither zero nor very small, i.e., ω must not be either a submultiple of ω_2 or very close to a submultiple of ω_2.

2. The "frequency equation" for the calculation of the "principal frequency" ω comes from the condition of nonzero α_n's of the infinite system (2). This condition in a determinatal form is

$$\Delta(n, \infty) = 0, \tag{4}$$

and the determinant $\Delta(n, \infty)$, for arbitrary integer n, is the following one-sided infinite determinant:

$$\Delta(n, \infty) = \lim_{\substack{n = \text{fixed integer} \\ m = 1, 2, 3, \ldots \\ m \to \infty.}} \Delta(n, m) \equiv \left\| \begin{array}{cccccccc} 1 & 0 & q_n & 0 & \cdots & & & 0 \\ 0 & 1 & 0 & q_{n+1} & 0 & \cdots & & 0 \\ p_{n+2} & 0 & 1 & 0 & q_{n+2} & 0 & \cdots & 0 \\ \cdots & \cdots & \cdots & \cdots & \cdots & \cdots & \cdots & \cdots \\ 0 & \cdots & & 0 & 1 & & & q_{n+m-2} \\ 0 & \cdots & & p_{n+m-1} & 0 & 1 & 0 & \\ 0 & \cdots & & 0 & p_{n+m} & 0 & 1 & \end{array} \right\| \tag{5}$$

Reprinted from the *Proceedings of the National Academy of Sciences* **47** (1961), 883–887.

Consider a weak nonlinearity, i.e., $\lambda_1 |x|^3 \ll \omega_1^2 |x|$. Then, since α_1 is the dominant coefficient of the series (3), $\max x^2 = 2\alpha_1^2$ when

$$\lambda_1 \ll \frac{K_1 + K_2}{2m_1\alpha_1^2}. \tag{6a}$$

For such a small λ_1, we can get

$$p_n = \lambda_1 \frac{3\alpha_1^2}{k_n} - \lambda_1^2 \frac{18\alpha_1^2 |\alpha_1|^2}{k_n^2} + \ldots, \quad q_n = \lambda_1 \frac{3\bar{\alpha}_1^2}{k_n} - \lambda_1^2 \frac{18\bar{\alpha}_1^2 |\alpha_1|^2}{k_n^2} + \ldots \tag{6}$$

Thus, the elements of the determinant (5) not in the main diagonal have the small λ_1 as a common factor, and we can get an expansion of the determinant in powers of λ_1. Following the appropriate way to get this expansion,[2] by taking the first terms of (6), we obtain

$$\Delta(n, \infty) = 1 - 9\lambda_1^2 |\alpha_1|^4 \sum_{m=0}^{\infty} \frac{1}{k_{n+m} \cdot k_{n+m+2}} + 0(\lambda_1^3) \tag{7}$$

For any ω, which is not a zero of the denominators of the elements of the series in (7), this series converges. Combining (4) and (7), if we confine ourselves to the first term of the series in (7), and putting $n = 1$, the frequency equation becomes approximately the following equation, quartic in ω^2:

$$\left(\omega_1^2 - \omega^2 - \frac{\lambda_2\lambda_3}{\omega_2^2 - \omega^2}\right)\left(\omega_1^2 - 9\omega^2 - \frac{\lambda_2\lambda_3}{\omega_2^2 - 9\omega^2}\right) - 9\lambda_1^2 |\alpha_1|^4 = 0, \tag{8}$$

which gives the principal frequency ω of order $0\ (\lambda_1^2)$. For values of the principal frequency of order higher than $0\ (\lambda_1^2)$, we find in the expansion of $\Delta(n, \infty)$ terms of order higher than $0\ (\lambda_1^2)$ and continue in the same way to find the corresponding algebraic equation.

3. The calculation of the coefficients α_n is based on the following formula of the previous paper:[1]

$$\alpha_n/\alpha_{n-2} = -p_n \cdot \cfrac{1}{1 + \cfrac{X_{n,1}}{1 + \cfrac{X_{n,2}}{1 + \ldots}}} \tag{9}$$

with

$$X_{n,m} = -p_{n+2m} \cdot q_{n+2(m-1)}$$

$$n = \text{any fixed integer}, \quad m = 1,2,3, \ldots \tag{9a}$$

By using the product of m linear transformations of the form

$$X_1(v) = 1, \quad X_m = \frac{1}{1 + X_m v}, \quad m = 2,3,4, \ldots$$

$$X_p \neq 0,\ p = 2, 3, \ldots m; \quad X_p = 0,\ p > m, \tag{10}$$

one can prove that the mth approximant of the continued fraction of (9) may be equal to the ratio of two determinants of which the denominator is

$$\bar{\Delta}(n,m) =$$

$$\left\|\begin{matrix} 1 & -1 & 0\ldots & 0 & 0 & 0 \\ -p_{n+2}q_n & 1 & -1\ldots & 0 & 0 & 0 \\ 0 & -p_{n+4}\cdot q_{n+2} & 1\ldots & 0 & 0 & 0 \\ \cdots & \cdots & \cdots & \cdots & \cdots & \cdots \\ 0 & 0 & 0\ldots & -p_{n+2(m-1)}\cdot q_{n+2(m-2)} & 1 & -1 \\ 0 & 0 & 0\ldots & 0 & -p_{n+2m}\cdot q_{n+2(m-1)} & 1 \end{matrix}\right\| \quad (11)$$

and the numerator, $\bar{\Delta}(n + 2,m)$, is obtained from (11) by omitting its first row and first column. For fixed n, and $m \to \infty$, we can write

$$\alpha_n/\alpha_{n-2} = -p_n \frac{\bar{\Delta}(n + 2, \infty)}{\bar{\Delta}(n, \infty)} \quad (12)$$

where the infinite determinants are of von Koch's type.

The induction procedure applied to (12) gives for the coefficients with even index:

$$\alpha_{2n} = (-1)^n \alpha_0 p_2 p_4 \ldots p_{2n}\cdot \frac{\bar{\Delta}(2n + 2, \infty)}{\bar{\Delta}(2, \infty)}, \; n = 1,2,3, \ldots \quad (13a)$$

For coefficients with odd index, starting from

$$\alpha_3 = -\lambda_1 \frac{\alpha_1^3}{k_3}, \quad (13b)$$

which can be found by using (3) and the Appendix, we have

$$\alpha_{2n+1} = (-1)^{n+1}\lambda_1 \frac{\alpha_1^3}{k_3}\cdot p_5 p_7 \ldots p_{2n+1} \frac{\bar{\Delta}(2n + 3, \infty)}{\bar{\Delta}(5, \infty)}, \quad n = 2,3, \ldots, \quad (13c)$$

α_0 and α_1 are arbitrary coefficients. By using the property $\alpha_{-n} = \bar{\alpha}_n$ and taking α_1 real, the solution (1) can be written as follows:

$$\chi(t) = \alpha_0 + 2\alpha_1 \cos \omega t - 2\alpha_3 \cos 3\omega t + 2 \sum_n \alpha_N \cos N\omega t,$$

$$\psi(t) = \frac{\lambda_3}{\omega_2^2} \alpha_0 + \frac{2\lambda_3\alpha_1}{\omega_2^2 - \omega^2} \cos \omega t - \frac{2\lambda_3\alpha_3}{\omega_2^2 - 2^2\omega^2} \cos 3\omega t + 2 \sum_n \frac{\lambda_3\alpha_n}{\omega_2^2 - N^2\omega^2} \cos N\omega t, \quad (14)$$

where α_3, $\alpha_{N=2n}$, and $\alpha_{N=2n+1}$ are given by (13*b*), (13*a*), and (13*c*), respectively. Formulas (14) give the solution in its final form, and by using (13*a*), (13*b*), and (13*c*), we can calculate as many coefficients of the solution (14) as we want, in terms of powers of λ_1. By expanding the determinants of (13*a*) and (13*c*) in powers of the small λ_1,[2] we find that their ratios are equal to $1 + 0\,(\lambda_1^2)$, and, as we can easily see, the calculation of the coefficients, in terms of powers of λ_1, needs only the unit as value of these ratios. The first two terms of (14) are independent of λ_1. For terms of (14) of order $O\,(\lambda_1)$, we take the first term of p_n of (6) and, combining it with (13*a*) for $n = 1$, we get the 2nd harmonic terms of the series (14), when the terms of order $O\,(\lambda_1)$ are

$$-2\lambda_1\left\{\frac{3\alpha_0\alpha_1^2}{k_2} \cos 2\omega t + \frac{\alpha_1^3}{k_3} \cos 3\omega t\right\}, \text{ of } \chi(t).$$

$$-2\lambda_1\left\{\frac{3\lambda_3\alpha_0\alpha_1^2}{k_2(\omega_2^2 - 2^2\omega^2)} \cos 2\omega t + \frac{\lambda_3\alpha_1^3}{k_3(\omega_2^2 - 3^2\omega^2)} \cos 3\omega t\right\}, \text{ of } \psi(t). \tag{14a}$$

If we take the terms of p_n of (6) up to the order $O(\lambda_1^2)$ and combine them with (13a) for $n = 2$, we obtain the 4th harmonic terms of order $O(\lambda_1^2)$:

$$18\lambda_1^2 \frac{\alpha_0\alpha_1^4}{k_2k_4} \cos 4\omega t, \quad 18\lambda_1^2 \frac{\alpha_0\alpha_1^4\lambda_3}{k_2k_4(\omega_2^2 - 4^2\omega^2)} \cos 4\omega t \tag{14b}$$

of $\chi(t)$ and $\psi(t)$, respectively. Following the same way, we can calculate the higher harmonics in terms of higher powers of λ_1. The solution (14), constructed as indicated, must be convergent, and its coefficient of the fundamental term must be much larger than any other coefficient. These requirements imply that the following conditions are satisfied:

$$\alpha_0 << \alpha_1, \quad \lambda_1 << \min\left\{\frac{K_1 + K_2}{2m_1\alpha_1^2}, \frac{k_2.}{2\alpha_0\alpha_1}, \frac{k_3}{\alpha_1^2}\right\}. \tag{15}$$

For the second condition, the inequality (6a) was taken into account. The 4th, 5th, ... harmonic terms are of order $O(\lambda_1^2)$, $O(\lambda_1^3)$, ..., and the convergence is guaranteed.

4. Another approach for the determination of principal modes is based on the manner in which the system is set into motion. This is equivalent to considering the principal modes as solutions of special initial-value problems. The solution (14), found in the previous section, gives an indication about the appropriate initial conditions. The initial displacements are, according to (14), the sum of the coefficients, and the initial velocities are zero.

Consider the differential equations of the "dual-mode" nonlinear system subject to the two sets of initial conditions:

$$\chi(0) = \chi_0, \quad \psi(0) = \psi_0, \quad \dot{\chi}(0) = \dot{\psi}(0) = 0 \tag{16a}$$

$$\chi(0) = \chi_0, \quad \psi(0) = -\psi_0, \quad \dot{\chi}(0) = \dot{\psi}(0) = 0. \tag{16b}$$

If the initial displacements χ_0 and ψ_0 are appropriately related, the initial velocities being zero, then each set of the initial conditions (16a) or (16b) gives rise to special vibration modes, which are, by definition, the principal modes of the system. If we assume the series

$$\chi(t) = \sum_{n=-\infty}^{\infty} \alpha_n e^{in\omega t}, \quad \psi(t) = \sum_{n=-\infty}^{\infty} A_n e^{in\omega t} \tag{17}$$

as solution of the differential equations of the system, by repeating the previous procedure, we will terminate to get the solution in the form (14), which gives the appropriate relationship between the initial displacements of the two oscillators. An approximation of this solution for very small λ_1, and α_1 as dominant coefficient, is:

$$\chi(t) = 2\alpha_1 \cos \omega t, \quad \psi(t) = 2\frac{\lambda_3}{\omega_2^2 - \omega^2}\alpha_1 \cos \omega t, \tag{18}$$

where the initial conditions are

$$\chi_0 = 2\alpha_1, \quad \psi_0 = \frac{\lambda_3}{\omega_2^2 - \omega^2}\chi_0, \quad \dot{\chi}_0 = \dot{\psi}_0 = 0. \tag{18a}$$

Formula (18*a*) give the initial conditions required for the solution to be of "principal modes" type.

Appendix.—If in the elements of the sequence $\{\alpha_n\}$, $n = 0, \pm 1, \pm 2, \ldots$, where $\alpha_{-n} = \bar{\alpha}_n$, the element α_1, and then α_{-1}, is dominant, it is easily seen that the dominant sum of the double series of (3) is given by

$$3\alpha_1^2\alpha_{n-2} + 6|\alpha_1|^2\alpha_n + 3\bar{\alpha}_1^2\alpha_{n+2}, \tag{A_1}$$

where n is any integer except $n = \pm 1, \pm 3$. For these exceptions, the dominant terms of the double series are

$$3\bar{\alpha}_1\alpha_1^2 \text{ for } n = 1, \quad 3\alpha_1\bar{\alpha}_1^2 \text{ for } n = -1, \quad \alpha_1^3 \text{ for } n = 3, \bar{\alpha}_1^3 \text{ for } n = -3. \tag{A_2}$$

[1] Magiros, D. G., these PROCEEDINGS, **46,** 1608 (1960).

[2] Brillouin, L., *Wave Propagation in Periodic Structures,* 2nd ed. (New York: Dover Publications, Inc., 1953), pp. 34, 35.

Method for Defining Principal Modes of Nonlinear Systems Utilizing Infinite Determinants

Demetrios G. Magiros
General Electric Company, Missile and Space Vehicle Department, Philadelphia, Pennsylvania
(Received April 21, 1961)

A method for calculation of "principal modes" of linear or nonlinear systems is discussed. The physical definition of "principal modes" is formulated mathematically in two ways. The trial solution of the differential equation of the motion of the system is taken in an appropriate structure. The calculation of principal modes leads to infinite determinants of Hill's and von Koch's type, which are analyzed. The above method yields the possibility of getting the "principal modes" in the form of a series, all the coefficients of which can be calculated.

1. INTRODUCTION

The concept of "principal modes" plays the predominant role in the analysis of the oscillatory systems, no matter what field the systems occur in.

The principal modes of linear systems are, by definition the fundamental set of solutions of which a linear combination gives the general solution of the linear differential equations, which govern the motion of the linear system. This means that any kind of oscillations in linear systems can be discussed in terms of some special modes of oscillation of the system, the "principal modes" of the system.

This definition of "principal modes" is meaningless in nonlinear systems, since the "principle of superposition" does not hold in those systems.

The study of the principal modes of systems, either linear or nonlinear, may be made by using two definitions, namely the "proportionality definition" of principal modes and their definition as solutions of "initial value problems" of special type. Calculations, based on these definitions, are shown for a nonlinear "dual-mode" system. If the solution of the differential equations of this system is taken as an exponential series with complex coefficients, the calculation of the frequency ω and the coefficients of the series leads to a recursion formula, which gives rise to "infinite determinants" of special type. The analysis of the infinite determinants involved, and the solution in its final form is discussed. The nonunit elements of the determinants contain the coefficient of the nonlinearity as a common factor, and, for a weak nonlinearity, we can get an expansion of these determinants in powers of the coefficient, then an appropriate approximation of them. Thus the "frequency equation," in an infinite determinantal form, is reduced to a quartic in ω^2, and the ratio of the determinants of the coefficients of the series to unit.

The solution given is in accordance with both definitions of "principal modes," and imposes a relation between the initial displacements of the masses, and this relation and the condition for the initial velocities to be zero distinguished the special initial value problems appropriate for the "principal modes."

A brief discussion of the present paper has been published in the Proceedings of National Academy of Sciences.[1a,b]

2. THE DEFINITION OF PRINCIPAL MODES, ITS APPLICATION TO A FREE NONLINEAR SYSTEM OF TWO DEGREES OF FREEDOM, AND THE RECURSION FORMULA FOR THE SOLUTION

By using the terminology of mechanical systems with s degrees of freedom, the principal modes of oscillations of the system are defined as those oscillations of the system for which the nonzero amplitudes of the fundamental and the corresponding harmonics of the displacements of any two of the oscillating masses have, separately, a constant ratio. For such motions, the masses all oscillate about their equilibrium positions, where they pass at the same time. Their common frequency is the "principal frequency" of the system.

If x_i, $i=1, 2, \cdots s$ are the displacements of the masses m_i, $i=1, 2, \cdots s$ from their equilibrium positions, and x_{in}, $x_{i'n}$ the corresponding amplitudes of the nth harmonic of the displacements x_i, $x_{i'}$, of the two masses m_i, $m_{i'}$, and if there exist constants c_n such that the conditions

$$x_{in}/x_{i'n}=c_n, \quad i\neq i'; \quad i, i'=1, 2, \cdots s, \quad c_n\neq 0, \infty \tag{1}$$

are satisfied, then these motions are, by definition, the "principal modes" of the system.

We restrict ourselves, without loss of generality, to a two-degrees-of-freedom nonlinear system; namely, we get—as a mechanical model—the "two-masses-three-springs" system with one of the anchor springs nonlinear and such that the corresponding restoring force is an odd-cubic function of the distance, or—as an electrical model—the "two-inductances-three-capacitances" system with one capacitance variable and the others constant, Fig. 1 (a), (b).

If m_1 and m_2 are the oscillating masses, $\bar{K}_1=K_1+\mu x^2$, K_3 and K_2 the stiffnesses of the first and second anchor springs and of the coupling, μ a constant which characterizes the nonlinearity, and x and y the displacements of m_1 and m_2 from their equilibrium positions, the equa-

[1] (a) D. G. Magiros, Proc. Natl. Acad. Sci. U. S., Dec. (1960). (b) D. G. Magiros, Proc. Natl. Acad. Sci. U. S., June (1961).

Reprinted from the *Journal of Mathematical Physics* 2 (1961), 869–875.

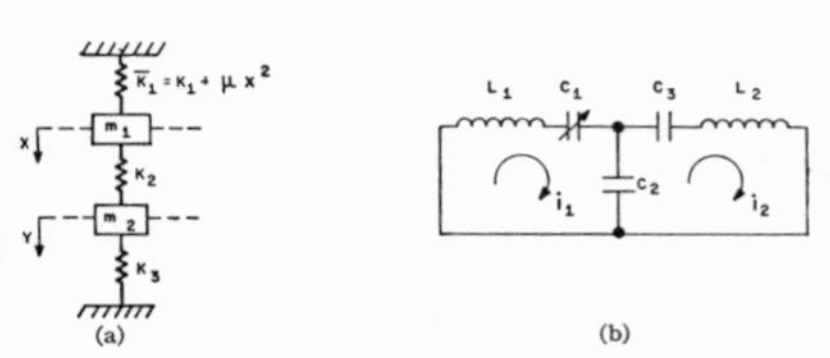

FIG. 1. (a) The mechanical model. (b) The electrical model.

tions of motion of this sytem are:

$$\ddot{x}+\omega_1^2x+\lambda_1x^3-\lambda_2y=0,\qquad \ddot{y}+\omega_2^2y-\lambda_3x=0, \tag{2}$$

where

$$\omega_1^2=\frac{K_1+K_2}{m_1},\quad \omega_2^2=\frac{K_2+K_3}{m_2},$$

$$\lambda_1=\frac{\mu}{m_1},\quad \lambda_2=\frac{K_2}{m_1},\quad \lambda_3=\frac{K_2}{m_2}. \tag{2a}$$

We proceed to find the solutions $x(t)$ and $y(t)$ of (2) with one fundamental frequency ω for both oscillators by imposing a certain organic structure for the functions $x(t)$ and $y(t)$. Assume solutions of (2) of the form

$$x(t)=\sum_{n=-\infty}^{\infty}\alpha_n e^{in\omega t},\quad y(t)=\sum_{n=-\infty}^{\infty}A_n e^{in\omega t}. \tag{3}$$

The coefficients α_n and A_n are complex, then they include the phase angle.

For the reality of the solution, one takes $\bar{\alpha}_n=\alpha_{-n}$, $\bar{A}_n=A_{-n}$, where $\bar{\alpha}_n$ and $\bar{A}_n$ are conjugates of α_n and A_n. Also assume that

$$\sum_{n=-\infty}^{\infty}|\alpha_n|<\infty,\quad \sum_{n=-\infty}^{\infty}|A_n|<\infty,$$

$$\sum_{n=-\infty}^{\infty}n^2|\alpha_n|<\infty,\quad \sum_{n=-\infty}^{\infty}n^2|A_n|<\infty.$$

The first two of these inequalities guarantee the convergence of the series (3), while the second ones the existence of the second derivatives $\ddot{x}$, $\ddot{y}$.

By using the first of (3), x^3 is given by

$$x^3=\sum_{\rho_1=-\infty}^{\infty}\sum_{\rho_2=-\infty}^{\infty}\sum_{\rho_3=-\infty}^{\infty}\alpha_{\rho_1}\alpha_{\rho_2}\alpha_{\rho_3}e^{i(\rho_1+\rho_2+\rho_3)\omega t},$$

which can be written as follows:

$$x^3=\sum_{n}\sum_{\rho_1}\sum_{\rho_2}\alpha_{\rho_1}\alpha_{\rho_2}\alpha_{n-\rho_1-\rho_2}e^{in\omega t}. \tag{3a}$$

Inserting (3) and (3a) into (2), the following system results:

$$\left(\omega_1^2-n^2\omega^2-\frac{\lambda_2\lambda_3}{\omega_2^2-n^2\omega^2}\right)\alpha_n+\lambda_1\sum_{\rho_1}\sum_{\rho_2}\alpha_{\rho_1}\alpha_{\rho_2}\alpha_{n-\rho_1-\rho_2}=0, \tag{4a}$$

$$A_n/\alpha_n=\lambda_3/(\omega_2^2-n^2\omega^2), \tag{4b}$$

where n, ρ_1, ρ_2 are integers. This nonlinear system consists of infinitely many nonlinear ($\lambda_1\neq 0$) homogeneous equations for the infinitely many unknown coefficients α_n and A_n.

The equation (4b) expresses the definition of principal modes applied to the above dual-mode system, and by (4b) the calculation of A_n is deduced from α_n, then the calculation of principal modes of the system is deduced from α_n, by using (4a). If α_1 is the dominant coefficient of the sequence $\{\alpha_n\}$, then the approximate value of the double series of (4a), see Appendix I, is given by

$$3\alpha_1^2\alpha_{n-2}+6|\alpha_1|^2\alpha_n+3\bar{\alpha}_1^2\alpha_{n+2}, \tag{5}$$

with an error $o(\alpha_1^2)$.

Inserting the expression (5) into the place of the double series of (4a), we can get the following recursion formula:

$$p_n\alpha_{n-2}+\alpha_n+q_n\alpha_{n+2}=0, \tag{6}$$

where

$$p_n=\frac{3\lambda_1\alpha_1^2}{k_n+6\lambda_1|\alpha_1|^2},\quad q_n=\frac{3\lambda_1\bar{\alpha}_1^2}{k_n+6\lambda_1|\alpha_1|^2},$$

$$k_n=\omega_1^2-n^2\omega^2-\frac{\lambda_2\lambda_3}{\omega_2^2-n^2\omega^2}. \tag{6a}$$

It is the recursion formula (6) which will be used for the calculation of the "principal frequency" ω and the coefficients α_n of the solution (3). We notice here that for the convergence of the series (3) it is necessary, according to (4b), that $(\omega_2^2-n^2\omega^2)$ is neither zero nor very small, i.e., ω must not be either a submultiple ω_2 or very close to a submultiple of ω_2.

3. CALCULATION OF THE PRINCIPAL FREQUENCY ω

The recursion formula (6) gives infinitely many homogeneous equations for the infinitely many unknowns α_n. The corresponding infinite matrix of the coefficients of these equations is

$$\begin{bmatrix} \cdots & \cdots & \cdots & \cdots & \cdots & \cdots & \cdots & \cdots & \cdots \\ \cdots 0 & p_{-1} & 0 & 1 & 0 & q_{-1} & 0 & \cdots & \\ & \cdots 0 & p_0 & 0 & 1 & 0 & q_0 & 0 \cdots & \\ & & \cdots 0 & p_1 & 0 & 1 & 0 & q_1 & 0\cdots \\ \cdots & \cdots & \cdots & \cdots & \cdots & \cdots & \cdots & \cdots & \cdots \end{bmatrix}.$$

For nonzero unknowns α_n, the corresponding infinite determinant must be zero. This doubly infinite determinant, by taking n arbitrary integer, becomes one-sided infinite determinant, and we can write

$$\Delta(n,\infty)=0, \tag{7}$$

where the infinite determinant $\Delta(n,\infty)$ is given by the limit

$$\begin{array}{l}\lim\Delta(n,m)= \\ n=\text{fixed integer} \\ m=0,1,2,3,\cdots \\ m\to\infty\end{array}\left\|\begin{array}{ccccccccc} 1 & 0 & q_n & 0 & 0 & \cdots & \cdots & 0 & 0 \\ 0 & 1 & 0 & q_{n+1} & 0 & \cdots & \cdots & 0 & 0 \\ p_{n+2} & 0 & 1 & 0 & q_{n+2} & \cdots & \cdots & 0 & 0 \\ \cdots & \cdots & \cdots & \cdots & \cdots & \cdots & \cdots & \cdots & \cdots \\ 0 & 0 & \cdots & \cdots & \cdots & 0 & 1 & 0 & q_{n+m-2} \\ 0 & 0 & \cdots & \cdots & \cdots & p_{n+m-1} & 0 & 1 & 0 \\ 0 & 0 & \cdots & \cdots & \cdots & 0 & p_{n+m} & 0 & 1 \end{array}\right\|. \tag{8}$$

Consider a weak nonlinearity, i.e., $\lambda_1|x|^3\ll\omega_1^2|x|$, or $\lambda_1x^2\ll(K_1+K_2)/m_1$. From the first of (3) we get

$$\max x^2=\left[\sum_{n=-\infty}^{\infty}\alpha_n\right]^2=\sum_{\rho_1=-\infty}^{\infty}\sum_{\rho_2=-\infty}^{\infty}\alpha_{\rho_1}\alpha_{\rho_2},$$

and since α_1 is considered as dominant element of that of the sequence $\{\alpha_n\}$, when $\max x^2=2\alpha_1^2$, then

$$\lambda_1\ll(K_1+K_2)/(2m_1\alpha_1^2). \tag{9a}$$

For a weak nonlinearity we can write

$$\rho_n=\frac{3\lambda_1\alpha_1^2}{k_n+6\lambda_1|\alpha_1|^2}=\frac{3\lambda_1\alpha_1^2}{k_n}\times\left\{1-\lambda_1\frac{6|\alpha_1|^2}{k_n}+\lambda_1^2\left(\frac{6|\alpha_1|^2}{k_n}\right)^2-\cdots\right\}$$
$$=\lambda_1\frac{3\alpha_1^2}{k_n}-\lambda_1^2\frac{18\alpha_1^2|\alpha_1|^2}{k_n^2}+\cdots, \tag{9}$$
$$q_n=\lambda_1\frac{3\bar{\alpha}_1^2}{k_n}-\lambda_1^2\frac{18\bar{\alpha}_1^2|\alpha_1|^2}{k_n^2}+\cdots.$$

All the elements not in the main diagonal of the determinant (8) have the small coefficient λ_1 as a common factor. Then, by applying formula (B) of Appendix II, and taking the first terms of ρ_n and q_n from formulas (9), the determinant (8) can be written approximately as

$$\Delta(n,\infty)=1-9\lambda_1^2|\alpha_1|^4\sum_{m=0}^{\infty}\frac{1}{k_{n+m}k_{n+m+2}}+0(\lambda_1^3). \tag{10}$$

$k_{\bar{n}}$, given by (6a), is an even function of $\bar{n}$. To examine the convergence of the series in (10) we confine ourselves to non-negative integers $\bar{n}$. Since $\{|k_{\bar{n}}|\}$ is a sequence with positive terms monotonically increasing with $\bar{n}$, and $|k_{\bar{n}}|\to\infty$, as $\bar{n}\to\infty$, the series in (10) converges. Convergence requirements of the series in (10) necessitates the ω^2 must not be a zero of $k_{\bar{n}}(\omega^2)$; hence

$$\omega^2\mp\omega_{0\pm}^2=\frac{1}{2\bar{n}^2}\{\omega_1^2+\omega_2^2\pm[(\omega_1^2-\omega_2^2)^2+4\lambda_2\lambda_3]^{\frac{1}{2}}\}. \tag{11}$$

The principal frequency ω can be determined from (7) by using (10).

Then, if we confine ourselves to the first term of the series in (10) and put $n=1$, the principal frequency ω is approximately a root of

$$k_1k_3=9\lambda_1^2|\alpha_1|^4,$$

or root of

$$\left(\omega_1^2-\omega^2-\frac{\lambda_2\lambda_3}{\omega_2^2-\omega^2}\right)\left(\omega_1^2-9\omega^2-\frac{\lambda_2\lambda_3}{\omega_2^2-9\omega^2}\right)-9\lambda_1^2|\alpha_1|^4=0, \tag{12}$$

which is quartic in ω^2. Formula (12) gives the principal frequency ω of order $O(\lambda_1^2)$. For values of the principal frequency of order higher than $O(\lambda_1^2)$, we find in the expansion of $\Delta(n,\infty)$ terms of order higher than $O(\lambda_1^2)$,[2] and continue in the same way to find the corresponding algebraic equation.

4. CALCULATION OF THE COEFFICIENTS

To calculate the coefficients α_n, we use in the recursion formula (6) the notation

$$u_n=\alpha_n/\alpha_{n-2}, \tag{13}$$

when we can get

$$u_n=-\frac{\rho_n}{1+q_nu_{n+2}},$$

which leads to the infinite continued fraction

$$\frac{\alpha_n}{\alpha_{n-2}}=-\cfrac{\rho_n}{1-\cfrac{\rho_{n+2}q_n}{1-\cfrac{\rho_{n+4}q_{n+2}}{1-\cdots}}}, \tag{14}$$

where the ρ's and q's are given by (9). The formula (14) is written as

$$\alpha_n/\alpha_{n-2}=-\rho_nZ_{n,\infty}, \tag{15}$$

[2] W. Magnus, "Infinite determinants in the theory of Mathieu's and Hill's equations," Research Report No. BR-1, Mathematical Research Group, Washington Square College of Arts and Science, New York University, 1953.

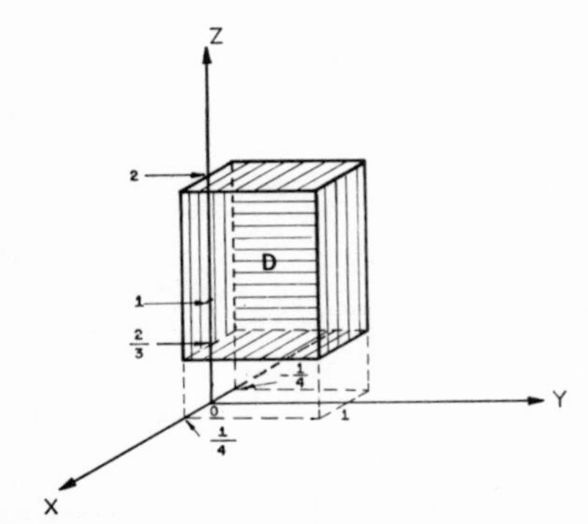

FIG. 2. The domain D in the X, Y, Z space is the appropriate one for the convergence of the continued fraction (16).

where

$$Z_{n,\infty}=\cfrac{1}{1+\cfrac{X_{n,1}}{1+\cfrac{X_{n,2}}{1+\cdots}}}. \tag{16}$$

$$X_{n,m}=-p_{n+2m}q_{n+2(m-1)}, \tag{16a}$$

n= fixed integer, $m=1,2,3,\cdots$.

Since the elements $X_{n,m}$ of (16) are functions of the real variable ω, the regions E, Y, V, defined in Appendix III, are segments of lines, and according to von Koch's[3] and Worpitzky's theorems, as stated in Appendix III,

$$\begin{aligned} &E:\ -\tfrac{1}{4}\leq X_{n,m}\leq\tfrac{1}{4}, \\ &Y:\ 0<Y_{n,m}=\sum_{m=2}|X_{n,m}|<1, \\ &V:\ \tfrac{2}{3}\leq Z_{n,m}\leq 2. \end{aligned} \tag{17}$$

Consider a Cartesian coordinate system in space, and let us take the region element on the X axis, the series region on the Y axis, and the value region on the Z axis. The inequalities (17) correspond then to the interior of the orthogonal parallelepiped D, Fig. 2, which is an open domain.

The domain

$$D:\ -\tfrac{1}{4}\leq X\leq\tfrac{1}{4},\quad 0<Y<1,\quad \tfrac{2}{3}\leq Z\leq 2$$

is the appropriate one for the continued fraction (16).

To evaluate the continued fraction (16) we apply the theory of Appendix III. The denominator of the ratio, which gives the mth approximant Z_m of the continued fraction (16) is

$$B_{n,m-1}=\left\|\begin{array}{cccccc} 1 & -1 & 0\ \cdots & 0 & 0 & 0 \\ -p_{n+2}q_n & 1 & -1\ \cdots & 0 & 0 & 0 \\ 0 & -p_{n+4}q_{n+2} & 1\ \cdots & 0 & 0 & 0 \\ \cdots & \cdots & \cdots & \cdots & \cdots & \cdots \\ 0 & 0\cdots & \cdots & -p_{n+2(m-1)}q_{n+2(m-2)} & 1 & -1 \\ 0 & 0\cdots & \cdots & 0 & -p_{n+2m}q_{n+2(m-1)} & 1 \end{array}\right\|. \tag{18}$$

The numerator $A_{n,m-1}$ of the ratio can be obtained from (18) if we omit its first row and first column. Taking the limit as $m\to\infty$, we obtain

$$\frac{\alpha_n}{\alpha_{n-2}}=-p_n\frac{\bar{\Delta}(n+2,\infty)}{\bar{\Delta}(n,\infty)}, \tag{19}$$

where

$$\bar{\Delta}(n,\infty)=\lim_{m\to\infty}B_{n,m-1},$$
$$\bar{\Delta}(n+2,\infty)=\lim_{m\to\infty}A_{n,m-1}. \tag{19a}$$

The determinants of (19) are of von Koch's type and they converge by von Koch's rule, that is, when the series $\sum_n|p_nq_{n-2}|$ converges, which happens here, as was pointed out at the discussion of the convergence of the series of (10).

Since $\alpha_n=\bar{\alpha}_n$, we may restrict ourselves to non-negative integers for the calculation of the coefficients α_n. Formula (19) suggests starting with $n=2$; then α_0 and α_1 are arbitrary. The induction procedure applied to (19) for the coefficients with even index gives

$$\alpha_{2n}=(-1)^n\alpha_0p_2p_4\cdots p_{2n}\frac{\bar{\Delta}(2n+2,\infty)}{\bar{\Delta}(2,\infty)},$$
$$n=1,2,3\cdots. \tag{20}$$

For the determination of the coefficients with odd index, we first calculate the coefficient α_3. This, according to Appendix I, is an exception.

If, according to Appendix I, we take $\alpha_1{}^3$ instead of the double series of the formula (4a), and we apply this formula for $n=3$, there results

$$\alpha_3=-(\lambda_1\alpha_1{}^3/k_3), \tag{21}$$

where

$$k_3=\omega_1{}^2-9\omega^2-[\lambda_2\lambda_3/(\omega_2{}^2-9\omega^2)]. \tag{21a}$$

[3] H. von Koch, Compt. rend., 120, 144 (1895).

Now, by applying the induction procedure to (19) starting from the coefficient α_5 and using the value of α_3 given by (21), one can get

$$\alpha_{2n+1}=(-1)^{n+1}\frac{\lambda_1\alpha_1^3}{k_3}\rho_5\rho_7\cdots\rho_{2n+1}\frac{\bar{\Delta}(2n+3,\infty)}{\bar{\Delta}(5,\infty)},\quad n=2,3,4\cdots. \tag{22}$$

By applying formula (B) of Appendix II to the determinants of (20) and (22), there is obtained

$$\frac{\bar{\Delta}(2n+2,\infty)}{\bar{\Delta}(2,\infty)}=1-9\lambda_1^2|\alpha_1|^4\sum_{m=0}^{\infty}\left(\frac{1}{k_{2n+2+m}k_{2n+4+m}}-\frac{1}{k_{2+m}k_{4+m}}\right)+O(\lambda_1^3),$$

$$\frac{\bar{\Delta}(2n+5,\infty)}{\bar{\Delta}(5,\infty)}=1-9\lambda_1^2|\alpha_1|^4\sum_{m=0}^{\infty}\left(\frac{1}{k_{2n+5+m}k_{2n+7+m}}-\frac{1}{k_{5+m}k_{7+m}}\right)+O(\lambda_1^3). \tag{23}$$

The formulas (21), (20), and (22) give the coefficients of the first of the series (3) for any positive n, with arbitrary α_0 and α_1. For the determination of the coefficients α_{-n} we use the property $\alpha_{-n}=\bar{\alpha}_n$. The coefficient α_0 is real, and α_1 in general complex, $\alpha_1=|\alpha_1|e_i^{\varphi_1}$. If we take α_1 real, the solution (3) can be written as follows:

$$x(t)=\alpha_0+2\alpha_1\cos\omega t-2\alpha_3\cos 3\omega t+2\sum_n\alpha_N\cos N\omega t,$$

$$y(t)=\frac{\lambda_3}{\omega_2^2}\alpha_0+\frac{2\lambda_3}{\omega_2^2-\omega^2}\alpha_1\cos\omega t-\frac{2\lambda_3}{\omega_2^2-3^2\omega^2}\alpha_3\cos 3\omega t+2\sum_n\frac{\lambda_3}{\omega_2^2-N^2\omega^2}\alpha_N\cos N\omega t, \tag{24}$$

where the coefficients α_3, $\alpha_{N=2n}$, $\alpha_{N=2n+1}$ are given by the formulas (21), (20), (22), respectively. The formulas (24) give the solution in its final form, and the formulas (20)–(23) can be used for the calculation of as many coefficients of the solution (24) as we want in terms of powers of λ_1.

We can easily see that for the calculation of the coefficients, in terms of powers of λ_1, does not need but only the unit as value of the ratios of the determinants (23). The first two terms of the solution (24) are independent of λ_1. For terms of order $O(\lambda_1)$, we take the first term of ρ_n of (9) and combining it with (20) for $n=1$ we get the 2nd harmonic, which, with (21), gives

$$x(t)=\alpha_0+2\alpha_1\cos\omega t-2\lambda_1\left\{\frac{3\alpha_0\alpha_1^2}{k_2}\cos 2\omega t+\frac{\alpha_1^3}{k_3}\cos 3\omega t\right\}+O(\lambda_1^2), \tag{25}$$

$$y(t)=\frac{\lambda_3}{\omega_2^2}\alpha_0+\frac{2\lambda_3}{\omega_2^2-\omega^2}\alpha_1\cos\omega t-2\lambda_1\left\{\frac{3\lambda_3\alpha_0\alpha_1^2}{k_2(\omega_2^2-2^2\omega^2)}\cos 2\omega t+\frac{\lambda_3\alpha_1^3}{k_3(\omega_2^2-3^2\omega^2)}\cos 3\omega t\right\}+O(\lambda_1^2). \tag{25}$$

If we take the terms of ρ_n of (9) up to the order $O(\lambda_1^2)$, and combine them with (22) for $n=2$ we obtain the 4th harmonic terms of order $O(\lambda_1^2)$:

$$18\lambda_1^2\frac{\alpha_0\alpha_1^4}{k_2k_4}\cos 4\omega t,\quad 18\lambda_1^2\frac{\alpha_0\alpha_1^4\lambda_3}{k_2k_4(\omega_2^2-4^2\omega^2)}\cos 4\omega t, \tag{25a}$$

of $x(t)$ and $y(t)$, respectively. The above procedure indicates how we can get higher harmonics in terms of higher powers of λ_1. The solution (24), constructed as indicated above, must be convergent and its coefficient of the fundamental term must be much larger than any other coefficient. These requirements imply that the following conditions are satisfied:

$$\alpha_0\ll\alpha_1,\quad \lambda_1\ll\min\left\{\frac{K_1+K_2}{2m_1\alpha_1^2},\frac{k_2}{2\alpha_0\alpha_1},\frac{k_3}{\alpha_1^2}\right\}. \tag{26}$$

For the second condition, the inequality (6a) was taken into account. The 4th, 5th, $\cdots$ harmonic terms are of order $O(\lambda_1^2)$, $O(\lambda_1^3)$, $\cdots$, and the convergence is guaranteed.

Since α_1 is much larger compared to α_0, the solution in the linear case is approximately

$$x(t)=2\alpha_1\cos\omega t,\qquad x(t)=2\frac{\lambda_3}{\omega_2^2-\omega^2}\alpha_1\cos\omega t, \tag{25b}$$

and the motions of the oscillators are "in phase" for $\omega_2<\omega$, and "180° out of phase" for $\omega_2<\omega$.

5. THE PRINCIPAL MODES AS SOLUTIONS OF INITIAL VALUE PROBLEMS

Another approach for the determination of principal modes may be based on the manner in which the system is set into motion. This is equivalent to considering the principal modes as solutions of special initial value problems.

The differential equations of the "dual-mode" system are considered subject to the restriction that the masses are displaced from their equilibrium positions either both up, or one up and the other down, by amounts x_0 and y_0, respectively, and released without velocity; i.e.,

$$x(0)=x_0,\quad y(0)=y_0,\quad \dot{x}(0)=\dot{y}(0)=0, \tag{27a}$$

$$x(0)=x_0,\quad y(0)=-y_0,\quad \dot{x}(0)=\dot{y}(0)=0. \tag{27b}$$

If the initial displacements x_0 and y_0 are appropriately related, then each one of these initial conditions gave rise to special vibration modes, which are, by definition, the "principal modes" of the system. To calculate the principal modes of our system utilizing infinite determinants and using the above definition, assume a solution in the form of complex exponential series (3), as in the previous case. The calculation of the principal frequency and the coefficients of the series has been completed throughout the preceding sections and the solution is found to be in the form given by (25). An approximation of this solution is given by (26), associated with the initial conditions

$$x_0=2\alpha_1,\quad y_0=\frac{\lambda_3 x_0}{\omega_2^2-\omega^2}=\frac{2\alpha_1\lambda_3}{\omega_2^2-\omega^2}. \tag{28}$$

Formulas (28) give the relations of the initial displacements required for the solution to be of "principal modes" type. These sinusoidal motions are "in phase" for the initial conditions (27a) if $\omega<\omega_2$.

Both definitions of principal modes lead to the same solution; they have the same physical interpretation and they are equivalent.

The discussion here is based on two definitions of principal modes and the final solution found by analyzing the infinite determinants involved. It may be mentioned that G. W. Hill, in his *Lunar Theory* brought into notice the infinite determinants, and H. Poincaré first gave conditions for their convergence.

ACKNOWLEDGMENT

The author is greatly indebted to Professor L. Brillouin for his unflagging interest and advice.

APPENDIX I. THE DOMINANT SUM OF A DOUBLE SERIES

Suppose in the sequence $\{\alpha_n\}$, $n=0, \pm1, \pm2, \cdots$ the complex elements have the property $\alpha_{-n}=\bar{\alpha}_n$. If α_1 and α_{-1} are dominant elements in the sequence, then it is easily seen that the dominant sum of the double series of (4a) is given by

$$3\alpha_1^2\alpha_{n-2}+6|\alpha_1|^2\alpha_n+3\bar{\alpha}_1^2\alpha_{n+2}, \tag{A1}$$

where n is any integer except $n=\pm1, \pm3$. For these exceptions the dominant terms of the double series are

$$3\bar{\alpha}_1\alpha_1^2 \text{ for } n=1,\quad 3\alpha_1\bar{\alpha}_1^2 \text{ for } n=-1,$$
$$\alpha_1^3 \text{ for } n=3,\quad \bar{\alpha}_1^3 \text{ for } n=-3. \tag{A2}$$

APPENDIX II. APPROXIMATE VALUE OF AN INFINITE DETERMINANT

If in an infinite convergent determinant,

$$\Delta=\|B_{m,n}\|_{-\infty}^{+\infty},$$

the elements in the main diagonal are equal to unity, and all the elements not in the main diagonal have a small common factor,[4] say ϵ, i.e., if $B_{m,m}=1$, $B_{m,n}=\epsilon\bar{B}_{m,n}$, $m\neq n$, we may get an expression of the determinant in powers of ϵ. The first term in this expression is independent of ϵ; it is the product of all the elements in the main diagonal, that is 1. The next terms in the expansion are in ϵ^2 and are obtained by replacing the product $\prod_m B_{m,m}$ the elements in the main diagonal (m,m) and (n,n) by the elements not in the main diagonal (m,n) and (n,m). These terms have a minus sign, according to the laws of determinants, and they are $-\epsilon^2\sum_m\sum_n \bar{B}_{m,n}\bar{B}_{n,m}$; the determinant Δ can be written in the form

$$\Delta=1-\epsilon^2\sum_m\sum_n \bar{B}_{m,n}\bar{B}_{n,m}+0(\epsilon^3). \tag{B}$$

APPENDIX III. A CONTINUED FRACTION AS A RATIO OF TWO INFINITE DETERMINANTS

Given the continued fraction

$$Z_\infty=\cfrac{1}{1+\cfrac{X_2}{1+\cfrac{X_3}{1+\cdots}}}, \tag{C1}$$

where the complex elements X are subject to specified conditions, its mth approximant Z_m, obtained by stopping with the mth partial quotient, can be estimated If the elements X of the sequence $\{X_m\}$ of (C1) have arbitrary values in a region, the "region element" E, then the correspondent series $\sum_{m=2}^{\rho}|X_m|$, $\rho=2, 3, 4, \cdots$ has its values in the "series region" Y, and the approximants Z_m have all their values in the value region V. The following theorems give relationships between the above regions.[5]

H. von Koch's Theorem

"If

$$Y:\ \sum_{m=2}^{\rho}|X_m|<1,\quad \rho=2, 3, 4, \cdots \tag{C2}$$

then: the continued fraction (C1) converges."

Worpitzky's Theorem

"If

$$E:\ |X_m|\leq\tfrac{1}{4},\quad m=2, 3, 4\cdots \tag{C3}$$

then

$$V:\ |Z_m-\tfrac{4}{3}|\leq\tfrac{2}{3}.\text{''} \tag{C4}$$

[4] L. Brillouin, *Wave Propagation in Periodic Structures* (Dover Publications, New York, 1953), 2nd ed., pp. 34, 35.

[5] H. Wall, *Analytic Theory of Continued Fractions* (D. Van Nostrand Company, Inc., Princeton, New Jersey, 1948), pp. 26, 42, and 51.

The mth approximant of (C1), Z_m, can have the form of a ratio of two determinants. To show that, one associates the continued fraction (C1) a sequence of linear transformations

$$X_1(v)=1, \quad X_m=1/(1+X_m v), \quad m=2, 3, 4, \cdots.$$

If:

$$X_\rho \neq 0, \quad \rho=2, 3, \cdots, m; \quad X_\rho=0, \quad \rho>m,$$

then the product of m of the above transformations is

$$Z_m=X_1X_2\cdots X_m(v)=\cfrac{1}{1+\cfrac{X_2}{1+\cdots+\cfrac{X_{m-1}}{1+X_m v}}}=\frac{X_m v A_{m-2}+A_{m-1}}{X_m v B_{m-2}+B_{m-1}}, \tag{C5}$$

where the A's and B's may be calculated by means of the recursion formulas

$$A_\rho=A_{\rho-1}+X_\rho A_{\rho-2}, \quad B_\rho=B_{\rho-1}+X_\rho B_{\rho-2}, \quad \rho=1, 2, 3, \cdots. \tag{C6}$$

For the above we require the initial values

$$A_{-1}=1, \quad A_0=0, \quad B_{-1}=0, \quad B_0=1, \quad X_1=1.$$

The mth approximant of (C1), Z_m, is given by (C5) if $v=0$, then it is equal to the ratio A_{m-1}/B_{m-1}.

The recursion formulas (C6) give two systems of homogeneous linear equations, one in the variables A, the other in the variables B. These systems give rise to two determinants, which give the values of the A's and B's. The B's are given by the determinant

$$B_{m-1}=\begin{Vmatrix} 0 & -1 & 0 & 0 & \cdots & & & \cdots & 0 \\ X_2 & 1 & -1 & 0 & \cdots & & & \cdots & 0 \\ 0 & X_3 & 1 & -1 & \cdots & & & \cdots & 0 \\ \cdots & \cdots & \cdots & \cdots & \cdots & \cdots & \cdots & \cdots & \cdots \\ 0 & & & & \cdots & X_{m-1} & 1 & & -1 \\ 0 & & & & \cdots & 0 & X_m & & 1 \end{Vmatrix}, \tag{C7}$$

$$m=2, 3, \cdots.$$

The determinant for the A's can be obtained from the above determinant by omitting its first row and its first column. These determinants are different from zero.

The value of the continued fraction (C1) is given, by definition, by the limit $\lim_{\rho\to\infty} (A_\rho/B_\rho)$.

ΠΡΑΚΤΙΚΑ
ΤΗΣ ΑΚΑΔΗΜΙΑΣ ΑΘΗΝΩΝ

ΕΤΟΣ 1963 : ΤΟΜΟΣ 38ος

ΑΝΑΤΥΠΟΝ
ΣΕΛ. 33 - 36

On the Convergence of Series Related to Principal Modes of Nonlinear Systems*.

by ***Demetrios G. Magiros*****.

'Ανεκοινώθη ὑπὸ τοῦ 'Ακαδημαϊκοῦ κ. 'Ιωάνν. Ξανθάκη.

I. INTRODUCTION

a. In previous papers [1], where the principal modes of a «dual mode» nonlinear system have been discussed, the solution is found in the form of series. The object of the present announcement is to give a brief discussion of the convergence of these series. Details of the discussion will appear in a forthcoming paper. The convergence is based on the «Abel's test» of convergence.

b. We state for reference that the solution found is the following series:

$$\text{(a).}\quad x(t)=a_0+2a_1\cos\omega t+\frac{2\lambda_1 a_1^3}{k_3}\cos 3\omega t+2\sum_N a_N \cos N\omega t \tag{1}$$

$$\text{(b).}\quad \psi(t)=\frac{\lambda_3 a_0}{\omega_2^2}+\frac{2\lambda_3 a_1}{\omega_2^2-\omega^2}\cos\omega t+\frac{2\lambda_1\lambda_3 a_1^3}{k_3(\omega_2^2-3^2\omega^2)}\cos 3\omega t+2\sum_N \frac{\lambda_3 a_N}{\omega_2^2-N^2\omega^2}\cos N\omega t$$

where:

$$\text{(a).}\quad a_{N=2n}=(-1)^n a_0 p_2 p_4 \ldots p_{2n}\,,\quad n=1,2,3,\ldots$$

$$\text{(b).}\quad a_{N=2n+1}=(-1)^{n+1}\frac{\lambda_1 a_1^3}{k_3}p_5 p_7\ldots p_{2n+1}\,,\quad n=2,3,4,\ldots$$

$$\text{(c).}\quad p_N=\frac{3\lambda_1 a_1^2}{k_N}\left[1-\frac{6\lambda_1|a_1|^2}{k_N}+\left(\frac{6\lambda_1|a_1|^2}{k_N}\right)^2-\ldots\right] \tag{2}$$

$$\text{(d).}\quad k_N=-N^2\omega^2+\omega_1^2+\frac{\lambda_2\lambda_3}{N^2\omega^2-\omega_2^2}$$

* Of the Valley Forge Space Technology Center of the General Electric Comp. M.S.D. Philadelphia, Pa.

** ΔΗΜ. ΜΑΓΕΙΡΟΥ, 'Επὶ τῆς συγκλίσεως σειρῶν τῶν πρωταρχικῶν ταλαντώσεων μὴ γραμμικῶν συστημάτων.

Reprinted from the *Proceedings of the Athens Academy of Sciences* **38** (1963), 33–36.

The values ω which are either submultiples of ω_2 or zeros of the k's are the sigularities of this solution.

2. THE CONVERGENCE OF THE SERIES

The convergence of the series (1) is deduced from that of the series $\sum_N a_N$. The proof of convergence of $\sum_N a_N$ can be done in two steps, namely by using either the first term of the series (2c) (first step), or all its terms (second step).

The formula (2d), for large values of N, shows that $|k_N|$ is of order N^2 and that the sign of k_N is negative. Then a value N_1 of N can be found such that k_N remains negative for any $N > N_1$ and $|k_N|$ increases with N.

First Step

By taking the integers $\bar{N}$, $\bar{n}$, n, σ such that $\bar{N}=2\bar{n}>N_1$, $n=\bar{n}+\sigma$, where N_1, $\bar{N}$, $\bar{n}$ are fixed, the coefficients a_N can be written as:

$$a_{N=2n} = C_{2\bar{n}} \cdot (3\lambda_1 a_1^2)^\sigma \cdot \frac{1}{k_2(\bar{n}+1) \cdots k_2(\bar{n}+\sigma)}$$

$$a_{N=2n+1} = C_{2\bar{n}+1} \cdot (3\lambda_1 a_1^2)^\sigma \cdot \frac{1}{k_2(\bar{n}+1)+1 \cdots k_2(\bar{n}+\sigma)+1} \tag{3}$$

Where $C_{2\bar{n}}$ and $C_{2\bar{n}+1}$ are constants, and the k's positive. The series $\sum_N a_N$ can be split as follows:

$$\sum_N a_N = \sum_{N=0}^{N_1-1} a_N + C_{2\bar{n}} \sum_{\sigma=1}^{\infty} (3\lambda_1 a_1^2)^\sigma \frac{1}{K_2(\bar{n}+1) \cdots K_2(\bar{n}+\sigma)} + C_{2\bar{n}+1} \sum_{\sigma=1}^{\infty} (3\lambda_1 a_1^2)^\sigma \frac{1}{k_2(\bar{n}+1)+1 \cdots K_2(\bar{n}+\sigma)+1} \tag{4}$$

By imposing the restriction $\lambda_1 < \frac{1}{3a_1^2}$, the Abel's test for the convergence [2b] of the infinite series of the right-hand member of (4) can be applied, since:

(a) the geometrical series $\sum_{\sigma=1}^{\infty} (3\lambda_1 a_1^2)^\sigma$ is convergent, and

(b) the sequences of the products of the inverse of k's are monotonic decreasing and bounded sequences.

Second Step

The coefficients a_N in this case can be expressed as follows:

$$a_{N=2n} = (-1)^n a_0 \frac{(3\lambda_1|a_1|^2)^n}{k_2 \dots k_{2n}} \left[1 - \frac{6\lambda_1|a_1|^2}{k_2} + \left(\frac{6\lambda_1|a_1|^2}{k_2}\right)^2 - \dots\right] \dots \left[1 - \frac{6\lambda_1|a_1|^2}{k_{2n}} + \left(\frac{6\lambda_1|a_1|^2}{k_{2n}}\right)^2 - \dots\right] \quad (5)$$

$$a_{N=2n+1} = (-1)^{n+1} \frac{a_1}{3} \frac{(3\lambda_1|a_1|^2)^n}{k_3 \dots k_{2n+1}} \left[1 - \frac{6\lambda_1|a_1|^2}{k_3} + \left(\frac{6\lambda_1|a_1|^2}{k_3}\right)^2 - \dots\right] \dots \left[1 - \frac{6\lambda_1|a_1|^2}{k_{2n+1}} + \left(\frac{6\lambda_1|a_1|^2}{k_{2n+1}}\right)^2 - \dots\right]$$

Then the series $\sum_N a_N$ can be written as:

$$\sum_N a_N = \sum_{N=2n} A_{2n} \prod_{2n} + \sum_{N=2n+1} A_{2n+1} \prod_{2n+1} \quad (6)$$

where A's are the factors of the right--hand members of (5) outside the brackets, and Π's the products of the brackets, namely:

$$\prod_r^N \left[1 - \frac{6\lambda_1|a_1|^2}{k_r} + \left(\frac{6\lambda_1|a_1|^2}{k_r}\right)^2 - \dots\right]$$

which for r either even or odd is either $\prod_{2n}$ or $\prod_{2n+1}$, respectively.

The series $\sum A_{2n}$ and $\sum A_{n+1}$ are convergent, according to the preceding step. In addition the Π's of (6) are sequences with monotone and bounded terms, because $\prod_r^N$ can be wiritten as:

$$\prod_r^N \left[1 - \frac{6\lambda_1|a_1|^2}{k_r} + \left(\frac{6\lambda_1|a_1|^2}{k_r}\right)^2 - \dots\right] = \prod_r^{N1-1} \left[1 - \frac{6\lambda_1|a_1|^2}{k_r} + \left(\frac{6\lambda_1|a_1|^2}{k_r}\right)^2 - \dots\right] \prod_{N=N1}^N \left[1 + \frac{6\lambda_1|a_1|^2}{k_r} + \left(\frac{6\lambda_1|a_1|^2}{k_r}\right)^2 + \dots\right] \quad (7)$$

where $\prod_r^{N1-1}$ is a finite fixed number, and $\prod_{N=N1}^N$ is convergent as $N \to \infty$, since the series $\sum_{r=N1}^{\infty} \left(\frac{6\lambda_1|a_1|^2}{k_r}\right)^2$ is convergent. [2α]

The above proves that the series (la) is convergent. For the proof of the convergence of the series (lb), we see that the Abel's test of convergence can be applied to the series.

$$\sum_{N=0}^{\infty} \frac{\lambda_s}{N^2\omega^2 - \omega_2^2} a_N .$$

The nature of the singularities of the series (1) and some subjects related to these singularities will be discussed in another paper.

REFERENCES

[1]. D. G. MAGIROS: (a) Nat. Ac. Sc. Vol. 46, 1608 (1960); Vol. 47, 883 (1961); (b) J. Math Phys. Vol 2, No 6, pp. 869-875, (Nov-Dec. 1961).

[2]. K. KNOPP: «Infinite Sequences and Series», Dover Publications, Inc. New York (1956), pg. (a) 94, (b) 137.

ΠΕΡΙΛΗΨΙΣ

Εἰς προηγουμένας ἐργασίας μας ἔχει εὑρεθῆ ὑπὸ μορφὴν σειρῶν ἡ λύσις προβλήματος τῶν πρωταρχικῶν ταλαντώσεων μὴ γραμμικῶν συστημάτων. Ἐνταῦθα δίδεται σύντομος ἐξέτασις τῆς συγκλίσεως τῶν σειρῶν αὐτῶν. Κατὰ τὴν πορείαν διὰ τὴν ἔρευναν τῆς συγκλίσεως γίνεται χρῆσις τοῦ θεωρήματος τοῦ Abel περὶ συγκλίσεως.

Ἡ πλήρης ἐξέτασις τῆς συγκλίσεως, καθὼς καὶ ἡ φύσις τῶν ἀνωμάλων σημείων τῆς λύσεως θὰ ἐκτεθοῦν εἰς ἐργασίαν, ἡ ὁποία συντόμως δημοσιεύεται ἀλλαχοῦ.

ΠΡΑΚΤΙΚΑ
ΤΗΣ ΑΚΑΔΗΜΙΑΣ ΑΘΗΝΩΝ

ΕΤΟΣ 1960: ΤΟΜΟΣ 35ος

ΑΝΑΤΥΠΟΝ
ΣΕΛ. 96 - 103

The motion of a projectile around the earth under the influence of the earth's gravitational atraction and a thrust*,

by ***Dem. G. Magiros*** ****.**

Ἀνεκοινώθη ὑπὸ τοῦ κ. Ἰωάνν. Ξανθάκη.

Abstract.

In this paper the motion of a projectile around the earth under the influence of the gravitational attraction of the earth and a thrust is discussed. The orbit of the projectile and its velocity along the orbit are found in two cases, namely when the thrust is suddenly applied to the projectile,

* Republic Aviation Corp., U.S.A.

** ΔΗΜ. ΜΑΓΕΙΡΟΥ, Ἡ κίνησις βλήματος πέριξ τῆς γῆς ὑπὸ τὴν ἐπίδρασιν τῆς ἑλκτικῆς δυνάμεως τῆς γῆς καὶ μιᾶς ὠστικῆς δυνάμεως.

Reprinted from the *Proceedings of the Athens Academy of Sciences* **35** (1960), 96–103.

and when it is gradually applied. The types of the Keplerian orbit, when the thrust is suddenly removed, are also discussed.

Introduction.

The following problem is discussed:

A projectile is moving around the earth in a Keplerian orbit, when a thrust is applied. Find the motion of the projectile during the action of the thrust.

This problem is solved for a general thrust vector, either suddenly or gradually applied to the projectile and acting continuously for non-infinitesimal time. The problem is specialized in the case of constant thrust vector, case which is identical with the problem of Quantum Mechanics in connection with «Stark Effect». Conditions are given in connection of the types of the Keplerian orbit, when the thrust is suddenly removed from the projectile.

I. *Mathematical formulation of the problem.*

If $\underline{T}$ is the thrust per unit mass, acting for time τ, the motion of the projectile during the time τ is governed by the differential equation:

$$\ddot{\underline{r}} = -\frac{G}{r^3}\underline{r} + \underline{T} \tag{1}$$

where r is the magnitude of $\underline{r}$, G a constant equal to $\frac{k(M+m)}{m}$, M and m are the masses of the earth and the projectile, respectively; k the constant of gravitation.

The time τ is split according to:

$$\tau = t_o + t_1, \tag{2}$$

where t_o is infinitesimal, and we take as initial conditions to the differential Eq. (1) the following conditions:

$$\underline{r}(t)_{t=t_o} = \underline{r}_o \;,\; \dot{\underline{r}}(t)_{t=t_o} = \dot{\underline{r}}_o + \underline{I}_o, \tag{3}$$

where $\underline{r}_o$ and $\dot{\underline{r}}_o$ are the position vector and velocity vector, respectively, at the point of the original orbit where the thrust is applied to the projectile; and $\underline{I}_o$ is the impulse per unit mass of the thrust $\underline{T}$ acting on the projectile in the infinitesimal time t_o. Such a selection of initial conditions leads to a solution which is very helpful to practical problems.

II. *Solution of the problem.*

By changing the time t according to:

$$\bar{t} = t - t_0 \tag{2.1}$$

the dif. Eq. (1) keeps its form, and the initial conditions (3) in the new scale time are:

$$\underline{r}(\bar{t})_{\bar{t}=0} = \underline{r}_0, \qquad \dot{\underline{r}}(\bar{t})_{\bar{t}=0} = \dot{\underline{r}}_0 + \underline{I}_0 \tag{3.1}$$

A solution of the system (1) and (3.1) of the form:

$$\underline{r}(\bar{t}) = a(\bar{t})\,\underline{r}_0 + b(\bar{t})\,(\dot{\underline{r}}_0 + \underline{I}_0) + c(\bar{t})\,\underline{T}_0 \tag{4}$$

is going to be established by calculating the functions $a(\bar{t})$, $b(\bar{t})$, $c(\bar{t})$ appropriately. $\underline{T}_0$ is the thrust at $\bar{t}=0$.

For $\bar{t}=0$ the function of Eq. (4) gives

$$a(0) = 1, \quad b(0) = 0, \quad c(0) = 0 \tag{4.1}$$

and its derivative:

$$\dot{a}(0) = 0, \quad \dot{b}(0) = 1, \quad \dot{c}(0) = 0 \tag{4.2}$$

The function of Eq. (4) must satisfy identically the differential Eq. (1); then:

$$\ddot{a}(\bar{t})\,\underline{r}_0 + \ddot{b}(\bar{t})\,(\dot{\underline{r}}_0 + \underline{I}_0) + \ddot{c}(\bar{t})\,\underline{T}_0 \equiv$$

$$-\frac{G}{r^3}\,a(\bar{t})\,\underline{r}_0 - \frac{G}{r^3}\,b(\bar{t})\,(\dot{\underline{r}}_0 + \underline{I}_0) - \frac{G}{r^3}\,c(\bar{t})\,\underline{T}_0 + \underline{T}(\bar{t}). \tag{5}$$

If the projections of $\underline{T}(\bar{t})$ on the constant vectors $\underline{r}_0$, $\dot{\underline{r}}_0 + \underline{I}_0$ and $\underline{T}_0$ are, by omitting constant factors, $T_1(\bar{t})$, $T_2(\bar{t})$, and $T_3(\bar{t})$, respectively from the identity (5) we can get:

$$\ddot{a}(\bar{t}) + \frac{G}{r^3}\,a(\bar{t}) = T_1(\bar{t}), \tag{6.1}$$

$$\ddot{b}(\bar{t}) + \frac{G}{r^3}\,b(\bar{t}) = T_2(\bar{t}) \tag{6.2}$$

$$\ddot{c}(\bar{t}) + \frac{G}{r^3}\,c(\bar{t}) = T_3(\bar{t}) \tag{6.3}$$

The determination of $a(\bar{t})$, $b(\bar{t})$ and $c(\bar{t})$ from Eqs. (6), subject to the initial conditions (4.1) and (4.2), requires r, T_1, and T_2, and T_3 to be known functions of time $\bar{t}$. For a prescribed thrust $\underline{T}$, the functions $T_1(\bar{t})$, $T_2(\bar{t})$ and $T_3(\bar{t})$ are known, but r is unknown, then we can not solve the Eqs. (6) for $a(\bar{t})$, $b(\bar{t})$, and $c(\bar{t})$.

In spite of that, approximations of these functions can be found in the following way.

We restrict ourselves without loss of generality to the case of any thrust constant in magnitude and direction. For such a thrust the Eqs. (6) may be replaced by:

$$\ddot{a}(\bar{t}) + \frac{G}{r^3} a(\bar{t}) = 0 \tag{7.1}$$

$$\ddot{b}(\bar{t}) + \frac{G}{r^3} b(\bar{t}) = 0 \tag{7.2}$$

$$\ddot{c}(\bar{t}) + \frac{G}{r^3} c(\bar{t}) = 1. \tag{7.3}$$

Let us assume for $a(\bar{t})$, $b(\bar{t})$, and $c(\bar{t})$ the Meclaurin's expensions:

$$a(\bar{t}) = a(0) + \dot{a}(0)\bar{t} + \ddot{a}(0)\frac{\bar{t}^2}{2} + \cdots a^{(n)}(0)\frac{\bar{t}^n}{n!} + \cdots \tag{8.1}$$

$$b(\bar{t}) = b(0) + \dot{b}(0)\bar{t} + \ddot{b}(0)\frac{\bar{t}^2}{2} + \ldots b^{(n)}(0)\frac{\bar{t}^n}{n!} + \ldots \tag{8.2}$$

$$c(\bar{t}) = c(0) + \dot{c}(0)t + \ddot{c}(0)\frac{\bar{t}^2}{2} + \ldots c^{(n)}(0)\frac{\bar{t}^n}{n!} + \ldots \tag{8.3}$$

We can calculate as many coefficients of these series as we want, by using Eqs. (7), (4.1), (4.2) and (1). The first two coefficients are known from the conditions (4.1) and (4.2). The third coefficients, found from Eqs. (7), are:

$$\ddot{a}(0) = -\frac{G}{r_0^3}, \quad \ddot{b}(0) = 0, \quad \ddot{c}(0) = 1 \tag{4.3}$$

The fourth coefficients can be obtained from Eqs. (7) by differentiation, and then using conditions (4.1), (4.2) and (3.1). We can get:

$$\dddot{a}(\bar{t}) = -G(r\dot{a} - 3a\dot{r})/r^4, \tag{4.4}$$

and similar formulae for $\dddot{b}(\bar{t})$ and $\dddot{c}(\bar{t})$, then:

$$\dddot{a}(0) = 3G(r_0 + \dot{I}_0)/r_0^4, \quad \dddot{b}(0) = -G/r_0^3, \quad \dddot{c}(0) = 0 \tag{4.5}$$

For the fifth coefficients we use the formula (1). From the Eq. (4.4) we get:

$$\ddddot{a}(\bar{t}) = -G\left\{\frac{\ddot{a}}{r^3} - 3\frac{2\dot{a}\dot{r} + a\ddot{r}}{r^4} + 12a\frac{\dot{r}^2}{r^5}\right\} \tag{4.6}$$

and similar formulae for $\ddddot{b}(\bar{t})$ and $\ddddot{c}(\bar{t})$, which for $\bar{t}=0$ give conditions where all the quantities of their right-hand members are known, the value

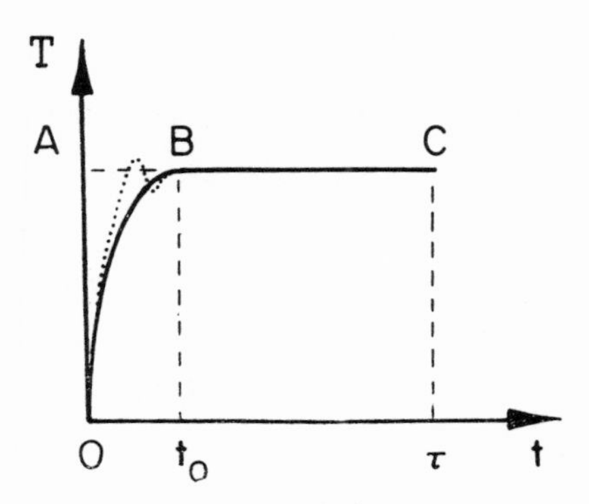

Fig. 1. The solid line is in accordance with the relation (13), the dotted line with practical problems where $t_0 = 0.02$ seconds, approximately.

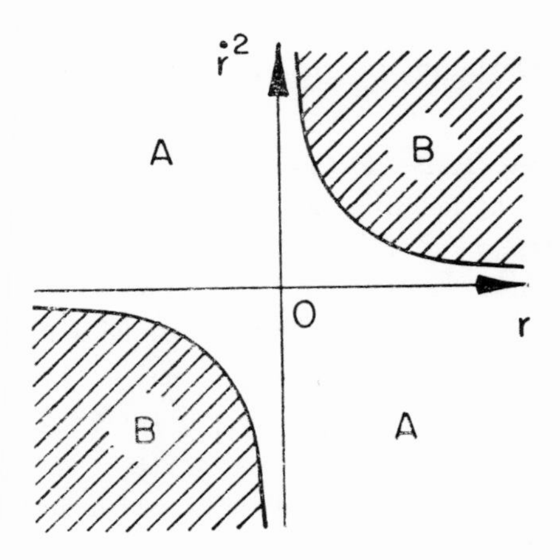

Fig. 2. The points of the rectangular hyperbola $r \cdot \dot{r}^2 = 2\mu$ in the $r \cdot \dot{r}^2$ — plane give a parabolic orbit; the points of the regions A give an elliptic orbit, and that of B a hyberbolic.

$\ddot{r}_o$ being given by the Eq. (1). For the other coefficients we follow the same procedure.

If we restrict ourselves to the approximations up to the order $\bar{t}^2$, we have:

$$a(\bar{t}) = 1 - \frac{G}{2r_o^3}\bar{t}^2,\ b(\bar{t}) = \bar{t},\ c(\bar{t}) = \frac{1}{2}\bar{t}^2 \tag{9}$$

then an approximation of the solution (4), if we come back to the original scale time by using (2,1), is:

$$\underline{r}(t) = \left\{1 - \frac{G}{2r_o^3}(t-t_o)^2\right\}\underline{r}_o + (t-t_o)(\dot{\underline{r}}_o + \underline{I}_o) + \frac{1}{2}(t-t_o)^2\underline{T}_o, \tag{10}$$

from which we get:

$$\dot{\underline{r}}(t) = -\frac{G}{r_o^3}(t-t_o)\underline{r}_o + (\dot{\underline{r}}_o + \underline{I}_o) + (t-t_o)\underline{T}_o. \tag{11}$$

III. *The solution when the thrust is either suddenly or gradually applied.*

The formulae (10) and (11) are valid in $t_o \leq t \leq \tau$, and the thrust $\underline{T}_o$ and its impulse $\underline{I}_o$ are the values of the thrust and the impulse at $t = t_o$. The impulse $\underline{I}_o$ during the time t_o is, by definition, given by:

$$\underline{I}_o = \int_0^{t_o} \underline{T}(t)\,dt. \tag{12.1}$$

If the thrust $\underline{T}(t)$ is, during the time t_o, a constant, $\underline{T}_o = T_o\underline{n}$ say, then:

$$\underline{I}_o = t_o\,\underline{T}_o, \tag{12.2}$$

and we speak about «*sudden application*» of the thrust. If the thrust is changed in $0 \leq t \leq t_o$, following any law, being zero at $t = 0$, we speak about «*gradual application*» of the thrust.

If the thrust is, say, constant in its direction but its magnitude varies accarding to parabolic law during the infinitesimal time t_o, being constant at the remainder time, i.e.

$$\underline{T}(t) = T(t)\underline{n},\quad \underline{n} = \text{constant},\quad T(t) = \begin{cases} at^{1/2}, & 0 \leq t \leq t_o \\ at_o^{1/2}, & t_o \leq t \leq \tau \end{cases} \tag{13}$$

where $\underline{n}$ is the unit vector, and the number a, which characterizes the parameter of the parabola, is a large positive constant, the impulse is:

$$\underline{I}_o = \frac{2}{3}t_o\,\underline{T}_o,\quad \underline{T}_o = at_o^{1/2}\underline{n}. \tag{12.3}$$

The above is an example of «gradual application», which can approximate the practical problems. In Fig. 1, the solid line is the graph of the thrust (13), while the dotted line is the thrust according to practical problems, t_0 being approximately equal to: 0.02 seconds.

The cases (12.2) and (12.3) can be written as:

$$\underline{I}_0 = m t_0 \underline{T}_0, \quad \underline{T}_0 = a t_0^{1/2} \underline{n} \tag{12.4}$$

where m=1 corresponds to sudden application, and $m = 2/3$ to gradual application, then the formulae (10) and (11) can be written as:

$$\underline{r}(t) = \left\{1 - \frac{G}{2r_0^3}(t-t_0)^2\right\}\underline{r}_0 + (t-t_0)\dot{\underline{r}}_0 + \left\{m t_0 (t-t_0) + \frac{1}{2}(t-t_0)^2\right\}\underline{T}_0, \tag{10.1}$$

$$\dot{\underline{r}}(t) = -\frac{G}{r_0^3}(t-t_0)\underline{r}_0 + \dot{r}_0 + \left\{m t_0 + (t-t_0)\right\}\underline{T}_0. \tag{11.1}$$

These formulae describe the motion of the projectile at time $t_0 \leq t \leq \tau$; m=1 corresponds to a sudden application of the constant thrust $\underline{T} = a t_0^{1/2} \underline{n}$, m=2/3 corresponds to a gradual application of a thrust according to (13).

IV. *The type of the Keplerian orbit if the thrust is suddenly removed.*

The position vector $\underline{r}(t)$ and the velocity vector $\dot{\underline{r}}(t)$, given above, determine the motion of the projectile on the non-Keplerian arc of its orbit during the action of the thrust. When the thrust is suddenly removed at time τ, the values $\underline{r}(\tau)$ and $\dot{\underline{r}}(\tau)$ determine completely the Keplerian orbit of the projectile after time τ.

The type of this Keplerian orbit depends upon the sign of the right-hand member of the relation:

$$\frac{1}{A_1} = \frac{2}{r} - \frac{\dot{r}^2}{\mu} \tag{14}$$

where A_1 is the semi-major axis of the orbit, $\mu = k(M+m)$, and if this member is bigger, equal to or less than zero, we have an ellipse, parabola or hyperbola, respectively, and this leads to the restrictions:

$$r \cdot \dot{r}^2 \lesseqgtr 2\mu \tag{15}$$

for an ellipse, parabola or hyperbola, respectively. In fig. 2 the graph of (15) in the r, $\dot{r}^2$—plane is given. The regious A correspond to an ellipse; the regious B to a hyperbola, and their boundary, the curve $r.\dot{r}^2 = 2\mu$, to a parabola.

The auther is indebted to Mr. S. Pines for fruitful discussion on some points of the paper, and to Dr. H. Wolf for suggesting the problem.

ΠΕΡΙΛΗΨΙΣ

Ἐνταῦθα σπουδάζεται ἡ κίνησις ὀχήματος πέριξ τῆς γῆς ὑπὸ τὴν ἐπίδρασιν τῆς ἑλκτικῆς δυνάμεως τῆς γῆς καὶ μιᾶς ὠστικῆς δυνάμεως. Ἡ τροχιὰ καὶ ἡ ταχύτης τοῦ ὀχήματος εὑρίσκονται εἰς δύο περιπτώσεις, ὅταν δηλαδὴ ἡ ὠστικὴ δύναμις ἐφαρμόζεται ἐπὶ τοῦ ὀχήματος εἴτε ἀκαριαίως εἴτε βραδέως. Δίδονται ἐπίσης καὶ αἱ συνθῆκαι ὑπὸ τὰς ὁποίας διακρίνομεν τὰ εἴδη τῆς Κεπλερείου τροχιᾶς, ὅταν ἡ ὠστικὴ δύναμις παύσῃ ἀκαριαίως νὰ δρᾷ.

ΑΝΑΤΥΠΟΝ
ΣΕΛ. 191 - 202

The Keplerian orbit of a projectile around the earth, after the thrust is suddenly removed**.

By ***Dem. G. Magiros********.

Ἀνεκοινώθη ὑπὸ τοῦ κ. Ἰωάνν. Ξανθάκη.

Introduction.

In the following we discuss the elements of the Keplerian orbit of a projectile around the earth, after the thrust is suddenly removed, in the cases of sudden or gradual application of the thrust, if the thrust acts continuously either for infinitesimal time t_0 or for non-infinitesimal time τ. Formulae are given for the elements of the Keplerian orbit in terms of the elements of the Keplerian orbit either the original or that which corresponds to time t_0. For the calculation of the elements of the Keplerian orbit when the thrust is removed, the position vector and the velocity vector at that time must be known. These vectors are given in a suitable form in a previous paper [1], «paper I», contained in the present volume. We treat first the case of infinitesimal time, then the case of non-infinitesimal time, if the thrust in both cases is suddenly or gradually applied. The numbers ε throughout the paper, if multiplied by 100, give the percentage of increment of the corresponding element.

I. *The case of infinitesimal time.*

If the thrust, acting for infinitesimal time t_0, ceases at the point M_0

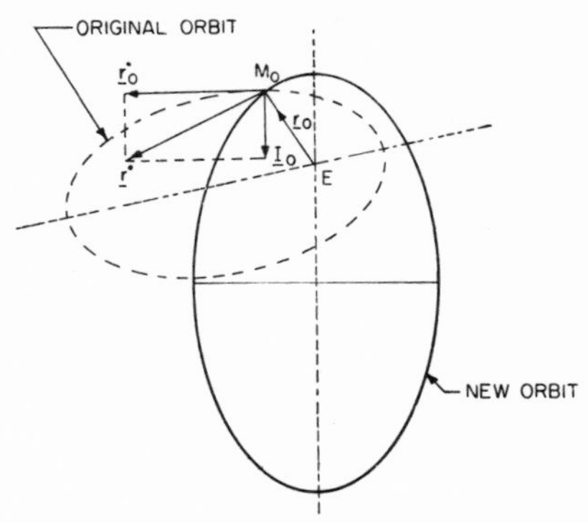

Fig. 1.— The velocity $\dot{\underline{r}}$, tangent to the new orbit at the point M_0, is the resultant of the initial velocity $\dot{\underline{r}}_0$ and the thrust $\underline{I}_0$. The original and the new orbits have the same focus E, the earth.

say, of the original orbit, Fig. 1, then the position vector and the velocity

* ΔΗΜ. ΜΑΓΕΙΡΟΥ, Τὰ στοιχεῖα τῆς Κεπλερίου τροχιᾶς ὀχήματος πέριξ τῆς γῆς, ὅταν ἀποτόμως παύσῃ ἡ ἐπ' αὐτοῦ ἐνεργοῦσα ὠστικὴ δύναμις.

** **Republic Aviation Corp., U.S.A.**

Reprinted from the *Proceedings of the Athens Academy of Sciences* **35** (1960), 191–202.

vector at M_o are given, according to the formulae (10) and (11) or (10.1) and (11.1), of the «paper I», by:

$$\underline{r}(t_o)=\underline{r}_o, \quad \dot{\underline{r}}(t_o)=\dot{\underline{r}}_o+\underline{I}_o. \tag{1}$$

After the remarks on the impulse $\underline{I}_o$ made in Chapter III of «Paper I», we proceed to calculate the elements of the orbit, after the thrust is suddenly removed at $t=t_o$, these elements being designated by the subscript I.

a. *The semi-major axis* a_I.

For the original and the new semi-major axis, a and a_I respectively, we have [2].

$$\frac{1}{\alpha}=\frac{2}{r_o}-\frac{(\dot{\underline{r}}_o)^2}{\mu} \tag{2.1}$$

$$\frac{1}{\alpha_I}=\frac{2}{r_o}-\frac{(\dot{\underline{r}}_o+\underline{I}_o)^2}{\mu} \tag{2.2}$$

which give:

$$\frac{1}{\alpha_I}=\frac{1}{\alpha}-\frac{1}{\mu}\left(2\,\dot{r}_o\,I_o\cos\vartheta+I_o^2\right) \tag{2.3}$$

then:

$$\frac{1}{\alpha_I}=\frac{1}{\alpha}\,\psi \tag{2.4}$$

with

$$\psi=1-\frac{\alpha}{\mu}\left(2\,\dot{r}_o\,I_o\cos\vartheta+I_o^2\right). \tag{2.5}$$

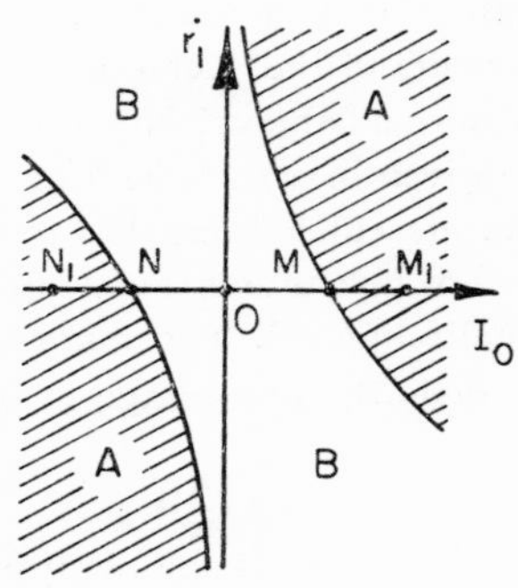

Fig. 2.— The points of the regions A give hyperbolic orbits, those of the region B elliptic orbits, and their boundary, the curve: $I_o^2+2\,I_o\,\dot{r}_1-\frac{\mu}{\alpha}=0$, *parabolic orbits.*

We also can have:

$$\frac{1}{\alpha_I}=\frac{1}{\alpha}\,(1+\varepsilon), \quad \varepsilon=\psi-1. \tag{2.6}$$

In the above r_o, $\dot{r}_o$, I_o are magnitudes of $\underline{r}_o$, $\dot{\underline{r}}_o$, $\underline{I}_o$; and ϑ the angle of $\dot{\underline{r}}_o$ and $\underline{I}_o$.

The conditions for the kind of the new orbit are:

$$I_o^2+2\,I_o\,\dot{r}_1-\frac{\mu}{\alpha}\lesseqgtr 0 \tag{3}$$

for elliptic, parabolic and hyperbolic ones, respectively, $\dot{r}_1$ is the projection of $\dot{\underline{r}}_o$ along $\underline{I}_o$, $\dot{r}_1=\dot{r}_o\cos\vartheta$. The graph of the conditions (3) is given in Fig. 2.

b. *The angular momentum vector $\underline{H}_{I}$, and the angle between the original and the new orbits.*

The original and the new angular momentum vectors, $\underline{H}$ and $\underline{H}_I$, are, by definition, given by the vector products:

$$\underline{H} = \underline{r}_o \times \underline{\dot{r}}_o, \quad \underline{H}_I = \underline{r}_o \times (\underline{\dot{r}}_o + \underline{I}_o). \tag{4.1}$$

If $\Delta\underline{H}$ is the increment vector, we can write:

$$\underline{H}_I = \underline{H} + \Delta\underline{H}, \quad \Delta\underline{H} = \underline{r}_o \times \underline{I}_o.$$

The length of $\underline{H}$ and $\Delta\underline{H}$ are:

$$H = r_o \dot{r}_o \sin \varphi_1, \quad \Delta H = r_o I_o \sin \varphi,$$

φ_1 being the angle of $\underline{r}_o$ and $\underline{\dot{r}}_o$, φ that of $\underline{r}_o$ and $\underline{I}_o$, Fig. 3α. The vectors $\underline{H}$ and $\Delta\underline{H}$ are perpendicular to the $\underline{r}_o$, $\underline{\dot{r}}_o$ -plane and to the $\underline{r}_o$, $\underline{I}_o$ -plane, respectively.

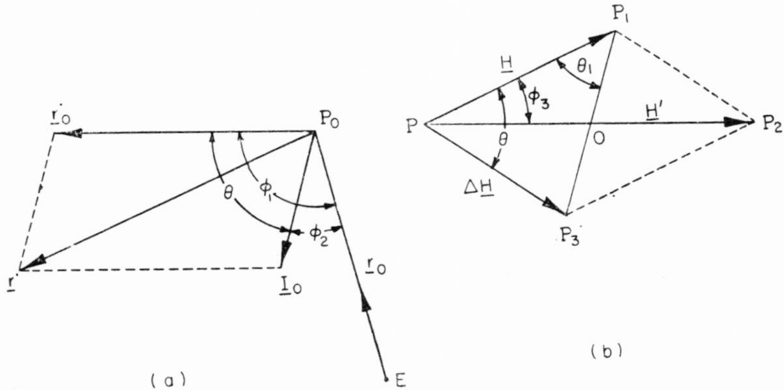

Fig. 3.—a) The vectors $\underline{r}_o$, $\underline{\dot{r}}_o$, $\underline{I}_o$ define a trihedral angle with vertex P_o. b) The new angular momentum $\underline{H}'$ is the resultant of the original angular momentum $\underline{H}$ and its increment $\Delta\underline{H}$.

If ϑ is the angle of these planes, ϑ must be the angle of $\underline{H}$ and $\Delta\underline{H}$. The new angular momentum vector $\underline{H}$ Is parallel to the $\underline{H}$, $\Delta\underline{H}$ -plane. Fig. 3b helps for calculation of the length H_I of $\underline{H}_I$ and its angle φ_3 with $\underline{H}$. From the triangle $P_1 P P_3$ we get:

$$(P_1 P_3) = \varrho = r_o \{(\dot{r}_o \sin \varphi_1)^2 + (I_o \sin \varphi_2)^2 - 2 \dot{r}_o I_o \sin \varphi_1 \sin \varphi_2 \cos \vartheta\}^{1/2},$$

$$\sin \vartheta_1 = (\Delta H/P) \sin \vartheta.$$

From the triangle P_1PO we get:

$$(PO) = \left\{H^2 + \left(\frac{\varrho}{2}\right)^2 + H\varrho \cos \vartheta_1\right\}^{1/2}, \quad \sin \varphi_3 = ((P_1O)/(PO)) \sin \vartheta_1.$$

Then

$$H_I = 2\,(PO) = 2\left\{H^2 + \left(\frac{\varrho}{2}\right)^2 - H\varrho\cos\vartheta_1\right\}^{1/2} \tag{4.2}$$

$$\sin\varphi_3 = \frac{1}{2}\left\{\frac{1}{4} + \left(\frac{H}{\varrho}\right)^5 + \frac{H}{\varrho}\cos\vartheta\right\}^{-1/2} . \sin\vartheta_1 . \tag{4.3}$$

c. *The period P_I, and the mean angular motion n_I.*

We have:

$$P = \frac{2\pi}{K\sqrt{\mu}}\,\alpha^{3/2}, \quad n = \frac{2\pi}{P} = K\sqrt{\mu}\left(\frac{1}{\alpha}\right)^{3/2}$$

$$P_I = \frac{2\pi}{K\sqrt{\mu}}\,\alpha_I^{3/2}, \quad n_I = \frac{2\pi}{P_1} = K\sqrt{\mu}\left(\frac{1}{\alpha_I}\right)^{3/2},$$

then

$$P_I = P\,(1 + \varepsilon_1), \quad n_I = n\,(1 + \bar{\varepsilon}_1) \tag{5.1}$$

with

$$\varepsilon_1 = \psi^{-3/2} - 1, \quad \bar{\varepsilon}_1 = \psi^{3/2} - 1. \tag{5.2}$$

ψ is given by (2.5).

d. *The eccentric anomaly E_o, and the eccentricity e_I.*

From the formulae for the original orbit [3]:

$$e\sin E_o = \frac{\underline{r_o}\,\dot{\underline{r_o}}}{\sqrt{\mu\alpha}} \tag{6.1}$$

$$e\cos E_o = \frac{\alpha - r_o}{\alpha} \tag{6.2}$$

we get:

$$\tan E_o = \sqrt{\frac{\alpha}{\mu}}\,.\,\frac{\underline{r_o}\,\dot{\underline{r_o}}}{\alpha - r_o}, \tag{6.3}$$

$$e^2 = \frac{(\underline{r_o}\,\dot{\underline{r_o}})^2}{\mu\alpha} + \frac{(\alpha - r_o)^2}{\alpha^2}. \tag{6.4}$$

Applying (6.3) we have for the eccentric anomaly E_{oI}:

$$\tan E_{oI} = \sqrt{\frac{\alpha}{\mu}}\,.\,r_o\,.\,\frac{\sqrt{\psi}}{\alpha - r_o\psi}\,(\dot{r}_o\cos\varphi_1 + I_o\cos\varphi_2), \tag{6.5}$$

φ_1 the angle of $\underline{r_o}$ and $\dot{\underline{r_o}}$, φ_2 that of $\underline{r_o}$ and $\underline{I_o}$.

Therefore:

$$\tan E_{oI} = (1 + \varepsilon_2)\tan E_o \tag{6.6}$$

with

$$\varepsilon_2 = (\alpha - r_o) \frac{\sqrt{\psi}}{\alpha - r_o \psi} \left(1 + \frac{\dot{I}_o}{r_o} \frac{\cos \varphi_2}{\cos \varphi_1}\right) - 1. \tag{6.7}$$

For the new eccentricity e_I we get:

$$e_I = \frac{1}{\alpha} X^{1/2} \tag{7.1}$$

with:

$$X = \frac{\alpha}{\mu} \psi r_o^2 (\dot{r}_o \cos \varphi_1 + I_o \cos \varphi_2)^2 + (\alpha - r_o \psi)^2, \tag{7.2}$$

then:

$$e_I = e(1 + \varepsilon_3) \tag{7.3}$$

with:

$$\varepsilon_3 = \sqrt{\mu X} \left\{ \alpha (r_o \dot{r}_o \cos \varphi_1)^2 + \mu (\alpha - r_o)^2 \right\}^{-1/2} - 1. \tag{7,4}$$

e. *The «Perigee» q_I, «parameter» or «latus rectum» p_I and «true anomaly V_{oI}.*

For the perigee we have:

$$q = \alpha(1 - e), \quad q_I = \alpha_I (1 - e_I) = \frac{\alpha}{\psi} \left(1 - \frac{1}{\alpha} X^{1/2}\right),$$

then:

$$q_I = (1 + \varepsilon_4) q \tag{8.1}$$

$$\varepsilon_4 = \frac{\alpha - X^{1/2}}{\alpha(1-e)\psi} - 1. \tag{8.2}$$

For the parameter:

$$p = \alpha(1 - e^2), \quad p_I = \alpha_I (1 - e_I^2)$$

then:

$$p_I = (1 + \varepsilon_5) p, \tag{9.1}$$

$$\varepsilon_5 = \frac{\alpha^2 - X}{\alpha^2 (1 - e^2) \psi} - 1. \tag{9.2}$$

For the true anomaly V_{oI} we use the formula [4].

$$\cos V_o = \frac{\cos E_o - e}{1 - e \cos E_o}$$

which, with the help of (6.4), can be written as:

$$\cos V_o = \frac{\alpha - \alpha e - r_o}{r_o e}$$

when:

$$\cos V_{oI} = \frac{\alpha_I - \alpha_I e_I^2 - r_o}{r_o e_I} = \frac{\alpha(\alpha - r_o \psi) - X}{r_o \psi X^{1/5}},$$

then:

$$\cos V_{oI} = (1 + \varepsilon_6) \cos V_o \tag{10.1}$$

$$\varepsilon_6 = \frac{e\{\alpha(\alpha - r_o\psi) - X\}}{\psi X^{1/2}(\alpha - \alpha e^2 - r_o} - 1. \tag{10.2}$$

f. *Orientation cosines:* $\underline{P}_I$, $\underline{Q}_I$, $\underline{W}_I$.

The orientation cosines $\underline{P}$, $\underline{Q}$, $\underline{W}$ are unit vector; $\underline{P}$ the «perigee vector» from the earth towards the perigee; $\underline{Q}$ vector directed along the lactus rectum; $\underline{W} = \underline{P} \times \underline{Q}$. The $\underline{P}$ and $\underline{Q}$ are given by: [5]

$$\underline{P} = \frac{\cos E_o}{r_o} \underline{r}_o - \sqrt{\frac{\alpha}{\mu}} \sin E_o . \dot{\underline{r}}_o \tag{11,1}$$

$$\underline{Q} = \frac{1}{\sqrt{1-e^2}} \left\{ \frac{\sin E_o}{r_o} \underline{r}_o + \sqrt{\frac{\alpha}{\mu}} (\cos E_o - e) . \dot{\underline{r}}_o \right\} . \tag{11.2}$$

For the new $\underline{P}_I$ and $\underline{Q}_I$, by taking into account (1) (2.5), (6.1), (6.2), (7.1), (7.2), we get:

$$\underline{P}_I = \frac{\alpha - r_o\psi}{r_o X^{1/2}} \underline{r}_o - \frac{\alpha}{\mu X^{1/2}} r_o \dot{r}_o \cos\varphi_1 (\dot{\underline{r}}_o + \underline{I}_o), \tag{12.1}$$

$$\underline{Q}_I = \frac{\alpha}{\sqrt{\alpha^2 - X}} \left\{ \sqrt{\frac{\alpha\psi}{\mu X}} \cos\varphi_1 . r_o . \underline{r}_o + \frac{1}{\sqrt{\alpha\mu\psi X}} \Big(\alpha(\alpha - r_o\psi) - X \Big) \Big(\dot{\underline{r}}_o + \underline{I}_o \Big) \tag{12.2}$$

We can omit the subscript o from the formulae (12.1) and (12.2) since the orientation cosines do not vary with time, then they are independent of the position of the point M_o on the orbit.

g. *Orientation angles:* i_I, ω_I, Ω_I.

The «inclination» i is the angle between the plane of the orbit and that of the equator or of the ecliptic. The «argument of perigee» ω is the angle between the nodal line (to the direction of the ascending node) and the semi-major axis of the orbit (to the direction of the perigee). The «longitude of node» Ω is the angle (on the equator) of the nodal line to the ascending node and the intersection of equator - ecliptic.

For the new inclination i_I we notice that the inclinations are measured from the plane of equator and the angle φ_s of the original and the new orbits from the plane of the original orbit. If i, i_I, φ_s are positive in the same rotation as shown in Fig. 4, we can get the relationship:

$$i_I = i \pm \varphi_3 \tag{13}$$

φ_3 being given by the formula (4.3).

Now, for the new angles ω_I and Ω_I, take the orthogonal xyz-system as it is shown in Fig. 4. If P_{Ix}, P_{Iy}, P_{Iz} are the components of the new perigee vector $\underline{P_I}$ along the axes of this system, we can get the following formula: [3]

$$\begin{aligned} P_{Ix} &= \cos \omega_I . \cos \Omega_I - \sin \omega_I . \sin \Omega_I . \cos i_I, \\ P_{Iy} &= \cos \omega_I \sin \Omega_I + \sin \omega_I \cos \Omega_I \cos i_I, \\ P_{Iz} &= \sin \omega_I . \sin i_I . \end{aligned} \tag{14}$$

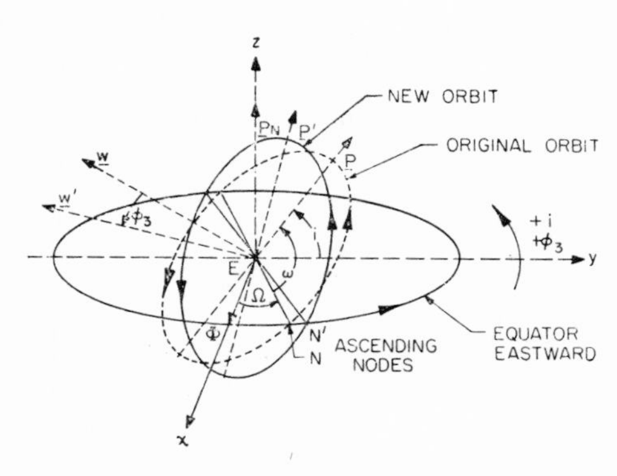

Fig. 4.—The location of the original and the new orbits with respect to the equator. The arrows in the orbits show the direction of the projectile.

We have the same formulae for the components P_x, P_y, P_z of the original perigee $\underline{P}$ by omitting the subscript I from (14).

By using the foamulae (12.1) and (13), the only unknowons in the system (14) are the angles ω_I and Ω_I. The last equation of (14) gives the new argument of perigee ω_I:

$$\sin \omega_I = P_{Iz} / \sin i_I . \tag{15}$$

Inserting the known value of ω_I into the first equation of (14), we can determine the new longitude of node Ω_I:

$$\cos \Omega_I = \frac{1}{\varrho^2 + \cos^2 \omega_I} \left\{ P_{Ix} \cos \omega_I \pm \varrho \left(\varrho^2 + \cos^2 \omega_I - P^2_{Ix} \right)^{1/2} \right\}, \tag{16.1}$$

with :

$$\varrho = \sin \omega_I . \cos i_I . \tag{16.2}$$

The second equation of (14) must be satisfied by the values found above, and this gives indication for the selection of plus or minus sign of the formula (16.1).

II. *The case of non-infinitesimal time.*

We consider in this section the case of a thrust of special type suddenly or gradually applied to the projectile and suddenly removed either after infinitesimal time t_o, case I, or after time τ, case II.

The formulae for the elements of the Keplerian orbit in case II are given in terms of that of the case I. The procedure can be used as a model to treat the calculation when other types of thrust are given.

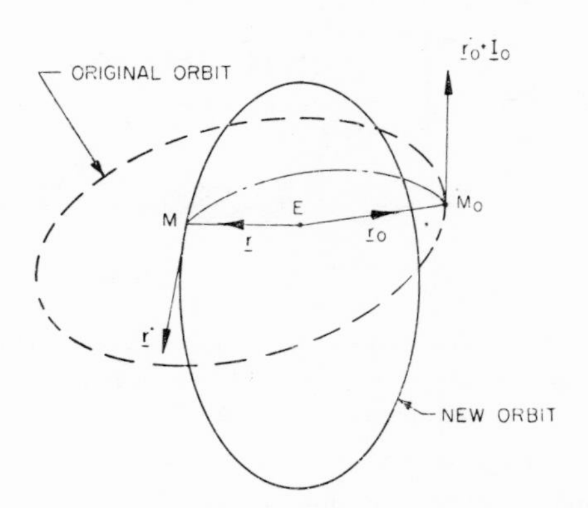

Fig 5.—The dotted are M_0 M shows the part of the orbit during the action of the thrust.

Take the direction of the thrust parallel to the initial velocity $\dot{\underline{r}}_o$, that is $\underline{T}(t) = \lambda(t)\, \dot{\underline{r}}_o$, λ being the factor of proportionality, and its magnitude constant for sudden application, or according to the law shown in Fig. 1 «Paper I», for gradual application.

The impulse $\underline{I}_o$ in the formulae (10) and (11) of «Paper I» is given by $\underline{I}_o = \frac{2}{3}\alpha t_o^{\,2/3}\, \dot{\underline{r}}_o$ in case of a gradual application with parabolic law in

$o \leqq t \leqq t_o$; and by $\underline{I}_o = \alpha t_o{}^{3/2} \dot{\underline{r}}_o$ in case of a sudden application of the constant thrust $\underline{T} = \alpha t_o{}^{1/2} \dot{\underline{r}}_o$. If we write: where $m=1$ for sudden application, and $m=2/3$ for gradual application of parabolic type, the formulae (10) and (11) of «paper I» can be written as:

$$\underline{r}(t) = \left\{ 1 - \frac{G}{2r_o{}^3}(t-t_o)^2 \right\} \underline{r}_o + (t-t_o) \left\{ 1 + mt_o\lambda + \frac{1}{2}(t-t_o)\lambda \right\} \dot{\underline{r}}_o. \quad (17)$$

$$\dot{\underline{r}}(t) = -\frac{G}{r_o{}^3}(t-t_o)\underline{r}_o + \left\{ 1 + mt_o\lambda + (t-t_o)\lambda \right\} \dot{\underline{r}}_o, \quad (18)$$

where $\lambda = \alpha t_o{}^{1/2}$. These formulae describe the motion of the projectile at time $t_o \leqq t \leqq \tau$, when the arc of the orbit is represented by the broken arc of Fig. 5.

a. *The semi-major axes a_I, a_{II}.*

From the formulae (2) we have:

$$\frac{1}{\alpha_I} = \frac{2\mu - (1+mt_o\lambda)^2 r_o \dot{r}_o{}^2}{\mu r_o}, \quad \frac{1}{\alpha_{II}} = \frac{2\mu - r_o \dot{r}^2}{\mu r}$$

then we can write:

$$\frac{1}{\alpha_{II}} = \frac{1}{\alpha_I}(1 + \varepsilon_\tau) \quad (18.1)$$

with:

$$\varepsilon_\tau = \frac{r_o}{r} \frac{2\mu - r\dot{r}_o}{2\mu - r_o \dot{r}_o{}^2 (1+mt_o\lambda)^2} - 1 \quad (19.2)$$

r and $\dot{r}$ are the magnitudes of $\underline{r}$ and $\dot{\underline{r}}$, given by (17) and (18).

b. *The angular momentum vectors $\underline{H}_I$, $\underline{H}_{II}$.*

For the angular momentum vectors $\underline{H}$, $\underline{H}_I$, $\underline{H}_{II}$ we have:

$$\underline{H} = \underline{r}_o \times \dot{\underline{r}}_o, \quad \underline{H}_I = (1+mt_o\lambda)^2 \underline{r}_o \times \dot{\underline{r}}_o, \quad \underline{H}_{II} = \underline{r} \times \dot{\underline{r}},$$

then for their magnitudes:

$$H = r_o \dot{r}_o \sin\varphi_1, \quad H_I = (1+mt_o\lambda)^2 r_o \dot{r}_o \sin\varphi_1, \quad H_{II} = r\dot{r} \sin\varphi,$$

φ_1 being the angle of r_o, $\dot{r}^o$ and φ that of r, $\dot{r}$.

We can write:

$$H_2 = H_1 (1 + \varepsilon_8) \tag{20.1}$$

with:

$$\varepsilon_8 = \frac{r \dot{r} \sin \varphi}{H (1 + m t_0 \lambda) \sin \varphi_1} - 1 \tag{20.2}$$

The plane of the original and the new orbits are the same, then we have the same direction for the original and the new angular momentum vectors, and the inclination remains constant.

c. *The periode P_I, P_{II} and the mean angular motions n_I, n_{II}.*

For the new periods we have:

$$P_{II} = \frac{2\pi}{K\sqrt{\mu}} \alpha_{II}^{3/2}, \quad P_I = \frac{2\pi}{K\sqrt{\mu}} \alpha_I^{3/2}$$

then:

$$P_{II} = P_I \left(\frac{\alpha_{II}}{\alpha_I}\right)^{3/2} = P_I (1 + \varepsilon_7)^{3/2} = P_I \left(1 - \frac{3}{2} \varepsilon_7\right) \tag{21}$$

for small ε_7, which is given by: (19.2).
In the same way for the new mean angular motion we can get:

$$n_{II} = n_I \left(1 + \frac{3}{2} \varepsilon_7\right). \tag{22}$$

d. *The eccentric anomalies E_{oI}, E_{oII}, and the eccentricities e_I, e_{II}.*

By using (6.3) we get:

$$\tan E_{oI} = \sqrt{\frac{\alpha_I}{\mu}} \quad \frac{r_o \dot{r}_o (1 + m t_o \lambda) \cos \varphi_1}{\alpha_I - r_o}$$

$$\tan E_{oII} = \sqrt{\frac{\alpha_{II}}{\mu}} \frac{r \dot{r} \cos \varphi}{\alpha_{II} - r} = \sqrt{\frac{\alpha_I (1 + \varepsilon_7)^{-1}}{\mu}} \cdot \frac{r \dot{r} \cos \varphi}{\alpha_I (1 + \varepsilon_7)^{-1} - r},$$

then:

$$\tan E_{oII} = (1 + \varepsilon_9) \tan E_{oI} \tag{23.1}$$

with:

$$\varepsilon_9 = \left(1 - \frac{1}{2} \varepsilon_7\right) \frac{r \dot{r} \cos \varphi}{r_o \dot{r}_o (1 + m t_o \lambda) \cos \varphi_1} \cdot \frac{\alpha_I - r_o}{\alpha_I (1 - \varepsilon_7) - r} - 1. \tag{23.2}$$

For the new eccentricities, by applying (6.4) we have:

$$e_I^2 = \frac{\{r_o \dot{r}_o (1 + m t_o \lambda) \cos \varphi_1\}^2}{\mu \alpha_I} + \frac{(\alpha_I - r_o)}{\alpha_I^2}$$

$$e_{II}^2 = \frac{(r \dot{r} \cos \varphi)^2}{\mu (1 - \varepsilon_7) \alpha_I} + \frac{\{\alpha_I (1 - \varepsilon_7) - r\}}{(1 - 2 \varepsilon_7) \alpha_I^2},$$

then:

$$e_{II}^2 = e_I^2 (1 + \varepsilon_{10}) \tag{24.1}$$

with:

$$\varepsilon_{10} = \frac{(1-\varepsilon_7)\,\alpha_I (r\dot{r}\cos\varphi)^2 + \mu\,[(1-\varepsilon_7)\,\alpha_I - r]^2}{(1-2\varepsilon_7)\,[\{r_o \dot{r}_o (1+mt_o\lambda)\cos\varphi_1\}^2 \alpha_I + \mu(\alpha_I - r_o)^2} - 1 \tag{24.2}$$

e. *The perigees q_I, q_{II}, and the parameter p_I, p_{II}.*

For the perigees we have:

$$q_I = \alpha_I (1 - e_I),\quad q_{II} = \alpha_{II}(1 - e_{II}) = \alpha_I (1 - \varepsilon_7) \left\{ 1 - e_I \left(1 + \frac{1}{2}\varepsilon_{10}\right) \right\},$$

then:

$$q_{II} = (1 - \varepsilon_7)\, q_I - \frac{1}{2}\varepsilon_{10}\, \alpha_I\, e_I; \tag{25}$$

and for the parameters:

$$p_I = \alpha_I (1 - e_I^2),\quad p_{II} = \alpha_{II}(1 - e_{II}^2) = \alpha_I (1 - \varepsilon_7) \left\{ 1 - e_I^2 (1 + e_{10}) \right\},$$

when, for small ε_7 and ε_{10}, we can write:

$$p_{II} = (1 - \varepsilon_7)\, p_I - \varepsilon_{10}\, \alpha_I\, e_I^2. \tag{26}$$

ΠΕΡΙΛΗΨΙΣ

Ἡ παροῦσα ἐργασία ἀποτελεῖ συμπλήρωμα καὶ συνέχειαν προηγουμένης ἐργασίας περιεχομένης εἰς τὸν παρόντα τόμον τῶν Πρακτικῶν σελ. 96 – 103. Εἰς τὴν προηγουμένην μελέτην μελετᾶται ἡ κίνησις ὀχήματος πέριξ τῆς γῆς ὑπὸ τὴν ἐπίδρασιν τῆς ἑλκτικῆς δυνάμεως τῆς γῆς καὶ μιᾶς ὠστικῆς δυνάμεως. Εἰς τὴν παροῦσαν ἐξετάζονται τὰ στοιχεῖα τῆς Κεπλερείου τροχιᾶς τοῦ ὀχήματος, ὅταν ἡ ὠστικὴ δύναμις παύσῃ ἀποτόμως νὰ ἐνεργῇ ἐπὶ τοῦ ὀχήματος. Δίδονται τύποι συνδέοντες τὰ στοιχεῖα τῆς Κεπλερείου τροχιᾶς πρὸς τὰ στοιχεῖα τῆς Κεπλερείου τροχιᾶς εἴτε τῆς ἀρχικῆς (ὅταν ἤρχισε ἐνεργοῦσα ἡ ὠστικὴ δύναμις), εἴτε τῆς ἀντιστοιχούσης εἰς τὸν χρόνον t_o. Οἱ ἀριθμοὶ ε εἰς τοὺς διδομένους τύπους, ἑκατονταπλασιαζόμενοι, δίδουν τὸ ποσοστὸν ἐπὶ τοῖς ἑκατὸν τῆς αὐξήσεως τῶν ἀντιστοιχούντων στοιχείων.

REFERENCES

[1] D. G. MAGIROS, «The motion of a projectile arount the earth under the influence of the earth's gravitational attraction and a thrust». this volume, Pag 96-103.

[2] W. SMART, «Celestial Mechanics», (1953). Longmans, Green and Company, p. 21, paragraph 2.12.

[3] S. HERRICK, «Formulas, Constants, Definitions, Notation for Geocentric and Heliocentric Orbits», Systems Laboratories Corporation, Spacenautics Division Report SN 1.

[4] C. Hilton and S. Herrick. «A Technical Note Concerning Coordinate Systems for Linear Vehicle Orbits». (1958), p. 13.

[5] J. Feyk and H. Karrenberg, «Equation Relating to the Trajectory of a Lunar Vehicle», (1958). Notes Systems Corporation of America.

ΠΡΑΚΤΙΚΑ
ΤΗΣ ΑΚΑΔΗΜΙΑΣ ΑΘΗΝΩΝ

ΕΤΟΣ 1963: ΤΟΜΟΣ 38ΟΣ

ΑΝΑΤΥΠΟΝ
ΣΕΛ. 36 - 39

On the convergence of the solution of a special two - body problem*.

by ***Demetrios G. Magiros (**)***.

Ἀνεκοινώθη ὑπὸ τοῦ Ἀκαδημαϊκοῦ κ. Ἰωάνν. Ξανθάκη.

1. INTRODUCTION

a. In previous papers [1], where the motion of a projectile a Newtonian center during the action of general thrust vector was investigated, a series solution for this problem was constructed. The purpose of the present note is to give a brief discussion of the determination of the time in-

(*) Of the Valley Forge Space Technology Center, of the General Electric Co., M.S.D. Philadelphia, Pa.

(**) ΔΗΜ. Γ. ΜΑΓΕΙΡΟΥ, Ἐπὶ τῆς συγκλίσεως τῆς λύσεως ἑνὸς εἰδικοῦ προβλήματος τῶν δύο σωμάτων.

Reprinted from the *Proceedings of the Athens Academy of Sciences* **38** (1963), 36–39.

terval for which the solution found is valid. Details of the present note and related subjects will appear elsewhere.

b. We state for reference that the differential equation and the initial conditions of the problem in vector form are:

$$\begin{aligned} \ddot{\underline{r}}(\tau) &= -\frac{\mu}{r^3(\tau)}\,\underline{r}(\tau) + \underline{T}(\tau) \\ \underline{r}(o) = \underline{r}_0, \quad \dot{\underline{r}}(o) &= \dot{\underline{r}}_0 + \underline{I}_0 \end{aligned} \tag{1}$$

valid in the region D: $|\underline{r}(\tau)| < M_1$, $|\dot{\underline{r}}(\tau)| < M_2$ for any value of time τ in D_1: $o \leq \tau \leq \tau'$. μ is a constant, $\underline{T}$ the thrust, $\underline{I}_0$ the impulse of the thrust for very small time, $\underline{r}$ and $\dot{\underline{r}}$ the displacement and velocity vectors.

If the reference coordinate system is: $(P;\ \underline{r}_0^*,\ \underline{s}_0^*,\ \underline{T}_0^*)$, where P is the position of the projectile when the thrust starts, $\underline{r}_0^*$, $\underline{s}_0^*$, $\underline{T}_0^*$ the unit vectors along $\underline{r}_0$, $\dot{\underline{r}}_0 + \underline{I}_0$, $\underline{I}_0$, respectively, a solution of the form

$$\underline{r}(\tau) = \alpha_1(\tau)\,\underline{r}_0^* + \alpha_2(\tau)\,\underline{s}_0^* + \alpha_3(\tau)\,\underline{T}_0^* \tag{2}$$

can be determined by calculating the scalar functions $\alpha_i(\tau)$, $i = 1,2,3$, in Mac Laurin's expansions at $\tau = o$.

$$\alpha_i(\tau) = \sum_{n=o}^{\infty} \frac{\alpha_i^{(n)}(o)}{n!}\,\tau^n \quad ; \; i = 1, 2, 3 \tag{3}$$

The functions $\alpha_i(\tau)$ satisfy the conditions

$$\begin{aligned} \ddot{\alpha}_i + \frac{\mu}{r^3}\,\alpha_i &= T_i \quad ; \; i = 1, 2, 3 \\ \alpha_1(o) = r_0,\ \alpha_2(o) &= \alpha_3(o) = 0 \\ \dot{\alpha}_1(o) = \dot{\alpha}_3(o) &= 0,\ \dot{\alpha}_2(o) = s_0 \end{aligned} \tag{4}$$

where $s_0 = |\dot{\underline{r}}_0 + \underline{I}_0|$ and T_1, T_2, T_3, projections of $\underline{T}$ on the $\underline{r}_0^*$, $\underline{s}_0^*$, $\underline{T}_0^*$ – axis. By using (1), (2), (4) we can determine the coefficients of the series (3).

2. THE RADIUS OF CONVERGENCE OF THE SERIES (3).

The radius of convergence of the series (3) is the reciprocal of the upper limit: [2]

$$\overline{\lim_{n\to\infty}} \left\{ |\alpha_i^{(n)}(o)| \,/\, n! \right\}^{1/n}, \; i = 1, 2, 3 \tag{5}$$

The nth derivative of $\alpha_i(\tau)$, found from the equation (4), is:

$$\alpha_i^{(n)}(\tau) = -\mu \left[\frac{\alpha_i(\tau)}{r^3(\tau)} \right]^{(n-2)} + T_i^{(n-2)}(\tau) \tag{6}$$

The general term $\left(\alpha_i^{(n)}(\tau) / n! \right)$, if we rake into account the formula (e) of the Appendix, becomes:

$$\frac{\alpha_i^{(n)}(\tau)}{n!} = -\sum_{m=0}^{n-2} \frac{\alpha_i^{(m)}(\tau)\left[V_1 r^{n-m-3} + V_2 r^{n-m-4} + \cdots + V_{n-m-2} \right]}{(n-1)n\, m!\,(n-m-2)!\, r^{n-m+1}} + \frac{T_i^{(n-2)}(\tau)}{n!} \tag{7}$$

where V's are polynomicals in the derivatives of r up to the order $(n-m-2)$ with coefficients smaller than $(n-m)!$

To find the limit of the nth root of the absolute value of the right-hand member of (7) for $\tau = o$ as $n \to \infty$, we consider each term of it as positive, and then take the nth root of each such term. Their sum Sn must satisfy the relation:

$$\left[\left| \alpha_i^{(n)}(o) \right| / n! \right]^{1/n} \leqq Sn \tag{8}$$

Since $r_o \neq o$ and all derivatives of r and T_i are bounded at $\tau = o$, the sum $Sn \to o$ as $n \to \infty$, then the limit of the left-hand member of (8) is zero, and, as a result, the radius of convergence is ∞, and the series (3) converge for any value of τ'.

APPENDIX:

The nth derivative of the product of the functions $\sigma(\tau)$, $\varphi(\tau)$ is given by:

$$[\sigma\varphi]^{(n)} = \sum_{m=0}^{n} \frac{n!}{m!\,(n-m)!}\, \sigma^{(m)}\, \varphi^{(n-m)} \tag{a}$$

and that of:

$$\varphi(\tau) = \frac{1}{r^3(\tau)} \tag{b}$$

by:

$$\varphi^{(n)} = \frac{P(r)}{r^{n+3}} \tag{c}$$

with:

$$P(r) = V_1 r^{n-1} + V_2 r^{n-2} + \cdots + V_n \tag{d}$$

where the V's are polynomials in the derivatives of r up to the order n^{th} with coefficients smaller than $(n+2)!$

Combining (a), (b), (c) we can get:

$$[\sigma / r^3]^{(n)} = \sum_{m=0}^{n} \frac{n!}{m!\,(n-m)!}\, \sigma^{(m)}\, \frac{P(r)}{r^{n-m+3}} \tag{e}$$

where:

$$P(r) = V_1 r^{n-m-1} + V_2 r^{n-m-2} \cdots + V_{n-m} \tag{f}$$

The V's polynomials have coefficients smaller than $(n-m+2)!$

REFERENCES

[1] D. G. Magiros: (a). Praktika of Athens Academy of Sciences, Vol. 35 (1960), 96-103, Athens, Greece. (b). «The motion of a Projectile under the influence of the Attractive Force of a Newtonian Center and a Thrust». General Electric Co., MSD, Technical Information Series, No. 625SD221, Nov. 15, 1962.

[2] K. Knopp: «Theory and Applications of Infinite Series», § 18, Section 94, 154-155 Hafner Publishing Co., New York.

ΠΕΡΙΛΗΨΙΣ

Εἰς προηγουμένας ἐργασίας κατεσκευάσθη ἡ λύσις τοῦ προβλήματος τῆς κινήσεως ὀχήματος ὑπὸ τὴν ἐπίδρασιν Νευτωνίου ἑλκτικοῦ κέντρου καὶ μιᾶς γενικῆς ὠστικῆς δυνάμεως. Ἐνταῦθα ἐκτίθεται ἐν συντομίᾳ τὸ ζήτημα τῆς συγκλίσεως τῆς λύσεως αὐτῆς. Λεπτομέρειαι, καθὼς καὶ ἄλλα συναφῆ ζητήματα, θὰ ἐκτεθοῦν εἰς ἐργασίαν, ἡ ὁποία θὰ δημοσιευθῇ ἀλλαχοῦ.

The Impulsive Force Required to Effectuate a New Orbit Through a Given Point in Space*

by

Demetrios G. Magiros[1]

ABSTRACT

A method for treatment of a special two-body problem is discussed in this paper. The problem is: "To calculate the impulse or impulsive force required to effectuate a new Keplerian orbit around a center through a given point T in space." The solution of this problem, according to the present method, is based on the solution of an auxiliary problem, treated here geometrically, on the "projection property" of the admissible impulse, and trigonometric calculations, which make the admissible impulse an appropriate one. Three groups of problems are discussed: one when the point T is on the plane of the original orbit, and two when T is out of this plane. In the case when the $(\dot{\mathbf{r}}_0, \mathbf{I})$-plane is not perpendicular to the line CM, use is made of projections on a plane perpendicular to CM. The calculation of the impulsive force in terms of the impulse is also given. Illustrated problems for each group complete the paper.

SYMBOLS

$\mathbf{r}_0, r_0$	initial position vector, its magnitude
$\dot{\mathbf{r}}_0, \dot{r}_0$	initial velocity vector, its magnitude
$\mathbf{F}, F$	impulsive force vector, its magnitude
$\mathbf{I}, I$	impulse vector, its magnitude
$\mathbf{I}_r, I_r$	resultant of $\mathbf{I}$ and $\dot{\mathbf{r}}_0$, its magnitude
t_0	time of duration of $\mathbf{I}$
μ	constant
C	center of attraction (first focus)
E, E_3	second focus
a, c	semi-major axis, semi- inter-focal distance of ellipse
τ, E	period and total energy related to ellipse
σ_1, σ_2	arcs of circumference (C, l)

* As General Electric Company, MSD, Technical Report 63SD256, April, 1963, appeared in slightly different form; also presented and discussed at the XIVth International Astronautical Congress, Paris, France, 1963.

[1] Consultant, General Electric Company, Valley Forge Space Technology Center, Philadelphia, Pa.

Reprinted from the *Journal of the Franklin Institute* **276** (1963), 475–489.

l, l', l'', L	distances defined in the text
$M, T, T_1, T_2, T_3, P, P_1, P_2, P', P_1', P_2', N$	points defined in the text
M_1M_2, MM_3	lines defined in the text
$\varphi, \varphi_1, \varphi_2, \varphi', \varphi_1', \varphi_2', \psi_1, \psi_1', \psi_2$	angles defined in the text
$\vartheta, \vartheta', \gamma, \tau, \tau_1, \tau_2, \tau_3, \tau_4$	
$\measuredangle$	angle
δ	Dirac δ-function

INTRODUCTION

In this paper a special two-body problem is treated. For the solution of the problem, without reference to general theory, an elementary method is proposed which may be used as a preliminary discussion for practical problems of current interest.

A vehicle of constant mass is in an elliptical orbit around a Newtownian center C and an impulsive force $\mathbf{F}$ is applied at a place M of the vehicle at time $t = 0$, when the position and velocity vectors are $\mathbf{r}_0$ and $\dot{\mathbf{r}}_0$. If the impulsive force is removed after acting for a short time t_0, the vehicle is placed in a new Keplerian orbit. The problem is: *To calculate the impulsive force required to effectuate a new orbit around the center through a given point T in space* (**1**).[2]

In **Part A** of the paper an auxiliary problem is treated geometrically. In **Part B**, the appropriate impulse is calculated by using the auxiliary problem. By the acceptance of the so-called "projection property" of the impulse, the impulse becomes an admissible one when, by trigonometric calculations, we get the appropriate impulse. In **Part C** the impulse force is calculated by using the corresponding impulse. A variety of problems in a single plane and in space is illustrated

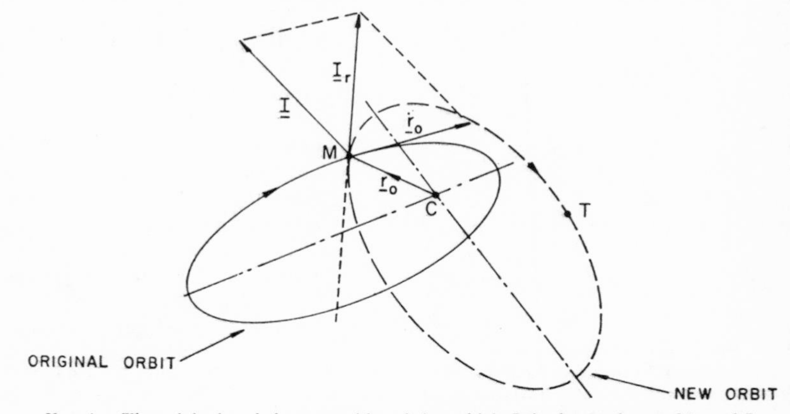

FIG. 1. The original and the new orbits of the vehicle $\mathbf{I}_r$ is the resultant of $\dot{\mathbf{r}}_0$ and $\mathbf{I}$.

[2] The boldface numbers in parentheses refer to the references appended to this paper.

PART A. THE AUXILIARY PROBLEM

The position and velocity vectors, when the force is removed, may be $\mathbf{r}_0$ and $\mathbf{I}_r = \dot{\mathbf{r}}_0 + \mathbf{I}$, respectively. $\mathbf{I}$ is the impulse of the impulsive force $\mathbf{F}$ during the time t_0 given by:

$$\mathbf{I} = \int_0^{t_0} \mathbf{F}(t)dt. \tag{A.1}$$

The plane of the new orbit must contain the vector $\mathbf{I}_r$, which must be tangent to the new orbit at the point M, Fig. 1. Then it is suggested physically to find the solution of the following auxiliary problem:

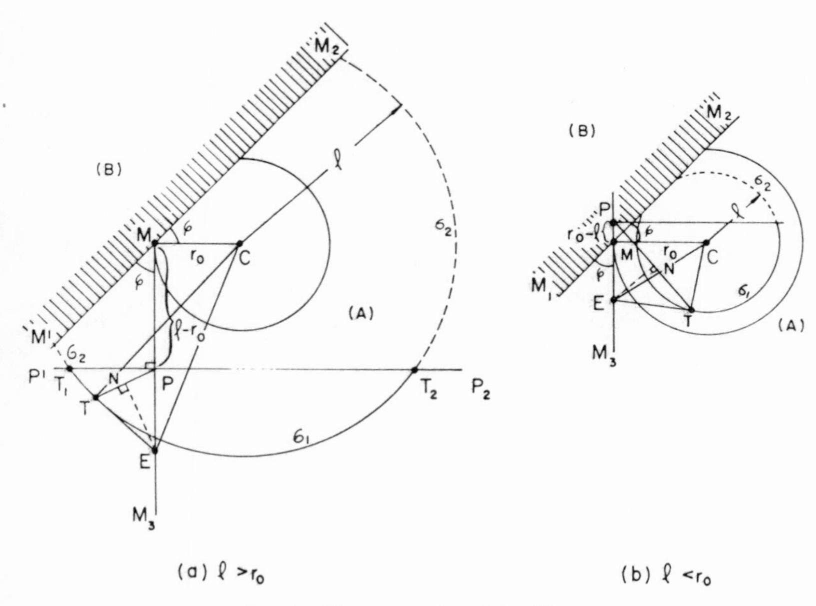

FIG. 2. The construction of the ellipse.

"*Construct an ellipse of which are known the focus C, the points M and T, and the tangent M_1M_2 through the point M.*"

The solution to the problem stated in the Introduction can be obtained by the solution of this auxiliary problem.

The ellipse of the auxiliary problem can be constructed by a unique construction of its second focus, when, based on this construction, one can calculate the major axis and the interfocal distance of the ellipse.

1. *Construction of the ellipse.*

For the construction of the ellipse, four conditions, independent of one another, are given. These conditions are such that the following property of an ellipse can be applied: "*The normal to an ellipse at any point of the ellipse bisects the angle included by the focal radii at that point.*" Application of this property in the problem at hand suggests the construction of the line MM_3, Fig. 2, such that $\sphericalangle M_1MM_3 = \sphericalangle CMM_3 = \varphi$. The second focus E must be a point of the line MM_3 and, by using this line, the focus E can be uniquely constructed as follows. Take first $TC = l$, $MC = r_0$, and draw the circle (C, l) with the center C and radius l. For the construction of E two cases can be distinguished.

(a) *The point T is not inside the circle (C, l), $l \geqq r_0$.*

We take a point P on the line MM_3, Fig. 2a, such that $MP = l - r_0$, draw the segment TP and construct the perpendicular to its midpoint N. The intersection point E with the line MM_3 is the second focus of the ellipse, for:

$$TE + TC = TE + l$$

$$ME + MC = MP + PE + r_0 = (l - r_0) + TE + r_0 = TE + l.$$

(b) *The point T is inside the circle (C, l), $l < r_0$.*

We take a point P on the extension of M_3M, Fig. 2b, such that $MP = r_0 - l$, draw the segment TP and the perpendicular to its midpoint, when the intersection point E is the second focus, for:

$$TE + TC = TE + l$$

$$ME + MC = PE - PM + r_0 = TE - (r_0 - l) + r_0 = TE + l.$$

2. *Calculation of the major axis and interfocal distance of the ellipse.*

The previous construction of the ellipse gives a procedure for the calculation of the major axis 2a and the interfocal distance 2c of the ellipse by using triangles of the construction. As shown in Fig. 3, we have $2a = ME + MC$, $2c = EC$. The points M, C, T we consider fixed, then r_0, l, γ, φ are known. We calculate from the triangle MCK the length MK and the $\sphericalangle MKC$, from triangle KTP the lengths KP and TP and the $\sphericalangle KPT$, and from the triangle TPE the length PE, when the calculation of the major axis 2a results. The triangle CEK can be used for the calculation of the interfocal distance CE.

3. *Discussion of the construction.*

The tangent M_1M_2 divides the CM_1M_2-plane into parts (A) and (B), when the ellipse and the second focus must lie on the same part with the given focus C, namely on the part (A), Fig. 2a & b.

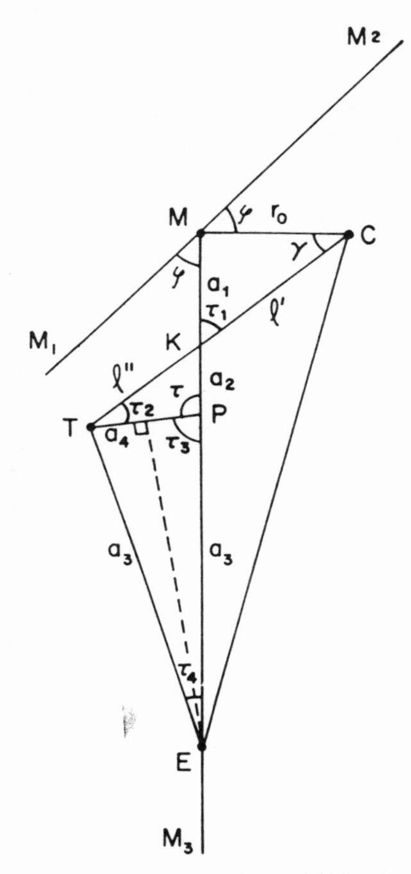

FIG. 3. Triangles for calculation of the major axis 2_a and the interfocal distance 2_c of the ellipse.

The line P_1P_2 perpendicular to MM_3 at P, which is fixed point for any place of the point T on the fixed circle (C, l), has important properties in connection with the above construction. P_1P_2 divides the circumference (C, l) into arcs σ_1 and σ_2, solid and dashed, respectively, in Fig. 2a & b, and we have the following cases, if $l \geqq r_0$.

(a) In case T is on P_1P_2, that is $T \equiv T_1$ or $T \equiv T_2$, the perpendicular to the midpoint N of the segment PT intersects MM_3 at infinity, $PE = \infty$, and the ellipse is a parabola.

(b) In case T is point of the arcs σ_2, the point E appears above P and in general above M in the part (B), when the construction is not valid, the places of T on the arcs σ_2 are not acceptable, and we have a domain of non-permissible places of T.

(c) In case T is point of the arc σ_1, the point E is in part (A) of the plane, the distance PE is finite, the construction is valid, the places of T on the arc σ_1 are acceptable, and we have ellipses.

We can get appropriate remarks if $l < r_0$.

4. *Special important case of the construction,* $\varphi = \pi/2$.

The case $\varphi = \pi/2$, that is when M is a vertex of the ellipse and CM contains the second focus E, is of special importance. We examine this case.

(a) *Case* $l \geqq r_0$.

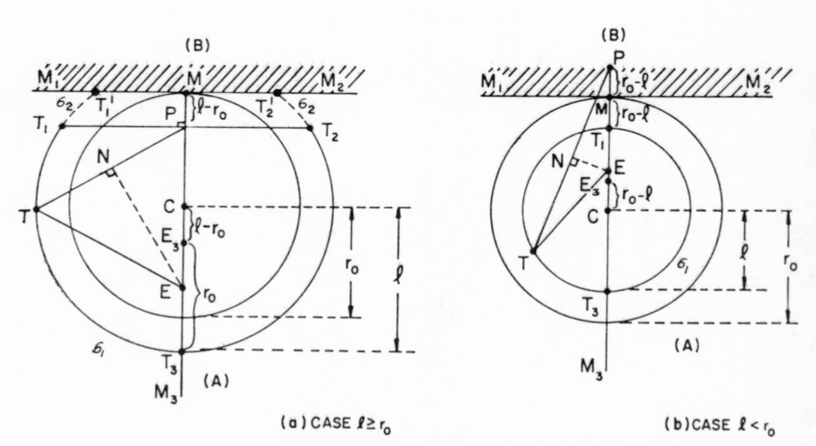

FIG. 4. Construction when MC perpendicular to M_1M_2.

The line T_1PT_2, Fig. 4a, which gives the arcs σ_1 and σ_2 of the circumference (C, l) is parallel to the tangent M_1M_2. If L is the distance of T from the tangent M_1M_2, we can have:

(i) The condition $L < l - r_0$ corresponds to non-permissible places of T.
(ii) The condition $L = l - r_0$ corresponds to parabolas ($T \equiv T_1$ or $T \equiv T_2$).
(iii) The condition $L > l - r_0$ corresponds to ellipses.
(iv) When $T \equiv T_3$, then $E \equiv E_3$, $2a = l + r_0$, $2c = l - r_0$.
(v) The condition $l = r_0$ corresponds to a circle.
(vi) The focus E of any ellipse is not above E_3, then for all ellipses the vertex M is pericenter.

(b) *Case* $l < r_0$.

From Fig. 4b we have:

(i) The condition $L = r_0 - l$ corresponds to non-permissible places of $T(T \equiv T_1$ only)
(ii) The condition $L > r_0 - l$ corresponds to ellipses.
(iii) When $T \equiv T_3$, then $E \equiv E_3$, $2a = l + r_0$, $2c = r_0 - l$.
(iv) The focus E of any ellipse is not below E_3, then for all ellipses the vertex M is apocenter.

The graph of a and c versus l, based on $a = \frac{1}{2}(l + r_0)$, $c = \frac{1}{2}|l - r_0|$, is given in Fig. 5.

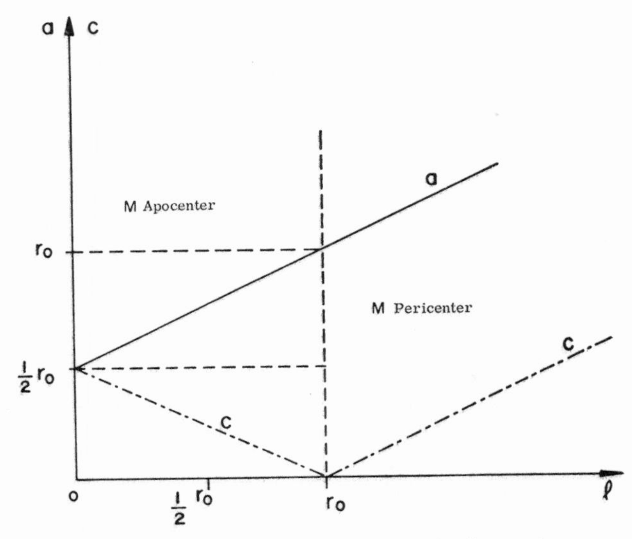

FIG. 5. The "a line" and "c line" versus the distance l.

PART B. THE APPROPRIATE IMPULSE

In the previous section we have discussed the orbit through the point T, the new orbit, in its plane. In this section we will discuss the impulse compatible with the new orbit, the appropriate impulse.

The central point C, the main point M and the target point T are taken fixed, when l, $\mathbf{r}_0$, $\dot{\mathbf{r}}_0$, γ, φ, Fig. 6, are known.

The point T may be either on the plane of the original orbit or out of it, and we consider the "single plane problems" and "space problems," separately.

1. *The single plane problems.*

When the target point T is on the plane of the original orbit, the impulse **I**, the angles ϑ and φ and the new orbit must lie on this plane. The angles ϑ, φ_1, φ, φ_2, ψ_1, ψ_2 are defined as shown in Fig. 6. We have:

$$\varphi = \varphi_1 + \psi_1, \qquad \vartheta = \psi_1 + \psi_2 \tag{B.1}$$

and from triangle MPP_1:

$$I_r = \{\dot{r}_0^2 + I^2 + 2\dot{r}_0 I \cos\vartheta\}^{1/2} \tag{B.2}$$

$$\tan\psi_1 = \frac{\sin\vartheta}{\cos\vartheta + (\dot{r}_0/I)}. \tag{B.3}$$

By using the above formulae and the construction of the new orbit according to the previous section, we can solve a variety of problems, the problems of "Group A."

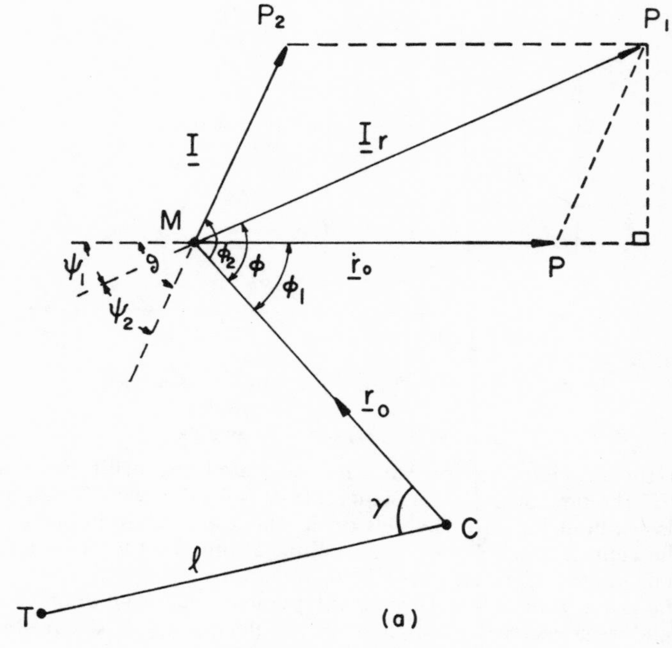

FIG. 6. The resultant $\mathbf{I}_r$ must be on the $(\mathbf{r}_0, T)$-plane.

Some of these problems are illustrated here.

Problem 1: "Given an impulse-vector, determine the new orbit through the target point, if this point is in a permissible place."

Solution: The problem accepts a unique solution. Since $\dot{\mathbf{r}}_0$, I, ϑ are known, tan ψ_1 can be calculated from Eq. B.3, when from the two angles ψ_1 we select the appropriate one to the directions of the known vectors $\dot{\mathbf{r}}_0$ and $\mathbf{I}$. The angle φ_1 is known, the angle ψ_1 is now calculated, when from Eq. B.1 we have the angle φ. The appropriate ellipse can now be constructed according to the instructions of the previous section and the construction is valid—that is, the solution is physically accepted—if the target point is in a permissible place.

Problem 2: "Given the direction of an impulse-vector and the angle ψ_1, determine the magnitude of the appropriate impulse and the orbit through the target point."

Solution: The problem accepts a unique solution. Since ψ_1, φ_1 are known, φ can be found, when the orbit through the target point can be determined. Since ψ_1, ϑ, $\dot{r}_0$ are known, we calculate the magnitude I of the appropriate impulse $\mathbf{I}$ by using Eq. B.3.

Problem 3: "Given the magnitude I of an impulse-vector and the angle ψ_1, determine the direction of the appropriate impulse-vector and the corresponding orbit through the target point."

Solution: The problem has a non-unique solution. Since ψ_1 and φ_1 are known, φ can be found and then the orbit can be constructed. For the direction of the impulse, we calculate sin ϑ by using Eq. B.3, when two angles ϑ are accepted, then two directions of the impulse-vector.

Problem 4: "Discuss the orbit through the target point and related subjects under the following restrictions:

(i) *The impulse is tangential to the original orbit;*
(ii) *The point M is vertex of the original orbit;*
(iii) *The point T is on the line MC.*

Solution: This problem is related to the special case of construction, paragraph 4 of the previous section, Fig. 4.

(i) Since the impulse is tangential, the directions of the impulse will be either $\vartheta = 0$ or $\vartheta = \pi$, then, for prescribed magnitude I of the impulse, we have extremes to the magnitude I_r of the resultant vector $\mathbf{I}_r = \dot{\mathbf{r}}_0 + \mathbf{I}$ at M, namely max $I_r = \dot{r}_0 + I$ when $\vartheta = 0$, min $I_r = \dot{r}_0 - I$ when $\vartheta = \pi$.

(ii) Since M is vertex of the original orbit and the impulse tangential at M, then M must necessarily be vertex of the new orbit.

(iii) Since T is on MC, the only permissible place of T is $T \equiv T_3$, Fig. 4, and we have:

$$a = \tfrac{1}{2}(l + r_0), \qquad c = \tfrac{1}{2}|l - r_0|.$$

As T_3 moves on MC, l is changed and we have the following cases:
As $l = r_0$, the ellipse becomes the circle (C, r_0); as $l > r_0$, M is pericenter; and as $l < r_0$, M is apocenter.

(iv) The family of the ellipses, corresponding to $l > r_0$ has an ellipse with minimum major axis; and the family of the ellipses, corresponding to $l < r_0$, has an ellipse with maximum major axis. For both families, and for these distinct extremal ellipses, the point T is the point $T \equiv T_3$ of line CM.

This remark, associated with the formula (2):

$$\tau = 2\pi\sqrt{a^3/\mu} = \tfrac{1}{2}\sqrt{\pi\mu/E^3}$$

where τ is the periodic time of the ellipse in the Newtonian two-body field, and E the total energy, gives the following result:
"From all ellipses of the family corresponding to $l > r_0$ (or $l < r_0$), the ellipse of the present problem ($T \equiv T_3$) has the minimum (maximum) period, and the maximum (minimum) total energy."

2. *The space problems.*

In case the point T is not on the plane of the original orbit, we have the following planes, Fig. 6.

The $(\mathbf{r}_0, \dot{\mathbf{r}}_0)$-plane, the original orbit plane,	"First-plane"
The $(\mathbf{r}_0, T)$-plane, the new orbit plane,	"Second-plane"
The $(\mathbf{r}_0, \mathbf{I}_r)$-plane, plane corresponding to resultant $\mathbf{I}_r$,	"Third-plane"
The $(\mathbf{r}_0, \mathbf{I})$-plane,	"Fourth-plane"
The $(\dot{\mathbf{r}}_0, \mathbf{I})$-plane, which contains $\mathbf{I}_r$.	"Fifth-plane"

All the above planes are identical with the plane of the original orbit if the target point T is in this plane.

(a) The impulse $\mathbf{I}$ must be such that the resultant $\mathbf{I}_r$ necessarily lies on the new orbit plane, that is the above "second" and "third" planes must coincide. In the general case when the above "fifth-plane" is not perpendicular to CM, by using a plane per-

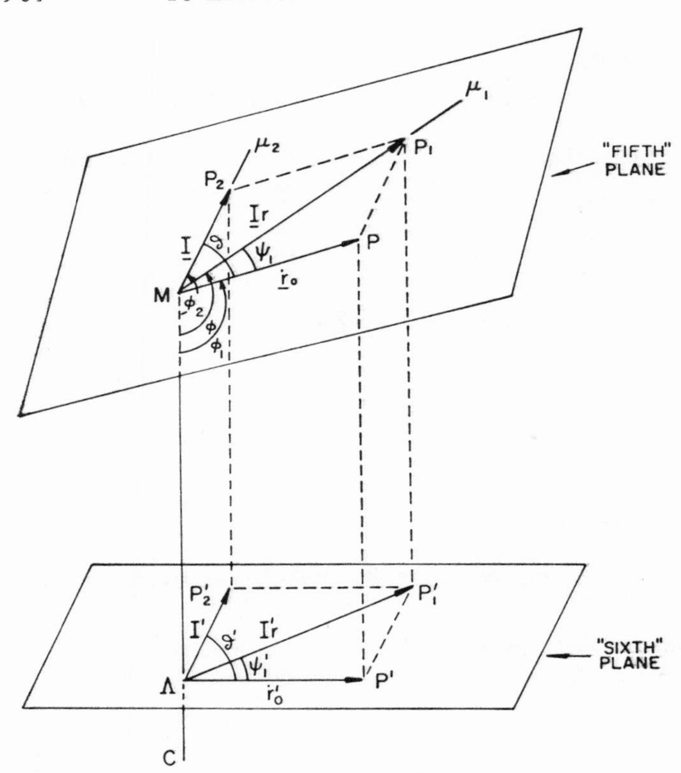

FIG. 7. The "fifth"-plane nonperpendicular to MC contains MPP_1P_2. The "sixth"-plane perpendicular to MC contains the projection of MPP_1P_2.

pendicular to CM, the "sixth-plane," the above restriction of the impulse can be stated as follows:

"Projection property of the impulse:" "The orthogonal projection of the angle ($\dot{\mathbf{r}}_0$, $\mathbf{I}_r$) *on the sixth plane must be equal to dihedral angle of the planes of the original and the new orbits."*

The impulse which obeys this "projection property" is a "permissible" impulse and has the "possibility" of making the impulse required. Trigonometric calculations can make the permissible impulse an appropriate one.

If we define the angles φ_1, φ, φ_2, ψ_1, ϑ as shown in Fig. 7, and the angles φ_1', φ', φ_2' as the inclination angles of $\dot{\mathbf{r}}_0$, $\mathbf{I}_r$, $\mathbf{I}$, respectively, to the "sixth-plane," we can have:

$$\varphi_1 = \frac{\pi}{2} + \varphi_1', \qquad \varphi = \frac{\pi}{2} + \varphi', \qquad \varphi_2 = \frac{\pi}{2} + \varphi_2'. \quad \text{(B.4)}$$

The orthogonal projection of the parallelogram MPP_1P_2 of the "fifth-plane" on the "sixth-plane" is a new parallelogram $\Lambda P'P_1'P_2'$, for which we can obtain:

$$\Lambda P' = \dot{r}_0 \cos \varphi_1', \qquad \Lambda P_1' = I_r \cos \varphi', \qquad \Lambda P_2' = I \cos \varphi_2', \qquad \vartheta' = \text{proj.}\, \vartheta, \qquad \psi_1' = \text{proj.}\, \psi_1 \quad \text{(B.5)}$$

$$I_r' = \{(\dot{r}_0 \cos \varphi_1')^2 + (I \cos \varphi_2')^2 + 2\dot{r}_0 I \cos \varphi_1' \cos \varphi_2' \cos \vartheta'\}^{1/2} \quad \text{(B.6)}$$

$$\tan \psi_1' = \frac{\sin \vartheta'}{\cos \vartheta' + \dfrac{\dot{r}_0}{I} \dfrac{\cos \varphi_1'}{\cos \varphi_2'}}. \quad \text{(B.7)}$$

The angle ψ_1' must be, according to the above restriction, equal to the dihedral angle of the original and new orbital planes.
A variety of problems, the problems of "*Group B*" can be solved by using the above formulae and the construction of the ellipse of the previous section. We give just two:

Problem 1: "Given the line of the action of an impulse, find the appropriate magnitude of the impulse required, and the new orbit through the target point, if it is in a permissible place."

Solution: The line of action of the impulse and that of $\dot{\mathbf{r}}_0$ determine the "fifth-plane" where the angle ϑ is given, Fig. 7. The "second-plane" which contains the target point, intersects the "fifth-plane" into the line $M\mu_1$, which must necessarily be the action line of the resultant $\mathbf{I}_r$ and tangent to the new orbit. The "first," "second," and "fifth" planes are known in this problem and from MP, ψ_1, ϑ we can construct the parallelogram MPP_1P_2. The construction leads to calculation of the magnitudes $MP_1 = I_r$, $MP_2 = I$. For the construction now of the new orbit we need to know the angle φ, and, since $\varphi = \frac{\pi}{2} + \varphi'$, we calculate φ'. For the calculation of φ', we use the orthogonal projection of the parallelogram MPP_1P_2 on the "sixth-plane," namely the parallelogram $\Lambda P'P_1'P_2'$. In the formula Eq. B.7 we know φ_1', φ_2', ψ_1', $\dot{r}_0$, I, then we calculate $\cos \vartheta'$, when from formula Eq. B.6 we callate I_r'. The angle φ' needed can be determined from $\cos \varphi' = I_r'/I_r$.

Problem 2: "Given the magnitude I of the impulse and the line of action of the resultant $\mathbf{I}_r$, *find the appropriate line of the action of the impulse re-*

quired, and the new orbit through the target point, if it is in a permissible place."

Solution: The line $M\mu_1$ of action of the unknown resultant $\mathbf{I}_r$ must be on the "second-plane," and the "fifth-plane" is the $(MP, M\mu_1)$-plane. The angle ψ_1 is known. We construct the parallelogram MPP_1P_2 on the "fifth-plane." The circle (P, I) in the "fifth-plane" intersects $M\mu_1$ at P_1, and the line of action of the impulse is the line $M\mu_2$ parallel to PP_1. Again, for the construction of the new orbit, we calculate φ'.

(b) Investigating the properties of the point M as a vertex of the original and/or the new orbit in connection with an impulse perpendicular or oblique to CM, another group of problems, the "Group C," may arise.
Properties of the inclination angles φ_1', φ', φ_2' of the coplanar vectors $\dot{\mathbf{r}}_0$, $\mathbf{I}_r$, $\mathbf{I}$ on the "sixth-plane" are used to distinguish this group of problems. We use the following statements of solid geometry:

(i) "A line is, by definition, perpendicular to a plane when it is perpendicular to every line on the plane which meets it;"
(ii) "If a line is perpendicular to each of two intersecting lines at their point of intersection, it is perpendicular to the plane determined by the two lines;"
(iii) "A line oblique to a plane is perpendicular to only one line in the plane through the point of intersection;"

We can have the following properties of φ_1', φ', φ_2', based on the above statements of solid geometry:

(i) "These angles may be altogether either zero or non-zero;"
(ii) "If one of them is zero, the other two must be either both zero or both non-zero;"
(iii) "If one of them is non-zero, the other two must be either both non-zero or one zero and the other non-zero, but never both zero;"
(iv) "If two of them are zero, the third one must be zero; if two of them are non-zero, the third one may be either zero or non-zero."

By using these properties of φ_1', φ', φ_2', and that the meaning of $\varphi_1' = 0$ (or $\varphi' = 0$) is that the point M is a vertex of the original (or new) orbit, and $\varphi_2' = 0$ is that the impulse is perpendicular to CM, we may have some of the problems of "Group C:"

Problem 1: "*If the impulse is perpendicular to CM, then:*

(i) *M being a vertex of the original (new) orbit implies that M is a vertex of the new (original) orbit;*
(ii) *M being a non-vertex of the original (new) orbit implies that M is non-vertex of the new (original) orbit;*"

Problem 2: "*If M is a vertex of the original (new) orbit, then:*

(i) *M being a vertex of the new (original) orbit, implies that the impulse is perpendicular to CM;*
(ii) *M being a non-vertex of the new (original) orbit implies that the impulse is oblique to CM.*"

Problem 3: "*If the impulse is oblique to CM and M non-vertex of the original (new) orbit, then M may or may not be a vertex of the new (original) orbit.*"

Problem 4: "*If the impulse is perpendicular to CM and M a vertex of the original (new) orbit, then M is a vertex of the new (original) orbit.*"

Problem 5: "*If M is non-vertex of the original and the new orbits, then the impulse may be either perpendicular or oblique to CM.*"

PART C. THE APPROPRIATE IMPULSIVE FORCE

The determination of the appropriate impulsive force, based on the integral definition Eq. A.1 of the impulse, in terms of the impulse, is a problem of special singular integral equations, subject to modern mathematical theories.

The impulsive forces are, by nature, not forces of ordinary type. Their convenient idealization is the improper "Dirac δ-function," of which the impulsive forces represent a simple physical interpretation. We can approximate the impulsive force by appropriate assumptions. If for example, we assume that it is constant during its short duration t_0, we can get from the formula Eq. A.1 that: $F = I/t_0$.

This formula gives an approximate transition from the impulse required to the corresponding impulsive force, valid approximately for any direction of the impulse, which direction is the same with that of the impulsive force.

* * * *

The author dedicates this paper to B. H. Caldwell and W. R. Becraft of General Electric Company, Missile and Space Division, for their encouragement and help in connection with this research work.

REFERENCES

(1) Transfer between elliptical orbits has been investigated by D. Lawden, P. Longe, G. Smith, R. Plimmer, J. Horner, etc. We may have another procedure for the solution of the present problem, without reference to the general theory, by using the D. Lawden article: "Impulse Transfer Between Elliptical Orbits," which is Chapter 11 of G. Leitmann's book: "Optimization Techniques with Applications to Aerospace Systems,', New York, Academic Press, 1962.

(2) J. Synge and B. Griffith, "Principles of Mechanics," 3rd Ed., New York, McGraw-Hill, 1959, Page 165.

Motion in a Newtonian Forced Field Modified by a General Force

by DEMETRIOS G. MAGIROS

General Electric Company

Valley Forge Space Technology Center, Philadelphia, Pa.

ABSTRACT: *The motion of a man-made celestial body under the influence of an attractive center according to the inverse square Newton's law and a general force is discussed. A solution of special form, referred to a specific inertial coordinate system, is established. For the time-dependent coefficients of the solution MacLaurin expansions are taken. It is proven that the formal solution constructed satisfies all the requirements needed for really being a solution of the physical problem; besides its region of convergence is found. The problem is specialized in the case of a constant force. The case of suddenly or gradually applied forces is also examined.*

1. Introduction

A variety of current problems, especially from the point of view of engineering, are related to the following problem:

> *"A man-made celestial body is moving under the influence of a 'centripetal force' obeying the inverse square Newton's law toward an attractive center, when an impulsive force is applied, it acts for an interval of time, then is removed. Find the motion of the body during the action of the force."*

A mathematical discussion of this particular two-body problem is given in this reference paper (1). A solution of special form, referred to a specific inertial coordinate system, is established. The time-dependent coefficients of this solution are taken in MacLaurin expansions. The solution is not only a formal one, but is actually accepted as a solution of the above physical problem since it satisfies all the appropriate requirements, besides the solution is valid for any interval of time, however large. In discussing the convergence of the series, by using the Leibnitz formula appropriately applied to this problem as shown in the Appendix, a specific form for their general term is given when the root test is successfully applied. This is discussed in Sections 3 and 4.

As an application, the problem is specialized in the case of a constant force, when the solution is simplified and taken in an approximation of the error found. A direct physical application of this is given in Section 5.[1]

The solution contains the impulse of the force for small time t_o, and this

[1] From the author's discussions with John Immel, Guidance and Advance Requirements Operation, G. E., Valley Forge Space Tech. Center.

Reprinted from the *Journal of the Franklin Institute* **278** (1964), 407–416.

impulse differs in the case when the force is either suddenly or gradually applied to the body. Further remarks on this subject are given in Section 6.

2. *Mathematical Formulation of the Problem*

The differential equation in vector form:

$$\ddot{\mathbf{r}}(t) = -\frac{\mu}{r^3(t)}\mathbf{r}(t) + \mathbf{T}(t), \qquad 0 \leq t \leq t' \tag{1}$$

governs the motion of the body during the action of the force. $\mathbf{T}$ is the force, r the magnitude of the position vector $\mathbf{r}$, the constant $\mu = K(M + m)$, M and m the masses of the attractive center and the body, respectively, K the constant of gravity.

The time t' of the action of the force is split according to

$$t' = t_o + \tau', \tag{2}$$

where t_o is very small time.

Initial conditions attached to Eq. 1 are:

$$\mathbf{r}(t_o) = \mathbf{r}_o, \qquad \dot{\mathbf{r}}(t_o) = \dot{\mathbf{r}}_o + \mathbf{I}_o, \tag{3}$$

where $\mathbf{r}_o$ and $\dot{\mathbf{r}}_o$ are the position vector and velocity vector of the body at time $t = 0$; $\mathbf{I}_o$, the impulse of the force $\mathbf{T}$ during the time t_o, is defined by the formula

$$\mathbf{I}_o = \int_o^{t_o} \mathbf{T}(t)dt. \tag{4}$$

Small terms of the order of the increments δr and $\delta \dot{r}$ of the vectors $\mathbf{r}$ and $\dot{\mathbf{r}}$, due to the action of an attractive center during the small time t_o, are omitted in Eq. 3.

The introduction of Eqs. 2 and 3 leads to a solution which is helpful in practical problems, as we can easily see.

By changing the time scale to

$$\tau = t - t_o \tag{5}$$

the system Eqs. 1 and 3 become

$$\ddot{\mathbf{r}}(\tau) = -\frac{\mu}{r^3(\tau)}\mathbf{r}(\tau) + \mathbf{T}(\tau), \qquad 0 \leq \tau \leq \tau' \tag{6}$$

$$\mathbf{r}(0) = \mathbf{r}_o, \qquad \dot{\mathbf{r}}(0) = \dot{\mathbf{r}}_o + \mathbf{I}_o. \tag{7}$$

Equation 6 is nonlinear because of the appearance of r^3. The time τ' of the action of the force is not necessarily considered small, as we cannot take the "average value" of r during the action of the force and work with linearized equations.

Every function $\mathbf{r}(\tau)$ which satisfies Eq. 6 must be twice differentiable, when this function and its first derivative $\dot{\mathbf{r}}(\tau)$ must necessarily be continuous functions and bounded for any τ in the interval D: $0 \leq \tau \leq \tau'$. Thus, if the constants M_1 and M_2 are the least upper bounds of $\mathbf{r}(\tau)$ and $\dot{\mathbf{r}}(\tau)$, respectively, then any solution $\mathbf{r}(\tau)$ of Eq. 6 as well as its derivative $\dot{\mathbf{r}}(\tau)$ are continuous in the domain:

$$D_1: \ |\mathbf{r}(\tau)| < M_1, \qquad |\dot{\mathbf{r}}(\tau)| < M_2$$

for all τ in D.

3. A Solution of the Problem

Let us select a reference coordinate system $(P; \mathbf{r}_o^*, \mathbf{s}_o^*, \mathbf{T}_o^*)$ defined as follows: P is the position of the body when the force starts acting; $\mathbf{r}_o^*, \mathbf{s}_o^*, \mathbf{T}_o^*$ are unit vectors of $\mathbf{r}_o$, $\dot{\mathbf{r}}_o + \mathbf{I}_o$, $\mathbf{I}_o$, respectively, see Fig. 1. We establish in the following a solution of the system Eqs. 6 and 7 valid in the domain D_1 for any τ in D, of the form:

$$\mathbf{r}(\tau) = a_1(\tau)\mathbf{r}_0^* + a_2(\tau)\mathbf{s}_o^* + a_3(\tau)\mathbf{T}_o^* \tag{8}$$

by an appropriate calculation of the scalar functions $a_i(\tau)$, $i = 1, 2, 3$. This function $\mathbf{r}(\tau)$, as solution of Eq. 6 must be differentiable twice, and we take this function differentiable as many times as needed. Then, the functions $a_i(\tau)$ of Eq. 8 are differentiable, continuous and bounded in the domain D_1, for any τ in D.

We can now insert the function $\mathbf{r}(\tau)$ of Eq. 8 into Eq. 6 when, if $T_1(\tau)$, $T_2(\tau)$, $T_3(\tau)$ are the projections of the force $\mathbf{T}(\tau)$ on the axis of the coordinate system selected, Eqs. 6 and 7 yield

$$\left.\begin{array}{llll} \text{(i)} & \ddot{a}_1(\tau) + \dfrac{\mu}{r^3(\tau)} a_1(\tau) = T_1(\tau), & a_1(0) = r_o, & \dot{a}_1(0) = 0 \\[2ex] \text{(ii)} & \ddot{a}_2(\tau) + \dfrac{\mu}{r^3(\tau)} a_2(\tau) = T_2(\tau), & a_2(0) = 0, & \dot{a}_2(0) = s_o \\[2ex] \text{(iii)} & \ddot{a}_3(\tau) + \dfrac{\mu}{r^3(\tau)} a_3(\tau) = T_3(\tau), & a_3(0) = 0, & \dot{a}_3(0) = 0 \end{array}\right\}. \tag{9}$$

By using the conditions Eqs. 9, satisfied by the functions $a_i(\tau)$, we can uniquely determine $a_i(\tau)$. To do that, let us take Eq. 9(i).

By changing the notation according to

$$a_1 = x_1, \qquad \dot{a}_1 = x_2, \tag{10}$$

Eq. 9(i) can be replaced by the equivalent normal system

$$\left.\begin{array}{l} \dot{x}_1(\tau) = x_2(\tau) \\[1ex] \dot{x}_2(\tau) = -\dfrac{\mu}{r^3(\tau)} x_1(\tau) + T_1(\tau) \end{array}\right\} \tag{11}$$

valid in the domain $D \times D_1$: $0 \le \tau \le \tau'$, $|x_1| < M_1$, $|x_2| < M_2$ with initial conditions $x_{10} = r_0$, $x_{20} = 0$. The right-hand members of Eq. 11:

$$f_1 \equiv x_2, \qquad f_2 \equiv -\frac{\mu}{r^3} x_1 + T_1, \tag{12}$$

Demetrios G. Magiros

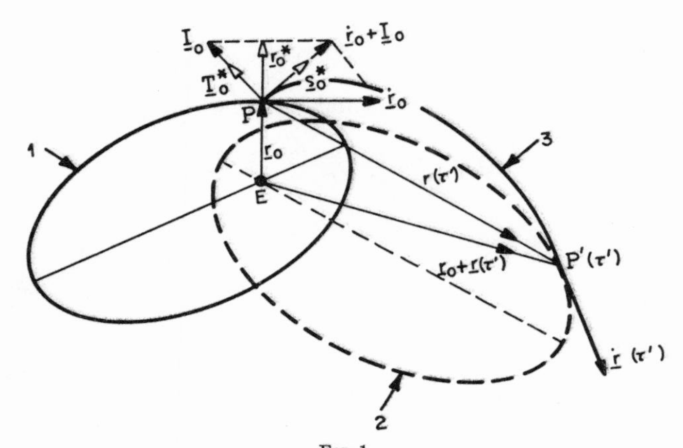

FIG. 1.
1. Original orbit; 2. New orbit; 3. Orbit during the action of the thrust.

if T_1 is taken as a differentiable function of τ in D, and the value $r = 0$ is excluded, are single-valued, continuous and bounded functions of all their arguments simultaneously, and satisfy a Lipschitz condition on x_1, x_2 in D for any value τ in D. Then, (2) solution of Eq. 11 will exist and be uniquely defined by given values x_{10}, x_{20} of x_1, x_2 in D_1, for any value τ_o of τ in D, say $x_{10} = r_o$, $x_{20} = 0$ for $\tau_0 = 0$. Therefore, Eq. 9i has only one solution $a_1(\tau)$.

By the same reasoning and using the other equations of (9), the functions $a_2(\tau)$, $a_3(\tau)$ can be uniquely defined.

For a prescribed force $\mathbf{T}(\tau)$, its components $T_i(\tau)$, $i = 1, 2, 3$ are known, but as $r(\tau)$ is unknown, it looks impossible to calculate $a_i(\tau)$ from Eq. 9. In spite of that, approximations of $a_i(\tau)$ can be found by considering their MacLaurin expansions at $\tau = 0$.

Due to the properties of $a_i(\tau)$ pointed out above, MacLaurin expansions of $a_i(\tau)$ at $\tau = 0$ convergent for any value τ in D_1, can be assumed:

$$\left.\begin{aligned} a_1(\tau) &= a_1(0) + \dot{a}_1(0)\tau + \frac{\ddot{a}_1(0)}{2}\tau^2 + \cdots + \frac{a_1^{(n)}(0)}{n!}\tau^n + \cdots \\ a_2(\tau) &= a_2(0) + \dot{a}_2(0)\tau + \frac{\ddot{a}_2(0)}{2}\tau^2 + \cdots + \frac{a_2^{(n)}(0)}{n!}\tau^n + \cdots \\ a_3(\tau) &= a_3(0) + \dot{a}_3(0)\tau + \frac{\ddot{a}_3(0)}{2}\tau^2 + \cdots + \frac{a_3^{(n)}(0)}{n!}\tau^n + \cdots \end{aligned}\right\} \quad (13)$$

We may calculate as many coefficients of these series as we want by successive differentiation of Eq. 9 and using Eqs. 6 and 7.

The first two coefficients are given in Eq. 9. For the third coefficients, by using Eq. 9, we get

$$\ddot{a}_1(0) = -\frac{\mu}{r_o^3} + T_1(0), \qquad \ddot{a}_2(0) = T_2(0), \qquad \ddot{a}_3(0) = T_3(0). \tag{14}$$

For the fourth coefficients, we differentiate Eq. 9, when

$$\left.\begin{aligned} \dddot{a}_1(\tau) &= -\mu \frac{d}{d\tau}\left\{\frac{a_1(\tau)}{r^3(\tau)}\right\} + \dot{T}_1(\tau) \\ \dddot{a}_2(\tau) &= -\mu \frac{d}{d\tau}\left\{\frac{a_2(\tau)}{r^3(\tau)}\right\} + \dot{T}_2(\tau) \\ \dddot{a}_3(\tau) &= -\mu \frac{d}{d\tau}\left\{\frac{a_3(\tau)}{r^3(\tau)}\right\} + \dot{T}_3(\tau) \end{aligned}\right\} \tag{15}$$

and take their values at $\tau = 0$. Applying the procedure of successive differentiation, we can get as many coefficients of the series 13 as we want when the series become known series, then the coefficients $a_i(\tau)$ of the function $\mathbf{r}(\tau)$ given by Eq. 8 are determined, and the solution $\mathbf{r}(\tau)$ of Eqs. 6 and 7 is formally constructed.

4. Further Discussion of the Solution

The solution constructed in the preceding section is a formal solution, which may or may not be an actual solution of the physical problem stated in the introduction.

The "existence" of the solution of the form Eq. 8, its "uniqueness" and "continuous dependence" with the data, which is a kind of stability of the solution, properties sometimes called "Hadamard's postulates," (**3**) decide about the acceptance of the solution physically.

Solution of Eq. 8 should satisfy all the above requirements, if the functions $a_i(\tau)$ exist uniquely and depend continuously on the data.

We prove that the functions $a_i(\tau)$ have these properties.

First, about the function $a_1(\tau)$. The functions f_1 and f_2 of Eq. 12 are differentiable in D_1, for all τ in D, the partial derivatives $\frac{\partial}{\partial x_j} f_i(x_1, x_2, \tau)$ of the functions f_1 and f_2 with respect to x_1 and x_2 exist and are continuous in $D \times D_1$. Then (**2**) the partial derivatives $\frac{\partial}{\partial x_{j0}} x_i(x_{10}, x_{20}, \tau)$ of the functions x_1 and x_2 with respect to the initial conditions x_{10} and x_{20} exist and are continuous functions of x_{10}, x_{20}, τ, when x_1 and x_2 themselves depend continuously on the initial conditions given in $D \times D_1$.

This proves that the function $a_1(\tau)$, as a solution of Eq. 9(i), exists, is unique and depends continually on the initial conditions.

In the same way we can prove also that the functions $a_2(\tau)$ and $a_3(\tau)$ have the above properties.

Therefore, as a result, the function $r(\tau)$ of Eq. 8 satisfies the requirements needed for being an actual solution of the series 13.

It remains to determine the domain of the validity of the solution found, and for that we can use the correspondent radius of convergence.

Demetrios G. Magiros

The radius of convergence of the series 13 is the reciprocal of the upper limit (4):

$$\overline{\lim_{n\to\infty}} [|a_i^{(n)}(0)|/n!]^{1/n}, \qquad i = 1, 2, 3. \tag{16}$$

To determine this limit, we first find the general term of the sequence

$$\{a_i^{(n)}(\tau)/n!\}, \qquad i = 1, 2, 3. \tag{17}$$

By using Eqs. 9, the nth derivative of $a_i(\tau)$ for any $n \geq 2$, is given by

$$a_i^{(n)}(\tau) = -\mu \left[\frac{a_i(\tau)}{r^3(\tau)}\right]^{(n-2)} + T_i^{(n-2)}(\tau), \qquad i = 1, 2, 3. \tag{18}$$

The general term of the sequence 17, if we take into account in Eq. 18 formula V of the Appendix, can be given by

$$\frac{a_i^{(n)}(\tau)}{n!} = -\sum_{m=0}^{n-2} \frac{a_i^{(n)}(\tau)[V_1 r^{n-m-3} + V_2 r^{n-m-4} + \cdots + V_{n-m-2}]}{(n-1)nm!(n-m-2)!r^{n-m+1}} + \frac{T_i^{(n-2)}(\tau)}{n!}, \tag{19}$$

where $V_1, V_2, \cdots$ are polynomials in the derivatives of r up to the order $(n - m - 2)$ with coefficients smaller than $(n - m)!$

Due to the convergence of the series 13, the ratio $|a_i^{(n)}(0)|/n!$ as $n \to \infty$ must tend to zero, and its nth root to a limit less than one.

To find the limit of the nth root of the above ratio, that is, to find the limit of the nth root of the absolute value of the right-hand member of Eq. 19 for $\tau = 0$ as $n \to \infty$, consider each term of it as positive, and then take the nth root of each such term. Their sum S_n must satisfy the relation

$$[|a_i^{(n)}(0)|/n!]^{1/n} \leq S_n. \tag{20}$$

Since $r_o \neq 0$ and all derivatives of r and T_i are bounded at $\tau = 0$, the sum S_n tends to zero as $n \to \infty$, then the limit of the left-hand member of Eq. 20 is zero as $n \to \infty$, and, as a result, the radius of convergence is infinite, and the series 13 converge for any value of τ'.

In case r_o is large compared to other quantities appearing in the coefficients of the series 13, and the time τ' of the action of the force is not large, the series 13 converge rapidly, then we can get in this case a good approximation of the coefficients of the solution Eq. 8, if we confine ourselves to a few terms of the series 13.

5. *The Case of a Constant Force*

In the previous discussion a solution of a specific form of the problem at hand is established. In the following an application is given.

Consider the case of a force constant in magnitude and direction with respect to the inertial coordinate system taken. In this case T_1, T_2, T_3 of

Eq. 9 are constants and the calculation of the coefficients of the series 13 is simplified. More specifically, take the force identical with the vector $\mathbf{T}_o^*$. In this specific case, $T_1 = T_2 = 0$, $T_3 = 1$, and the coefficients of the series 13 are

$$\left.\begin{array}{llll} a_1(0) = r_o, & \dot{a}_1(0) = 0, & \ddot{a}_1(0) = -\dfrac{\mu}{r_o^2}, & \dddot{a}_1(0) = \dfrac{3\mu s_o}{r_o^3}, \cdots \\ a_2(0) = 0, & \dot{a}_2(0) = s_o, & \ddot{a}_2(0) = 0 & \dddot{a}_2(0) = -\dfrac{\mu s_o}{r_o^3}, \cdots \\ a_3(0) = 0, & \dot{a}_3(0) = 0, & \ddot{a}_3(0) = 1 & \dddot{a}_3(0) = 0, \cdots \end{array}\right\}. \tag{21}$$

For large r_o but small τ, we can get a good approximation for a_1, a_2, a_3 if we take a few terms of the corresponding series 13; we can write

$$\begin{aligned} a_1(\tau) &= r_o - \frac{\mu}{2r_o^2}\tau^2 + 0(\tau^3/r_o^4) \\ a_2(\tau) &= s_o\tau + 0(\tau^3/r_o^4) \\ a_3(\tau) &= \tfrac{1}{2}\tau^2 + 0(\tau^4/r_o^5). \end{aligned} \tag{22}$$

Therefore, by using the original time scale, the solution Eq. 8 and its derivative become

$$\left.\begin{aligned} \mathbf{r}(t) &= \left(r_o - \frac{\mu}{2r_o^2}(t - t_o)^2\right)\mathbf{r}_o^* + (t - t_o)s_o\mathbf{s}_o^* + \tfrac{1}{2}(t - t_o)^2\mathbf{T}_o^* \\ \dot{\mathbf{r}}(t) &= -\frac{\mu}{r_o^2}(t - t_o)\mathbf{r}_o^* + s_o\mathbf{s}_o^* + (t - t_o)\mathbf{T}_o^* \end{aligned}\right\}. \tag{23}$$

Equations 23 give the position vector and the velocity vector of the body in this special case during the action of the force.

When the force is removed at time t', say, the values $\mathbf{r}_o + \mathbf{r}(t')$ and $\dot{\mathbf{r}}(t')$ calculated from Eq. 23, can be used for the determination of the elements of the new Keplerian orbit of the body (**5**) which has the same focus, the attractive center E, Fig. 1, as the original Keplerian orbit.

A Physical Application

A direct physical application of the above special case can be given.

Consider a projectile that is moving around the earth in a Keplerian orbit and it is desired to change the initial Keplerian orbit into a new one that intersects the surface of the earth for the purpose of recovering the projectile. The thrust required for the desired change in orbit is produced by a rocket attached to the projectile, and the direction of the thrust vector is maintained in a specified initial direction by rotating the projectile at a high angular velocity ω about the thrust axis. The gyroscopic action of the rotating projectile resists changes in the direction of the thrust vector; thus, if the rocket provides a

Demetrios G. Magiros

constant magnitude of the thrust after time t_o, and if no changes occur in the direction of the thrust vector, it can be seen that the conditions $T_1 = T_2 = 0$, $T_3 = 1$ will be realized. Also, because the projectile is rotating prior to ignition of the rocket, $\mathbf{I}_o$ is in the same direction as $\mathbf{T}_o^*$. To further illustrate the physical application of the solution consider a typical situation encountered in practice. In many cases the orbiting object is composed of two parts, one being especially adapted for injection into orbit and attitude control while in orbit, and the other for re-entry into the earth's atmosphere. Figure 2 illustrates the above situation. It should be pointed out that the thrust direction shown is typical, chosen to provide the most favorable re-entry conditions for a given rocket impulse capacity. Also, the re-entering portion of the projectile is designed to be aerodynamically stable so that atmospheric forces align the nose of the projectile with the velocity vector as re-entry occurs. Figure 2 shows (a)

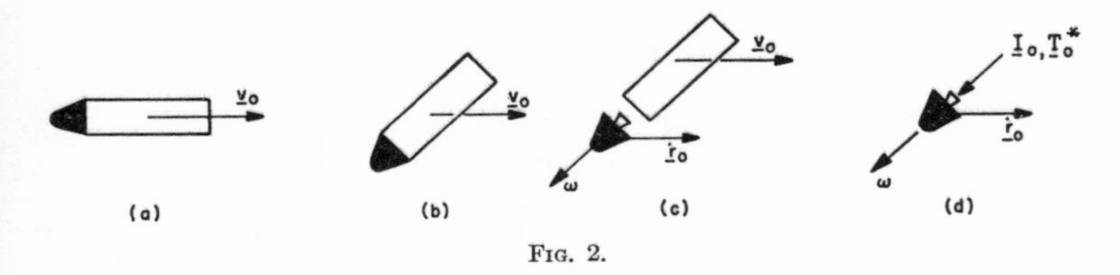

FIG. 2.

projectile in orbit with re-entry portion shaded; (b) projectile in attitude for thrust application; (c) re-entry portion separated and rotating at angular velocity ω; the unshaded portion remains moving around the earth in Keplerian orbit; and (d) enlarged view of the re-entry portion during thrust application.

6. The Impulse of the Force in Two Cases

The solution found is valid in $t_o \leq t \leq t'$ and contains the impulse $\mathbf{I}_o$, which is given by the formula Eq. 4, it then depends on the function $\mathbf{T}(t)$ in $0 \leq t \leq t_o$. We have two cases.

(a) For a force $\mathbf{T}(t)$ constant in magnitude and direction during time t_o with the same direction as $\mathbf{T}_o^*$, $\mathbf{T} = T_o\mathbf{T}_o^*$, the impulse is

$$\mathbf{I}_o' = t_o T_o \mathbf{T}_o^* \tag{24}$$

and in this case we can speak about a "sudden application" of force. The broken line $00_10_20_3$ in the t, T-plane, Fig. 3, corresponds to this situation.

(b) Let us take in this case as the direction of the force in $0 \leq t \leq t_o$ the direction of $\mathbf{T}_o^*$ and its magnitude varied according to a parabolic law during the time $0 \leq t \leq t_o$, being constant at the remainder time, i.e.,

$$\mathbf{T}(t) = T(t)\mathbf{T}_o^*, \qquad T(t) = \begin{cases} at^{1/2}, & \text{for} \quad 0 \leq t \leq t_0 \\ at_o^{1/2}, & \text{for} \quad t_0 \leq t \leq t', \end{cases} \tag{25}$$

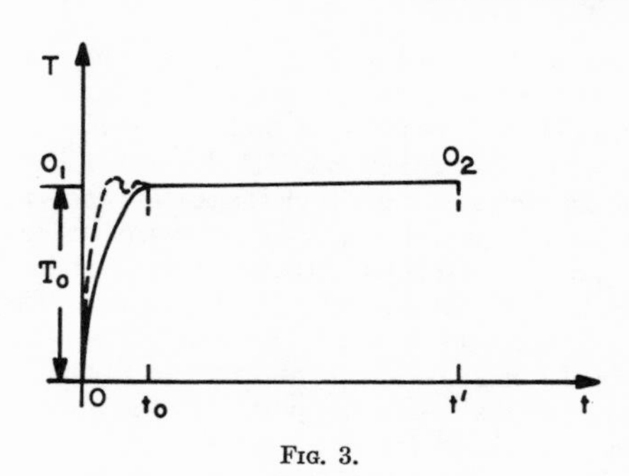

FIG. 3.

where the number a, the parameter of the parabola, is a large positive number. The impulse in this case is

$$\mathbf{I}_o'' = \tfrac{2}{3}at_o^{3/2}\mathbf{T}_o^*$$

or if, as a comparison, we take $T_o = at_o^{1/2}$, then

$$\mathbf{I}_o'' = \tfrac{2}{3}t_oT_o\mathbf{T}_o^* = \tfrac{2}{3}\mathbf{I}_o'. \tag{26}$$

The above is an example of a "gradual application," which approximates practical problems. In Fig. 3 the solid line is the graph of the magnitude of the force according to formula 25, while the dotted curve is the force according to practical experience, t_o being approximately equal to 0.02 seconds.

Appendix

The nth Derivative of the Function $\sigma(t)/r^3(t)$

We find upon examination that the nth derivative of the product of the functions $\sigma(t)$ and $\varphi(t)$ is given by the Leibnitz formula:

$$(\sigma \cdot \varphi)^{(n)} = \sum_{m=0}^{n} \frac{n!}{m!(n-m)!}\sigma^{(m)} \cdot \varphi^{(n-m)} \tag{I}$$

and that, if

$$\varphi(t) = \frac{1}{r^3(t)} \tag{II}$$

its nth derivative is

$$\varphi^{(n)} = \frac{P(r)}{r^{n+3}}, \tag{III}$$

where

$$P(r) = V_1r^{n-1} + V_2r^{n-2} + \cdots + V_n \tag{IV}$$

V_1, V_2, $\cdots$ are polynomials in the derivatives of r up to the order n^{th} with coefficients smaller than $(n+2)!$ Combining I, II, III we derive

$$\left(\frac{\sigma}{r^3}\right)^{(n)} = \sum_{m=0}^{n} \frac{n!}{m!(n-m)!}\sigma^{(m)}\frac{P(r)}{r^{n-m+3}} \tag{V}$$

Demetrios G. Magiros

where

$$P(r) = V_1 r^{n-m-1} + V_2 r^{n-m-2} + \cdots + V_{n-m} \quad \text{(VI)}$$

and the polynomials V_1, V_2, $\cdots$ have coefficients smaller than $(n - m + 2)!$

* * *

This paper appeared in a slightly different form as a General Electric M. S. D. technical report, No. 625D256. It was also presented and discussed at the XVth International Astronautical Congress, Warsaw, Poland, 1964.

References

(1) D. G. Magiros, "On the Convergence of the Solution of a Special Two-Body Problem," Athens, Greece, Practica of Athens Academy of Sciences, Vol. 38, pp. 36–39, 1963.
(2) W. Hurewicz, "Lectures on Ordinary Differential Equations," New York, M.I.T. Technical Press and John Wiley and Sons, 1958.
(3) R. Courant, "Boundary Value Problems in Modern Dynamics," *Proc. Intern. Congress of Mathematicians*, II, 1952.
(4) K. Knopp, "Theory and Applications of Infinite Series" New York, Hafner Publishing Company, Sec. Ed., §18, Sect. 94, pp. 154–155, 1948.
(5) W. Smart, "Celestial Mechanics," New York, Longmans, Green and Company, p. 21, paragraph 2.12, 1953.

MOTION IN A NEWTONIAN FORCE FIELD MODIFIED BY A GENERAL FORCE. II

by

D. G. MAGIROS*

General Electric Company, Philadelphia, Pennsylvania (U.S.A.)

1. Introduction

In a preceding paper published under a similar title,** the motion of an artificial celestial body P, which is moving under the influence of a Newtonian force $\underline{N}$ from an attractive center E and a general force $\underline{T}$, was discussed.

The forces $\underline{T}$ may be such that the moving body either collides or not with the attractive center E, that is the paths of the body pass or not through the center E and they are either "collision paths" or "non-collision paths", and the forces $\underline{T}$ are either "collision forces" or "non-collision forces", respectively, when one has two distinct cases of the problem, the "collision case" and the "non-collision case".

(a) In the preceding paper the "non-collision case" of the problem was discussed. It is established for this case an actual mathematical solution of specific form in terms of power series in time, found the general term of the series, proved its convergence, determined the region of convergence, and indicated the conditions for a rapid convergence, which permits an approximation of the solution within a given error. The procedure was based on the hypothesis of "non-zero distance r" between the attractive center E and the moving body P during its motion. $r = 0$ is a singularity of the series found, and this singularity is associated with important physical significance, which makes the subject very worth studying.

(b) In the present paper the "collision case" of the problem is investigated. Conditions are discussed for the forces to be "collision forces", and calculation is given of the "collision time" τ, that is the finite time needed

* Consulting Mathematician, Valley Forge Space Technology Center.

** MAGIROS, D. G., Motion in a Newtonian force field modified by a general force, *Journal of Franklin Institute* (December 1964).

Reprinted from the *Journal of the Franklin Institute* 278 (1964), 349–355.

for the moving body P to collide with the attractive center E. The series solution of the problem, found in the preceding paper, which is valid for very large time in the "non-collision case", is now valid for the "collision case" for time smaller than the finite collision time τ, which corresponds to the particular collision problem one has.

2. Useful Remarks

The following two remarks are useful for the collision problem.

(a) In a Newtonian force field, the attractive force $\underline{N}$ to the body P of distance r from the attractive center E is: $\underline{N} = -(\mu/r^2)\underline{r}^*$, where $\underline{r}^*$ is a unit vector, Figure 1(a), $\mu = K(M+m)$, K the constant of gravity, M and m the masses of the attractive center and the body, respectively.

If $\underline{r}(t)$ is the position vector in this field,

$$\underline{r}(t) = r(t) \cdot \underline{r}^*(t) \tag{1}$$

the corresponding linear velocity $\underline{V}(t)$ is

$$\begin{aligned} \dot{\underline{r}} &= \dot{r}\underline{r}^* + r\dot{\underline{r}}^* \\ &= \dot{r}\underline{r}^* + r\omega\underline{r}_1^* \\ &= \underline{V}(t) = V(t)\cdot\underline{V}^*(t) \end{aligned} \tag{2}$$

where ω is the angular velocity $\dot{\varphi}$. The vectors $\underline{r}^*$ and $\dot{\underline{r}}_1^*$ are perpendicular.

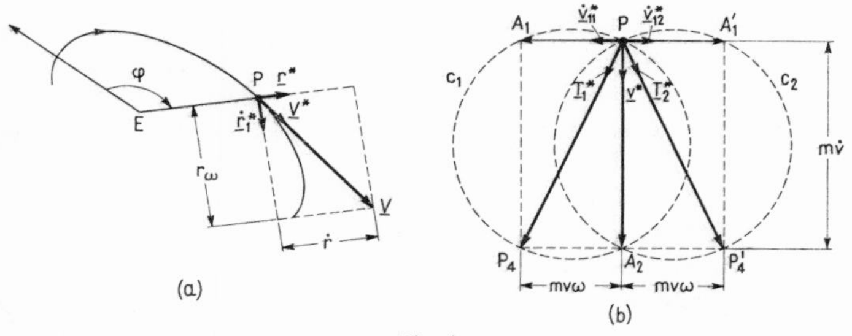

Fig. 1

The speed V of the velocity $\underline{V}$ is given by

$$V = \begin{cases} \sqrt{\dfrac{\mu}{a}} \cdot \sqrt{\dfrac{2a-r}{r}}, & \text{for ellipse } (0 \leqslant e < 1) \\ \sqrt{\dfrac{\mu}{a}} \cdot \sqrt{\dfrac{2a+r}{r}}, & \text{for hyperbola } (e > 1) \\ 2\mu \cdot \dfrac{1}{\sqrt{r}}, & \text{for parabola } (e = 1), \end{cases} \tag{3}$$

where e is the orbital eccentricity and α the semi-major axis of the orbit. The Eqs. (3), if $V_1 = \sqrt{2\mu/r}$ and V_P the speed at P, give

$$V_P < V_1, \quad V_P > V_1, \quad V_P = V_1, \tag{4}$$

when P is on an ellipse, hyperbola or parabola, respectively, with the same focus E, and V_P, speed at P, is elliptic, hyperbolic or parabolic speed, respectively.

(b) The general force $\underline{T}$ and its corresponding velocity $\underline{v}$ are such that

$$\begin{aligned} \underline{T}(t) &= T(t)\cdot\underline{T}^*(t) \\ &= m\cdot\dot{\underline{v}}(t) \\ &= m\frac{d}{dt}\{v(t)\cdot\underline{v}^*(t)\} \\ &= m\dot{v}\underline{v}^* + mv\dot{\underline{v}}^* \\ &= m\dot{v}\underline{v}^* + mv\omega\dot{\underline{v}}_1^* \end{aligned} \tag{5}$$

$\underline{v}^*$ and $\dot{\underline{v}}_1^*$ are perpendicular. Given $\underline{v}^*$ one may have two directions for the perpendicular unit vector $\dot{\underline{v}}_1^*$, namely $\dot{\underline{v}}_{11}^*$ and $\dot{\underline{v}}_{12}^*$, then two forces symmetric to $\underline{v}^*$ and of the same magnitude, Fig. 1b.

3. Collision Problem

The body P is under the action of the Newtonian force $\underline{N}$ and the general force $\underline{T}$, and its velocity $\underline{R}$ due to these forces is tangent to the path of the body. For the following we assume that the velocity $\underline{R}$ of the body is considered as resultant of the velocities $\underline{V}$ and $\underline{v}$, associated with the forces $\underline{N}$ and $\underline{T}$, respectively.

One may have collision or non-collision cases of the problem, by restricting the forces $\underline{T}$ to have certain properties of rate of change.

In the following we discuss the:

Collision Problem: "*Determine the force $\underline{T}$ which is such that the resultant $\underline{R}$ of the velocities $\underline{V}$ and $\underline{v}$ is through the center E for all time t in* $0 \leqslant t < \tau$, *τ being appropriate time.*" Forces $\underline{T}$, which satisfy the requirement of this problem, are "collision forces", and the time τ is the "*collision time*" of the problem. The series solution of the preceding paper is now valid for time $t < \tau$.

The discussion of the above problem is based on the following two auxiliary problems:

Auxiliary Problem A: "*Given the vector $\underline{V}$, determine a vector $\underline{v}$ such that their resultant $\underline{R}$ is through a constant point E.*" One has two cases, accordingly as the direction or the magnitude of $\underline{v}$ is given.

(a) Given the direction of $\underline{v}$, that is the angle φ_2, Fig. 2, from the triangle PP_1P_2 of the parallelogram $PP_1P_2P_3$ one has

$$\left.\begin{aligned} &\text{(i)} \qquad R = \frac{\sin\varphi_2}{\sin(\varphi_2-\varphi_1)}\cdot V \\ &\text{(ii)} \qquad v = \frac{\sin\varphi_1}{\sin(\varphi_2-\varphi_1)}\cdot V \end{aligned}\right\}. \tag{6}$$

The graphical construction shows that the vector $\underline{v}$ must be in the angle EPP_1', then for the above auxiliary problem the direction of $\underline{v}$ is restricted according to

$$0 < \varphi_1 < \varphi_2 < \pi. \tag{7}$$

Remark. The restrictions (7) hold in the regular construction. In an extreme case, that is in case either $\varphi_1 = 0$ or $\varphi_2 = 0, \pi$, one of these conditions implies the other.

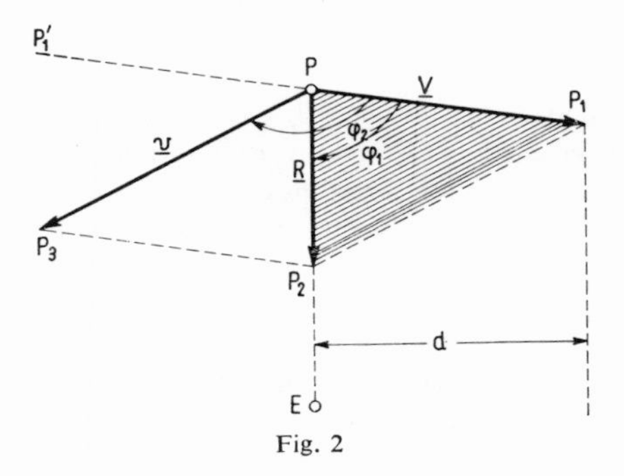

Fig. 2

(b) Given the magnitude of $\underline{v}$, one determines the appropriate direction of $\underline{v}$ as follows. With center the end-point P_1 of $\underline{V}$, Fig. 2, and radius the magnitude v of $\underline{v}$ one draws a circumference which intersects the line PE in either one or two points, or does not intersect it. If P_2 is such a point of intersection, then the direction of $\underline{v}$ is parallel to P_1P_2 from P. Therefore, one may have either one or two solutions, or no solution, if

$$d = v, \qquad d < v, \qquad d > v, \tag{8}$$

respectively. d is the distance of P_1 from the line PE.

Auxiliary Problem B: "*Determine the force $\underline{T}$ in terms of its corresponding velocity $\underline{v}$, if $\underline{v}$ satisfies the auxiliary problem A.*"

Given v, the force $\underline{T}$ can be determined by using formulae (5) and Fig. 1b. Either the diagonal PP_4 or PP_4' is the force $\underline{T}$ of magnitude

$$T = m(v^2\omega^2 + \dot{v}^2)^{\frac{1}{2}}. \tag{9}$$

In case the velocity $\underline{v}$ is of constant direction, the force $\underline{T}$ is along this direction and its magnitude is: $T = m\dot{v}$.

In case $\underline{v}$ is of constant magnitude, the force is perpendicular to the velocity $\underline{v}$ and its magnitude is: $T = mv\omega$.

Combining the above auxiliary problems, one can have the following problems equivalent to the collision problem:

Problem A: "*Given the Newtonian velocity $\underline{V}$ and the direction of the velocity $\underline{v}$, which satisfies the requirement of the collision problem, find the appropriate collision forces $\underline{T}$.*"

Problem B: "*Given $\underline{V}$ and the magnitude of $\underline{v}$, which satisfies the requirement of the collision problem, find the appropriate collision forces $\underline{T}$.*"

For the solution of the problem A, if the direction of $\underline{v}$ satisfies the requirement (7) the magnitude v of $\underline{v}$ is given by the formula (6ii), then the component PA_2, Fig. 1b, of the force is known. There are two directions for the perpendicular to $\underline{v}^*$, namely $\dot{\underline{v}}_{11}^*$ and $\dot{\underline{v}}_{12}^*$, Fig. 1b, and the second component of the force is either PA_1, or PA_1', both of magnitude $mv\omega$, according to auxiliary problem B. Then, the problem A has two solutions.

For the solution of the problem B, one first must check which of the conditions (7) is satisfied by v. Then, by using formula (6ii), one determines the angle φ_2, which gives the direction of $\underline{v}$, when, Fig. 1b, one sees that for each accepted v two forces can be found. Therefore one has:

(i) Case $v < d$, when there exist no collision forces $\underline{T}$ satisfying the collision problem;

(ii) Case $v = d$, when there exist two solutions of the collision problem;

(iii) Case $v > d$, when there exist four solutions of the collision problem.

Application: "*Motion in case velocity $\underline{v}$ is parallel to velocity $\underline{V}$*. Combining the above remark of the auxiliary problem A and the requirements of the collision problem, one can distinguish two cases:

(a) The regular case of Keplerian orbit. In this case $\varphi_1 \neq 0$, and $\varphi_2 = \varphi_1$ or $\varphi_2 = \varphi_1 + \pi$, then the forces $\underline{T}$, which produce $\underline{v}$, can not be collision forces, and one has a non-collision case of the problem, when the series solution is valid for large time.

(b) The degenerate case of Keplerian orbit. In this case $\varphi_1 = 0$, and $\varphi_2 = 0$ or $\varphi_2 = \pi$, then the forces $\underline{T}$ are collision forces and one has a collision case of the problem, when the series solution is valid for time smaller than the collision time corresponding to this collision case.

4. The Collision Time

The collision time τ determines the time-interval for convergence of the series solution, $0 \leqslant t < \tau$.

We calculate the collision time in the case of the present collision problem.

In this problem, the motion of the body P is along the line PE towards E, and its speed R, according to formula (6i) and (3) is known function of the distance r of P from E, $R = R(r)$. Then, along the r-axis one has

$$\dot{r} = R(r)$$

when

$$t - t_0 = \int_{r_0}^{r} \frac{dr}{R(r)}. \tag{10}$$

The limits r_0 and r are values of r at time t_0 and t, respectively. One can get the "collision time" τ from formula (10) by taking $t_0 = 0$ and $r = 0$, when

$$\tau = \int_{r_0}^{0} \frac{dr}{R(r)} = -\int_{0}^{r_0} \frac{dr}{R(r)}. \tag{11}$$

The integrand of (11), according to formulae (6i) and (3), is given by

$$-\frac{1}{R(r)} = \begin{cases} \dfrac{\sin(\varphi_1 - \varphi_2)}{\sin\varphi_2} \sqrt{\dfrac{a}{\mu}} \sqrt{\dfrac{r}{2a-r}}, & \text{for ellipse} \\ \dfrac{\sin(\varphi_1 - \varphi_2)}{\sin\varphi_2} \sqrt{\dfrac{a}{\mu}} \sqrt{\dfrac{r}{2a+r}}, & \text{for hyperbola} \\ \dfrac{\sin(\varphi_1 - \varphi_2)}{\sin\varphi_2} \dfrac{\sqrt{r}}{\sqrt{2\mu}}, & \text{for parabola} \end{cases} \tag{12}$$

We remark that: for any place of the body in the line PE, a speed R is associated with the body, and this R depends on the distance r of the body from the center, on the number a and on the angles φ_1, φ_2.

If at that place of the body one takes specified values for a, φ_1, φ_2, then R, and therefore the integrand of (11) given by (12), is a function of r. If we assume a, φ_1, φ_2 as parameters, from (11) and (12) one can get

$$\left.\begin{aligned} &\text{(i)} \quad \tau_e = \sqrt{\frac{a}{\mu}} \cdot \frac{\sin(\varphi_1 - \varphi_2)}{\sin\varphi_2} \cdot I_e \\ &\text{(ii)} \quad \tau_h = \sqrt{\frac{a}{\mu}} \cdot \frac{\sin(\varphi_1 - \varphi_2)}{\sin\varphi_2} \cdot I_h \\ &\text{(iii)} \quad \tau_p = \frac{1}{\sqrt{2\mu}} \cdot \frac{\sin(\varphi_1 - \varphi_2)}{\sin\varphi_2} \cdot I_p \end{aligned}\right\} \tag{13}$$

where

$$\text{(i) } I_e = \int_0^{r_0} \sqrt{\frac{r}{2a-r}}\, dr, \quad \text{(ii) } I_h = \int_0^{r_0} \sqrt{\frac{r}{2a+r}}\, dr, \quad \text{(iii) } I_p = \int_0^{r_0} \sqrt{r}\, dr. \tag{14}$$

The indices e, h, p in (13) and (14) correspond to elliptic, hyperbolic, parabolic Keplerian velocity $\underline{V}$, respectively.

By using the transformation $2a-r = y^2$ in the integral (14i), and $2a+r = y^2$ in the integral (14ii), the calculation of the integrals (14) gives

$$\left.\begin{aligned} &\text{(i)} \quad I_e = a\pi - \sqrt{r_0(2a-r_0)} - 2a\sin^{-1}\left(\sqrt{\frac{2a-r_0}{2a}}\right) \\ &\text{(ii)} \quad I_h = \sqrt{r_0(2a+r_0)} + 2a\log\frac{\sqrt{2a}}{\sqrt{r_0}+\sqrt{2a+r_0}} \\ &\text{(iii)} \quad I_p = \tfrac{2}{3}r_0\sqrt{r_0} \end{aligned}\right\} \tag{15}$$

Inserting (15) into (13) we have the collision time in our case

$$\left.\begin{aligned} &\text{(i)} \quad \tau_e = \sqrt{\frac{a}{\mu}}\,\frac{\sin(\varphi_1-\varphi_2)}{\sin\varphi_2}\times \\ &\qquad\qquad \times\left\{a\pi - \sqrt{r_0(2a-r_0)} - 2a\sin^{-1}\left(\sqrt{\frac{2a-r_0}{2a}}\right)\right\} \\ &\text{(ii)} \quad \tau_h = \sqrt{\frac{a}{\mu}}\,\frac{\sin(\varphi_1-\varphi_2)}{\sin\varphi_2}\times \\ &\qquad\qquad \times\left\{\sqrt{r_0(2a+r_0)} + 2a\log\frac{\sqrt{2a}}{\sqrt{r_0}+\sqrt{2a+r_0}}\right\} \\ &\text{(iii)} \quad \tau_p = \frac{1}{3}\sqrt{\frac{2}{\mu}}\,\frac{\sin(\varphi_1-\varphi_2)}{\sin\varphi_2}\,r_0\sqrt{r_0}. \end{aligned}\right\} \tag{16}$$

Remark. The calculation of the collision time in the present collision problem was based on the fact that the integration line is the segment PE, and on the assumption that a, φ_1, φ_2 are parameters. This assumption needs a special consideration.

MOTION IN A NEWTONIAN FORCE FIELD MODIFIED BY A GENERAL FORCE III APPLICATION: THE ENTRY PROBLEM

by

D. MAGIROS and G. REEHL

General Electric Company, Valley Forge Space Technology Center, Philadelphia, Pennsylvania (U.S.A.)

In the present paper the entry problem is discussed as an application of papers published by one of the authors during the last years.[1(a)(b)(c)]

The discussion is of mathematical type, but it can help for a treatment of physical entry problems.

1. The Problem

The problem is the following: "An artificial celestial body is moving in an elliptical Keplerian orbit around the Earth. A force $\boldsymbol{T}$ is applied to the body, acts for some time, then stops, when the body starts moving in a new Keplerian orbit around the Earth. Conditions are required under which the new Keplerian orbit intersects the surface of the Earth."

2. Solution of the Problem

We discuss first three kinds of orbits separately.

The original orbit

The position vector $\boldsymbol{r}_0$ and the velocity vector $\dot{\boldsymbol{r}}_0 = \boldsymbol{V}_0$ at a point P_0, due only to the Newtonian force, determine the Keplerian orbit through P_0 with focus the attractive center E of the Earth. If these vectors are known, the distance $r_0 = EP_0$, the true anomaly ϑ_0, the angle φ_0, and the speed V_0, associated with the point P_0, will be known.

The Keplerian orbit is an ellipse, parabola or hyperbola if V_0 is smaller than, equal to or bigger than $\sqrt{2\mu/r_0}$, respectively.

Reprinted from the *Proceedings of the XVIIth Intl. Astron. Congress*, Madrid (1966), 149–154.

$\mu = k(m_1+m_2)$, $k =$ gravitational constant, m_1 and m_2 masses of the Earth and the moving body, respectively.

The semi-major axis a_0, the eccentricity e_0, the perigee distance p_0 of the orbit, and the angle φ_0 can be calculated in terms of the known quantities by using appropriate formulae, which, in the case of an elliptical orbit, are:[2]

$$\left.\begin{array}{ll} a_0 = \dfrac{\mu}{(2\mu/r_0)-V_0^2}, & e_0 = \left\{1-\dfrac{r_0}{a_0^2}(2a_0-r_0)\sin^2\varphi_0\right\}^{1/2} \\ p_0 = a_0(1-e_0), & \sin\varphi_0 = \dfrac{1+e_0\cos\vartheta_0}{\{1+e_0^2+2e_0\cos\vartheta_0\}^{1/2}} \end{array}\right\} \quad (1)$$

The transfer orbit

If the time is split according to $t = t_0+\tau$, where t_0 is very small, and if the reference coordinate system is $(P_0; \boldsymbol{r}_0^*, \boldsymbol{s}_0^*, \boldsymbol{T}_0^*)$ as indicated in Fig. 1, the conditions at P_0 can be:

$$\left.\begin{array}{ll} \text{(i)} & \boldsymbol{r}(0) = \boldsymbol{r}(t_0) = \boldsymbol{r}_0 = r_0\cdot\boldsymbol{r}_0^* \\ \text{(ii)} & \dot{\boldsymbol{r}}(t_0) = \dot{\boldsymbol{r}}(0)+\boldsymbol{I}_0 = \boldsymbol{V}_0+\boldsymbol{I}_0 = s_0\cdot\boldsymbol{s}_0^* \\ \text{(iii)} & \boldsymbol{I}_0 = \displaystyle\int_0^{t_0}\boldsymbol{T}(t)\,dt, \quad \text{the impulse.} \end{array}\right\} \quad (2)$$

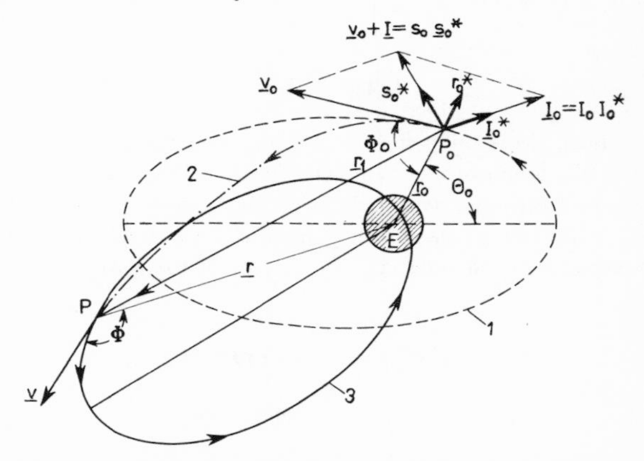

Fig. 1 *1* is the "original orbit" (departure); *2* is the "transfer orbit" (non keplerian); *3* is the "new orbit" (destination). The force $\boldsymbol{T}$ starts at P_0 and stops at P with duration t'.

The motion of the body during the action of the force on the non-Keplerian transfer orbit is given by the position vector $\boldsymbol{r}(\tau)$ and the velocity $\dot{\boldsymbol{r}}(\tau)$ according to the formulae:

$$\left.\begin{array}{l} \boldsymbol{r}(\tau) = \boldsymbol{r}_0+\boldsymbol{r}_1(\tau) = \alpha_1(\tau)\boldsymbol{r}_0^*+\alpha_2(\tau)\boldsymbol{s}_0^*+\alpha_3(\tau)\boldsymbol{T}_0^* \\ \dot{\boldsymbol{r}}(\tau) = \boldsymbol{V}(\tau) = \dot{\alpha}_1(\tau)\boldsymbol{r}_0^*+\dot{\alpha}_2(\tau)\boldsymbol{s}_0^*+\dot{\alpha}_3(\tau)\boldsymbol{T}_0^* \end{array}\right\} \quad (3)$$

where:

$$\left.\begin{aligned}\alpha_1(\tau) &= \alpha_1(0)+\dot{\alpha}_1(0)\tau+\frac{1}{2}\ddot{\alpha}_1(0)\tau^2+\ldots+\frac{1}{n!}\alpha_1^{(n)}(0)\tau^n+\ldots\\ \alpha_2(\tau) &= \alpha_2(0)+\dot{\alpha}_2(0)\tau+\frac{1}{2}\ddot{\alpha}_2(0)\tau^2+\ldots+\frac{1}{n!}\alpha_2^{(n)}(0)\tau^n+\ldots\\ \alpha_3(\tau) &= \alpha_3(0)+\dot{\alpha}_3(0)\tau+\frac{1}{2}\ddot{\alpha}_3(0)\tau^2+\ldots+\frac{1}{n!}\alpha_3^{(n)}(0)\tau^n+\ldots\end{aligned}\right\} \tag{4}$$

In the series (4), which is proved to be convergent, the first two coefficients are known, while the others can be calculated by using the formulae:

$$\left.\begin{aligned}\ddot{\alpha}_1(\tau) &= -\frac{\mu}{r^3(\tau)}\alpha_1(\tau)+T_1, \quad \alpha_1(0)=r_0, \quad \dot{\alpha}_1(0)=0\\ \ddot{\alpha}_2(\tau) &= -\frac{\mu}{r^3(\tau)}\alpha_2(\tau)+T_2, \quad \alpha_2(0)=0, \quad \dot{\alpha}_2(0)=s_0\\ \ddot{\alpha}_3(\tau) &= -\frac{\mu}{r^3(\tau)}\alpha_3(\tau)+T_3, \quad \alpha_3(0)=0, \quad \ddot{\alpha}_3(0)=0\end{aligned}\right\} \tag{4a}$$

T_1, T_2, T_3 are projections of the force $\boldsymbol{T}$ on the coordinate axis.
The three-dimensional entry problem is characterized by the fact that the force $\boldsymbol{T}$ and the unit-vector $\boldsymbol{T}_0^*$ are not on the $(\boldsymbol{V}_0, \boldsymbol{r}_0^*)$-plane of the original Keplerian orbit. Having the impulse $\boldsymbol{I}_0$ at P_0 we calculate the vector $\boldsymbol{V}_0+\boldsymbol{I}_0 = s_0 \cdot \boldsymbol{s}_0^*$. During the action of the force $\boldsymbol{T}$, the body is moving on the transfer orbit, when the position vector and the velocity vector of the body are given by the formulae (3) indicated above.

When $\boldsymbol{T}$ stops acting, say at the point P, the position vector $\boldsymbol{r} = \boldsymbol{EP}$ and the velocity vector $\dot{\boldsymbol{r}} = \boldsymbol{V}$ at P, that is the magnitudes $r = EP$ and V as well as the angles which determine the direction of the vectors $\boldsymbol{r}$ and $\boldsymbol{V}$, are known by the above formulae.

The destination orbit

The destination orbit is determined by the position vector and the velocity vector at the last point P of the transfer orbit, that is, by using r, V, φ corresponding to P in the $(E, \boldsymbol{V})$-plane. The destination orbit is an ellipse in case: $\sqrt{2\mu/r} > V$. The destination orbit intersects the surface of the Earth, when the entry problem is realized, if its perigee distance is smaller than or equal to the radius ϱ of the Earth taken spherical, that is if, in case of elliptical destination orbit, the following condition is satisfied:

$$\alpha(1-e) \leqslant \varrho. \tag{5}$$

We notice that α and e are dependent upon known and calculated angles and lengths and upon the time of the duration of the force $\boldsymbol{T}$.

3. The Solution of Some Restricted Problems

By imposing restrictions to magnitude and/or to direction of the force $\boldsymbol{T}$ one can have a variety of entry problems. We distinguish here two most important cases, namely the cases of constant force $\boldsymbol{T}$ in three-dimensional and in two-dimensional entry problems.

The force $\boldsymbol{T}$ constant in space

Let us take:

$$T_1 = 0 \quad \text{on} \quad \boldsymbol{r}_0^*, \qquad T_2 = 0 \quad \text{on} \quad \boldsymbol{s}_0^*, \qquad T_3 = T \quad \text{on} \quad \boldsymbol{T}_0^* \tag{6}$$

From formulae (4a) one can get:

$$\left.\begin{aligned}
&\alpha_1(0) = r_0,\ \dot{\alpha}_1(0) = 0,\ \ddot{\alpha}_1(0) = -\frac{\mu}{r_0^2},\ \dddot{\alpha}_1(0) = 0,\ \ddddot{\alpha}_1(0) = \frac{\mu^2}{r_0^5},\ \ldots\\
&\alpha_2(0) = 0,\ \dot{\alpha}_2(0) = s_0,\ \ddot{\alpha}_2(0) = 0,\ \dddot{\alpha}_2(0) = -\frac{\mu s_0}{r_0^3},\ \ddddot{\alpha}_2(0) = 0,\ \ldots\\
&\alpha_3(0) = 0,\ \dot{\alpha}_3(0) = 0,\ \ddot{\alpha}_3(0) = T,\ \dddot{\alpha}_3(0) = 0,\ \ddddot{\alpha}_3(0) = -\frac{\mu T}{r_0^3},\ \ldots
\end{aligned}\right\} \tag{6a}$$

For large r_0 and small τ, if we restrict ourselves to the first four terms of the expressions (4), the solution (3) becomes:

$$\left.\begin{aligned}
\boldsymbol{r}(\tau) &= \left(r_0 - \frac{\mu\tau^2}{2r_0^2}\right)\boldsymbol{r}_0^* + \left(s_0\tau - \frac{\mu s_0\tau^3}{6r_0^3}\right)\boldsymbol{s}_0^* + \tfrac{1}{2}T\tau^2\boldsymbol{T}_0^*\\
\dot{\boldsymbol{r}}(\tau) &= -\frac{\mu\tau}{r_0^2}\boldsymbol{r}_0^* + \left(s_0 - \frac{\mu s_0\tau^2}{2r_0^3}\right)\boldsymbol{s}_0^* + T\tau\,\boldsymbol{T}_0^*
\end{aligned}\right\} \tag{6b}$$

The errors of the coefficients in the solution (6b) are known.

The constant force $\boldsymbol{T}$ on the original orbit plane

We take:

$$T_1 = 0 \quad \text{on} \quad \boldsymbol{r}_0^*, \qquad T_2 = 0 \quad \text{on} \quad \boldsymbol{s}_0^*,\ T_3 = T \text{ on } \boldsymbol{T}_0^* = -\boldsymbol{s}_0^* \tag{7}$$

when the solution (6b) becomes:

$$\left.\begin{aligned}
\boldsymbol{r}(\tau) &= \left(r_0 - \frac{\mu\tau^2}{2r_0^2}\right)\boldsymbol{r}_0^* + \left(s_0\tau - \tfrac{1}{2}T\tau^2 - \frac{\mu s_0\tau^3}{6r_0^3}\right)\cdot\boldsymbol{s}_0^*\\
\dot{\boldsymbol{r}}(\tau) &= -\frac{\mu\tau}{r_0^2}\boldsymbol{r}_0^* + \left(s_0 - T\tau - \frac{\mu s_0\tau^2}{2r_0^3}\right)\cdot\boldsymbol{s}_0^*
\end{aligned}\right\} \tag{7a}$$

4. Numerical Examples

We now apply equations (7a) to specific numerical examples, taking the coordinate system as indicated in Fig. 2 and the original orbit circular with 100 miles altitude. Thus we have:

radius of spherical Earth: $\varrho = 2.090290 \times 10^7$ ft

radius of circular original orbit: $r_0 = 2.143090 \times 10^7$ ft

the satellite velocity: $\boldsymbol{V}_0 = 2.562762 \times 10^{14} \dfrac{\text{ft}}{\text{sec}} \cdot \boldsymbol{s}_0^*$

the constant coefficient: $\mu = 1.407528 \times 10^{16} \dfrac{\text{ft}^3}{\text{sec}^2}$.

Assume $t_0 = 10^{-4}$ sec and a retro-engine starting at $t = 0$ and acting for time t' giving a constant thrust $\boldsymbol{T} = -10 \dfrac{\text{ft}}{\text{sec}^2} \cdot \boldsymbol{s}_0^*$.

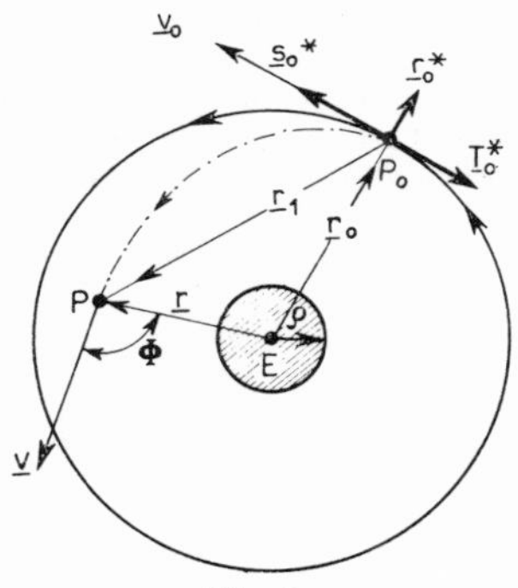

Fig. 2

At time t_0 we get:

$$\left.\begin{aligned} \boldsymbol{r}(0) = \boldsymbol{r}(t_0) &= 2.143090 \times 10^7 \cdot \boldsymbol{r}_0^* \\ \dot{\boldsymbol{r}}(t_0) &= 2.562762 \times 10^4 \cdot \boldsymbol{s}_0^* \end{aligned}\right\} \tag{8}$$

(i) For time $t' = 100$ sec, Eqs. (7a) and the values (8) give:

$$r(100) = 2.127767 \times 10^7 \cdot r_0^* + 2.506654 \times 10^5 \cdot s_0^*$$

$$\dot{r}(100) = -3.064617 \times 10^2 \cdot r_0^* + 2.444438 \times 10^4 \cdot s_0^*$$

Initial conditions for the new Keplerian orbit are:

$$r = 2.142481 \times 10^7 \text{ ft}$$

$$V = 2.463574 \times 10^4 \frac{\text{ft}}{\text{sec}}$$

$$\varphi = 90° - \tan^{-1} \frac{2.506654 \times 10^5}{2.127767 \times 10^7} + \tan^{-1} \frac{3.064617 \times 10^2}{2.444438 \times 10^4} = 90°25'37.5''$$

when:

$$\alpha = 1.990836\times 10^7 \text{ ft}, \quad e = 0.076534, \quad p = 1.838471\times 10^7 \text{ ft}.$$

and since $p < \varrho = 2.090290\times 10^7$ ft, the entry problem has been achieved in the above case.

(ii) For time $t' = 10$ sec, we get:

$$r(10) = 2.142937\times 10^7 \cdot \boldsymbol{r}_0^* + 2.557701\times 10^5 \cdot \boldsymbol{s}_0^*,$$

$$\dot{r}(10) = -3.064617\times 10^2 \cdot \boldsymbol{r}_0^* + 2.552579\times 10^4 \cdot \boldsymbol{s}_0^*$$

$$\varphi = 90° \; 0' \; 14.5''$$

$$\alpha = 2.126527\times 10^7 \text{ ft}, \quad e = 0.0077698, \quad p = 2.110004\times 10^7 \text{ ft}$$

and since $p > \varrho$ is this case, the entry problem has not been achieved.

References

[1] MAGIROS. D. G., Proc. International Astronautical Congress: (a) XIV 1963 (Paris), (b) XV 1964 (Warsaw), (c) XVI 1965 (Athens)

[2] ROY. A. E., The Foundations of Astrodynamics, pp. 80, 83, The MacMillan Co., New York (1965)

ΜΑΘΗΜΑΤΙΚΑ.— **The Entry Problem***, *by* ***Demetrios G. Magiros and George Reehl*****. Ἀνεκοινώθη ὑπὸ τοῦ Ἀκαδημαϊκοῦ κ. Ἰ. Ξανθάκη.

I. INTRODUCTION

(a) The Entry Problem.

«An artificial celestial body is moving in an elliptical Keplerian orbit around the earth. A force T is applied to the body, acts for some time, then stops, when the body starts moving in a new Keplerian orbit around the earth. Condition are required under which the new Keplerian orbit intersects the surface of the earth».

The conditions required will be found from the restriction of the perigee distance of the new Keplerian orbit to be smaller than or equal to the radius of the earth, taken spherical.
Use will be made of the results of one of the author's papers in connection with the motion of the body on the non-Keplerian orbit during the action of the general force T.

(b) The Motion of the Body During the Action of the Force.

We state for reference [1(b)] that if the time is split according to: $t = t_o + \tau$, where t_o is very small, the reference coordinate system: $(P_o; r^*_o, S^*_o, T^*_o)$, which is in general non-orthogonal and three-dimensional, is defined as shown in Figure 1, the conditions at P_o are:

$$\left.\begin{array}{ll} \text{(i)} & r(0) = r(t_o) = r_o = \mathrm{r}_o \cdot r^*_o \\ \text{(ii)} & \dot{r}(t_o) = \dot{r}(0) + I_o = V_o + I_o = \mathrm{s}_o \cdot s^*_o \\ \text{(iii)} & I_o = \int_0^{t_o} T(t)\,dt \end{array}\right\} \quad (1)$$

the motion of the body during the action of the force on the non-Keplerian arc of the orbit is given by the position vector $r(\tau)$ and the velocity $\dot{r}(\tau)$ according to the formulae:

$$r(\tau) = r_o + r_1(\tau) = \mathrm{a}_1(\tau) \cdot r^*_o + \mathrm{a}_2(\tau) \cdot s^*_o + \mathrm{a}_3(\tau)\, T^*_o \quad (2)$$

$$\dot{r}(\tau) = V(\tau) = \dot{\mathrm{a}}_1(\tau)\, r^*_o + \dot{\mathrm{a}}_2(\tau) \cdot s^*_o + \dot{\mathrm{a}}_3(\tau)\, T^*_o \quad (3)$$

---

* DEMETRIOS S. MAGIROS καὶ GEORGE REEHL, Τὸ πρόβλημα τῆς εἰσόδου τεχνητοῦ δορυφόρου.

** General Electric Co, Missile and Space Division, Philadelphia, PA.

Reprinted from the *Proceedings of the Athens Academy of Sciences* **41** (1966), 246–251.

where:

$$\left.\begin{aligned}
a_1(\tau) &= a_1(o) + \dot{a}_1(o)\tau + \frac{1}{2}\ddot{a}_1(o)\tau^2 + \dots + \frac{1}{n!}a_1^{(n)}(o)\tau^n + \dots \\
a_2(\tau) &= a_2(o) + \dot{a}_2(o)\tau + \frac{1}{2}\ddot{a}_2(o)\tau^2 + \dots + \frac{1}{n!}a_2^{(n)}(o)\tau^n + \dots \\
a_3(\tau) &= a_3(o) + \dot{a}_3(o)\tau + \frac{1}{2}\ddot{a}_3(o)\tau^2 + \dots + \frac{1}{n!}a_3^{(n)}(o)\tau^n + \dots
\end{aligned}\right\} \quad (4)$$

The first two coefficients of these series are given, and the others can be calculated by using:

$$\left.\begin{aligned}
\ddot{a}_1(\tau) &= -\frac{\mu}{r^3(\tau)}a_1(\tau) + T_1, \qquad a_1(o) = r_o, \qquad \dot{a}_1(o) = 0 \\
\ddot{a}_2(\tau) &= -\frac{\mu}{r^3(\tau)}a_2(\tau) + T_2, \qquad a_2(o) = o, \qquad \dot{a}_2(o) = s_o \\
\ddot{a}_3(\tau) &= -\frac{\mu}{r^3(\tau)}a_3(\tau) + T_3, \qquad a_3(o) = o, \qquad \dot{a}_3(o) = 0
\end{aligned}\right\} \quad (4a)$$

T_1, T_2, T_3 are projections of the force T on the coordinate axis, $\mu = K(m_1 + m_2)$, K the gravitational constant, and m_1, m_2 the masses of the earth and the body, respectively.
The convergence of the series (4a) is proved.[1b]

II. THE SOLUTION OF THE ENTRY PROBLEM

By the following procedure we can solve the three-dimensional entry problem.

(a) The Original Keplerian Orbit.

The position vector r_o and the velocity vector $\dot{r}_o = V_o$ at a point P_o, due to the Newtonian force only, determine the Keplerian orbit through P_o with focus the attractive center E of the earth. If these vectors are known, the distance $r_o = EP_o$, the true anomaly d_o, the angle φ_o, and the speed V_o, associated with the point P_o, will be known.
The Keplerian orbit is an ellipse, parabola, or hyperbola if V_o is smaller than, equal to, or bigger than $\sqrt{2\mu/r_o}$, respectively.[1(c)]
The semi-major axis a_o, the eccentricity e_o and the perigee distance p_o of the orbit can be calculated in terms of the known quantities by using appropriate formulae, which in case of elliptic orbit are:[2]

$$\left.\begin{aligned} a_o &= \frac{\mu}{\frac{2\mu}{r_o} - V_o}, & e_o &= \left\{1 - \frac{r_o}{a_o^2}(2a_o - r_o)\sin^2\varphi_o\right\} \\ P_o &= a_o(1 - e_o), & \sin\varphi_o &= \frac{1 + e_o\cos d_o}{(1 + e_o^2 + 2e_o\cos d_o)^{1/2}} \end{aligned}\right\} \quad (5)$$

(b) The Non-Keplerian Arc of the Orbit.

The three-dimensional entry problem is characterized by the fact that the force T and the unit vector T_o^* are not on the plane (V_o, r_o^*) of the original Keplerian orbit, Figure 1.

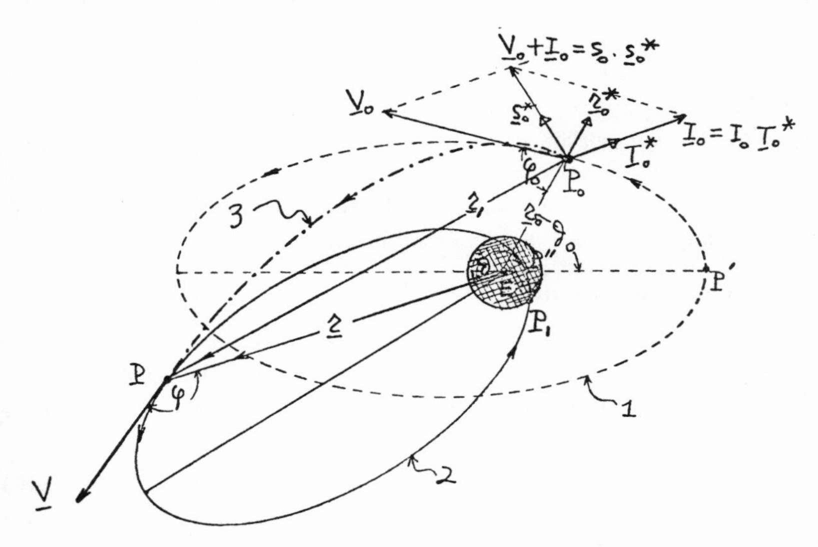

1 : Original orbit (Departure). P_o point where the force starts.
2 : New orbit (Destination). P point where the force stops.
3 : Non-Keplerian arc (Transfer). P_o P during the action of the force.

Figure 1

Having the impulse I_o at P_o, given by the formula (1,iii) we calculate the vector: $V_o + I_o = s_o \cdot s_o^*$.

During the action of the force T, the body is moving on the non-Keplerian arc of the orbit, which is a curve in space, and the position vector and the velocity vector of the body are given by the formulae (2) and (3).

When the force T stops acting, say at the point P, formulae (2) and (3) give the position vector $r = EP$ and the velocity vector $\dot{r} = V$ at P, that is the magnitudes r = EP and V, as well as the angles which determine the direction of the vectors r and V.
We notice that the plane (r_1, V) in general does not contain the point E.

(c) The New Keplerian Orbit.

To determine the new Keplerian orbit, which corresponds to the body at P, Figure 1, where the force T stops, one must know the position vector $r = EP$ and the velocity vector V at P, that is, one must know the magnitudes r and V, and the angle φ on the plane (E, V) of the new orbit, which are now known.
The new orbit is an ellipse, parabola or hyperbola, if V is smaller than, equal to, or bigger than $\sqrt{2\mu/r}$, respectively.
By using the appropriate formulae the Keplerian orbit can be calculated. In case of elliptic new Keplerian orbit formulae (5) can be used, if, instead of the quantities r_o, V_o, a_o, φ_o, e_o, p_o, d_o, the quantities r, V, a, φ, e, p, d, corresponding to the new Keplerian orbit, are used.

(b) The Condition for Entry.

The new Keplerian orbit, constructed as above, intersects the surface of the earth, (when one has a realization of the entry problem), if its perigee distance is smaller than or equal to the radius ρ of the earth, that is, in the case of elliptical orbit, if the condition:

$$EP'' = a\,(1 - e) \leqq \rho \tag{6}$$

is satisfied, Figure 1. The quantities a and e are dependent, as we saw, upon known and calculated angles and lengths, and upon the time t' of the duration of the force T.

III. THE SOLUTION IN RESTRICTED PROBLEMS

By restricting the force T, that is by imposing restrictions to its magnitude and direction, which are valid during its action on the body, a variety of entry problems may occur, when the above general procedure for their solution is simplified.
We consider two restricted cases of the entry problem of special interest.

(a) Force T Constant.

If the magnitude and the direction of T are constant, the problem is simplified. As an example, in case $T_1 = 0$, $T_2 = 0$, $T_3 \doteq T$, and large r_0 and small τ, formulae (2) and (3) become: [1(b)]

$$r(\tau) = \left(r_0 - \frac{\mu\tau^2}{2r_0^2}\right) r_0^* + \left(s_0\tau - \frac{\mu s_0 \tau^3}{6\, r_0^3}\right) s_0^* + \frac{1}{2} T\tau^2 T_0^* \tag{7}$$

$$\dot{r}(\tau) = V = -\frac{\mu\tau}{r_0^2} r_0^* + \left(s_0 - \frac{\mu s_0 \tau^2}{2r_0^3}\right) s_0^* + T\tau T_0^* \tag{8}$$

(b) Two-dimensional Entry Problem.

The two-dimensional entry problem occurs in case the force T is on the plane of the original orbit. In this case the non-Keplerian arc and the new orbit are on the plane of the original orbit, and the coordinate system (P_0; r_0^* , s_0^*) is suggested. All calculations of the general procedure of the previous section are simplified, especially in case of large r_0 and small τ, when the following formulae can be used:

$$r(\tau) = \left(r_0 - \frac{\mu\tau^2}{2r_0^2}\right) r_0^* + \left(s_0\tau - \frac{T\tau^2}{2} - \frac{\mu s_0 \tau^3}{6\, r_0^3}\right) s_0^* \tag{9}$$

$$\dot{r}(\tau) = V = -\frac{\mu\,\tau}{r_0^2} r_0^* + \left(s_0 - T\tau - \frac{\mu s_0 \tau^2}{2\, r_0^3}\right) s_0^* \tag{10}$$

REMARK. The present mathematical discussion of the entry problem can be taken as basic discussion of «physical entry problems».

REFERENCES

1. D.G. Magiros : «Proceedings of International Astronautical Congress», (a) XIV 1963 (Paris), (b) XV 1964 (Warsaw), (c) XVI 1965 (Athens).

2. A.E. Roy : «The foundations of Astrodynamics», The MacMillan Company, New York, (1965), pg. 80, 83.

ΠΕΡΙΛΗΨΙΣ

Τὸ πρόβλημα τῆς εἰσόδου ἑνὸς τεχνητοῦ δορυφόρου ἐξετάζεται εἰς τὴν παροῦσαν ἐργασίαν ὡς ἐφαρμογὴ γενικωτέρας ἐρεύνης δημοσιευθείσης ὑφ᾽ ἑνὸς τῶν συγγραφέων. Ἡ καμπύλη μεταφορᾶς λαμβάνεται συμφώνως πρὸς τὴν ἐργασίαν 1 (b). Ἡ τροχιὰ προορισμοῦ πρέπει νὰ πληροῖ τὴν συνθήκην (6). Δίδονται δύο εἰδικαὶ περιπτώσεις.

*

Ὁ Ἀκαδημαϊκὸς κ. Ἰω. **Ξανθάκης** κατὰ τὴν ἀνακοίνωσιν τῆς ἀνωτέρω ἐργασίας εἶπε τὰ κάτωθι:

Εἰς τὴν παροῦσαν ἐργασίαν ἐξετάζεται τὸ «Πρόβλημα τῆς Εἰσόδου» ὑπὸ τὴν ἀκόλουθον μορφήν:

«Τεχνητὸς δορυφόρος τῆς Γῆς κινεῖται πέριξ αὐτῆς εἰς ἐλλειπτικὴν τροχιάν. Μία γενικὴ δύναμις ἐφαρμόζεται ἐπὶ τοῦ δορυφόρου, δρᾷ ἐπὶ χρονικόν τι διάστημα καὶ μετὰ παύει δρῶσα, ὁπότε ὁ δορυφόρος ἀρχίζει νὰ κινῆται ἐπὶ νέας Κεπλερείου τροχιᾶς. Ζητοῦνται αἱ συνθῆκαι καὶ οἱ περιορισμοὶ ὑπὸ τοὺς ὁποίους ἡ νέα τροχιὰ δύναται νὰ συναντήσῃ τὴν ἐπιφάνειαν τῆς Γῆς».

Ἡ λύσις τοῦ παρόντος προβλήματος ἐπιτυγχάνεται δι' ἐφαρμογῆς τῶν πορισμάτων προηγηθεισῶν ἐργασιῶν ἑνὸς ἐκ τῶν συγγραφέων τῆς παρούσης ἐρεύνης. Οἱ ζητούμενοι περιορισμοὶ προκύπτουν ἐκ τοῦ γεγονότος ὅτι «ἡ ἐκ τοῦ περιγείου ἀπόστασις τοῦ δορυφόρου τῆς νέας Κεπλερείου τροχιᾶς δέον νὰ εἶναι μικροτέρα ἢ ἴση τῆς ἀκτῖνος τῆς Γῆς, θεωρουμένης σφαιρικῆς».

Εἶτα ἐξετάζεται τὸ πρόβλημα τοῦτο τῆς εἰσόδου εἰς τὰς μερικὰς περιπτώσεις κατὰ τὰς ὁποίας:

α) Ἡ δρῶσα δύναμις ἔχει σταθερὸν μέγεθος καὶ διεύθυνσιν καὶ

β) Αὕτη κεῖται ἐπὶ τοῦ ἐπιπέδου τῆς ἀρχικῆς τροχιᾶς.

PART III

DYNAMICAL SYSTEMS ANALYSIS

In this Part, Magiros' important papers on stability analysis, precessional phenomena and separatrices of dynamical systems were selected. In papers 30 and 31 a unified formulation of the three basic stability concepts in the sense of Lyapunov, Poincaré and Lagrange is given, on the basis of an appropriate interpretation of momentary or permanent effects of the perturbations acting on the system. Specialized stability concepts, as well as some relationships between the general stability concepts are indicated. By appropriate geometrical interpretations, the stability concepts are clarified. Illustrative and application examples are also included. Paper 32 is referred to the "in-orbital-plane" angular motion, and its stability characteristics of a spherical shape satellite, acted upon by aerodynamic and gravitational torques. Different stability concepts are employed and conditions are determined for the stability of the equilibrium attitudes. Paper 33 deals with the stability concepts of the class of DE with deviating arguments, and Paper 34 provides some remarks on stability concepts, by which a better understanding of current stability problems can be obtained. Finally paper 35 is a unification of the concepts and results involved in the previous papers. Paper 42 is a brief version of paper 35, published in Russian.

In paper 36 the stability of the precessions defined and studied in a previous paper of Magiros (see [32] in the full list of publications) is treated on the basis of remarks given in paper 31. The main results are: (i) the "regular" and "non-regular periodic" precessions are stable, but the helicoid precession is unstable in the Lagrange sense, (ii) The "regular" and "non-regular periodic" precessions are stable, but not asymptotically stable in the Poincaré sense. The "helicoid" precession is asymptotically stable in the

Poincaré sense, and (iii) The “non-regular periodic” and the “helicoid” precessions are unstable in the Lyapunov sense, but the “regular” precession is either unstable or stable in the Lyapunov sense, depending on the sudden perturbations which affect or do not affect, respectively, the initial angular velocity component along the roll axis. In paper 37 a re-entry problem is solved through the concept of the helicoid precession. This is the problem of determining the orientation of a spin-stabilized axi-symmetric re-entry vehicle. In paper 38 the “helicoid precession concept” is used as a mean to treat the problem of finding the error of the orientation of the angular momentum vector of a re-entry vehicle at the end of spin-up, and in paper 39 the stability of a class of helicoid precessions is studied. It is found that all members of the class are Lyapunov unstable, but Poincaré (orbitaly) stable, or asymptotically stable, or unstable, if the limiting value of the “pitch distance” of the member at hand is either constant, or zero, or infinite, respectively.

Finally, paper 40 presents new results for the determination of separatrices of dynamical systems and their usefulness in the analysis of physical systems. On the basis of the new definition of separatrices, theorems are formulated for the existence, the number, and the determination of separatrices. The basic concepts used are the concepts of “nonlinearities” and “sectors” with apex singular points, of the system.

Paper 41 is also devoted to the study of separatrices of dynamical systems. Using a supplementery definition of separatrices, some theorems are proved concerning separatrices at the neighborhood of elementary and nonelementary critical points in the plane.

ON STABILITY DEFINITIONS OF DYNAMICAL SYSTEMS

By Demetrios G. Magiros

GENERAL ELECTRIC COMPANY, VALLEY FORGE SPACE TECHNOLOGY CENTER,
PHILADELPHIA, PENNSYLVANIA

Communicated by Leon Brillouin, March 25, 1965

1. *Introduction.*—The dominant subject in the analysis of dynamical systems is the stability of the insuing motion. Many fields of interest under current research involve dynamical systems. These range from fields such as astrodynamics, meteorology, and control systems to biology, chemistry, medicine, and economics. In each of these fields dynamical systems can be formulated, and their stability can be examined as special cases of stability problems in dynamics. During recent years the stability concepts of dynamical systems have been advanced, either by modifying old ideas or by creating new stability concepts. These advances permit a deeper penetration into the more profound stability problems.

In this note, stability concepts of physically realistic dynamical systems, which are mathematically consistent, are discussed. Based on appropriate interpretation of the effects of perturbations, acting on the system either momentarily or permanently, a unified formulation of the three basic stability concepts in the sense of Liapunov, Poincaré, and Lagrange is given. Specialized stability concepts, introduced within the basic concepts, are indicated by appropriate remarks, and some relationships between the stability concepts are emphasized, and the stability concepts are clarified by geometrical interpretations.

2. *Physical Stability Considerations.*—Physically, we may have three basic stability aspects, depending on stability considerations of a motion on its orbit, of the orbit of a motion, and of the boundedness of the motion and its orbit.

A motion or its orbit is considered as stable, if, by giving a small disturbance to the motion or to its orbit, the disturbed motion or its orbit, initially near to the unperturbed motion or to its orbit, remains near to it for all time. More specifically,

If for small disturbances the effect on the motion or on its orbit is small, we say that the motion or its orbit is in a "stable" situation.

If for small disturbances the effect is considerable, the situation is "unstable."

If for small disturbances the effect tends to disappear, the situation is "asymptotically stable."

If, regardless of the magnitude of the disturbances, the effect tends to disappear, the situation is "asymptotically stable in the large."

Reprinted from the *Proceedings of the National Academy of Sciences* **53** (1965), 1288–1294.

The stability in the sense of Liapunov and Poincaré is based on the above physical stability definitions.

The boundedness of all motions and orbits of a system in connection with bounded disturbances is another physical stability aspect, on which the stability in the sense of Lagrange is based.

These three different stability aspects are of a qualitative type. In the following paragraphs a unified quantitative discussion and a geometrical interpretation of these stability aspects are given.

The disturbances are due to perturbations, which are considered as minor disturbing forces acting on the system either momentarily or constantly.

3. *The Effects of Perturbations.*—The equations of the dynamical system are

$$\dot{x}_i(t) = X_i(t, x_1, \ldots x_n), \qquad i = 1, \ldots n$$

$$X_i(t, 0, \ldots, 0) = 0, \; x_i(t_0) = x_{i0}. \tag{1}$$

Let $x_i(t)$ be a solution of the system (1), that is, a motion with orbit L (Fig. 1). The effect of perturbations is a change of certain quantities dependent on the motion, that is, a change of the motion into the perturbed motion $\bar{x}_i(t)$ with orbit $\bar{L}$.

If the distance ρ between a point P of L and a point $\bar{P}$ of $\bar{L}$ measures the effect of perturbations at the point P, the points P and $\bar{P}$ are correspondent points of the unperturbed motion $x_i(t)$ and the perturbed one $\bar{x}_i(t)$.

We may have different kinds of correspondence between points P of L and points $\bar{P}$ of $\bar{L}$, and this correspondence characterizes different stability concepts of the motion.

We distinguish here two such correspondences, most physical, on which two important stability concepts are based, namely, the stability in the sense of Liapunov and Poincaré.

(*a*) The points P and $\bar{P}$ are points of L and $\bar{L}$, respectively, at the "same time" (Fig. 1), either in the phase space or in the parameter space of the system, and the distance $\rho = P\bar{P}$ is "time-dependent," namely, we have

$$\rho = \left\{ \sum_{i=1}^{n} [x_i(t) - \bar{x}_i(t)]^2 \right\}^{1/2} \quad \text{in the phase space, and}$$

$$\rho = \left\{ \sum_{i=1}^{n} [x_i(t,a_j) - x_i(t,\bar{a}_j)]^2 \right\}^{1/2}, \quad \text{in the parameter space.} \tag{2}$$

This kind of correspondence is an appropriate one for stability discussion of a motion on its orbit. We call "Liapunov distance" the distance ρ defined with the above correspondence between points P and $\bar{P}$.

(*b*) The points P and $\bar{P}$ on L and $\bar{L}$ correspond to each other in such a way that the distance $\rho = P\bar{P}$ is the minimum from the distances of $\bar{P}$ from all points of L:

$$\rho = \min \left\{ \sum_{i=1}^{n} (x_i - \bar{x}_i)^2 \right\}^{1/2}, \tag{3}$$

When $\bar{P}$ is sometimes on the normal to L at P (Fig. 2), and the distance ρ, defined in this way, is "time-independent".

This kind of correspondence is an appropriate one for stability discussion of the

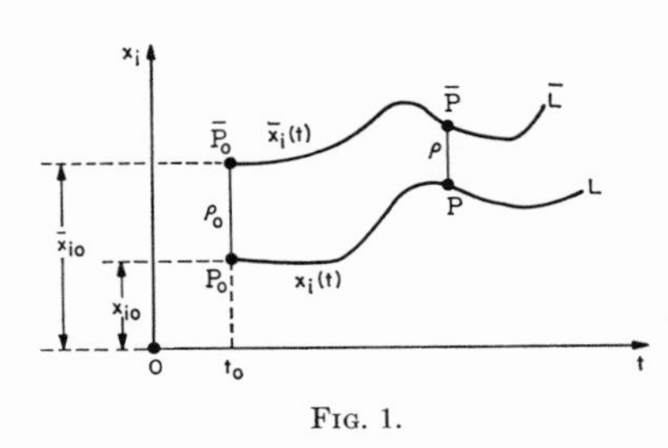

FIG. 1.

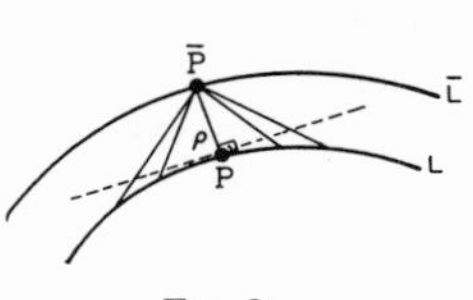

FIG. 2.

orbit of a motion. We call "Poincaré distance" the distance ρ defined in the above manner.

4. *Stability Definitions of a General Trajectory.*—The distance ρ defined in the preceding can be used for finding stability definitions of an analytic type in agreement with the physical stability definitions.

The motion $x_i(t)$ of the system (1) is said to be "stable," if, for any given positive number ϵ, however small, a positive number δ, depending on ϵ and, in general, on the initial time t_0, can be chosen such that, for any perturbed motion $\bar{x}_i(t)$, the inequality

$$\rho_0 < \delta, \tag{4}$$

where ρ_0 is the distance initially, implies

$$\rho < \epsilon \tag{5}$$

for all time $t \geq t_0$.

The motion is "unstable," if, given ϵ as above, for sufficiently small positive δ, the inequality (5) is not satisfied even for at least one perturbed motion.

The motion is "asymptotically stable," if it is stable, and, in addition, a positive number $\delta_1 \geq \delta$ exists such that, starting from any ρ_0, which is

$$\rho_0 < \delta_1, \tag{4a}$$

the limiting condition

$$\lim_{t \to \infty} \rho = 0 \tag{6}$$

holds uniformly relative to the quantities t_0, x_{t0}.[1a]

The motion is "quasi-asymptotically stable," if, from the above stability conditions, only the limiting condition (6) holds.

The motion is "asymptotically stable in the large," or "globally stable," or "completely stable," if it is asymptotically stable and the number δ_1 is very large.

Remarks: (*a*) The above definitions are "stability definitions in Liapunov sense," either in the phase space or in the parameter space, if the distances ρ_0 and ρ are "Liapunov distances" either in phase space or in parameter space, respectively.

The definitions are "stability definitions in Poincaré sense," or "orbital definitions," if the distances ρ_0 and ρ are "Poincaré distances."

(*b*) In case the orbit L shrinks to an equilibrium point of the system, the distances in Liapunov and Poincaré sense are identical, and therefore the stability definitions in Liapunov and Poincaré sense are equivalent at the singular points of dynamical systems.

(*c*) A nonlinear autonomous system, of which no eigenvalue has positive real part, may have, under some appropriate restrictions of the nonlinearities, the origin as an equilibrium point asymptotically stable in the large. Such nonlinear systems are called "absolutely stable," and there are methods to find classes of nonlinearities for absolute stability of the systems. The above kind of stability of the systems plays an important role in modern technology.[2]

(*d*) The stability definitions in Liapunov sense in periodic motions are narrow definitions, because they classify as unstable motions some motions which can be practically considered as stable. The definitions in Poincaré sense are the appropriate ones in periodic motions.

(*e*) In case δ is independent of t_0 as in autonomous systems, the definitions are "uniform"; so we may have "uniform stability," "uniform asymptotic stability," etc., in Liapunov or in Poincaré sense. In nonautonomous systems the selection of the initial time t_0 is not free; there is in those systems a value τ of time such that $t_0 \geq \tau$.[3]

(*f*) The stability conditions (4) and (5), in case of Liapunov stability in parameter space of initial conditions, become conditions of continuity of the solution $x_i\,(t, t_0, x_{10}, \ldots, x_{n0})$ of (1) on the initial conditions. The continuity of the solution on the initial conditions, which is one of the three "Hadamard's postulates" in order that the solution be accepted physically, is then a special stability case in Liapunov sense.[4]

(*g*) If the stability situation of the motion $x_i(t)$ is invariant in a parameter space S of the system, we speak about a "structural stability" of the system in the parameter space S, which is the "domain of structural stability" of the system. This property of the system gives information about the "insensitivity" of the system to perturbations.

5. *Lagrange Stability of a System.*—The boundedness of all motions and orbits of a system is the third basic stability concept of the system.

If ρ is the distance of the origin of the system from any point of any solution $x_i(t, t_0, x_{10}, \ldots, x_{n0})$ of the system, and if, given any finite positive number ϵ, we can find another finite positive number δ such that condition (4) implies condition (5) for all $t \geq t_0$, the system is "stable in Lagrange sense." The system is "unstable in Lagrange sense," if its response has the tendency to grow without bounds.

Remarks: (*a*) The distance ρ used in stability conditions in Lagrange sense is identical to the distance in Liapunov or in Poincaré sense applied to origin.

(*b*) The three basic stability concepts are given in the preceding, independently of each other. We justify this by the following examples:

The solution $x = t + c$, where c is arbitrary constant, of the equation $\dot{x} = 1$, is unstable in Lagrange sense, but it is stable in Liapunov sense.

The family of solutions $x = a \sin\,(at + b)$, where a and b are parameters, of the equation

$$\ddot{x} = -\, {}^1/_2\, x\{x^2 + (x^4 + 4\dot{x}^2)^{1/2}\} \tag{7}$$

is stable in Poincaré and Lagrange sense, but unstable, except the zero solution, in Liapunov sense, when the ratio of any two values of a is irrational.

(c) Nevertheless, the stability concepts are in many cases connected. For example:[5]

In linear homogeneous systems, $\dot{x} = \sum_{j=1}^{n} a_{ij}(t)x_j$ with $a_{ij}(t)$ continuous in $t \geq t_0$, the Lagrange stability implies the Liapunov stability, and conversely.

In linear inhomogeneous systems, $\dot{x} = \sum_{j=1}^{n} a_{ij}(t)\, x_j + f_i(t)$ with $a_{ij}(t)$ and $f_i(t)$ continuous in $t \geq t_0$, the Lagrange stability implies the Liapunov stability, but for the converse the boundedness of one solution is needed.

In nonlinear systems, connections between the stability concepts either are, in general, hard to find or do not exist.

6. *Stability under Persistent Perturbations.*—If $p_i(t, x_1, \ldots, x_n)$ are persistent perturbations, the equations of the dynamical system are

$$\dot{x}_i(t) = X_i(t, x_1, \ldots, x_n) + p_i(t, x_1, \ldots, x_n), \tag{8}$$

where p_i must be such that (8) has unique solutions corresponding to given initial conditions. The solution $x_i(t)$ of (1) is "stable under persistent perturbations," if, in addition to conditions (4) and (5), some appropriate restrictions for the magnitude of p_i are accepted, and which hold for all accepted x_i and $t \geq t_0$.

In case the magnitude of the perturbations is according to

$$\max|p_i(t, x_1, \ldots, x_n)| < \eta, \tag{9}$$

where η is an appropriate positive number, we speak about "total stability" of the solution $x_i(t)$.

If the condition (9) is replaced by

$$\int_0^{\infty} \max|p_i(t, x_1, \ldots, x_n)|\, dt < \eta, \tag{9a}$$

we speak about "integral stability."

In case of total stability the perturbations must be small, but in case of integral stability they may be large in a small interval of time.

The properties of both these types are possessed by stability under persistent perturbations "bounded in the mean," when we have "stability in the mean." The last type of stability is obtained, if, instead of condition (9), we have

$$\int_t^{t+\mathrm{T}} \max|p_i(t, x_1, \ldots, x_n)|\, dt < \eta, \tag{9b}$$

where η and δ of (4) depend on T, in general.

If, in addition to the above, condition (6) holds, we have "stability of asymptotic type."

The above stabilities are in Liapunov or in Poincaré sense, if the distances ρ_0 and ρ in (4) and (5) are Liapunov or Poincaré distances.

There are relationships between the above stability under persistent perturbations. For example:[6]

Asymptotic integral stability implies asymptotic stability in the mean, and conversely.

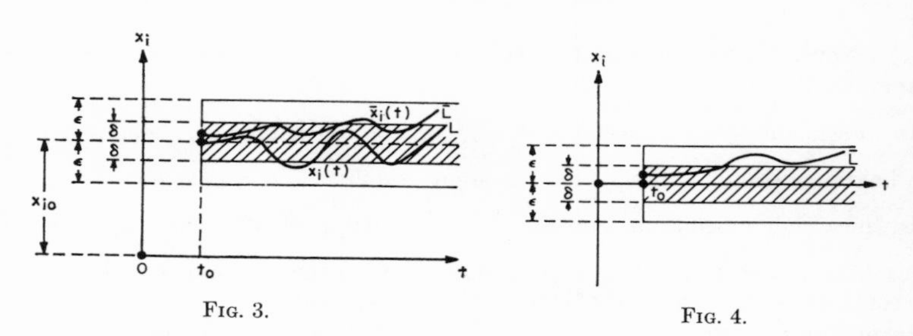

FIG. 3. FIG. 4.

Asymptotic integral stability implies total asymptotic stability, and this implies total stability.

Asymptotic stability in the mean implies stability in the mean, and this implies integral stability.

Remarks: (*a*) The preceding analytical stability definitions need appropriate modifications, changes, supplements in order to be useful practically, that is, in order to meet the requirements for a "practical stability." These requirements are:[1b, 7] "to know the size of the acceptable deviations of a state, of the region of the appropriate initial conditions, of the acceptable perturbations, and the finite time for which the above happens."

(*b*) The character of the stability, by using any of the previous concepts, is not invariant with a general transformation of the coordinates of the system, that is, the stability depends on the coordinate system with reference to which the variables are to be considered.

7. *Geometric Interpretation of Stability Definitions.*—We clarify the stability concepts by their geometrical interpretation. We call "r-tube" around a curve l in n-dimensional space the set of all points of which the distance ρ in Poincaré sense from l is smaller than r; r is the width and l the central curve of the tube. For finite r the tube is bounded.

The above definitions can appropriately be used for a geometrical interpretation of the stability concepts.

For stability in the Liapunov sense of a general motion $x_i(t)$, we consider the "δ-tube" and "ϵ-tube" around the straight line $x_i = x_{i0}$ which is parallel to t-axis through the point (t_0, x_{i0}).

A motion $x_i(t, t_0, x_{10}, \ldots, x_{n0})$ of orbit L is stable in Liapunov sense, if given the "ϵ-tube" around the line $x_i = x_{i0}$ (Fig. 3), we can find a "δ-tube" around the line $x_i = x_{i0}$ such that any perturbed motion of orbit $\bar{L}$, starting in the "δ-tube." remains for all $t \geq t_0$ in the "ϵ-tube."

In case of equilibrium points, that is, the origin, the above tubes are around the t-axis (Fig. 4).

For the asymptotic stability we need, in addition, another tube, "δ_1-tube," which will be unbounded for asymptotic stability in the large.

For stability of a general motion in Poincaré sense, that is, for "orbital stability" of the motion, the "δ-tube" and "ϵ-tube" around the orbit L can be used. A motion is "orbitally stable" if, given the "ϵ-tube" around L, we can find a "δ-tube" around L such that any perturbed orbit $\bar{L}$ starting in the "δ-tube" remains always in the "ϵ-tube" (Fig. 5).

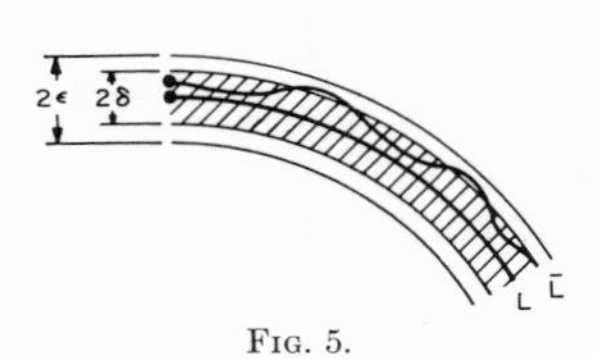

FIG. 5.

For stability of equilibrium points, the orbit L is the t-axis, then the tubes are around the t-axis (Fig. 4), and the discussion is the same as in Liapunov case. For the periodic motions, the tubes around the closed orbit L are tori with L as central curve.

For Lagrange stability, bounded tubes around the t-axis (Fig. 4) can be used.

The system is "Lagrange stable," if for every bounded "ϵ-tube" around the t-axis another bounded "δ-tube" around the t-axis can be found such that every orbit of the system starting in the "δ-tube" does not leave the "ϵ-tube" for all time $t \geq t_0$.

For stability under persistent perturbations the "δ-tube" and "ϵ-tube" must be around the t-axis, but, in addition, an appropriate "η-tube" around the t-axis is needed.

The author wishes to express his thanks to O. Klima, R. Chapman, and J. Weinstein of the General Electric Company, Missile and Space Division, Valley Forge Space Technology Center, the first two for their encouragement and help in connection with this research work, and the third for his assistance in editing this paper.

[1] Zubov, V. I., *Mathematical Methods for the Study of Automatic Control Systems* (New York: The Macmillan Co., 1963): (*a*) chap. I, §1; (*b*) p. 13.

[2] Aizerman, M., and F. Gantmaher, *Absolute Stability of Regulator Systems* (San Francisco: Holden-Day, Inc., 1964).

[3] LaSalle, J., and R. Rath, "Eventual stability," in *Proceedings of the Fourth Joint Automatic Control Conference*, University of Minnesota, Minneapolis (1963), pp. 468–470.

[4] Magiros, D. G., *Physical Problems Discussed Mathematically*, General Electric Co., Valley Forge Space Technology Center, Technical Report No. 64SD286, October 1964.

[5] Cesari, L., *Asymptotic Behavior and Stability Problems in Ordinary Differential Equations* (Berlin: Springer-Verlag, 1959), pp. 7–13.

[6] Vrkŏv, I., "On some stability problems," "Differential equations and their applications," in *Proceedings of a Conference in Prague*, September 1962 (New York: Academic Press, 1963), pp. 217–221.

[7] LaSalle, J., and S. Lefshetz, *Stability by Liapunov's Direct Method* (New York: Academic Press, 1961), p. 121.

Stability Concepts of Dynamical Systems

DEMETRIOS G. MAGIROS*

General Electric Company, Missile and Space Division, Philadelphia, Pennsylvania

I. INTRODUCTION

Many fields of interest under current research involve dynamical systems. In these fields dynamical systems can be formulated, and their stability examined as a special case of problems of stability in dynamics.

One can distinguish classes of concepts of stability depending on the nature of the dynamical systems, the manner in which the system approaches a given state or deviates from it, the properties of the perturbations of the system, and the space variables selected.

During recent years the concepts of stability of dynamical systems have been advanced, either by modifying old ideas or by creating new ones, and these advances permit a deeper penetration into the more profound problems of stability.

In this paper, of which a first draft is published in the Proceedings of National Academy of Sciences (Magiros, 1965a), we discuss the three basic concepts of stability in the sense of Liapunov, Poincaré, and Lagrange, and some specialized ones. Based on an appropriate interpretation of the effects of perturbations, acting on the systems either momentarily or permanently, a unified formulation of the basic concepts of stability is given. Some relationships between the concepts of stability are emphasized and appropriate examples and geometrical interpretations of the concepts clarify the discussion.

II. PHYSICAL STABILITY CONSIDERATIONS

Physically, one may have three basic aspects of stability depending on stability considerations of the motion in a given orbit, of the orbit of a given motion and of the boundedness of the motion and its orbit.

* The author wishes to express his thanks to O. Klima and R. Chapman of General Electric Company, Missile and Space Division, for their encouragement and help with this research work.

Reprinted from *Information and Control* 9 (1966), 531–548.

A motion or its orbit is considered stable, if, by giving a small disturbance to the motion or to its orbit, the disturbed motion or its orbit, remains close to the unperturbed one for all time.

More specifically.

a. If for small disturbances the effect on the motion or on its orbit is small, one says that the motion or its orbit is in a "stable" situation.

b. If for small disturbances the effect is considerable, the situation is "unstable."

c. If for small disturbances the effect tends to disappear, the situation is "asymptotically stable."

d. If, regardless of the magnitude of the disturbances, the effect tends to disappear, the situation is "asymptotically stable in the large." The stability in the sense of Liapunov and Poincaré is based on the above physical stability definitions.

The boundedness of motions and orbits of a system in connection with bounded disturbances is another physical aspect of stability, on which the stability in the sense of Lagrange is based.

These three different aspects of stability are of a qualitative type. In the following a unified quantitative discussion of these aspects of stability is given.

III. THE EFFECTS OF PERTURBATIONS

The disturbances of the systems are due to perturbations, which are considered as minor disturbing forces acting on the system either momentarily or constantly.

The equations of a nonautonomous dynamical system in case of sudden perturbations are:

$$\begin{gathered} \dot{x}_i(t) = X_i(t, x_1, \cdots, x_n), \qquad x_i(t_0) = x_{i0} \\ X_i(t, 0, \cdots, 0) \equiv 0; \qquad i = 1, \cdots, n \end{gathered} \tag{1}$$

Let $x_i(t)$ be a solution of the system (1), that is, a motion with orbit L, Fig. 1. The effect of perturbations is a change of certain quantities dependent on the motion, that is, a change of the motion into the perturbed motion $\bar{x}_i(t)$ with orbit $\bar{L}$.

If the distance ρ between a point P of L and a point $\bar{P}$ of $\bar{L}$ measures the effect of perturbations at the point P, the points P and $\bar{P}$ are corresponding points of the unperturbed motion $x_i(t)$ and the perturbed one $\bar{x}_i(t)$.

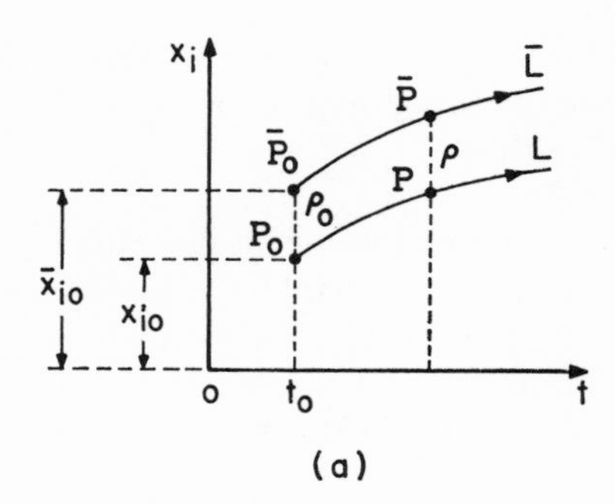

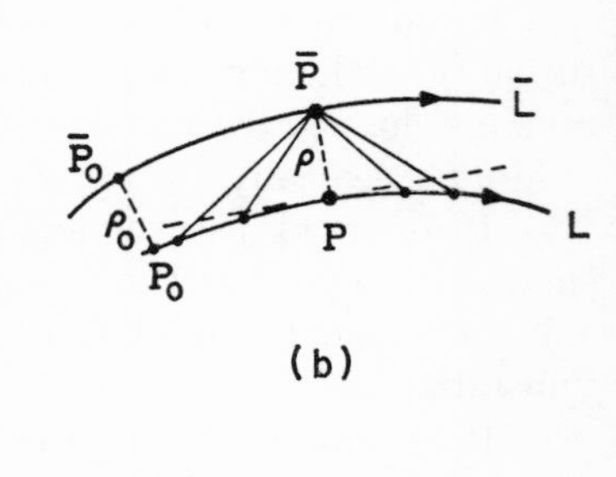

FIG. 1

One may have different kinds of correspondence between points P of L and points $\bar{P}$ of $\bar{L}$, and this correspondence characterizes different stability concepts of the motion.

One can distinguish two such correspondences, which are mostly physical, on which two important concepts of stability are based, namely, the stability in the sense of Liapunov and Poincaré:

a. The points P and $\bar{P}$ are points of L and $\bar{L}$, respectively, at the "same time," Fig. 1(a), either in phase space or in parameter space of the system, and the distance $\rho = P\bar{P}$ is "time-dependent," namely, one has:

$$\rho = \left\{\sum_{i=0}^{n} [x_i(t) - \bar{x}_i(t)]^2\right\}^{1/2}, \quad \text{in the phase space, and}$$

$$\rho = \left\{\sum_{i=0}^{n} [x_i(t, a_j) - x_i(t, \bar{a}_j)]^2\right\}^{1/2}, \quad \text{in the parameter space.}$$

This kind of correspondence is an appropriate one for stability discussion of a motion in its orbit.

We call "Liapunov distance" the distance ρ defined with the above correspondence between points P and $\bar{P}$.

b. The points P and $\bar{P}$ on L and $\bar{L}$ correspond to each other in such a way that the distance $\rho = P\bar{P}$, Fig. 1(b), is the minimum from the distances of $\bar{P}$ from all points of L:

$$\rho = \min\left\{\sum_{i=1}^{n} (x_i - \bar{x}_i)^2\right\}^{1/2}$$

The distance ρ defined in the above manner is "time-independent."

This correspondence is an appropriate one for stability discussion of the orbit of a motion.

We call "Poincaré distance" the distance ρ defined in the above manner.

IV. STABILITY DEFINITIONS OF A GENERAL MOTION

The distance ρ defined in the preceding can be used for finding stability definitions of a general motion of analytical type in agreement with the physical stability definitions.

a. A motion $x_i(t)$ of the system (1) is said to be "stable," if, for any given positive number ϵ and initial time $t_0 \geqq 0$, a positive number δ, depending on ϵ and, in general, on t_0, can be found such that, for any perturbed motion $\bar{x}_i(t)$, the inequality

$$\rho_0 < \delta \tag{2}$$

where ρ_0 is the distance initially, implies, for all time $t \geqq t_0$,

$$\rho < \epsilon \tag{3}$$

b. The motion is "unstable", if, given ϵ as above, for sufficiently small positive δ the inequality (3) is not satisfied even for at least one perturbed motion.

c. The motion is "asymptotically stable", if it is stable and in addition a positive number $\delta_1 \geqq \delta$ exists such that, starting from any ρ_0, which is:

$$\rho_0 < \delta_1 \tag{2a}$$

the limiting condition:

$$\lim \rho = 0 \tag{4}$$

holds uniformly to the quantities t_0, x_{i0} (Žubov, 1963a).

d. The motion is "quasi-asymptotically stable", if, from the above stability conditions, only the limiting condition (4) holds.

e. The motion is "asymptotically stable in the large", if it is asymptotically stable and the number δ_1 is very large.

The above definitions express the stability definitions in a unifying way, since in the above conditions the kind of the distance is not yet specified.

V. REMARKS

Remark 1: Stability Definitions in the Liapunov and Poincaré Senses

The above definitions are "stability definitions in the Liapunov sense," either in phase space or in parameter space, if the distances ρ_0 and ρ are "Liapunov distances" either in phase space or in parameter space, respectively.

The definitions are "stability definitions in the Poincaré sense," or "orbital definitions," if the distances ρ_0 and ρ are "Poincaré distances."

Remark 2: Equivalence of Liapunov and Poincaré Definitions in Equilibrium Points

In case the orbit L shrinks to an equilibrium point of the system, the distances in the Liapunov and Poincaré senses are identical, and therefore the stability definitions in the Liapunov and Poincaré senses are equivalent at the singular points of dynamical systems.

Remark 3: Stability Definition in the Lagrange Sense

The Lagrange stability of a system, namely, the stability concept in connection with the boundedness of motions and orbits of the system, is given by the preceding general stability conditions.

If ρ is the distance of the origin of the system from a point of any solution $x_i(t, t_0, x_{i0}, \cdots, x_{n0})$ of the system, and if, given any finite positive number ϵ, one can find another finite positive number δ such that condition (2) implies condition (3) for all time $t \geqq t_0$, the system is "stable in the Lagrange sense" (Hahn, 1963a).

The system is "unstable in Lagrange sense," if its response can grow without bounds.

The distance ρ used in stability conditions in Lagrange sense is identical with the distance in Liapunov or in Poincaré sense applied to origin.

Remark 4: Stability at Right and/or at Left

If the above stability definitions hold for $t \geqq t_0$, one can speak about "stability at right"; if they hold for $t \leqq t_0$, one can speak about "stability at left." One may have "stability at right and left" in case the conditions hold for all time (Cesari, 1959a).

REMARK 5: STABILITY DEFINITIONS IN AUTONOMOUS AND NONAUTONOMOUS SYSTEMS

a. If δ is independent of t_0, as in the case of autonomous systems or nonautonomous but periodic ones, the definitions are called "uniform"; so one may have "uniform stability," "uniform asymptotic stability," etc., in the Liapunov or in the Poincaré sense. In nonautonomous systems the selection of the initial time t_0 is not free, as, e.g., for stability of equilibrium points there is a value τ of time such that $\tau \leqq t_0$ (La Salle and Rath, 1963).

b. There is a one-to-one correspondence between the definitions of stability expressed by the previous simple stability conditions where the distances are either Liapunov or Poincaré distances, and the known definitions of stability of invariant sets (Zubov, 1964) applied to equilibrium states or to periodic motions.

REMARK 6: CONTINUOUS DEPENDENCE OF A SOLUTION ON INITIAL CONDITIONS

In case a solution $x(t, x_0)$ of the system is stable in the special parameter region of the initial conditions, for any two sets of values x_{01} and x_{02} of the initial conditions and for appropriate numbers ϵ and δ, the two inequalities:

$$\rho_0 = \{\sum(x_{01} - x_{02})^2\}^{1/2} < \delta, \qquad \rho = \{\sum [x(t, x_{01}) - x(t, x_{02})]^2\}^{1/2} < \epsilon$$

are compatible.

These two inequalities are the conditions for a continuous dependence of the solution $x(t, x_0)$ on the initial conditions x_0 uniformly in t, and this property of the solution is a Hadamard's postulate of the solution for being physically acceptable (Magiros, 1965b). Therefore, "the continuous dependence of a solution on the initial conditions corresponds to a stability situation of the solution in the Liapunov sense in a special parameter space, namely, in the initial conditions space."

REMARK 7: STABILITY OF PERIODIC MOTIONS

The stability definitions in Liapunov sense of periodic motions are narrow definitions, because they classify as unstable motions some motions which can be practically considered as stable.

The definitions in Poincaré sense are the appropriate ones for periodic motions.

REMARK 8: STABILITY OF THE ORIGIN WITH RESPECT TO ONLY CERTAIN COORDINATES (Zubov, 1963b)

If $x_i(t; x_{10}, \cdots, x_{n0})$ is a perturbed solution of a system through the point $(x_{10}, \cdots, x_{n0})$, and the distances ρ, ρ_k, $\rho_{\bar{k}}$ from the origin are given by the expressions:

$$\rho = \left\{\sum_{i=1}^{k} x_i^2(t, t_0, x_{10}, \cdots, x_{n0})\right\}^{1/2}, \rho_k = \left\{\sum_{i=1}^{k} x_{i0}^2\right\}^{1/2}, \rho_{\bar{k}} = \left\{\sum_{i=k+1}^{n} x_{i0}^2\right\}^{1/2}$$

the origin of the system is "stable with respect to the coordinates $x_1, \cdots, x_k$," $k < n$, if, for every small positive number ϵ, there exist two positive numbers δ_1 and δ_2, $\delta_1 < \epsilon$, such that:

$$\rho_k < \delta_1, \qquad \rho_{\bar{k}} < \delta_2$$

imply, for all $t \geqq t_0$,

$$\rho < \epsilon$$

If, in addition,

$$\lim_{t\to\infty} \rho = 0$$

the origin is "asymptotically stable with respect to the coordinates $x_1, \cdots, x_k$.

Example: The system (Zubov, 1963c)

$$\dot{x}_1 = x_1 + x_2, \qquad \dot{x}_2 = -(4x_1 + x_2 + x_3), \qquad \dot{x}_3 = -(2x_1 + x_2 + x_3)$$

has a solution which is "asymptotically stable at the origin" in the third component of the solution, but "asymptotically unsatable at the origin" in the first two components.

REMARK 9: ABSOLUTE STABILITY

A nonlinear autonomous system of which not any eigenvalue has positive real part, may have, under some restrictions of the nonlinearities, the origin as an equilibrium point "asymptotically stable in the large." Such nonlinear systems are called "absolutely stable systems," and there are methods for finding classes of nonlinearities for the absolute stability of the systems (Aizerman and Gantmaher, 1964).

The above kind of stability of the systems plays an important role in modern technology.

REMARK 10: STRUCTURAL STABILITY

If the stability situation of a motion $x_i(t)$ of a system is invariant in a parameter space S of the system, one speaks about a "structural stability" of the system in this space which is the "domain of structural stability" of the system. This property of the system gives information about the "insensitivity" of the system to perturbations (Lefschetz, 1957).

VI. STABILITY UNDER PERSISTENT PERTURBATIONS

In the preceding, the perturbations were considered momentarily acting to the system, and consequently the perturbations do not appear in the formulation of the equations of the system.

In the case of constantly acting perturbations $p_i(t; x_1, \cdots, x_n)$, the equations of the system must contain the perturbations, for which the equations of the system during the action of persistent perturbations are:

$$\begin{aligned} \dot{x}_i(t) &= X_i(t; x_1, \cdots, x_n) + p_i(t; x_1, \cdots, x_n) \\ x_i(t_0) &= x_{i0}; \qquad i = 1, \cdots, n \end{aligned} \tag{5}$$

where p_i must be such that the system (5) has unique solutions corresponding to the initial conditions.

A solution $x_i(t)$ of the system (1) is "stable under persistent perturbations," if, in addition to conditions (2) and (3), some appropriate restrictions on the magnitude of p_i are accepted and which must hold for all accepted x_i and $t \geqq t_0$.

In case the magnitude of the perturbations is according to (Vrkŏv, 1963):

$$\max_{x_i} | p_i(t; x_1, \cdots, x_n)| < \eta \tag{6}$$

where η is an appropriate positive number, one speaks about a "total stability" of the solution $x_i(t)$.

If the condition (6) is replaced by:

$$\int_0^\infty \max_{x_i} | p_i(t; x_1, \cdots, x_n)| \, dt < n \tag{6a}$$

one speaks about "integral stability."

In case of total stability the perturbations must be small in magnitude, but in case of integral stability they may be large in a small interval of time.

These magnitude restrictions of the perturbations of the above stabilities are possessed by stability under perturbations which are "bounded in the mean," and in this case one has "stability in the mean," which corresponds to perturbations satisfying the condition:

$$\int_t^{t+T} \max_{x_i} | p_i(t; x_1, \cdots, x_n)| \, dt < \eta \tag{6b}$$

where η of 6(b) and δ of (2) depend on T in general.

If, in addition to the above, condition (4) holds, one has "stability of asymptotic type" under persistent perturbations.

The above stability of persistent perturbations are in the Liapunov or Poincaré sense, if the distances ρ_0 and ρ in (2) and (3) are Liapunov or Poincaré distances.

VII. COMPARISON AND CONNECTIONS OF STABILITY CONCEPTS—EXAMPLES

The three basic concepts of stability in case of sudden or persistent perturbations, although expressed by the same mathematical conditions, are independent of one another. It is therefore possible for a solution to have different situations of stability under different definitions of stability. Nevertheless, the concepts of stability are in many cases connected.

In the following we clarify these statements.

1. *Liapunov and Poincaré Definitions of Stability Applied to Periodic Motions*

We apply the Liapunov and Poincaré definitions of stability to the following physical problem:

"*Discuss the stability of the motion of a mass in an elliptic orbit under the inverse square Newton's law of attraction of an attractive center.*"

The elliptic orbit of the moving mass in this Newtonian field can be determined by knowing the initial conditions, that is, the distance of the moving mass from the attractive center E and its velocity at time t_0. Let us take L the orbit by specifying the initial conditions, and T the corresponding period of the motion on L. If these initial conditions are changed, then the new initial conditions correspond to a new orbit $\bar{L}$ of the perturbed motion with new period $\bar{T}$.

a. *The motion of the mass on the orbit L is "orbitally stable", but not "asymptotically orbitally stable".*

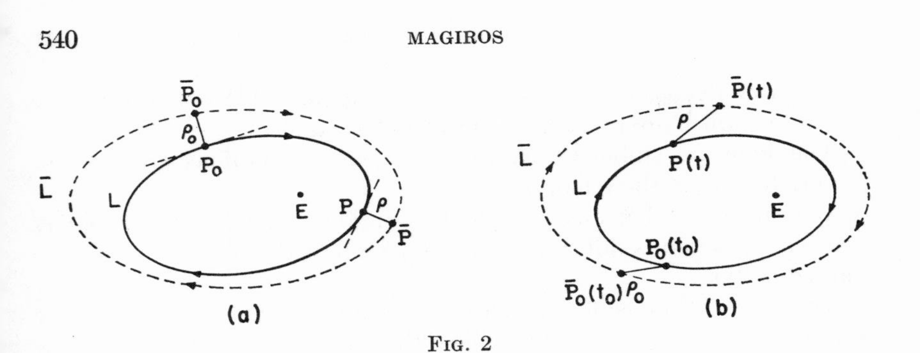

FIG. 2

For any point P on L, Fig. 2(a), the corresponding point $\bar{P}$ on $\bar{L}$ must be such that the distance $\rho = P\bar{P}$ is the minimum distance from $\bar{P}$ to L.

If one wants this distance ρ to be smaller than a given number $\epsilon > 0$ for all time, one needs for orbital stability to find a $\delta > 0$ such that, if the distance initially is smaller than δ, $\rho_0 < \delta$, one has $\rho < \epsilon$. This is possible. Given a perturbation p, the perturbed orbit $\bar{L}$ is known, then the deviation ρ, corresponding to any point P of L, is known. This deviation for all points of L has a maximum ρ_M, and a minimum ρ_m; and the smaller the perturbation p, the smaller ρ_M and ρ_m. Given now a small positive ϵ, if one wants $\rho < \epsilon$ for all points P of L, one must use a very small perturbation p such that the corresponding ρ_M is smaller than ϵ, when the appropriate δ must be smaller than ρ_m. Therefore, the motion is "orbitally stable."

Since the distance ρ does not tend to zero as t changes, the motion is not "asymptotically orbitally stable," that is, the ellipse is not a "limit cycle."

b. *The above motion on L is "unstable in Liapunov sense.*

The distance $\rho = P\bar{P}$, Fig. 2(b), in Liapunov definition is the distance of the points P and $\bar{P}$ on L and $\bar{L}$, which are places of the mass at the "same time." The periods T and $\bar{T}$ of the motion on L and $\bar{L}$ are dependent on the length of the corresponding major axes, thus the points P and $\bar{P}$, very close initially but traveling on different ellipses, may find themselves at opposition and thus at great distance from each other in due course of time. Then given small positive ϵ, one can not find a δ such that $\rho_0 < \delta$ implies $\rho < \epsilon$, and this is instability of the motion in Liapunov sense.

Motions as of the present problem are practically considered as stable, and by the above example one sees that the Liapunov stability definitions are narrow definitions, since they classify as unstable situations some situations which are practically considered as stable.

The Poincaré stability definitions in periodic motions are the appropriate ones in these motions.

We can prove that any periodic motion is unstable in the Liapunov sense if its period is different from the period of the corresponding perturbed motion.

The free vibrations of a simple pendulum are stable in the Liapunov sense if they are linear and unstable if they are nonlinear, because in the case of linearity the isochronous phenomenon exists, and in the case of nonlinearity the isochronous phenomenon is violated.

2. *Examples of Motions Stable or Unstable Under Different Definitions of Stability*

a. The rectilinear motion:

$$x(t) = c_1(t - t_0) + c_2$$

is the general solution of:

$$\ddot{x} = 0, \qquad \dot{x}(t_0) = c_1, \qquad x(t_0) = c_2.$$

The above motion is stable or unstable depending on the initial conditions and the definitions of stability used. We have the following cases:

(i) $c_1 = 0$, $c_2 =$ *arbitrary*. The motion can be represented in the "t, x-plane" by lines parallel to "t-axis," the Liapunov distance ρ_L, the Poincaré distance ρ_P and the Lagrange distance ρ_{La} are, Fig. 3(a), $\rho_L = \rho_P = P\bar{P}$, $\rho_{La} = OP'$, which are constants, then the motion is "stable" in the Liapunov, Poincaré, and Lagrange senses.

(ii) $c_1 \neq 0$ *fixed*, $c_2 =$ *arbitrary*. The motion can be represented by parallel lines with nonzero slope c_1, Fig. 3(b), the Liapunov distance $\rho_L = P\bar{P} =$ constant, the Poincaré distance $\rho_P = P\bar{\bar{P}} =$ constant, and the Lagrange distance $\rho_{La} = OP'$ increasing to infinity as t, and then P, tends to infinity. Then, the motion is "stable" in Liapunov and Poincaré sense, but "unstable" in Lagrange sense.

(iii) $c_1 \neq 0$ *arbitrary*, c_2 *either fixed or arbitrary*. The trajectories L and $\bar{L}$ are, in general, as shown in Fig. 3(c). The Liapunov, Poincaré and Lagrange distances are: $\rho_L = P\bar{P}$, $\rho_P = P\bar{\bar{P}}$, $\rho_{La} = OP'$, and all of them increase to infinity with t, and then P goes to infinity when the motion is "unstable" in all the three senses.

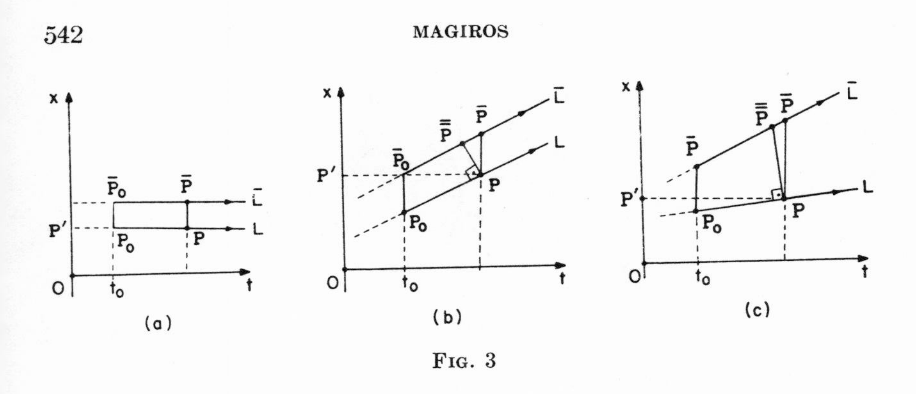

FIG. 3

b. The motion $x(t) = a \sin(at + b)$, where a and b are arbitrary parameters is a solution of the equation:

$$\ddot{x} = -\tfrac{1}{2}x\{x^2 + (x^4 + 4\dot{x}^2)^{1/2}\}$$

it is stable in Lagrange and Poincaré sense, but, with the exception of the origin, it is unstable in Liapunov sense.

(c) The motion $x = c_1 \cos(\mu t + c_2)$, $y = c_1 \sin(\mu t + c_2)$ of the system: $\dot{x} = -\mu y$, $\dot{y} = \mu x$, $\mu \neq 0$ where c_1 and c_2 are arbitrary constants, is a periodic motion with frequency μ in the concentric circles $x^2 + y^2 = c_1^2$.

This motion is Liapunov stable in case μ is a constant, it is Liapunov unstable in case μ is an arbitrary parameter, say in $\mu_1 \leqq \mu \leqq \mu_2$.

d. The equilibrium point of the system

$$\dot{x} = 2xy, \qquad \dot{y} = y^2 - x^2$$

that is the origin of the system is "quasi-asymptotically stable." The solutions of this system are the members of the one parameter family of curves: $(x - \rho)^2 + y^2 = \rho^2$, then circles through the origin with radius ρ and the center on the x-axis, Fig. 4. Starting from any point of the x, y-plane, the circle through this point ultimately terminates to origin, then the limiting condition (4) is satisfied. All solutions of the system starting from points of the circle $(0, \delta)$ can not be included in the circle $(0, \epsilon)$, then the conditions (2) and (3) are not compatible, even if δ is small and ϵ large.

e. The motion $x = ae^{-t} \sin t$, where a is an arbitrary parameter, is "asymptotically stable" in case $t_0 \neq k\pi$, k an integer, and it is "quasi-asymptotically stable" in case $t_0 = k\pi$ (Hahn, 1963b).

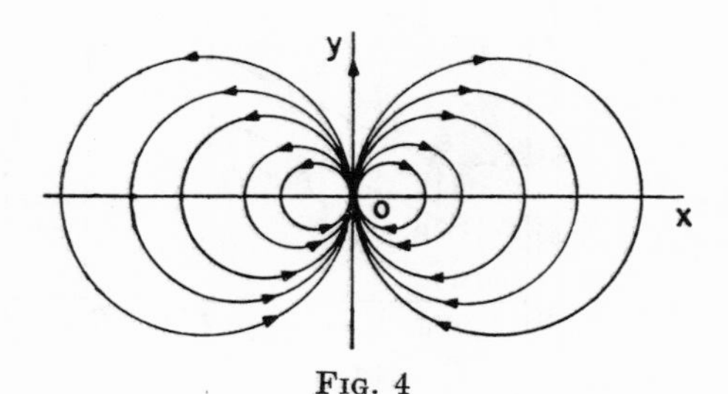

FIG. 4

3. Connections Between Stability Concepts

a. In linear homogeneous systems:

$$\dot{x}_i(t) = \sum_{j=1}^{n} d_{ij}(t)x_j ; \qquad i = 1, \cdots, n$$

where $d_{ij}(t)$ are continuous functions of t in $t \geqq t_0$:

"*The Lagrange stability of its solutions implies their Liapunov stability*", *and conversely* (Cesari, 1959b).

b. In linear nonhomogeneous systems:

$$\dot{x}_i(t) = \sum_{j=1}^{n} d_{ij}(t)x_j + f_i(t); \qquad i = 1, \cdots, n$$

where $d_{ij}(t)$ and $fi(t)$ are continuous functions of t in $t \geqq t_0$: "*The Lagrange stability of its solutions implies their Liapunov stability, but for the converse the boundedness of one solution is needed as an additional requirement*" (Cesari, 1959b).

In nonlinear systems, connections between the above concepts of stability either is in general hard to be found or they do not exist.

In case of stabilities of persistent perturbations connections can be introduced by the following statements (Vrkǒv, 1963):

c. "*Asymptotic integral stability implies asymptotic stability in the mean, and conversely*";

d. "*Asymptotic integral stability implies total asymptotic stability, and this implies total stability*";

e. "*Asymptotic stability in the mean implies stability in the mean, and this implies either integral stability or total stability*";

f. "*Integral stability does not imply total stability*";

g. "*Stability in the mean does not imply total asymptotic stability*";

h. "*Total asymptotic stability does not imply integral stability*".

VIII. GEOMETRICAL INTERPRETATION OF THE STABILITIES

The concepts of stability may be clarified by their geometrical interpretation.

One may define as "r-tube" around a curve l in n-dimensional space the set of all points of which the distance ρ in the Poincaré sense from l is smaller than r. The quantity r is the width and l the central curve of the tube. For finite r the tube is bounded.

These definitions can appropriately be used for a geometrical interpretation of the stability concepts.

A motion $x_i(t, t_0, x_{10}, \cdots, x_{n0})$ of orbit L is stable in Liapunov sense, if given the "ϵ-tube" around the line $x_i = x_{i0}$, Fig. 5(a), one can find a "δ-tube" around the line $x_i = x_{i0}$ such that any perturbed motion of orbit $\bar{L}$, starting in the "δ-tube," remains for all $t \geqq t_0$ in the "ϵ-tube."

In case of equilibrium points, that is, the origin, the above tubes are around the t-axis, Fig. 5(b).

For asymptotic stability one needs, in addition, another tube, a "δ_1-tube," which will be unbounded for asymptotic stability in the large.

For stability of a general motion in Poincaré sense, that is, for "orbital stability" of the motion, the "δ-tube" and "ϵ-tube" around the orbit L can be used.

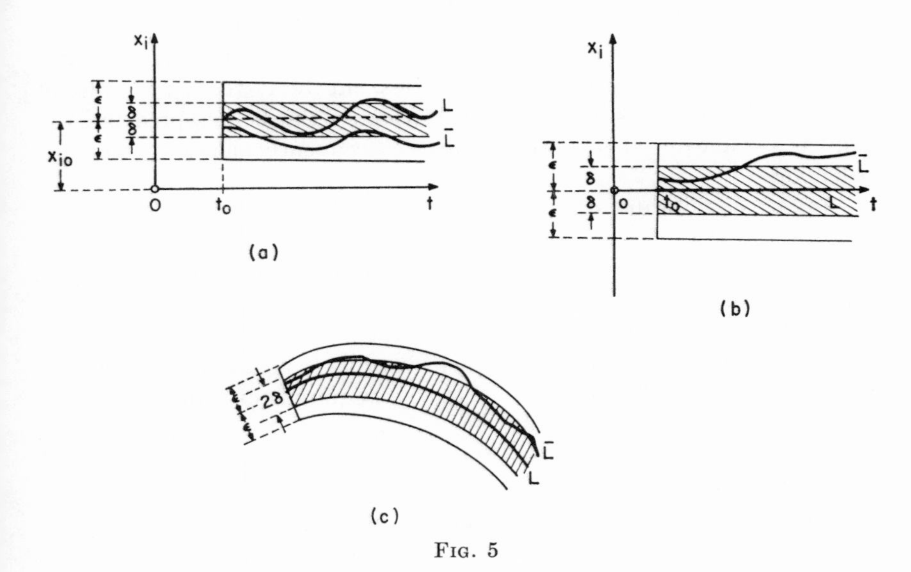

FIG. 5

A motion is "orbitally stable," if given the "ϵ-tube" around L, one can find a "δ-tube" around L such that any perturbed orbit $\bar{L}$ starting in the "δ-tube" remains always in the "ϵ-tube", Fig. 5(c).

For stability of equilibrium points, the orbit L is the t-axis, then the tubes are around the t-axis, Fig. 5(b), and the discussion is the same as in the Liapunov case.

For periodic motions, the tubes around the closed orbit L are tori with L as a central curve.

For Lagrange stability bounded tubes around the t-axis, Fig. 5(b) can be used.

The system is "Lagrange stable," if, for every bounded "ϵ-tube" around the t-axis another bounded "δ-tube" around the t-axis can be found such that every orbit of the system starting in the "δ-tube" does not leave the "ϵ-tube" for all time $t \geqq t_0$.

For stability under persistent perturbations the "η-tube" around the t-axis is needed, in addition. The "δ-tube" and "ϵ-tube" must be around the t-axis in the Liapunov case, or around the orbit L in Poincaré case.

IX. PRACTICAL STABILITY (Zubov, 1963(d); La Salle-Lefschetz, 1961)

All the previous discussion on concepts and definitions of stability, although modified in many points for the purpose of their application to practical problems, are of mathematical type.

One can see that a state of a system may be unstable mathematically, that is, under one of the previous definitions, but the system may oscillate sufficiently near this state and its performance can be accepted practically as a stable one. The motion of missiles frequently shows this kind of behavior.

Also, an equilibrium state of a system may be stable mathematically in a small region, but in practice the perturbations expected may cause the system to go far from the equilibrium state; then the system is practically unstable at the equilibrium state.

The definitions of stability discussed in the preceding need appropriate modifications, changes, and supplements in order to be useful for practical problems, that is, in order to meet the requirements for "practical stability".

For practical stability one needs to know:

(a) the size of the region of deviations, that is, the width of the "ϵ-tube," which gives the "acceptable states" and a "satisfactory operation" of the system;

(b) the size of the region of initial data, that is, the width of the "δ-tube," which gives the permitted size of the initial conditions which can be controlled;

(c) the size of the region of perturbations, that is, the width of the "η-tube"; and

(d) a finite time T over which the stability is valid, that is, the lengths of the above tubes, in other words, the time $t_0 \leqq t \leqq t_0 + T$, for which the solution $x_i(t)$ satisfies the above requirements.

In determining practical stability, the linearization of the system is not, in general, permitted, since practical stability depends on the nonlinearities of the system.

X. STABILITY AND THE SYSTEM OF COORDINATES

The character of the stability, under any of the previous concepts, is not invariant with a general transformation of the coordinates of the system. In other words, the stability is dependent on the coordinate system with reference to which the variables are to be considered; then for a discussion of stability the selection of the appropriate space variables of the system is suggested.

By means of the following examples, we clarify the dependence of the stability on the coordinate system (Cesari, 1959c).

Example 1. The dynamical system:

$$\dot{x} = -y(x^2 + y^2)^{1/2}, \qquad \dot{y} = x(x^2 + y^2)^{1/2} \tag{7}$$

accepts the two parameter family of solutions:

$$x = a \cos (at + b), \qquad y = a \sin (at + b) \tag{8}$$

where a and b are arbitrary.

This family of solutions is stable in Poincaré and Lagrange sense, but it is unstable in Liapunov sense.

By introducing new variables, one can make the solution Liapunov stable.

In case the new variables r and b are introduced by means of the relations:

$$x = r \cos \vartheta, \qquad y = r \sin \vartheta, \qquad \vartheta = at + b \tag{9}$$

the original system (7) is transformed into the new one:

$$\dot{r} = 0, \qquad \dot{b} = 0 \tag{10}$$

for which the solution

$$r = c_1 , \qquad b = c_2 \tag{11}$$

is Liapunov stable, where c_1 and c_2 arbitrary constants.

Example 2. The equation of pendulum:

$$\ddot{x} + \sin x = 0 \tag{12}$$

accepts the two parameter family of solutions:

$$x = a \sin \{\varphi(a)t + b\} \tag{13}$$

where a and b are arbitrary, and $\varphi(a)$ can be expressed in terms of elliptic functions.

The solutions (13) are, except for the origin, unstable in Liapunov sense, but by using new coordinates, r and b, according to the transformation formulas:

$$x = r \sin \{\varphi(r)t + b\}, \qquad y = r \cos \{\varphi(r)t + b\} \tag{14}$$

the system (12) leads to the system:

$$\dot{r} = 0, \qquad \dot{b} = 0 \tag{15}$$

of which the solutions are Liapunov stable.

Example 3. Let us take the nonautonomous system

$$\dot{x}_i(t) = X_i(t, x_1 , \cdots , x_n); \qquad i = 1, \cdots , n \tag{16}$$

of which the solution in its implicit form is:

$$\phi_i(t, x_1 , \cdots , x_n) = c_i \tag{17}$$

where $c_1 , \cdots , c_n$ are constants. By using the transformation:

$$y_i = \phi_i(t, x_1 , \cdots , x_n) \tag{18}$$

the original system (16) is reduced to:

$$\dot{y}_i = 0 \tag{19}$$

of which the solution:

$$y_i = c_i \tag{20}$$

is Liapunov stable, but this is not the case, in general for the solution (17) of the system (16).

RECEIVED: October 8, 1965

REFERENCES

AIZERMAN, M. AND GANTMAHER, F. (1964), "Absolute Stability of Regulator Systems." Holden-Day, San Francisco.

CESARI, L. (1959), "Asymptotic Behavior and Stability Problems in Ordinary Differential Equations," (a) p. 6, (b) pp. 7-8, (c) pp. 12-13. Springer, Berlin.

HAHN, W. (1963), "Theory and Applications of Liapunov's Direct Method," Prentice-Hall, Englewood Cliffs, New Jersey. (a) p. 129, (b) p. 6.

LA SALLE, J. AND RATH, R. (1963), Eventual stability. *Proc. 1963 IFAC.*

LA SALLE, J. AND LEFSCHETZ, S. (1961), "Stability by Liapunov's Direct Method," p. 121. Academic Press, New York.

LEFSCHETZ, S. (1957), "Differential Equations: Geometric Theory," p. 239. Interscience, New York.

MAGIROS, D. (1965a), On stability definitions of dynamical systems. *Proc. Natl. Acad. Sci., U.S.A.* **53**, 1288–1294.

MAGIROS, D. (1965b), Physical problems discussed mathematically. *Bull. Greek Math. Soc., New Ser. II*, **6**, 143–156.

VRKŐV, I. (1963), On some stability problems. *Proc. Conf. Prague, September, 1962*, pp. 217–221 (Academic Press, New York).

ZUBOV, V. (1963), "Mathematical Methods for the Study of Automatic Control Systems," Macmillan, New York. (a) Chap. I, Sect. 1, (b) Sect. 4, (c) Section 8, (d) p. 13.

ZUBOV, V. (1964), "Methods of A. M. Liapunov and Their Application," pp. 22–26. Noordhoff, Groningen, The Netherlands.

*Attitude Stability of a Spherical Satellite**

by D. G. MAGIROS *and* A. J. DENNISON

General Electric Company
Valley Forge Space Technology Center, Philadelphia, Pa.

ABSTRACT: *The in-orbit-plane angular motion of a spherical satellite with a mass eccentricity is studied, and both aerodynamic and gravitational torques along with large angle motions are considered. The equilibrium attitudes of the motion are investigated and the conditions for stability presented. The stability concepts on which the study is based are delineated in an Appendix.*

I. Introduction

Problems associated with the angular motion and attitude stability of satellites acted upon by aerodynamic and gravitational torques have received much attention in recent years. Of the many papers published, a few have considered the problem of particular interest here, *viz.*, the attitude stability of a spherical satellite with a mass eccentricity.

D. M. Schrello (**2**), as part of a comprehensive work considering many aerodynamic shapes, has studied the attitude stability of the sphere employing small angle approximations. Nam Tum Po (**7**), while admitting large angular motions, considers only the aerodynamic rate damping effects in studying the attitude stability of a spherical satellite. Finally, the form of the basic differential equation employed in this paper, excluding the dynamic damping term, appears in other works, e.g., Minorsky (**6**) considers the problem of a rotating pendulum and discusses the stability characteristics of that system.

In general, the admission of large angular motion and the inclusion of dynamic damping effects can significantly alter conclusions regarding the stability characteristics of the physical system being studied. Accordingly, this paper considers both gravitational and aerodynamic torques, including the aerodynamic rate damping term, and admits large angular motion in determining equilibrium attitudes and establishing stability conditions.

For the purposes of this paper the motion of the satellite is assumed to be confined to the orbital plane. The authors intend to discuss the out-of-plane rotational problems at a later time.

II. Development of the Equation Governing the Angular Motion

Rigid Body Equations of Motion. The satellite is considered as a rigid body. In general, the motion of a rigid body consists of two motions in coexistence, a translational motion of its mass center o_1 with respect to a fixed orthogonal coordinate system $O_1X_1X_2X_3$, and a rotational motion about axes through

* This is part of the senior author's paper discussed at the 18th International Astronautical Congress, Belgrade, Yugoslavia, Sept. 1967.

Reprinted from the *Journal of the Franklin Institute* **286** (1968), 193–203.

D. G. Magiros and A. J. Dennison

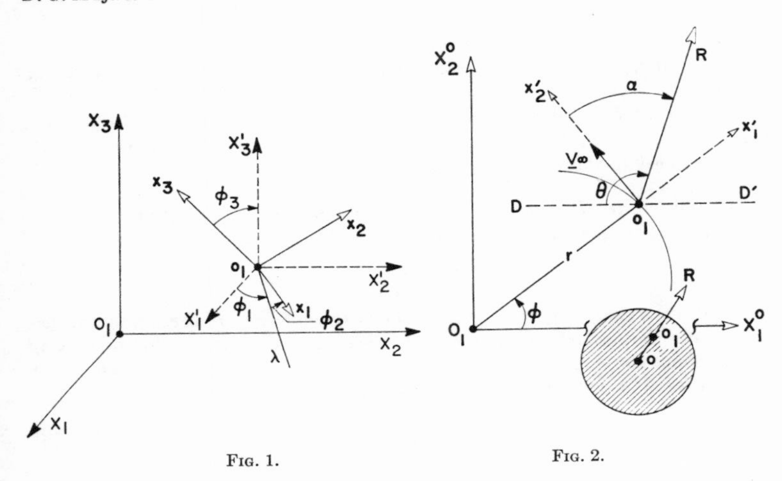

Fig. 1. Fig. 2.

the mass center referred to a moving orthogonal coordinate system $o_1x_1x_2x_3$ rigidly fixed in the body, see Fig. 1.

If the resultant **F** of all external forces acting on the body is displaced parallel to itself to act on the mass center, a resultant torque **L** of the external forces results, and if the x_1, x_2, x_3-axes are oriented as the principal axes of inertia of the body, the motion of the body in scalar form is given by:

$$m\dot{v}_i = F_i, \qquad i = 1, 2, 3 \tag{1a}$$
$$I_i\dot{w}_i = L_i, \qquad i = 1, 2, 3 \tag{1b}$$

where m is the constant mass of the body, v_i the translational velocities, w_i the angular velocities, and I_i the principal moments of inertia. Equations 1a refer to the fixed system $O_1X_1X_2X_3$, and Eqs. 1b, the moving system $O_1x_1x_2x_3$.

Equations 1 are assumed to be uncoupled, i.e., the force is independent of the angular velocity and the torque is independent of the translational velocity. Accordingly, the *translational* motion problem and the *rotational* motion problem are considered separately.

Planar Equations of Motion. We assume that the translational motion of the satellite is confined to an inertially-fixed orbit plane and that the motion in this plane consists of a circular orbit with the earth at its center, see Fig. 2. Based on this assumption, Eqs. 1a are not considered further.

Furthermore, it is assumed that the rotational motion of the satellite is confined to the orbital plane and for this case the equation of motion is

$$I_P\ddot{\theta} = L \tag{2}$$

where I_P is the moment of inertia about the axis (of rotation) through the center of mass o_1 perpendicular to the orbital plane, θ the inertial angle, shown in Fig. 2, and L the component of the resultant torque on the axis of rotation. The resultant torque L is taken as the sum of the applied gravitational torque, L_G, and the applied aerodynamic torque, L_A. Expressions for these applied torques are developed in the next two sections.

Gravitational Torque. For an inverse square central force field the gravitational torque, L_G, can be approximated by (**2**)

$$L_G = \tfrac{3}{2}(K/r^3)(I_P - I_R)\sin 2\alpha \tag{3}$$

where r is the radius of the circular orbit of the satellite, K the product of the universal gravitational constant and the mass of the earth, I_R the moment of inertia about the satellite reference axis, and α the angle of attack shown in Fig. 2. Equation 3 is written, for convenience, in the form

$$L_G = I_P G \sin 2\alpha \tag{4}$$

where the constant G is defined by

$$G = \tfrac{3}{2}(K/r^3)[1 - (I_R/I_P)]. \tag{5}$$

Aerodynamic Torque. The aerodynamic torque, L_A, for a passive satellite is given by (**2, 3**)

$$L_A = [C_m + C_{m_q}(l_q/2V_\infty)]\tfrac{1}{2}\rho_\infty V_\infty{}^2 A l \tag{6}$$

where C_m and C_{m_q} are the static and dynamic aerodynamic moment coefficients, respectively; l and A are the aerodynamic reference length and area, respectively; ρ_∞ is the local atmospheric density; V_∞ is the satellite velocity relative to the atmosphere; and q is the total angular rate of the satellite in the orbital plane.

In Eq. 6, A, l, ρ_∞ and V_∞ can all be taken as constants; A and l are, of course, constants dictated by the size of the satellite. Since the satellite orbit is assumed to be circular, ρ_∞ can be taken as constant. Finally, V_∞ can also be considered as a constant since, for a circular orbit, its deviation from a constant is a second-order effect even when the atmosphere is considered to be rotating.

Owing to the restricted nature of the orbit, the time-rate of change of angle-of-attack, $\dot{\alpha}$, is the prime contributor to the total angular rate, q, and, accordingly, it is assumed that

$$q = \dot{\alpha} \tag{7}$$

for the purposes of Eq. 6. The static and dynamic aerodynamic coefficients, C_m and C_{m_q}, for a spherical satellite moving near the earth, are given by (**3**)

$$C_m = -k_1 \sin\alpha \tag{8a}$$
$$C_{m_q} = -(k_2 + k_3\cos\alpha) \tag{8b}$$

where k_1, k_2, k_3 are positive constants depending upon the free stream molecular speed ratio which is assumed constant, upon the radius a of the spherical satellite, and upon the distance $x = |oo_1|$, shown in Fig. 2.

Employing Eqs. 7 and 8, Eq. 6 can be put in the following convenient form

$$L_A = -I_P[A_1 \sin \alpha + (A_2 + A_3 \cos \alpha)\dot{\alpha}] \tag{9}$$

where the constants A_1, A_2 and A_3 are defined by

$$A_1 = k_1[(\tfrac{1}{2}\rho_\infty V_\infty^2)Al/I_P] \tag{10a}$$
$$A_2 = k_2[(\tfrac{1}{2}\rho_\infty V_\infty^2)Al/I_P] \cdot (l/2V_\infty) \tag{10b}$$
$$A_3 = k_3[(\tfrac{1}{2}\rho_\infty V_\infty^2)Al/I_P] \cdot (l/2V_\infty) \tag{10c}$$

the magnitudes of which satisfy the inequalities

$$A_1 \gg / A_3 > A_2 > 0. \tag{11}$$

Equation Governing Angular Motion. From Fig. 2 it is clear that the angle θ can be expressed in terms of α and ϕ, the polar angle defining the position of the satellite in its orbit, according to the relation

$$\theta = \tfrac{1}{2}\pi + \alpha - \phi. \tag{12}$$

Since the satellite orbit is assumed to be circular, $\dot{\phi}$ is a constant and, from Eq. 12

$$\ddot{\theta} = \ddot{\alpha}. \tag{13}$$

Now, substitution of Eqs. 4, 9 and 13 into Eq. 2, gives

$$\ddot{\alpha} + (A_2 + A_3 \cos \alpha)\dot{\alpha} + A_1 \sin \alpha - G \sin 2\alpha = 0 \tag{14}$$

the differential equation describing the in-orbit-plane angular motion of a spherical satellite. Finally, Eq. 14 is put into the equivalent normal form of an autonomous non-linear system

$$\dot{\alpha} = \omega \tag{15a}$$
$$\dot{\omega} = -(A_2 + A_3 \cos \alpha)\omega + (2G \cos \alpha - A_1) \sin \alpha \tag{15b}$$

thereby facilitating the investigation of the satellite's attitude stability.

III. Equilibrium Points and their Stability

Equilibrium Points of the System (15), are defined by the condition $\dot{\alpha} = 0$, $\dot{\omega} = 0$ and the points satisfying this condition are given by

$$\omega = 0, \qquad \alpha = m\pi \tag{16a}$$
$$\omega = 0, \qquad \alpha = \cos^{-1}(A_1/2G) + 2m\pi \tag{16b}$$

where $m = 0, \pm 1, \pm 2 \cdots$ and the point defined by (16b), of course, exists only on the range $-1 \leq (A_1/2G) \leq +1$.

Since the physical restriction of the angle α is $-\pi < \alpha \leq \pi$ and, since the values $(A_1/2G) = +1$ and $(A_1/2G) = -1$ correspond to $\alpha = 0$ and $\alpha = \pi$, respectively, the equilibrium points studied here are confined to

$$\alpha_1 = 0 \tag{17a}$$
$$\alpha_2 = \pi \tag{17b}$$
$$\alpha_3 = \pm \cos^{-1}(A_1/2G), \quad \text{where} \quad -1 < (A_1/2G) < 1. \tag{17c}$$

We remark here that the stability situation at each of the equilibrium points is a local behavior because of the nonlinearities in the system (15).

Stability of the Equilibrium Points (17), is investigated here on the basis of the definitions and theorems given in the Appendix.

Stability of $\alpha_1 = 0$. Expanding $\sin\alpha$ and $\cos\alpha$ by use of McLaurin series, the system (15) becomes

$$\dot{\alpha} = \omega \tag{18a}$$
$$\dot{\omega} = (2G - A_1)\alpha - (A_2 + A_3)\omega + F_1(\alpha, \omega) \tag{18b}$$

where $F_1(\alpha, \omega)$ is a known power series starting with terms of the second degree.

It can be shown that $F_1(\alpha, \omega)$ satisfies the requirement of Theorem I so that it is only necessary to consider the linear terms in Eqs. 18 in studying the stability of this equilibrium point. Accordingly, Theorem I, which defines the stability condition in terms of the characteristic roots of the linearized system, is applied in this case.

The characteristic roots of system (18), if linearized, are

$$\lambda_{1,2} = -\tfrac{1}{2}(A_2 + A_3) \pm \tfrac{1}{2}[(A_2 + A_3)^2 + 4(2G - A_1)]^{1/2}. \tag{19}$$

In view of the relative magnitudes of A_1, A_2 and A_3, as shown by (11), the signs of the real parts of $\lambda_{1,2}$ are dictated, for practical cases, by the term $4(2G - A_1)$. This being the case, it is clear that:

(a) If $G < A_1/2$, then both $\lambda_{1,2}$ have negative real parts and the equilibrium point is asymptotically stable.

(b) If $G > A_1/2$, then one of the $\lambda_{1,2}$ has a positive real part and the equilibrium point is unstable.

(c) If $G = A_1/2$, then one of the $\lambda_{1,2}$ has a negative real part and the other has a zero real part. This is a critical case and must be studied by other means.

In order to study the stability of the critical case where $G = A_1/2$, an appropriate Liapunov function $V(\alpha, \omega)$ is developed and Theorem II is applied.

Such a function can be determined, according to (4), by

$$V(\alpha, \omega) = \tfrac{1}{2}\omega^2 - \int_0^\alpha \dot{\omega}(\alpha, 0)\, d\alpha \tag{20}$$

which for Eqs. 15, with $G = A_1/2$, gives

$$V(\alpha, \omega) = \tfrac{1}{2}\omega^2 + \tfrac{1}{2}A_1(1 - \cos\alpha)^2 \tag{21}$$

which has the time derivative

$$\dot{V}(\alpha, \omega) = -(A_2 + A_3 \cos\alpha)\omega^2. \tag{22}$$

Clearly, $V(\alpha, \omega)$ is *positive* definite and $\dot{V}(\alpha, \omega)$ is *negative* definite in the neighborhood of the equilibrium point $(\alpha, \omega) = (0, 0)$. Hence, according to Theorem II, this equilibrium point is asymptotically stable when $G = A_1/2$.

In summary, then, it can be stated that the equilibrium point $(\alpha, \omega) = (0, 0)$ is: (a) asymptotically stable if $G \leq A_1/2$; and (b) unstable if $G > A_1/2$.

Stability of $\alpha_2 = \pi$. If in system (15) α is transformed into x according to relation $x = \alpha - \pi$ and if, in addition, $\sin x$ and $\cos x$ appearing in the resulting expression are expanded about $x = 0$, the following system of equations results:

$$\dot{x} = \omega \tag{23a}$$
$$\dot{\omega} = (A_1 + 2G)x + (A_3 - A_2)\omega + F_2(x, \omega). \tag{23b}$$

Note that the equilibrium point $(\alpha, \omega) = (\pi, c)$ for system (15) corresponds to the equilibrium point $(x, \omega) = (0, 0)$ for system (23).

As before, it can be shown that $F_2(x, \omega)$ satisfies the condition of Theorem I so that only the linear terms in Eqs. 23 need be considered in studying the stability. Also, as with the previous case, the stability conditions can be determined by examining the characteristic roots of the linearized equations.

For this case the characteristic roots are

$$\lambda_{1,2} = \tfrac{1}{2}(A_3 - A_2) \pm \tfrac{1}{2}[(A_3 - A_2)^2 + 4(2G + A_1)]^{1/2}. \tag{24}$$

Examination of Eq. 24, in view of Eq. 11, shows that one of the $\lambda_{1,2}$ always has a positive real part. Hence, according to Theorem I, the equilibrium point $(\alpha, \omega) = (\pi, 0)$ is always unstable.

Stability of $\alpha_3 = \pm\cos^{-1}(A_1/2G)$. As with the previously discussed equilibrium point, if in the system (15) α is transformed into x according to the relation $x = \alpha - \alpha_3$ and if, in addition, the $\sin x$ and $\cos x$ appearing in the resulting expression are expanded about $x = 0$, the following system of equations results

$$\dot{x} = \omega \tag{25a}$$
$$\dot{\omega} = -2G[1 - (A_1/2G)^2]x - A_3[(A_2/A_3) + (A_1/2G)]\omega + F_3(x, \omega) \tag{25b}$$

Again $F_3(x, \omega)$ satisfies the condition of Theorem I, and this Theorem can be applied. The characteristic roots of the system (25) are

$$\lambda_{1,2} = -\tfrac{1}{2}A_3\left(\frac{A_2}{A_3} + \frac{A_1}{2G}\right) \pm \frac{1}{2}\left[A_3^2\left(\frac{A_2}{A_3} + \frac{A_1}{2G}\right)^2 - 8G\left[1 - \left(\frac{A_1}{2G}\right)^2\right]\right]^{1/2}. \qquad (26)$$

The existence of the equilibrium point $(\alpha, \omega) = (\alpha_3, 0)$ implies that $-1 < (A_1/2G) < 1$, the equal signs being omitted as indicated earlier, so that the term $[1 - (A_1/2G)^2]$ is always positive and only two cases of practical interest result:

(a) If $G < -A_1/2$, then one of the $\lambda_{1,2}$ has a positive real part and the equilibrium point $(\alpha, \omega) = (\alpha_3, 0)$ is unstable.

(b) If $G > A_1/2$, then both of the $\lambda_{1,2}$ have negative real parts and the equilibrium point $(\alpha, \omega) = (\alpha_3, 0)$ is asymptotically stable.

Remarks on Dynamic Damping Effects

The above analysis and conclusions rest heavily on the existence and magnitude of the dynamic damping term, C_{m_q}, in the aerodynamic torque as defined by Eq. 6. In practical cases this term is quite small so that the satellite motion considered here is only lightly damped in those situations where asymptotic stability has been previously indicated. Indeed, numerical computations show that the time required to damp a given oscillation to half of its initial magnitude is, at best, of the same order of magnitude as the time required for the orbit to decay to the point where re-entry occurs.

In any case, it can be concluded that the system is stable and, in fact, lightly damped for those cases where asymptotic stability is indicated.

Stability and Instability Regions in Coefficient Space

The results concerning the stability of the equilibrium points α_1, α_2 and α_3, expressed in terms of inequalities in the preceding, can be summarized and clarified by geometrical means, namely, by giving the regions of stability and instability for the equilibrium positions α_1, α_2 and α_3 in a coefficient space.

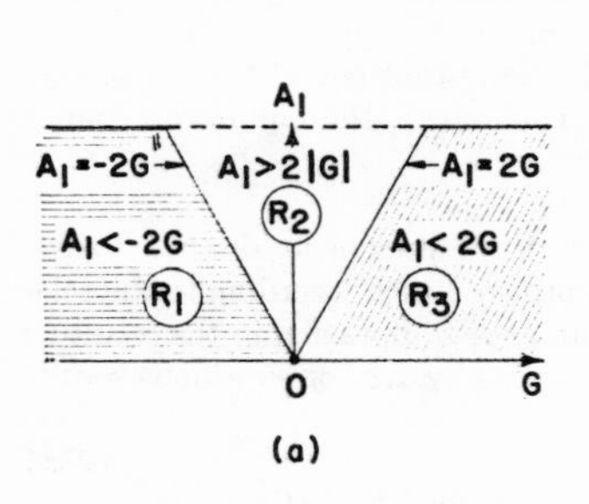

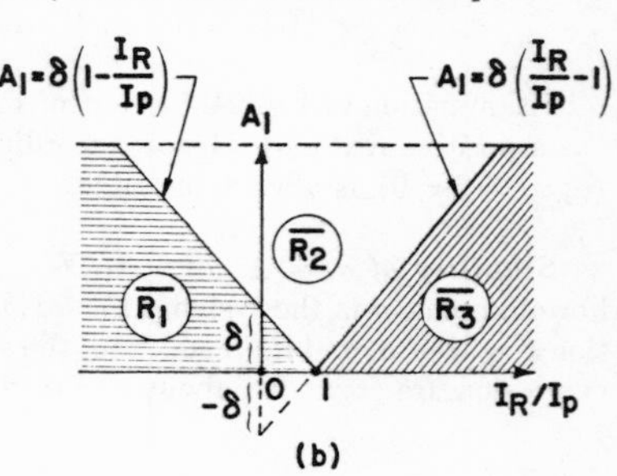

FIG. 3.

Figure 3(a) shows the regions of asymptotic stability and instability as well as the non-permissible regions for the equilibrium positions $\alpha_1 = 0$, $\alpha_2 = \pi$ and $\alpha_3 = \pm \cos^{-1}(A_1/2G)$ in the G, A_1-plane. Table I summarizes the results in terms of the regions shown in Fig. 3(a). The regions R_1, R_2, R_3 of Fig. 3(a) are defined in the parameter space (G, A_1), where G and A_1 are given by Eqs 5

TABLE I

	Regions		
Equilibrium points	R_1	R_2	R_3
$\alpha_1 = 0$	Asymptotically stable	Asymptotically stable	Unstable
$\alpha_2 = \pi$	Unstable	Unstable	Unstable
$\alpha j = \pm \cos^{-1}(A_1/2G)$	Unstable	Non-permissible	Asymptotically stable

and 10a, respectively. By using these formulae, the above regions can be mapped into the regions $\bar{R}_1$, $\bar{R}_2$, $\bar{R}_3$ in the parameter space $[(I_R/I_P), A_1]$, as shown in Fig. 3(b). The lines $A_1 = \pm 2G$ in the (G, A_1)-plane can be mapped into the lines $A_1 = \pm\delta[(I_R/I_P) - 1]$ of the $[(I_R/I_P), A_1]$-plane, where $\delta = 3K/r^3$. The quantity I_P can be taken either as unity or as a magnification factor. Table II summarizes the results in terms of the regions shown in Fig. 3(b).

TABLE II

	Regions		
Equilibrium points	$\bar{R}_1$	$\bar{R}_2$	$\bar{R}_3$
$\alpha_1 = 0$	Unstable	Asymptotically stable	Asymptotically stable
$\alpha_2 = \pi$	Unstable	Unstable	Unstable
$\alpha_3 = \pm \cos^{-1}\{A_1/\delta[1 - (I_R/I_P)]\}$	Asymptotically stable	Non-permissible	Unstable

The above Figures and Tables show that there always exists an asymptotically stable situation, *viz.*, $\alpha_1 = 0$ in the regions R_1 and R_2 (or $\bar{R}_2$ and $\bar{R}_3$), and $\alpha_3 = \pm \cos^{-1}(A_1/2G)$ in region R_3 (or $\bar{R}_1$). Furthermore, when α_1 is asymptotically stable all other equilibrium situations are unstable, and when α_3 is asymptotically stable, the remaining equilibrium situations are unstable. The equilibrium situation $\alpha_2 = \pi$ is always unstable.

Appendix: Definitions and Theorems Relating to Stability

Stability Definitions (5): Classes of stability concepts can be distinguished according to the nature of the dynamical systems, the manner in which the system approaches a given state or deviates from it, the properties of the perturbations of the system, and the system of coordinates selected.

Physically, there exist three basic stability concepts depending upon considerations of stability of the motion in a given orbit, of the orbit of a given motion, and of the boundedness of the motion and its orbit.

An analytical description of the stability concepts can be achieved by using the Liapunov, Poincaré and Lagrange distances ρ, shown in Fig. 4. These concepts interpret the correspondence between a point P of the unperturbed orbit L and a point $\bar{P}$ of the perturbed one $\bar{L}$ in different ways; the distance ρ measures the effect of the perturbations at P. The three stability concepts in the sense of Liapunov, Poincaré and Lagrange are of basic type and all others are regarded as special cases of them.

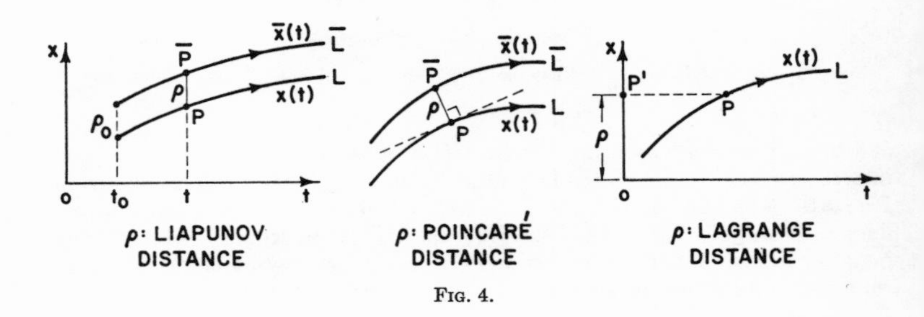

FIG. 4.

All stability concepts of dynamical systems are included in the same stability conditions:

$$\rho_0 < \delta \tag{27a}$$

$$\rho < \epsilon \tag{27b}$$

$$\lim_{t\to\infty} \rho = 0 \tag{27c}$$

$$|| p || < \eta \tag{27d}$$

by appropriately interpreting the distances ρ_0, the initial value of ρ, and ρ. These conditions express a "unification" of the stability concepts.

In case the perturbations affect the initial conditions only, the first two conditions (27) guarantee the stability of the motion $x(t)$, and the first three conditions the asymptotic stability of the motion. For the case where the whole system is perturbed by persistent perturbations $p(x, t)$, the definition of stability requires a restriction on the perturbations p, say, a restriction of its norm $|| p ||$, so that the fourth condition, (27), is needed for the stability definition under persistent perturbations.

The above remarks and unification of the stability concepts lead to important results concerning stability, of which we mention just two needed in the present paper:

(a) For the equilibrium states of a system, the stability concepts in the Liapunov and Poincaré sense are equivalent. For periodic motions, the Liapunov stability concept is narrow compared to the Poincaré stability concept. The isochronism characterizes the use of the Liapunov stability concept in periodic motions.

(b) The Lagrange stability concept, applied either to equilibrium states of to periodic motions in finite distance, always classifies these special motions of a system as stable, so that it is useless.

Stability Theorems (5, 6): Consider the nonlinear autonomous system

$$\dot{x}_i = \sum_{i=1}^{n} a_{ij}x_j + X_i(x_1, x_2, \cdots, x_n) \quad \text{with} \quad i = 1, 2, \cdots, n \tag{28}$$

where the a_{ij} are constants and the $X_i(x_1, x_2, \cdots, x_n)$ are power series beginning with terms of at least the second degree.

Theorem I: If the nonlinear terms, $X_i(x_1, x_2, \cdots, x_n)$, in Eqs. 28 are such that

$$\lim_{x_1\to 0 \ldots x_n \to 0} \left[\sum_{j=1}^{n} | X_j(x_1, x_2, \cdots x_n) | / \sum_{j=1}^{n} | x_j |\right] = 0 \tag{29}$$

then the nature of the singularity at the point of equilibrium $x_i = 0$ $(i = 1, \cdots, n)$ may be determined by considering only the linear terms in Eqs. 28.

If the system (28) is such that the condition (29) is satisfied, then the stability of the equilibrium point $x_i = 0$ $(i = 1, \cdots, n)$ is specified by the roots λ_i $(i = 1, \cdots, n)$ of the characteristic equation for the linear terms in Eq. 28 as follows:

(a) If all of the distinct roots have negative real parts, then the equilibrium point is asymptotically stable.

(b) If at least one of the non-zero roots has a positive real part then the equilibrium point is unstable.

(c) If none of the roots have positive real parts but some roots have zero real parts, then a so-called critical case exists and the question of stability cannot be resolved on the basis of the characteristic equation roots. Other means must be employed to study this case.

In general, if the nonlinearities in the system (28) are such that they must be considered in studying the stability characteristics of the system, then the following theorem is applicable.

Theorem II: Given the system (28) with the equilibrium point $x_i = 0$ $(i = 1, \cdots, n)$, the equilibrium is

(a) asymptotically *stable*, if it is possible to determine a positive (negative) definite function $V(x_1, x_2, \cdots x_n)$, defined on a domain D including the origin, whose time derivative $\dot{V}$ is negative (positive) definite; and

(b) *unstable*, if it is possible to determine a positive (negative) definite function $V(x_1, x_2, \cdots, x_n)$ whose time derivative $\dot{V}$ is also positive (negative) definite.

References

(1) H. Goldstein, "Classical Mechanics," Reading, Mass., Addison-Wesley Pub. Co., Chap. 5, 1959.

(2) D. M. Schrello, "Passive Aerodynamic Attitude Stabilization of Near-Earth Satellites, Vol. I, Librations due to Combined Aerodynamic and Gravitational Torques," WADD Tech. Rep. 61–133, Wright Air Dev. Div., Wright-Patterson A.F.B., Ohio, July 1961, *Unclassified.*

(3) P. H. Davison, "Passive Aerodynamic Attitude Stabilization of Near-Earth Satellites, Vol. II, Aerodynamic Analysis," WADD Tech. Rep. 61–133, Wright Air Dev. Div., Wright-Patterson A.F.B., Ohio, July 1961, *Unclassified.*

(4) W. Leighton, "On the Construction of Liapunov Functions for Certain Autonomous Nonlinear Differential Equations," Contributions to Differential Equations, New York, Interscience Pub. Co., p. 368, 1963.

(5) D. G. Magiros, "On Stability Definitions of Dynamical Systems," *Proc. Nat. Acad. Sci.*, Vol. 53, No. 6, pp. 1288–1294, June 1965.

(6) N. Minorsky, "Nonlinear Oscillations," Princeton, N. J., D. Van Nostrand Co. Inc., 1962

(7) Nam Tum Po, "On the Rotational Motion of a Spherical Satellite Under the Action of Retarding Aerodynamical Moments," Leningrad Univ., Math., Mech. and Astronomical Ser. No. 19, Vol. 4, pp. 129–134, 1964.

ΕΦΗΡΜΟΣΜΕΝΑ ΜΑΘΗΜΑΤΙΚΑ.— **Stability concepts of solutions of differential equations with deviating arguments,** *by Demetrios G. Magiros* *. Ἀνεκοινώθη ὑπὸ τοῦ Ἀκαδημαϊκοῦ κ. Ἰ. Ξανθάκη.

1. Introduction

The majority of physical and social systems can be expressed by relations between quantities under investigation and their rate of change, that is by differential equations.

The duration of the transmission of the action or signal can not in many cases be neglected, and the processes are governed by «differential equations with deviating arguments», as, e. g., by equations with «retarded arguments», which characterize situations with «delay periods» or «aftereffects». Modern needs of automatic regluation, or probability, of biology, of medicine, and of certain other fields, lead to such equations.

The study of the stability properties of the solutions of these equations is a fundamental problem, and this problem can be treated if, and only if, the «concepts of stability» are clarified and selected appropriately.

In this note we formulate stability concepts of the class of differential equations with deviating arguments, and give somme remarks concerning these stability concepts. The stability concepts depend especially upon the nature of the perturbations, the way the perturbations act on the system, their magnitude, the magnitude of their effect, etc. Definition of stability are given in case of sudden and permanent perturbations, and in case of perturbations of the deviating arguments. By the remarks, we clarify questions on the stability concepts.

2. Definitions

A system of differential equations with retarded arguments, that is variables containing «delay periods», can be given by:

---

* ΔΗΜΗΤΡΙΟΥ ΜΑΓΕΙΡΟΥ, **Αἱ ἔννοιαι τῆς εὐσταθείας τῶν λύσεων διαφορικῶν ἐξισώσεων μὲ ἀποκλινούσας μεταβλητάς.**

Consulting Scientist, General Electric Co., RESD Philadelphia, Pa., U.S.A.:

Reprinted from the *Proceedings of the Athens Academy of Sciences* **46** (1971), 273–278.

$$\dot{x}_i(t) = f_i[t, x_j(t - \tau_{jk}(t))]$$
$$i, j = 1, 2, \ldots, n; \quad k = 1, 2, \ldots, m, \quad m \leqslant n \tag{1}$$

where the retardations τ_{jk} are positive.

By «solution» of (1) we mean continuous functions $x_i(t)$ which satisfy (1) on $t \geqslant t_0$, and which, on the initial interval: $E_{t_0}: t_0 - \tau_{jk}(t) \leqslant t \leqslant t_0$ become $x_i(t) = \varphi_i(t)$, and $\varphi_i(t)$ are given continuous functions, called «initial functions» of (1).

By «stationary point» of (1) we mean a constant solution x_{i_0} of (1) on $t \geqslant t_0$, which is also constant on the initial set E_{t_0}.

The solution of (1) depends on the given arbitrary functions $\varphi_i(t)$. $\varphi_i(t)$ are extensions of the solutions in the interval E_{t_0}, and the solutions are uniquely determined by the initial functions and appropriate properties of f_i.

3. Stability in case of sudden perturbations

«Sudden perturbations» are perturbations «momentarily» applied to the initial functions $\varphi_i(t)$ attached to (1). Let x_φ be the solution of (1) with orbit L corresponding to the initial functions $\varphi_i(t)$ and x_ψ be the solution of (1), after the perturbation, with orbit $\bar{L}$ corresponding to the initial functions $\psi_i(t)$, Fig. (a).

If the points $\bar{P}_{01}$, $\bar{P}_0$, $\bar{P}$ of the perturbed curve $\bar{L}$ correspond to the points P_{01}, P_0, P of the unperturbed curve L, the following distances can be defined, corresponding to time indicated:

$$\begin{aligned} \varrho_{01} &= P_{01}\bar{P}_{01} = |\psi_i(t) - \varphi_i(t) && \text{on} \quad E_{t_0}: t_0 - \tau \leqslant t \leqslant t_0 \\ \varrho_0 &= P_0\bar{P}_0 = |\psi_i(t_0) - \varphi_i(t_0) && \text{at} \quad t = t_0 \\ \varrho &= P\bar{P} = |x_\psi(t) - x_\varphi(t) && \text{on} \quad t \geqslant t_0 \end{aligned} \tag{2}$$

These distances give the effect of the perturbation at the points P_{01}, P_0, P of L, Fig. (a).

The solution x_φ of (1) is «stable» if, given $\varepsilon > 0$ and $t_0 \geqslant 0$, there exists $\delta = \delta(t_0, \varepsilon) > 0$ such that the inequality $\varrho_{01} < \delta$ implies $\varrho < \varepsilon$.

This solutions is «asymptotically stable», if it is stable and, in addition, $\lim_{t \to \infty} \varrho = 0$.

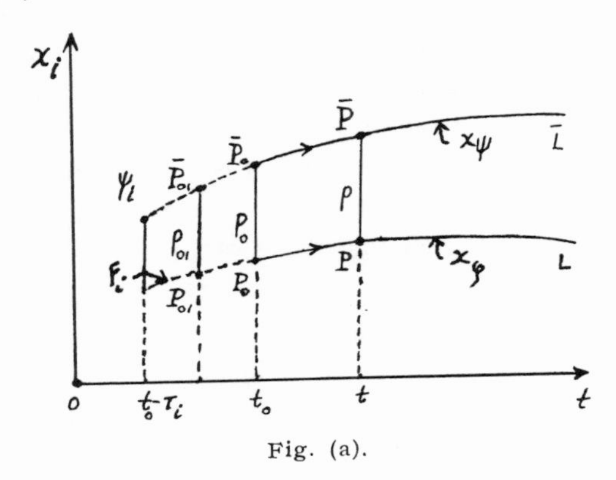

Fig. (a).

These definitions can be accepted for other types of differential equations with deviating arguments, say of neutral types.

4. Stability in case of persistent perturbations

If the system (1) is continuously perturbed by the perturbations R_i, the perturbed system, corresponding to (1), is :

$$\dot{x}_i(t) = f_i\big(t, x_j(t - \tau_{jk}(t))\big) + R_i\big(t, x_j(t - \tau_{jk}(t))\big) \tag{3}$$

R_i are sufficiently small in absolute value. Given $\varepsilon > 0$ and $t_0 \geqslant 0$, if there exist two positive quantities $\delta_1 = \delta_1(t_0, \varepsilon)$, $\delta_2 = \delta_2(t_0, \varepsilon)$ such that: $\varrho_{01} < \delta_1$ valid on E_{t_0}, and $\| R_i \| < \delta_2$ valid on $t \geqslant t_0$, imply $\varrho < \varepsilon$ valid on $t \geqslant t_0$, we say that the solution x_φ of (1) is «stable» with respect to persistent perturbations R_i.

We remark that many basic theorems on stability can be carried out without essential alteration to the case of differential equations with deviating argument, but, up to now, the stability theory for these equations is essential for stationary equations of the first approximation in noncritical cases.

5. Stability in case of perturbations of deviating arguments

In processes with after effects, described by differential equations with deviating arguments, the deviations themselves can not be prescribed exactly, that is deviations themselves may have small disturbances, when the question of stability of the equations with respect to small perturbations of the deviating arguments arises. The important stability problem arises when the perturbations of the deviating arguments have a continuing character, as, e.g., when in some processes with after effects the retardation or delay period is not precisely defined.

Instead of the system (1), we now take the system:

$$\dot{x}_i(t) = f_i\left(t,\ x_j(t - \bar{\tau}_{jk}(t))\right) \tag{4}$$

where $\bar{\tau}_{jk}(t)$ are the perturbed deviations. The initial interval is: $E_{\bar{t}_0} : \bar{t}_0 - \bar{\tau}_{jk}(t) \leqslant t \leqslant \bar{t}_0,\ \ t_0 \leqslant \bar{t}_0$.

Now, the solution of (1), x_φ, defined by $\varphi_i(t)$ on the set E_{t_0}, it said to be «stable with respect to perturbations of deviating arguments», if, for $\varepsilon > 0$, $t_0 \geqslant 0$, we can find $\delta_1 > 0$, $\delta_2 > 0$ such that, if $\varrho_{01} < \delta_1$, for t on E_{t_0}, and if $/\bar{\tau}_{jk}(t) - \tau_{jk}(t)/ < \delta_2$ for t on $t \geqslant t_0$, then $\varrho < \varepsilon$ for t on $t \geqslant t_0$.

In case τ_{jk} and $\bar{\tau}_{jk}$ are constant, we have «stability with respect to a continuously perturbed deviating argument».

In case the difference $(\bar{\tau}_{jk} - \tau_{jk})$ is either positive or negative, we have a «one sided perturbation of a deviating argument» (**1**).

REMARKS (2)

The following remarks may give to the reader an opportunity to think more deeply about the difficulties of the subject.

1. The above stability definitions are in the sense of Liapunov or Poincaré, if ϱ_{01}, ϱ_0, ϱ are interpreted as Liapunov or Poincaré distances.

The stability definitions in the sense of Liapunov or Poincaré are equivalent in case of equilibrium solutions, but for any other kind of solutions the stability situation may be different, if we employ different stability concepts, and naturally appears the important subject of the selection of the appropriate stability definition for the stability problem at hand.

2. The above classes of stability concepts can give any subclass of stability concepts by appropriate restrictions of the distances, the time, and other quantities. So, e.g., one can speak about «eventually uniform stability», if δ_1 is independent of t_0, that is $\delta_1 = \delta_1(\varepsilon)$, and t_0 has a minimum $\alpha(\varepsilon)$, that is $\alpha(\varepsilon) \leqslant t_0 \leqslant t$. Nonexistence of $\alpha(\varepsilon)$, that is $\alpha(\varepsilon) \equiv 0$, corresponds to «uniform stability».

3. In case of persistent perturbations, the selection of the kind of the norm of the perturbations, specifies the stability concept. So, one may have «total stability», or «integral stability» or «stability in the mean» under suitable norm of the perturbations (3).

4. The stability, as defined above, is a property of the solution different from its boundedness property, although in some cases there may exist regions where these two properties are equivalent and one implies the other.

5. All the above stability concepts are of mathematical type and the results, theorems or criteria, based on them, may not interpret the reality. In order that these results have a practical usefulness, which must be the ultimate purpose of the investigations, the investigation must be accompanied by some additional requirements, as, e.g., to know the region of the practically permitted deviations of the solutions, that is the «ε-region», the corresponding region of the initial conditions, that is the «δ_1-region», and the «δ_2-region» of the norm of the perturbation R_i.

REFERENCES

1. L. È. Èl'sqol'c: «Qualitative Methods of Mathematical Analysis» American Mathematical Society, Providence, R. I. (1964), pg. 167 - 199.
2. D. G. Magiros: «Stability Concepts of Dynamical Systems», Journal of Information and Control, Vol. 9, No. 9 (Oct. 1966), pg. 531 - 548.
3. I. Vrkov: «On Some Stability Problems», Academic Press, Proc. Conf. Prague, (Sept. 1962), pg. 217 - 221.

★

Ὁ Ἀκαδημαϊκὸς κ. **Ἰω. Ξανθάκης** κατὰ τὴν ἀνακοίνωσιν τῆς ἀνωτέρω ἐργασίας εἶπε τὰ κάτωθι:

«Ἡ ἀνακοίνωσις τοῦ κ. Μαγείρου ἀναφέρεται εἰς τὰς ἐννοίας τῆς εὐσταθείας τῶν λύσεων διαφορικῶν ἐξισώσεων μὲ ἀποκλινούσας μεταβλητάς.

Μία μεγάλη κατηγορία φυσικῶν καὶ κοινωνικῶν φαινομένων ἐκφράζεται μαθηματικῶς διὰ διαφορικῶν ἐξισώσεων μή - γραμμικῶν μὲ ἀποκλινούσας μεταβλητάς. Ἀφ᾽ ἑτέρου, ὡρισμένα προβλήματα αὐτομάτου ἐλέγχου, προβλήματα πιθανοτήτων, ἡ ἀνάπτυξις εἰς τὸν τομέα τῆς βιολογίας καὶ ἰατρικῆς ὁδηγοῦν εἰς ἐξισώσεις τοῦ ἐν λόγῳ τύπου.

Τὸ ζήτημα τῆς εὐσταθείας τῶν λύσεων τῶν ἐξισώσεων αὐτῶν εἶναι θεμελιῶδες. Ἡ σπουδὴ δὲ τῆς εὐσταθείας βασίζεται ἐπὶ διαφόρων ἀντιλήψεων καὶ ὑποθέσεων περὶ εὐσταθείας, αἱ ὁποῖαι ἐξαρτῶνται ἐκ τοῦ τρόπου τῆς δράσεως τῶν «διαταράξεων» καὶ ἐκ τοῦ εἴδους μετρήσεως τοῦ μεγέθους τῶν διαταράξεων.

Εἰς τὴν παροῦσαν ἀνακοίνωσιν ἐκτίθενται αἱ ἔννοιαι εὐσταθείας τῶν λύσεων τῶν ἐν λόγῳ ἐξισώσεων καὶ διατυποῦνται παρατηρήσεις τινὲς ἐπ᾽ αὐτῶν. Διατυποῦνται ἐπίσης οἱ ὁρισμοὶ ὅταν αἱ διαταράξεις εἶναι «αἰφνίδιαι» ἢ «συνεχῶς δρῶσαι», καθὼς καὶ ὅταν, αἱ ἴδιαι αἱ ἀποκλίσεις τῶν μεταβλητῶν ὑπόκεινται εἰς διαταράξεις.

Ἡ εἰσαγωγὴ τῆς ἐννοίας τῆς εὐσταθείας κατὰ Liapounov καὶ Poincaré παρέχει δύο διακεκριμένας γενικὰς κατηγορίας τῶν ἀντιλήψεων εὐσταθείας».

408

ΕΦΗΡΜΟΣΜΕΝΑ ΜΑΘΗΜΑΤΙΚΑ.— **Remarks on stability concepts of solutions of dynamical systems,** *by Demetrios G. Magiros**.
'Ανεκοινώθη ὑπὸ τοῦ 'Ακαδημαϊκοῦ κ. 'Ιω. Ξανθάκη.

INTRODUCTION

The study of the stability situation of physical and social phenomena, which are modeled as dynamical systems, is based on a variety of stability concepts, and this variety makes the study complicated and the stability results questionable in many cases.

In this paper, we will give a set of remarks on stability concepts, which may permit a better understanding of the difficulties of current stability problems.

The stability concepts may come from sources of different nature.

Examining the stability of a motion in its orbit and of the orbit of a motion, one can distinguish two basic stability concepts, which contain many other specialized concepts as special cases.

The manner in which a state of a system approaches another state, or deviates from it, the way the perturbations act on a system, or the way one measures their norm and their effect on the system, the type of the mathematical model of the system, etc., are sources for stability concepts of different nature.

All these different stability concepts can be «unified» into the same «stability relationships», and this «unification» of the stability concepts brings a natural simplicity in the understanding of subjects concerning stability, and gives rise to new results.

1. REMARKS ON PERTURBATIONS

It is necessary to make remarks concerning the purturbations of the systems and their solutions. A variety of stability concepts comes from the manner the purturbations act on the system, the way the norm

---

* **ΔΗΜΗΤΡΙΟΥ Γ. ΜΑΓΕΙΡΟΥ, Παρατηρήσεις ἐπὶ τῶν ἀντιλήψεων εὐσταθείας εἰς δυναμικὰ συστήματα.** Scientific Consultant, General Electric Co., RESD, Philadelphia, Pa. U. S. A.

Reprinted from the *Proceedings of the Athens Academy of Sciences* **49** (1974), 408–416.

of perturbations is taken, and the way their effect on the system is measured.

The perturbations, which can be considered as minor disturbing forces, may act on a system either «momentarily», when only the initial conditions are perturbed, or «permanently», when the system itself is disturbed and the perturbations, during their action, must enter the equations of the system explicitly. The «sudden» and «persistent» perturbations characterize two different classes of stability concepts.

The «norm» of the perturbations can be taken in different ways, and each of them characterizes a special stability concept under persistent perturbations. The perturbations may depend on deviating arguments, when the stability concepts will be related to the deviations of the arguments.

The «effect» of the perturbations is a change of certain quantities pertaining to the original motion and/or to its orbit, and this effect can be visualized by the change of the orbit S of the unperturbed motion $x_i(t)$ into the orbit $\overline{S}$ of the perturbed motion $\overline{x}_i(t)$.

Given a state of a dynamical system, that is a point P on S, if, as a result of the perturbations, $\overline{S}$ is the new orbit, and the point $\overline{P}$ on $\overline{S}$ corresponds to the point P of S, the distance $\varrho = P\overline{P}$ can be taken as the magnitude of the effect of the perturbations at P, Fig. 1.

To a given point P of S, one may make to correspond different points $\overline{P}$ on $\overline{S}$, and each correspondence characterizes a specific stability concept of the motion. Any such correspondence presuposes an assumption, and each assumption comes from a physical reason.

One can distinguish two, the most physical, correspondences between P and $\overline{P}$, when two important stability concepts result, namely, the stability concept in the sence of Liapunov and that in the sense of Poincaré, by using the stability distances $\varrho = P\overline{P}$, called Liapunov and Poincaré distances. In Fig. 2 «Liapunov distances» are shown P and $\overline{P}$ correspond to each other at the «same time». In Fig. 3, P and $\overline{P}$ correspond in such a way that :

$$\varrho = \varrho(P, \overline{S}) = P\overline{P} = \min \left\{ \sum_{i=1}^{n} |\overline{x}_i - x_i|^2 \right\}^{1/2}$$

and $\varrho = P\overline{P}$ is «Poincaré distance».

If the model of the system is expressed by differential equations with deviating arguments, e.g. by differential equations with retarded arguments of retardation $\tau_i(t)$, the «initial functions» $\varphi_i(t)$ of S and $\bar{\varphi}_i(t)$ of $\bar{S}$ define the «initial distance» $\varrho_{01} = P_{01}\bar{P}_{01} = |\bar{\varphi}_i(t) - \varphi_i(t)|$ taken over the interval $t_0 - \tau_i \leqslant t \leqslant t_0$, Fig. 2.

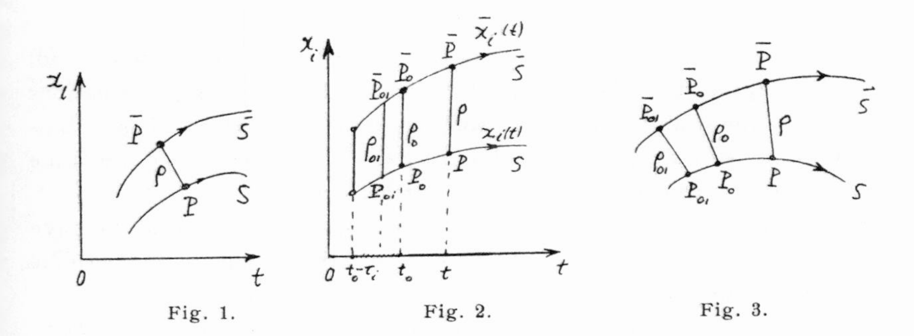

Fig. 1. Fig. 2. Fig. 3.

The above includes the ordinary differential equations, where $\tau_i = 0$, $\varrho_{01} = \varrho_0 = P_0\bar{P}_0$ at $t = t_0$, Fig. 2.

In case the retardations $\tau_i(t)$ are perturbed, one has a new stability concept characterized by the distance: $\varrho_\tau = |\bar{\tau}_i(t) - \tau_i(t)|$, where $\bar{\tau}_i$ the perturbed regardations (1).

2. REMARKS ON THE RELATIONSHIPS OF THE STABILITY CONCEPTS

Following considerations on «physical stability» and using the precending remarks and notations, any stability concept can be described quantitatively by the «stability conditions»:

$$(a): \varrho_{01} < \delta_1, \quad (b): \varrho < \varepsilon, \quad (c): \lim_{t \to \infty} \varrho = 0, \quad (d): \|p_i\| < \delta_2, \quad (e): \varrho_\tau < \delta_3 \quad (1)$$

which, then, «unify» all the stability concepts. ε, δ_1, δ_2, δ_3 are positive constants, p_i the perturbations, and, in general, δ_1 and δ_2 depend on t_0 and ε.

By a suitable combination of these relationships, by an appropriate interpretation of the distances involved, and by some restrictions of some quantities of these relationships, one can express any stability concept.

The following remarks may help to clarify the above statements.

The inequality 1(a) is valid for t in the initial inverval $E_{t_0}: t_0 - \tau_i \leqslant t \leqslant t_0$, while the inequalities 1(b), (d), (e) for t in $t \geqslant t_0$, $t_0 \geqslant 0$.

In case of «sudden perturbations», the inequality 1(d) is meaningless, and in this case 1(a), (b) express the «stability», while 1(a), (b), (c) the «asymptotic stability».

In case of «permanent perturbations», when 1(d) is meaningful, the selection of the kind of the norm of perturbations specifies the stability concept, so, one may have «total stability» or «integral stability», or «stability in the mean», under suitable norm of perturbations.

By restricting δ_1 and δ_2 to depend only on ε and not on t_0, we have the «uniform stabilities», as this happens in periodic systems, or in autonomous systems.

Restriction on t_0 to have a minimum, $\min t_0 = \alpha$, implies «eventual stabilities».

If in (1) the distances ϱ are interpreted as Liapunov or Poincaré distances, we have stabilities in the sence of Liapunov or Poincaré.

In case of equilibrium points of a system, when the orbit S shrinks to a point, the distinction between stabilities in the sence of Liapunov and Poincaré is meaningless.

The inequality 1(e) has a meaning in case of perturbed retardations.

In case of periodic motions, when S is a closed curve, the stability concept in Liapunov sense is a narrow concept compared to the stability concept in Poincaré sence. The «isochronism», that is the «constancy of the frequency», characterizes the Liapunov stability, while this notion does not enter in the Poincaré stability. Further we may have that :

. A motion stable in Liapunov sense is also stable in Poincaré sense ;

. A motion unstable in Poincaré sense is also unstable in Liapunov sense ;

. A motion unstable in Liapunov sense may be stable or unstable in Poincaré sense, and

. A motion stable in Poincaré sense may be stable or unstable in Liapunov sense.

The stability is a property of the solution different from its boundedness property, although in some cases there may exist regions where these properties are equivalent and one implies the other. The boundedness of the solution is characterized by the boundedness of $OP = \varrho_1 = |x_i|$, where O the origin of the coordinate system and P point of the orbit; but the stability is characterized by the boundedness of $\varrho = |\bar{x}_i - x_i|$, and it is possible for the orbits x_i and $\bar{x}_i$ to be unbounded as $t \to \infty$, when the stability distance ϱ gets the form $(\infty - \infty)$, when ϱ will be either infinite, or constant, or zero, and the unbounded solution x_i will be either unstable or stable.

All the stability concepts included in the relationships (1) are of mathematical type, and the results, theorems or criteria, based on them, may not interpret the reality. Also, one and the same phenomenon may be, mathematically speaking, stable or unstable depending on the stability concept employed in the discussion of the stability of the phenomenon when the selection of the stability concept, appropriate for the phenomenon, arises.

The mathematical stability concepts and the stability criteria based on them, represent a possible functionig of the physical system, and in order all these to have a practical usefulness, and to agree with «practical stability», which must be the ultimate purpuse of the stability investigations, appropriate modifications, changes, supplements of the mathematical stability concepts must accompany the investigations.

The notion of «practical stability» is a subject not yet completely studied, but in many cases is characterized by the knowledge of : (2)

. The size of deviation of the state acceptable for a satisfactory operation of the system ;

. The size of permitted initial conditions that can be controlled ;

. The size of permitted perturbations ;

. The finite time T for the stability investigation.

The nolinearities of the system play a decisive role for practical stability.

3. AN EXAMPLE

We terminate the discussion on stability remarks by using as an example the investigation of the «stability of some precessional phenomena», by which some of the above statements may be clarified.

The rotational motion of a rigid body around its axis of symmetry is governed by the Euler's (ordinary differential) equations, of which the stability of the solutions has been examined under sudden perturbations, then by employing the first three stability relationships (1).

If $\underline{L} = (L_1, L_2, L_3)$ is the external torque vector acting on the body, $\underline{\omega} = (\omega_1, \omega_2, \omega_3)$ the angular velocity vector, which characterizes the precession of the body, $\underline{\omega_0} = (\omega_{01}, \omega_{02}, \omega_{03})$ the initial angular velocity vector, and I_1, $I_2 = I_3 = I$ the moments of inertia, the precessional motion of the body in the following two cases are given by:

$$\left.\begin{array}{l} \text{(a)}\ \omega_1 = \omega_{10} = \text{constant}, \quad \omega_2 = A\cos Q_1, \quad \omega_3 = A\sin Q_1, \quad Q_1 = \dfrac{(I_1 - I)\,\omega_{10}}{I}t \\ \quad \text{(in case,}\ L_1 = L_2 = L_3 = 0; \quad I_1,\ I,\ A = (\omega_{02}^2 + \omega_{03}^2)^{1/2}\ \text{constants)} \\ \text{(b)}\ \omega_1 = \dfrac{L_1}{I_1}t, \quad \omega_2 = A\cos Q_2, \quad \omega_3 = A\sin Q_2, \quad Q_2 = Q_1 + \dfrac{(I_1 - I)\,L_1}{2\,I_1\,I}t^2 \\ \quad \text{(in case:}\ L_1 = \text{constant}, \quad L_2 = L_3 = 0) \end{array}\right\} \quad (2)$$

The «regular precession» 2(a) is bounded, while the «helicoid precession» 2(b) is unbounded as $t \to \infty$. The results for their stability situation are the following:

(i) The regular precession 2(a) is «stable» but «not asymptotically stable» in Poincaré sense (orbitally). In Liapunov sense it is «stable» but «not asymptotically stable», if ω_{01} is not affected by the perturbation; and it is «unstable», if ω_{01} is affected by the perturbations.

(ii) The helicoid precession 2(b) is «asymptotically stable» in Poincaré sense, but it is «unstable» in Liapunov sense.

(iii) The stability situation of the above example in Poincaré sense is preferred, because in this case it is proved that the requirements for «practical stability» are satisfied.

We remark that the stability of a system, by using any of the previous concepts, depends, in general, on the selection of the major variables of the system, and on the transformation of the variables. This will be subject of a next paper.

ΠΕΡΙΛΗΨΙΣ

1. Εἰς τὴν παροῦσαν ἐργασίαν δίδονται παρατηρήσεις ἐπὶ τῶν ἀντιλήψεων περὶ εὐσταθείας εἰς φυσικὰ καὶ κοινωνικὰ φαινόμενα, τὰ ὁποῖα μαθηματικοποιοῦνται ὡς δυναμικὰ προβλήματα. Ἡ ποικιλία τῶν ἀντιλήψεων εὐσταθείας κάμνει τὰ προβλήματα εὐσταθείας πολὺ πεπλεγμένα καὶ τ' ἀποτελέσματα τῆς ἐρεύνης βάσει αὐτῶν ὄχι δεκτὰ ἐνίοτε.

2. Ἀναφέρομεν μερικὰς πηγάς, ἀπὸ τὰς ὁποίας δυνάμεθα νὰ ἔχωμεν ποικιλίαν διαφόρου φύσεως ἀντιλήψεων περὶ εὐσταθείας:

—Ὁ τρόπος μὲ τὸν ὁποῖον μία κατάστασις ἑνὸς συστήματος πλησιάζει μίαν ἄλλην κατάστασιν ἢ ἀπομακρύνεται ἀπὸ αὐτήν.

—Ὁ τρόπος μὲ τὸν ὁποῖον προκαλοῦνται καὶ δροῦν αἱ διαταραχαὶ ἑνὸς συστήματος, ὁ τρόπος μὲ τὸν ὁποῖον μετροῦμεν τὴν ἔντασιν τῶν διαταραχῶν, καθὼς καὶ τὴν τῶν ἀποτελεσμάτων των ἐπὶ τοῦ συστήματος.

—Ἡ δομὴ τοῦ μαθηματικοῦ μοντέλου τοῦ συστήματος κλπ.

3. Ὅλαι αἱ ἀντιλήψεις εὐσταθείας, ἂν καὶ διαφόρου φύσεως, δύνανται ν' ἀναχθοῦν εἰς τὰς αὐτὰς μαθηματικὰς «σχέσεις εὐσταθείας», αἱ ὁποῖαι δίδουν μίαν «ἑνοποίησιν» τῶν ἀντιλήψεων εὐσταθείας, αὐτὴ δὲ ἡ ἑνοποίησις ὑποβοηθεῖ τὴν κατανόησιν τῶν ἀντιστοίχων προβλημάτων, δύναται δὲ νὰ ὁδηγήσῃ εἰς νέα συμπεράσματα. Μὲ κατάλληλον συνδυασμὸν τῶν «σχέσεων εὐσταθείας», καὶ κατάλληλον ἑρμηνείαν ἢ περιορισμὸν τῶν ποσοτήτων τῶν σχέσεων αὐτῶν, δύναται νὰ προκύψῃ ὁποιαδήποτε ἀντίληψις εὐσταθείας.

4. Αἱ ἀντιλήψεις εὐσταθείας, ποὺ περικλείονται εἰς τὰς «σχέσεις εὐσταθείας» ὑποδεικνύουν ἐνδεχομένην λειτουργίαν τοῦ συστήματος, τὰ δὲ συμπεράσματα, θεωρήματα ἢ κριτήρια, ποὺ βασίζονται ἐπ' αὐτῶν, ἐνδέχεται νὰ μὴ ἑρμηνεύουν τὴν πραγματικότητα κατὰ ἱκανοποιητικὸν τρόπον, ἢ ἐνδέχεται νὰ ἔχωμεν διὰ τὸ αὐτὸ φαινόμενον διαφόρους καταστάσεις εὐσταθείας, ὁπότε γεννᾶται τὸ πρόβλημα τῆς ἐκλογῆς τῆς καταλλήλου καταστάσεως διὰ τὸ φαινόμενον.

5. Αἱ μαθηματικαὶ ἀντιλήψεις περὶ εὐσταθείας, καθὼς καὶ τὰ βάσει αὐτῶν συμπεράσματα, διὰ νὰ ἑρμηνεύουν τὴν πραγματικότητα κατὰ ἱκανοποιητικὸν τρόπον, πρέπει νὰ συμφωνοῦν μὲ τὰ πορίσματα τῆς «πρακτικῆς εὐσταθείας», πρὸς τοῦτο δὲ χρειάζονται κατάλληλον τροποποίησιν καὶ συμπλήρωσιν. Ἡ πρακτικὴ εὐστάθεια δὲν ἔχει πλήρως σπουδασθῆ, ὅμως κύρια χαρακτηριστικά της δύναται νὰ εἶναι ἡ γνῶσις :

— τοῦ μεγέθους τῆς ἀποκλίσεως δεκτῶν καταστάσεων τοῦ συστήματος πρὸς ἱκανοποιητικὴν λειτουργίαν τοῦ συστήματος,

— τοῦ μεγέθους τῶν ἀρχικῶν συνθηκῶν, αἱ ὁποῖαι δύνανται νὰ ἐλεγχθοῦν,

— τοῦ μεγέθους τῶν ἐπιτρεπομένων διαταραχῶν,

— τοῦ χρόνου, ποὺ μελετῶμεν τὴν εὐστάθειαν.

— Αἱ μὴ γραμμικότητες τοῦ συστήματος παίζουν ἀποφασιστικὸν ρόλον διὰ τὴν πρακτικὴν εὐστάθειαν, καὶ δὲν δύνανται νὰ ἀμεληθοῦν.

REFERENCES

1. L. Èl'sgol'c, Qualitative methods in mathematical analysis, Amer. Math. Soc. (1964), 167 - 199, Providence, R.I., U.S.A.
2. J. La Salle and S. Lefshetz, Stability by Liapunov direct method with applications, Academic Press. New York, U.S.A. (1961).
3. D. Magiros, (a) Nat. Acad. Sci., Proc. 53, No. 6 (June 1965), U.S.A.
4. ——, (b) J. of Information and Control, 9, No. 9 (Oct. 1966), U.S.A.
5. ——, (c) 5th Intern. Conf. of Nonlinear Oscillations, Proc. 2, pp. 346 - 357, (Sept. 1969), Kiev, U.S.S.R.
6. ——, (d) Athens Acad. Sci., Proc. (May 25, 1972), Athens, Greece.
7. ——, (e) C. R. Acad. Sci., 266 A, (Apr. 8. 1968), 770 - 773. Paris, France.
8. ——, (f) C. R. Acad. Sci., 268 A, (March 1969), 652 - 655 Paris, France.
9. ——, (g) Athens Acad. Sci., Proc. (Dec. 9, 1971) Athens, Greece.

*

Ὁ Ἀκαδημαϊκὸς κ. **Ἰωάννης Ξανθάκης,** παρουσιάζων τὴν ἀνωτέρω ἀνακοίνωσιν, εἶπε τὰ ἑξῆς :

Εἰς τὴν ἐργασίαν ταύτην τοῦ κ. Μαγείρου, τὴν ὁποίαν ἔχω τὴν τιμὴν νὰ παρουσιάσω εἰς τὴν Ἀκαδημίαν, ἐκτίθενται ὡρισμέναι ἐνδιαφέρουσαι παρατηρήσεις ἐπὶ τῶν ἀντιλήψεων περὶ εὐσταθείας εἰς φυσικὰ καὶ κοινωνικὰ φαινόμενα, ποὺ ἐκφράζονται μαθηματικῶς ὡς δυναμικὰ προβλήματα.

Αἱ ἀντιλήψεις περὶ εὐσταθείας προέρχονται ἀπὸ πηγὰς διαφόρου φύσεως. Ἡ μελέτη τῆς σταθερότητος μιᾶς κινήσεως ἐπὶ τῆς τροχιᾶς της ἢ ἡ μελέτη τῆς τροχιᾶς μιᾶς κινήσεως μᾶς παρέχει δύο διακεκριμένας βασικὰς ἀντιλήψεις περὶ εὐσταθείας, αἱ ὁποῖαι περιέχουν πλῆθος ἄλλων εἰδικευμένων ἀντιλήψεων ὡς εἰδικὰς περιπτώσεις. Ὁ τρόπος μὲ τὸν ὁποῖον ἡ κατάστασις ἑνὸς συστήματος πλησιάζει πρὸς μίαν ἄλλην, ἢ παρεκκλίνει ἐξ αὐτῆς, ὁ τρόπος μετρήσεως τῆς ἐντάσεως τῶν διαταραχῶν ἑνὸς συστήματος, ἡ δομὴ τοῦ μαθηματικοῦ μοντέλου ἑνὸς συστήματος καὶ ἄλλα εἶναι πηγαὶ ἀντιλήψεων εὐσταθείας διαφόρου φύσεως. Ἡ ποικιλία αὕτη τῶν ἀντιλήψεων εὐσταθείας κάμνει τὰ προβλήματα λίαν πολύπλοκα καὶ τὰ ἀποτελέσματα τῶν ἐρευνῶν βάσει αὐτῶν ἐνίοτε δὲν εἶναι γενικῶς ἀποδεκτά.

Ὅλαι αἱ ἀντιλήψεις περὶ εὐσταθείας, ἂν καὶ διαφόρου φύσεως, δύνανται νὰ ἀναχθοῦν, κατὰ τὸν κ. Μάγειρον, εἰς τὰς αὐτὰς μαθηματικὰς «σχέσεις εὐσταθείας», αἱ ὁποῖαι παρέχουν μίαν ἑνοποίησιν τῶν διαφόρων περιστάσεων. Ἡ ἑνοποίησις αὕτη ὑποβοηθεῖ εἰς τὴν πληρεστέραν κατανόησιν τῶν ἀντιστοίχων προβλημάτων.

Τὰ συμπεράσματα, θεωρήματα ἢ κριτήρια, τὰ στηριζόμενα εἰς τὰς μαθηματικὰς σχέσεις ἀντιλήψεων εὐσταθείας, εἶναι δυνατὸν νὰ μὴ ἑρμηνεύουν κατὰ ἱκανοποιητικὸν τρόπον τὴν πραγματικότητα, ὁπότε παρίσταται ἀνάγκη τροποποιήσεων ἢ συμπληρώσεων τοῦ μαθηματικοῦ προτύπου, οὕτως ὥστε ἡ μαθηματικὴ διατύπωσις τῆς ἀντιλήψεως εὐσταθείας νὰ πλησιάζῃ, ὅσον τὸ δυνατὸν περισσότερον, πρὸς τὴν λεγομένην «Πρακτικὴν Εὐστάθειαν», ἥτις ὅμως δὲν ἔχει ἀκόμη πλήρως μελετηθῆ.

STABILITY CONCEPTS OF DYNAMICAL SYSTEMS

By: Demetrios G. Magiros

January, 1980

Reprinted from a *Technical Report, Genl. Electric Co.*, RSD, Philadelphia (Jan. 1980), 1–28.

-2-

TABLE OF CONTENTS

-3-

INTRODUCTION

Basic problems in many fields of interest under current research in physical, technological and life sciences can be formulated and solved as problems of stability of dynamical systems.
In such problems the investigation faces many difficulties, which make the problems complicated and the results questionable in many cases. There are difficulties in understanding the stability concept, difficulties in its definitions that are reasonable from the physical point of view and consistent from a mathematical point of view, difficulties in developing workable criteria for making a decision regarding stability or instability of the states of the system. During recent years the concepts of stability of dynamical systems and the corresponding stability criteria have been advanced either by modifying old ideas or by creating new ones, and these advances permit a deeper penetration into the more profound problems of stability, very important both in theory and practice.

In the present paper attention is given to the stability concepts.
The variety of the stability concepts is due to sources of different nature, as, e.g., to the type of the mathematical model of the physical system, the nature of the variables of the system, the manner in which a state of the system approaches to, or deviates from, another state, the way the perturbations act on the system, the way the norm of the perturbations and their effect are measured, etc.
The stability concepts are classified into two classes, of which one is characterized by the stability concept in the sense of Liapunov, and the other by the stability concept in the sense of Poincaré.

-4-

The stability concepts are expressed and unified by the same mathematical relationships, and such a unification of the stability concepts brings a natural simplicity in the understanding of problems concerning stability and gives rise to new results. These unification relationships, by specialization or restrictions of the quantities involved, give a variety of special stability concepts and some relations between them.

Appropriate remarks and examples clarify the discussion.

1. PHYSICAL STABILITY CONCEPTS

The notion of stability was originated in physical problems as a characterization of specific situations. A state of a phenomenon or system is said to be "stable" if small disturbances to the state have as an effect small changes to the state. If the effect is considerable, the state is "unstable", and if, for small disturbances of the state, the effect tends to disappear, the state is "asymptotically stable"; and if, regardless of the magnitude of the disturbances, the effect tends to disappear, the state is "asymptotically stable in the large".

These physical stability concepts are of "qualitative" nature, and by using them no stability problems can, in general, be solved. The discussion of stability problems necessitates a "quantitative" knowledge of the stability concepts, for which mathematical relationships are needed between the magnitude of the disturbances of the state of the system and the magnitude of their effect on the state.
The subject in the following is a quantitative discussion of the stability concepts, which is essentially based on remarks on perturbations and their effect.

2. SYSTEMS UNDER PERTURBATIONS ACTING MOMENTARILY OR PERMANENTLY.

The disturbances to a state of a system are due to perturbations, which can be considered as "disturbing forces" acting on the system either momentarily or permanently.

The "momentarily" acting perturbations give disturbances to the initial states of the system but they do not appear in the formulation of the equations

-6-

of the motion of the system. We take as the model of the physical system in case of momentarily acting perturbations the n-dimensional system of non-autonomous differential equations in its normal form:

$$\left.\begin{array}{l} \dot{x}_i(t) = X_i(t, x_1, \ldots, x_n) \\ x_i(t_0) = x_{i0}, \quad X_i(t, 0, \ldots, 0), \quad t_0 \leq t \end{array}\right\} \qquad (1)$$

where $x_1, \ldots, x_n$ are the "state variables", and x_{i0} the initial conditions.

The "permanently" acting perturbations give disturbances to the system itself, X_i will be changed, and the equations of the motion of the system must contain these persistent perturbations, and, in this case, we take the model of the system in the form:

$$\dot{x}_i(t) = X_i(t, x_1, \ldots, x_n) + p_i(t, x_1, \ldots, x_n) \qquad (2)$$

where p_i are the persistent perturbations. For the existence of unique solutions of the systems (1) or (2), the functions X_i and p_i must satisfy appropriate conditions. We remark that the distinction of the perturbations into sudden and persistent leads to a classification of the stability concepts into two important categories.

3. THE EFFECT OF THE PERTURBATIONS. SOME DEFINITIONS.

The effect of the perturbations on a system is a change of certain quantities pertaining to the system, and this effect can be visualized by the change of the trajectory S of the original (the unperturbed) system into the trajectory $\bar{S}$ of the new (the perturbed) system.
A state of a system is designated by a point of the trajectory, and if a point M of the unperturbed trajectory S corresponds to the point $\bar{M}$ of the perturbed trajectory $\bar{S}$, the distance $\rho = M\bar{M}$, Figure 1(a), can be considered as measuring the magnitude of the effect of the perturbations on M.

-7-

This distance ρ of which the unperturbed end M corresponds to the perturbed end $\bar{M}$, plays a decisive role in the stability concepts and in the formulation of the stability relationships. Let us call it a "stability distance".[7(a)(c)] To a given point M on S one may make to correspond different points $\bar{M}$ on $\bar{S}$. Any such correspondence presupposes an assumption and characterizes a specific stability concept of the trajectory S . Each assumption must be related to physical reality.

One can distinguish two the most physical, correspondences between M and $\bar{M}$, when one has two important and basic stability concepts.

(a): One correspondence between M on S and $\bar{M}$ on $\bar{S}$ is when M and $\bar{M}$ are taken at the "same time", Figure 1(b). In this case we call the distance $\rho=\rho_\ell = M\bar{M}$ a "Liapunov distance", which characterizes a basic stability concept, the "stability in Liapunov sense". This distance in the n-dimensional Euclidian space is:

$$\rho_\ell=\left\{\sum_{i=1}^{n}[\bar{x}_i(t)-x_i(t)]^2\right\}^{1/2}, \quad \text{or:} \quad \rho_\ell=\left\{\sum_{i=1}^{n}[x_i(t,\bar{a}_j)-x_i(t,a_j)]^2\right\}^{1/2}$$

in the state space and in the parameter space, respectively.

$\bar{x}_i$ and $\bar{a}_j$ are the perturbed state variables and parameters corresponding to the unperturbed ones x_i and a_j respectively.

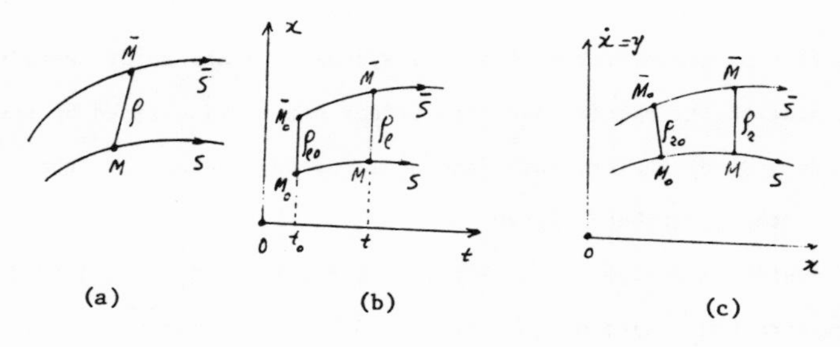

(a) (b) (c)

Figure 1

-8-

(b): In the second correspondence, the distance $\rho = \rho_2 = M\bar{M}$, Figure 1(c), is taken as the "minimum" of the distances from M to the points of $\bar{S}$, that is:

$$\rho = \rho(M, \bar{S}) = \rho_2 = min \left\{ \sum_{i=1}^{n} (\bar{x}_i - x_i)^2 \right\}^{1/2}$$

in the "phase space" (x, $\dot{x} = y$), parametrized by the time variable t.
The equations in the phase space can be found by eliminating t from the equations of the model, thus reducing them to differential equations containing x and $\dot{x} = y$.
The distance ρ_2, called a "Poincaré distance", characterizes a basic special stability concept, the "stability in Poincaré sense", also called an "orbital stability".

4. UNIFICATION OF STABILITY CONCEPTS

A variety of the stability concepts can be quantitatively described by the "stability relationships":

$$\text{(a): } \rho_0 < \delta_1 \,,\quad \text{(b): } \rho < \varepsilon \,,\quad \text{(c): } \lim_{t\to\infty} \rho = 0 \,,\quad \text{(d): } \|p\| < \delta_2 \,,\quad \text{(e): } \rho_\tau < \delta_3 \qquad (3)$$

which, then, "unify" the stability concepts. In these relationships $\varepsilon, \delta_1, \delta_2, \delta_3$ are positive constants, δ_1 and δ_2 depend on ε and on the initial time t_0, $t_0 \geq 0$, ρ is the stability distance, and $\|p\|$ designates the norm of the perturbations.
By a suitable combination of these relationships and an appropriate interpretation and restriction of the quantities involved, one can define specific stability concepts.
In the following we analyze the above statement.

-9-

5. STABILITY CONCEPTS IN CASE OF MOMENTARILY ACTING PERTURBATIONS.

In case the perturbations act on the system momentarily we have variations only of initial conditions, and we take into account the first three relations of (3).

- Let a solution of the system (1) be $x_i(t)$ and its perturbed $\bar{x}_i(t)$. Given a positive number ε however small, if it is possible to find a positive number $\delta_1 = \delta_1(t_0, \varepsilon)$ such that $\rho_0 < \delta_1$, where ρ_0 is the distance initially, implies, for all $t \geq t_0$, $\rho < \varepsilon$, then $x_i(t)$ is said to be "stable".

- The motion $x_i(t)$ is "unstable", if, for given small number ε, and for sufficiently small number δ_1 for which $\rho_0 < \delta_1$, the inequality $\rho < \varepsilon$ is not satisfied.

- $x_i(t)$ is "asymptotically stable", if it is stable and, in addition, a number $\delta \geq \delta_1$ exists such that starting from any ρ_0, with $\rho_0 < \delta$, the limiting condition $\lim_{t \to \infty} \rho = 0$ holds uniformly to x_{i0} and t_0.

- $x_i(t)$ is "asymptotically stable in the large", if it is asymptotically stable and δ is very large.

6. REMARKS. SPECIAL STABILITIES.

In this section we make a few remarks and, by using the relationships (3), we give a variety of special stability concepts all of which have a practical usefulness.

- Remark 1. Liapunov and Poincaré stabilities.

-10-

If the ρ's in the relationships (3) are interpreted as Liapunov or Poincaré distances, the stability concepts are in Liapunov or Poincaré sense. The Liapunov stability concept is an appropriate one for the discussion of the stability of the motion of a system, while the Poincaré stability concept is appropriate for the stability of the orbit (itself) of the motion.

In periodic motions, the Liapunov stability concept classifies as unstable situations which are practically considered as stable, so this stability concept is a narrow one in periodic motions.
The Poincaré stability concept is the appropriate one for investigations of periodic motions.

\- Remark 2. Stability of equilibrium points.

In case the orbit S shrinks to an equilibrium point of the system, the Liapunov and Poincaré stability distances can be considered as identical, when, for the stability of the singular points of dynamical systems, the stabilities in Liapunov and in Poincaré sense are equivalent.

\- Remark 3. Uniform, eventual and finite time stabilities.

If δ_1 in 3(a) is independent of the initial time t_0, then we speak about "uniform stabilities".

If the solution $x(t)$ is stable for $t \geq t_0$, then the solution will be stable for initial time bigger than t_0, but not necessarily for initial time smaller than t_0. There are cases of stability in which the initial time has a minimum value τ, which depends on ε, when $\tau(\varepsilon) \leq t_0$. In these cases we speak about "eventual stabilities".[5]

-11-

In the investigation of the stability of a solution $x(t)$ of the model of a system, $x(t)$ may become infinite either when $t \to \infty$, or when time takes a maximum value which is a finite number T, when $0 \le t_0 \le t \le t_0 + T$. This value T of time is called a "finite escape time".[3(c),10]

The so-called "finite time stability" is characterized by a finite time T, which is not, in general, the "finite escape time". This kind of stability characterizes real systems expressed in general by nonautonomous differential equations, especially if these differential equations contain (structural) parameters. The "practical stability", as we will see, presupposes finite time.

- Remark 4. Continuous dependence of a solution on the initial conditions and its stability.

If the solution $x(t, x_0)$ of a system is stable in the special parameter space of the initial conditions, then, for any two sets x_{10} and x_{20} of the initial conditions, and for appropriate numbers ε and δ_1, the two inequalities:

$$\rho_0 = \{\Sigma (x_{20} - x_{10})^2\}^{1/2} < \delta_1 \quad , \quad \rho = \{\Sigma [x(t, x_{20}) - x(t, x_{10})]^2\}^{1/2} < \varepsilon$$

are compatible. But these inequalities are the conditions for the solution $x(t, x_0)$ to be continuously dependent on the initial conditions x_0 uniformly in t, and this property of the solution is a Hadamard postulate for the solution to have a physical meaning.

Therefore, the continuous dependence of a solution on the initial conditions is equivalent to the stability situation in a special parameter space, which is the initial conditions space.

- Remark 5. Structural stability.

A real structure differs from the idealized structure designed by an engineer this difference being connected with small imperfections and defects. In order for the real structure to behave approximately as the idealized scheme, the structure must be postulated as stable with respect to small perturbations of the structure during its life.

We call "structural stability" of a system its stability, if it is "invariant" under small perturbations, either perturbations of the system itself or perturbations of its parameters.
The appropriate space of the perturbations of the system or of the parameters, where the system is structurally stable, is the "domain of structural stability" of the system.
The structural stability of a system in its domain implies a variety of important properties of the system.[6] Properties of structural stability suggest methods for investigation of physical problems with well accepted results, as, e.g., in problems of morphological processes or catastrophies in biology, etc.[8]
The structural stability of a system fails at points of "bifurcation" in the space of parameters of the system, where the topological structure of the system changes abruptly.

- *Remark 6*. *The boundedness of a solution and its stability*.

The boundedness of a solution $x(t)$ of a system is, by definition, characterized by the boundedness of the distance $\rho = \rho_{\ell} = OM = |x(t)|$, Figure 2, of any point M of the orbit S of the solution from the origin O of the coordinate system.

-13-

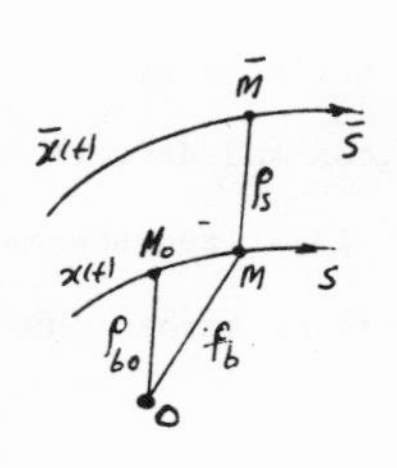

Figure 2

We remark that the relationships (3) contain the concept of boundness as a special case, by interpreting ρ as the distance $\rho_b = OM$, Figure 2.

If, given a point M_0 of S, corresponding to the distance $\rho_{b0} = OM_0$, such that $\rho_{b0} < \delta_1$, where δ_1 is a positive finite number, and if, for any point M of S, there is a positive finite number ε, however large, and $\rho_{b0} < \delta_1$ implies $\rho_b < \varepsilon$, the orbit S is bounded.

To different types of stability there, in general, correspond different types of boundedness.[11]

The concept of boundedness of a solution has been taken, from the time of Lagrange, as a stability concept, especially in celestial mechanics, called "stability in Lagrange sense".

The solar system, for example, is considered as "Lagrange stable" in the sense that none of the members of this system excapes to infinity, and not any two of its members collide. The velocities of the bodies at their collision are unbounded, and the Lagrange stability means that the coordinates and velocities of the bodies of the system are bounded.

The boundedness property of a solution of a system is different from its stability property. In some cases these two properties are identical and one implies the other, and in some other cases methods in studying one of these properties can often be so modified as to be of use also in problems involving the other.[1,3(b),11]

The Lagrange stability classifies as stable any situation of a state in finite distance, and as unstable any situation in a large distance, and such concepts are not, in general, accepted for investigation of stability problems on earth, say.

-14-

We may have unstable situations in bounded distances, and stable situations in unbounded distances. The "boundedness distance" $\rho_b = OM$, Figure 2, may be unbounded as $t \to \infty$, that is it may be: $\lim_{t\to\infty} \rho_b = \lim_{t\to\infty} |x| = \infty$, but the "stability distance" $\rho_s = M\bar{M} = |\bar{x} - x|$ for $t \to \infty$, may be of the form $(\infty - \infty)$, when the stability distance" ρ_s may be either finite or infinite, and the unperturbed situation may be either stable or unstable.

7. STABILITY CONCEPTS IN CASE OF PERSISTENT PERTURBATIONS.

In case the perturbations acting on a system are permanent, we have a variation of the second members X_i of (1), when we take the model of the system in the form (2), where X_i and p_i are such that (2) has a unique solution $x_i(t)$. We do not, in general, assume that $p_i(t, o) = 0$, then the origin of the perturbed system is not, in general, a solution of (2).

The stability concepts in the case of persistent perturbations come from the stability relationships (3), where 3(d) is meaningful, by interpreting or restricting appropriately the stability distances ρ and accepting special forms for the norm of the perturbations p_i depending on their magnitude.

We may have the following stability concepts.[9]

- In case the magnitude of p_i is small, the norm is taken in the form: $\|p_i\| = \max_{x_i} |p_i|$, when we have the concept of "total stability".

- If the magnitude of p_i is large in a small interval of time, the norm is taken in the form: $\|p_i\| = \int_0^\infty \max_{x_i} |p_i| dx$, and we have the concept of "integral stability".

- For both the above restrictions of the magnitude of p_i , the norm is taken as $\|p_i\| = \int_t^{t+T} \max_{x_i} |p_i| dx$, and we have the "mean stability". All these stability concepts are in Liapunov or in Poincaré sense, if the distances ρ in the relationships (3) are appropriately interpreted.

Some connections between stability concepts can be introduced by the following statements: [9]

- Asymptotic integral stability implies total asymptotic stability in the mean, and conversely

- Asymptotic integral stability implies total asymptotic stability and this implies total stability

- Asymptotic stability in the mean implies stability in the mean, and this implies either integral stability or total stability

- Integral stability does not imply total stability

- Stability in the mean does not imply total asymptotic stability

- Total asymptotic stability does not imply integral, stability.

8. STABILITY CONCEPTS IN CASE OF MODELS WITH DEVIATING ARGUMENTS.

If the model of the system is expressed by differential equations with deviating arguments, as, e.g., by differential equations with retarded arguments of retardations $\tau(t)$, we distinguish two cases, accordingly as the perturbations act either momentarily or permanently.[7(b)]

-16-

(a): In the first case, the model does not contain the perturbations, when the model may be of the form:

$$\left.\begin{array}{l} \dot{x}_i(t) = X_i[t, x_j(t, \tau_{jk}(t))], \quad \tau_{jk} > 0 \\ i, j = 1, 2, \ldots, n \; ; \; k = 1, 2, \ldots, m, \; m \leq n \end{array}\right\} \tag{4}$$

The initial functions $\varphi(t)$ are defined in the initial interval: $E_{t_0}: t_0 - \tau_{jk} \leq t \leq t_0$, and the corresponding solution $x_\varphi(t)$ of (4) holds for $t \geq t_0$.

For the Liapunov distances, Figure 3(a), we have $\rho_{10} = M_{01}\bar{M}_{01} = |\bar{\varphi} - \varphi|$ in the interval E_{t_0}, while ρ_ℓ in the interval $t \geq t_0$.

For Poincaré stability appropriate remarks hold, Figure 3(b).

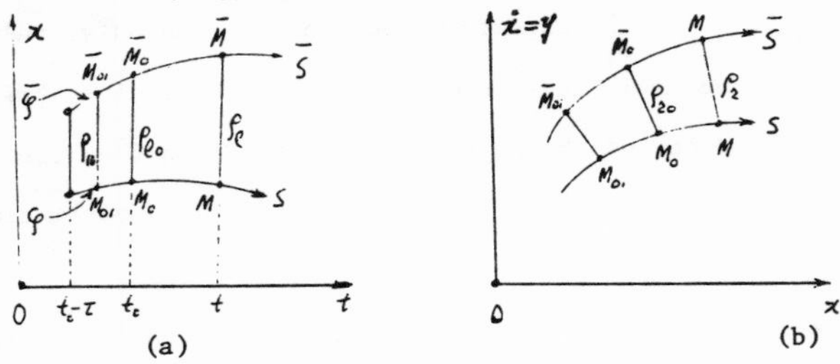

Figure 3

(b): In the second case, the model contains the perturbations, and the model may be of the form:

$$\dot{x}_i(t) = X_i\,[t, x_j(t - \tau_{jk}(t))] + p_i[t, x_j(t - \tau_{jk}(t))] \tag{5}$$

The solutions x_φ of (5), corresponding to the initial functions $\varphi(t)$, is "stable" with respect to perturbations p, if, given $\varepsilon > 0$ and $t_0 \geq 0$, there exist two positive constants $\delta_1 = \delta_1(t_0, \varepsilon)$, $\delta_2 = \delta_2(t_0, \varepsilon)$ such that for $\rho_{10} < \delta_1$ valid in the interval E_{t_0}, and $\|p\| < \delta_2$ valid in $t \geq t_0$, the inequality $\rho < \varepsilon$, in $t \geq t_0$, is implied.

(c): In case the retardations τ are perturbed, we have a new stability concept characterized by the distance $\rho_\tau = |\bar{\tau}(t) - \tau(t)|$, where $\bar{\tau}$ is the perturbed retardation corresponding to the unperturbed τ, which holds in $t \geq t_0$

9. THE STABILITY OF A SYSTEM AND THE COORDINATE SYSTEM.

For the study of a phenomenon or for investigation of a physical system, one selects some of its quantities, called "space variables" of the system, which give the "coordinate system" to which the investigation is referred.[7(d)] The nature of the stability of a physical system is, in general, "not invariant" with the selection of the space variables, or with a general transformation of the coordinate system. In other words, the coordinate system selected and its transformations affect, in general, the nature of the stability. Also, we may have stability with respect to some coordinates only, that is, the coordinates may not be equivalent from the point of view of stability. The problem of the dependence of the stability of a solution from the coordinate system is a very important problem, but no specific and systematic investigation upon it exists.
By means of some examples in the following, an idea is given about this problem.

10. GEOMETRICAL INTERPRETATION OF STABILITIES.

The stability concepts may be clarified by a geometrical interpretation that follows.

We define as a half "η-cylinder" around a curve Q in n-dimensional space the set of points of which the distance ρ from the curve Q is smaller than η. The curve Q is the axis of the cylinder.[3(d),12(b)]

Given the motion $x_i(t, t_0, x_{10}, \ldots, x_{n0})$ of orbit S, if we consider S as the axis of "δ_1-cylinder" and "ε-cylinder", where δ_1 and ε are the constants of the inequalities (3), and we interpret ρ either as a "Liapunov distance",

-18-

or as a "Poincaré distance", we can have a geometrical interpretation either of "Liapunov stability", or of "orbital stability".

The motion x_i with orbit S starting from an initial point x_{io} in the (t, x_i)-coordinates, is "Liapunov stable", if, given an "ε-cylinder" around S, one can find a "δ_1-cylinder" such that any perturbed orbit $\bar{S}$ starting from $\bar{x}_{io}$ in "δ_1-cylinder" remains in the "ε-cylinder for all $t \geq t_o$, Figure 4. ρ is taken as "Liapunov distance".
If, in addition, S and $\bar{S}$ tend to coincide as $t \to \infty$, S is "asymptotically stable".
For "orbital stability" of S, ρ must be interpreted as a "Poincaré distance", and S and $\bar{S}$ are referred in the phase space $(x_i, \dot{x}_i = y_i)$.

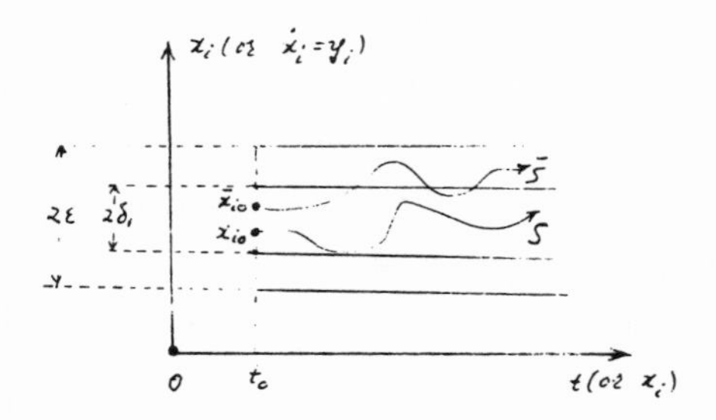

Figure 4

For stability under persistent perturbations, the "δ_2-cylinder" must be used.

We remark that the widths and lengths of the above cylinders, as well as the nature of the nonlinearities of the model of the system, play, as we will see, an important role for the so-called "practical stability".

11. PRACTICAL STABILITY.

The previous discussion on stability concepts is of theoretical or mathematical type. It represents a possible function of a physical system, but it is, in general, of small practical usefulness.
One can see that a state of a system may be unstable under the previous stability definitions, but the system may oscillate sufficiently near this state and its performance can be accepted practically as a stable one.
Missiles many times have this kind of behavior.
Also, an equilibrium state of a system may be stable mathematically in a very small region, but in practice the perturbations expected may cause the system to go far from the equilibrium state, when the system is practically unstable at the equilibrium state.

In order that the theoretical stability concepts interpret the reality adequately and have a practical usefulness, which must be the ultimate purpose of the investigator, they must be accompanied by appropriate modifications, changes and supplements, when, by all these, we will have the motion of the so-called "practical stability". Practical stability is defined as follows:[4,12(a)]

Given an interval of time $[t_0, t_0+T]$ and two sets: R_{t_0} and R_t in the n-dimensional space ($x_1, \ldots, x_n$), if there exists a set R_a in the space of admissible values of the parameters such that the integral curves $x(t)$ of the system, starting from t_0 of the set R_{t_0}, remain in the set R_t for all time of the interval $[t_0, t_0+T]$, then the solution $x(t)$ is said to be "practically stable".

Practical stability is characterized by the knowledge of:

(a): The size of the region R_t of deviations of the state acceptable for a

-20-

satisfactory operation of the system, that is the "width" of the ε-cylinder. This region is formulated from the definite requirement demanded of the design system.

(b): The size of the region R_{t_0} of the permitted initial conditions that can be controlled, that is the "width" of the δ_1-cylinder. This region is defined by the technical conditions of the functioning of the real system.

(c): The size of the region R_p of the permitted perturbations, that is the "width" of the δ_2-cylinder.

(d): The finite time T over which the stability is investigated, which corresponds to the "lengths" of the cylinders, that is the time interval $[t_0, t_0+T]$ for which the above requirements are satisfied for the solution $x(t)$ of which the stability is investigated.

(e): The size of the region R_2 in the space of admissible values of the parameters of the system such that $x(t)$ starting at $t=t_0$ in the δ_1-cylinder fall within ε-cylinder for $t\in[t_0,t_0+T]$ for perturbations in the δ_2-cylinder.

For a perfect practical stability the sizes of the above regions must be large.

For the computation of the above regions the use of the "nonlinearities" of the model of the system is necessary, and, therefore, investigations based on linear approximations of the models must not be taken too seriously.[7(a)(e)]

12. EXAMPLES.

In the following examples we discuss the stability situations by using only the stability definitions, and make appropriate remarks.

-21-

Example 1. Stability of rectilinear motions.

The rectilinear motion of a mass is given in the t,x-plane by:

$$x(t) = v(t-t_0) + d \tag{1}$$

where d is distance and v the velocity of the moving mass.
If at the initial time t_0 the distance is $d=d_0$ and the velocity $v=v_0$, the orbit of the motion is a straight line S , Figure 5(a), starting from the point $M_0(t_0, d_0)$ with velocity v_0 (slope of S). We consider two cases.

(a): Uniform rectilinear motion.

For this motion the velocity v is considered as a constant, when the perturbation of the motion will affect only the distance d in the equation (1), then the perturbed orbit $\bar{S}$ will have the same slope with S , and $S, \bar{S}$ are parallel, Figure 5(a). For any place M of the mass on S the Liapunov distance $\rho_\ell = M\bar{M}$ is constant, also the Poincaré distance $\rho_2 = M\bar{\bar{M}}$ is constant, then the uniform rectilinear motion is stable both in Liapunov and Poincaré sense.

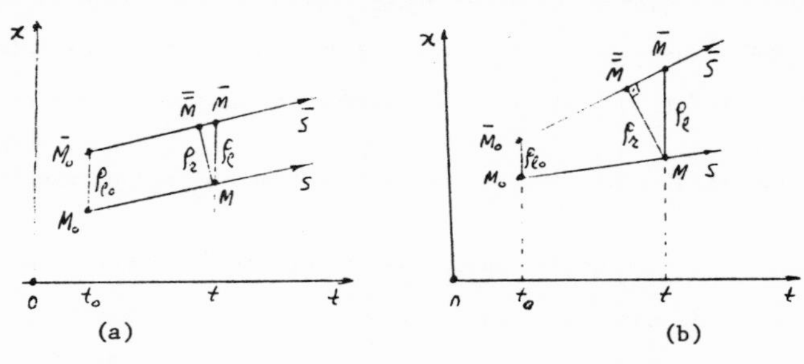

Figure 5

(b): Nonuniform rectilinear motion.

Since during this motion the velocity v is not constant, the perturbation of S will affect both the distance d and the slope (v) of the orbit S , and S and its perturbed $\bar{S}$ are not parallel. The stability distances ρ_ℓ and ρ_2 ,

Figure 5(b), increase to infinity as $t \to \infty$ and the motion is unstable both in Liapunov and Poincaré sense.
We remark that the orbits in these motions are unbounded as $t \to \infty$, then these motions are unstable in Lagrange stable.

Example 2. Linear systems with or without forcing terms.

(a):
$$\dot{x}_i(t) = \sum_{j=1}^{n} a_{ij}(t)\, x_j \;;\quad i = 1, \dots, n \tag{2.a}$$

the coefficients $a_{ij}(t)$ are continuous functions of t in $t \geq t_0$. We see that:[2] "The boundedness of the solutions of the system (2.a) implies the Liapunov stability, and conversely. If a_{ij} are constants and all solutions bounded, we have uniform stability".

(b):
$$\dot{x}_i(t) = \sum_{j=1}^{n} a_{ij}(t)\, x_j + f_i(t) \tag{2.b}$$

$a_{ij}(t)$ and $f_i(t)$ are continuous functions of t in $t \geq t_0$. We see that:
(i): "The boundedness of the solutions of (2.b) implies their Liapunov stability, but the converse is not true". For the converse we have:
(ii): "The Liapunov stability of the solutions of (2.b) implies the boundedness of all solutions, if, in addition, there is at least one solution bounded".
We remark that connections between stability concepts and boundedness of the solutions in nonlinear systems are found in some specific cases.

Example 3. Stability in a problem of astronomy.

We examine: "The stability of the motion of a mass moving in an orbit under the inverse square Newton's law of attraction of an attractive center".[2]

The orbit S of the mass M is in this case an ellipse which can be determined by using initial conditions. The initial conditions are the distance of the moving mass from the attractive center E and its velocity at an initial time t_0. The motion of M on S is periodic with period T. Perturbations acting momentarily affect the initial conditions, so we have a perturbed elliptic orbit $\bar{S}$ with a new period $\bar{T}$.

(a): <u>Stability in Liapunov sense</u>.

The Liapunov distance is $\rho_\ell = M\bar{M}$, Figure 6(a), where M and $\bar{M}$ are places of the mass on the orbits S and $\bar{S}$ at the "same time". The periods T and $\bar{T}$ depend on the length of the major axis of the correspondent ellipse. M and $\bar{M}$ being very close initially and traveling on different ellipses may find themselves at opposition and then at great distance from each other in due course of time. Then, given a small positive number ε, one can not find a δ_1 such that $\rho_0 < \delta_1$ implies $\rho < \varepsilon$, when the motion is "Liapunov unstable".

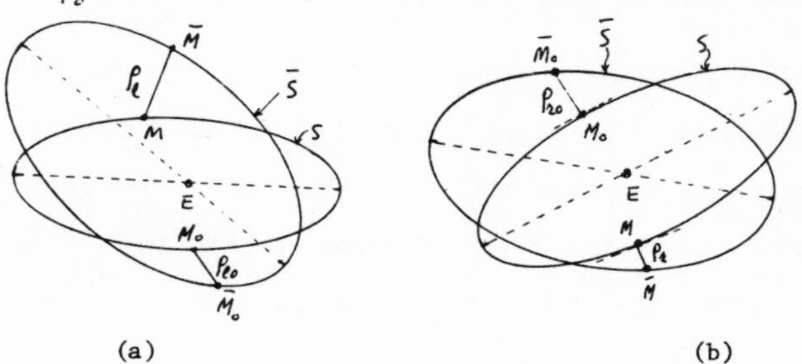

(a) (b)

<u>Figure 6</u>

We can prove that in general a periodic motion of which the corresponding perturbed motion has different period is Liapunov unstable.

The free vibrations of a simple pendulum are Liapunov stable if they are linear, and Liapunov unstable if they are nonlinear. In the case of linearity the isochronous phenomenon exists, while in the case of nonlinearity the isochronous phenomenon is violated.

(b): Stability in Poincaré sense (orbital).

The Poincare distance $\rho_2 = M\bar{M}$, Figure 6(b), must be the "minimum" distance from M to points of $\bar{S}$. If one wants ρ_2 to be smaller than a given positive number ε for all time, a positive number δ_1 must be found such that, if the Poincaré distance initially is $\rho_{20} < \delta_1$, the inequality $\rho_2 < \varepsilon$ must be implied. This is possible. Given a perturbation p, the perturbed orbit $\bar{S}$ is known, when ρ_2, corresponding to M on S, is known. The deviations for all points of S have a maximum ρ_M and a minimum ρ_m, and the smaller the perturbation p the smaller ρ_M and ρ_m. Given ε small, if we want $\rho_2 < \varepsilon$ for all M on S, we must use a very small perturbation p such that $\rho_M < \varepsilon$, when δ_1 must be $\delta_1 < \rho_m$, and this means that the motion is orbitally stable. The distance ρ_2 never tends to zero as t changes, then the orbital stability is not an asymptotic one.

According to the above the same phenomenon may be stable or unstable depending on the stability definition used. For the stability situation of the phenomenon, the appropriate stability definition must be selected according to practical usefulness.

In the above example practical reasons suggest to accept the stability situation coming from stability concept in Poincaré sense.

Example 4. Stability and the change of coordinates.

(a): The motion: $$x = e^{-\alpha t} \cdot \cos \omega t \tag{4.a}$$

for ω in $-\infty < \omega < +\infty$, and α a parameter, is "unstable" in Liapunov sense for $\alpha \leq 0$, but "quasi-asymptotically stable" for $\alpha > 0$.

(b): The nonlinear differential system:

$$\dot{x} = -\mu y \, , \quad \dot{y} = \mu x \, , \quad \mu \neq 0 \tag{4.b}$$

-25-

has as a solution the motion:

$$x = a\cos(\mu t + b), \quad y = a\sin(\mu t + b) \tag{4.b.1}$$

with orbit

$$x^2 + y^2 = a^2 \tag{4.b.2}$$

a and b are constant parameters, and μ characterizes the frequency or the period $(\frac{2\pi}{\mu})$ of the motion (4.b.1) on the orbit (4.b.2).

The motion (4.b.1) is "Liapunov stable" if μ is a constant, but it is "Liapunov unstable" if μ is changed, say in the interval $\mu_1 \leq \mu \leq \mu_2$.

For a constant, the "orbital stability" is meaningless; and for a changed, the motion is "orbitally stable".

The motions all are "Lagrange stable" for any a and μ.

(c): The nonlinear differential system:

$$\dot{x} = -y(x^2+y^2)^{1/2}, \quad \dot{y} = x(x^2+y^2)^{1/2} \tag{4.c}$$

accepts as a solution the motion:

$$x = a\cos(at + b), \quad y = a\sin(at + b) \tag{4.c.1}$$

with orbit

$$x^2 + y^2 = a^2 \tag{4.c.2}$$

a and b are constant parameters.

For a constant, the motion (4.c.1) is "Liapunov stable", but the orbital stability is meaningless. For a changed, the motion is "Liapunov unstable", but "orbitally stable". The motion (4.c.1) is "Lagrange stable".

— The solution (4.c.1), which is "Liapunov unstable" for a changed, can become "Liapunov stable" by changing the variables x and y in (4.c) according to appropriate relations.[2] If the new variables r and b are related to x and y according to:

$$x = r\cos\vartheta, \quad y = r\sin\vartheta, \quad \vartheta = at + b \tag{4.c.3}$$

-26-

the equations (4.1) are transformed into the new ones:

$$\dot{a}=0, \quad \dot{b}=0 \tag{4.c.4}$$

of which the solutions:

$$a=c_1 \quad , \quad b=c_2 \tag{4.c.5}$$

c_1 and c_2 arbitrary constants, are "Liapunov stable".

(d): The nonlinear equation:

$$\ddot{x}+\sin x=0 \tag{4.d}$$

accepts as a solution the motion:

$$x=a \sin\{\varphi(a)t+b\} \tag{4.d.1}$$

where a and b are arbitrary constants, and the frequency $\varphi(a)$ can be expressed in terms of elliptic functions.

The solutions (4.d.1) are, except the origin, "Liapunov unstable", but by using new coordinates a and b, according to the transformations:[2]

$$x=a\sin\{\varphi(a)t+b\},\ y=a\cos\{\varphi(a)t+b\} \tag{4.d.2}$$

the equation (4.d) leads to the equations

$$\dot{a}=0 \quad , \quad \dot{b}=0 \tag{4.d.3}$$

of which the solutions $a=c_1$, $b=c_2$ are "Liapunov stable".

(e): The nonautonomous normal nonlinear system:[2]

$$\dot{x}_i=X_i(t,x_1,\ldots,x_n);\ i=1,\ldots,n \tag{4.e}$$

has the functions:

$$\phi_i(t,x_1,\ldots,x_n)=c_i \tag{4.e.1}$$

as solutions in implicit form. $c_1,\ldots,c_n$ are arbitrary constants.

By the transformation:

$$y_i=\phi_i(t,x_1,\ldots,x_n) \tag{4.e.2}$$

(4.e) are reduced to

$$\dot{y}_i=0 \tag{4.e.3}$$

-27-

of which the solutions

$$y_i = c_i \qquad (4.e.4)$$

are "Liapunov stable", and this is not, in general, the case for the solutions (4.e.1) of (4.e).

III. REFERENCES

1. H. Antosiewicz: "Boundedness and Stability", in "Nonlinear differential equations and nonlinear mechanics", edited by: J. LaSalle and S. Lefschetz, Academic Press, New York (1963), pg. 259-267.
2. L. Cesari: "Asymptotic behavior and stability problems in ordinary differential equations", Springer-Verlag, Berlin (1959), pg. 6-13.
3. W. Hahn: "Theory and application of Liapunov direct method", Prentice-Hall, Englewood Cliffs, N.J. (1963), (a) pg. 10, (b) pg. 129, (c) pg. 126, (d) pg. 8.
4. J. LaSalle and S. Lefschetz: "Stability by Liapunov's direct method", Academic Press, New York (1961), pg. 121-126.
5. J. LaSalle and R. Ruth: "Eventual stability", Proc., 4th Joint Automatic control conference , Univ. of Minnesota, Minneapolis (1963), pg. 468-470.
6. S. Lefschetz: "Differential equations: Geometric theory", Interscience Publ., New York (1957), pg. 239-245.
7. D. Magiros: (a): "On stability definitions of dynamical systems", proc., Nat. Acad. 5c., Washington, D.C. (U.S.A), Vol. 53, No. 6 (1965), pg. 1288-1294.

-28-

(b): "Stability concepts of solutions of differential equations with deviating arguments", Practica, Athens Acad. SC, Athens (1971).

(c): "Remarks on stability concepts od dynamical systems", Practica, Athens Acad. SC (1974).

(d): "Mathematical models of physical and social systems", General Electric Co., RESD, Phila., PA. (Nov. 1976).

(e): "Characteristic properties of linear and nonlinear systems", Practica, Athens Acad. SC. (1976).

8. R. Thom: "Structural stability and morphogenesis", W. Benjamin, Reading, Mass, U.S.A. (1975).

9. I. Vrkŏv "On some stability problems", Proc. of a conference in Prague, Acad. Press, New York (1963), pg. 217-221.

10. L. Weis and E. Infante: "Finite time stability under perturbations and on product spaces", in "Differential equations and dynamical systems", ed. J. Hale and J. LaSalle, Acad. Press, New York (1967), pg. 341-350.

11. T. Yoshizawa: "Stability theory by Liapunov second method", Math. Soc. of Japan (1966).

12. V. Zubov: "Mathematical methods for the study of automatic control systems", The Mac Millan Co., New York (1963), (a): pg. 13, 123, (b): pg. 9-12, (c): pg. 96-97.

ON A CLASS OF PRECESSIONAL PHENOMENA AND THEIR STABILITY IN THE SENSE OF LIAPUNOV, POINCARÉ AND LAGRANGE

Demetrios G. MAGIROS *

Abstract

The stability of a class of precessions in the sense of Liapunov, Poincare' and Lagrange is investigated in this paper.

1. Introduction

In this paper we discuss the stability of a class of precessional phenomena of the rotational motion of a rigid body around its axis of symmetry. The stability is examined in the three basic stability concepts, that is stability in the sense of Liapunov, Poincare' and Lagrange. In section 2 we specify the class of precessions, of which the stability is discussed in section 4. The discussion of the stability is based on stability remarks and results of section 3. Ref. 1 and 2 are used for sections 2 and 3, respectively.

2. A Class of Precessional Phenomena

If a rigid body has an axis of symmetry and rotates around it, precessional phenomena occur depending upon the nature of the resultant external torque $\underline{L}$ acting on the body, the torque taken with respect to its center o_1 of the mass of the body. The rotational motion of our symmetric body is governed by the Euler's equations:

$$\dot{\omega}_1 = L_1/I_1, \ \dot{\omega}_2 = L_2/I - \omega_1\omega_3 (I_1 - I)/I, \ \dot{\omega}_3 = L_3/I + \omega_1\omega_2 (I_1 - I)/I \tag{1}$$

where ω_1, ω_2, ω_3 are components of the angular velocity $\underline{\omega}$, and L_1, L_2, L_3 components of the resultant torque $\underline{L}$ along the axes of the coordinate system $o_1x_1x_2x_3$ which is fixed in the moving body. The constants I_1, I_2, I_3 are moments of inertia about the coordinate axes, and if o_1x_1 is the axis of rotation, the last two moments of inertia are equal, $I_2=I_3=I$.

Each solution $(\omega_1, \omega_2, \omega_3)$ of (1) characterizes a precessional motion of the body dependent on the nature of the torque components L_1, L_2, L_3.

We distinguish here the class of precessional motions which correspond to a time-dependent torque, and especially the class corresponding to the torque with components:

$$L_1 = L_1(t), \ L_2 = L_3 = 0 \tag{2}$$

In this case the solution of (1) is given by:

$$\omega_1 = (1/I_1)\int L_1(t)\,dt + c, \ \omega_2 = A\cos Q(t), \ \omega_3 = A\sin Q(t), \ Q(t) = [(I_1 - I)/I]\int\omega_1 dt \tag{3}$$

c and A are constants determined by specializing the initial conditions.

The locus of the endpoint P of the angular velocity $\underline{\omega} = \underline{o_1 P}$ is a curve, the "precessional curve", of which the equations can be found from (3) by eliminating the time t. The precessional curves characterize the precessional phenomena corresponding to the torque component $L_1(t)$.

We specialize $L_1(t)$ as in the following cases:

$$L_1 = 0, \quad L_2 = L_3 = 0 \tag{4.1}$$

$$L_1 = \overline{L_1} = \text{const}, \quad L_2 = L_3 = 0 \tag{4.2}$$

$$L_1 = \sin t, \quad L_2 = L_3 = 0 \tag{4.3}$$

* Consulting Mathematician, General Electric Co., Re-entry and Environmental Systems Division, Philadelphia, Pa., U.S.A.

Reprinted from the *Proceedings of the VIIIth Intl. Symp. Space Tech. Sci.*, Tokyo (1969), 1163–1170.

By using (4), the formulae (3) give:

$$\omega_1 = c = \text{const.}, \ \omega_2 = A \cos Q_1(t), \ \omega_3 = A \sin Q_1(t), \ Q_1(t) = \frac{I_1 - I}{I} ct \tag{5.1}$$

$$\omega_1 = \frac{\overline{L_1}}{I_1} t + c, \ \omega_2 = A \cos Q_2(t), \ \omega_3 = A \sin Q_2(t), \ Q_2(t) = \frac{I_1 - I}{I} ct + \frac{(I_1 - I)\overline{L_1}}{2 I_1 I} t^2 \tag{5.2}$$

$$\omega_1 = -\frac{1}{I_1} \cos t + c, \ \omega_2 = A \cos Q_3(t), \ \omega_3 = A \sin Q_3(t), \ Q_3(t) = \frac{I_1 - I}{I} ct - \frac{I_1 - I}{I_1 I} \sin t \tag{5.3}$$

Equations (5) give three members of the class of precessions (3). (5.1) give the well-known "regular precession", (5.2) the "helicoid precession", and (5.3) a "non-regular periodic precession". Eliminating t in each case of equations (5), one can get the corresponding precessional curves, shown in Fig. 1, 2, 3, respectively.

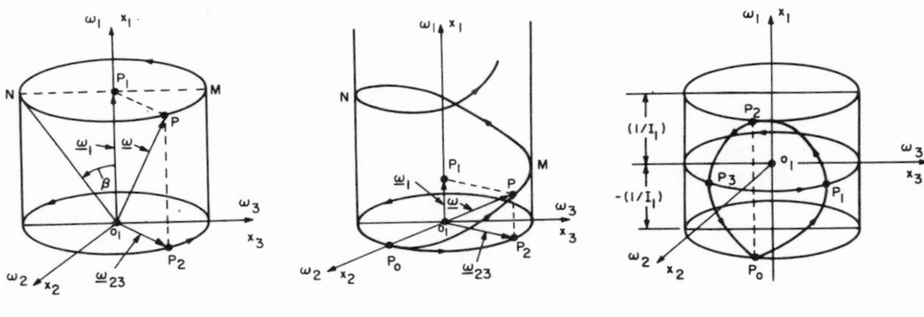

Figure 1 Figure 2 Figure 3

3. Remarks on Stability Concepts

Physically, there are three different basic stability concepts of the motions of dynamical systems depending upon stability considerations of the motion in a given orbit, of the orbit of a given motion, and of the boundedness of the motion and its orbit.

Analytically, a unified description of these stability concepts can be achieved by using the same stability conditions and utilizing the distance ρ in the sense of Liapunov, Poincare' and Lagrange, shown in Fig. 4.

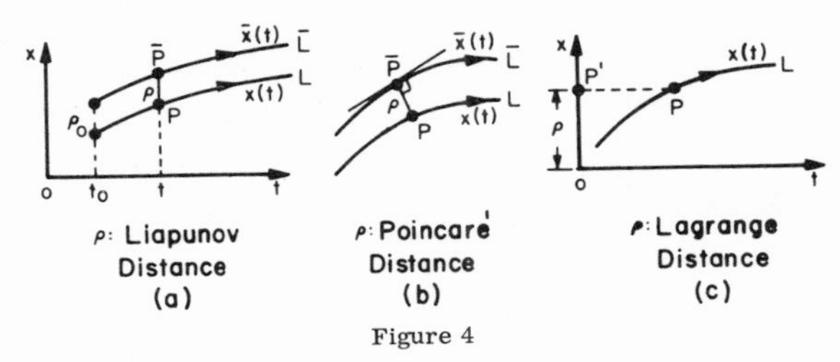

Figure 4

The first two cases of these distances interpret in different ways the correspondence between a point P of the unperturbed orbit L of which the stability is required and a point $\overline{P}$ of the perturbed orbit $\overline{L}$, and these distances measure the effect of a perturbation at P. In Liapunov sense, $\overline{P}$ of $\overline{L}$ corresponds to P of L at the "same-time", Figure 4(a), and in Poincare' sense, $\overline{P}$ of $\overline{L}$ corresponds to P of L in such a way that the distance $\rho = P\overline{P}$ is the minimum of the distances $(P, \overline{L})$ of P from the points of $\overline{L}$, Figure 4(b). In Lagrange sense, ρ is the distance of P from origin in the x-space. The conditions for stability definitions are:

$$\rho_0 < \delta, \quad \rho < \epsilon, \quad \lim_{t \to \infty} \rho = 0, \quad \| p \| < \eta \tag{6}$$

By an appropriate interpretation of the distance ρ and its initial value ρ_0 in these conditions, one can get the three stability concepts in the sense of Liapunov, Poincare' (orbital) and Lagrange (boundedness).

In case the perturbations affect only the initial conditions, that is in case of "sudden perturbations", the first two of the conditions (6) define the "stability" of the motion in L, and the first three of these conditions the "asymptotic stability" of the motion in L.

In case the perturbations $p(x,t)$ affect the whole dynamical system, that is in case of "persistent perturbations", the definitions of stability require a restriction of these perturbations, say a restriction of their norm, $\|p\|$, so that for the stability definitions under persistent perturbations the fourth condition of (6), in addition to all the others, is needed.

The above remarks and the unification of the stability concepts in the conditions (6) lead to important results concerning stability, and of these results we mention just three needed in the present paper:

(a) For the equilibrium states of a system, the stability concepts in the sense of Liapunov and Poincare' are equivalent. For periodic states, the Liapunov stability concept is a narrow one compared to the Poincare' stability concept (orbital). The isochronism, that is the constancy of the frequency, characterizes and suggests the use of the Liapunov concept in periodic states.

(b) A state stable in Liapunov sense is stable in Poincare' sense, and a state unstable in Poincare' sense is unstable in Liapunov sense. But a state stable in Poincare' sense may be either stable or unstable in Liapunov sense.

(c) The Lagrange stability concept, applied either to equilibrium states or to periodic motions in finite distance, classifies these special motions as stable, so the Lagrange stability concept is useless in these motions.

According to the above, one and the same physical phenomenon may be, mathematically speaking, stable or unstable depending on the stability concept which is used for the discussion of the phenomenon. It then becomes necessary that for any stability problem one must select beforehand the stability concept on which the stability discussion will be based, and also know which stability concept is the appropriate one for the problem.

The previous stability concepts and definitions are of mathematical type and in order to be physically accepted they must meet practical needs and agree with "practical stability" and for these appropriate modifications, changes and supplements are to be found, as, e.g., to know:

(a) the size of deviations, the "ϵ-region", that is the "acceptable states", for a satisfactory operation of the system;

(b) the size of initial conditions, the "δ-region", that is the "permitted size of the initial conditions", which can be controlled;

(c) the size of the perturbations, the "p-region".

For practical stability of a solution of a system the nonlinearities of the system must be used, and the linearization of the system is not, in general, permitted.

4. The Stability of the Precessions (5) in the Sense of Liapunov, Poincare' and Lagrange

Let $\omega_{10}, \omega_{20}, \omega_{30}$ be initial values of the precessions (5), and the perturbations affecting only these values, when the stability investigation needs only the first three of the stability conditions (6). We discuss the stability of the above precessions separately.

4.1 The Regular Precession (Eq. 5.1)

(a) Liapunov stability. By using formulae (5.1), the amplitude and frequency of the regular precession are given by:

$$A = (\omega_{20}^2 + \omega_{30}^2)^{1/2}, \quad f = (I_1 - I)\,\omega_{10}/I$$

If the perturbations affect ω_{10}, the constancy of the frequency, in going from the unperturbed precessional curve to the perturbed one, is violated, and the precession is: "Liapunov - unstable".

If the perturbations do not affect ω_{10}, the frequency remains constant and the precession is "Liapunov - stable", but not "Liapunov - asymptotically stable".

(b) Poincare' (or orbital) stability.

Theorem 1. "The regular precession is "orbitally stable", but not "orbitally asymptotically stable".

Proof. The unperturbed and perturbed precessional curves of the above precession are circumferences on parallel planes perpendicular to the ω_1 - axis, which contains their centers, then the Poincare' distance ρ at any point P of the unperturbed precessional curve is a constant, $\rho = \rho_o$.

According to the first two conditions (6), the unperturbed precession is orbitally stable if, for a given small positive number ϵ, it is possible to find a small positive number δ such that if the above distance initially is smaller than δ, $\rho_o < \delta$, then the distance ρ at any point P is smaller than ϵ, $\rho < \epsilon$.

Since $\rho = \rho_o$, the above two inequalities are satisfied by selecting $\delta = \epsilon$, when the precession is "orbitally-stable".

The limiting condition $\rho \to 0$ as $t \to \infty$ is not satisfied, and then the precession is not "orbitally-asymptotically stable".

As a result of the above the precessional curve is a "cycle" but not a "limit cycle".

(c) Lagrange stability (boundedness). The precessional curve of the above precession has, for any time, all its points in finite distance from the origin, then the precession is "Lagrange-stable".

4.2 The Helicoid Precession (Eq. 5.2)

(a) Liapunov stability. The vector $\underline{\omega}_{23}$ on the ω_2, ω_3-plane with components ω_2, ω_3, given by (5.2), is periodic with frequency depending on time, when the motion of its endpoint on the circumference with radius A and center at o_1, is "Liapunov-unstable", and, as a consequence, the precession is "Liapunov-unstable".

(b) Poincare' stability. The Poincare' (or orbital) stability situation of the helicoil precession can be successfully discussed by using the following definitions and theorems.

A generator of the cylinder surface intersects any two particular helicoid precessional curves L and $\overline{L}$ into two infinite sets of points, s: $P_0, P_1, P_2, \ldots . P_n, \ldots$ on the curve L, and $\overline{s}$: $\overline{P}_0, \overline{P}_1, \overline{P}_2, \ldots, \overline{P}_n \ldots$ on $\overline{L}$, Fig. 5(a). At any point P_n of L we can define three different distances, shown in Fig. 5(b),

$$D_n = P_n P_{n+1}, \quad d_n = P_n \overline{P}_n, \quad \rho_n = P_n \overline{P}'_n$$

P_n and P_{n+1} are consecutive points of L, $\overline{P}_n$ of $\overline{L}$ is consecutive point of P_n of L, and $\overline{P}'_n$ is the intersection point of $\overline{L}$ and the plane through P_n of L perpendicular to $\overline{L}$ at $\overline{P}'_n$. D_n is the pitch distance of L at P_n, d_n the pitch distance of L and $\overline{L}$ at P_n, and ρ_n the Poincare' distance of L at P_n, if $\overline{L}$ is the perturbed of L.

The distances D_n, d_n, ρ_n defined as above, have a common limiting property, which is related to the orbited stability of the helicoid precession.

Theorem 2. "The distances D_n, d_n, ρ_n at a point P_n of any particular helicoid precessional curve L of (5.2) decrease to zero, as the point P_n goes to infinity".

Proof. (i) Let us consider the particular helicoid precessional curve corresponding to initial conditions $t_o = 0$, $\omega_{10} = 0$, $\omega_{20} \neq 0$, $\omega_{30} = 0$, then starting from the point P_o $(0, \omega_{20}, 0)$, Fig. 5(a). From equations (5.2) we get:

$$c = 0, \quad A = \omega_{20}, \quad Q_2(t) = \frac{(I_1 - I)\,\overline{L}_1}{2I_1 I}\, t^2$$

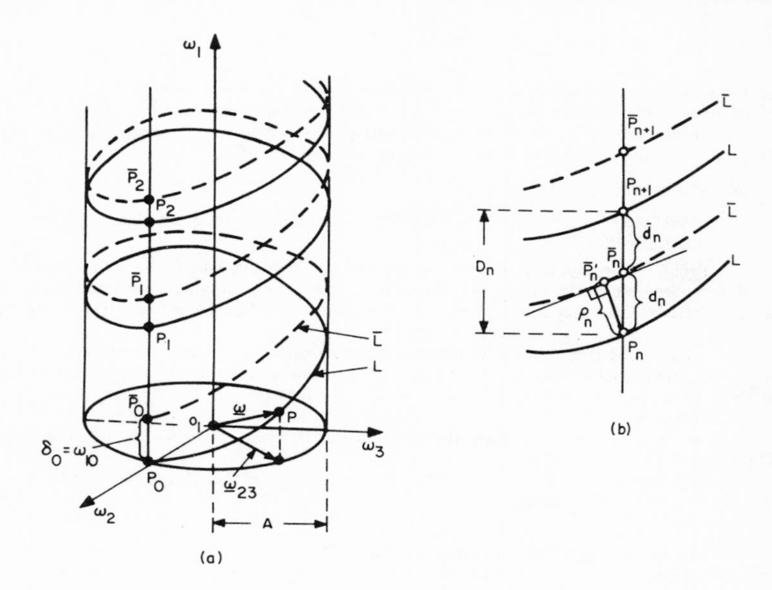

Figure 5

and the equations of L are:

$$\omega_1(t) = \frac{\bar{L}_1}{I_1} t, \; \omega_2(t) = \omega_{20} \cos \left\{ \frac{(I_1 - I)\bar{L}_1 t^2}{2 I_1 I} \right\}, \; \omega_3(t) = \omega_{20} \sin \left\{ \frac{(I_1 - I)\bar{L}_1 t^2}{2 I_1 I} \right\} \tag{7}$$

The generator through P_o intersects the curve (7) into points of which the projection on the ω_2, ω_3 - plane is the point P_o, then for all intersection points we have $\omega_2(t) = \omega_{20}$, $\omega_{30} = 0$, when from equations (7) we have:

$$\frac{(I_1 - I)\bar{L}_1 t^2}{2 I_1 I} = 2\pi n; \; n = 0, \pm 1. \pm 2, \ldots$$

and the time t for the intersection points is:

$$t = \left(\frac{4\pi n I_1 I}{(I_1 - I)\bar{L}_1} \right)^{1/2} \tag{8}$$

We notice that for the reality of t we take positive values of the integer n if $I_1 > I$, and negative if $I_1 < I$.
Inserting (8) into the first of (7) we get the values of ω_1 at the intersection points:

$$(\omega_1)_n = P_o P_n = a\sqrt{n} \tag{9}$$

where the constant a is given by:

$$a = \left(\frac{4 \pi I \bar{L}_1}{I_1 - I) I_1} \right)^{1/2} \tag{9.1}$$

and the integer n is $n \geq 0$.

From (9) we can get the distance D_n at P_n:

$$D_n = P_n P_{n+1} = a\,(\sqrt{n+1} - \sqrt{n}) \tag{10}$$

The sequence $\{\sqrt{n+1} - \sqrt{n}\}$, $n = 0, 1, 2, \ldots$ is monotone decreasing with zero limit as $n \to \infty$, and, since a is a constant, the sequence D_n, $n = 0, 1, 2, \ldots$ has the same property, and then, as the point P_n goes to infinity, we have:

$$\lim_{n \to \infty} D_n = 0 \tag{11}$$

We remark that this property of D_n of the curve (7) is property of D_n corresponding to any curve of (5.2) starting from any point.

(ii) From Figure 5(b) we see that: $D_n = d_n + \bar{d}_n$, when, according to (11):

$$\lim_{n \to \infty} d_n + \lim_{n \to \infty} \bar{d}_n = 0$$

and, since d_n and $\bar{d}_n$ are positive constants,

$$\lim_{n \to \infty} d_n = 0 \tag{12}$$

(iii) The Poincare' distance ρ_n at P_n, Fig. 5(b), is smaller than the distance d_n at P_n, $\rho_n < d_n$, then

$$\lim_{n \to \infty} \rho_n = 0 \tag{13}$$

Theorem 3. "The helicoid precession (5.2) is "orbitally-asymptotically stable".

Proof. Taking any helicoid precessional curve L and its perturbed one $\bar{L}$, we can see that, given an $\epsilon > 0$, we can find a $\delta > 0$ such that $\rho_0 < \delta$ implies $\rho_n < \epsilon$ for any ρ_n. Indeed, such a δ exists, say $\delta = \epsilon$, when if we take initially $d_0 = \delta$, when $\rho_0 < \delta$, we will have $\rho_n < \epsilon$, and the curve L is "orbitally stable"; and since, in addition, $\rho_n \to 0$ as $n \to \infty$, the curve L is "orbitally asymptotically stable".

The δ-region and ϵ-region of an helicoid precessional curve.

If an helicoid precessional curve L starts from the point with coordinates (Ω_{10}, Ω_{20}, Ω_{30}), we can determine the corresponding δ-region and ϵ-region which are needed in order that the above orbital stability of L is of practical importance.

Let $L_0, L_1, \ldots, L_n, \ldots$ be the arcs of L between the points P_0 and P_1, P_1 and P_2, ..., P_n and P_{n+1}, ...

The coordinates ω_1, ω_2, ω_3 of any point of L are related to the coordinates of the initial point of L according to:

$$\left.\begin{aligned} \omega_2^2 + \omega_3^2 &= \Omega_{20}^2 + \Omega_{30}^2 = \text{const.} \\ (\omega_1)_n &\leq \Omega_{10} \leq (\omega_1)_{n+1} \end{aligned}\right\}$$

By using (9), the last inequality reads:

$$a\sqrt{n} \leq \Omega_{10} \leq a\sqrt{n+1}$$

when

$$n \leq \left(\frac{\Omega_{10}}{a}\right)^2 \leq n+1$$

and the value of n is calculated, and therefore:

$$D_n = a(\sqrt{n+1} - \sqrt{n})$$

is known. This distance D_n is the upper limit of δ and ϵ.

(c) Lagrange stability. Since the endpoint P_1 of $\omega_1 = o_1P_1$, Fig. 2, according to the first of (5.2), goes to infinity with the time t, the above precession is "Lagrange-unstable".

4.3 The Non-Regular Periodic Precession (Eq. 5.3)

(a) Liapunov stability. From formulae (5.3) we can see that this precession is "Liapunov - unstable".

(b) Poincare' stability.

Theorem 4. "The above precession is "orbitally - stable", but not "orbitally - asymptotically stable".

Proof. For given initial conditions and known sudden perturbations, the above precession and the corresponding perturbed one are known, also is known the Poincare' distance ρ, associated with any point P of the unperturbed precessional curve. Let ρ_M and ρ_m of these distances be the maximum and minimum, respectively. The smaller perturbations the smaller ρ_M and ρ_m. Now, given a small positive number ϵ and using the Poincare' distance ρ, if one wants $\rho < \epsilon$ for all points P of the unperturbed precessional curve, one must use small perturbations such that $\rho_M < \epsilon$, when an appropriate δ must be $\delta < \rho_m$, a condition which can be satisfied, when the above precession is "orbitally - stable".

The conditions $\rho \to 0$ as $t \to \infty$ cannot be satisfied, when the precession is not "orbitally - asymptotically stable".

(c) Lagrange stability. All the points of the precessional curves of the above precession are in finite distance from the origin, then the precession is "Lagrange - stable".

The above results are summarized in the following table.

TABLE OF RESULTS

Kinds of Stability \ Kinds of Precession	Regular Precession (Eq. 5.1)	Helicoid Precession (Eq. 5.2)	Non-Regular Periodic Precession (Eq. 5.3)
Liapunov Stability	Stable - if the perturbations do not change ω_1 Unstable - if the perturbations do change ω_1	Unstable	Unstable
Poincare' Stability (Orbital)	Stable	Asymptotically Stable	Stable
Lagrange Stability (Boundedness)	Stable	Unstable	Stable

References

1. (a) Magiros, D. G.: The Study of the Orientation of an Earth Satellite, General Electric Company, TIS, 68SD249, RSD, April 19, 1968, Philadelphia, Pa.

(b) Magiros, D. G., and Reehl, G.: Sur quelques sortes de précession dans le mouvement de rotation d'un corps ridige ayant un axe de symétrie, Comptes Rendus, Academie des Sciences, Paris, 266, April 8, 1968.

2. (a) Magiros, D. G.: "On Stability Definitions of Dynamical Systems", National Academy of Sciences (U.S.A.), Proceedings, Vol. 53, No. 6, pp, 1288-1294, June 1965.

(b) Magiros, D. G.: Stability Concepts of Dynamical Systems, Information and Control (U.S.A.) Vol. 9, No. 5, October 1966.

§2.1. Flight Trajectories

ON THE HELICOID PRECESSION: ITS STABILITY AND AN APPLICATION TO A RE-ENTRY PROBLEM

by

D. G. Magiros and G. Reehl

General Electric Company, Re-entry System Organization,
Philadelphia, Pennsylvania (U.S.A.)

Introduction

This investigation deals with problems of helicoid precession, that is its stability and an application to a re-entry problem.

It is found that the helicoid precession is "unstable" in the sense of Liapunov and Lagrange, but "asymptotically stable" in the sense of Poincaré.

The re-entry problem, which is treated by using the concept of the helicoid precession, is the problem of determining the error in the orientation of a spin-stabilized axi-symmetric re-entry vehicle (RV).

In Sec. 1 we give the equations of the helicoid precession, in Sec. 2 the stability of the helicoid precession very briefly, and in Sec. 3 we state a re-entry problem of which the solution is discussed as an application of the concept of the helicoid precession.

1. The Helicoid Precession

If a rigid body has an axis of symmetry and rotates around it, and the resultant torque vector, acting on the body, is, in the body axes, a constant vector with magnitude L, and direction along the symmetry axis, the components w_1, w_2, w_3 of the angular velocity w, in the body axes, are given by

$$\begin{aligned} w_1 &= (L_1/I_1)t + c_1 \\ w_2 &= A\cos Q(t) \\ w_3 &= A\sin Q(t) \\ Q(t) &= \frac{I_1 - I}{I}\, ct + \frac{(I_1 - I)L_1}{2II_1}\, t^2 \end{aligned} \qquad (1)$$

Reprinted from the *Proceedings of the XXth Intl. Astronatucial Congress*, Buenos Aires (1969), 491–496.

c_1 and A are constants determined by the initial conditions; I_1 and I moments of inertia about the symmetry axis and a perpendicular to it, respectively.[1]

Equations (1) give the "helicoid precession", and the curve in space coming from these equations by elimination of time is the "helicoid precessional curve", that is the locus of the endpoint P of the vector $\boldsymbol{w} = \boldsymbol{o}_1 \boldsymbol{P}$, where o_1 is the mass center of the body (Fig. 1*a*). An individual helicoid precessional curve corresponds to given initial conditions w_{10}, w_{20}, w_{30}.

2. The Stability of the Helicoid Precession

By using sudden perturbations, which affect only the initial conditions, the three basic stability concepts in the sense of Liapunov, Poincaré and Lagrange are unified in the three stability conditions:

$$\varrho_0 < \delta, \quad \varrho < \varepsilon, \quad \lim_{t \to \infty} \varrho = 0 \tag{2}$$

where δ and ε are positive constants, and ϱ_0 and ϱ can be interpreted as distances in the sense of Liapunov, or Poincaré or Lagrange.[2]

The nature of the stability of the precession (1) can be found by using the conditions (2).[3]

We will have:

(a) *The helicoid precession* (1) *is unstable in Lagrange sense*, since the component w_1 is unbounded as $t \to \infty$.

(b) *This precession is unstable in Liapunov sense*, since the vector w_{23} with components w_2 and w_3 in the w_2, w_3-plane Fig. 1*a* is periodic with time-dependent frequency, when the motion of the end-point of w_{23} on the circumference with radius A and center at o_1 is unstable in Liapunov sense.

(c) *The above precession is asymptotically stable in Pioncaré sense* (*orbitally*). The proof of this statement comes from the property that the distances D_n, d_n, ϱ_n' defined in the following, decrease to zero as $n \to \infty$.

A generator of the surface of the cylinder, on which the helicoid precessional curves lie, intersects any two such curves L and $\bar{L}$ into two sets of points: s: $P_0, P_1, \ldots, P_n, \ldots$ on L, and $\bar{s}$: $\bar{P}_0, \bar{P}_1, \ldots, \bar{P}_n, \ldots$ on $\bar{L}$, Fig. 1*a*.

The distances D_n, d_n, ϱ_n at any point P_n of L, Fig. 1*b*, are defined as:

$$D_n = P_n P_{n+1}, \quad d_n = P_n \bar{P}_n, \quad \varrho_n = P_n \bar{P}_n' \tag{3}$$

where $\bar{P}_n'$ is the intersection point of $\bar{L}$ by the plane through the point P_n perpendicular to $\bar{L}$.

D_n and d_n are different kinds of pitch distances, and ϱ_n the Poincaré distance at P_n of L.

We can prove that:

$$D_n = \alpha(\sqrt{n+1} - \sqrt{n}), \quad d_n < D_n, \quad \varrho_n < d_n \tag{4}$$

where the constant a is dependent on L_1, I, I_1.

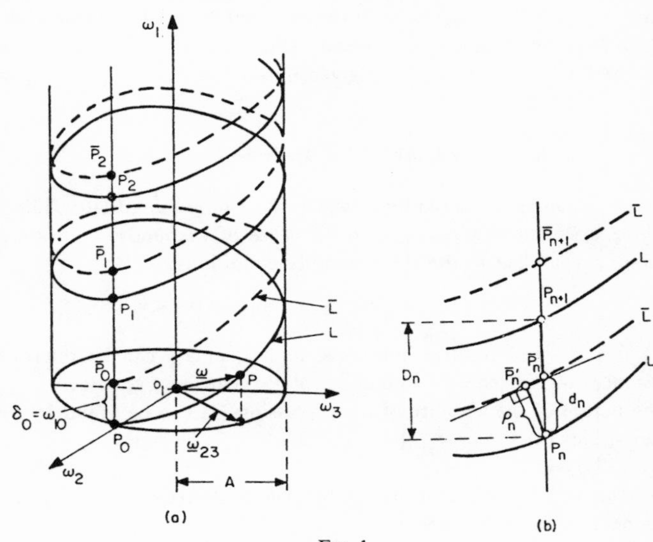

FIG. 1

Since the limit of $\{\sqrt{n+1} - \sqrt{n}\}$, as $n \to \infty$, is zero, we have:

$$\lim_{n\to\infty} D_n = 0, \quad \lim_{n\to\infty} \varrho_n = 0. \tag{5}$$

Combining (2) and (5) we can prove the asymptotic stability of the helicoid precession (1).

3. Application of the Concept of the Helicoid Precession to a Re-entry Problem

3.1. The problem

Shortly after it separates from its booster, an axi-symmetric RV is spun-up about its roll axis by means of small rockets. The reason for spinning it is to maintain the roll axis in the desired orientation during exospheric flight and to minimize dispersion. During firing of constant thrust rockets the motion is a helicoid precession, but after the spin rockets have fired the RV is spinning with regular precession. "*The problem is to determine the error in the orientation of the RV by finding the values of the coordinate angles at the time of the constant thrust termination* (that is at the time when the helicoid precession of the RV stops and the regular precession starts) *with respect to the initial attitude of the RV*".

The above problem has been studied by numerical methods on a linear basis, but the method presented here provides a quantitative assessment of the general behavior of non-linear type.

3.2. The solution

We consider the orthogonal coordinate system $OX_1X_2X_3$ (Fig. 2*a*), fixed in space, the system $o_1X_1'X_2'X_3'$ parallel to $OX_1X_2X_3$, o_1 the mass center of the RV, and the orthogonal system $o_1x_1x_2x_3$, fixed in the body, as a system of principal axes of inertia of the RV. The orientation of the RV moving in space can be specified by the "angular coordinates" θ, ψ, φ defined as follows (Fig. 2*b*):

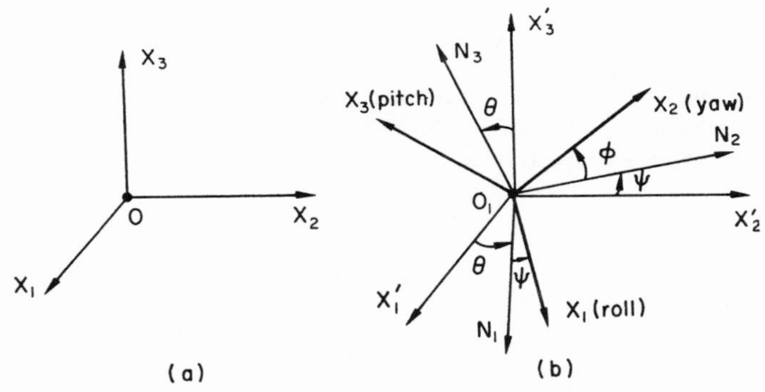

Fig. 2. Angle θ: rotate $X_1'X_3'$-plane about X_2'-axis through angle θ when X_1' comes to N_1 and X_3' to N_3. Angle ψ: rotate N_1X_2'-plane about N_3-axis through angle ψ when N_1 comes to x_1-axis and $X_2'$to N_2. Angle φ: rotate N_2N_3-plane about x_1-axis through angle φ when N_2 comes to x_2-axis and N_3 to x_3-axis

If we select the x_1, x_2, x_3-axes as the roll, yaw, pitch axes of the RV, and along these axes, respectively, we take the body coordinates w_1, w_2, w_3 of the angular velocity vector $\boldsymbol{w}$ of the rotation of the RV, these angular velocity coordinates are given by the following formulae:

$$\left.\begin{aligned} w_1 &= \dot{\varphi}+\dot{\theta}\sin\psi \\ w_2 &= \dot{\psi}\sin\varphi+\dot{\theta}\cos\varphi\cos\psi \\ w_3 &= \dot{\psi}\cos\varphi-\dot{\theta}\sin\varphi\cos\psi \end{aligned}\right\}. \tag{6}$$

By the assumption that the RV has the x_1-axis as the symmetry axis, which is a principal axis, the two moments of inertia I_2, I_3 at o_1 are equal,

$I_2 = I_3 = I \neq I_1$, where I_1 is the axial moment of inertia, and I the transverse moment of inertia. The angle ψ will normally be small, and then formulae (6) can be approximated by:

$$\left.\begin{aligned} w_1 &= \dot{\varphi} \\ w_2 &= \dot{\psi}\sin\varphi+\dot{\theta}\cos\varphi \\ w_3 &= \dot{\psi}\cos\varphi-\dot{\theta}\sin\varphi \end{aligned}\right\}. \tag{7}$$

from which we get:

$$\left.\begin{aligned} \dot{\varphi} &= w_1 \\ \dot{\delta} &= w_2\cos\varphi-w_3\sin\varphi \\ \dot{\psi} &= w_2\sin\varphi+w_3\cos\varphi \end{aligned}\right\}. \tag{8}$$

If $t = 0$ is the time of application of the spin-up on the RV, and at that time the coordinate system $X_1'X_2'X_3'$ and $x_1x_2x_3$ are in coincidence, then the initial conditions, associated with equations (8), are:

$$\varphi_0 = \theta_0 = \psi_0 = 0 \tag{8.1}$$

If the spin-up torque is along the roll axis with the constant magnitude L_1, the rotational motion of the RV during the action of this torque, is a helicoid precession, when Eqs. (1) read:

$$\left.\begin{aligned} w_1 &= \frac{L_1}{I_1}t \\ w_2 &= A\cos\left\{\frac{(I_1-I)L_1}{2II_1}t^2\right\} \\ w_3 &= A\sin\left\{\frac{(I_1-I)L_1}{2II_1}t^2\right\} \end{aligned}\right\}. \tag{9}$$

Inserting the first of (9) into the first of (8) and integrating we have:

$$\varphi = \frac{L_1}{2I_1}t^2 \tag{10}$$

Inserting now the second and third of (9) into the second and third of (8) and using appropriate trigonometric identities and formula (10) we get:

$$\left.\begin{aligned} \dot{\theta} &= A\cos\left(\frac{1}{2}\frac{L_1}{I}t^2\right) \\ \dot{\psi} &= A\sin\left(\frac{1}{2}\frac{L_1}{I}t^2\right) \end{aligned}\right\}. \tag{11}$$

To integrate (11), since the angle is very small compared to unity, we expand the functions "cosine" and "sine" in Taylor Series and take a few terms, when we have the following approximate formulae:

$$\left.\begin{aligned}\dot{\theta} &= A\left\{1-\frac{1}{2}\left(\frac{L_1}{2I}\right)^2 t^4+\frac{1}{24}\left(\frac{L_1}{2I}\right)^4 t^8+O(t^{12})\right.\\ \dot{\psi} &= A\left\{\frac{L_1}{2I}t^2-\frac{1}{6}\left(\frac{L_1}{2I}\right)^3 t^6+O(t^{10})\right.\end{aligned}\right\} \tag{12}$$

The terms omitted in the series expansion are of order $O(t^{12})$ and $O(t^{10})$ in the first and second of Eqs. (12), respectively. By integration of (12) we get:

$$\left.\begin{aligned}\theta &= A\left\{t-\frac{1}{10}\left(\frac{L_1}{2I}\right)^2 t^5+\frac{1}{216}\left(\frac{L_1}{2I}\right)^4 t^9\right\}+O(t^{13})\\ \psi &= A\left\{\frac{L_1}{6I}t^3-\frac{1}{42}\left(\frac{L_1}{2I}\right)^3 t^7\right\}+O(t^{11})\end{aligned}\right\} \tag{13}$$

If $t = t_s$ is the time of constant thrust termination, and $\varphi_s, \theta_s, \psi_s$ the values of the angular coordinates φ, θ, ψ at that time, we can get:

$$\left.\begin{aligned}\varphi_s &= \frac{1}{2} w_s t_s\\ \theta_s &= At_s\left\{1-\frac{1}{10}\left(\frac{I_1}{2I}w_s t_s\right)^2+\frac{1}{216}\left(\frac{I_1}{2I}w_s t_s\right)^4\right\}\\ \psi_s &= \frac{1}{3}At_s\left\{\frac{I_1}{2I}w_s t_s\right\}\left\{1-\frac{1}{14}\left(\frac{I_1}{2I}w_s t_s\right)^2\right\}\end{aligned}\right\} \tag{14}$$

where:

$$w_s = (w)_{t=t_s} = \frac{L_1}{I_1} t_s. \tag{14.1}$$

After the thrust termination the motion is a regular precession with the roll axis rotating around the angular momentum vector $\boldsymbol{H}_s$ at $t = t_s$ and precession angle β_s given by:

$$\tan\beta_s = \frac{A}{w_s} = \frac{AI_1}{L_1 t_s} \tag{15}$$

References

[1] MAGIROS, D. G. and REEHL, G., Sur quelques sortes de précession dans le mouvement de rotation d'un corps ayant un axe de symétrie, Comptes Rendus, Academy of Sciences in Paris, France, t. 266, pp. 770–773 (8 April 1968)

[2] MAGIROS, D. G., Stability concepts of dynamical systems, *J. Information and Control, U.S.A.* **9**, No. 5 (Oct. 1966)

[3] MAGIROS, D. G., La stabilité de la précession hélicoidale dans le sens de Liapunov, Poincaré et Lagrange, Comptes Rendus, Academy of Sciences in Paris, France, t. 268, pp. 652–654 (24 March 1969)

22nd INTERNATIONAL ASTRONAUTICAL CONGRESS

Brussels, Belgium

September, 1971

ORIENTATION OF THE ANGULAR MOMENTUM VECTOR OF A SPACE VEHICLE AT THE END OF SPIN-UP

ABSTRACT

In this paper, the angular momentum vector of a space vehicle at the end of spin-up is determined.

Use is made of the concept of the helicoid precession, which characterizes the rotational motion of the vehicle during the action of a constant thrust acting on the vehicle.

Reprinted from the *XXIInd Intl. Astronautical Congress*, Brussels (1971), 1–8.

ORIENTATION OF THE ANGULAR MOMENTUM VECTOR OF A SPACE VEHICLE AT THE END OF SPIN-UP(*)

BY: Demetrios G. Magiros(**)

1. INTRODUCTION

This investigation deals with the determination of the angular momentum vector of a space vehicle (S.V.) at the end of spin-up. The concept of the "helicoid precession", introduced and used in the papers of reference 1 and 2, is applied in this investigation.

The procedure in the paper gives a method, which seems to be a unique one, to treat such a subject.

Shortly after it separates from its booster, an axi-symmetric re-entry vehicle is spun-up about its roll axis by means of small rockets of constant thrust. During the action of the constant thrust, the "helicoid precession" characterizes the motion of the SV; but, just before and after the action of the constant thrust, the motion of the SV is characterized by different "regular precessions". As a result of this, the magnitude and orientation of the angular momentum vector H of the SV at the time of the termination of the constant thrust have different values than that before the action of the constant thrust.

(*)Communicated to the "22nd International Astronautical Congress, Brussels, Belgium, September, 1971.

(**) Consulting mathematician, General Electric Co., Re - Entry and Environmental Division, Philadelphia,Pa.,U.S.A.

The determination of the difference of these values, which gives the "error in the angular momentum vector" is the subject of this paper.

Mr. G. Reehl, an author's collaborator, helped in the calculations.

2. THE ANGULAR COORDINATES AND THEIR DERIVATIVES AS TIME FUNCTIONS

Following the reference paper (2), we give here the formulae for the angular coordinates and their derivatives as time functions, which will help the calculation of the angular momentum vector at the time of the constant thrust termination.

We consider the orthogonal coordinate system $OX_1X_2X_3$, Figure 1 (a), fixed in space, the system $oX_1'X_2'X_3'$ parallel to $OX_1X_2X_3$; o the mass center of the SV and the orthogonal system $ox_1x_2x_3$, fixed in the body, as a system with principal axes of inertia of the SV.

The orientation of the SV moving in space can be specified by the "angular coordinates" θ, ψ, φ , defined as follows, Figure 1(b).

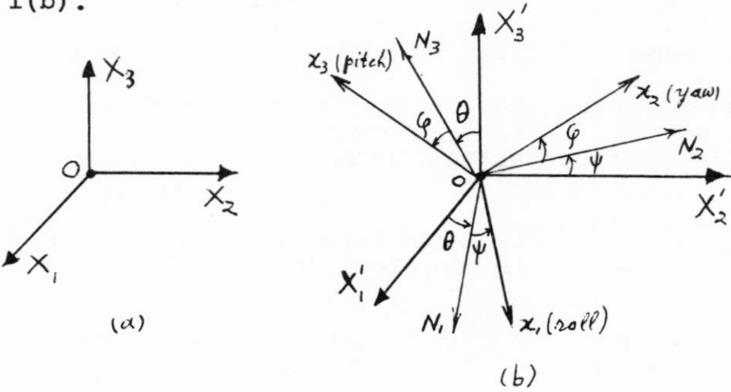

FIGURE 1

Angle θ : Rotate $X_1' X_3'$ - plane about X_2'- axis through angle θ , when X_1' comes to N_1 and X_3' to N_3 ;

Angle ψ : Rotate $N_1 X_2'$ - plane about N_3 - axis through angle ψ , when N_1 comes to x_1 - axis and X_2' to N_2 ;

Angle φ : Rotate $N_2 N_3$ - plane about x_1- axis through angle φ , when N_2 comes to x_2- axis and N_3 to x_3- axis.

If we select the x_1, x_2, x_3-axes as the roll, yaw, pitch-axes of the SV, and if $t=0$ is the time of the application of the spin-up on the SV, and if at that time the coordinate systems $X_1' X_2' X_3'$ and $x_1 x_2 x_3$ are in coincidence, when $\theta=\psi=\varphi=0$, these angles at time t , according to the reference paper 2, are approximately given by:

$$\left.\begin{aligned} \varphi &= \frac{L_1}{2I_1} t^2 \\ \theta &= A\left[t - \frac{1}{10}\left(\frac{L_1}{2I}\right)^2 t^5 + \frac{1}{216}\left(\frac{L_1}{2I}\right)^4 t^9\right] + O(t^{13}) \\ \psi &= A\left[\frac{L_1}{6I} t^3 - \frac{1}{42}\left(\frac{L_1}{2I}\right)^3 t^7\right] + O(t^{11}) \end{aligned}\right\} \quad (1)$$

where L_1 is the applied roll torque, I_1 the roll moment of inertia, I the moment of inertia about the transverse axis and $A = (\omega_{20}^2 + \omega_{30}^2)^{1/2}$; ω_{20}, ω_{30} initial angular velocity components along the x_2, x_3 - axes.

The symbol $O(t^m)$ denotes a function $k(t^m)$ such that $\frac{\|k(t^m)\|}{t^m} \leq c$ = constant as $t \to 0$. From formulae (1) we have:

$$\left.\begin{aligned} \dot{\varphi} &= \frac{L_1}{I_1} t \\ \dot{\theta} &= A\left[1 - \frac{1}{2}\left(\frac{L_1}{2I}\right)^2 t^4 + \frac{1}{24}\left(\frac{L_1}{2I}\right)^4 t^8\right] + O(t^{12}) \\ \dot{\psi} &= A\left[\frac{L_1}{2I} t^2 - \frac{1}{6}\left(\frac{L_1}{2I}\right)^3 t^6\right] + O(t^{10}) \end{aligned}\right\} \quad (2)$$

If $t=t_s$ is the time of the constant thrust termination and $\theta_s, \varphi_s, \psi_s, \dot{\theta}_s, \dot{\varphi}_s, \dot{\psi}_s$ are the values of the coordinate angles and their time rates at $t = t_s$, we have:

$$\left.\begin{aligned}
\varphi_s &= \tfrac{1}{2}\omega_s t_s \\
\theta_s &= A t_s [1-\tfrac{1}{10}(\tfrac{I_1}{2I}\omega_s t_s)^2+\tfrac{1}{216}(\tfrac{I_1}{2I}\omega_s t_s)^4 \\
\psi_s &= \tfrac{1}{3} A t_s [\tfrac{I_1}{2I}\omega_s t_s][1-\tfrac{1}{14}(\tfrac{I_1}{2I}\omega_s t_s)^2] \\
\dot{\varphi}_s &= \tfrac{L_1}{I_1} t_s = \omega_s \\
\dot{\theta}_s &= A [1-\tfrac{1}{2}(\tfrac{I_1}{2I}\omega_s t_s)^2+\tfrac{1}{24}(\tfrac{I_1}{2I}\omega_s t_s)^4] \\
\dot{\psi}_s &= A [\tfrac{I_1}{2I}\omega_s t_s][1-\tfrac{1}{6}(\tfrac{I_1}{2I}\omega_s t_s)^2]
\end{aligned}\right\} \qquad (3)$$

The first three formulae of (3) give the orientation of the SV at the t_s when the constant thrust stops acting on the SV.

3. FORMULAE FOR THE ANGULAR MOMENTUM

The angular momentum vector $\underline{H}_s$ at $t=t_s$ in vehicle coordinates is given by:

$$\underline{H}_s = \underline{x}_1 I_1 \omega_s + \underline{x}_2 I \omega_{2s} + \underline{x}_3 I \omega_{3s} \qquad (4)$$

where $\omega_s, \omega_{2s}, \omega_{3s}$ are the angular velocity components about the x_1, x_2, x_3- axes at $t=t_s$. From reference (2) we have:

$$\left.\begin{aligned}
\omega_{2s} &= \dot{\psi}_s \sin\varphi_s + \dot{\theta}_s \cos\varphi_s \\
\omega_{3s} &= \dot{\psi}_s \cos\varphi_s - \dot{\theta}_s \sin\varphi_s
\end{aligned}\right\} \qquad (5)$$

The body x_1, x_2, x_3 - axes are related to the inertial X_1, X_2, X_3 - axes by the matrix equation:

$$\begin{bmatrix} X_1 \\ X_2 \\ X_3 \end{bmatrix} = \begin{bmatrix} \cos\theta\cos\psi & \begin{Bmatrix} \sin\theta\sin\varphi - \\ -\cos\theta\cos\varphi\sin\psi \end{Bmatrix} & \begin{Bmatrix} \sin\theta\cos\varphi + \\ +\cos\theta\sin\varphi\sin\psi \end{Bmatrix} \\ \sin\psi & \cos\varphi\cos\psi & -\sin\varphi\cos\psi \\ -\sin\theta\cos\psi & \begin{Bmatrix} \cos\theta\sin\varphi + \\ +\sin\theta\cos\varphi\sin\psi \end{Bmatrix} & \begin{Bmatrix} \cos\theta\cos\varphi - \\ -\sin\theta\sin\varphi\sin\psi \end{Bmatrix} \end{bmatrix} \cdot \begin{bmatrix} x_1 \\ x_2 \\ x_3 \end{bmatrix} \quad (6)$$

Since θ_s and ψ_s are normally small, equation (6) at $t=t_s$ can be approximated by:

$$\begin{bmatrix} X_1 \\ X_2 \\ X_3 \end{bmatrix} = \begin{bmatrix} 1 & (\theta_s\sin\varphi_s - \psi_s\cos\varphi_s) & (\theta_s\cos\varphi_s + \psi_s\sin\varphi_s) \\ \psi_s & \cos\varphi_s & -\sin\varphi_s \\ -\theta_s & \sin\varphi_s & \cos\varphi_s \end{bmatrix} \cdot \begin{bmatrix} x_1 \\ x_2 \\ x_3 \end{bmatrix} \quad (7)$$

Using (4) and (7), the component of $\underline{H}_s$ along X_1 is given by:

$$H_{1s} = I_1\omega_s + I\,[\omega_{2s}(\theta_s\sin\varphi_s - \psi_s\cos\varphi_s) + \omega_{3s}(\theta_s\cos\varphi_s + \psi_s\sin\varphi_s)] \quad (8)$$

and substituting from equation (5):

$$H_{1s} = I_1\omega_s + I(\theta_s\dot{\psi}_s - \psi_s\dot{\theta}_s) \quad (9)$$

For values of the parameters normally encountered in practice, the second term is negligible compared to the first. Thus, approximately, we have:

$$H_{1s} = I_1 \omega_s \tag{10}$$

Also, from equations (4) and (7), the components of angular momentum along the inertial axes X_2 and X_3 are given by:

$$\left.\begin{aligned} H_{2s} &= I_1 \omega_s \psi_s + I\,(\omega_{2s}\cos\varphi_s - \omega_{3s}\sin\varphi_s) \\ H_{3s} &= -I\,\omega_s \theta_s + I\,(\omega_{2s}\sin\varphi_s + \omega_{3s}\cos\varphi_s) \end{aligned}\right\} \tag{11}$$

and substituting from equation (5):

$$\begin{aligned} H_{2s} &= I\,\dot{\theta}_s + I_1\,\omega_s\,\psi_s \\ H_{3s} &= I\,\dot{\psi}_s - I_1\,\omega_s\,\theta_s \end{aligned} \tag{12}$$

Since the final orientation of the angular vector will not deviate very much from X_1' in practical cases, it can be specified by two small angles $\Delta\theta$ about X_2' and $\Delta\psi$ about X_3', defined as follows:

$$\Delta\theta = -\frac{H_{3s}}{H_{1s}} = \theta_s - \frac{I\,\dot{\psi}_s}{I_1\,\omega_s} \tag{13}$$

$$\Delta\psi = \frac{H_{2s}}{H_{1s}} = \psi_s + \frac{I\,\dot{\theta}_s}{I_1\,\omega_s} \tag{14}$$

when substituting from equations (2) and (3), become:

$$\Delta\theta = \frac{At_s}{2}\left[1 - \frac{1}{30}\left(\frac{I_1}{2I}\omega_s t_s\right)^2 + \frac{1}{108}\left(\frac{I_1}{2I}\omega_s t_s\right)^4\right. \tag{15}$$

$$\Delta\psi = \frac{IA}{I_1\omega_s} + \frac{At_s}{12}\left(\frac{I_1}{2I}\omega_s t_s\right)\left[1 - \frac{1}{108}\left(\frac{I_1}{2I}\omega_s t_s\right)^2\right] \tag{16}$$

The first term of equation (16) is the precession angle of the regular precession, that is, the angle between the vehicle roll axis and the regular momentum vector.

We remark that the procedure and the results of the present paper may be useful to some research area of which here we mention the following two:

a. For manned planetary missions, any spacecraft should be designated to create artificial gravity, which could be accomplished by rotating the vehicle continuously during the flight. The rotation produces precession and special angular momentum vector.

b. If an atom is placed into a magnetic field in a direction, the field exerts a torque on the magmetic dipole moments, trying to align it with the direction. The atom has, in general, an angular momentum, that is an orbital and an intrinsic (spin) one, and the atom in the above situation becomes like any gyroscopic.

REFERENCES

1. D. G. Magiros and G. Reehl: "Sur quelques sortes de précession dans le movement de rotation d'un corps ayant an axe de symétrie," Comptes Rendus, Academie de Sciences, Paris, France, V. 266, 770-776, April 8, 1968.

2. D. G. Magiros and G. Reehl: "On the helicoid precession: Its Stability and an application to a re-entry problem," Proceedings, XX International Astronautical Congress, Buenos Aires, Argentina, October, 1969.

ΠΡΑΚΤΙΚΑ ΤΗΣ ΑΚΑΔΗΜΙΑΣ ΑΘΗΝΩΝ

ΣΥΝΕΔΡΙΑ ΤΗΣ 25ΗΣ ΜΑΪΟΥ 1972

ΠΡΟΕΔΡΙΑ ΓΡΗΓ. ΚΑΣΙΜΑΤΗ

ΑΝΑΚΟΙΝΩΣΕΙΣ ΜΗ ΜΕΛΩΝ

ΕΦΗΡΜΟΣΜΕΝΑ ΜΑΘΗΜΑΤΙΚΑ.— **The stability of a class of helicoid precessions in the sense of Liapunov and Poincaré,** *by Demetrios G. Magiros** Ἀνεκοινώθη ὑπὸ τοῦ Ἀκαδημαϊκοῦ κ. Ἰω. Ξανθάκη.

Introduction

In a previous paper, Ref. 1, we discussed the stability of a helicoid precession in case of constant torque, whereby, by employing different stability concepts, we found for this precession different stability situations. The concept of this helicoid precession was successfully applied in problems of current interest in Astrodynamics, treated in papers Ref. 2, 3.

In the present paper, we discuss the stability of a «class of helicoid precessions», of which the helicoid precession of the paper Ref. 1 is only a member.

The concepts of stability in the sense of Liapunov and Poincaré, Ref. 4, are employed.

We found that all the members of the class of precessions are

* ΔΗΜΗΤΡΙΟΥ Γ. ΜΑΓΕΙΡΟΥ, Ἐπιστημονικοῦ Συμβούλου τῆς General Electric Co, RESD Philadelphia, PA, U.S.A. : **Ἡ εὐστάθεια μιᾶς κλάσεως ἑλικοειδῶν μεταπτώσεων κατὰ τὴν ἔννοιαν τῶν Liapunov καὶ Poincaré.**

Reprinted from the *Proceedings of the Athens Academy of Sciences* **47** (1972), 102–110.

unstable in Liapunov sense; but in Poincaré sense the stability of a member S of the class is either stable, or asymptotically stable, or unstable, if the limit value of the pitch distance of S is either a constant, or zero, or infinite, respectively.

There are reasons which suggest that the stability situation of the above class of the helicoid precessions in the sense of Poincaré is close to practical stability, then it is preferred.

1. The class of the helicoid precessions

The rotational motion of a rigid body around its symmetry axis, is governed by the Euler's equations:

$$\dot{\omega}_1 = \frac{L_1}{I_1}, \quad \dot{\omega}_2 = \frac{L_2}{I} - \frac{I_1 - I}{I}\omega_1\omega_3, \qquad \dot{\omega}_3 = \frac{L_3}{I} + \frac{I_1 - I}{I}\omega_1\omega_3 \quad (1)$$

where $\underline{\omega} = (\omega_1, \omega_2, \omega_3)$ is the angular velocity; $\underline{L} = (L_1, L_2, L_3)$ the external resultant torque acting on the body, I_1, $I_2 = I_3 = I$ the moments of inertia about the coordinate axis $0_1\omega_1\omega_2\omega_3$, 0_1 the center of the mass of the boby.

In the case where the torque is:

$$L_1 = L_1(t), \qquad L_2 = 0, \qquad L_3 = 0 \qquad (2)$$

the solution of (1) is:

$$\left.\begin{aligned} &\omega_1(t) = \frac{1}{I_1}\int L_1(t) + c, \quad \omega_2(t) = A\cos Q(t), \quad \omega_3(t) = A\sin Q(t) \\ &\qquad Q(t) = \frac{I_1 - I}{I}\int \omega_1(t)\,dt \end{aligned}\right\} \quad (3)$$

c and A are constants to be determined from the initial angular velocity:

$$\underline{\omega_0} = (\omega_{10}, \omega_{20}, \omega_{30}), \quad \text{and} \quad A = (\omega_{20}^2 + \omega_{30}^2)^{1/2}.$$

In case $\omega_1(t)$ is increasing function of and tends to infinity with time t, the solution (3) is a helicoid curve S on the surface of an orthogonal circular cylinder of radius A and gives a helicoid precession corresponding to the specified function $L_1(t)$, and so (3) gives a «class of helicoid precessions», each member of which is determined by the specification of $L_1(t)$. We remark that if $\omega_1(t)$ does not satisfy the above requirement, the corresponding precession is not helicoid, as, e.g., for $L_1 = 0$, when we have the «regular precession» and its «precessional curve» is circumference on the surface of the cylinder; or for $L_1 = \sin t$, when the precessional curve is closed curve on the surface of the cylinder. But, for $L_1 =$ constant, $\omega_1(t) \rightarrow \infty$ as $t \rightarrow \infty$, and we have a helicoid precession, the simplest one of the class (3).

We discuss here the stability of the class of helicoid precessions (3) in the sense of Liapunov and Poincaré.

2. Stability in Liapunov sense

The vector $\underline{\omega}_{23} = (\omega_2, \omega_3)$ on the ω_2, ω_3 — plane, Fig. 1(a), of which the components are given by (3), is periodic in t, but with period dependent on t, then the motion of the end point of this vector on the circumference with radius A and center 0_1, is unstable in Liapunov sense, Ref. 4, and, as a consequence, *«the class of helicoid precessions is «unstable» in Liapunov sense»*.

3. Stability in Poincaré sense (orbital stability)

The orbital stability of any member of the above class of precessions depends upon the structure of the corresponding function $L_1(t)$. Some auxiliary distances and their properties, shown below, will help to create a criterion for the orbital stability.

3.1 Some auxiliary distances and their properties.

Let us take two helicoid precessional curves S and $\overline{S}$ belonging to the same family, that is corresponding to the same function $L_1(t)$, but starting from different points $P_0(0, \omega_{.0}, 0)$ and $\overline{P}_0(\omega_{10}, \omega_{20}, 0)$, respectively, Fig. 1(a).

The generator of the cylinder through P_0 intersects S into the points: $P_0, P_1, P_2, \ldots, P_n, \ldots$, and $\overline{S}$ into the points: $\overline{P}_0, \overline{P}_1, \overline{P}_2, \ldots, \overline{P}_n, \ldots, S_0$, at the point P_n of S, we can define, as shown in Fig. 1(b), the pitch distances: $D_n = P_n P_{n+1}$, $d_n = P_n \bar{P}_n$. A third distance at P_n is defined by the plane through P_n perpendicular to $\bar{S}$ at $\bar{P}'_n$, the distance $\varrho_n = P_n \bar{P}'_n$.

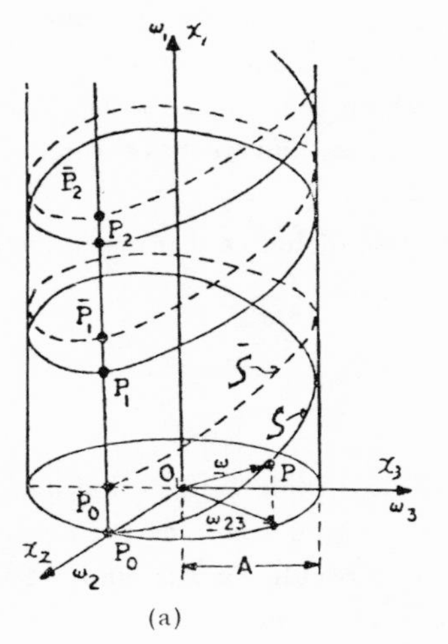

(a)

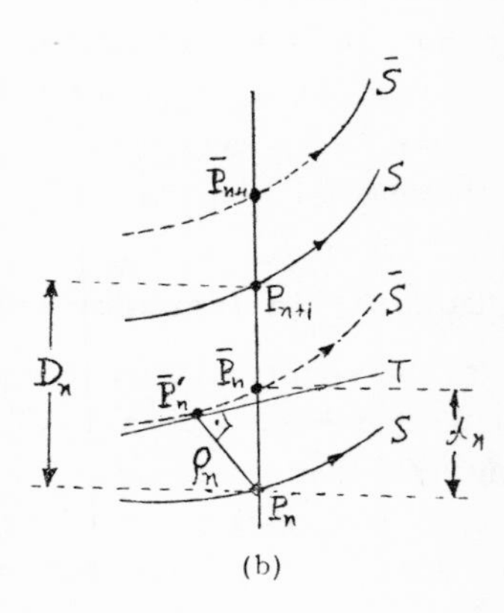

(b)

Fig. 1.

The above distances have properties very useful for the discussion of the orbital stability of the helicoid precessions. These properties are given by the following theorem.

Theorem 1. *The distances* D_n, d_n, ϱ_n *have the properties*:

. (a): $D_n > d_n > \varrho_n$

. (b): *The limit distance:* $\lim_{n \to \infty} D_n = \bar{D}$ *is either a constant,* or zero, *or infinite, when the* $\lim_{n \to \infty} d_n = \bar{d}$, $\lim_{n \to \infty} \varrho_n = \bar{\varrho}$ *are either constant, or zero, or infinite, respectively.*

Proof . (a): The point $\bar{P}_n$ is a point of the segment $P_n P_{n+1}$, Fig. 1 (b), which means that: $D_n > d_n$.

The plane through P_n and perpendicular to $\bar{S}$ at $\bar{P}'_n$ intersects perpendicularly the tangent $\bar{P}'_n T$ of $\bar{S}$ at $\bar{P}'_n$, so this tangent is perpendicular to the distance ϱ_n. The plane through $\bar{P}'_n$ and perpendicular to ϱ_n contains the tangent $\bar{P}'_n T$ and divides the whole space into two parts, one of which contains all the points of $\overline{P'_n}\,\overline{S}$, and the other contains the point P_n, so the distance $P_n \overline{P}_n$ is bigger than the distance $P_n \overline{P'_n}$, that is $d_n > \varrho_n$.

(b): For the second part of the theorem the calculation of the limit $\overline{D}$ is needed.

The curve S starts from $P_0\,(0, \omega_{20}, 0)$ at $t_0 = 0$, and corresponds to $c = 0$ in the formulae (3). Its equations are:

$$\left.\begin{aligned} \omega_1(t) = \frac{1}{I_1}\int L_1(t)\,dt, \quad \omega_2(t) = \omega_{20}\cos Q(t), \quad \omega_3(t) = \omega_{20}\sin Q(t) \\ Q(t) = \frac{I_1 - I}{I}\int dt \int L_1(t)\,dt \end{aligned}\right\} \quad (4)$$

For the points $P_0, P_1, P_2, \ldots, P_n$, we have $\omega_2(t) = \omega_{20}$, $\omega_3(t) = 0$, then the quantity $Q(t)$ of (4) for these points must be:

$$Q(t_n) = 2\pi n, \qquad n = 0, \pm 1, \pm 2, \ldots \quad (5)$$

and, if we take into account some restrictions of Q and the nature of n, we can solve (5) for t_n, when:

$$t_n = \overline{Q}\,(n) \ldots \quad (6)$$

Inserting (6) into the first of (4), we can get the value of $\omega_1(t)$ corresponding to the point P_n: $\omega_1(t_n) = P_0 P_n = (\omega_1)_n$, when the distance D_n at P_n is:

$$D_n = P_0 P_{n+1} - P_0 P_n = P_n P_{n+1} = (\omega_1)_{n+1} - (\omega_1)_n \quad (7)$$

As $n \to \infty$, the points P_n and P_{n+1} go to infinity, the distances $P_0 P_n$ and $P_0 P_{n+1}$ tend to infinity, and the $\lim_{n\to\infty} D_n = \overline{D}$ tends to get the undetermined form $(\infty - \infty)$, which may be either a constant, or zero or infinite, and, then the limits $\bar{d}$ and $\bar{\varrho}$ may be either constants, or zero, or infinite, respectively.

3.2 Orbital stability criterion

Based on the above properties of the distances D_n, d_n, ϱ_n, we can formulate a criterion for the orbital stability of the helicoid precessions (3), expressed by the:

Theorem 2. *«Any member of the class of the helicoid precessions (3), corresponding to a given function $L(t)$, is orbitally either stable, or asymptoti-*

cally stable, or unstable, if, respectively, the limit distance $\bar{D}$ is either a constant, or zero, or infinite».

Proof : We first see that, for sudden perturbation when the initial conditions are only perturbed, the curve $\bar{S}$ can be considered as the perturbed of S and the distance ϱ_n, defined above, is the «Poincaré distance» of S at P_n, Ref. 4.

If the number $\bar{D}$ is a constant, given $\varepsilon > 0$, we can really find a $\delta > 0$ such that, if the Poincaré distance initially is $\varrho_0 < \delta$, then inequality $\varrho_n < \varepsilon$, for all n, can be implied, since we can select $\delta = \varepsilon$, and $d_0 = \delta$, when $\varrho_0 < \delta$ implies $\varrho_n < \varepsilon$, and S is «orbitally stable».

If $\bar{D} = 0$, then $\bar{\varrho} = 0$, and S is «orbitally asymptotically stable».

If $\bar{D} = \infty$, S is «orbitally unstable».

3. 3 Example. As an example, we mention the case $L = \bar{L}_1 =$ constant, treated in Ref. 1.

The corresponding helicoid precession is in this case given by:

$$\left.\begin{array}{c} \omega_1(t) = \dfrac{\bar{L}_1}{I_1} t, \quad \omega_2(t) = \omega_{20} \cos Q_1(t), \quad \omega_3(t) = \omega_{20} \sin Q_1(t) \\ Q_1(t) = (I_1 - I) \bar{L}_1 t^2 / 2 I I_1 \end{array}\right\} \quad (8)$$

This precession, due to the form of $Q_1(t)$, is «Liapunov unstable»; but it is «orbitally asymptotically stable», since the distance D_n is given by: $D_n = \alpha(\sqrt{n+1} - \sqrt{n})$, $\alpha =$ constant, and of which the limit, as $t \to \infty$, is $\bar{D} = 0$.

For this example, we can determine the region of the permitted deviations of the precessional curve, the «ε - region», and the corresponding region of the initial points, the «δ - region», for which regions the helicoid precession is orbitally asymptotically stable, when this stability situation of (8) has a practical importance.

Given a point $(\bar{\omega}_{10}, \bar{\omega}_{20}, \bar{\omega}_{30})$ on the surface of the cylinder as a starting point of a helicoid precessional curve (8), the coordinates $\omega_1, \omega_2, \omega_3$ of any point of this curve are related to $\omega_{10}, \omega_{20}, \omega_{30}$ by:

$$\omega_2^2 + \omega_3^2 = \omega_{20}^2 + \omega_{30}^2 = \text{constant} \quad (8.1)$$

$$(\omega_1)_n \leqslant \bar{\omega}_{10} \leqslant (\omega_1)_{n+1} \quad (8.2)$$

The inequality (8. 2), by using $(\omega_1)_n = \alpha\sqrt{n}$, leads to:

$$n \leqslant \bar{\omega}_{10}/\alpha \leqslant n+1 \tag{8.3}$$

from which the integer n can be determined, when, as a result, $D_n = \alpha(\sqrt{n+1}-\sqrt{n})$ is known. This distance D_n is the upper limit of δ and ε.

We remark that we can calculate the «ε, δ-regions» of any member S of the helicoid precessions (3), if the distance $\bar{D}$ of S is zero or finite, when the orbital stability situation of S, and not its Liapunov stability situation, has a practical meaning.

3. 4 Remarks. We saw above that, for the same phenomenon, we have different stability situations, if we apply different stability concepts.

There arises the problem of the selection of the stability concept appropriate to the phenomenon, that is of the selection of the stability situation, which interprets the reality in an adequate way, and it is more close to «practical stability» of the phenomenon.

The possibility of the determination, by using a physical situation, of the region of the permitted deviations of motion and orbit, of the corresponding region of the initial points, and of the appropriate region of the perturbation, in case of persistent perturbations, Ref. 5, makes the stability results practically important and physically accepted.

Stability investigations, which may satisfy mathematical curiosities or needs, will become useful if they are oriented towards «practical usefulness».

REFERENCES

1. MAGIROS, D. G.: Comptes Rendus, Academy of Sciences, Paris, France 268, No. 12 (1969), Series A, 652-654.
2. » Proceedings XX International Astronautical Congress, Buenos Aires, Argentina (1969).
3. » Proceedings, XXII International Astronautical Congress, Brussels, Belgium (1971).
4. » J. Information and Control, U.S.A. 9, No. 5 (Oct. 1966), 531-548.
5. LA SALLE, J. and LEFSCHETZ, S.: «Stability by Liapunov's direct method with Applications», Academic Press (1961), New York.

ΠΕΡΙΛΗΨΙΣ

Εἰς προηγουμένην ἐργασίαν, ἀνακοινωθεῖσαν εἰς τὴν Ἀκαδημίαν τῶν Παρισίων (1969), ἐμελετήθη ἡ εὐστάθεια μιᾶς ἑλικοειδοῦς μεταπτώσεως εἰς τὴν περίπτωσιν σταθερᾶς ἐξωτερικῆς ροπῆς μὲ χρησιμοποίησιν διαφόρων ὁρισμῶν εὐσταθείας, καὶ εὑρέθησαν διαφορετικαὶ καταστάσεις εὐσταθείας, ὑπεδείχθη δὲ ποία ἐκ τῶν καταστάσεων εὐσταθείας τῆς ἑλικοειδοῦς ἔχει πρακτικὴν ἀξίαν.

Εἰς τὴν παροῦσαν ἐργασίαν μελετᾶται ἡ κατάστασις εὐσταθείας μιᾶς κλάσεως ἑλικοειδῶν μεταπτώσεων, ποὺ περιέχει ὡς ἕνα μέλος της τὴν ἑλικοειδῆ μετάπτωσιν τῆς προηγουμένης ἐργασίας.

Χρησιμοποιοῦνται δύο ὁρισμοὶ εὐσταθείας, οἱ κατὰ Liapunov καὶ Poincaré. Τὰ συμπεράσματα τῆς παρούσης ἐργασίας εἶναι :

α. Ὅλα τὰ μέλη τῆς κλάσεως τῶν ἑλικοειδῶν μεταπτώσεων εἶναι εἰς ἀσταθῆ κατὰ Liapunov κατάστασιν.

β. Ἡ κατὰ Poincaré κατάστασις εὐσταθείας οἱουδήποτε μέλους S τῆς κλάσεως ἐξαρτᾶται ἀπὸ τὴν ὁριακὴν τιμὴν τοῦ βήματος τῆς ἑλικοειδοῦς S, καὶ ὅταν ἡ ὁριακὴ τιμὴ εἶναι σταθερὰ ἢ μηδὲν ἢ ἄπειρον, τότε ἡ ἑλικοειδὴς εἶναι εὐσταθής, ἢ ἀσυμπτωτικὰ εὐσταθὴς ἢ ἀσταθής, ἀντιστοίχως.

γ. Εἰς τὴν περίπτωσιν ποὺ ἡ ἑλικοειδὴς S εἶναι ἀσυμπτωτικὰ εὐσταθής, τότε ἡ κατάστασις αὐτὴ καὶ μόνον ἔχει πρακτικὴν ἀξίαν.

★

Ἔχω τὴν τιμὴν νὰ παρουσιάσω εἰς τὴν Ἀκαδημίαν Ἀθηνῶν τὴν ἐργασίαν τοῦ κ. Δημητρίου Μαγείρου, Ἐπιστημονικοῦ Συμβούλου τῆς General Electric τῶν Η.Π.Α. ὑπὸ τὸν τίτλον «Ἡ Εὐστάθεια μιᾶς Κλάσεως Ἑλικοειδῶν Μεταπτώσεων κατὰ Liapunov καί Poincaré».

Ὁ κ. Μάγειρος εἰς προηγουμένην ἐργασίαν του, ἀνακοινωθεῖσαν εἰς τὴν Ἀκαδημίαν Παρισίων, ἐμελέτησε τὴν εὐστάθειαν μιᾶς Ἑλικοειδοῦς Μεταπτώσεως εἰς τὴν περίπτωσιν σταθερᾶς ἐξωτερικῆς ροπῆς.

Εἰς τὴν παροῦσαν ἐργασίαν μελετᾶται ἡ κατάστασις εὐσταθείας μιᾶς κλάσεως ἑλικοειδῶν μεταπτώσεων εἰς τὴν ὁποίαν ἓν ἐκ τῶν μελῶν της εἶναι καὶ ἡ ἑλικοειδὴς μετάπτωσις τῆς ἀναφερθείσης ἤδη προηγουμένης ἐργασίας.

Χρησιμοποιοῦνται πρὸς τοῦτο οἱ ὁρισμοὶ εὐσταθείας κατὰ Liapunov καὶ Poincaré, τὰ δὲ ἀντίστοιχα πορίσματα τῆς ἐρεύνης εἶναι τὰ κάτωθι :

α) Ὅλα τὰ μέλη τῆς κλάσεως τῶν ἑλικοειδῶν μεταπτώσεων εὑρίσκονται εἰς ἀσταθῆ κατὰ Liapunov κατάστασιν.

β) Ἡ κατὰ Poincaré κατάστασις εὐσταθείας οἱουδήποτε μέλους τῆς ἐξεταζομένης κλάσεως ἐξαρτᾶται ἀπὸ τὴν ὁριακὴν τιμὴν τοῦ βήματος τῆς ἑλικοειδοῦς. Οὕτω, ὅταν ἡ ὁριακὴ τιμὴ εἶναι σταθερὰ ἢ μηδὲν ἢ ἄπειρος, τότε ἡ ἑλικοειδὴς εἶναι ἀντιστοίχως εὐσταθής, ἀσυμπτωματικὰ εὐσταθής, ἢ ἀσταθής.

γ) Εἰς τὴν περίπτωσιν ὅπου ἡ ὁριακὴ τιμὴ εἶναι μηδέν, ὁπότε ἡ ἑλικοειδὴς εἶναι ἀσυμπτωματικὰ εὐσταθής, τότε καὶ μόνον τότε ἡ κατάστασις αὕτη ἔχει πρακτικὴν ἀξίαν.

ΕΦΗΡΜΟΣΜΕΝΑ ΜΑΘΗΜΑΤΙΚΑ.— **On the separatrices of dynamical systems,** *by Demetrios G. Magiros**. Ἀνεκοινώθη ὑπὸ τοῦ Ἀκαδημαϊκοῦ κ. Φ. Βασιλείου.

INTRODUCTION

Separatrices are special trajectories or motions of dynamical systems, and play an important role in the study of problems of the systems of current interest, especially when quantitative aspects enter the problems. But there is no general and systematic discussion on the use of the properties and on the determination of the separatrices on nonlinear dynamical systems, this determination being by itself an important problem.

In this paper we will see remarks and results concerning the properties of separatrices and their use for the study of physical problems.

The definition of separatrices given in topology is supplemented as needed in physical problems, some theorems on separatrices are stated, a list of useful properties of separatrices is given, and by selected examples we see the usefulness of the separatrices in the study of physical problems.

1. DEFINITION OF SEPARATRICES

We give the definition of separatrices both from a topological and a dynamical point of view.

We can say that a space W is filled by a collection S of solution curves of a dynamical system, if each solution curve of S lies in W, and each point in W is on exactly one solution curve of S.

The whole space W of the validity of a differential system may be decomposed into subspaces of which the corresponding collection of the solution curves has common properties which characterize each space.

These subspaces are called «canonical regions» of the space W, and

---

* Δ. Γ. ΜΑΓΕΙΡΟΥ, **Ἐπὶ τῶν διαχωριστικῶν καμπυλῶν τῶν δυναμικῶν συστημάτων.**

Reprinted from the *Proceedings of the Athens Academy of Sciences* **54** (1979), 264–287.

the paths of the solution curves of the system, which bound these canonical regions, are called «separatrices» of the system [**7**, **10**].

In this topological definition of separatrices the solution curves are considered only as paths, that is as locus of point sets. In the reality, the solution curves of the dynamical systems are time-parametrized curves, that is paths on which the law of the motion of the system is known, when the topological definition of the separatrices, although it helps the investigation in some aspects, is unrealistic.

The time must be included in the concept of separatrices of physical problems. This can be succeeded by accepting the separatrices as «special "limiting" trajectories through special equilibrium states». Supplemented by this property, the topological definition of separatrices satisfies physical requirements and acquires a «physical validity».

By examples which will follow we clarify concepts related to separatrices and emphasize the usefulness of their properties in the investigation of physical problems.

2. THEOREMS RELATED TO SEPARATRICES

The separatrices are intimately related to the singular points of the system. It is the nature of the trajectories at the neighborhood of a singular point which guarantees the existence of separatrices through the singular point. We give, without analysis or proof, statements of theorems concerning singular points and corresponding separatrices, and the formulation of these theorems is given as needed in applications.

Theorem 1. Given a «noncritical linear dynamical system» in its normal form, if m is the number of the solution curves through a point of the space W of its validity, we may have the following cases:

a) For $m = 1$, the point is «regular», but for any other value of m the point is «singular»;

b) For $m = 0$ the singular point is a «center»;

c) For m a finite even integer, the singular point is a «saddle» point, and all the solution curves through this point are «branches of a separatrix» ;

d) For $m = \infty$, the singular point is either a «node» or a «spiral» point, and among the infinitely many solution curves through the point some of them may be separatrices.

The above singular points are «elementary singular points» and characterize the linear noncritical systems.

Theorem 2. In «noncritical nonlinear systems» it is the order of magnitude of the nonlinearities of the system which decides on the nature of the singular point and of the corresponding separatrices, and we have the following cases:

a) If the order of magnitude of the nonlinearities is appropriately small, the singular point is «elementary» and the situation of separatrices is as in Theorem A.

b) If the smallness of the magnitude of the nonlinearities can not be restricted appropriately, the singular point is «nonelementary» and the situation of separatrices is a complicated matter.

Theorem 3. We distinguish two cases:

a) In «critical linear systems» the singular point may be elementary or nonelementary, and the separatrices will be in a complicated situation, especially if the system has many singular points.

b) In «critical nonlinear systems», or in «nonlinear systems without linear part», the phase portrait near the nonelementary singular point is very complicated. A small neighborhood around such a point may be devided by separatrices into sectors with this point as the apex. These sectors may be of «nodal» (parabolic), or «elliptic», or «saddle» (hyperbolic) type.

We remark that there are cases very complicated, and only a few results are known today for highly nonelementary singular points and the corresponding separatrices. The following theorem is due to Bendixson [**3**].

T h e o r e m 4. If a system is the x, y-phase plane is given by

$$y' = x^{-m}[\alpha y + bx + B(x, y)] \qquad (12.1)$$

where $B(x, y)$ is a polynomial of degree at least two, and $\alpha \neq 0$, we have the following four cases:

a) If $\alpha > 0$ and $m =$ even integer, then there is only one branch of integral curves tending to the origin on the left side of y-axis, $x < 0$, while integral curves on the other side, $x > 0$, constitute a nodal distribution; that is, there is a coalescence of «saddle-nodal» points.

In this case there exists a separatrix through the origin.

b) If $\alpha < 0$ and $m =$ even integer, this case can be transformed to the previous case, and we have a coalescence of «nodal-saddle» points (node at $x < 0$, and saddle at $x > 0$).

A separatrix exists through the origin in this case.

c. If $\alpha > 0$ and $m =$ odd integer, the origin is a nodal point, when a separatrix may exist.

d. If $\alpha < 0$ and $m =$ odd integer, the origin is a saddle point and a separatrix exists.

The analysis of the above statements is based on the definition of the separatrices, on the concepts of the «α-limiting» and «ω-limiting» properties of the separatrices, and on other concepts.

3. SOME PROPERTIES OF SEPARATRICES

Combining the definition of separatrices and results coming from the theorems, one can find properties of separatrices, which are very useful in applications. In the following we list some of these properties.

— The separatrices may be points, lines, surfaces, depending on the dimensions and the structure of the dynamical system.

— There is no separatrix through a center.

— A separatrix through a singular point may be either a «α-limiting» or a «ω-limiting» trajectory, when, starting from a point of the separatrix, the time to reach the terminating point is «infinite».

— Separatrices starting from a singular point may terminate to the same singular point, when they are «closed» separatrices, and they have a finite length. «Non-closed» or «open» separatrices do not start and terminate at the same singular point. They start from a singular point and they may terminate either to another singular point or to infinity. Some of these open separatrices may have finite length.

— Separatrices through a node have at this point a definite tangent, and separatrices terminating to a spiral point move around it spirally and they do not have a definite tangent at this point.

— An «isolated closed path» of a dynamical system, in case all its points are regular, is a «limit-cycle» of the system, when it corresponds to a periodic phonomenon with a fixed period. But, if this closed path is through a singular point, the periodicity disappears and the closed path is not a limit cycle, but it is a «closed separatrix». The limit cycle is a separatrix according to the topological definition; but it is not a separatrice according to the supplemented definition of separatrices.

— The separatrices have important physical significance. We indicate some of them.

They may determine the whole region of the validity of a dynamical system and separate it from the «empty regions» which are without real solutions of the system.

They may be boundary curves of the regions in each of which the solutions are characterized by different stability situations, when the separatrices are in a «neutral stability situation». This property is of paramount importance in contemporary nonlinear control problems.

They may have some other physical meaning.

4. DETERMINATION OF SEPARATRICES

For the determination of the separatrices we see two cases. In case we know the general solution of the mathematical model of the physical problem, the determination of separatrices is identical to the determination of special particular solutions through appropriate singular points of the system.

In case the general solution is not known, approximate methods, either geometrical, or numerical, or analytical, map help the investigation towards the determination of the separatrices.

Remark. We remark that the concept of «index» of singular points, of trajectories in general and of separatrices in particular, introduced by Poincaré, plays an important role for investigation of their nature [**5**, **13**].

5. EXAMPLES

In each of the following examples appropriate remarks are given related to properties of separatrices.

Example 1. The separatrices in this example determine exactly the boundary of the canonical regions, which are regions of the validity of the system where real trajectories exist, and empty regions.

The dynamical system [**4** (a)] :

$$\dot{x}^2 = 1 - x^2, \qquad \dot{y}^2 = 1 - y^2 \tag{1.1}$$

has four singular points, the points $(\pm 1, \pm 1)$, which are points of intersection of the lines $x = \pm 1$, $y = \pm 1$.

The system (1.1) corresponds in the x, y-phase plane to the DE:

$$y'^2 = \frac{1 - y^2}{1 - x^2}. \tag{1.2}$$

For the reality of the solutions of (1.2), the $(1 - x^2)$ and $(1 - y^2)$ must have the same sign, and this restriction helps to determine the real regions of the validity of (1.2). By separating the variables and integrating, one can find the general solution of (1.2):

$$\left.\begin{array}{lll} \text{arc sin}\, y \pm \text{arc sin}\, x = c\,; & |x| < 1, & |y| < 1 \\ \text{arc cos h}\, y \pm \text{arc cos h}\, y = c\,; & |x| > 1, & |y| > 1 \end{array}\right\} \tag{1.3}$$

c is the arbitrary constant. Figure 1 shows the phase-portrait of (1. 3).

The separatrices are the lines $x = \pm 1$ and $y = \pm 1$, which separate the whole x, y - plane into nine regions five of which have families of real solutions and four are empty regions.

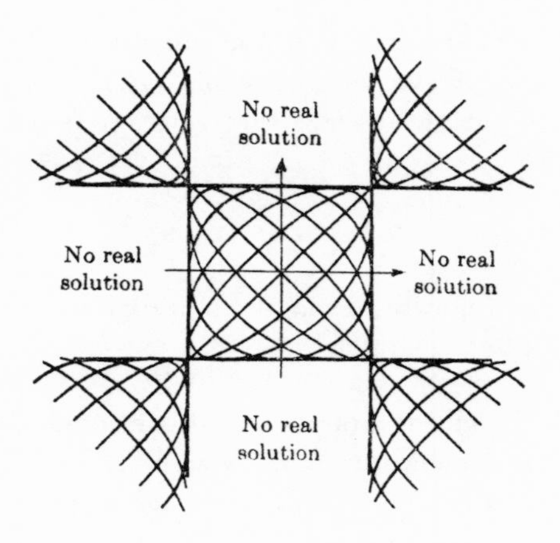

Fig. 1.

E x a m p l e 2. In this example we see that the separatrices, with the help of some other curves (which are not separatrices), determine the boundaries of the canonical regions, and that the common property of each family of trajectories of the regions is a special stability situation.

The dynamical system:

$$\dot{x} = x(\varepsilon x - 1), \quad x(0) = x_0, \quad t \geqslant 0 \tag{2. 1}$$

has as singular points the points $x = 0$ and $x = \frac{1}{\varepsilon}$. The general solution of (2. 1) is:

$$x(t) = \frac{x_0}{\varepsilon x_0 - (\varepsilon x_0 - 1) e^t} \tag{2. 2}$$

of which the portrait is shown in Figure 2.

The separatrices in the t, x-plane are the lines $x = 0$, $x = \frac{1}{\varepsilon}$, $t = 0$. These separatrices, with the help of the line:

$$t = \log \frac{\varepsilon x_0}{\varepsilon x_0 - 1} \tag{2.3}$$

separate the half t, x-plane, $t \geqslant 0$, into four regions of which I, II, III are real regions of the validity of (2.1), and IV is the empty region.

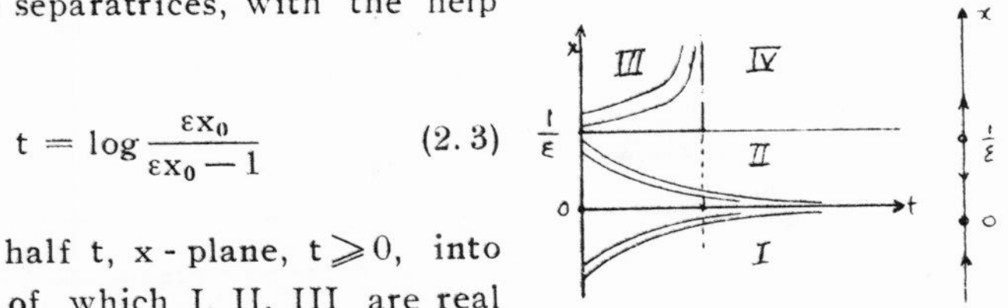

Fig. 2.

E x a m p l e 3. There exist dynamical systems without singular points, then without separatrices, of which the canonical regions are separated by curves of nature different than the separatrices. In this example these separating curves are «asymptotes».

The dynamical system:

$$\dot{x} = 2, \qquad \dot{y} = y^2 - 1 \tag{3.1}$$

is without singular points, then without separatrices. This system corresponds to the DE:

$$y' = \frac{1}{2}(y^2 - 1) \tag{3.2}$$

of which the general solution is:

$$y = \frac{1 + ce^x}{1 - ce^x}. \tag{3.3}$$

The phase-portrait of (3.3) is shown in Figure 3.

The lines $y = \pm 1$ separate the x, y-plane into three canonical regions, and these lines are «asymptotes» of the families of the solutions of the regions. We remark that in the previous example, Figure 2, the separatrix $x = 0$ is an asymptote for the trajectories of the regions I and II.

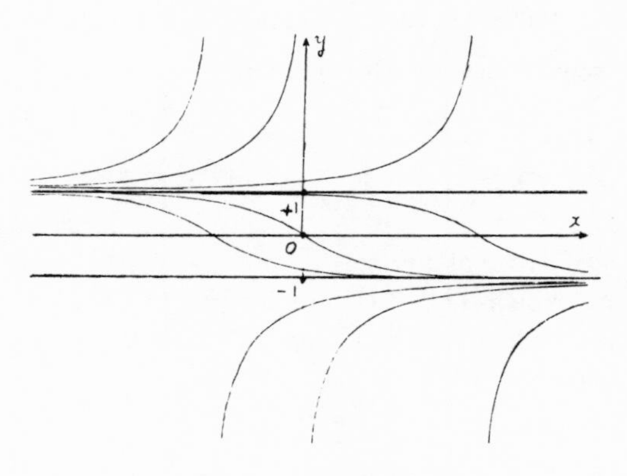

Fig. 3.

E x a m p l e 4. Here we distinguish the concept of a separatrix from the concept of an envelope of a family of solution curves. Both have the property to separate the region of the validity of the system into canonical regions, but the separatrices are special members of the families of the solutions, while the envelopes have not, in general, this property but they are special singular solutions of the system, tangent to all members of the families of the solutions.

We give appropriate examples.

a. The motion of a projectile in a vacuo.

1. We imagine all trajectories described by projectiles fired from the same point 0 with the same initial velocity $\bar{v}_0$ on a x, y - plane, each trajectory corresponding to a different direction of firing φ, Figure 4. All these trajectories belong to a family of parabolas given by [**6**]:

$$y = mx - \frac{g(1+m^2)x}{2v_0} \tag{4. 1}$$

where $m = \tan\varphi$ is the parameter, and g serves as a retardation.

For the envelope of the family (4. 1), we eliminate the parameter m between (4.1) and its derivative with respect to m, when the result:

$$y = \frac{v_0^2}{g^2} - \frac{gx^2}{2v_0^2} \tag{4.2}$$

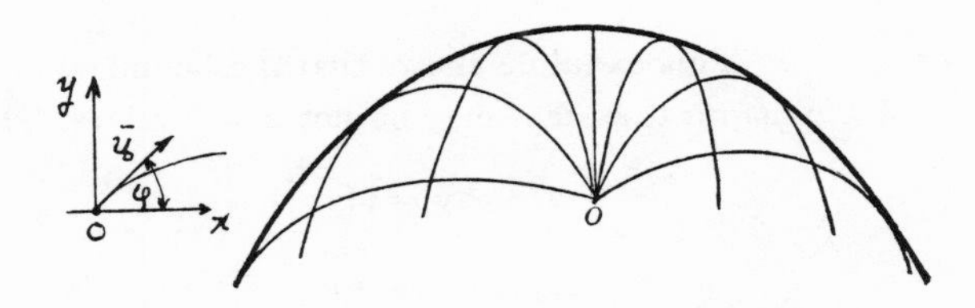

Fig. 4.

a parabola, is the envelope. This envelope separates the points which can be reached from those which can not be reached, and it is not a number of the family (4.1), Figure 4.

The family (4.1) has no separatrix.

2. There are other physical problems of the same nature, e.g., the «caustics» in optics are envelopes of light rays reflected by a mirror.

Let us see another example.

b. Consider the DE:

$$y'^2 = \frac{x^2}{1 - x^2} \tag{4.3}$$

which is equivalent to:

$$y' = \pm \frac{x}{\sqrt{1 - x^2}} \tag{4.4}$$

valid in the strip $|x| \leqslant 1$. Its general solution is:

$$f \equiv x^2 + (y + c)^2 - 1 = 0. \tag{4.5}$$

That is, a family of circumferences with centers on the y-axis and tangent to the lines $x = \pm 1$.

For every «regular» point of the strip two circumferences pass, but this is not the case for the points of the lines $x = 0$, $x = \pm 1$, which are «singular lines». The lines $x = \pm 1$ are boundaries of the strip and they are tangents to every member of the family (4.5), and these lines are «singular solutions» of (4. 4), envelopes of the family (4. 5). There is no separatrix in the family (4. 5).

We remark that in some systems, as in the Example 11, Figure 12, envelope and separatrix exist and are identical.

Example 5. This example shows that the boundary of the stability regions of a dynamical system may be not a separatrix. The system:

$$\dot{x} = -y, \qquad \dot{y} = f(x, y) \tag{5. 1}$$

with

$$f(x, y) = \begin{cases} -x + 2x^3y^3, & \text{if: } x^2y^2 < 1 \\ -x, & \text{if: } x^2y^2 > 1 \end{cases} \tag{5. 2}$$

has the origin as the singular point, which is in the region $x^2y^2 < 1$, when the appropriate equation in the x, y - plane is:

$$y' = \frac{-x + 2x^3y^2}{-y}. \tag{5. 3}$$

The eigenvalues of (5. 3) are both real and negative, then the origin is a «node», a «regular attractor», when starting from any point of the

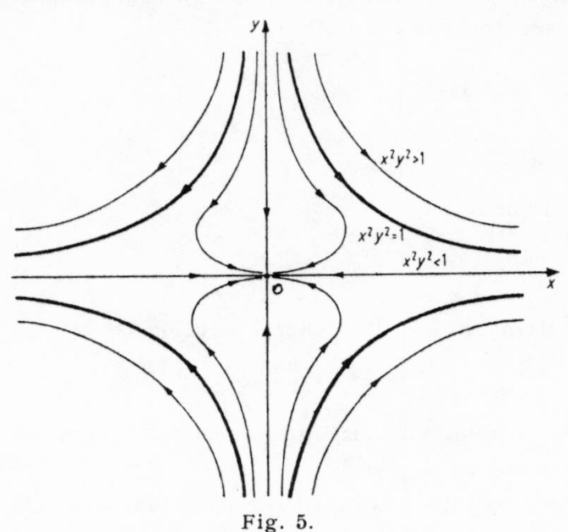

Fig. 5.

region $x^2y^2 < 1$ and following the corresponding trajectory we will terminate to the origin. The order of the magnitude of the nonlinearity of

(5. 3) agrees with this result. The phase-portrait of the solutions of (5. 3) is shown in Figure 5. The curve $x^2y^2=1$, which consists of four branches, determines the region of attractiveness of the origin. Outside of this region the stability situation is different. The curve $x^2y^2=1$, which is the boundary of different stability situations of the regions, is not a separatrix of the system.

Example 6. By this example we see changes in the nature of the separatrices by putting restrictions to the constants of the system [**9**].

Take the system:

$$\ddot{x} = x(\alpha^2 - x^2) + b\dot{x} \tag{6. 1}$$

which has the normal form:

$$\dot{x} = y, \quad \dot{y} = x(\alpha^2 - x^2) + by, \quad \alpha b \neq 0. \tag{6. 2}$$

The singular points are $O(o, 0)$, $A_1(\alpha, 0)$, $A_2(-\alpha, o)$.

For the nature of the origin we find the characteristic equation of (6. 2):

$$\lambda^2 - b\lambda - \alpha^2 = 0$$

when the eigenvalues are:

$$\lambda_{1,2} = \frac{1}{2}\left(b \pm \sqrt{b^2 + 4\alpha^2}\right) \tag{6. 3}$$

and since these eigenvalues are real and of opposite sign, the origin is a saddle point.

For the nature of the points A_1 and A_2, we use the transformations $x = \bar{x} + \alpha$, $y = \bar{y}$, when (6. 2) is reduced to a perturbed system of which the origin corresponds to A_1 and A_2, and the characteristic equation of the perturbed system is $\lambda^2 - b\lambda + 2\alpha^2 = 0$, and the eigenvalues are:

$$\lambda_{1,2} = \frac{1}{2}\left(b \pm \sqrt{b^2 - 8\alpha^2}\right). \tag{6. 4}$$

We have two cases.

a. If $b^2 < 8\alpha^2$, λ_1 and λ_2 are complex numbers with real part of sign of b, when A_1 and A_2 are spirals, stable for $b<0$ unstable for $b>0$.

Figure 6(a) has been drawn for A_1, A_2 stable spirals. For unstable spirals one merely reverses the arrows in this Figure.

The separatrices in this case connect saddle and spiral points and have infinite length.

b. If $b^2 > 8\alpha^2$, λ_1 and λ_2 are real and of the same sign as b, when A_1, A_2 are nodes, stable for $b<0$, unstable for $b>0$. Figure 6(b) shows the case of stable nodes A_1, A_2, and for unstable nodes we reverse the arrows in this Figure. Two of the separatrices are of finite length and two of infinite length.

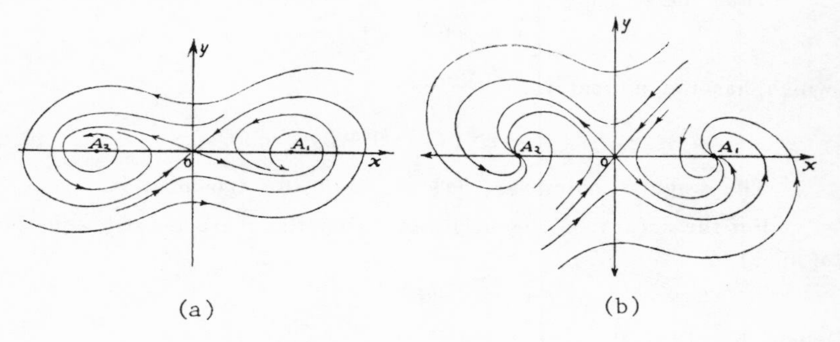

(a) (b)

Fig. 6.

E x a m p l e 7. Here we have a physical problem of biology or economics in which the separatrices are calculated as special particular solutions of the general solution of the model of the problem. In addition we see a property of separatrices which is very important in interpreting theoretical results.

There are many assemblies around us of which the elements influence each other through competition and cooperation.

The «p r o b l e m o f p o p u l a t i o n g r o w t h» is a problem of this nature.

We discuss this problem as a biological problem, but, by appropriate changes in the meaning of the variables and the constants involved, the problem can become a problem in other fields, as, e.g., in economics.

We consider two coexisting species of population numbers x and y at time t, both hunters, that is one species kills members of the other species. By using appropriate assumptions, the correspondent mathematical model is the nonlinear differential system [**8**] :

$$\dot{x} = \alpha x - cxy, \qquad \dot{y} = by - dxy \tag{7.1}$$

x and y are positive integers, but they can be considered as positive continuous functions of time. The coefficients α, b, c, d have a physical meaning and here are taken as positive integers.

The equilibrium points of (7.1) are the origin and the point $A\left(\frac{b}{c}, \frac{\alpha}{c}\right)$, and we can check that the origin is a «node», and A a «saddle» point.

The system (7.1) corresponds in the x, y-phase plane to the DE:

$$y' = \frac{b - dx}{x} \cdot \frac{y}{a - cy} \tag{7.2}$$

of which the general solution is :

$$y^{\alpha} \cdot e^{-cy} = k \cdot x^{b} \cdot e^{-dx}. \tag{7.3}$$

The constant k in (7.3), which corresponds to the point A, is :

$$k = \left(\frac{\alpha}{c}\right)^{\alpha} \cdot \left(\frac{d}{b}\right)^{b} \cdot e^{b-\alpha}. \tag{7.4}$$

Inserting (7.4) into (7.3) one gets the equation of the separatrix through A. For a specific case, let us take $\alpha = 4$, $b = 3$, $c = 2$, $d = 1$, when the point A and the constant k are: A(3, 2), $k \simeq .218$, and the equation of the separatrix through A is:

$$(y^2/x^2) \cdot e^{x-2y} = .218\,. \tag{7.5}$$

An investigation of (7.5) leads to the Figure 7, in which the four branches of the separatrix are the courves through the point A, and these branches separate the first quadrant into the four regions I, II, III, and IV.

Starting from any point of any of these regions, we see that, as $t \to \infty$, one of the species tends to vanish asymptotically, while the other species tends to become infinite. In addition, we see that the species y eventually dissapears if the corresponding (x, y) - point is in the region

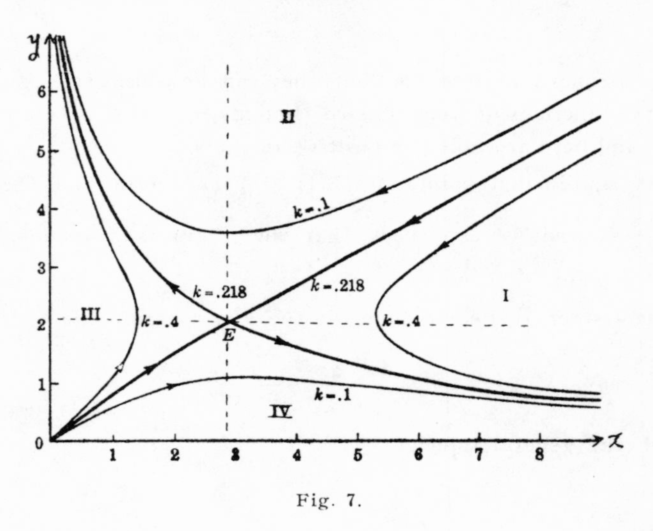

Fig. 7.

I or IV, and we see the opposite situation if the (x, y) - point is in the regions II and III. These results indicate a new property of separatrices, and show how important it is to know the location of the separatrices in the x, y - plane.

The origin 0 is a repulsor in the regions III and VI, and in III the species $x \to 0$, while in IV $y \to 0$.

Of course, due to the over-simplification of the model (7.1) of the problem, the above results are somehow unrealistic. For better results, the model of the problem must be modified by taking into account other influences for the growth of the species, e.g., the food supply, etc.

We remark that the previous discussion, modified by suitable changes to the problem and appropriate specification on the competitive species and the limiting resources, might be useful for an investigation of a problem of nature different than the above. E. G., one can have a problem in the field of economics if the variables denote the size or extent of commercial enterprizes for a common source and for a common market.

Example 8. By this example we see how the separatrices can be calculated in case of coexistence of many singular points, and also that the spearactrices, either closed or open, may be of finite length.

If the system is expressed by the DE [**11**]

$$\ddot{x} + 3x - 4x^3 + x^5 = 0 \tag{8.1}$$

by using $\dot{x} = y$, this equation can be reduced to

$$y' = \frac{1}{y}\{x(x^2-1)(x^2-3)\} \tag{8.2}$$

valid in the (x, y) - phase plane. The singular points are the origin and the points $(\pm 1, 0)$ and $(\pm\sqrt{3}, 0)$, and we can check that the origin and $(\pm\sqrt{3}, 0)$ are «centres», while $(\pm 1, 0)$ are «saddle» points.

The general solution of (8.1) in the (x, y) - phase plane is:

$$y^2 = c - 3x^2 + 2x^4 - \frac{1}{3}x^6. \tag{8.3}$$

The value of the arbitrary constant c of (8.3) corresponding to the saddle points $(\pm 1, 0)$ is $c = \frac{4}{3}$, when the separatrix through the points $(\pm 1, 0)$ is:

$$y^2 = \frac{1}{3}(4 - 9x^2 + 6x^4 - x^6) \tag{8.4}$$

of which the graph is shown in Figure 8.

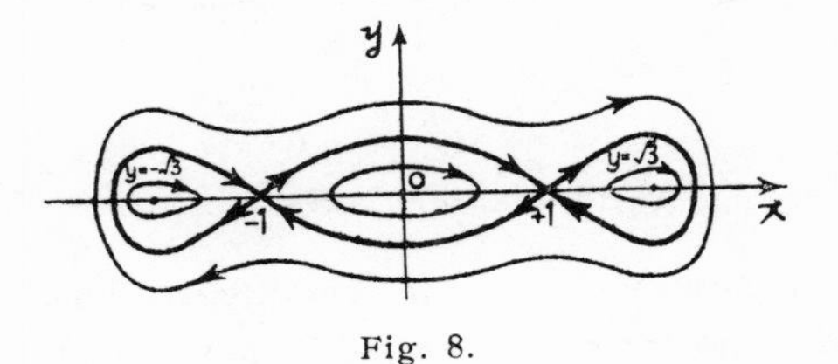

Fig. 8.

The separatrix (8.4) has four branches all of which have a finite lenght. The branches around the points $(\pm\sqrt{3}, 0)$ are «closed», and that around the origin are «open».

From this and the previous example we see a procedure for the determination of separatrices through saddle points of the system, of which the existence guarantees the existence of the separatrices.

Example 9. In these examples we see systems which have infinitely many «open» separatrices with finite or infinite lengths.

a. $$\ddot{x} + \omega^2 \sin x = 0. \qquad (9.1)$$

This is the pendulum equation and it is equivalent to the system:

$$\dot{x} = y, \quad \dot{y} = -\omega^2 \sin x, \qquad (9.2)$$

ω is the proper frequency. The singular points are infinitely may and they are the points $x = n\pi$, $n =$ integer, of the x-axis.

For even n the singular points are centers, and for odd n saddle points. There are infinitely many canonical regions and infinitely many separatrices running from a saddle point to the nearest saddle point. Figure 9 gives the corresponding phase-portrait. The separatrices are open and have finite length.

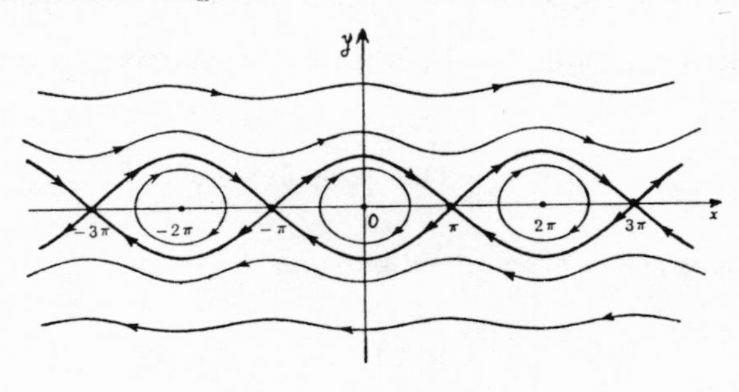

Fig. 9.

b. $$\ddot{x} + k\dot{x}|\dot{x}| + \omega^2 \sin x = 0. \qquad (9.3)$$

The singular points are $x = n\pi$, $n =$ integer, on the x-axis. For even n are spirals, and for odd n are saddle points. The infinitely many separatrices are of infinite length and run from a saddle point to the nearest spiral points, or they run from infinity to saddle points. Figure 10 shows the corresponding phase-portrait.

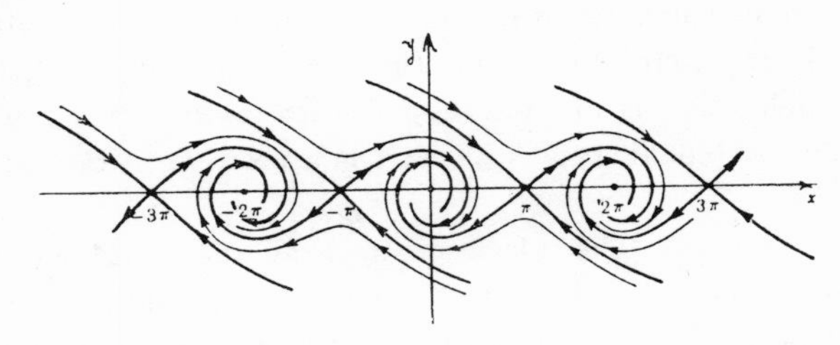

Fig. 10.

Example 10. In this example we see an nonelementary singular poit of complicated nature.

We take the system in polar coordinates

$$\dot{r} = r(1-r), \qquad \dot{\vartheta} = \sin^2\left(\frac{\vartheta}{2}\right). \tag{10.1}$$

Its singular points are $O(r=0,\ \vartheta=0)$, $O_1(r=1,\ \vartheta=0)$.

The DE in the phase-plane, corresponding to (10.1) is:

$$\frac{dr}{d\vartheta} = \frac{r(1-r)}{\sin^2\left(\frac{\vartheta}{2}\right)} \tag{10.2}$$

which can be integrated, and the family of the solutions $r = r(\vartheta) + c$ is shown in Figure 11.

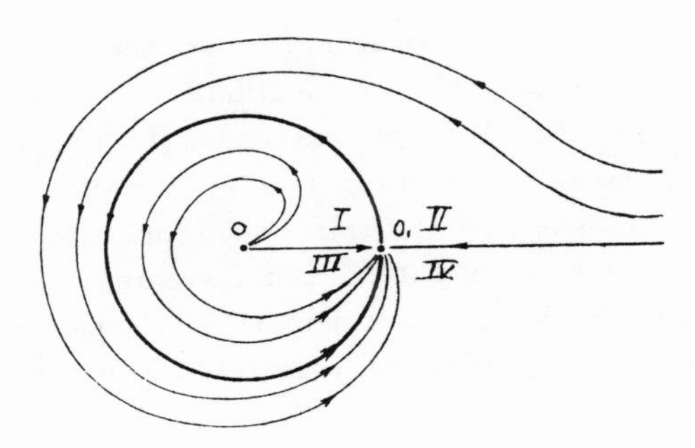

Fig. 11.

The separatrix is the circumference with center O and radius $OO_1 = 1$. The origin is a repulsor or a negative attractor (unstable). The point O_1 is a nonelementary singular point of complicated nature. The line OO_1 and the separatrix divide the neighborhood of O_1 into four sectors of which I and II are of «hyperbolic» (saddle) type, and III and IV of «parabolic» (nodal) type where O_1 is a positive attractor.

Example 11. The phase-portrait at the neighborhood of a nonelementary singular point may be complicated, but the separatrices through this point may be very simple; in addition «separatrices» and «envelopes» may be identical.

This is shown by the present example [**4**(b)].

The system:

$$\dot{x} = x(2y^3 - x^3), \qquad \dot{y} = -y(2x^3 - y^3) \tag{11.1}$$

corresponds to the DE:

$$y' = -\frac{y(2x^3 - y^3)}{x(2y^3 - x^3)}. \tag{11.2}$$

The origin is the only singular point and, since the system is without linear part, this point is nonelementary.

The right hand member of (11.2) is a function of the ratio (y/x), when by using the transformation $y = x \cdot u(x)$ one can separate the variables and integrate. The general solution of (11.2) can be found to be

$$x^3 + y^3 - 2cxy = 0, \tag{11.3}$$

c is the arbitrary constant. (11.3) is a one parameter family of curves known as «Folia of Descartes». The equation (11.3) is satisfied at the origin for any value of c, then all curves of (11.3) are through the origin. Figure 12 shows the graph of (11.3). The axes of coordinates are the separatrices. The first and third quadrants are elliptic sectors, the second and fourth are negative nodal sectors. In this example, although the origin is highly nonelementary, the separatrices are very simple lines, the axes of coordinates. Figure 12 shows the phase-portrait of (11.3).

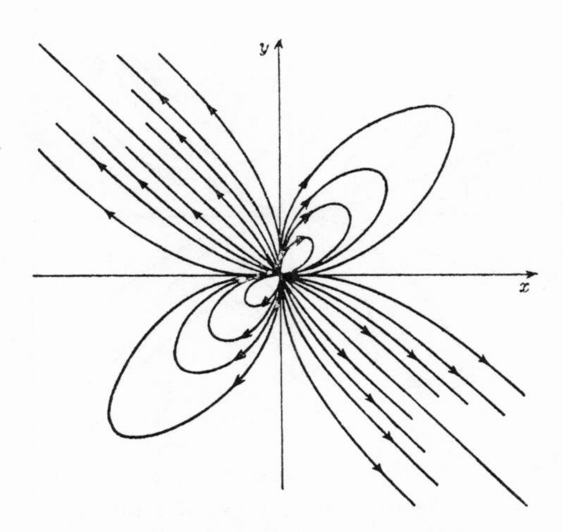

Fig. 12. (Folia of Descartes).

We remark that

The derivative of (11.3) with respect to the arbitrary constant c gives $x = 0$, $y = 0$, then the axes of coordinates of the above system are «envelopes» of the families of the solutions of the system and are identical with the separatrices of the system.

Example 12. In this example the nonelementary singular point is of «nodal-saddle» type, and we have three separatrices.

The system:

$$\dot{x} = -x^6, \qquad \dot{y} = y^3 - yx^4 \tag{12.1}$$

corresponds to:

$$y' = \frac{yx^4 - y^3}{x^6}. \tag{12.2}$$

The origin is the only singular point which is nonelementary.

The phase portrait, shown in Figure 13, can be found approximately by, say, geometrical methods.

There are four sectors I, II, III, IV and three separatrices which are the y-axis and the curve OO_1 and OO_2. The 180° sector I has negative nodal trajectories.

The origin is a positive attractor in the sector II. The sectors III and IV are of saddle type.

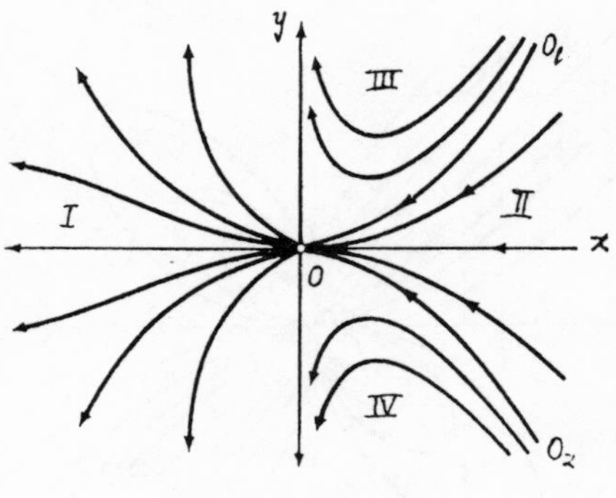

Fig. 13.

Example 13. By this example we give a highly nonelementary singular point, where the separatrices is difficult to be calculated. If the system in its phase-plane is given by [5]:

$$y' = x\left\{\frac{x^2 - xy - xy^2}{x^2 - y^2 - x^2y^2} + \frac{y}{x^2}\right\}. \tag{13. 1}$$

the origin is the nonelementary singular point. Figure 14 shows the graph of the solutions of (13. 1) at the neighborhood of the origin found

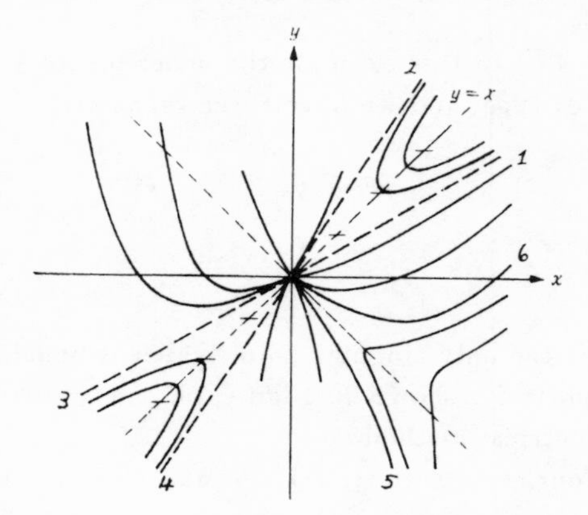

Fig. 14.

by approximate methods. We have six sectors with apex the origin, of which three are of hyperbolic type and three of parabolic type. Two hyperbolic sectors contain the line y=x in the first and third quadrants, and the third hyperbolic sector contains the line y=−x in the fourth quadrant. The six branches of the separatrix are the curves 01, 02, 03, 04, 05, 06 shown in Figure 14.

E x a m p l e 14. At the neighborhood of the saddle points of the previous examples, either elementary or nonelementary saddle points, the behavior of the trajectories were characterized by the property that these trajectories do not intersect the correspondent separatrices.

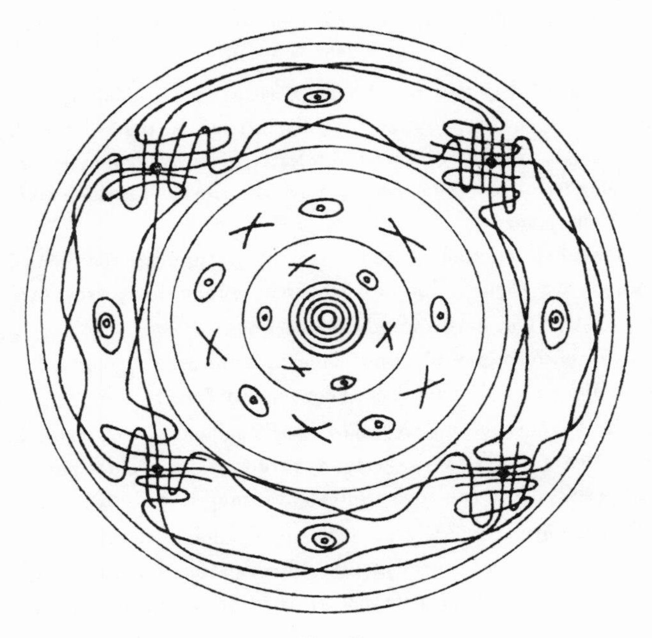

Fig. 15.

There are systems with saddle points at the neighborhood of which the behavior of the trajectories is very complicated. The three-body problem and some problems of dynamics show this complexity, Figure 15 [**1**, **2**, **12**].

ΠΕΡΙΛΗΨΙΣ

Αἱ διαχωριστικαὶ καμπύλαι (ΔΚ) τῶν δυναμικῶν συστημάτων παίζουν σπουδαῖον ρόλον εἰς τὴν ἔρευναν φυσικῶν προβλημάτων τρέχοντος ἐνδιαφέροντος. Ὅμως δὲν ὑπάρχουν σήμερον εἰδικὰ δημοσιεύματα ἀναφερόμενα εἰς τὸν γενικὸν προσδιορισμὸν τῶν ΔΚ, ὅπως καὶ εἰς τὴν χρῆσιν τῶν ἰδιοτήτων των εἰς τὴν ἔρευναν φυσικῶν προβλημάτων ποσοτικοῦ τύπου, ὁ δὲ προσδιορισμός των εἶναι ἀφ' ἑαυτοῦ ἕνα σπουδαῖον πρόβλημα.

Εἰς τὴν παροῦσαν ἐργασίαν δίδονται παρατηρήσεις καὶ εὑρίσκονται συμπεράσματα σχετικὰ μὲ τὴν ὕπαρξιν καὶ τὸν προσδιορισμὸν τῶν ΔΚ, ὅπως καὶ μὲ τὴν χρῆσιν τῶν ἰδιοτήτων των εἰς τὴν ἔρευναν καὶ ἑρμηνείαν φυσικῶν προβλημάτων.

Ὁ τοπολογικὸς ὁρισμὸς τῶν ΔΚ συμπληρώνεται καταλλήλως ὥστε νὰ γίνῃ χρήσιμος εἰς τὴν ἔρευναν πρακτικῶν προβλημάτων, διατυπώνονται θεωρήματα σχετικὰ μὲ τὰς ΔΚ χωρὶς ἀνάλυσιν ἢ ἀπόδειξιν, δίδονται παρατηρήσεις διὰ τῶν ὁποίων ὑποβοηθεῖται ὁ προσδιορισμὸς τῶν ΔΚ, τονίζονται ἰδιότητες τῶν ΔΚ χρήσιμοι διὰ τὴν ἔρευναν.

Διὰ τῶν παραδειγμάτων καταφαίνεται ἡ χρησιμότης τῶν ἰδιοτήτων τῶν ΔΚ εἰς ἐφαρμογὰς κυρίως. Ἐκ τῶν ἰδιοτήτων αὐτῶν τονίζονται δύο κυρίως:

(α): αἱ ΔΚ εἶναι δυνατόν, μόναι των ἢ καὶ μὲ τὴν βοήθειαν καὶ ἄλλων καμπυλῶν, νὰ προσδιορίζουν τὰ χωρία ὅπου τὰ δυναμικὰ συστήματα ἔχουν λύσεις πραγματικὰς ἀπὸ τὰ χωρία ὅπου δὲν ὑπάρχουν κἂν λύσεις,

(β): αἱ ΔΚ διαχωρίζουν τὸ χωρίον προσδιορισμοῦ τῶν συστημάτων εἰς χωρία, ὅπου αἱ λύσεις ἔχουν διαφόρους καταστάσεις εὐσταθείας ἕκαστον χωρίον, ὁπότε αἱ ΔΚ εὑρίσκονται ὑπὸ «οὐδετέραν» κατάστασιν εὐσταθείας.

Ἡ ἰδιότης αὕτη τῶν ΔΚ εἶναι μεγάλης σημασίας εἰς προβλήματα εὐσταθείας τῆς νεωτέρας θεωρίας «μὴ γραμμικῶν συστημάτων ἐλέγχου».

Ἡ παροῦσα ἐργασία θὰ συμπληρωθῇ καταλλήλως πολὺ σύντομα.

REFERENCES

1. V. Arnold, «Small denominators and problems of stability of motion in classical and celestial mechanics», Transl. Division, Foreign Technical Div., WP-AFB, Ohio (May 5, 1964), U. S. A.
2. V. Arnold and A. Avez, «Ergodic problems of classical mechanics», Benjamin Inc., New York (1968), pg. 91, U. S. A.

3. I. Bendixson, «Sur les courbes définies par des équations differentielles», Acta Math. 24 (1899), Sweden.
4. G. Birkhoff and G. Carlo, «Ordinary differential equations», Ginn and Co., New York (1962), (a): pg. 15 - 16, (b): pg. 138 - 142, U. S. A.
5. E. Davis and E. James, «Nonlinear differential equations», Addison-Wesley Publ. Co., New York (1966), pg. 28 - 39, U. S. A.
6. J. Jeans, «Theoretical mechanics», Ginn & Co., New York (1907), pg. 211, U. S. A.
7. W. Kaplan, «Regular curve families filling the plane», Duke Math. J., I. 7 (1940), II. 8 (1941), U. S. A.
8. J. Kemeny and J. Snell, «Mathematical models of the social sciences», MIT - Press, Cambridge, Mass. (1972), pg. 24 - 33, U. S. A.
9. S. Lefschetz, «Differential equations, geometric theory», Interscience Publishers, Inc., New York (1957), pg. 223 - 231, U. S. A.
10. L. Markus, «Global structure of ordinary differential equations in the plane», Trans. Am. Math. Soc. 76 (1954), pg. 127 - 148, U. S. A.
11. N. McLachlan, «Ordinary nonlinear differential equations in engineering and physical sciences», 2nd Ed. (1958), Oxford, Clarendon Press, pg. 196, England.
12. V. Mel'nicov, «On the stability of the center for time-periodic perturbation», Math. Proc., Moscow Math. Soc., U. S. S. R.
13. H. Poincaré, «Sur les courbes definies par des équations differentielles», Journal de Mathematiques, 1881, 1882, . . . , Paris, France.

UDC 517.9

D. G. Magiros

Philadelphia, USA

SEPARATRICES OF DYNAMICAL SYSTEMS

Separatrices are special trajectories or motions of dynamical systems, they have a fundamental geometrical and physical significance and play an important role in the study of physical, engineering and social problems of current interest.

There is no systematic discussion on the existence and determination of separatrices, on their properties and their use. Also, their known definitions are somehow defective, they lead to confusion and need a clarification and modification.

In this paper, the known definitions of separatrices are appropriately supplemented, and, based on the supplemented definition, theorems are discussed related to separatrices at the neighborhood of elementary or nonelementary critical points of dynamical analytical systems in the plane. The systems are taken in general as homogeneous, and three cases are distinguished, namely the cases: $\Delta \neq 0$, $\Delta \equiv 0$, $\Delta = 0$, where Δ is the discriminant of the system.

By an appropriate transformation of the given system, an «auxiliary system» results in the study, which facilitates the investigation. The concepts of «nonlinearities», «eigenvalues», «sectors» are used in this paper.

We intend in another paper to use the concept of the «index» and give a discussion of the properties of separatrices and their use in physical applications.

On the Definitions of Separatrices. We can say that a space W is filled by a collection S of solution curves of a system of differential equations, if each solution curve of S lies in W and each point in W is on exactly one solution curve of S.

The whole space W of the validity of the differential system may be decomposed into subspaces in each of which the corresponding collection of the solution curves has common properties which characterize each subspace.

These subspaces are called «cononical regions» (or cells) in the space W and the special solution curves of the differential system, which have the property to bound the canonical regions, are called «separatrices» of the system [6, 9].

In this topological definition of separatrices the solution curves are considered only as geometrical entities, as a locus of point sets, as paths which are just carriers of the motion.

By using this definition one cannot distinguish the concept of a separatrix from the concepts of limit cycles, or of asymptotes and envelopes of a family of curves. In the reality, the solution curves of dynamical systems are time-parameterized curves, they are trajectories, that is paths on which the law of the motion is known. The above «topological property» of, separatrices helps the investigation in some aspects; it is necessary for the definition of separatrices, but it is not sufficient. The time must be included in the definition of the concept of separatrices.

Another definition of separatrices is based on stability concepts [1, 2, 7, 10]. But some of the so defined separatrices are stable or unstable in Liapunov or in Poincare' sense, or stable or unstable structurally [1, 2] when this definition is accompanied by confusion. The definition of separatrices must be given independently of the stability concepts.

Reprinted from the *Proceedings of the 9th Intl. Conf. on Nonlinear Oscillations* (Kiev, 1981) v. 2, *Qualitative Methods of the Theory of Nonlinear Oscillations*, Kiev Naukova Dumka (1984), 280–287.

We may define as separatrices of a dynamical system its trajectories which are characterized by the two properties [8]: (a) The «topological property» according to which these trajectories separate the whole region of the validity of the system into canonical subregions, and, (b) The «dynamical property» according to which they are «limiting trajectories», either «ω-limiting», or/and α-limiting» trajectories, through special critical points of the system, with definite tangents there.

By «ω-limiting» trajectory we mean trajectory tending to critical point as $t \to \infty$, and by «α-limiting» as $t \to -\infty$.

Such a definition of separatrices elucidates the concept of separatrices, satisfies physical requirements and acquires a physical validity.

A relation Between Figenvalues and Slopes at the Origin. We consider the dynamic system

$$\dot{x} = a_1 x + b_1 y + X_2(x, y), \quad \dot{y} = a_2 x + b_2 y + Y_2(x, y) \tag{1}$$

where a_1, a_2, b_1, b_2 are real constants, and X_2, Y_2 series convergent in a region of the validity of (1), which have terms of degree at least two in x, y. The linearization of (1) is

$$\dot{x} = a_1 x + b_1 y, \quad \dot{y} = a_2 x + b_2 y. \tag{1.1}$$

The origin is a critical point of these systems, and if their discriminant $\Delta = \begin{vmatrix} a_1 b_1 \\ a_2 b_2 \end{vmatrix}$ is different than zero the origin is an «elementary critical point», while if $\Delta \equiv 0$, or $\Delta = 0$, the origin is a «nonelementary critical point». The «eigenvalues» λ of the systems are given by

$$\lambda = \frac{1}{2}\{a_1 + b_2 \pm \sqrt{(a_1 + b_2)^2 - 4\Delta}\}. \tag{1.2}$$

For both λ real, the elementary critical points are either «nodals» (if both λ of the same sign) or «saddles» (if the λ are of opposite sign); and when λ are complex the critical points are «spirals», while for λ purely imaginary they are «centres».

The concept of the «slope» of the trajectories through the origin taken as a critical point gives important help to the investigation of the separatrices through the origin.

For system (1), in which X_2, Y_3 are infenitesimals stronger than x, y at the neighborhood of the origin, the slope K of the tangents of the trajectories at the origin can be given by

$$K = \frac{1}{2b_1}\{a_1 - b_2 \pm \sqrt{(a_1 - b_2)^2 + 4a_2 b_1}\}. \tag{1.3}$$

Since the discriminants of (1.2) and (1.3) are equal, K can be given in terms of λ by

$$K = (\lambda - b_2)/b_1. \tag{1.4}$$

K must be real, and it is real for real λ, according to (1.4), then for the investigation of separatrices spirals and centres must be excluded.

Separatrices of Elementary Critical Points. The Case $\Delta \neq 0$. We examine the problem of separatrices in case of a nodal or saddle at the origin.

We distinguish three kinds of nodal points: the «ordinary», the «degenerate», and the «dicritical» nodal points. The ordinary nodal is characterized by real distinct eigenvalues both of which are either negative or positive; in the degenerate nodal both eigenvalues are equal, and in the dicritical nodal $a_1 = b_2$, $a_2 = b_1 = 0$ in (1.1). In the system (1), in which the linear part dominates its nonlinear part at the neighborhood of the origin, the separatrices through the origin are as follows.

Theorem 1. (a) «All the trajectories in nodal points either stop at or start from these points, and some of them are separatrices.

There are four separatrices in an ordinary nodal point, two in a degenerate nodal, and four in a dicritical nodal.

The separatrices in the ordinary and degerate nodals are all either ω-separatrices» or «α-separatrices». The separatrices in the dicritical nodals are lines of «neutral» type (neutral in the sense of stability) (Fig. 1). (b) In a saddle point there are four separatrices, two «ω-separatrices» and two «α-separatrices», which are the only trajectories of the system through the saddle point.

A «ω-separatrix» and its consecutive «α-separatrix» considered together can be taken as one separatrix, when in a saddle point one can have four separatrices transversing the saddle point».

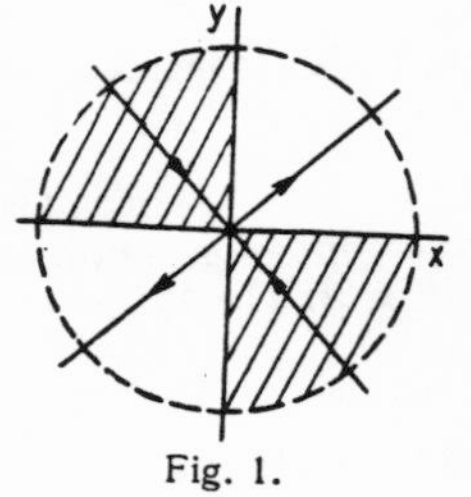

Fig. 1.

Separatrices of Nonelementary Critical Points. The Case $\Delta \equiv 0$. We consider a nonlinear system without linear part, ($\Delta = 0$), in the form

$$\dot{x} = X_m(x, y) + f(x, y), \quad \dot{y} = Y_m(x, y) + f_2(x, y) \tag{2}$$

where X_m, Y_m are homogeneous polynomials without common factors, and $m > 1$; f_1 and f_2 are convergent series in x, y with terms of degree bigger than m. Omitting f_1, f_2 in (2) and using the transformation $y = ux$ one gets the «auxiliary, equation»

$$\frac{du}{dx} = \frac{F(u)}{xX_m(1, u)}, \tag{2.1}$$

where the «auxiliary function» $F(u)$ is

$$F(u) = Y_m(1, u) - uX_m(1, u). \tag{2.2}$$

The nature of the critical points of (2.1), which are the real roots of $F(u)$ on u — axis, plays, a decisive role for the investigation of the nature of the critical point at the origin of x, y-plane and of the nature of the sector and separatrices through the origin. There exist criteria [3, 4] to check the nature of these critical points in x, u-plane which are of «nodal» (N), or «saddle» (S), or «nodal-saddle» (NS) type. If $F(u)$ is without real roots, the critical points in x, u-plane are of «spiral» or «centre» type, and this case is of no interest to us. To each real root u_i of $E(u)$ on u-axis a «critical line», perpendicular to u-axis, corresponds, and, by the transformation $y = ux$, to each real root u_i the straight line $y = u_i x$ corresponds in x, y-plane, which is the carrier of two separatrices through the origin in x, y-plane. The number l of the real roots u_i, $i = 1, 2, \ldots, l$ depends on the degree of $F(u)$ which the maximum is $m + 1$, and $l \leqslant m + 1$, when the number of the sectors and separatrices through the origin in x, y-plane is $2l \leqslant 2(m + 1)$. As for the nature of the sectors and separatrices through the origin of x, y-plane, we remark the following[3,4]:

Two consecutive nodals u_i and u_{i+1} on u-axis correspond to two elliptic sectors at the origin of x, y-plane;

Two consecutive saddles correspond to two hyperbolic sectors;

A saddle and a consecutive nodal correspond to two parabolic sectors;

Two consecutive nodal-saddles give two elliptic sectors, if the sides of nodals are consecutive, or give two hyperbolic sectors if the sides of saddles are consecutive, or give two parabolic sectors, if the side of nodal follows the side of saddle.

The same happens if the consecutive of nodal-saddle is a nodal or a saddle. If the degree of $F(u)$ is its maximum $m + 1$, $F(u)$ has $m + 1$ real distinct finite roots, corresponding to $2(m + 1)$ straight line separatrices through the origin in x, y-plane. In case the degree of $F(u)$ is smaller than $m + 1$ the point u_∞ at the infinity of u-axis is a critical point in x, u — plane, and the number of the separatrices depends on the multiplicity of u_∞ as a root of $F(u)$ and this number is $< 2(m + 1)$

The nature of the portraits and of the separatrices through the origin of x, y-plane depend on the nature of the real roots of $F(u)$ on u-axis, their number and their multiplicity.

In the following we specialize the above statements in cases of systems of «quadratic» and «cubic» homogeneous forms [3].

Quadratic Homogeneous Systems. In systems of quadratic homogeneous forms, one can distinguish four cases depending on the degree of $F(u)$, which may be three (maximum), or two, or one, or $F(u)$ may be a constant, in which cases the number of separatrices is six, four, or two.

Theorem 2. (a) **If the degree of F (u) is three, we have three cases.**

(i) ***F* (u) has three real simple roots.** The number of the separatrices through the origin of x, y-plane is six. The nature of the sectors and of the separatrices depend on the nature of the three roots, which may be positive, or negative and of N, or S or (NS) type. We distinguish three combinations of the roots: S, S, S, or S, S, N, or S, N, N. Fig. 2 corresponds to N, S, S.

(ii) **One root simple and one double.** The separatrices are four. We have two combinations of the roots: S, (NS), or N, (NS). Fig. 3 (combination N, NS)).

(iii) **One real root either simple or triple,** which will be of N or S type. The separatrices are two. Fig. 4 (N).

(b) ***F* (u) of degree two.** $F(u)$ has a simple root u_∞ at infinity of u — axis, and the finite roots may be either two simple or one double. This case can be deduced to the previous cases, and y-axis is a trajectory of N or S type.

(c) ***F* (u) of degree one.** u_∞ is a double root of $F(u)$, the separatrices are four, and the situation is similar to the above case (a, ii), Fig. 5 (S, (NS)).

(d) ***F* (u) is a constant.** u_∞ is a triple root of Fu the separatrices two, and the situation is similar to the case (a, iii), Fig. 6 (S).

Cubic Homogeneous Systems. In cubic homogeneous systems the degree of $F(u)$ is $\leqslant 4$, and the separatrices are in an even number $\leqslant 8$. We consider some cases.

Theorem 3. (a) F (u) **of degree four.** We have five cases.

(i) ***F* (u) with four roots simple.** The number of separatrices through the origin of x, y- plane is eight. We distinguish five combinations of roots: S, S, S, S, or S, S, S, N, or S, S, N, N, or S, N, S, N, or S, N, N, N. Fig. 7 (combination S, S, S, S).

(ii) ***F* (u) with two simple and one double roots.** We have six separatrices, and four root combinations are considered: S, S, (NS), or S, N, (NS), or S, N, (SN), or N, N, (NS), Fig. 8 (S, N, (NS).

(iii) ***F* (u) with two real roots and two complex, or one simple and one triple.** There are four separatrices, and the root combinations: S, S, or S, N, or N, N, Fig. 9 (N, N).

(iv) ***F* (u) with two double roots.** The separatrices are four, and the root combinations: (NS), (NS), or (NS), (SN). Fig. 10 (SN), (NS)).

(V) ***F* (u) with one real double root and two complex, or one real of multiplicity four.** The separatrices are two. Fig. 11 ((SN)).

(b) ***F* (u) of degree < 4.** The y-axis in x, y-plane is a trajectory the nature of which depends on the order of multiplicity of the root u_∞. The separatrices are in a number < 8.

The case of a Multiple Critical Point at the Origin of x, u- plane. The separatrices in this case are given according to the following [3].

Theorem 4. Let the auxiliary system be of the form

$$\frac{du}{dx} = \frac{U_m(x, u)}{X_m(x, u)}, \tag{3}$$

with X_m, U_m homogeneous in x, u without common factors, and $m > 1$. The origin in x, u-plane is supposed to be an isolated critical multiple point from which

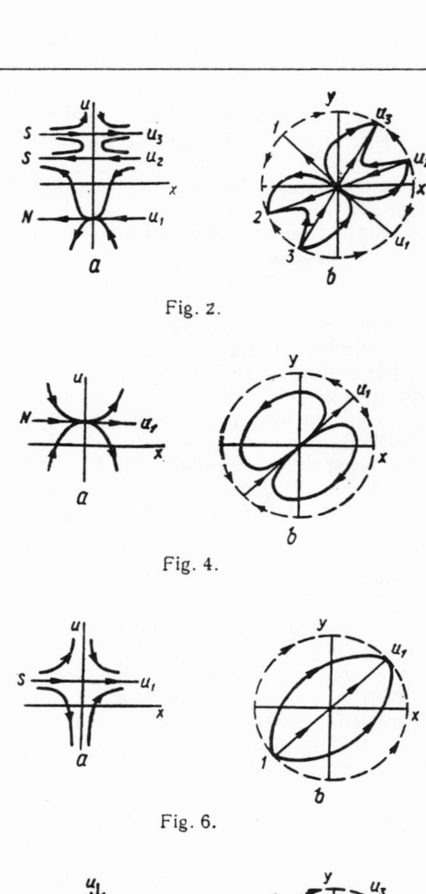

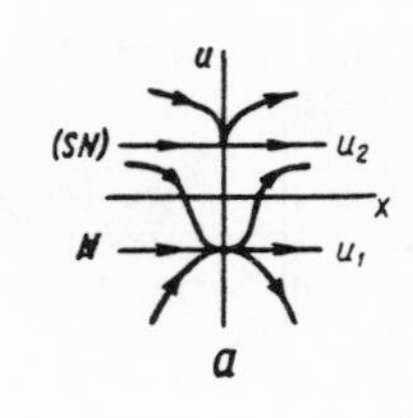

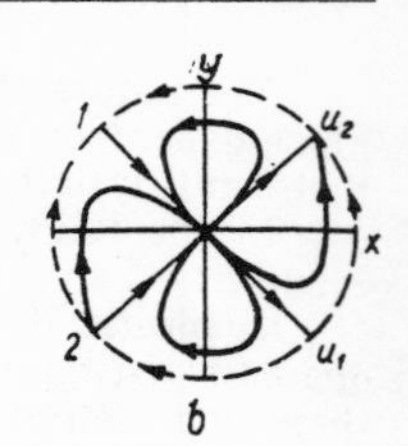

Fig. 2.

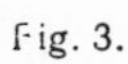

Fig. 3.

Fig. 4.

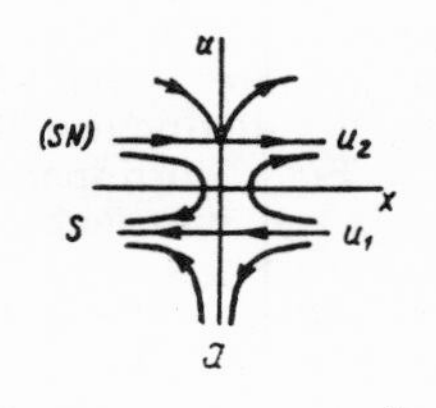

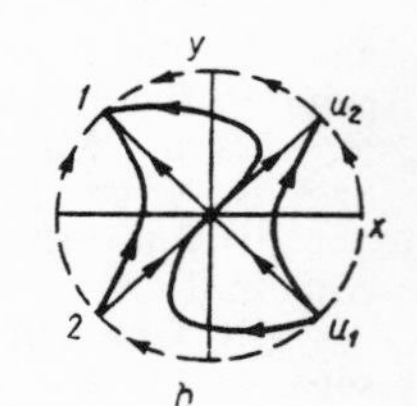

Fig. 5.

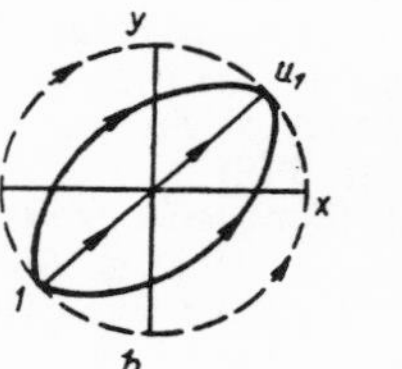

Fig. 6.

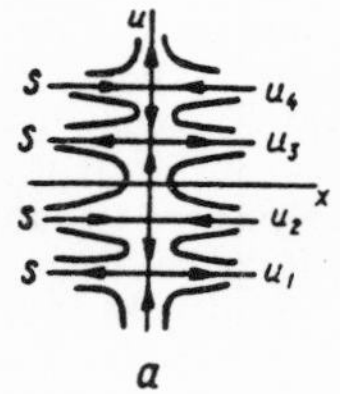

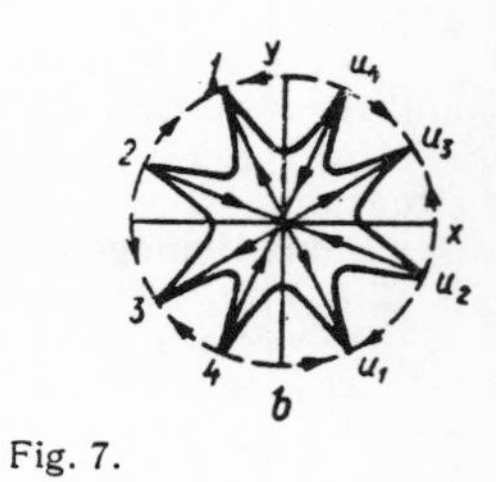

Fig. 7.

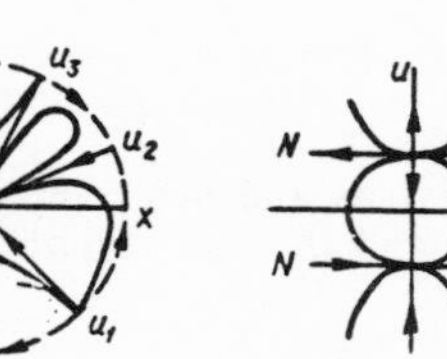

Fig. 8.

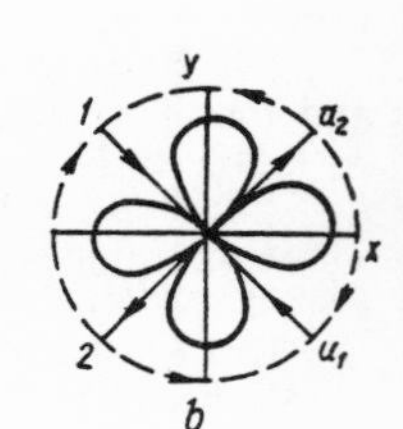

Fig. 9.

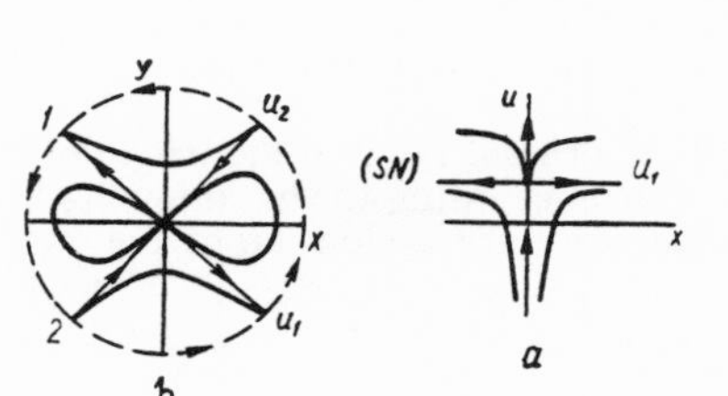

Fig. 10.

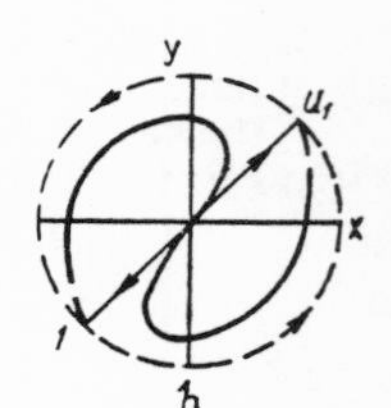

Fig. 11.

rectilinear trajectories of (3) pass with finite slopes. The auxiliary function is

$$F(\alpha) = U_m(1, \alpha) - \alpha X_m(1, \alpha) \tag{3.1}$$

and has finite roots α_i. By the transformation $y = xu$ the equation (3) leads to

$$\frac{dy}{dx} = \frac{yX_m(x^2, y) + x^2U_m(x^2, y)}{xX_m(x^2, y)} \tag{3.2}$$

If u-axis is not a trajectory of (3) and α_i are real, then the rectilinear critical trajectories $u = \alpha_i x$ of (3) are transformed in x, y-plane into the parabolas $y = \alpha_i x^2$ each of which is the carrier of two separatrices (Fig. 12).

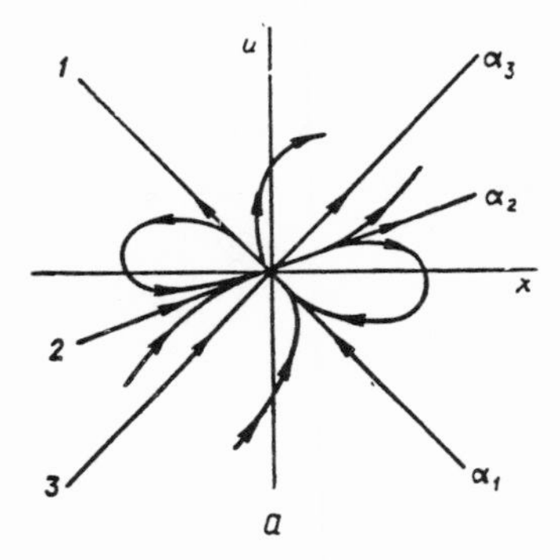

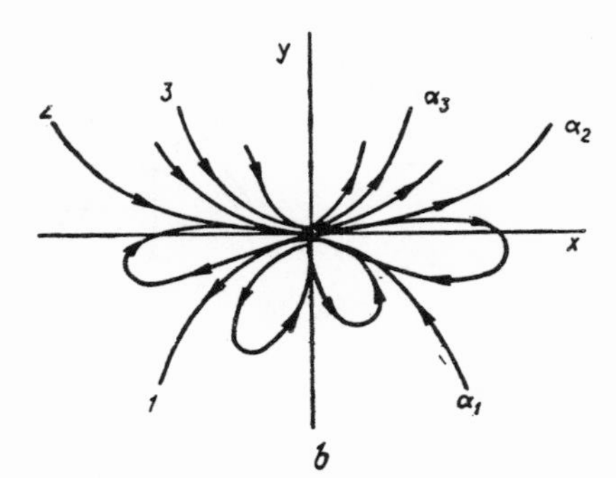

Fig. 12.

Separatrices Through Nonelementary Critical Points. The case $\Delta = 0$. We consider the system [3, 11]

$$\dot{x} = y + \sum_{i=2}^{p} X_i(x, y), \quad \dot{y} = \sum_{i=2}^{q} Y_i(x, y), \tag{4}$$

where X_i, Y_i are homogeneous polynomials with real coefficients of degree i in x, y. Transforming by $y = xu$ and taking terms of degree smaller than three the auxiliary system is

$$\dot{x} = xu + ax^2, \quad \dot{y} = bx - u^2 + (c - a)xu + dx^2, \tag{4.1}$$

where a, b, c, d are real coefficients. The nature of the trajectories at the neighborhood of the origin in x, u-plane depend on the values of the quantities: a, b, c, d, E, α_1, α_2, which may be positive, negative, or zero. Important cases are when $b \geq 0$, $E \gtrless 0$. In case $b = 0$, if $A(x, u)$ and $B(x, u)$ are the members of (4.1) at right, respectively, the auxiliary equation $F(\alpha) = B(1, \alpha) - \alpha A(1, \alpha) = 0$ leads to

$$-2\alpha + (c - 2a)\alpha + d = 0, \quad \alpha = u/x, \tag{4.2}$$

with roots

$$\alpha_\infty, \alpha_i = \frac{1}{4}\{(c - 2a) + (-1)^i\sqrt{E}\}, \quad i = 1, 2, \tag{4.3}$$

$$E = (c - 2a)^2 + 8a. \tag{4.4}$$

The separatrices through the origin of x, y-plane in the case of the system (4.1) are given by the following theorems.

Theorem 5. (a) If in the auxiliary system (4.1): $b = 0$, $E > 0$, and one of the real and distinct roots α_1, α_2 is zero, then the x-axis is a trajectory in both

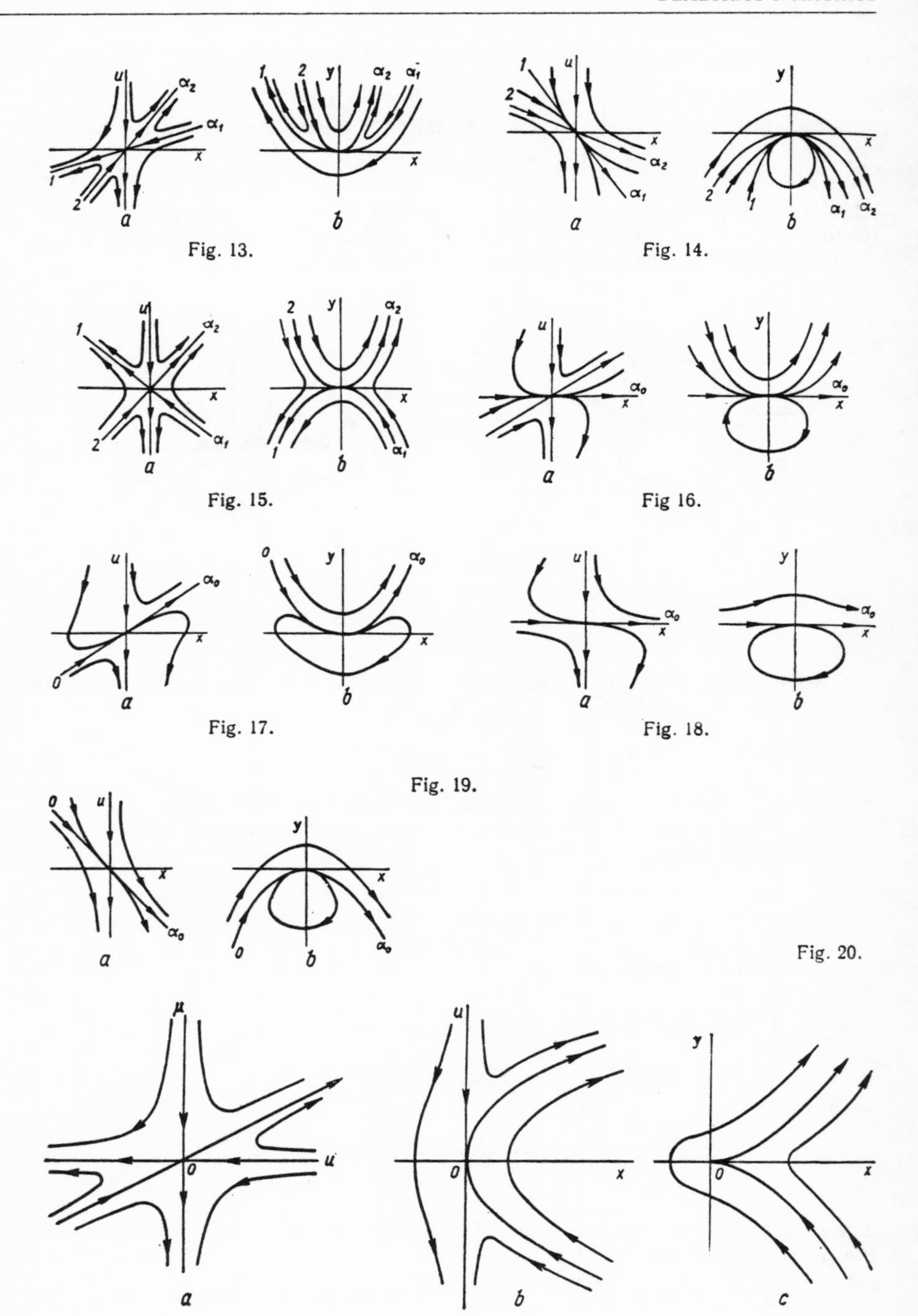

Fig. 13.

Fig. 14.

Fig. 15.

Fig 16.

Fig. 17.

Fig. 18.

Fig. 19.

Fig. 20.

the x, u-plane and the x, y-plane, and the straight lines $u = \alpha_i x$, $i = 1, 2$ in x, u-plane correspond to the parabolas $y = \alpha_i x^2$ in x, y-plane with x-axis tangent to these parabolas at the origin. Each of these two parabolas is the carrier of two separatrices, when the separatrices in x, y-plane are four. We can distinguish four cases according to the sign of α_1 and α_2 (Fig. 13—16).

(b) If in the system (4.1) $b = 0$, $E = 0$, $c + 2a > 0$, the finite roots coincide: $\alpha_1 \text{-} \alpha_2 \text{-} \alpha_0 \text{-} \frac{1}{2}(c - 2a)$, and the corresponding parabola $y = \alpha_0 x^2$ in x, y-plane is the carrier of two separatrices. We can distinguish three cases (Fig. 17—19).

Theorem 6. If in (4.1) $b > 0$, by using the transformation $x = u\mu$ and taking the quadratic terms only, one can get

$$\dot{u} = u(b\mu - u) = C(u, \mu), \quad \dot{\mu} = \mu(2u - b\mu) = D(u, \mu) \tag{4.5}$$

The critical points of (4.5) are given by

$$K(\alpha) = D(1, \alpha) - \alpha C(1, \alpha), \quad \alpha = \mu/u \tag{4.6}$$

and they are

$$\alpha_\infty, \ \alpha_0 = 0, \quad \alpha_1 = 3/2b. \tag{4.7}$$

The portraits are symbolically represented by

$$(u, \mu) \xrightarrow[x=u\mu]{} (x, u) \xrightarrow[y=ux]{} (x, y)$$

the origin in x, y-plane is of «cusp» type, and the separatrices are two (Fig. 20).

General remark. In the preceding the separatrices of some dynamical systems in the plane were examined locally at the neighborhood of a critical point (the origin). The global investigation of separatrices is also very important. Various physical problems have solution — curves connecting two critical points, and there are conditions which assume the existence of such solutions, which are physically realistic if they are locally unique and (in some sense) persistent to small perturbations.

The separatrices connecting two critical points can be thought of as being formed by intersection of two manifolds each of which is associated with a critical point [5]. The separatrices may be closed or open. In the closed separatrices the starting and terminal critical points are identical (the separatrices are loops). In the open separatrices the terminal point is different then the starting one they may have finite or infinite length and if they have infinite length either go to infinity or go to a spiral point, when they become regular spiral curves and they lose the property of being separatrices.

1. *Андронов А. А., Леонтович Е. А., Гордон И. И. и др.* Качественная теория динамических систем второго порядка.— М. : Наука, 1966.— 568 с.
2. *Андронов А. А., Леонтович Е. А., Гордон И. И. и др.* Теория бифуркаций динамических систем на плоскости.— М. : Наука, 1967.— 488 с.
3. *Argemi J.* Sur les points singuliers multiples de systèmes dunamiques dans R^2.—Theses, Université d'Aix—Marseille, France, 1967.
4. *Bendixson I.* Sur les courbes définies par les équation différentielles.—Acta Math., 1899, 24, Sweden.
5. *Gordon P.* Paths connecting elementary critical points of dynamical systems.— SIAM, J. Appl. Math., 1974, **26**, N 1.
6. *Kaplan W.* Regular curve families filling the plane.— Duke Math. J., 1940, **1**, 7; 1941, **11**, N 8, U.S.A. p. 155—185, p. 11—46.
7. *Lefschetz S.* Differential equations: Geometric theory.— Interscience Publishers, Inc., New York, 1957, p. 212.
8. *Magiros D.* Separatrices of dynamical systems.— Practica. Athens Academy June 1979, Greece.
9. *Markus L.* Global structure of ordinary differential equations in the plane.— Trans. Amer. Math. Soc., 1954, **76**, p. 127—149.
10. *McLachlan N.* Ordinary non linear differential equations in engineering and physical sciences, 2nd Ed. Oxford, the Clarencon Press, 1950, p. 188.
11. *Sansone G., Conti R.* Nonlinear differential equations. The McMillan Co., New York, 1964.

APPENDIX: PAPERS IN RUSSIAN

Ukrainian Academy of sciences,
Ukrainian Mathematical journal
Vol. 30, No. 2, 1978

D. G. Magiros: NONLINEAR DIFFERENTIAL EQUATIONS WITH SEVERAL GENERAL SOLUTIONS.

A nonlinear differential equation may have several general solutions. The investigation of the existence and construction of such solutions may be succeeded by "restricting quantities of the equation", or by "factorizing the equation".

УДК 517.93

Д. Г. Мажирос

Нелинейные дифференциальные уравнения с несколькими общими решениями

Нелинейное обыкновенное дифференциальное уравнение может иметь несколько общих решений. Их существование и определение может быть рассмотрено путем «ограничения решений» дифференциального уравнения или путем «факторизации дифференциального уравнения».

1. Введение. Общие решения нелинейных обыкновенных дифференциальных уравнений — наиболее желательные решения, особенно для приложений, однако только для некоторых классов уравнений можно найти их в замкнутом виде.

В то время как линейное обыкновенное дифференциальное уравнение имеет одно общее решение, нелинейное обыкновенное дифференциальное уравнение может иметь одно или несколько общих решений.

В этой статье рассматриваются классы нелинейных обыкновенных дифференциальных уравнений, имеющие несколько общих решений. Для этого класса уравнений можно получить хорошие результаты либо путем ограничения величин дифференциального уравнения, либо путем его факторизации.

Уточним некоторые понятия, связанные с понятием общих решений дифференциальных уравнений.

Рассмотрим нелинейное дифференциальное уравнение (F), справедливое в области (R) пространства своих переменных, и рассмотрим функцию Φ, зависящую от переменных дифференциального уравнения (F), содержащую ряд произвольных постоянных и принимающую значения в области (C)

Reprinted from the *Ukrainian Mathematical Journal*, Ukrainian Academy of Sciences **30** (1978), 238–241.

пространства этих постоянных. Функция, возникающая из (Φ) при определении всех ее произвольных постоянных, и такая, что удовлетворяет (F), является частным решением уравнения (F).

Функция (Φ), рассматриваемая как совокупность всех частных решений уравнения (F), возникающих из (Φ), является общим решением этого уравнения (F).

Любая функция, возникающая из общего решения (Φ) нелинейного дифференциального уравнения (F) при определении некоторых произвольных постоянных в (Φ), которая содержит оставшиеся неопределенными постоянные в качестве произвольных параметров, является «частью» общего решения уравнения (F).

Если имеются функции (Φ_i), $i = 1, 2, \ldots, n$, обладающие свойствами, аналогичными свойствам (Φ) относительно (F), и, кроме того, частные решения любых двух функций (Φ_i) не тождественны, то их следует рассматривать как «разные общие решения» дифференциального уравнения (F); тогда говорим, что (F) имеет «несколько общих решений». Другими словами, рассматривая (Φ_i) как множество решений дифференциального уравнения (F) таких, что ни одна из любых двух функций (Φ_i) не включает другую, но они могут иметь пересечение, мы считаем их разными общими решениями нелинейного дифференциального уравнения (F).

2. Метод ограничения величин дифференциального уравнения. Ограничение решений или других величин дифференциального уравнения может дать несколько общих решений дифференциального уравнения. Приведем примеры.

Пример 1.

$$y'y''' - y''^2 = 0. \tag{1}$$

Ограничим решение y и получим 3 случая:

а) если y ограничено посредством $y' = 0$, откуда следует $y'' = 0$, тогда (1) удовлетворяется посредством $y' = 0$, $y'' = 0$, так что $y = c$ удовлетворяет (1); c — произвольная постоянная;

б) если y ограничено посредством $y'' = 0$, откуда следует $y''' = 0$, тогда (1) удовлетворяется посредством $y'' = 0$, $y''' = 0$, так что

$$y = a_1 x + a_2, \tag{2}$$

с произвольными постоянными a_1, a_2, является общим решением уравнения (1);

в) если y ограничено посредством $y' \neq 0$ и $y'' \neq 0$, дифференциальное уравнение (1) можно записать в виде

$$\frac{y'''}{y''} = \frac{y''}{y'},$$

откуда посредством интегрирования можно получить

$$\frac{y''}{y'} = c_1, \tag{3}$$

c_1 — произвольная постоянная. Интегрирование (3) дает

$$y' = c_2 e^{c_1 x}, \tag{4}$$

c_2 — вторая произвольная постоянная. Интегрирование (4) дает

$$y = \frac{c_2}{c_1} e^{c_1 x} + c_3, \tag{5}$$

которое содержит три произвольных постоянных.

Функции (2) и (5) — два различных общих решения дифференциального уравнения (1), содержащие прямые линии, параллельные оси x-ов, как общую часть.

Пример 2.

$$y'''(1+y'^2)-2y'y''=0. \tag{6}$$

Ограничим решение y двумя способами:

а) если y ограничить посредством $y''=0$, откуда следует $y'''=0$, тогда дифференциальное уравнение (6) будет одновременно удовлетворяться посредством $y''=0$, $y'''=0$, и функция

$$y=a_1x+a_2 \tag{7}$$

будет являться общим решением уравнения (6);

б) если y ограничивается посредством $y''\neq 0$, тогда (6) может быть точно проинтегрировано [1], и функция

$$x^2+y^2+c_1x+c_2y+c_3=0, \tag{8}$$

т. е. семейство окружностей в плоскости (x, y) будет общим решением уравнения (6).

Так как определение произвольной постоянной (7) и (8), которое отождествляет частное решение (7) с частным решением (8) не осуществлено, то в таком случае функции (7) и (8) являются различными общими решениями уравнения (6).

Пример 3.

$$y''+\frac{2}{x}y'+y^n=0. \tag{9}$$

В этом примере ограничим скаляр n. Это уравнение хорошо известно в астродинамике как дифференциальное уравнение «Емдена» [2]. Его решение в замкнутом виде известно в случаях $n=0$, $n=1$, когда (9) линейно; в случае $n=5$, функция, содержащая произвольную постоянную, удовлетворяет этому дифференциальному уравнению, и рассматривается как «часть» «неизвестного» общего решения. Ни в каких других случаях скаляра n дифференциальное уравнение (9) не имеет общего решения.

Хотя общие решения и наиболее желательны, но встречаются случаи, когда они не в состоянии отразить всей сути рассматриваемых физических явлений и любая попытка найти общие решения должна побуждаться только теоретическим любопытством. Емденовское дифференциальное уравнение и является таким примером.

В своих исследованиях Емден нашел, что решения уравнения (9) имеют физический смысл в случае, когда n находится между 0 и 5. Он также нашел, что в физических ситуациях исследуются граничные условия: $x=0$, $y=1$, $y'=0$, и при этих условиях частные решения уравнения (9), известные в случаях $n=0, 1, 5$, не приемлемы. Емден и его последователи нашли решение, удовлетворяющее физическим требованиям, и этим решением было разложение Тейлора относительно $x=0$ с последующим применением аналитического продолжения ряда.

Заметим, что, используя ограничения величин дифференциального уравнения, можно получить несколько общих решений нелинейного дифференциального уравнения, отличных от приведенных примеров.

3. Метод факторизации дифференциального уравнения. Факторизация дифференциального уравнения может привести к нескольким общим решениям дифференциального уравнения. Проиллюстрируем этот метод на нескольких примерах.

Пример 4.

$$x^3y''y'''+x^2y''^2-2xy'y''+2yy''=0, \tag{10}$$

y'' — общий множитель всех членов уравнения (10), так что либо

$$\text{а)}\ y''=0, \quad \text{либо б)}\ x^3y'''+x^2y''-2xy'+2y=0,$$

откуда имеем

$$\text{а)}\ y = a_1 x + a_2; \quad \text{б)}\ y = c_1 x + c_2 x^2 + c_3 x^{-1}. \tag{11}$$

Функции (11)— два общих решения уравнения (10). Видим, что семейство прямых линий, проходящих через начало координат,— общая часть этих общих решений.

Пример 5. $F(x, y, y') = 0$, когда это выражение есть полином по y' степени m.

В этом случае имеем

$$F(x, y, y') \equiv y'^m + P_1(x, y)\, y'^{m-1} + \ldots + P_{m-1}(x, y)\, y' + P_m(x, y) = 0. \tag{12}$$

Мы можем решить (12) относительно y', если $F_{y'}(x, y, y') \neq 0$; если $y_1', y_2', \ldots, y_m'$ являются m простыми корнями, то можно записать

$$F(x, y, y') \equiv [y_1' - P_1(x, y)]\,[y_2' - P_2(x, y)] \ldots [y_m' - P_m(x, y)] = 0. \tag{13}$$

Приравнивая к нулю каждый множитель, из соответствующих дифференциальных уравнений получаем m функций:

$$\varphi_1(x, y, c_1) = 0, \ \varphi_2(x, y, c_2) = 0, \ \ldots, \ \varphi_m(x, y, c_m) = 0, \tag{14}$$

являющихся m разными общими решениями выражения (12). Из них выбирают такие, которые являются действительными функциями от действительных переменных. Например, в дифференциальном уравнении

$$y'^2 y''' + y^2 y''' + y''' = 0 \tag{15}$$

факторизация дает: $y'''(y'^2 + y^2 + 1) = 0$ и, так как $y'^2 + y^2 + 1 \neq 0$, это дифференциальное уравнение (15) эквивалентно $y''' = 0$, так что функция

$$y = c_1 x^2 + c_2 x + c_3 \tag{16}$$

является общим решением дифференциального уравнения (15). Это общее решение состоит из двух частей разного характера — семейства «парабол» ($c_1 \neq 0$) и семейства «прямых линий» ($c_1 = 0$).

ЛИТЕРАТУРА

1. Davis H. Introduction to Nonlinear Differential and Integral Equations. Dover Publications Inc., New York, 1962, p 371—377.
2. Goursat E. Differential Equations. Ginn and Company, Boston, 1917, pg. 5.

США

Поступила в редакцию
19.V. 1977 г.

Magiros D.G., UNIFICATION OF STABILITY CONCEPTS. -Mat. Fizika, V. 33 (1983): 16-21.

A great number of concepts of stability are illustrated and unified in the present paper by means of the same mathematical relations, which give the possibility to obtain concrete definitions of stability and some correlations between them.

The author considers the systems under the conditions of instantly and constantly effecting perturbations, and unifies the concepts of stability on the basis of metrics. Relations between stability according to Lyapunov, Poincaré and Lagrange are discussed.

УДК 517.91

Д. Г. МАЖИРОС

УНИФИКАЦИЯ ПОНЯТИЙ УСТОЙЧИВОСТИ

Исследования основных физических, технологических и социальных проблем приводят к формулировке проблем устойчивости динамических систем. Эти исследования наталкиваются на трудности различной природы — в понимании концепции устойчивости, приемлемых с физической точки зрения и совместных с математической; в разработке эффективных критериев, с помощью которых можно было бы сделать заключение об устойчивости или неустойчивости состояния системы. Разнообразие понятий устойчивости возникает по различным причинам, обусловленным формой математической модели системы; природой выбора системы переменных; способами описания состояния системы; способами приближения или отталкивания от другого состояния системы; характером возмущений, действующих на систему; способом измерения нормы возмущений и их действия на систему. В данной работе множество понятий устойчивости выражено и унифицировано с помощью одних и тех же математических соотношений, которые посредством конкретизации параметров дают возможность получить конкретные понятия устойчивости и некоторые соотношения между ними. Эта унификация приносит естественное упрощение в понимании проблем, касающихся устойчивости, и является источником для получения новых результатов.

Системы при мгновенно или постоянно действующих возмущениях. Опишем математические модели физических систем, устойчивость решений которых будем исследовать. Нарушения состояния системы зависят от возмущений, которые можно рассматривать, как возмущающие силы, действующие на систему

Reprinted from *Matematich. Fizika*, Ukrainian Academy of Sciences **33** (1983), 16–21.

мгновенно или постоянно. Мгновенно действующие возмущения нарушают начальные состояния, но не появляются в математической модели, т. е. они не проявляются в уравнениях, описывающих движение системы. В случае мгновенных возмущений модель можно взять в векторной форме

$$\dot{x}(t) = X(t, x), \quad x(t_0) = x_0, \quad X(t, 0) = 0, \quad t \geqslant t_0, \tag{1}$$

где $X = (X_1, \ldots, X_n)$, $x = (x_1, \ldots, x_n)$ — переменные состояния; $x_0 = (x_{10}, \ldots, x_{n0})$ — начальные условия. Эти уравнения рассматриваются в (t, x)-пространстве. Постоянно действующие возмущения изменяют состояние системы, X изменяется в зависимости от возмущений, а математическая модель имеет вид

$$\dot{x}(t) = X(t, x) + p(t, x). \tag{2}$$

Здесь p — постоянно действующие возмущения. Для существования единственного решения уравнений (1), (2) величины X и p должны удовлетворять определенным условиям.

Когда модель выражается дифференциальными уравнениями с отклоняющимися аргументами, например дифференциальными уравнениями с запаздывающим аргументом $\tau(t)$, различаем два типа моделей:

1) при мгновенно действующих возмущениях модель имеет вид

$$x(t) = X[t, x(t - \tau(t))], \quad \tau > 0, \tag{3}$$

при этом начальная функция $\varphi(t)$ определена на начальном интервале $E_{t_0}: t_0 - \tau \leqslant t \leqslant t_0$, а соответствующее решение $x_\varphi(t)$ (3) при $t \geqslant t_0$;

2) при постоянно действующих возмущениях —

$$x(t) = X[t, x(t - \tau(t))] + p[t, x(t - \tau(t))]. \tag{4}$$

Действие возмущений. Некоторые определения. Следующие соображения и определения дадут возможность унифицировать понятия устойчивости.

Результатом действия возмущений на систему является изменение определенных величин, принадлежащих системе, и это воздействие может быть наглядно выражено путем изменения траектории S начального (невозмущенного) решения системы в траекторию $\bar{S}$ нового (возмущенного).

Состояние системы обозначается точкой траектории, и если точка M невозмущенной траектории S соответствует точке $\bar{M}$ возмущенной траектории $\bar{S}$, расстояние $\rho = M\bar{M}$ (рис. 1) может рассматриваться как мера величины действия возмущений на M.

Это расстояние играет важную роль в понимании устойчивости и формулирования ее соотношений. Назовем его расстоянием устойчивости.

Данной точке M траектории S могут соответствовать разные точки $\bar{M}$ траектории $\bar{S}$. Такое соответствие допускает предположение о физической пригодности и характеризует конкретное понятие устойчивости траектории S. Если имеется два основных понятия устойчивости, то можно выделить два наиболее физически осмысленных соответствия между M и $\bar{M}$:

1) одно соответствие между M и $\bar{M}$ на S и $\bar{S}$ справедливо тогда, когда эти точки взяты в один и тот же момент времени (рис. 2); в этом случае расстояние $\rho = \rho_l = M\bar{M}$ называется ляпуновским и характеризует одно из основных понятий устойчивости — устойчивость в смысле Ляпунова, а математическая модель относится к пространству (t, x);

2) при другом соответствии расстояние $\rho = \rho_r = M\bar{M}$ (рис. 3) берется как минимальное расстояние от точки M до точек траектории $\bar{S}$ и называется расстоянием Пуанкаре, характеризующим другое основное понятие устойчивости — устойчивость в смысле Пуанкаре, или орбитальную устойчивость, в этом случае модель относится к пространству $(x, y = \dot{x})$, где t является параметром.

Унификация понятий устойчивости. Разнообразие понятий устойчивости может быть количественно описано следующими соотношениями:

$$\rho_0 < \delta_1, \tag{5}$$

$$\rho < \varepsilon, \tag{6}$$

$$\lim_{t\to\infty} \rho = 0, \tag{7}$$

$$\| p \| < \delta_2, \tag{8}$$

$$\rho_\tau < \delta_3, \tag{9}$$

которые унифицируют понятия устойчивости. Здесь ε, δ_1, δ_2, δ_3 — положительные постоянные; δ_1, δ_2 зависят от ε и начального момента $t_0 \geqslant 0$; ρ_0, ρ — расстояния устойчивости; $\| p \|$ обозначает норму возмущений; ρ_τ является расстоянием между невозмущенным и возмущенным запаздываниями.

Укажем некоторые практически полезные понятия устойчивости, вытекающие из соотношений (5) — (9) путем их комбинаций, конкретизации ρ и из ограничений других величин.

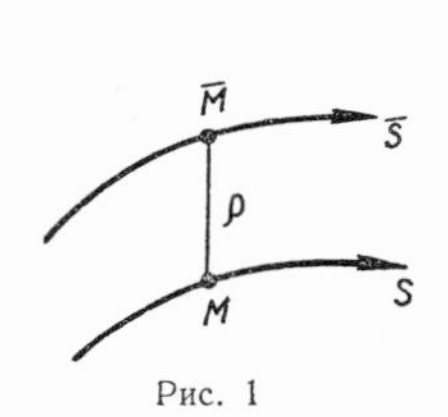

Рис. 1

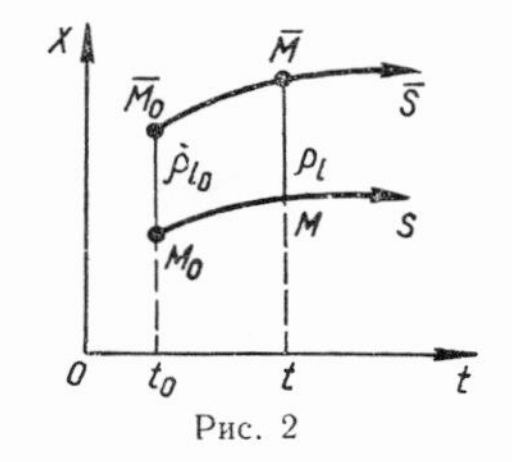

Рис. 2

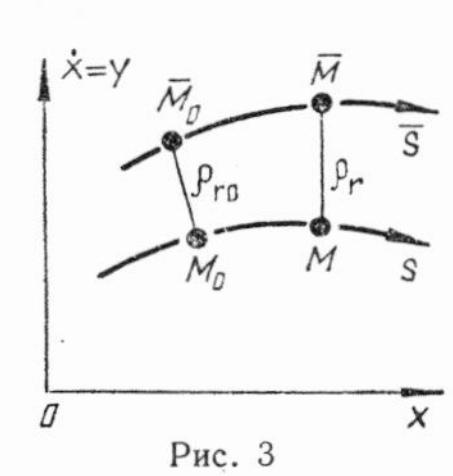

Рис. 3

Понятия устойчивости, вытекающие из соотношения (5). Для понятий устойчивости в случае модели (1) используем соотношения (5) — (7). Неравенство (8) используется в случае модели (2), а неравенство (9) — если в моделях (3), (4) возмущается запаздывание.

1. Решение $x(t)$ системы (1) называется устойчивым, если для данного сколь угодно малого ε можно найти $\delta_1 = \delta_1(t_0, \varepsilon)$ такое, что из неравенства $\rho_0 < \delta_1$ следует, что $\rho < \varepsilon$ для $t \geqslant t_0$. Расстояние $\rho = |\bar{x} - x|$, где $\bar{x}$ — возмущение x; ρ_0 — начальное расстояние. Если ρ является ляпуновским расстоянием, то $x(t)$ устойчиво в смысле Ляпунова, а если ρ — расстояние Пуанкаре, то имеет место устойчивость в смысле Пуанкаре (орбитальная устойчивость). Если к тому же выполняется условие (7), то имеем асимптотическую устойчивость.

2. Для устойчивости решения уравнения (2) является важным неравенство (8). Различным нормам p соответствуют различные типы устойчивости. Укажем три из них [10]: траектория S является устойчивой в целом, если p мало и его норма определяется как $\| p \| = \max_x | p |$; S интегрально устойчива, если p может быть большим на малых интервалах времени, а его норма определяется как $\| p \| = \int_0^\infty \max_x | p | \, dx$; S устойчива в среднем, если $\| p \| = \int_t^{t+T} \max_x | p | \, dx$.

Существуют утверждения, дающие соотношения между этими понятиями устойчивости. Указанная выше устойчивость может пониматься как в смысле Ляпунова, так и в смысле Пуанкаре.

Некоторые замечания по понятиям устойчивости. 1. Понятие устойчивости по Ляпунову соответствует устойчивости движения по траектории, а понятие устойчивости Пуанкаре — устойчивости самой траектории. В то время как понятие устойчивости Ляпунова является слишком ограничительным для периодических движений, понятие устойчивости Пуанкаре является именно тем, которое нужно для периодических движений.

2. При исследовании устойчивости точек равновесия системы расстояния по Ляпунову и Пуанкаре идентичны, так как устойчивость точек равновесия в смысле Ляпунова и Пуанкаре эквивалентна.

3. Если δ_1 не зависит от t_0, $\delta_1 = \delta_1 \varepsilon$, обе устойчивости являются равномерными.

4. Непрерывная зависимость решения от начальных условий эквивалентна устойчивости решения по отношению к начальным условиям, рассматриваемым как параметры решений.

5. Понятие ограниченности решений, которое может быть дано как частный случай соотношений (5) — (9), есть частный случай понятия устойчивости, называемый устойчивостью по Лагранжу. Но свойство ограниченности решения является отличным от свойства устойчивости, хотя в некоторых случаях одно вытекает из другого [1; 3, с. 129; 11].

6. Природа устойчивости физических систем является неинвариантной по отношению к выбору пространственных переменных, используемых как координаты системы, или же по отношению к преобразованию этой системы координат [6]. Можно также иметь устойчивость только по отношению к части координат и неустойчивость по отношению к другим координатам [3, с. 10; 12, с. 96—97]. Зависимость устойчивости решения от системы координат и ее преобразований является весьма важной проблемой, но по ней нет никаких систематических исследований.

7. Система определяется как структурно устойчивая, если ее устойчивость инвариантна при малых возмущениях параметров либо формы правых частей самой системы. Из структурной устойчивости системы вытекает много важных свойств [5]. Свойства структурной устойчивости подсказывают методы исследования физических проблем с хорошо понимаемыми результатами, как в задачах, связанных с морфологическими процессами, катастрофами в биологии и т. д. [9].

8. Обсуждаемые выше понятия устойчивости носят теоретический или математический характер, и чтобы адекватно интерпретировать реальные соотношения, необходимы соответствующие модификации, изменения, дополнения [3, с. 126; 4; 12, с. 13; 123].

Устойчивость, имеющая практическую пригодность, называется практической устойчивостью. Для нее требуется знание следующих параметров:

(*i*) размера области отклонений системы, пригодной для удовлетворительной работы;

(*ii*) размера области допустимых начальных условий, которые могут быть управляемы;

(*iii*) размера области допустимых возмущений;

(*iv*) размера области допустимых значений параметров;

(*v*) конечного времени, в течение которого исследуется устойчивость.

Для безупречной практической устойчивости размеры указанных выше областей должны быть большими. Вычисляются эти области с использованием нелинейностей в математической модели системы, и в этом случае метод линеаризации модели не допускается.

Рассмотрим примеры, являющиеся приложением определений устойчивости.

1. Прямолинейное движение: $x(t) = v(t - t_0) + a$ в плоскости (t, x), где a — расстояние, v — скорость движущейся массы (наклон прямолинейной траектории движения):

i является устойчивым в смысле Ляпунова и Пуанкаре, если движение равномерное;

ii является неустойчивым в смысле Ляпунова и Пуанкаре, если движение неравномерное.

Упомянутое выше движение в общих случаях неустойчиво по Лагранжу.

2. Движение массы, притягиваемой к центру по обратно квадратическому закону Ньютона, является периодическим с эллиптической орбитой. Это движение по Ляпунову неустойчиво, а по Пуанкаре и Лагранжу устойчиво.

3. Система нелинейных дифференциальных уравнений

$$\dot{x} = -y(x^2 + y^2)^{\frac{1}{2}}, \quad \dot{y} = x(x^2 + y^2)^{\frac{1}{2}} \tag{10}$$

в качестве решения имеет

$$x = a\cos(at + b), \quad y = a\sin(at + b) \tag{11}$$

с орбитой

$$x^2 + y^2 = a^2, \qquad (12)$$

где a, b — параметры. Имеем (i), если a — постоянная, решение (11) устойчиво по Ляпунову, но не имеет смысла по Пуанкаре;

(ii) если a — переменная, решение (11) неустойчиво по Ляпунову, но устойчиво по Пуанкаре. В обоих случаях движение устойчиво по Лагранжу.

Отметим, что ляпуновская неустойчивость в случае переменного a может быть устранена посредством замены переменных x, y новыми r, b по формулам перехода [2]

$$x = r\cos v, \quad y = r\sin v, \quad v = at + b, \qquad (13)$$

когда неравенство (6) переходит в уравнения $\dot{r} = 0$, $\dot{b} = 0$, решением которых является $r = c_1$, $b = c_2$, где c_1, c_2 — произвольные постоянные. Это решение устойчиво по Ляпунову.

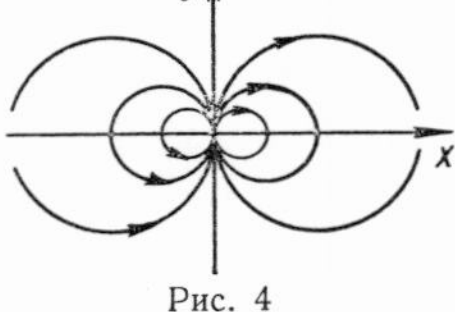

Рис. 4

4. Система нелинейных дифференциальных уравнений

$$\dot{x} = 2xy, \quad \dot{y} = x^2 - x^2 \qquad (14)$$

в качестве решения имеет однопараметрическое семейство кривых (циклов)

$$(x - r)^2 + y^2 = r^2, \qquad (15)$$

единственной точкой равновесия которого является начало координат (рис. 4). Этот пример напоминает картину магнитного силового поля вокруг электрона. Решение, начинающееся в любой точке на любой интегральной кривой, стремится при $t \to \infty$ к предельной точке — началу координат, т. е. $\lim\limits_{t\to 0} \rho = 0$. Видно, что для данного $\varepsilon > 0$ не существует δ_1 такого, что для любого начального $\rho < \delta_1$ все точки соответствующей кривой решения будут находиться внутри окружности радиуса ε, т. е. соотношение $\rho < \varepsilon$ не может выполняться. Так как выполняется только условие (7), то начало координат является «аттрактором» специального типа и псевдоасимптотически устойчиво.

5. Если твердое тело вращается вокруг своей оси симметрии, то возникает явление процессии, которое зависит от природы внешнего момента вращения, действующего на тело. В частном случае момент вращения процессии соответствует некоторой кривой на поверхности ортогонального кругового цилиндра и эта кривая стремится к бесконечности при $t \to \infty$, когда процессия относится к геликоидному типу. Этот вид процессии неустойчив по Ляпунову, но асимптотически устойчив по Пуанкаре. Из практических соображений в случае геликоидной процессии для описания устойчивости следует принять асимптотическую устойчивость по Пуанкаре [6].

6. Исследование численных процессов с использованием быстродействующих ЭВМ дало начало математическим моделям с возмущающими членами, устойчивыми в смысле Ляпунова и Пуанкаре.

1. *Antosiewicz H.* Boundedness and stability.— In: Nonlinear differential equations and nonlinear mechanics. New York : Acad. press, 1963.— 270 p.
2. *Cesari L.* Asymptotic behavior and stability problems in ordinary differential equations. — Berlin : Springer, 1959.— 271 p.
3. *Hahn W.* Theory and application of Liapunov direct method.— Englewood Cliffs : Prentice-Hall, 1963.
4. *LaSalle J., Lefschetz S.* Stability by Liapunov's direct method.— New York : Acad. press, 1961.
5. *Lefschetz S.* Differential equations : geometric theory.— New York : Intersci. publ., 1957.— 364 p.
6. *Magiros D.* Mathematical models of physical and social systems.— Practica, Athens Academy of sciences, 1976.
7. *Magiros D.* Characteristic properties of linear and nonlinear systems.— Ibid.
8. *Magiros D.* On the stability of a special class of precessions.— In: Yolume in memorium D. Eginitis. Athens, s. a., p. 189—198.
9. *Thom R.* Structural stability and morphogenesis.— Reading. (Mass.) : Benjamin, 1975.— 362 p.
10. *Vrkoc I.* On some stability problems.— In: Proc. conf. Prague. New York : Acad. press, 1963, p. 217—221.

11. *Yoshizawa T.* Stability theory by Liapunov second method.— Tokyo : Math. Soc. Jap., 1966.— 223 p.
12. *Zubov V.* Mathematical methods for the study of automatic control systems.— New York : MacMillan, 1963.— 324 p.

Корпорация «Дженерэл Электрик»

Поступила в редколлегию
24.10.80

УДК 62.502

BIOGRAPHICAL NOTE OF D.G. MAGIROS

Demetrios G. Magiros was born in Euboia Island, Greece, December 29,1912. He died in Philadelphia, Pennsylvania, January 19, 1982.

Magiros studied pure mathematics at the University of Athens, Greece, where he prepared his doctor's thesis in 1940. He begun his teaching career at the National Metsoveion Polytechneion (National Technical University) of Athens as a lecturer of mechanics and geodesy, and later as a professor of mathematics of the same institute. In 1949 he left for the United States with the purpose of carrying out further studies and advanced research in applied mathematics. Magiros showed early in his career his enormous interest in the applications of mathematics and his conviction that pure mathematics had important applications that mathematicians had a duty to seek to facilitate. In this attitude he differed sharply from many mathematicians who praised pure mathematics for its beauty and logical rigor alone, while disdaining any thought of application.

In the United States he continued his studies in applied mathematics at Brown University, the Courant Institute of Mathematical Sciences of New York University, and the Massachusetts Institute of Technology. He was first engaged in research at the IBM Watson Laboratory of Columbia University, and continued at the Republic Aviation Corporation and the Courant Institute (NYU). At Hofstra University he was appointed professor of mathematics and mechanics, while at the same time he accepted the duties of scientific consultant at the Space and Missile Center of General Electric Company in Philadelphia.

By 1960, when Magiros assumes full-time responsibilities as a researcher and mathematical consultant of the G.E. Space and Missile Center (presently, Re-Entry and Environmental Systems Division), he has developed a remarkable multi-dimensional scientific activity. He keeps in constant communication with distinguished scientists and academicians of various countries, such as, Leon Brillouin, Henry Villat, S. Lefschetz, C. Friedricks, L.H. Thomas, Y. Mitropolsky, G. Duboshin, C. Kodradiyev, who influenced to a degree his research orientation.

It was in this scientific center, one of the most advanced and dynamic space centers of the world where, with incomparable enthusiasm and dedication, he prepared most of his important work at a time when the knowledge on space vehicles and their orbits was limited while the world competition was highly intense.

His contribution to mathematical formulation and solution of high research space problems, such as, long-range ballistic missiles, ballistic re-entry, hypersonic aerothermodynamics with high speed re-entry, planetary entry, absorb energies in certain spectra, theories for improvement of the system effectiveness, etc., is of fundamental significance. At the same time his influence as a consultant expanded in many G.E. scientific centers in U.S. and his seminars and lectures were attended intensively by numerous scientists from all over the world.

The Space Center of General Electric Co. and the Athens Academy of Sciences repeatedly honored Magiros' distinguished service to mathematical advancement.

Magiros' remarkable scope of interests and his important contribution to mathematical research and technical progress rendered him an internationally recognized scientist. An intense, driving yet sensitive and modest man, a prolific mathematician and an inspiring teacher Magiros will continue for a long time to be an influence on mathematical research through his own works and through those he will be inspiring.

COMPLETE CHRONOLOGICAL LIST OF MAGIROS' PUBLICATIONS

[1] 1946 Ο τόπος κορυφής μεταβλητής αλυσοειδούς {On the locus of the vertex of a variable catenary}. Tech. Chron., Sc.J.Tech.Chamber,Greece,V.23,no. 265-266 (1946):99-102.

[2] 1946 Περί της τάσεως εις μεταβλητήν αλυσοειδή {The study of the tension of a variable catenary} .Proc. Athens Acad.Sci.,V.21 (1946):41-46;V.23,no.273-274 (1947): 3-7.

[3] 1946 Περί τασικών διαγραμμάτων των διατομών πακτώσεως εις ανηρτημένα σύρματα. {Diagrams of the tension of the cross-section in the suspension of hanging wires} . Proc.Athens Acad.Sci.,V.21 (1946):46-50; Tech. Chamber, Greece, V.23,no.273-274 (1947): 8-10.

[4] 1956 Ελεύθεραι πλευρικαί ταλαντώσεις απλής ράβδου υπό πλαστικήν εξαίτησιν, {I. Lateral free vibrations of simple prismatic bar in plasticity} . Proc. Athens Acad.Sci.,V.31 (1956):16-21.

[5] 1956 Ελεύθεραι πλευρικαί ταλαντώσεις απλής ράβδου υπο πλαστικήν εξαίτησιν, II. {Lateral free vibrations of simple prismatic bar in plasticity,II} .Proc. Athens Acad.Sci.,V.31 (1956):168-171.

[6] 1957 Subharmonics of any order in case of nonlinear restoring force, pt.I.Proc.Athens Acad.Sci.,V.32 (1957):77-85.

[7] 1957 Subharmonics of order one third in the case of cubic restoring force. Proc.Athens Acad.Sci.,V.32 (1957):101-108.

[8] 1957 Remarks on a problem of subharmonics. Proc.Athens Acad.Sci.,V.32. (1957):143-146.

[9] 1957 On the singularities of differential equations, where the time figures explicity. Proc.Athens Acad. Sci.,V.32 (1957):448-451.

[10] 1958 Subharmonics of any order in nonlinear systems of one degree of freedom:application to subharmonics of order 1/3. Inf.and Control,V.1,no.3 (1958): 198-227.

[11] 1959 On a problem of nonlinear mechanics. Inf.and Control,V.2,no.3 (1959):297-309;Proc.Athens Acad.Sci., V.34 (1959):238-242.

[12] 1960 The motion of a projectile around the earth under the influence of the earth's gravitational attraction and a thrust. Proc.Athens Acad.Sci.,V.35 (1960):96-103.

[13] 1960 The Keplerian orbit of a projectile around the earth, after the thrust is suddenly removed. Proc. Athens Acad.Sci.,V.35 (1960):191-202.

[14] 1960 A method for defining principal modes of nonlinear systems utilizing infinite determinants, I . Proc. Natl. Acad. Sci.,U.S.,V. 46,no.12 (1960):1608-1611.

[15] 1961 A method for defining principal modes of nonlinear systems utilizing infinite determinants, II . Proc. Natl.Acad.Sci.,U.S.,V.47, no.6 (1961):883-887.

[16] 1961 Diffraction by a semi-infinite screen with a rounded end (with Joseph B. Keller). Comm.Pure Appl. Math.,v.14,no.3 (1961):457-471.

[17] 1961 Method for defining principal modes of nonlinear systems utilizing infinite determinants. J.Math. Phys.,V.2,no.6 (1961):869-875.

[18] 1961 Remarks on Rosenberg's paper "The normal modes of nonlinear n-degree-of-freedom systems". J.Appl. Mech.,Trans.ASME,V.30,ser.E,no.1 (1963):151.

[19] 1963 On the convergence of series related to principal modes of nonlinear systems. Proc.Athens Acad.Sci., V.38 (1963):33-36.

[20] 1963 On the convergence of the solution of special two-body problem. Proc.Athens Acad.Sci.,V.38 (1963): 36-39.

[21] 1963 The impulsive force required to effectuate a new orbit through a given point in space. J.Frank.Inst., V.276, no.6 (1963):475-489;Proc.XIVth Intl.Astron. Congress, Paris, 1963.

[22] 1964 Motion in a Newtonian forced field modified by a general force, { I }. J. Frank.Inst.,V.278,no.6 (1964):407-416;Proc.XVth Intl.Astron.Congress, Warsaw, 1964.

[23] 1965 Motion in a Newtonian force field modified by a general force,{ II}. XVIth Intl.Astron.Congress, Athens, Greece (1965):349-355.

[24] 1965 On the stability definitions of dynamical systems, Proc.Natl.Acad.Sci. (U.S.),V.53,no.6 (1965):1288-1294.

[25] 1965 Physical problems discussed mathematically.Bull. Soc.Math.Greece,nouv.ser.,t.6II,fasc.I (1965): 143-156.

[26] 1966 Motion in a Newtonian force field modified by a general force, III.Application:the entry problem (with G. Reehl). XVIIth Intl.Astron.Congress,Madrid (1966):149-154.

[27] 1966 The entry problem (with George Reehl) Proc.Athens Acad.Sci.,V.41 (1966):246-251.

[28] 1966 Stability concepts of dynamical systems,Inf.and Control,V.9,no.5 (1966):531-548.

[29] 1967 Unification and classification of stability concepts of dynamical systems. Proc.VIIth Intl.Symp. Space,Tech. and Sci.,Tokyo, 1967.

[30] 1967 Attitude stability of a spherical satellite. Remarks on stability.Proc.XVIIIth Intl.Astron.Congress,Belgrade, Yugoslavia, 1967.

[31] 1967 A class of nonautonomous differential equations reducible to autonomous ones by an exact method. Proc.Natl.Acad.Sci.,V.58,no.2 (1967):412-419.

[32] 1968 Sur quelques sortes de précession dans le mouvement de rotation d'un corps rigide ayant un axe de symetrie. Comptes Rendus, Acad. Sci.,Paris, ser.A, t. 266 (1968):770-773.

[33] 1968 Attitude stability of a spherical satellite (with A.J.Dennison).J.Frank.Inst.,V.286,no.3 (1968):193-203;Bull.Amer.Phys.Soc.,ser.2,V.12,no.3 (1967): p.288 (Abstract).

[34] 1969 La stabilité de la précession hélicoidale dans le sens de Liapunov,Poincaré et Lagrange. Comptes Rendus, Acad.Sci.,Paris, sér.A, t. 268 (1969): 652-654.

[35] 1969 On a class of precessional phenomena and their stability in the sense of Liapunov, Poincare and Lagrange. Proc.VIIIth Intl.Symp.on Space Tech. Sci., Tokyo (1969): 1163-1170.

[36] 1969 The stability in the sense of Liapunov, Poincaré and Lagrange of some precessional phenomena. Proc. Vth Intl.Conf.Nonl.Oscill.,Kiev (1969):347-357

[37] 1969 On the helicoid precession:its stability and an application to a re-entry problem (with G. Reehl). Proc. XXth Intl.Astron.Congress,Buenos Aires, Argentina (1969):491-496.

[38] 1969 Stability concepts of dynamical systems. Applications to flight dynamics and numerical analysis. Proc.Athens Acad.Sci.,V.44 (1971):270-284.

[39] 1970 Actual mathematical solutions of problems posed by reality, I:a classical procedure. Proc.Athens Acad.Sci.,V.45 (1971):179-187.

[40] 1971 Actual mathematical solutions of problems posed by reality, II:applications. Proc.Athens Acad. Sci.,V.46 (1971):21-31.

[41] 1971 Orientation of the angular momentum vector of a space vehicle at the end of spin-up. Proc.XXIIth Intl.Astron.Congress, Brussels, Belgium, 1971.

[42] 1971 Stability concepts of solutions of differential equations with deviating arguments. Proc. Athens Acad.Sci.,V.46 (1971):273-278.

[43] 1972 The stability of a class of helicoid precessions in the sense of Liapunov and Poincare. Proc.Athens Acad.,Sci.,V.47 (1972):102-110.

[44] 1974 Remarks on stability concepts of solutions of dynamical systems. Proc.Athens Acad.Sci.,V.49 (1975): 408-416.

[45] 1974 On the stability of a special class of precessions. In memoriam of "Demetrios Eginitis". Athens Acad. Sci.(1975);Proc.XXVth Intl.Astron.Congress,

[46] 1975 (a) Μαθηματικά μοντέλα φυσικών και κοινωνικών συστημάτων (Mathematical models of physical and social systems). Math.Epith.,Greek Math.Soc., V.3 (1975):81-124.

(b) Mathematical models of physical and social systems. Genl.Electric Co.,R.S.D.,Philadelphia, 1980.

[47] 1977 On the linearization of nonlinear models of the phenomena,pt.I:linearization by exact methods. Proc.Athens Acad.Sci.,V.51 (1977):659-668.

[48] 1977 On the linearization of nonlinear models of the phenomena,pt.II:linearization by approximate methods. Proc.Athens Acad.Sci.,V.51 (1977): 669-683.

[49] 1977 Characteristic properties of linear and nonlinear systems. Proc.Athens Acad.Sci.,V.51 (1977): 907-935.

[50] 1977 (a) Nonlinear differential equations with several general solutions. Proc.Athens Acad.Sci.,V. 52 (1977):221-229.

(b) Nonlinear differential equations with several general solutions (In Russian). Ukrainian Math. J.,Ukrainian Acad.Sci.,V.30,no.2 (1978):238-241.

[51] 1978 The general solutions of nonlinear differential equations as functions of their arbitrary constants. Proc.Athens Acad.Sci.,V.52 (1978):524-532.

[52] 1979 On the separatrices of dynamical systems.Proc. Athens Acad.Sci.,V.54 (1980):264-287;Proc.IXth.Intl. Congress on Nonl.Oscill.,Kiev, 1981.

[53] 1980 Unification of stability concepts. Philadelphia: Genl.Electric.Co.,R.S.D.,1980. Mathematich. Fizika, Ukrainian Acad.Sci. (April) 1983.

[54] 1980 Stability concepts of dynamical systems. Philadelphia:Genl.Electric Co.,R.S.D.,1980.

MAGIROS' UNPUBLISHED WORKS

LECTURE NOTES

1963 Special functions and applications.
1963 Integral transforms.
1964 On some topics of nonlinear mechanics.
1965 Selected topics in applied mathematics.
1966 Stability concepts and criteria of dynamical systems.
1967 Nonlinear ordinary differential equations with closed form solutions.
1968 Methods for solutions of nonlinear differential equations.
1972 Linear, nonlinear and linearized phenomena.
1973 Exact solutions on nonlinear ordinary differential equations.
1974 Geometrical methods for solution of nonlinear ordinary differential equations.
1974 Numerical methods for solution of Nonlinear ordinary differential equations.
1975 Calculus of variations and applications, I.
1975 Optimal control systems. Applications.
1975 Matrices and transformations.

MONOGRAPHS

1976 Differential equations with one argument.
1976 Linearization of nonlinear models of the phenomena.
1977 Mathematics and real world.
1977 General and singular solution of nonlinear differential equations.
1977 Approximate methods for solutions of nonlinear differential equations: an introduction with some remarks.
1978 Approximate analytical methods for solution of nonlinear differential equations.
1979 Differential equations with one argument. Applications.
1979 Mathematics in economics.

REGULAR REPORTS

1961 Problems related to the motion of a projectile under the influence of a Newtonian center and a thrust (Genl. Electric Co., TIS no. 61SD36).

1961 Motion of a projectile under the influence of the attractive force and a Newtonian center and a thrust (Genl. Electric Co., TIS no. 61SD142).
1962 An actual solution of a problem of nonlinear mechanics (Genl. Electric Co., TIS no. 62SD144).
1963 Problems of classical celestial mechanics and astrodynamics (Genl. Electric Co., TIS no. 63SD957) {Rev. ed. 1978}
1964 Linear and nonlinear phenomena (Genl. Electric Co., TIS no. 64SD291).
1968 The study of orientation of an earth satellite (Genl. Electric Co., TIS no. 68SD249).
1968 Groups of problems of practical importance in numerical analysis, flight dynamics and nonlinear mechanics (Genl.Electric Co., TIS no. 68SD273).
1971 Remarks on linearization of models of nonlinear phenomena (Genl. Electric Co., TIS no. FM#71-1).
1972 Linear, nonlinear and linearized phenomena (Genl. Electric Co., TIS no. 72SD 225).
1972 The study of mathematical models of physical and real life phenomena.
1972 The role of mathematics in the investigation of physical and real life phenomena (Genl. Electric Co., TIS no. 72SD228).
1973 Exact solutions of nonlinear ordinary differential equations (Genl. Electric Co., TIS no. 73SD240).
1974 Geometrical methods for solution of nonlinear ordinary differential equations (Genl. Electric Co., TIS no. 74SD204).
1974 Numerical methods for solution of nonlinear ordinary differential equations (Genl. Electric Co., TIS no. 74SD20).
1976 Optimal control systems: Theory and applications, pt. I. (Genl. Electric Co., TIS no. 76SDR007).
1977 Matrices and transformations. Applications (Genl. Electric Co., TIS no. 76SDR042).

INTERIM REPORTS

1961 Periodic solutions of the general linear inhomogeneous system with periodic coefficients (NAC Analysis and synthesis, no. 131-096).
1961 On a system of differential equations of which the coefficients are functions of many parameters (OAO Stability contract).
1963 The nature of the stability of nonlinear control systems.
1963 Criteria of stability of nonlinear control systems.
1965 On the blast problem.
1969 Linearization of nonlinear models of physical and social phenomena.
1970 A problem of evolution.
1970 Competition between two species, I: Mathematical remarks.
1970 Competition between two species, II: The stability of equilibrium.
1970 Competition between two species, III: Equilibrium point of the model with depletion terms.

1970 Competition between two species, IV-V: The stability of the equilibrium state of the model with completion terms (2 v.)

oasis

supersoni

THE COMPLETE, AUTHORIS
AND UNCUT INTERVI

CURATED BY SIMON

Contents

Introduction

'How busy are you?' That was Noel's first question as he sat in our kitchen while I poured him a cup of strong tea, some time in the early months of 2014.

By that time, it had been over twenty years since he and I first met in Los Angeles, where I was living at the time. Oasis were shooting a video for a US version of their single 'Supersonic'. Back then, I was designing record sleeves, including for Paul Weller, and happened to mention to him on the phone beforehand that I was popping down to the set, as the director Nick Egan – another ex-pat – was a friend. Paul told me that I should introduce myself, not forgetting to tell Noel that he and I were mates. It was really that connection that served as a gold seal, firstly at the shoot and then that same evening at the Hollywood Roosevelt Hotel where I'd been invited to join the band for a light ale or two. This was where my friendship with Noel and Liam began and over the next couple of years I'd see them on a fairly regular basis when they'd breeze through town either for a gig or another video shoot.

To fill in the gaps very briefly between that first meeting and that cuppa with Noel twenty years on. By the middle of 1996 I was back living in London – a stone's throw from both Noel and Liam's houses in north-west London and we'd see a fair bit of each other.

Towards the end of the nineties I began designing LP covers for Oasis, and at the same time had the privilege of taking photos of the band at close quarters – whether it be at the studio, in Paris, Milan, and even on a return trip with them to the West Coast. The boys were always extremely welcoming and their own studio, Wheeler End, out in Buckinghamshire was a regular meeting spot. It felt like being a member of a gentlemen's club of sorts, where the time was split – for the most part – between chatting, laughing and listening to great music.

So there we were, 2014, in my kitchen. Noel tells me that August 2016 will be the twentieth anniversary of Oasis playing their two landmark gigs at Knebworth and asks if I might be interested in producing a short documentary to celebrate that. Would I? Needless to say, I immediately replied with a very enthusiastic, 'Yes.' I was unbelievably flattered to be asked, to be honest.

It wasn't until later that evening that I really had a chance to think about the potential and the scale of this idea. It seemed to me that this was not a short TV 'walk down memory lane' documentary. It really felt like those first three years of the Oasis story, from being signed by Alan McGee to their landmark shows at Knebworth, were nothing short of a phenomenon. I had recently seen Asif Kapadia's excellent documentary *Senna* and I felt that this Knebworth celebration could be elevated to a theatrical feature-length documentary in very much the same way, and like *Senna*, utilise archive footage alongside contemporary audio interviews to tell the story.

With this in mind I got in touch with James Gay Rees who produced *Senna* to see if they'd be interested in collaborating on this feature idea. James was in from the get-go. Asif came on board as an executive producer but wasn't available to direct the film as he was in the middle of making *Amy* and was committed to another project right after. We needed a director.

Prior to all this, I had already had some dealings with Mat Whitecross on another film project that, like most film projects, sadly didn't come to anything. I remembered being really impressed by his passion, understanding and commitment. Mat is a super-smart filmmaker. I also knew that because he had worked with, and been friends with, Coldplay since their student days, he'd have a real understanding of the dynamics of a

rock band. We approached Mat and thankfully he, like me, couldn't say yes quick enough. He and his producing partner Fiona Neilson were now firmly on board.

All we needed now was for Liam to sign off on the idea. So Mat, James and myself headed up to a restaurant in Highgate to meet with him and his partner Debbie to talk them through the idea. He was keen but had two questions, 'Who is the hero and who is the villain?'

That was a key question and one that made it our resolve to be certain this film would be an absolutely fair and frank telling of the band's story. It's a tale that goes beyond the phenomenon to tell the story of a group that captured the zeitgeist, but with a tale of two brothers at its heart.

Throughout 2015, with Mat Whitecross at the helm, we were able to interview all the key players at great length and this enabled us to make Supersonic the film we had hoped it would be. We were very fortunate to have access to so many of those who were close to the band, including family members. All very kindly and candidly recounted their stories in great depth, over many, many hours.

Mat did an outstanding job weaving the stories together, creating a film that resonated on such an amazing level with the band, the fans and critics

alike, even picking up some awards along the way. It wasn't just Liam who claimed that the film was 'biblical'. Fans of the film reached far and wide from the late George Michael to Chris Martin to Paul Thomas Anderson to Ewan McGregor. Lena Dunham even hosted a screening in Los Angeles with Brad Pitt introducing the film. Yes, Brad Pitt. What was really surprising though, was that the film really struck a chord with a whole new generation of kids that weren't even born when Oasis played Knebworth.

As the twenty-fifth anniversary of Knebworth approached, I thought it the perfect opportunity to revisit the transcripts of all the interviews recorded for the film. We had over sixteen hours in the can with Noel and twelve with Liam. It had not been physically possible to include it all in a two-hour documentary film, so the idea for a book, uncut, made perfect sense. The good sense in this idea has been confirmed over the last few months, during which I have had the opportunity to revisit and dig deep into the full transcripts with all the key players.

As a result this book genuinely feels like the last word – a complete and true bookend to a documentary that I'm very proud of, and a real testament to a band that genuinely mattered, and still does to this day.

Simon Halfon, June 2021

Liam Oasis was definitely like a fucking Ferrari: great to look at, great to drive. It would spin out of control every now and again when you go too fast. I'd have that any day over the old Volvo – being in a band like that didn't interest me one bit. We just wanted to fucking take it to the max, every day, have a great time, every fucking day, and then if it ended tomorrow, it ended tomorrow.

Noel I would go back and fucking do that journey again in a heartbeat.

Liam I just wanted it all fucking there and then. The beginning, the middle bit and the end, right now. I just wanted it. Whatever the ending was I wanted it now, and whatever the middle bit was I wanted it now, and whatever the beginning was . . . I just wanted it all to happen in one big fuck-off explosion of madness.

Noel Everybody used to say, 'We are the best band in the world', I am not sure anybody truly believed it. I actually fucking believed it. And I still believe it to this day. There was a period where we were fucking untouchable. It was only short, eighteen months, maybe two years, but we were up there with the greats.

Growing up

Noel I was born in Longsight in 1967 on 29th May. *Sgt Pepper* came out on 1st June and I do believe on hospital radio they were playing *Sgt Pepper* as I was born into the world and if that's not fucking true, that's the story I've been sticking to for the past forty-eight years.

Peggie Gallagher Paul arrived ten months after I was married, then Noel arrived a year after that again and everything was fine, it was going alright, or so I thought it was.

Paul Gallagher We had bowl heads, knitted jumpers, little shorts. She used to knit our clothes. Imagine, you've got no choice, you couldn't say, 'I don't want that, Mam,' 'cause she's just spent four years knitting you a jumper. We were dressed identical, maybe it was cheaper to do two instead of one.

Peggie Liam arrived five and a half years after. I don't know, I think there was always that bit of jealousy with Liam and Noel, because there was only Paul and Noel for so long. I idolised Paul and Noel because I had only the two of them. Noel was absolutely beautiful when he was a baby. Then of course, Liam comes along, takes the limelight. You could tell the disagreement was there with them.

Noel We lived off a busy main road; Longsight Market was across the road from our house. It was a busy Mancunian suburb maybe two miles from the city centre. I don't romanticise my childhood, but it was alright. We had an outside toilet. It's grim up north and all that.

Peggie It was great, that was the best time with the kids. Everybody knew everybody, I knew hundreds of Irish people. If you went out shopping with the kids on Stockport Road, you'd always run into friends that you knew when you were young and their families. Going to church, going to school, everybody knew everybody. It was a great community for the Irish people then. I was down Longsight in a two-bedroom house – two up and two down – and that was demolished. We got rehoused to a place called Burnage which was further south. Compared to Longsight, this was like fucking the Cotswolds. You had a front garden and a back garden instead of an alleyway and a front door that opened onto the street. But if you'd seen my rooms upstairs – you couldn't swing a bloody cat in them.

Paul You've got a three-bedroom council house, there's not a lot of space. You're going to get friction. That's what people don't understand, you know, since Liam was ten till seventeen, he shared a bedroom with Noel.

Noel Our Paul's always had his own room, *bastard.* It's something I've never quite forgiven him for. I got to share with Liam, which wasn't really a problem until we were teenagers and then he was just a pain in the arse . . . and has remained that way ever since.

Liam Paul had his own room, I shared one with Noel. Yeah, it was alright.

Noel I didn't hang around with Liam until I joined the band. Although I shared a room with him, five years is a generation apart. All his mates, when I was fifteen, they are all ten, you know, so that's massive. I'm leaving school at fifteen, he's ten years old. You're fucking smoking weed at fifteen, he's just out of short pants, so there is no relationship.

Liam I think we got on, I think. He was a bit of a stoner, a bit of a loner, one of them people that you'd throw stones at. He had a guitar so he was copping for it, you know what I mean? One of them, walking round with his guitar with his weird mates. Our Paul was a mod so he had a bit more about him, he wasn't so weird like Noel was.

Peggie Always very quiet, Noel. He would go upstairs and bury himself. Always strumming a guitar. Many's a time I went up and I knocked at the door and said, 'Bloody guitar, you get on my bloody nerves.' 'You leave my

guitar alone,' he said. I used to do an awful lot of knitting and he'd be always drumming with the needles when he was young. It was in him. I used to go, 'He'll put that bloody knitting needle in his eye yet.'

Paul Noel was quiet, moody, skinny, withdrawn, he kept himself to himself. Then you've got Liam who is a livewire. Imagine Zebedee from the Magic Roundabout versus, I don't know, Mickey Mouse? You got a lot of noise.

Liam Noel's definitely a bit cagey, I'd call it shifty.

Noel I don't know why me and Liam would be so different, we both had the same childhood, do you know what I mean? I am more of a loner. I genuinely like my own company. I'm not a shy, can't-really-speak-to-anyone person . . . I'm very outgoing and fucking love my circle of friends and all that, but I don't need them . . . A few years in nick, in solitary confinement, that wouldn't bother me in the slightest.

Liam Anyone who could do a stretch, I would watch out for them. Anyone who says, 'Yeah, I could do fucking time,' that's scary. I couldn't do that.

He definitely keeps himself to himself. He's a funny cunt without a doubt. He's great and all that, but he's a bit of a dick as well.

Noel Liam is a fucking major pain in the arse . . . there isn't actually a word that could adequately sum up his fucking buffoonery.

Liam Our kid's more a thinker, I wasn't a thinker, I didn't have no time to think. It was just, 'Let's get straight into this fucking day.' I'd never sit there and scratch my chin and go, 'How am I going to play this Tuesday afternoon?' There was none of that. It was like, 'Let's fucking have Tuesday, full on, let's freak Tuesday out so it doesn't come back again. Let Tuesday tell Wednesday, "Fucking hell, you're in for it tomorrow, mate."'

Peggie He was a devil, Liam, full of it. Many's a time, I swear to God, if I caught him I would have broke the brush across his back, because he used to torment me in the kitchen.

Liam I used to have too much Weetabix, a lot of energy. Three in the morning, three when I get in from school and three before I go to bed. I was just fucking bouncing off the walls, man.

He was a devil, Liam, full of it.

Peggie Liam was a mammy's boy, always with me. He'd always come looking for you to see where you were. They were all close, but I suppose Liam was always closer to me, because he was the youngest. Everybody loved him round here because he had a lot of time for older people. You'd be coming back from the shops there and he'd say, 'Let me carry that bag for you.' He used to go up and down Burnage Lane waving to everyone. The big wave. Everybody loved him round here.

Noel Liam, I would suggest, needs an audience. He is an ideal frontman.

Paul Total attention seeker, robbing your clothes, robbing your records, robbing your this, robbing your that, robbing your money.

Liam I was pretty confident all the way through life, you know what I mean – fuck knows, I guess just looking in the mirror and seeing what your reflection is, you kind of, if you look like a knob you're going to act like a knob, I guess, you know what I mean. I kind of dig the way I look so went for it. I'm definitely a bit of a show-off and that, always loved the attention and that and still do, but not to the point where it's like fucking Bonnie Langford or one of them little fucking brats with jazz hands.

Peggie I used to go to the school plays and he'd always look to see if he could spot you in the audience. Liam always wanted to be top of everything, from a small little boy. If he didn't get the main part, he didn't want to be in it. Noel was a different kettle of fish altogether.

Noel I didn't hate school, I wasn't like an anarchist – fuck school and fuck all the teachers and all that. When I joined there was a mix-up with another kid called Gallagher from another school. There was three top classes and two classes for headcases in our year. Due to a mix-up, I got put in the top class for a term. I was thinking, there is something not right here; I don't understand what that teacher is saying; I don't know what she is fucking going on about. Who are these fucking eggheads? Then I walked into another classroom one day and was like, 'Aha! Okay, right.' Took my rightful seat at the back of the class.

Liam School was alright, I enjoyed it, I mean I didn't learn anything, but I enjoyed the fucking about part of it. I guess that set me up for being in a band and being on tour. I liked being in a gang, I liked hanging out with me mates, you know what I mean. It was a good crack. I wasn't one of them, sitting on me own, thinking about stuff, I was definitely always up to some kind of shit.

We would just do fuck all, all day, apart from listen to music, get into mischief.

Noel I don't look back on it with any fondness or any hate, do you know what I mean? It was just somewhere you had to go until you grew up, and then they fucking let you go and that was it. I didn't play truant, or 'wag it' as it used to be called, because I hated school, I did it because it was fucking boring. One of my mates lived right by the school and his mam and dad worked all day so you could stay at his house. We would just do fuck all, all day, apart from listen to music, get into mischief. Me mam got a job as a dinner lady and then that made things a bit trickier because you would always have to be around school at dinner time, but not in the morning or the afternoon.

Peggie Noel would go to school as brazen as you like. I was a dinner lady then and he'd come in and stand in the queue with these big Doc Martens on him. He'd get whatever he wanted, because it was free dinners, and then he'd always came back to me at the end with a biscuit that he didn't want

and give it to me, as much as to say, 'This will get me out of trouble.' Then the teacher would say, 'Where is Noel?'
'I've just seen him, here; he was in front of me.'
'No, Mrs Gallagher, he's not been in school for three weeks.'
'I've just been talking to him; he's come up here with his tray and everything.' He was skipping over the wall, he was trouble.

Liam I got me hair shaved, ears pierced, told a couple of teachers to fuck off – like you do. Didn't do me homework, it got ate by an alien one day, all that nonsense. The usual. Didn't bother turning up because we were sitting at home playing pool, getting stoned, wagging it or whatever they call it these days. But yeah, but when I was there, it was a laugh.

Peggie I remember the teacher saying once to me, 'I don't know how you put up with him, Mrs Gallagher. I only have him a couple of hours a day, I have to go home and take a tablet. You've got to put up with him all the time.' I still meet that lady sometimes. She said, 'He was full of energy.' He was a devil, Liam.

Somebody hammered the music into him . . .

Liam I enjoyed going to school, like I say, didn't learn much, but I liked it because I just fucked about. I remember just having a laugh. I weren't that clever at school, didn't pass any exams, wasn't interested, always looking out the window, just fucking flicking people on the ear and robbing shit and stuff. Just being the class clown and shit. Obviously at home there was a bit of tension and a bit of shit going down, so it was a bit of relief getting into school and hanging out with your mates and just fucking about. We were always running around fighting with schools and shit like that, it was a good crack. I wouldn't say I went round looking for trouble, but if someone started then I certainly wouldn't do a runner, you know what I mean. I wasn't like tough guy or that, there were a lot harder people than me. I had a few mates that knew how to knock people out.

Paul You have the cocks of the school; Liam was the cock of the school and then, one day, some rival school turned up and boshed him on the head with an 'ammer.

Liam With a hammer, yeah, that was in St Mark's. We were stood there having a cig in one of the fucking little areas round the back. The girls used to come over to our school to use our facilities, because ours were better, and I think we were talking to our mate's sister, she was about twelve so we were about fifteen. I remember all these lads coming down, I don't know what school they were from. They've come down, hoods up, and one of them's fucking punched her. So we got stuck in. I've hit this kid and this kid pulled out this little fucking hammer and went whack on me head. Then I've ended up in hospital, blood everywhere. Got out of fucking double maths though, so that was alright. That was it. From that day on, it was like as if something had fucking clicked. I started hearing music, it started making sense. Up until then I wasn't into music. I know it sounds stupid, I probably said it before, but whoever he is, thank you.

Noel Somebody hammered the music into him . . . he's got a lot to answer for, hasn't he? I've got a perfect alibi for that – so it's nowt to do with me.

Proper jobs

Noel I left school in 1983. There wasn't a great deal to do in Manchester in 1983 if you were fifteen. What would we do? Nothing, is the answer to that. Sign on. Work for six, seven months until you got laid off. Or I would go and work with my old fella. When I didn't have shitty labouring jobs I would be hanging out with my mates in the local park, either picking magic mushrooms or fucking trying to get drugs – there is fuck all to do.

Peggie Noel didn't want to do any job, to be quite honest with you, Noel just wanted to laze around. It was all too much bother for Noel to do any job.

Liam I think I got kicked out of school when I was about fifteen. They found me a job working up the road, in a garden centre, and I done that for a bit, which was alright. It was a nice peaceful job, you'd sit there on this wood and it was pretty chilled and that. I can't remember what I was getting, I think I was getting 50 quid a week or something. I'd go to work, come down to my old school on my little BMX, and wave me 50 quid across the railings going, 'Knobheads!' to me mates who were still at school. So it was good. Then the bloke says, 'You've got to clean these toilets out.'
'I ain't fucking cleaning the toilets out, not a chance.'
So then he sacked me.

Peggie Liam didn't like work at all. He walked out, he wasn't cleaning no toilet for nobody. That was that.

Liam Never had a proper job, it was just like bits of shit. There was a guy from Levenshulme. He'd have a contract for us painting the fucking satellites at Jodrell Bank out in Cheshire. Or cleaning the shit up around these ICI factories. So we'd do that, they were alright. We'd just get stoned and fucking go and sit in the toilets for about an hour. Skin up and he'd be, 'Where's Gallagher?' We'd be in the toilet, crashed out with a brush between our legs, fucking snoring.

Noel I got a job for a sub-contractor to British Gas, putting in gas mains all over Manchester. I used to work out in the trenches, literally digging fucking holes for a living. Then for some reason I was put in the stockyard. Maybe I'd done my back in or something, which fucking plagues me to this very fucking day. No one would ever come to this fucking place and you could sit in there all fucking day listening to the radio. Or I would bring a ghetto blaster in and listen, just bring tapes in and stuff, you know, playing along to them. There was a load of – let me get technical about this – big fucking iron caps they would put on the end of gas mains. Some of them needed shifting and one of them fell and broke my right foot.

Liam I done another job doing neon signs. I had to do the Granada TV sign in fucking November. You'd have to get harnessed over it – someone would fucking hold your legs and drop you down – and you'd be fucking changing light bulbs with no gloves on, freezing, for a tenner a day. I can't even change a light bulb. I'd just hang about there going, 'I can't work it,' so I got fucked off from that one.

Noel My old fella was a sub-contractor who would put concrete floors in flats. He was a proper grafter, a labourer and instilled in us a work ethic. And he was a City supporter which is a relief because we could have been fucking dirty reds and let's face it, nobody would have wanted that.

Liam We'd do a bit of work on the buildings and that and he was still a dick, never pay you on time. You do big hours on this building site, grafting like fuck – proper graft, with grown big fucking Irishmen – and then you go down to get the wages off him the next weekend and he'd never be in.

Noel Family life was a bit like *The Royle Family* but with added violence and tension. Not all the time. It was the seventies and then leading into the eighties and Thatcher, unemployment and all that. My mam was working two jobs to try and make ends meet. They were bleak times. There wasn't a great deal to do apart from get into trouble. Smoke weed, sniff glue, listen to music, go to football. That was it. I don't look back on them with any romance whatsoever, but then again when I look back on them they don't fill me with any dread either, it was just growing up. I was arrested a few times, for shoplifting, robbery, all of which fucking haunt me to this day. I've still got a police record. If you're caught three times you pretty much know that you're not destined for a life of crime. If I'd have got away with it a bit more I might have been like a proper gangster, but I was shit at it.

Peggie I will tell you a funny thing about Noel, once he got up to some mischief and went to court. There was a hole in the bottom of his shoe, it was a big hole in the bottom and he had a piece of cardboard in it. I said, 'I tell you what, Noel, if you get half off your father, I'll give you the other half.' He said, 'Okay.' So I said to himself, 'Noel wants a pair of shoes.'
'Well, our Lord didn't have shoes, what does he want shoes for?'
And then I thought to myself, 'You are not right in the head.'

Liam If it wasn't for music, who fucking knows, I could either be dead or I could be in nick, simple. I don't think I'd have got my head together to do a normal job.

Noel I guess music is a form of escapism for me. I've always enjoyed i
and loved playing it and I guess as I've got older it means everything tc
me now. Everything I've got that is good in my life has come through my
ove of music. Of all things, I think the thing I value the most is being able
to just shut the fucking living-room door, pick up the guitar and just sit and
strum it for an hour, an hour and a half
and not even know what I am playing
It makes me feel great and it takes me
somewhere else, whether I am playing
it, writing it, doing it, talking about it.
do often wonder what do other people
do? Some people actually play golf tc
unwind and fucking go to the gym, you
know, or play video games, imagine tha

erything
e got that
good in my
e has come
rough my
e of music.

Liam I wasn't into it until later on,
was just into being out all night and
playing football. Anyone with a guitar
or in a band I thought was a bit suspect, you know what I mean. I would
throw stones at them and shit like that. Hurl abuse at them when they were
walking down the road. All these people go, 'Oh yeah, man, I was into like
George Formby at age three.' Really? Well, I was into just shitting in me
nappy, you know what I mean? I didn't get The Beatles at five, I got them
at nineteen.

Noel I can tell you the first record that was bought for me. My old fella
bought me, 'The Show Must Go On' by Leo Sayer because I'd seen him
dressed as a clown on *Top of the Pops*. I don't know why I was fascinated
with the clown, but anyway he bought me that on seven inch. Dad is a shi
dad and a shit husband but I must kind of credit him for the musical side
of us. He was a DJ and he had a record collection and all that, so there was
always music in the house. Everyone in the Irish community knew who my
old man was in Manchester, everyone. He was a DJ in the Irish social clubs
and that's where I first heard Elvis, Motown and all that kind of thing.

Peggie He was a DJ for years and he was always into Irish music, country
and western.

Paul We had to roadie for him, me and Noel. We got good at playing
pool from doing that. He'd have an amp, two decks, speakers, records and

he would be playing in some holy name club in Manchester, you'd get out, 'You two, carry that gear.' Fucking roadies at seven . . . 'There you go, there's a pound for some Coke, go and play pool.'

Liam Me old fella was playing the Chieftains, the Dubliners, Daniel O'Donnell and all that nonsense. He'd be playing that, Mam wasn't really playing music. I guess Noel was into The Smiths, he was into a bit of the Pistols. Paul was more a Jam fan, he was more of a mod. So there was always music round the house, but it just wasn't my time. Then when the Roses popped up, I sort of got into music.

Noel *Never Mind the Bollocks* as an album is probably the first record that I wished to own, because all the older lads on the estate where I lived had it and it had swearing in it. It said fuck and shit and it was like, 'Oh my God, swearing on a record.' I remember that being the first really important record.

Liam Our kid had the Roses albums and The Smiths and stuff, a band called The Bodines and all that Manchester stuff. Then our Bod had The Jam, The Style Council, UB40, he was more of a soul boy, more of a mod kind of thing.

Paul I thought Weller was amazing, even though he was from a different part of the country. My dad was against anything British, he wouldn't allow me to have a green parka. 'You can't be walking round with a target on your back.' He didn't understand, so I had to get a brown snorkel jacket and try and sneak patches on the arms, I must've looked a right dickhead.

Noel Once I discovered weed and guitars and The Jam and The Smiths and all that, what would you want to go out for? Everything I ever wanted in life was coming out of the speakers and I was playing along to it and it was going somewhere I felt, it was great.

Liam Obviously I was sharing a room with Noel so I would hear it all, I never really was a big Smiths head, our kid got heavily into them. I like Johnny Marr, I like the tunes and that, but Morrissey always just rubbed me up the wrong way. As I've got older I've kind of liked him, he's a funny cunt, but as a kid I couldn't really fucking stomach him, you know what I mean? So like I said, when the Roses popped up, that was where it was at for me. They were talking about their influences – Simon and Garfunkel, Hendrix, The Yardbirds and stuff – so I started listening through them, kind of. Then I made my own way around the musical world and made my own thing. It's like an advent calendar, once you open that door you open another door, and it's just like this big long chain. Before you know it I was in Beatle world, and still am, and it's fucking amazing.

Liam Me old fella had a guitar, yeah. He didn't play it, used to just whack us with it, I think it was more of a fucking weapon than an instrument, you know what I mean?

Noel The guitar that I learnt on, the guitar that lit the fire, was behind our living-room door, stood against the wall. How it got there is anyone's guess. My dad could never play the guitar, my mam certainly couldn't, and my two brothers can't play it so I don't know why it was there.

I didn't really take the guitar seriously until Liam asked me to join Oasis.

Liam Our kid had a guitar and he'd be playing his guitar in his bedroom. Me and my mates, we'd be like, 'Fucking weird brother upstairs playing that guitar, shall we go out and smash some cunt's head in?'
'Yeah, let's do that then.'
Then we would come back and he would still be fucking playing the guitar. So my mates would be thinking he's a bit weird. But he wasn't super weird; he wasn't like the geezer out The Cure or anything.

Paul He'd just sit in his room, music on, he'd have a little amp, headphones, didn't see him for years.

Noel I didn't pick it up to become anything, I used to get grounded – well, they call it grounded now – I used to have to stay in a lot because I was a bit of a wild man at school, in the sense that I never fucking went. I was getting in trouble so I got to stay in a lot. I remember kind of strumming along, not strumming the full six strings but the odd string, to Joy Division basslines, I'd while away the hours of being stuck in your room. But I didn't really take the guitar seriously until Liam asked me to join Oasis. I always had one, I was a bit into it, but not really that fucking arsed . . . It was something I did for pleasure, it was never going to be the career. I was just so lucky that that guitar was there and I was made to fucking stay in on curfew. Thank God for that.

Liam I played violin at one point, very badly, like everyone, I don't know anyone that plays violin well. Noel got guitar lessons, I think he was playing guitar in school and me mam decided to get me a fucking violin for some bizarre reason, which got booted to school, there and back. It wasn't cool, I looked pretty gangsterish with it, but it was shit when you whipped it out.

Peggie I put myself in debt for a violin, it was £30 at the time, and as soon as I paid it off he said, 'I'm not carrying that violin up to school, people will think I'm a sissy.'

Noel The first gig I ever went to was The Damned at the Apollo in 1980 in Manchester. The second gig I ever went to was Stiff Little Fingers, then I went to see Public Image and U2 – all at the Apollo. The reason for that was Shaun Dolan did the door at the Apollo, and he was married to one of my mam's family friends from the area where they were all born in Ireland. I was underage so we would go down there and he would wait till all the crowd had gone in, then he would let me and whoever I was with in to stand at the back. I think I was twelve or thirteen and it was in fucking Ardwick which is in town, miles away from where we lived. I am going to these punk gigs, and they are pretty hairy things, you know. We'd go and stand at the back getting blown away.

Liam I think the first gig I went to see was The Stone Roses in the International One in Longsight. I was on the dole at the time so I borrowed like £15 off me mam. The ticket cost a tenner, bought it off some tout. I know a lot of people, you know, don't condone touts, but if it wasn't for that tout that day I wouldn't have seen The Stone Roses and I wouldn't have joined Oasis. So viva la tout. They weren't dressed in leather kecks or like The Cure or The Smiths and that. I always found them bands a little bit fucking odd

because they weren't dressed like me. Then when I saw the Roses and the Mondays, I thought, 'You know what? I'll stop throwing stones and gobbing at them,' because they look a little bit like us. That made more sense to me. Seeing The Stone Roses, I just thought, 'You know what, you don't have to have a perm, you don't have to have leather kecks on, you don't have to be this fucking Jim Morrison poet dude to be in a band.' They weren't farting about on stage, they were just stood there playing the music – he didn't have flowers hanging out of his arse, you know what I mean? They just didn't seem pretentious to me, it seemed like it was just normal. There's four guys up there on that stage that looked like our kid's mates, that looked like my mates. Obviously they are older and have just sat and learned their instruments and are doing it right, you know what I mean. I thought I could do that, it doesn't look hard, it was that kind of thing. That ain't a diss on the Roses or anything, they just looked effortless and I think that's what it should be. A lot of people make too much out of this rock-and-roll thing. If it's cool, it looks effortless. There's no need for make-up and painting your fingernails and all that nonsense. At that gig, I had two spliffs, bought two shitty beers, went down to the front, checked it out and thought, 'I'm having a bit of that.' Went home, Mam said, 'What was the gig like?'
I said, 'Mega, I'm going to be in a band.'

I had two spliffs, bought two shitty beers, went down to the front, checked it out and thought, 'I'm having a bit of that.'

Peggie Actually, I never thought Liam was interested in music, until he used to sit out there in the kitchen and say, 'I'm going to be famous one day, Mam.'
I said, 'Are you? Get off your bloody arse and get out and get a job.'
'Oh no, Mam' he says, 'I'm telling you,' he says, 'I'm going to be really famous one day and you're going to be really proud of me.'
I said, 'Am I? Well, I hope it's before I start pushing up daisies. because we need the money now. What are you going to do, Liam?'
'I'm a singer.'
I said, 'I never heard you singing.'
'I'm going to be in a band' he says, 'I'm telling you.'
That's the first I knew about it. 'Never mind talking a load of bull there, get out and get a job for yourself.'

The Inspirals

Noel The gig where I met Graham Lambert was an Anti-Clause 28 gig. The Mondays, the Roses and James played this gig. You put that gig on now it would be 175,000 people at it; then it was in the International Two, which was previously called the Carousel, the place where my mam and dad met, funnily enough. It was a huge, old, Irish dance hall and I remember being stood up in the balcony, watching The Stone Roses. I saw this guy just leaning on the balcony. There was a little red light where he was stood and I noticed he had a tape recorder. I went up to him and said, 'Are you taping this gig?' and he said, 'Yeah.'

So I said, 'If I give you my address can you send me a copy of the tape?'

I was fucking obsessed with this band.

He said, 'Yeah.'

Lovely guy, we got chatting about music and he said, 'Who else are you into?' and I said I'd been to see Inspiral Carpets a few times. He said, 'I'm in Inspiral Carpets, and I was kind of like, 'Really? Fucking hell!' Pretty soon after that their lead singer left and Graham said, 'Do you want to come and audition for the singer's job because you know all the songs?' I went up and it soon became clear that my singing chops hadn't arrived just yet. Fair play to them though, they didn't like fuck us off, they said, 'Well, fuck it, you can look after the gear then.' Brilliant.

Clint Boon* My first memories of Noel would be midway through 1988; he was a fan of the band and he used to come and see the Inspirals a lot. I remember a period, probably for a month or two, he had a broken leg so he was coming to gigs on crutches. It made an impact on us. He could sing, we all know that, but he just wasn't the kind of vocalist we wanted. We needed somebody with a bigger set of lungs really.

Noel So I became their roadie, worked in the office and all that, worked for them for years, it was fucking great.

* *Keyboard player and occasional vocalist of Inspiral Carpets.*

Liam He was just buzzing, you know what I mean, he was around the music, around the Inspirals. At that time all this Manchester thing was going down, it was a good time, I guess, making a few quid, seeing the world. It was great.

Paul Noel just came back with all weird-looking clothes on. 'Where have you been?'
'I've been on tour.'
'Okay, who with?'
'Inspiral Carpets.'
If your brother goes away working on the other side of the world you aren't going to go, 'Well done, brother, fantastic!' You're going to say, 'And? . . . You're missing the City games, mate, give a shit.'

Noel I thought, this is it, this'll do for me, this is fucking great, set up these guitars, play them before the band get up, fucking excellent. Get three hundred quid a week, fucking brilliant. What can be better? Drugs, women, booze, travelling the world, it doesn't get any better than that. In fact in many ways it is better than being in a fucking band.

Peggie I was excited because I was glad he'd got what he wanted to do, but God, I was worried about him – I'm an awful worrier, me, you know. You thought of drugs and I was always well clued-up to that. If they ever tried to pull the wool, I said, 'I know all about it, you don't have to tell me.'

Liam I used to do a bit with him, when they'd go down to London and they'd need a helping hand or something, someone to lug it about. I'd kind of jump in every now and again. I went to Konk Studios once, which is The Kinks' place. That was good.

Clint Boon Liam had often come along to gigs. He'd be in the dressing room with his mates and it always struck me how quiet he was, always so quiet and polite and appreciative. He'd never ask for anything, he'd never gone in the dressing room and helped himself to beer, which is what a lot of kids did, he would wait until he was asked if he wanted something, just dead polite.

Noel I met Mark Coyle the first day of their first major British tour, I was working for them on my own for quite a while and then the whole thing blew up. They were going on this tour and of course they had to have a huge fucking tour bus and a crew and all that. I met Mark Coyle sat at the back of the tour bus smoking weed and me and him got on like a fucking house on fire.

Mark Coyle That's how I met Noel, just walking into a room and he was on that side, pulling this fucking face, because the organisation that he was in, had just grown. This is his thing, he's the roadie for that band. Didn't speak to him for, probably a couple of days, and as soon as we started talking, well, we were fucking best mates straight away. There is one thing I remember about Noel, he never had a fucking dirty hand. I'd load the truck, take charge of all the boxes going on, a fucking hard job. But he was very committed to the cause, he loved that band. He was a big fan of the Inspiral Carpets. They were all terrified of him, he was just the boss all the way, from the minute I met him.

Clint Boon You'd never see him sweating, he is famous for not sweating. He was good at getting the local crew to do all the hard work and he'd stand there saying, 'Put that there, plug that in there.' During the gig you'd often look over and he'd be sat having a beer or he'd be looking bored, giving that knobhead sign to you.

Drugs, women, booze, travelling the world, it doesn't get any better than that.

Noel Me and Mark used to share a room, we had some great, funny fucking times, doing mushrooms and all sorts of mad shit . . . fucking proper laugh.

We were the archetypal roadies, fucking hell, man, we went everywhere, all over the world, loads of fucking times. We loved it.

Clint Boon Touring the world, getting decent money, getting free booze every night, not having to worry about record sales and all that kind of stuff. He was just a passenger on this amazing vehicle that took him round the world a few times.

Noel The Inspirals were a fucking good laugh, really were. Clint Boon and Graham Lambert, they used to make me belly laugh all the time, we always had a good laugh with them.

Clint Boon He was confident, cocky, but in a good way, like a lot of Mancunians are. Not arrogant, just confident, very cool and infectious. He was always dead funny. On the night that he auditioned to be the singer he was singing one of the songs I'd written and in the middle of it he stopped and said, 'Who the fuck's written this? This is shite.' Never been one to bite his lip really, was he? We grew to love him, we came to see him as a brother.

Noel Me and Mark – lazy fuckers. I thought we were great. I mean, clearly we weren't because we did get fired, separately got fired, by the band for being unprofessional and somewhat unapproachable by various tour managers.

I thought we were great. I mean, clearly we weren't because we did get fired . . .

Clint Boon My favourite memories of Noel are when I'd go to see him in his flat at India House in Manchester. It's just like a little one-, maybe two-, bedroom flat that he had there, and he'd sit on his bed and he'd play us these songs. You never felt like, this is genius, you know what I mean? I think we took the piss a bit at the time. It's probably the only time in Noel's career that somebody's took the piss out of his songwriting, and he got the last laugh, didn't he?

Noel I wasn't sitting at the side of the stage tuning guitars thinking, 'One day this is going to be me,' you know what I mean? Not in the slightest. I thought that was it. I never thought, 'That could be me,' or 'I deserve to be there.' That never entered into my sphere of thinking at all, at all. I wasn't as egotistical as that. The ego thing came, coincidentally, accompanied by the massive bag of fucking cocaine, you know. There's a funny thing, isn't it?

The Rain

Liam My life then was going to sign on, getting me dole, cashing me cheque, going into Sifters, getting a record if the cover looked great, getting all me classics built up and that, buying some weed, going into Greggs, going back home and fucking blasting it out. I had to play it loud. I wanted it so fucking bad, man, and I was just obsessed with being in a band, just ob-fucking-sessed, man. Right up for it, thought I could do it and, in fact, knew I could do it, just needed to get people to write the songs, I guess, or to get a band and that. Definitely, I knew I could do it – without a doubt. That's all I wanted to do.

I wanted it so fucking bad, man, and I was just obsessed with being in a band . . .

Peggie Liam was sure he was going to be famous. Then of course, Bonehead started coming round and Tony McCarroll and Guigsy, that's when Liam formed the band.

Paul Bonehead used to play keyboards and I used to see him round Westpoint when he had hair and this little Bontempi or whatever under his arm. I didn't see him for years and then he popped up with Liam.

Bonehead I probably got the name Bonehead when I was eight. In 1973, most people in my primary school did have long hair, which was the style of the time, and I didn't. It was just like straight down the barber's every Friday, you know, typical Irish Catholic parents and they're having none of it, no long hair in our house. They gave me my fifty pence and I had to go down, 'What is it? Usual short back and sides?' No long hair on our street. So, that was me nickname from the age of eight, you know, some kid, 'Ah, look at the bonehead!' and that was it. When I got to secondary school even the teachers referred to me as Bonehead, even when they were angry. *Bonehead.* It wasn't 'Arthurs'. That's when you knew you were in trouble, when you got the surname. It was Bonehead and has been ever since. When we got really well known we used to do interviews in France and they'd want to know why I was called Bonehead. I always used to tell them that if you went right back in my family history my real name is *Bonaparte Headimus*; and Bonehead was for short. People fell for it. The people that I grew up with weren't massive music fans. It was either football or music that you got into, and the circle of friends I associated with, it was football. It was United

or City and nothing in between, that was it. They were big football fans and as we got older and we got to the age where we could go to pubs, it was football and drinking. But luckily my older brother was really into music. I shared a bedroom with him as we were growing up and he had a record player, an amp and two speakers, a really good, cool stereo system. He had a massive vinyl collection which was brilliant. He played guitar, he'd bought himself an amp and he was twanging that around the house, I used to have a go of it when he wasn't around. Friday night was pub night so I got to know Chris and Guigs just through people in the pub, who knew people who knew people. Everybody was into the Roses and the other bands coming out of Manchester.

'We're going to form a band, you know, do you want to be in it?'

Guigs was like, 'What shall I do?' Play bass, one note, dead easy. Chris can sing, get a microphone. Plugged it in in his garage and did it, badly, but it was great because all of a sudden Friday nights were Friday night in the garage making a racket through one amp and it was really good, it was cool. Guigs had never seen a bass guitar in his life. Chris had never sang in his life, we did a couple of gigs – we did one in the pub, which didn't even put gigs on, we just said to the landlord, 'We want to do a gig,' and he was, 'What do you mean, gig?' We used a drum machine, the same beat for every song, and somebody just said, 'Look, I know a guy from Levenshulme, Tony McCarroll.' I didn't know Tony, but he played drums, so we had a drummer then. It was a racket, it was crap, but it was brilliant as well.

Liam I knew Bonehead and I knew Guigs and they were in a band called The Rain that were playing pubs and stuff. I was impressed because they were in a band and I wanted to be in a band. They'd heard I was cool and was looking at being in a band, that I'd had my epiphany and all that bollocks. So then some lad come over to me and said, 'They've sacked Chris' – who was their singer – and then they've asked, 'Do you want to go down and meet them?'
I said, 'Yeah, go on, I'll give it a go.' Put me money where me mouth is kind of thing, so I did it and then they liked it.

Bonehead I knew who Liam was, and someone mentioned that Liam wanted to be in a band, he wanted to sing and that he actually could sing, you know.

Liam I think I went to Bonehead's house and just fucking sang, with him on guitar. I don't know what we fucking sang. So long ago. But he was like, 'Cool' and his missus Kate was going, 'Better than the last one anyway.' I mean it wouldn't have been fucking hard to beat him, he was shit.

Bonehead His look . . . I mean, he looked like Liam's always looked, you know? He had a great topcoat on and great haircut, a great walk, great voice.

I looked like a rock star even when I was digging holes in Manchester.

Noel He had the haircut for sure and the walk and all that.

Liam I was fucking cool. I looked like a rock star even when I was digging holes in Manchester. I was cool then. People would clock my head even when I was wearing fucking overalls and had a fucking shovel in my hand. Full of shit with a pneumatic drill, I still looked cool. A lot of people look like that in Manchester and dress like that anyway, so I wouldn't say it was my look, it was a shared look with everyone.

Bonehead His voice was just like, whoa . . . Wasn't the Liam we all know; a bit softer and a lot more melodic.

Liam So they liked it and then they went, 'Right, look, do you want to be in this band?'
I says, 'Yeah, but we'll have to change that fucking name though 'cause it's terrible,' and we changed it to Oasis and that was it. So, for the record, I was never in a band called The Rain.

THE OASIS BAR

Bonehead The Rain. I think he thought it was a crap name so Liam, being Liam, says, 'No, that's shit, we need to change the name.' I think he got it off an Inspiral Carpets tour poster, Swindon Oasis, the venue, I think it was. I was just like, 'Whatever, yeah, that'll do.' A name's a name; once people get familiar with your name, it's just the name of the band, isn't it?

Noel Halfway down Market Street in Manchester there was the Underground Market. There was this stall called Oasis – which sold Adidas trainers – where the Manchester look came from. We used to shop there and all that. What a great name, and what a fucking great story that would have been for the band. Unfortunately the fucking singer decides to tell somebody that I had an Inspiral Carpets tour poster up on my wall and on it was a place they played in Swindon called the Oasis Centre.

Liam It certainly wasn't about Swindon Oasis. There was a stall, I think it was in Affleck's Palace, that sold cool clothes. There was a song by the Mondays called Oasis, there was a kebab shop called Oasis, there was a taxi rank called Oasis . . . you're looking around and that name keeps coming up, all the time in my head. Oasis, just sounded good. I know a lot of people think it's shit, and it probably is a shit name, but everything's shit, innit? Until you make it good. The fucking Jam, that's a shit name, The Jam. But it isn't because they are good and they write good music. The Who's a shit name as well; The Beatles is a fucking shit name.
We wrote a couple of shit tunes, a tune called 'Take Me', which we even knew were shit at the time and still are to this day, but it was something. I wasn't interested in writing songs to be fair. We did it because Noel wasn't in the band at that time. It was like as soon as someone comes who knows what they are doing with melody and guitars, then they can have that gig, man. I was not arsed about that, I just wanted to be a fucking frontman. Bonehead wrote the music, I wouldn't even know how to look at a guitar. He'd just write it, send it over or whatever. I can't remember how it worked. I'd listen to it back and get a melody and write some shitty words down, still doing the same thing today, you know what I mean?

Bonehead Me and Liam together, we were never going to write an album for certain. I'm not a songwriter.

Liam We were nowhere near what it had to be. We could still be fucking there now doing it if we were doing it. Never, ever had the desire to want to write songs. I was not interested in that at all. We knew we were shit, never said we were good. I don't think the confidence came until we had the songs. We were just doing it, that's the main thing. We were just, you know, opening the door.

And then there were five

Liam I thought the main thing was getting Noel in the band. He was a songwriter, he'd been doing it for quite a bit, he was a lad. He was touring with the Carpets so he'd seen a bit. No one wants to be setting people's gear up all their life, that's like work, innit? That's graft, especially if you've got a talent. I thought, when he comes back and he sees that we're doing it instead of talking about it, he'll want to join our band and he'll do the business. Whether we are good or not. We weren't fucking great, but we were better than nothing. Get him in the band: he'll be the songwriter and I can just fart about. Then when he did get in the band and his songs were great I just thought all we need to do now is get a deal and then we'll tell them what we are about.

Noel I was doing a gig in Munich and I used to phone home to speak to me mam on a Sunday. I'd be at the phone box at the venue, somebody soundchecking a bass drum – boom, boom, 'one, two, sibilance, sibilance, one, two' – and I'm going, 'How you doing? I'm in Munich.'
'Where's Munich?'
'It's in Germany.'
'Where's Germany?' Fucking hell . . .
Chatting away about family stuff and then, 'How is Liam?'
'Oh, he's out rehearsing.'
'What for? Fucking hell, he's not joined the Shakespeare fucking group, has he?'
'No, he's in a band.'
'What?! Doing what?'
'He's the singer.'
'He can't fucking sing.'
'Oh, I don't know now. Well, he said he's the singer.' And that was it.
Got home and our Paul is going, 'Oh yeah, he's in this band. They are pretty fucking good, you know.'

Liam I think we were rehearsing in Longsight at the time, in some Irish centre. At that point it was good as it was going to be. Let's do a gig. It's like a footballer, you've got to go and get some time on the pitch under your belt.

Noel They were playing at the local band night. I went down to see them and I remember being pretty impressed. They had their own songs and Liam didn't look that out of place and I thought, 'Fucking hell, that's pretty good.'

Liam It was shit. I don't think there was anyone there, but at least we did it . . . It can't have been that fucking bad if he wanted to join.

Noel I'm not in Oasis at this point, I'm not in a band, I have no intention of being in a band, I'm just trying not to get sacked from this job because I fucking blagged my way into it and I'm an absolute chancer. It didn't dawn on me at all that I was going to join that band. They came up and said, 'What do you think?'
I said, 'It's fucking great.'
'We were thinking, would you fancy being our manager?'
'What? What the fucking hell are you talking about? No.'
'Because you know loads of people and all that.'
'No, I'm not sure about that, I think you can get a better manager than me.'

Get him in the band: he'll be the songwriter and I can just fart about.

Liam We were asking him to manage us at first. I was thinking he might have a connection through the Inspiral Carpets and he was like, 'Fucking manage you, I'll write the songs for you.' He wanted to do it; we didn't put a gun to his head . . . The truth is he got on his hands and knees and said, 'Listen, I'll do anything, anything, just please let me be in your band.'
I said, 'Get up off your hands and knees, son, you're alright, you can do it.' That's the main event. I think I've got a picture of it somewhere.

Noel A couple of weeks later Liam said, 'Come and fucking jam with us.' So I went and sat in with them playing their tunes and it was great. I think the second time I went I had a riff of something, I don't know what it was. Liam was going, 'Play them that fucking song that you played us.' It just went from there really. Once everybody joins in and you hear this thing that you've 'written' being played back to you in this room it's like, 'Wow! Fucking hell!' To hear your own stuff and saying, 'If you play that, why don't you play that there and I'll play that there.' It was really a mega moment. And then it went nowhere for two years.

Paul It was perfect for Noel because he'd seen Liam, he'd seen potential. Someone young, good looking; I believe I have got the songs, or I will have, put them together: magic. But there is no genius involved, it just happened.

Liam He'd gone from being the loner, the stoner, the little weirdo . . . He'd sort of come out of himself and I guess that's through drugs and pills and that. He started getting a bit more swagger. I just knew he'd be the one, he'd be great. Once he got sacked, he went, 'Fuck it, I'm going to have it with Oasis.' I'd like to thank the Inspiral Carpets for sacking him, well done.

Noel The actual reason why I got sacked is amazing. We were travelling all on the same bus, the band and the crew, and it's back in the day before open borders and all that. The band would sit at the front of the bus and all the fucking potheads would sit at the back. Somebody would come up and say, 'We are coming to the Spanish border.' The singer was asleep in his bunk. We had to clean the fucking bus up, there were drugs everywhere. The sniffer dogs got on and all that and there's a fucking scene. As the bus is pulling off, the lead singer wakes up, I'm stood up by his bunk – he says, 'Fucking hell, have they gone?' and I said, 'Yeah, they've gone,' and he said, 'What about the fucking drugs on the bus?' and I said, 'Fucking hell, there was shitloads of it everywhere,' and he said, 'Where did you hide them?' and I put my hands under his pillow and said, 'Under your pillow.' I got the sack about two weeks after that. One of my finest moments.

Clint Boon I think towards the end of '92 his heart wasn't in it. He'd already started with Oasis, he was writing with them, he was rehearsing with them when he had the chance. He was phoning home every night, talking to Liam about his ideas . . . I think it dawned on us that his heart was no longer with the Inspirals so we literally let him go. We give him £2,000 as a golden handshake, which was a lot of money back then.

Noel There is the myth that I kicked open the rehearsal room door to the theme tune from *The Good, the Bad and the Ugly* and said, 'Everybody stop what they're doing, I am here to make us all millionaires.' There wasn't that at all, I kind of fell into the whole thing by accident really. I never had a clear vision of anything until Liam asked me to join and then I turned into a megalomaniac. I always thought that we were greater than the sum of our parts, do you know what I mean?

I never had a clear vision of anything until Liam asked me to join and then I turned into a megalomaniac.

Noel Somebody who'd worked for one of the three biggest bands in Manchester, Inspiral Carpets, really should have had a head start, but we didn't. But what we did have was the tunes and when the music was shit for that first eighteen months, two years, we never gave up.

Liam We weren't the best musicians and I obviously weren't the best singer and that, but we just, I don't know, we just got our heads down and got on with it. I don't think there was any fucking plan, it was let's get some gigs under our belt and just learn our craft. We were very, very fucking serious about it.

Paul If there was a master plan there would be no Tony McCarroll in the band, and there probably would have been no Guigsy, and there probably would have been no Bonehead because he's balding. If you are going to do it that way, you are going to pick people with hair, aren't you?

Noel Imagine a band coming along now with a geezer like Bonehead, what is the first thing anybody would say to you? He's got to fucking go. I remember Alan McGee saying to us once, 'Can we no get him a fucking hat?' No, he is what he is; he's not even going to wear shades. In fact we should shave that bald patch, fucking paint a Union Jack on it.

Bonehead was kind of the glue that held all togeth

Liam He was like the geezer, you know, the doctor off *Back to the Future*, the mad fucking doctor. He was always in mad bands and that, just always had some instrument on him. Fixing it or breaking it or trying to play it. He was kind of like that.

Noel He would lose his fucking mind on red wine. He'd also lose the use of his neck muscles when he was drunk. I swear to God, when he'd get pissed it was like his head was a medicine ball.

Liam I like Bonehead's style of playing, there's no frills. Even though now he thinks he is fucking Jimi Hendrix. Chill, man, that fucking straight rhythm, it's a dying art, man. That was our sound pretty much. Just a wall of sound and Noel over the top and Guigs just doing the root notes.

Noel Bonehead was kind of the glue that held it all together. I would say he was the most forward thinking, initially, because he had been in bands before Oasis. If anything, I would say Bonehead was probably, in the early days, the spirit of Oasis. He didn't give a fuck and why would you?

Bonehead What does he mean by the spirit of Oasis? I don't know. Maybe he said that because whenever Noel and Liam had a bit of a fight and it got more than heated, to the point that someone is going to have to come between them, I would always, always wait and watch and jump in between them. You can guarantee it would be me who would try and lighten it up the next morning, who'd come falling out of the lift doing a *Morecambe and Wise* stunt, half-drunk still, to lighten the atmosphere to get them all laughing so we could get on the bus. Maybe I was in that sense, yes.

Noel When it would be going off between me and Liam, he was the only one person that would try and get in between us and we would just completely ignore him. There would be a lull in the argument while he said his bit and then we would just go, 'Whatever,' and carry on fucking rowing about – I don't know, fur collars on jackets and shit.

Liam Bonehead. Mental fucker. Cannot describe how mental he is. Another brother, love him, still in touch with him. He's a top musician and all, he can play anything. But he's a mental cunt and I loved his mad side, so me and him would get up to mad shit together, you know what I mean?

Noel Guigs brought a calmness to it all, he never got flustered. Just a real, quiet fucking dude. Guigs didn't care about playing on records. Loved cricket and *Doctor Who* and weed and Man City. I'd say fifth after that was being in Oasis. I'd say that came a lowly fifth. I think he's got two prized possessions in his life, none of them are like fucking Ivor Novellos and shit, I think one of them is a Dalek that he bought off the BBC, another one is a fucking Ian Botham signed cricket bat or something. Just a recluse. We were in the band with him till 1998, so that was what, seven years. I reckon if I wrote down everything he said to me it would fill about a paragraph and a half. Bass players are a weird bunch though, aren't they? They are always the quiet ones.

Liam Guigsy, chilled-out motherfucker, man. Stoner. He is like fucking Bob Marley. Lovely lad, adore him, but just a complete and utter fucking stoner. He was into cricket, which I found dead weird. Fucking cricket, what the fuck's that about? He loved the music, man, loved it, he was deep into it, but maybe a bit too fucking deep. Sometimes you've got to come out of it and go, 'It's only music.'

Bonehead He is 'Lee Perry Guigs', he should have been a Rasta. How he ended up in a rock-and-roll band and not a reggae band I don't know.

Liam Tony McCarroll was good, had some good times with Tony, man. Tony was more into the Irish scene. We would go to the Irish pubs, he would hang about with all these Irish musicians. But he was cool, man.

Tony McCarroll I was knackered, you know, I was absolutely knackered, I never forget that, that moulded me, I had a different outlook on life. None of them lot, initially, had kids, none of them had that responsibility, tough on the shoulders of an eighteen-year-old. I don't think they appreciated it, they couldn't see why I wasn't as loud and brash as them, which I could never be. I wasn't with Paula for so long, the mother of Gemma, me child, she became pregnant quite fast, accidentally. I had to go and earn more money. At the beginning, the band thing wasn't being lucrative, bringing any money in, I ended up on building sites, but I needed to earn that money.

Bonehead How to describe Noel? We didn't call him the chief for nothing. He is a natural born leader. He's also got a real sense of humour that has got to be seen and heard to be believed. He is one of the funniest people you could possibly meet. That aside, don't step on his toes the wrong way or you will know about it; but then he won't hold a grudge.

Liam I didn't call Noel the chief. But I know people that get called chiefs and it's not a good thing.

Noel Liam? We wouldn't have been what we were without him, that's for sure. As important and as vital as those songs were and still are, I think the two elements that made Oasis was his thing and them songs. If it wasn't for him we might have been just another band. I couldn't imagine anybody else being the singer.

If it wasn't for him we might have been just another band.

Bonehead Liam is Liam. I suppose everybody thinks they know Liam. Everybody knows Liam's walk, everyone knows Liam's clothes, everyone knows Liam's voice, everybody. Because of the press and the interviews whatever, you'd feel as if you knew him and you'd probably be three-quarters right because he is honest; he is a really, really honest person. Everybody's read a million interviews with Liam and that is the Liam you are going to meet, he doesn't lie. I love him to bits for that, his total honesty. But if Liam starts growing a beard, watch out: something is going to go wrong. The longer the beard gets the more you should worry.

Liam Me? Beautiful, mental, the best one out of the lot of them. I could have multiple personalities, but they are all fucking amazing. Whoever they are they are all fucking great. Every one of them is good.

Home from home

Bonehead There was a club in Manchester called the Boardwalk, it's gone now, sadly. It was a great venue round the corner from the Haçienda. Looked like an old church and it was a cool place, used to have good club nights at the weekends, great gigs during the week. Any band from around the country would roll into the Boardwalk and play at some point. It was a cool place to be, cool crowd, just people out, loving music. They had a big loading bay and these stairs down into the basement where a few Manchester bands used to rehearse. I think the Mondays used to rehearse in there, the Inspirals used to go in there . . . We got this room that we actually could leave our gear – all the amps, the PA – in. We painted it, put pictures up and draped things on the walls. We made it our own.

Liam I painted that Union Jack on that wall, that's my work. Stuck a Beatles poster next to it. That was our home for a bit. I loved it. I wasn't into dance music and the whole fucking town, the whole city, was like just immersed in this fucking music that didn't make sense to me. Everyone was in the Haçienda popping pills. Everyone was fucking stood there fucking off their head, dancing like idiots. All me mates would be knocking on the door going, 'What you doing tonight? Going down to fucking . . .? Come out, man, and fucking let's go out.'

'No, I'm not going out listening to that fucking stupid music. I'm staying down here and listening to the Faces. Fuck that, mate, you're only fucking dancing to it because you've got some drugs inside you. You try to tell me

if you put that tune on there without nothing inside you, you'll be buzzing like you are now? That's bollocks.' I can listen to Led Zeppelin and I can listen to The Beatles, fucking clean as a whistle, and it still gets me off.

Bonehead There was always a fight when it got to 11 o'clock at night. Everyone would just leg it, get your guitar in the case and just run. Last one out of the door had to bring the key up and you knew you were going to get battered for rent. But the people who ran it were cool and we got away without paying rent for months in there, which was great.

Liam 'We'll give it you next week, come on, chill the fuck out, man, have a spliff.'

Noel We were always getting in trouble with the people from the Boardwalk: for smoking drugs, for any fucking wrong 'uns hanging out. If anything ever got stolen they said it was our fault, but it never was.

If anything ever got stolen they said it was our fault, but it never was.

Bonehead We were pretty committed and dedicated to rehearsing. It didn't matter what night it was or who was doing what on what night. If we were rehearsing, we were rehearsing and that was it: 'Sorry, boys, we are not going there, we are not doing that.' People couldn't understand it, but that's how we developed our sound, playing his songs, singing his words, as that unit. Every night, you know? I was working, Guigs was working; Guigs had a great job, he worked for British Telecom with a pension, it was like, job for life. So, he would just finish work and walk straight down the Boardwalk. Every single night. We didn't go in there and just plug in and say, 'Alright, let's have an hour's jamming, like, give us a beat, Tony, let's come up with some ideas.' We actually went in there, get all the chat out of the way, laughs and jokes, plug in and away we go. We'd have a set list of songs that we'd play, again, and again, and again, night after night. We didn't dick about jamming and messing about. I don't think it was, we've got to make this happen, it was, we're *going* to make this happen.

Noel It would have been very easy to say, 'Me bird's getting on me case,' or, 'I'm fucking skint,' or whatever, you know what I mean? Then, when the music arrived, when you've got the songs, you don't need anything else. We were fucking dedicated if nothing else, and that is the thing we carried on right to the end really. For us, rehearsing was sacred, we would never go out on tour without two months of solid rehearsing. Day after day after day, because people are going to come and see you and you have got to blow them away.

LG: I read something in the Bible, yeah.
What are them two guys called?
Abel and fucking...
NG: Abel and Cable
LG: There was Adam and Eve, yeah, and they
had two sons, yeah. These two sons had a
fight one day. One fucking stabbed the
other and that...
That's fucking you two...
LG: This is in the Bible man.
NG: Get to fuck...
LG: That's what it's all about. That's why we'll be
the best band in the world, because I fucking
hate that twat.
NG: It's not hate, it's love.
I don't hate him, I love him.

LG: True hate.
NG: It's not hate. I don't hate him.
He doesn't hate me. I know he doesn't.
He loves me.
LG: What you're saying is complete and utter fucking
blah-ski-boo !

Liam We never jammed, man, never, ever jammed. We weren't like Led Zeppelin or anything. Just rehearsed and rehearsed and worked it out.

Noel I fucking love rehearsing, what could be better than rehearsing in a room with no pressure on you. Playing as loud as fuck, playing guitar and working out bits, it's great.

Bonehead Coyley was always there. Coyley would bring a little four-track recorder and a microphone. He'd fix the mixer because, 'We need Liam louder, but it's feeding back.'

Coyley They were there all the time, they were very committed. At that time, they sounded very powerful, moody, belligerent, dark; it just had a devilish, filthy little sound to it and I was immediately drawn to that, I like that kind of thing, you know.

Playing as loud as fuck, playing guitar and working out bits, it's great.

Bonehead I remember it being a great sound because, all of a sudden, there was two guitars. The noise we were making was outrageous. Everyone always says, 'Bonehead, how did you get that sound, the driving guitars?' I just turned the fucker up, every dial, turn it up, crank it. It didn't take long until that noise we were making, in that room, was going to do something. I knew it was going to reach out to people, I knew if we could bring that onto a stage and people would come and see us they would get it immediately. I never went into rehearsals thinking, 'Just rehearsals,' half-heartedly strumming. I was in, I was doing it, I was nailing it, I was getting it how it should be. I had to do it right every time. Had to. We got it just right where you could hear Liam, you could hear Noel, you could hear me, you could hear Guigs, you could hear drums . . . but massively loud. It sounded incredible in that room. We could all hear each other, Coyley had sorted that out for us, and it was like doing this massive gig every night and it got tighter and tighter. There came a real confidence after a time. You really start to believe that you are going to do something with this. We didn't do a lot of gigs, it was all rehearse, rehearse, rehearse. We played the Boardwalk more often than not, because it was a case of going upstairs and knocking on the office door. 'Can we play next Friday, is there a slot?' – 'Yeah, go on.' We'd do it that way.

Liam I think we just went in there and just made a fucking racket, you know what I mean, I think we were the loudest thing going.

I was rehearsing with them and then Bonehead said, 'We've got a gig next Tuesday.' The first one – and it hadn't dawned on me until that momen hat I'd never played guitar standing up. I was like, 'I haven't even got a strap or me guitar . . .' I had to borrow a strap from someone and stand in the bedroom – I should have done all this with the tennis racquet years ago because it's a crucial thing, where you are going to hold that guitar and how ong that strap's going to be. Too long and you're heading into punk-rock erritory. Too high and it's a bit Haircut 100. Luckily for me the place where t felt most comfortable was the place where it fucking looked the best But I have got to say it was a big thing learning to play that set of songs ir about five days standing up. It's a really kind of innocuous thing, but it was ike such a big deal for me.

Liam Bonehead had been in bands so he'd done fucking lots of gigs. He knew his way around the stage and venues and shit like that.

Clint Boon I remember seeing that gig at the Boardwalk, Noel's first gig with the band. There weren't many people there. It was like probably twenty of us and five of those were the Inspiral Carpets. It wasn't the Oasis that we came to know, and Liam was very much under the spell of Ian Brown, think he was in his mind, thinking, I'll do it the way that Ian does it and that will get me by.

idn't atter to whether ople liked r not, that sn't what were ng it for.

Liam I don't know if there was anyone in there. I mean there might have been a few, but I think it was just mainly ou mates, just sort of going, 'What the fuck are they doing?' I don't know, I can't really remember. It didn't matter to me whethe people liked it or not, that wasn't what we were doing it for. We were doing it just to fucking do it.

Tony Our friends, of course, would be supportive, but you know, you'd hear calls of 'Stone Roses', but we knew we were something different from The Stone Roses, without a doubt

12
Flitcroft Street

Bonehead Nobody clapped, it was just silence, it was like, fuck. Then you got the odd clap, it was like, shit, that went down a storm, didn't it?

Noel I've never really had stage fright or nerves because we rehearsed so much, we knew it backwards. We were very, very well-rehearsed and that went right through the whole thing with Oasis, the band were fucking on it, drilled, do you know what I mean? There was no stone left unturned as to make it the best that it could be.

Liam We were good live, I can tell you that, we were fucking shit hot. Individually we weren't great musicians or great singers or particularly great songwriters but we were great together.

We were good live, I can tell you that, we were fucking shit hot.

Noel The Boardwalk was very small. From street level, there was two stairs, one leading down into the basement where the rehearsal rooms were and a staircase leading up to the door where they took the money. You walked in through the door and there was a bar to your left, then there was another bar to your right then there was the stage kind of in front of you. Small, but a pretty fucking good gig. We did it at least half a dozen times.

Coyley Sharp as fuck, loud as fuck, you couldn't help but notice there was something very different and very special about this band. Not that there was a lot there, some of these early gigs are fucking half empty, most of them are half empty.

Liam At the time we weren't there for anyone else except for us. We'd walk off and we'd be, 'Fucking great, all we need now is a crowd and we'll be happy,' but in our heads we were just waiting for everyone else to catch up I guess.

Noel I was kind of awkward playing these things. I couldn't play the guitar stood up so I was really concentrating. That's when this 'stillism' became this thing that we'd invented.

Liam It felt natural for me to do nothing, it was like I never, ever broke into a sweat about should I be jumping around like Mick Jagger, should I be fucking wearing make-up like Bowie or should I be doing this or that? It just felt natural to do fuck all. Except for rip that mic apart and to sing as best as you can, as powerful as you can. And that was it. I thought, 'I can do that.

Who needs make-up and fucking leather kecks and fucking Cuban heels. Just stand there and fucking have it. I can do that.'

Who needs make-up and fucking leather kecks and fucking Cuban heels. Just stand there and fucking have it.

Paul There would be no audience interaction, it would just be Liam going, 'This one's called . . .' and then just play it. Kind of cocky Manc arrogance, just go on stage, play your tunes and get off.

Liam I've tried dancing, done it in the mirror once and I looked like a fucking cock, so not happening, mate. I can stand still really good.

Noel The singer wasn't going, 'Fucking love me!' you know what I mean? Mick Jagger showmanship was fucking beneath us. We don't have to. You are going to come to us because these songs are brilliant.

Liam I've never really held a microphone, only in the early days, I guess. All the thought of taking that microphone off and walking around like Ian Brown just didn't give me any power. I just don't know how people can fucking sing like that to be fair. I had to be right up on it, hands behind me back and just fucking pouring everything into it. The more people went on about it the more we played on it. Not that we could have done anything else because we are not dancers or anything like that. It was getting to the point where you'd kind of get fined if you moved, like James Brown would fine people for fucking a lyric up. It would be like that.
'Did I fucking see you move in the middle eight of "Supersonic"?'
'No.'
'Better not, bitch. Did I fucking see you tap your foot in the middle of fucking "Wonderwall"?'
'No, no.'
'Okay, must be seeing things. If I see any of that nonsense there will be a fine going down. Static.'
I can stand still really good. You're never going to outdo 'Jumping Jack Flash', you're never going to dance better than James Brown, so don't fucking bother. I don't know many other people that stand still on stage so we were owning that.

National Flat Cap Day

Noel Bonehead's brother must have worked for Granada Television and, as I'm thinking of it now, it is amazing. What on earth were we doing? Was it Red Nose Day? It definitely wasn't Children in Need. Children in Need is fucking massive. It must have been something though. I've got a feeling it was a bit more local than that, it was like Save the Hot Pot or something; National Flat Cap Day or Where's Me Whippet?

Tony We were told, you're going to be filmed, you're going to get exposure, TV exposure, this could be it, this could be the one we'd been waiting for.

Bonehead It was an all-day thing outdoors. Somehow my brother knew somebody who was involved in it so I was, 'Yeah, alright, cool, it's going to be on TV.' I don't know what it was all about. I think it was a bit of a shambles really, from what I remember. It was one of them gigs, what was the point of that? A couple of mums and dads and four-year-old kids in prams sat on the grass watching us . . .

Noel We went on after Alvin Stardust, *the* Alvin Stardust, and we mimed to one of our demos. It's like grannies, a Sunday-afternoon thing. The guy introduced us as a rave band, unbelievable. It was one of Liam's and Bonehead's songs called 'Take Me' which they steadfastly refused to record. Fucking great song and all, the *one* great song they wrote. How that footage has never come out is amazing. Obviously someone's flipped the off switch and gone, 'Fuck these cunts,' do you know what I mean?

We went on after Alvin Stardust, *the* Alvin Stardust, and we mimed to one of our demos.

Liam I remember we borrowed Alvin Stardust's drum kit. We'd done this tune with a sound of a big fucking air-raid siren going down, very strange, man. We sounded good I think, I don't think there was anyone there, it was like a village fete kind of thing, it was just pretty odd.

Tony Gutted, we were devastated.

All the things that I wanna be

Noel The way that I write songs is I have songs that are in cold storage for years and I know they are great. I know I am going to finish it one day, but I'm not going to rush it, it is going to fall out of the sky, I'm not going to just write it for the sake of it. That is why I have always got a backlog of forty-five slightly unfinished tunes that need a first verse or a middle eight, but they are essentially fully formed.

Liam I didn't really get involved in the music – I was like, 'That's your thing, you do that, I'll just be cool as fuck over there.'

Bonehead Me and Liam would often go round his flat and just strum acoustic guitars. Liam would sing, no microphone, nothing, just sit on the couch and we'd strum through Noel's songs. It was great to do it that way sometimes. But he'd bring them into the rehearsal room and then they would really come to life; there would be drums, bass, amplifier involved and Liam would be singing through a PA with the delay on his voice. For Noel to hear them like that, he must have thought 'wow'. To see them come alive in that way with a band, that's how you hear them in your head when you write them.

The first real song that I completed that was good was 'Live Forever'.

Noel The first real song I wrote was 'Live Forever'. The first real song that I completed that was good was 'Live Forever'. Oasis was going for about maybe six months or a year before that song and we were kind of alright, we could all play a little bit. I remember one night being in a rehearsal room and we were one kind of band – like a shit indie band from Manchester – then the next night going down with 'Live Forever' and everything changed. I knew enough about music and about songs to know that that was a great song. Then one followed another and it's like this is happening now. Even if nobody else takes any notice of it, this is happening. Bonehead was saying, 'You've not just written that, there is no way that is your song.' He still to this day fucking debates whether I'd written it or not.

Bonehead I wouldn't believe it, you know? You heard songs like that on the radio and you didn't hear them coming from someone you know in the same room, claiming that they had wrote it. So, for someone to come in and just go, 'I'll play you one of my songs,' and play you 'Live Forever', fuck off. You didn't write that. He was like, 'Why didn't I?' Because listen to it, you know? Wow, what a song.

'You didn't write it.'

'Yeah, I did.'

'No, you didn't.'

We're still in denial now. He didn't write any of it in fact; in fact, he's bluffed it, he's winged it, someone's wrote it, wasn't him.

Liam I don't remember that one specifically. I know it's a great tune that, but all of them were good. It wasn't like a eureka moment, all of it was kind of like, 'Fucking hell, we've got some good tunes now, man. These are fucking great, this is it.' We'd got our sound, because that is the hardest thing about being a band. We'd just sounded like a really shit Stone Roses at the beginning, really shit.

Noel I kind of write without writing if you like. I'll be strumming three chords on a guitar. The opening chord sequence on 'Live Forever' I must have strummed that for weeks and weeks on the guitar until a melody came, the words come later, right at the end. So I will be strumming the chords that are saying something to me musically, but in its early stages I won't know what it is. Then a melody will arrive and so I'll have the music and the melody and then I'll arrange the whole song without any words. It will just be me singing whatever is coming into my head and then, over a period, little bits of words will stick. So the first line, I'll have that straight away, and then maybe another couple of lines after that and then I will get to a point where I will write down the bits that have fallen out of the sky and then I will try and fill in the gaps. That has never changed from the minute I started writing songs; it has always been about the melody and the music. I don't really put too much stock into the words if I am being honest.

Bonehead Every night he'd come in with something, he'd just go, 'Hang on a minute, I'll just play you this song,' and you'd just be all blubbering wrecks and he just kept churning them out. They weren't half formed, 'I've got this idea'. It was, 'Here's me latest song.' Done. I was doing a riff in rehearsals one night and he was, 'That's tops that, what is it? It sounds like "Taxman" or whatever.' He come in the next day and just went, 'Check this out.' And he played 'Up in the Sky' start to finish, done, arranged.

'What's that?'

'I did it last night after rehearsal, in bed.'

It felt like the songs weren’t his and they weren’t mine, they were ours.

Liam The songs were uplifting, they were definitely escapism, but they were definitely real. For me, when I was singing them, they weren't like something that couldn't be attained, it was real fucking life. 'Rock 'n' Roll Star', that's me, and I've not even fucking done a gig yet. I can go with that.

Noel I like music to be up, I think it comes from listening to a lot of acid house music in the eighties, which was so uplifting; I always wanted my music to be euphoric. I was at the Haçienda when acid house was invented and that music is still really important to me. I never write words down particularly. I find that if I've written something it stays there and it can never be changed. But if I just let the words come, and if they are good enough, then you remember them, you don't need to write them down. I only used to ever write them down to give to Liam and then, as I'm writing them down, I would be thinking, 'What is this song about?' And you'd have to make up some meaning for Liam because he would get freaked about things like that.

Liam It felt like the songs weren't his and they weren't mine, they were ours. So it felt pretty natural, all the words he was singing, they weren't alien to me, it was like 'Yeah, I get this'. Coyley had some kind of recording equipment up in his bedroom, we'd go round there, everyone would be downstairs, smoking dope, talking about pyramids and fucking aliens and all that nonsense, like you do. We'd go up and we'd record some of these tunes with him. I think it was 'Married With Children', 'D'Yer Wanna be a Spaceman' and stuff like that. Then go back down, get stoned and walk home. They were good times, innocent times, not knowing what's going on; you are kind of just hoping that it takes off. Yeah, good times.

Noel We were just hanging out at Coyley's house, you know, he had a four-track in his bedroom, and that's where I met Phil Smith.

Phil We'd be up late, but Coyley, he'd be up all night, we used to call him 'The Prince of Darkness'. We just started hearing noise when we were going to bed, usually it was some mate's band, and they were crap. Tell him to go and turn it down. But I remember going up to bed and knocking on the door and asking, 'Who's that?' and he went, 'Oh, it's one of Noel's tunes,' and it was kind of like, 'Ah, well, that's pretty good.' And then a few days later, going to bed and it's a different tune, 'Oh, that's another one of Noel's.' 'Oh, right.' That's something else, that.

Liam I'd hear him playing his songs and I'd be thinking, 'I can't believe my brother's wrote them songs.' You just go, 'That's fucking cool, you make that all by yourself?' Writing songs, words and melodies and getting it to a guitar was, 'You're fucking amazing,' totally.

UMBRO
MANCHESTER CITY F.C.
brother

UBLIC
BAR

First demos

Liam It was serious and it was fucking intense, but it was never desperate. It was never, ever desperate. We just knew it would happen and we never licked arse and we never begged. We were never running around, banging people's doors down going, 'Please give us a deal,' or, 'Please listen to us.'

Noel The original demo tape, which is the one we gave to Alan McGee, we did that in the Real People's practice room. I was on tour with Inspiral Carpets and the Real People were supporting them, a band from the arse-end of the eighties who I really fucking love and still like to this day. Typical Scousers, two brothers, Tony and Chris Griffiths, headcases, brilliant, just talk shit all night, brilliant, brilliant bullshit all night. Me and Liam loved them. I got to know them on that tour and then when that tour finished the Real People came to Manchester and did a gig at the Boardwalk where we were rehearsing and we all went down to see them. Then we all went back to my flat in India House on Whitworth Street and there were guitars out and boozing and smoking weed and all that. They said that they had this eight-track recorder in their practice room in Liverpool. So off we went and recorded what became the famous demo with the Union Jack on the front – the demo tape that became sought after, the holy grail demo tape.

Liam Yeah, I can't remember much about that. There was acid involved, lots of weed, going back and forth. It was great. They were out there, man, they were top, top musicians definitely. They were miles better than us, they were great players and they were great songwriters.

Noel We had some great times. We were doing acid and just sitting up till all hours talking about music and they were happy days. We ended up driving over every few days to record 'Rock 'n' Roll Star', 'Columbia' and 'Bring It On Down' . . . we did about ten or twelve songs and we got to know them quite well. I loved them, I loved being there and hanging out in their rehearsal room. They had a real enthusiasm not only for music but for *our* music. We were doing 'Columbia' as an instrumental and I think either Tony or Chris said, 'That is a boss tune that, mate, why don't you write some words for it?' It is still a bone of contention to this day as to who wrote what. I definitely wrote most of the words, I think Liam wrote a line or two, but as we were all tripping the night we did it . . .

Liam We were up all night and I wouldn't like to say what it was like, but it was pretty out there. I remember everyone was playing and I remember waking up with fucking blisters on my fingers and I was going, 'What the fuck is all that?' I'd been on the drum kit all night, just tripping me head off. So I don't remember doing much of the singing, or recording it to be fair. Or the writing. But I knew who'd written it.

Noel We had ten copies pressed up and gave them all away to our friends, didn't send them to any record labels. We only went to one record label and that was Factory and we went to see Phil Sax who was head of A&R because we had this demo tape and that's what you do. It was shit, it wasn't any of the stuff that became the famous songs.

Bonehead So we go upstairs with the demo tape for him to have a listen and have a chat through it. Me, Noel and Liam, three of us. I had a copy of this tape obviously and I put it in my system at home and I would turn it up and it sounded pretty fucking good to me. He put it in his and I remember he had the bass turned right down to the left and he had the treble up there and he had the balance there and he played it sort of half-heartedly, dead quiet. He sort of, skimmed through it. We were looking at each other going, 'He's got all the bass turned off there, turn the fucker up, have a listen!' I remember coming out and thinking, 'He didn't have a clue, man.' Didn't seem arsed. Thank God, eh?

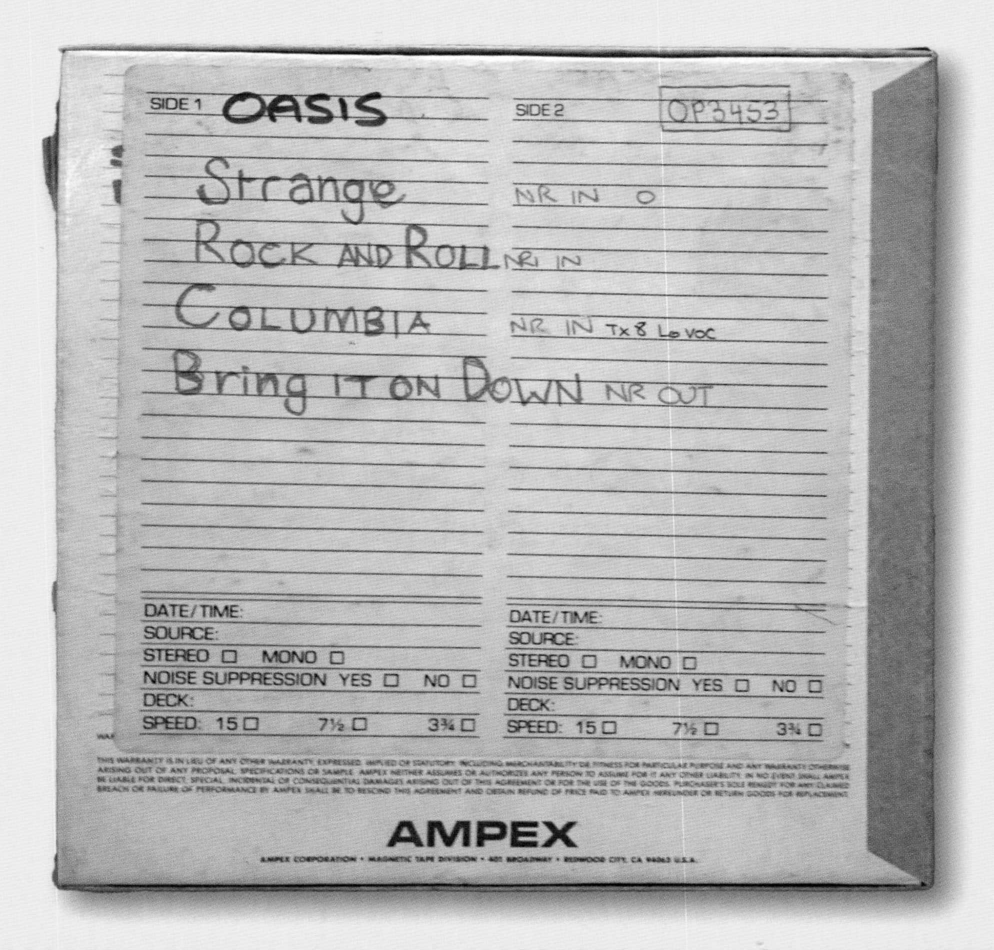

Noel He said, 'We'll play it to Tony Wilson and see what Tony thinks,' and we never heard from them again. Fair enough.

Liam Maybe he just didn't get it and that's fucking cool, I listen to lots of shit that I don't get. He just didn't see it, which means, fair enough, we weren't meant to be on that label. McGee seen it and got it and the rest is history. I wasn't arsed whether he got it or not. The funny thing is that he turned round and said, 'You sound a bit too Manchester for my liking.' That's a bit fucking strange considering we are from Manchester, you lunatic . . . As opposed to what, being from Wakefield?

Noel I remember feeling if it's going to happen they are going to come to us, I'm not going to go to anybody else.

Bonehead There were a lot of people going, 'Why are you dicking about, man?' There were a lot of people questioning it. Why? I don't think I even bothered answering them. Because, that's why. It's like asking George Best, 'Why are you always kicking a football?' Because I'm George Best, I'll show you why. That's what we thought. We knew.

Noel We'd done this festival in Manchester called In The City where all the record labels come, it was an unsigned band night. We did two of those. Fuck all, nothing, not even a drink, nothing. No one even said we were shit. We were completely and utterly fucking ignored. I'd already written the first three albums. I'd already written the melody and the chords, I was only waiting on the words. To me that is a finished song because I would quite frankly just get fucking pissed and make the words up and freestyle it. A lot of *Morning Glory* was in my head as we were recording *Definitely Maybe* for sure. And I knew, for instance, that 'All Around The World' was going to be the last track on our third album before I had a record deal. Bonehead used to laugh and say, 'What third album? We've not even got a manager.' I truly believed that it would come to pass. I don't know where that belief came from.

Liam I weren't the best singer, Noel weren't the best guitar player, Bonehead weren't the best guitar player, but we had spirit, man, and that was lacking, massively, at that point in music. I kind of knew to be honest and I know people think you are being big headed, but I had a feeling it would happen, man. Just because, if you want something so much, you know, it happens. I fucking wanted it, man, and I wanted it for the right reasons. I actually fucking needed it. It was like, this shit has to happen. And I don't know what the consequences would have been without it happening, that's how much it had to happen. It needed to happen because if it didn't happen then the world would be black, that's how mad it was.

King Tut's

Noel No one in the outside world is taking a fucking blind bit of notice, we didn't have one single paragraph written about us, ever. Then we met McGee and of course this is the guy who signed Primal Scream, Jesus and Mary Chain, Teenage Fan Club and all that, and he's coming up to you and offering you a record deal saying, 'This is fucking great, you are fucking great.' It is like thank you very much, I know, I thought as much. We were sharing a rehearsal room with an all-girl band called Sister Lovers; there is no need to google it, they didn't go anywhere, I am not even sure whether they put any records out. Unbeknown to us, one of the girls in the band, Debbie Turner, God bless Debbie, was an ex-girlfriend of Alan McGee. We were asking them what they were up to and they are saying, 'We are going up to Glasgow to do this gig.' We'd never heard of King Tut's Wah Wah Hut – what a stupid name for a place anyway.

Debbie Turner I think I was probably being quite cocky actually, saying, 'Noel, we're playing in Glasgow.' For an unsigned band to get a gig outside of Manchester then was like playing Glastonbury.

Noel She said, 'Why don't you come with us – be on the bill?'
We were like, 'Yes – fucking Glasgow – let's do it.'

Liam 'How the fuck have you got a gig and not us?'
She said, 'Oh, we know this band, why don't you just jib along with it?'
We were thinking, 'We'll get there and they will tell us to fuck off.'

Coyley Noel decides we're going to do this the proper way, we're not driving up, we're going to get a fucking driver and a bus.

Noel We kind of got a few headcases together. If we all put in £25 we can hire a splitter van and get up there, kip in the van, do the gig, get back, fucking Friday night, yeah, let's do it. I remember it was a gold van which I thought was amazing.

Liam Beautiful day, smoking weed, all the nonsense, drinking. One of them days where you just go, 'It's fucking going to be the day today.' Maybe it was the drugs or something.

Noel We get there really early of course and we go to the gig before anyone's turned up and said we were Oasis from Manchester, we'd come to play tonight. The guy said . . .

GLASGOW !!!
PHOTO
Glasgow
Manchester
MANCHESTE
Glasgow
Manchester

Liam 'Never fucking heard of you, you are not down.'

Noel We said, 'Yeah, yeah, it's alright, we're with Debbie,' and he's like . . .

Liam 'Fuck you, you're not fucking booked.'

Noel Turns out Debbie and Sister Lovers are supporting this band called 18 Wheeler that McGee had either signed or was about to sign, so it wasn't even their gig. Why the fuck they were even inviting us up is . . . well, who knows. There is this story that we threatened to smash the gaff up and all that, that's bollocks. If anybody's fucking been to a nightclub in Glasgow, you don't go throwing your weight around up there and live to tell the tale.

Liam You would have got fucking rolled up in a little ball and fucking slung in the corner. There was none of that.

Noel We said to the guy, 'Come on, man, we come all the way from Manchester, all these guys here, our mates, we are alright.' Debbie's turned up and the rest of 18 Wheeler. 'Well, if you don't mind they can go on first.' In the middle of the afternoon or something.

There is this story that we threatened to smash the gaff up and all that, that's bollocks.

Debbie Turner We said that if they're not playing then we're not going to play. We'll do a really short set, which wasn't hard because we only had five songs anyway. Then Oasis were allowed to play.

Noel So, we got to do four songs. We used to finish with 'I Am the Walrus' and we never really worked out an ending, so what Liam used to do is he would finish the 'coo, coo, cachoo' bit and he'd walk off. Then I would go round it maybe three or four times with the band and think, 'Fuck it, I don't know how to finish it,' so I would leave the guitar feeding back and I would walk off and leave. Then we would go and stand on the dance floor, watch them and take the piss and, where circumstances would allow it, start slinging shit out of the dressing room like bags of crisps and oranges and apples. Some nights that song used to go on for ages because the three of them didn't have a clue how to fucking end it. People thought this was performance art going on but invariably it would be a shit ending. At King Tut's I think there was loads of feedback, a smattering of applause.

Coyley The set is very, very short, they squeezed us on. Very grateful, because all we want to do is go and play King Tut's, that's fucking big time at that level. King Tut's, a very famous little gig in its own right. So, the gig finishes, I walk over to the bar and order a drink and I swung me head to me right and fucking Alan McGee. So, we walk upstairs and I introduce them.

Bonehead I was well aware of who Alan McGee was, I loved the Mary Chain, loved Primals, was a big fan, so I knew who he was. Didn't know he was in the audience.

Alan McGee Debbie doesn't even know that I was coming up to that gig and being an evil twisted fucker, I thought I would just show up to put her on edge. I really do believe that some things are meant to fucking be. I'm standing there with my kid sister Susan. She immediately says, 'You should sign these.'
'Let's hear the second song.'
Then it was like, I am signing these; third song, I'm *definitely* signing these.

Noel I'm up by the mixing desk with Coyley, Alan McGee walks up and says, 'What's your band called?' and I said, 'Oasis,' and, these were his exact words, he said, 'Do you want a record deal?'
'Who with?'
'Creation Records.'

We kind of went back that night, I don't remember anybody high-fivin
the back of the van thinking, 'Yeah, this is it!' I don't remember us be
nonchalant about it either. We were going back to Manchester and get
into where I was living at 6 o'clock in the morning. My then missus, Lou
was getting ready to go to work. I said, 'Creation Records have offerec
a record deal,' and she started crying. I was like, fucking hell, I expec
a bit more than that. She knew then that, you know, that was going tc
the end of us.

Coyley For me, says it all about Alan McGee, he knew, immediately, there is no negotiation to be done, no fucking about and he just come and said, 'I want to sign you,' just like that.

I often thin what would have happe if he'd have not been th that night.

Noel I often think what would have happened if he'd have not been there that night. The arrogant side of me thinks we would have definitely got a record deal and changed the fucking world and blah, blah, blah because the songs were too good. really, if it had been someone else, I don't know.

fuck, this isn't a record company. You know what, I thought we were being taken for a ride, I thought that guy had thought it well worth his while to shell out £120 of free train tickets, just to sit on a roof with a pair of binoculars, watching us, pissing himself, going, 'Dickheads.' But there was a staircase like that, and a little sign, you know, Creation, and we were like, 'It's got to be up there.'

Noel Inside the front door of Creation was scrawled on the wall, in big felt-pen letters, 'Northern Ignorance'. It was in massive letters, scrawled in felt pen and I thought it was fucking brilliant. As soon as I'd seen that I thought, I fucking love it here. I feel comfortable here already.

Liam There was loads of people, loads of like record-business types, but not like the knobs that are in it today, just cool people. I can't remember any of the names and that, but just cool people. Just doing the business.

Noel The first person that we bump into is Tim Abbott and he had a mullet and really strange clothes on. His brother Chris was running the dance music side of things for Creation, and Tim had his own little corner in the office. It was Tim that had written 'Northern Ignorance' on the wall. McGee had an office downstairs called the bunker and on his desk he had this sign that said 'President of Pop'. I remember it being very laid-back. Somebody sent out to the pub for some booze and we just sat in the office for quite a while, just talking about music.

Liam Who is this fucking dude, Willy Wonka of fucking Creation? He looked like Gene Wilder. He had a bit of a weird head, curly hair and ginger . . . Tim was like a David Essex lookalike. So there's fucking Willy Wonka and David Essex and I'm stoned out of me head and thinking, 'Yeah, let's fucking sign to these fuckers.'

Noel McGee just got it. What can you say when the head of the record label, the guy that owns it, is offering you a record deal and he's into it and you become friends and fucking drug buddies and all that. That's it, what more do you want out of life? There's nothing else a record company can give you. We didn't know it then, but he was the last of a dying breed in the music business. Never got involved in anything other than making sure you had everything you wanted to go and do the records. He would come down and say, 'This is fucking great.' Alan McGee was brilliant. He is up there, one of the most important people

JESUS AND MARY CHAIN

TEENAGE FAN CLUB

RIDE

MY BLOODY VALENTINE

PRIMAL SCREAM

THE BOO RADLEYS

in my life that's made a fucking real difference. Being on Creation is as important as Oasis being from Manchester, I feel. Major record labels are now set up for commerce and not for the art. It's all about the numbers and hits on YouTube and all that.

Liam I was impressed, not that I would have known what another label was like. McGee was top, really enthusiastic about music and about us and just buzzing off the vibes, you know what I mean?

Bonehead I think if we did go into the big glass building with the palm trees and the receptionist and the guy manning the lift and the fella behind the big leather chair, I'd have come out thinking, 'Fuck, do I really want to do this? I don't like this music business, record company, office thing.' But I came out of there and it was like, 'Really like them, cool people.' Absolutely. I do remember the feeling – what was a record deal? Do you want a record deal? Fucking right I do. What's one of them, what happens, what do you do next? Dunno. I'd never had one. I didn't have a clue what it was, it was just like, 'Yeah, fucking great.'

Noel That is what happened, it was that straightforward. There was no doing the dance and all that. In any case I would have signed to Creation for nothing, you know, I gave them my word and that was it. The fact that I never asked for a record deal or I never asked to be in this band, has served me well down the years, do you know what I mean? Because, hey man, you fucking asked me. When the shit was hitting the fan with Creation I would say, 'You asked me to be on your record label, not the other fucking way round.'

I would have signed to Creation for nothing you know, I gave them my word and that was it.

Peggie I remember the day Liam came back and told me he had the record deal with Alan McGee. They come up here and said, 'We got a record deal, Mam!' And of course I thought, this is great, they'll get a bit of money and maybe buy a house. I never really thought it would go the way it did go. All Liam wanted was new clothes. Then it went haywire. I never, ever thought that it was going to go like that, never. I found it very hard to deal with actually at the beginning.

Paul I remember Liam being excited he didn't have to go to work any more. He got paid to be Liam, which was sing a bit, smoke loads of weed, hang out with your mates, get drunk . . .

Liam The main thing was just getting a deal, that's the best thing. Better than Knebworth, better than the best . . . When people go, 'What's the best thing about Oasis?' it was just having someone have the faith in us and letting us show them what we could do, getting that chance to record our album, those songs. Because we knew we weren't going to fuck it up.

Peggie

Noel Me mam was fucking fearless.

Liam Mam was an angel, still is. Me mam's cool, she's the coolest woman that's ever walked this fucking planet, in my eyes. Me mam's an absolute diamond. I know a lot of people have fucking dickheads as mams, but my mam's cool as fuck, absolute cool as fuck. Everything that is good about me is definitely from her.

Peggie I'm one of eleven, there are seven girls and four boys. We were the poorest of the poor, you could say. We had nothing, but we were happy enough. School was about a mile, we used to go down across the fields. with no shoes on us, bare feet, in the summertime. And if we got a pair of wellingtons, we'd share, we couldn't all get them. We used to love them wellingtons. I suppose we all brought each other up, you know, a lot of fighting and arguing going on. I left school when I was thirteen and a half and went out to work. I was working all my life really. My dad was no good, he was always coming and going. He'd go, then he'd come back, she used to get pregnant again, then he'd go and then he'd come back. So, we never really knew him. Eventually, he just went off and never came back at all. I was nearly eighteen when I left. We were very naïve. What I thought was going to happen, I don't really know. We just came over here to Manchester because everybody else did, but I hated it when I came over here first. I must have cried for about six months. I used to think, I wish I could get on a train and go back home. It was a big shock, I'm telling you, to come to a big city out of a small village. Oh, it was massive, I couldn't get over the size of the buildings. The Carousel was a big Irish club that was just up the road, it used to be packed. I would go there, but I was never really one for going out all that much, I'd only go out maybe one night a week whereas my sisters would go out Friday, Saturday and Sunday. I used to think, 'I can't be bothered going out,' one night was enough. I used to go on the Saturday night and that's all we did. Go to church then go to the Carousel. You'd be there and you'd dance and jive and do all this old-time waltzing and all that. Never really let themselves go that much, if you know what I mean. It was after the New Year, I think it was 1963, I went to the Carousel and I met him there. Tommy Gallagher. It was lovely when I met him but he was always a dodgy one from day one, did exactly what he

Everything that is good about me is definitely from her.

wanted. He was always disappearing as well. But anyways, we got married after about nine, ten months and that was it. People used to think he was lovely, 'But your husband would do anything for anyone, he'd be a true friend.' I said, 'Yeah, but you don't know what he's done in the house.' That's why I always said street angel and the house devil.

Liam Mam's not impressed by all this rock-and-roll business. She is proud and all that, but I guess there are some bits where she's completely and utterly just gone, 'You fucking knobheads,' cringed and rightly so. There is some shit that come out of our mouths that was just fucking ridiculous.

Peggie I remember the first time he'd come back and he walked in to the house and he said, 'I'm used to big hotels, Mam,' and I said, 'I don't care what you're used to, you're back here now. If you want a cup of tea, get out and put on the kettle and make it.'

Noel Thinking back on it now, as long as we were happy and out of her hair, she didn't really give a fuck.

Liam I wanted to make her proud, wanted to get her from having to do three jobs, I wanted her to put her feet up and have nice things.

Peggie I was cleaning five houses a day, me. I was forever on the bloody roads, walking, going from one house to the next. Then I worked in an old people's home for two years. I used to be shattered. I used to get up in the morning, clean, tidy up everything here, then go out and clean maybe two houses in the day then go on at night time and work all night.

Liam At this point we still weren't making any money,
so she was like, 'You're still fucking scrounging off me,
you little shit, I don't give a shit what band you're in.
When are you going to start fucking pay me some rent?'

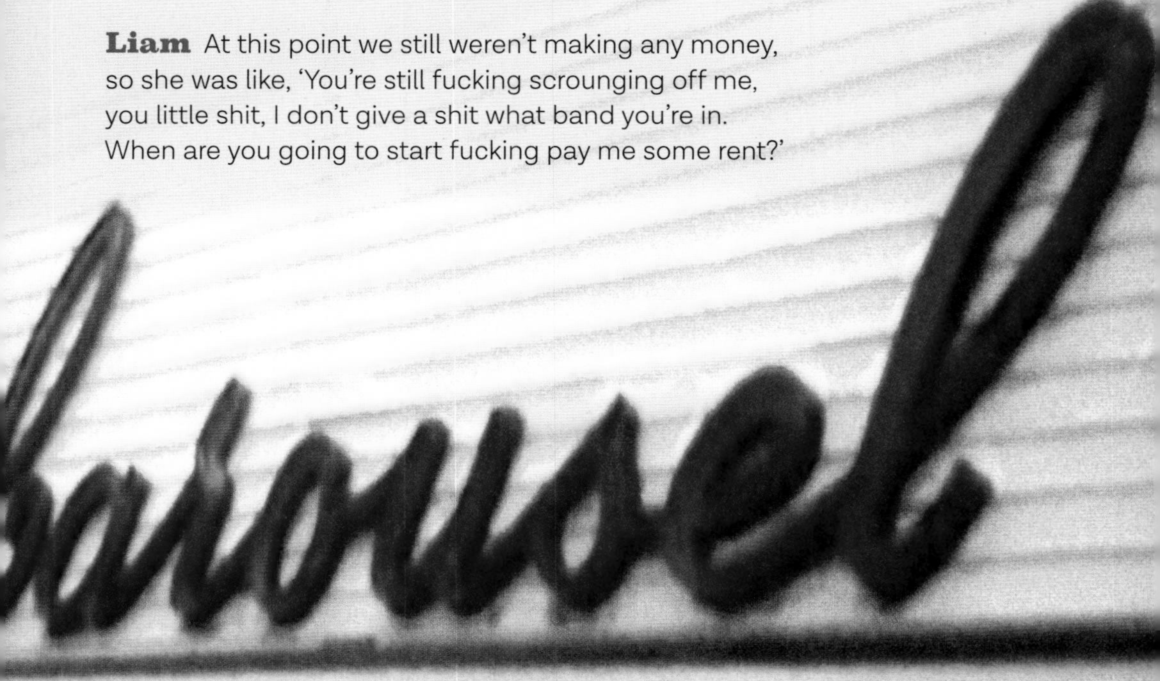

Peggie Do you know what, it never came into my mind that they would make it. I just thought, well, they are in a band, hopefully, they will go out and they will make money for themselves and they will have a better life. I was glad they were together in a band. I would not have wanted Liam in a band without Noel and I wouldn't have wanted Noel in a band without Liam. I always thought when they started out, so long as the two of them are together I didn't need to worry as much. I thought Noel will look out for Liam so they are alright, but . . . and they did look out for each other for years like, but they grew up.

Liam Then I bought her a house round the corner up in Eaton, beautiful, 120 grand which is a lot around that time. I said, 'There is your keys, I'm out of here,' and she says, 'I don't fucking want it.'
I was like, 'I've bought it now, you can have it.'
She was adamant, 'No, I don't want it.' She didn't want her life to change; 'What do I want to live in a big posh house for when all my mates are down here? Your life's changing and good luck to you, but mine's staying the same. Get me a garden gate if you want.' So that was basically it, she's a cheap date. I bought her a garden gate, I think Noel bought her a fence, and she's not asked for anything ever since.

> **She could hear in my voice that I wasn't turning into a fucking idiot . . .**

Peggie I did worry, I've done nothing but worry about them. When they'd be away I'd worry about them. I spend my whole life worrying about them.

Liam I guess she thought, 'One of them is going to end up dead or one of them is going to end up with a big drug habit.' There's been a few scrapes but we're still here, man, I've still got my fucking head screwed completely on. I'm not a casualty and I never will be. There are a lot of dickheads out there going, 'Oh yeah, I'm going to get fucked up in the first year.' Well, good luck to you, mate. I was fucked up at fifteen before I even heard a note of music. I didn't join a band to be a casualty. I speak to me mam three or four times a day. So I'd be always reassuring her. She could hear in my voice that I wasn't turning into a fucking idiot, she'd get it. I think she's just going, 'Jesus, I wished they've never joined that bloody band,' because of the shit. But she's well fucking into it. She fucking loves it.

Peggie Never in a million years did I think that they were going to be as big as what they were.

Enter Marcus

Noel Alan had handed me a list of managers that he knew and at the top of that list was Marcus Russell. Funnily enough I'd just become friends with Johnny Marr and he was Johnny's manager; I didn't know any of this. Johnny just said, 'He's a straight-up guy and fucking great.' That was it.

Liam When we met Marcus, I was impressed. With every other person that we'd met, they were a bit fucking tin pot, fucking two bob, you know what I mean? This geezer had a Beemer – I was impressed with his car. So I thought, if he's got a fucking motor like that and he manages Johnny Marr, he's got to know what he's doing.

Noel When we got back to Manchester and I was speaking to Marcus I told him we'd already got an offer of a record deal. I think he initially said, 'Hang on a minute, you don't just take the fucking first record deal that you can get.' But I had already given McGee my word and I wanted to be on Creation, not because of Creation records, but because of Alan. The way that they loved us and they loved our music and they got it – and they got us. Where the businessman thing came from is fucking beyond me. I sat back into a life of chemicals and fucking expensive champagne after that, thank you very much.

Liam I never really got involved with what we were signed for and whatever. You don't think like that, do you? I was thinking about the bigger picture, man. I mean a contract didn't mean jack shit. If you are thinking about contracts at nineteen, then there is something fucking wrong with you. I don't think I ever signed a contract with anyone. It was like 'Look, you fuck me over and I'll burn your fucking house down' and I still will to this day, that contract still fucking stands.

I never really got involved with what we were signed for and whatever. You don't think like that, do you?

Bonehead There was no contracts, no lawyers, no nothing from Marcus, that is the way Marcus worked which was old school, it was really cool. Probably stupid, you would be advised not to, but just, 'Yeah, right, I will be your manager,' spit on his hand, shake your hand. It was done on a handshake. Cool by us.

Noel There has been this misconception that I'm some character walking round with a clipboard, with a graph, a pie chart, fucking spreadsheets while Liam is this caged panther. Our *Bo Selecta* puppets summed it up perfectly: Liam was this dribbling 'duh, duh' and I was this posh guy going, 'Oh hello, fucking hell, dreadful, what's the merch score? How much did we take on the merchandise?' I haven't got a fucking clue about anything, for all I know I've been ripped off fucking left, right and centre for the last twenty fucking years, I really don't know. As long as there's money in the hole in the wall when I go there I couldn't give a fuck really, you know what I mean?

Good evening, Great Britain

Bonehead Not long after signing we did our first radio broadcast, *Hit the North*. It was a big deal but I don't think we got mad excited; I think we were pretty calm about it actually. I remember Peter Hook being on it and Liam made some comment about Hooky's leather trousers. He was like, 'Are you wearing them leather trousers up there, you dick?' Or something to that effect.

Noel The story goes that it was Mark Radcliffe's show and he fucking hated us, like most people in Manchester did. He was away for a week so Peter Hook stood in for him with Mark Riley; they liked us so they got us on while he was away. The reason it sounds a bit odd is because there was only one vocal mic in the studio, which Liam was using, but he wouldn't speak to them. They came in and said, 'Right, we are going to do a little tune and we are going to do a little chat,' and Liam was, 'I'm not fucking speaking.' He couldn't get his head around an interview, so he said to me, 'We'll do the song and you come and do the talking.' So we had to quickly swap. Peter Hook was having a little pop. Him in leather trousers. Fucking hell, mate, come on, it's the nineties.

It was a big deal but I don't think we got mad excited; I think we were pretty calm about it actually.

Liam Being on the radio, it was fucking good, yeah. It was like, 'We are fucking on the radio, who wants it?' I remember telling him, putting him in his place, yeah. He was saying something about 'You're banned from my club,' and it was like, 'Look, I don't fucking go in it anyway.' He obviously thought that everyone in Manchester was in his club and I'm sure they were, but I wasn't.

Bonehead We played a blinder. We did 'Bring It On Down', I can't remember if we played anything else. I remember it sounding great.

Noel Afterwards me, Coyley and Phil all bowled into a van and went up to the moors because Halley's Comet was passing. We all sat on the top of this van smoking weed, going, 'Hey man, cosmic shit, man.'

Fifty quid on the door

Bonehead Going out on stage, we just gave it our all. In rehearsals we gave it our all, we didn't do it by halves, it didn't matter who we were playing to. We did a gig in Leeds to two people.

Noel We had a gig booked at one of the nights of In The City and maybe two nights before that we did a warm-up for it in Leeds at the Duchess of York. We soundchecked in the afternoon, the guy came in and said, 'How much do you want to charge for this gig tonight?'

Liam We were like, 'What are you on about?' and he's going, 'How much do you want to put outside?' Fucking joking we said, 'Fifty quid to come and see us.'

Noel So we played the gig to a barman who was just cleaning glasses. I think towards the end a boy and a girl might have come in and were just fucking snogging on one of the benches as we were bravely going through 'Live Forever'.

Bonehead The only applause we were getting were off the bar staff who were just sat there, chewing their nails, watching us.

Liam Between songs you could hear the bar lady just squeaking the fucking glasses. But we were mega, man, every gig we did at that time was like we were fucking steaming it, man. A couple of months later I guess, we were selling places out.

Bonehead I remember breaking a string on 'Rock 'n' Roll Star' and fixing it dead quick as if there was a crowd there. There was nobody, but we played a blinder, we played it like we meant it, you know, top gig as well, it was brilliant, one of the best.

Noel We went back there six months later to play, couldn't even get onto the street, it was fucking bedlam.

Bonehead It was like, 'Oh God, we're going back to that venue again, we're doing Duchess of York again.' I remember we did the soundcheck sitting in the dressing room looking out the window and it was just like 'Fucking hell!' The door was directly below and there was just a queue of people coming out of the front, turning left, going right down the street and round the corner.

ORANGE

Feeling supersonic

Noel Creation were only really aware that we had about six songs; I knew I had about thirty-six. At this point they'd never heard 'Slide Away', never heard 'Married with Children', they had never heard 'Shakermaker'. I'm not the kind of person that would sit in a dressing room and sing fucking fifty songs in a row, that's fucking boring to me, let's wait till we record them. So they never knew what *Definitely Maybe* would eventually sound like even though I did.

Liam He always had more, he always had tunes coming, but they were like planes, you've got to wait for the fucker to land. My vision was like when those planes go round and round above Heathrow Airport, you've got to land one of them first for the others to land. Otherwise it all just becomes silly, you can't land them all at once.

Noel When somebody at Creation said they were going to put out 'Columbia' as a one-sided 12-inch white label, great. Then Radio 1 put it on the playlist and I remember being a little bit horrified thinking, 'Oh no, that is the first thing anyone's going to associate us with.' It sounded a bit tinny, there is no bass on it, and it was clearly fucking recorded on an eight-track. At the same time, thinking, 'Wow, we are on the radio, they've said the name Oasis from Manchester.'

I remember thinking, we can't be in a studio for three days and go back with nothing.

Noel So we've got to the point where we've signed the record deal, 'Columbia' has been played on the radio, it's now time to record a single. This is what's great about Creation Records, they just gave us some money, they didn't give us a producer or anything, they just booked us a studio. We thought we would do it in Liverpool because if we are in Manchester it will be fucking chaos at the studio. So we'll do it in Liverpool, at a place called The Pink Museum. McGee suggested that 'Bring It On Down' should be our first single and I was like, 'Great, I fucking love that song, it's like the Pistols, like the Stooges, that will do for me.'

McGee 'Bring It On Down', for me, it's just punk; 'You're the outcast, you're the under-class, but you don't care, because you're living fast.' I just loved it.

Liam That's a tune, mate, that was what Oasis was about before they got caught up in this Beatle web. We were rocking, steaming, it was like the Pistols with melodies.

Coyley We were all very inexperienced and I think we booked in for two days. The first day is just horrible and it gets worse and worse, and the session starts degenerating.

Noel We couldn't get it right, whatever we had in our rehearsal room and on stage wasn't translating into the studio.

Liam I don't remember not being happy . . . it certainly wasn't my vocals, they were rocking.

Noel It would become apparent that session didn't work because our drummer wasn't the most consistent from one fucking bar to the next, never mind one day to the next.

Tony 'Bring It On Down' just wasn't coming together, the tempo, the speed, I don't know what the fuck was wrong with it like. I can't say that I understood why it was chosen, it was out of my hands really.

Noel I remember thinking, we can't be in a studio for three days and go back with nothing. What if they tell us to fuck off? It was either Tony or Chris Griffiths, who were with us in the studio, who said, 'Well, if it's not happening, just do something else.'

Coyley Noel's got a riff, but that's all he's got, so they're just kind of knocking this little thing around, which all bands do, a little riff.

Noel Why I chose to write a new song as opposed to record any of the others off *Definitely Maybe*, is still a little bit of a mystery. Because it's quite a mad thing to do. I remember someone sent out for Chinese or fish and chips or Chinese fish and chips, and I went in the back room and, as bizarre as it sounds, wrote 'Supersonic' in about however long it takes six other fucking guys to eat a Chinese meal. It was a brilliant moment in time because I would never be able to do that again.

Bonehead Noel's just sat there with the guitar and he just wrote the music, that will do, and then he wrote the words, any old fucking words and he came back in the room with us, with his guitar and he said, 'Look, I've just written another song.' He started singing it and we nailed it and mixed it that night, rapid, because that's what we'd been doing every night in the Boardwalk, you know, and it sounded massive, absolutely massive.

Noel We all stood in the same room and it was like, 'Well, this is how it goes.' It's really slow because we are all in the same room looking at each other, nodding for the changes. We could never get it that slow playing it live. We recorded it and mixed it in that night and I've got to say, listening to it on the way home on the cassette deck in Mark Coyle's Renault, I thought it was fucking brilliant. As good as 'Bring It On Down' and 'Columbia' – I thought it was fucking great.

Tony It just unfolded, within minutes maybe, done. Some of the greatest things that have ever been recorded, just happen like this, bang, there it is. Eat that.

Noel What really set it apart was Tony Griffiths had come up with those backing vocals – the 'aahs' in the bridge – and it was just a fucking brilliant moment in time. It showcased everybody's talent: Liam singing is fucking brilliant, the drumming's great, Bonehead's bits are great, Guigsy's bits are great, the guitar solo was great – even though everyone thinks it's ripped off something or other – but it was all good. Lyrics are fucking bananas . . .

Liam The producer had this big fucking Rottweiler called Elsa, and she would be there all the time sniffing shit and getting off her tits. I think Noel just wrote lyrics about her.

Noel The way I generally write is the first few lines will form a story and then it kind of gets confused and muddled up a little bit. Then you might have a line that might be true, in the sense that that's what might have happened, but it doesn't work in a song. So you have got to fill in the gaps . . . but it's the overall thing that matters. It's why, when I'm asked about my songwriting I find it quite difficult to talk about because 90 per cent of every line I've ever written has got some kind of meaning to me, but then you are kind of just filling in gaps with lines that are quite vague but get you to the next place. I'm a songwriter at the end of the day, I'm not writing novels here, I am not writing reportage; they are pop songs and a great deal of pop lyrics are fucking nonsense. It's all about the tune, but you do have hit songs from time to time that really get to sum up an experience that you had.

Liam Didn't care what they meant, still don't care what they mean. They mean something different to everyone, don't they, so . . .

Noel One night a girl came up to me and asked if 'Supersonic' was about prostitution . . . I don't think I've been speechless many times in my life, but that was one.
'The whole thing about Elsa doing it with the doctor in a helicopter . . .'

NG: If we're gonna get rid of Phil Collins
and Sting... junk-food music, McDonald's
music...
We've got to get in the charts and stamp
them out.
I want the severed head of Phil Collins
in my fridge by the end of this decade.
And if I haven't...I'll be a failure.

Elsa's a fucking Rottweiler, it is a fat stinky dog. She was convinced it was about prostitution and I told her she was right as well. I would never spoil it for anybody. It's usually lyricists that will tell you that the words are fucking everything. They are not, the words don't mean shit to anybody. It's the melody is what you remember. We all whistle tunes; it's always about the melody, and that's what I do. Roger Waters for instance is very fucking willing to tell you what every single line of his songs mean. When I first heard *The Wall* I was gone . . . and then you find out it was all about his dad. Don't fucking tell me that, I'm not fucking interested. I thought it was all about me. I like to think that all my favourite songs are somehow about me. Which is why you love them all. I leave it up to people to interpret those songs. When journalists come after the meaning I always dismiss it and say they don't mean anything. I don't want anybody to know about me and my life, my songs. A great deal of odd lines and even half sentences have real proper relevance to my life and growing up but it is not something I'm interested in telling you or anyone else because then it spoils it. That is why Oasis was so brilliant because they were so inclusive. They weren't about anybody on the stage, they were about – if anybody – the people in the crowd. And I think that is why it was so immense because it appealed to so many people, right across the board: the middle classes, the working classes, the toffs; the older generation, the sixties generation and still kids to this day. There was something magical going on, and still is, in those tunes, I don't know what it is.

There was something magical going on, and still is, in those tunes . . .

Liam I've never spoke to our kid about what any of them mean because they mean the world to me, but I don't know what. I could ramble on, but I would never do them injustice by just saying something flippant. I could never go, 'It means this,' because they just mean everything. I don't want to know what they are about. That's the beauty of it.

Coyley I'm not a big word man, me, I don't really listen to the words, but it's how they make me feel and if I can feel it, it turns me on, I don't give a fuck what the singer's saying. I'm on it, I'm with you, Noel, I didn't understand what 'feeling supersonic' meant, but whatever it is, I'm fucking there. I'd die for you right now. That's how I felt about Oasis.

Noel 'Supersonic' was never even remixed. The mix that's on *Definitely Maybe* is the rough mix from that night. A brilliant moment in time and if push comes to shove it's probably still my favourite ever recording because of the flash of inspiration; something was going on that night, bang and it was there.

Bonehead I remember going down to Maida Vale and doing a BBC session straight after and McGee turning up.

McGee Noel came in and said in true Noel fashion, 'The recording session was rubbish, it never worked out, but I've written a smash.'

Bonehead Noel put this cassette in, McGee was expecting to hear 'Bring It On Down' and instead 'Supersonic' comes fucking blurting out of the speakers. Blew his head off. Blew mine off.

Noel We played it and he went fucking mental, he loved it.

McGee But I was nuts, I was fucking nuts. I mean, most people would have gone, wait a minute, but I was on my own trip. Then they gave me a line of coke and four or five seconds into a line of cocaine, fucking everything seems like a fucking great idea. I did wake up the next day going, 'What the fuck happened, how did they do that?'

Coyley It just showed the potential that was just bubbling underneath at that time. The explosion was just imminent, this is going to happen any second and here is a little taster of how it's going to be.

Monnow Valley

Liam Beautiful studio, Monnow Valley.

Phil Everyone used to go and do a debut album there. It's an old farmhouse with a barn attached to it that's been converted into a studio. You're just holed up in the country, windswept and rainy, with a massive log fire.

Noel It's not even the best studio in Monmouth, Rockfield is four minutes up the road; The Stone Roses were there recording *The Second Coming*.

Liam We met them on the high street. I was star-struck by them. Ian Brown had a big beard, he looked like Jesus, he looked cool as fuck.

Tony God, it was a massive adventure, a holiday we'd never had. I just didn't know what to expect. A fantastic place but quite daunting and nerve-wracking. It's not a demo, we're recording for real, paying thousands of pounds to be here.

Noel It wasn't the most glamorous place to be, but it was nice. I'd been in recording sessions with Inspiral Carpets – I've loaded gear in and out of studios and sat in on their sessions for years – so I am already a bit underwhelmed by it all. I'm not running around like, 'Yay, we are in the studio! Fucking hell, look at all these teabags. It's unbelievable. Crisps!' I just wanted to get cracking.

Liam I'm not having that. He'd set up an amplifier, that's about it . . .

Bonehead We had never been in a recording studio in our lives. People were putting a lot of faith and money into us so we just strolled in there like we knew what we were doing. Did we fuck, didn't have a clue. Everyone was pretty green, but that probably helped it work. No one came in with preconceived ideas, no one came in lording it: 'I've done this before; I know what I'm doing.' Everyone was winging it to an extent.

Noel Johnny Marr sent down some guitars because I only had one guitar and Bonehead had one guitar.

Bonehead We didn't have that much in the way of equipment. Because Marcus was managing Johnny Marr, Johnny was really good. I remember going up to Johnny's house in Cheshire, with Noel, and we went downstairs – he had a studio in the basement – and he had a whole collection of guitars. He was just like, 'Take your pick.'

Noel He sent down this Les Paul and I took it out of the case and immediately wrote 'Slide Away' on it.

Liam Obviously I get a decent room, but they're all playing tricks, going, 'Oh look, it's fucking haunted.'

Bonehead One day, one of the staff casually mentioned, 'So, who's sleeping in room number three?'
Liam says, 'That's me that, I'm in room number three.'
'Alright, you got the haunted one then.'
We were just like, 'Whoa! What are you on about? A ghost?'
And she was like, 'Yeah, real ghosts; some dark figure's been seen.'
So Liam, of course, is just shitting it.

Liam I go to bed one night, wake up in the morning, all the furniture is moved round. I've gone downstairs going, 'Look, it must be fucking haunted, I've got to change room.'

Bonehead The next night I'm thinking about this ghost thing. By the front door there are a load of fishing rods with hooks and fishing line so, if you're into that sort of thing, you can go fishing by the river that goes through Monnow Valley when you've got a break. I got a load of fishing line off one of these rods and looped it through a little hole in the corner of the *Daily Mirror*, right in the top right-hand corner. Put a bit of fishing line through the laces of somebody's pair of trainers as well.

Phil The line went along the picture rail, round the back of the chairs, all around this massive living room, round the back of the TV.

Bonehead Everyone else knew about it, but Liam didn't. We're all watching football or something on TV and I just started pulling this line.

Liam So we are sitting that night round the fire, there was lots of fucking about, and the chandelier tilts and then a paper goes over.

Bonehead I'm doing the newspaper; I just started pulling it so the page started curling open, held it vertical and then let the line go so it dropped.
'Did you fucking see that – did you see that?'
And we were like, 'See what?'
And he's like, 'That fucking newspaper just opened and shut.'
I say, 'I'm sure it did, it's the wind.'
And he's saying, 'No, it did, it did.'
'It can't have done.' But you could see him watching it so I did it again.
He's like, 'It fucking moved again!'
Then I really tugged it so the whole newspaper just went fuck off across the table. So, of course, we all jumped up like, 'Wow!' Liam was shitting it, because he didn't know anything about it. So he really jumped up. Someone tugged the trainer so, as he headed for the door, a pair of Adidas trainers, just scooted across the floor, chasing him. He was done. He was off, down the fields. See you later, Liam.

Noel Liam is a great believer. For a man that actually has no spiritual content whatsoever, he is a great believer in the spirit world, which I find fascinating.

Liam I don't remember doing much recording, just shitting me pants because it was haunted.

Noel This guy called Dave Batchelor got to produce the first incarnation; that was all my fault. He was the Skids sound engineer and he'd done a festival run with Inspiral Carpets four or five years before. I remember travelling around Europe with him talking about music and punk rock and all that. I thought he was a fucking cool guy, I used to like the Skids and blah, blah, blah . . . When it came to, 'Who do you want to produce your first album?', it was like, 'I want this guy to do it.'
Everybody went, 'What? He's not done anything for fucking . . . what are you talking about?'
I was like, 'No, this guy is going to do it.'

McGee Noel was a real road dog, he loved going out on tour. He'd bonded with the guy and he promised him he could do his album. Because he'd promised it, I gave Dave a chance and it was fucking wrong.

Noel I remember it becoming frustrating, but I don't remember entering into it with any sense of trepidation. I remember being pretty relaxed; all we have to do is play. It's fucking easy. I've written all the songs, I've come up with all the parts for everybody, fucking hell what else do you want me to do, make the tea? We had a producer in the room, it's his job to make it sound good. I know what I'm doing. If everyone else knows what they're doing, what could possibly go wrong?

Bonehead I presumed that we were going to get in and someone was going to say, 'Right, okay, we're recording, one, two, three, four, away we go.' But it wasn't like that, it was take, after take, after take.

I don't remember doing much recording, just shitting me pants because it was haunted.

Noel We were all in separate booths and it just wasn't fucking happening.

Liam I didn't like that, that did my head in. Felt like you were on the naughty step all the time. We were a good band, we played a lot of gigs and we were tight. I don't think all that doing it separately worked. It should have been done all at the same time. I think that captures the vibe, that's the glue. If you're doing it separately you're taking away all the glue . . .

Noel This is where the thing with Tony started. He was good live and all that but at that very moment of trying to record a fucking album, it was a pain in the fucking arse because he was very inconsistent.

Tony That wasn't a comfortable session, we quickly recognised that this isn't the right guy for us.

Bonehead We would go in the control room and listen back, and Dave just wasn't getting it. At some points it would get pretty tense.

The idea of the song diminishes a little bit in the studio because you can never recreate what is in your head . . .

Noel We did one version of the album which we mixed a fucking thousand times. Trying to replicate the sound that you have is the ongoing battle you have as an artist anyway. The idea of the song diminishes a little bit in the studio because you can never recreate what is in your head – my fucking head space is massive and the speakers are this big. Every time I go into the studio I'm trying to get something out that is in my head and it can become like a dog chasing its tail.

Coyley They all knew it was no good.

Noel We went away on tour, and the tapes were sent off to be mixed somewhere by someone. You are getting ready to go on stage in a fucking pub in Southampton, you're expecting your manager to turn up with a cassette of the album and it's like, 'It's fucking awful'. The feedback coming back is, 'Sounds shit.'
That to me is always the worst thing anyone can ever say about anything: 'It's shit.'
'Why?'
'Don't know, it just is.'
That doesn't mean fuck all to me. If you can't articulate what's wrong with it, how are you supposed to put it right next time? It would be great if somebody said, 'It sounds shit because of this, and this is what we've got to do to put it right.'
Then you'd just go, 'Right, well, let's do it then.'

Liam I'm sure it wasn't quite right, but when you start picking at these things you can fucking twist your head. I guess Noel carried the weight of the band. I was definitely not stressed; I was fucking having the time of my life.

Definitely plan B

Noel This is when the discussion about the sound started: 'It doesn't sound like you lot live.' I would be thinking, I don't know what anybody's going on about here because it fucking sounds like me when I'm stood beside my amp. Every time we would do a track and you'd think, well, it kind of sounds alright to me, McGee or Marcus or Coyley would say, 'Doesn't sound like you lot live.'

And I would be like, 'What do we fucking sound like live? I don't know, I'm on stage?'

'It just doesn't sound like you lot live.'

'Who *does* it fucking sound like? Spandau Ballet?'

McGee I knew what Oasis should sound like because I'd seen it live and I just knew the performances were wrong. I was at the point of going, 'Let's put the fucking demos out.' The demos were brilliant – better than the fucking recordings. We wouldn't have been as big, but on an indie level, we'd have got really big.

Noel We only came to the conclusion that Mark Coyle should go in with the band and record it live by drunken accident one night. If what you lot want is a live sound, who does the live sound? Mark does. So why doesn't Mark do the record?

'Fucking hell, that's a good idea!'

And I am sitting there thinking, is this how easy this shit is?

'Man, you are a fucking genius.'

'I know, thank you.'

Fuck me, if he doesn't know what it's supposed to sound like we are fucked.

Coyley Time for plan B on *Definitely Maybe*. I would be interested to know why they took that risk, because it's some risk, putting me in charge of that, no track record whatsoever. But somebody recognised a situation the band were comfortable with and just said, you do it.

Bonehead It's like, 'Right, we're going to redo it, we're going to take it down a studio in Cornwall, Sawmills Studio. We're going to get Coyley in.' Thank God for that, because we all knew that Coyley understood us, he got us.

Noel Thank God for McGee, you know. He's just like, 'Here's some more money, let's do it again, let's get it right.'

I’ve writter
I’ve come up
parts for
What else do
you want me
make

all the songs,
with all the
everybody.
to do,
the tea?

Liam People go, 'It weren't quite right.' Well, I'm fucking having it was that good, we did it again.

Noel I remember it taking forever to get there. We drove down, it took fucking ninety hours . . . And no one had told us that Sawmills was only accessible by a fucking canoe and if you missed the tide you had to walk up the train tracks. So we get to wherever it is in Cornwall and they say, 'We've got to load the gear onto this boat and then you've got to sail to this studio.' It was fucking freezing. We were like, 'What fucking studio is this?'

It was really enjoyable as I remember. Liam was on top form, everybody was good.

Bonehead It's mad really. We threw the amps on and drums and guitars and that sails over, and then they come back and get the band. There is no way out of that studio, once the tide disappears, you're stuck.

Coyley It's just idyllic, it was beautiful. We spent the week having chips and egg.

Bonehead Coyley used to have a little DAT player and a microphone, and he would go recording crickets, or bumblebees buzzing, so he could sample them on music at home. I caught him one day by the river and he was just recording the river rippling, you know, just being a weirdo. So, I sneaked up and stood behind a tree. I was making weird noises, hid behind this tree, and you could see him thinking, 'Oh man, I've just captured the sound of the greater crested woodpecker, this is going to be incredible when I put it on a dance record.' The dick's just sat there, you know, being Coyley. Then I started giving it all that and you could see him thinking, 'That sounded like "fuck off".' And you've got to own up, haven't you, so I stuck me head round the tree. I was like, 'Got you there, didn't I, Coyley?' He got really angry, absolutely lost the plot. He chased me all over the place, the guy was purple, he was going to kill me. Hippies, man.

Liam You've got to get a boat or walk through the woods and along a railway track to the one pub. Coming back was pretty funny some nights, just having a laugh, man. It was like fucking camping, it was great.

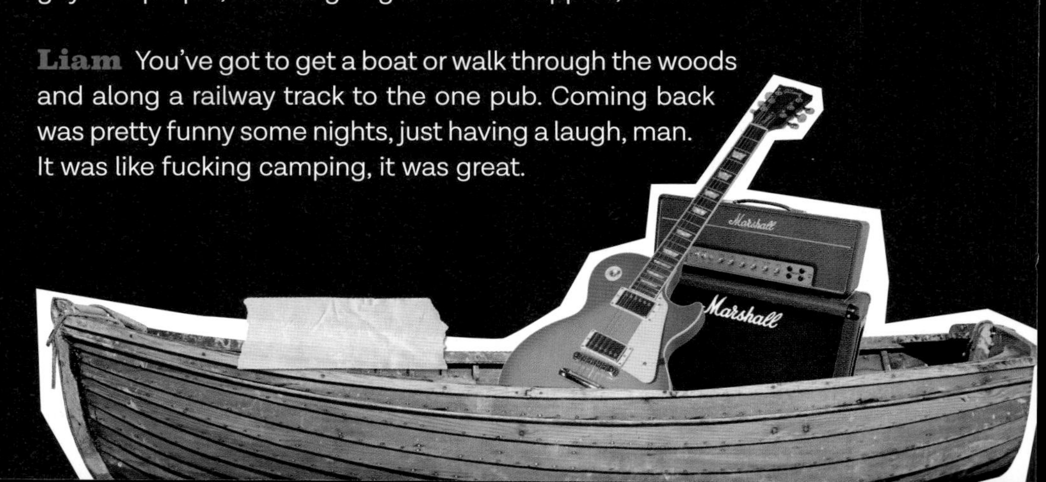

Coyley We started recording *Definitely Maybe* and we got results immediately. It was the easiest thing that any of us ever did.

Noel We hit upon this system where we would start in the morning when everybody had had breakfast and cigs and got their shit together. We all stood in the same room with no headphones, just like a rehearsal and just went through it. We would do three takes of each song and then move on. No going into the control room listening back to anything. Three versions of 'Rock 'n' Roll Star', tune up, three versions of 'Live Forever', tune up . . . That's how we used to rehearse. So the vibe was like it used to be in the Boardwalk. At night we would sit and listen and say, 'I like that version.' We would work on that, put some overdubs on and that would be it. It was really enjoyable as I remember. Liam was on top form, everybody was good.

Coyley I think there is a lot of nonsense written and spoken about making records. You can take your technical bumph and your know-how and your knowledge and you can shove it right up your arse, because I made *Definitely Maybe*, where a technically minded man couldn't do it. So, it must be about the musician just feeling good and feeling comfortable and feeling confident. It was so relaxed, it was the way it should have been the first time round. I would have thought that album was recorded in three days, maximum.

Noel I remember mixing it with Mark and Anjali Dutt in Olympic Studios and struggling with it, a lot. I don't know why. The lesson it taught me is that I am not a studio technician; I shouldn't have to fucking know about stuff like that. Have you looked at a mixing desk? It's a thousand fucking knobs. Do they all do something? Fuck that. I write the songs, I play the guitar, I'll even do some backing vocals, I'm not getting involved in all that, I'm not interested.

Coyley Me and Anjali were there for a couple of weeks. It didn't really sound very good. I'd never mixed a record before, I think I was just a little out of me depth. I wish that I'd had more experience because I could have truly made that my own record, but I think the bottom line is that I just didn't know what I was doing.

Bonehead But it was there, it was all played right, that much was obvious. It sounded good, but it didn't sound great.

Noel It turns out even then it was shit so I don't fucking know what was going on.

Have you looked
at a mixing desk?
It’s a thousand
fucking knobs.

Enter Owen

Noel You are talking to these people who make records for a living and they are telling you it is not right, but they can't articulate why. I don't know what I was supposed to do about it. I remember taking a cassette of a mix that we did to McGee's flat and sensing that he was completely underwhelmed by it. Me thinking, 'Fuck it, I give up. I really give up.' I couldn't understand why 'Supersonic' sounded so amazing and everyone thought these songs sounded so flat. Maybe it's because 'Supersonic' was written that night and these other songs had been on the verge of being three years old.

McGee At one point, Noel was so frustrated with the whole thing, he went, 'Look, why don't we just put it out and we'll get it right on the second album,' and I went, 'You'll never get to the second album, we've got to get it right on the first album.'

Coyley That record had to be made, whatever cost it was going to be, it would have been made three times. Alan McGee knew there was a good record there.

Noel We were never put under pressure by Creation, ever, once, ever. Alan was such a fucking dude; that record label was never started for commercial reasons, it was for the love of music. I knew we'd spent everything, we'd definitely spent the money that we were given to do it, that was gone. As I remember it Alan had to go to Sony to get some more money and, because I was signing a publishing deal at the time, I paid for some of it with some of the advance.

Bonehead Alan McGee . . . a believer, I suppose; an absolute belief in what he does. A visionary. He is the type of person who, if he believes in something, will go out and do it and he won't stop. He will get that thing and he will take it to wherever it goes. He doesn't care if it just goes to there and falls. He'll still believe in it and still love it.

We were never put under pressure by Creation, ever . . .

McGee Marcus announces that he's going to get Owen Morris to do a mix on spec and I'm thinking, 'Why?' I don't get it. He did one mix – I think it was 'Rock 'n' Roll Star' – and we went, 'Right, he's doing the album.' So, it was Marcus's relationship with Owen that was the final element in the whole jigsaw puzzle.

Owen Morris I was kind of angry and pissed off at music and the music business, so it was a relief to be given free rein of Oasis. I thought, this is my one crack: they've got a deal, said in the *NME* they are on the fucking go, Radio 1 all over them. I might get some more work off this. And they let me do exactly what I wanted.

Liam Mental Welshman; biggest laugh in the world. I loved him. He was fucking off his head, completely off his head. He is trouble, but he was great, because I'd be going, 'I'll do some shit and blame it on him,' and vice versa. Owen come along and just turned everything fucking up. He kept it simple whereas maybe other people were trying to get a bit flash with it. But Owen come along and went, 'Right, you lot just need to turn the fucking thing up,' and that's basically what he did. He was always setting fire to shit and things were always blowing up because he had it too loud, which was great, it was fucking great. With Owen, those recordings are the nearest it got to the gigs.

Noel Owen would get really excited and start throwing beer over everyone. He used to get very, very boisterous.

Tim Abbott The thing is with Owen, he's got a wall of sound with guitars but he can isolate the lead vocal. That's his trick. I think Coyley could do that on a sound desk, but he could never do it on a recording desk.

Noel Me and Liam were always arguing about the mixes, always, always, always. This went on all the time: 'The vocals should be louder.' My thinking is the vocals shouldn't be louder than anything else, it's all about the soup. The great records don't have loud vocals. He was always, always trying to get the vocals up. I can understand it if you've written these great lyrics and you're a singer and you need them to be heard. He didn't write a thing, not a fucking thing. He didn't even write the gaps in between the singing. Even I wrote those. So why he was wanting the vocals to be loud I don't know.

Owen Marcus sent me cassettes of the mixes they had so far, and he said, 'What do you think we should do with it?' Oasis were touring then, and Marcus was like, 'Fucking sort it out.' So I had a listen and pretty much said, 'We'd better re-record every vocal.' We went down to Loco in Wales for a weekend, to try me out and to try a couple of mixes. That's when I met Liam, I had already been briefed that he was a John Lennon freak and all that so recording Liam was really easy: I told him he sounded like John Lennon and he was like, 'Fucking . . . you got it.'
There was no vocal on 'Rock 'n' Roll Star', so that's the first thing we did, Noel just left me and Liam to it. Liam's lead vocal fucking blasted through it. Piece of piss.

Liam I loved the way Owen recorded. It was very simple to me and I think that is the way it should be done. A lot of musos will sit there and go, 'Blah, blah . . .' Well, I tell you fucking what, you go back to twiddling with your guitar over there, and your bass and all that nonsense. Your record sounds naff compared to ours because you've got the rule book out and you are on page sixty-two and your music sounds shite. Listen to ours, we are not on any page, we have not even got the book out. Ours will eat yours alive, and your mate's and your other mate's. So that is what I liked about it, it was just straightforward, simple and that is what our band was about. It was just solid rock and roll with great songs and great melody. There was no Slash doing his diddly dee, there was no drum fills, there was none of that. I would never go in there and sing it light. I found it hard singing less. I always had to give it 120 per cent. No one would ever have to ask me, 'Look, can you give it a bit more fucking bollocks?' I would just go in and just fucking rip the arse out of it and people would go, 'Stop singing it so hard, you are not at a gig now, back off a bit.'

I remember one fucking producer going, 'Can you get into a role?' I was, 'Get to fuck, mate, I ain't playing a role, I'm going to sing it.' I'm going to belt it out and fucking turn me down.

Owen They were good, the mixes were good, really fucking good actually. Noel comes in and was like, 'That sounds great, that'll do.'

Noel I don't like to spend too much time fucking about in studios, they are kind of in and out, put the kettle on, that's me done, let's go and do something else. If you've not got it in the third take you've not got it. That's still one of my golden rules to this day. If I'm doing a song and I haven't got it the third time I will go and do something else. I'm not one for sitting around all day fannying about, moving a mic an eighth of an inch.

Liam We'd do two takes and just go, 'Right, let's go to the fucking pub.' Take hundred is going to be absolutely fucking shit. End of. Take five is going to be fucking utter shit as well. Who gives a fuck about take seven? If it's not done in three takes, then you shouldn't be in a fucking band or you need to go and have a word with yourself.

If you've not got it in the third take you've not got it. That's still one of my golden rules to this day.

Owen They went away, and everyone liked it. I think a cassette went to Noel, he listened on the tour bus. And that's when they said to finish the rest. I mastered it at Johnny Marr's studio. Noel came over and I just got on with it. There was this new box that could make things twice as loud without distortion. So, I thought, if I make it twice as loud as everything else, that will mask the fact that my mixes are not technically very fucking good. My theory was quantity, rather than quality, at the time really. When it went out, it was brilliant – in every fucking jukebox in the country you could go in and put Oasis on and it would be twice as loud. As funny as fuck, man.

Noel We mixed fucking shitloads of times and then Owen got hold of it and we all went, 'Fucking hell, at last.' I don't know what he did to it, but fucking hell that's what you hear today.

Owen come along and just

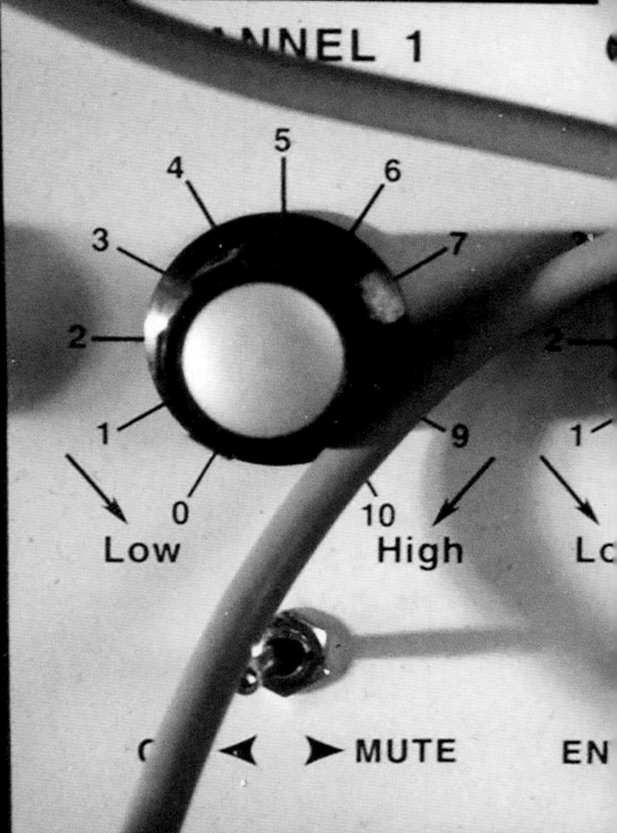

turned everything
fucking up.
EL 2
11
High
CH2

The Dam

Noel We were doing this gig with The Verve in Amsterdam. What the fuck for, who knows why? It is not even a tour, it is one gig. It was going to be our first international gig, which even saying it like that sounds ridiculous. Of course it soon descends into chaos. We didn't have any roadies at this point, so we get rounded up in Manchester and this guy picked us up in a van – Jason Rhodes – he's going to be our roadie. He knew Marcus because he worked for New Order and he'd done Bernard's guitars.

Jason Rhodes Marcus got in touch: 'I've got these three bands. I've got Oasis, I've got Push, I've got this, I've got that.' I know Noel because he's a roadie, I've met him through Inspirals, so I'll go with them. 'Right,' he says, 'do you want to go to Amsterdam?'
I went, 'Yeah.'

Liam The first time I met Jason Rhodes, he picks us up at me mam's house in this fucking van and he looked mental. He looked like a fucking psycho killer.

Noel I got in the van, 'Jason, nice to meet you' and all that, 'Did they give you a float? You need to give that to me.'
And he's saying, 'No, no, no, that is for petrol.'
I said, 'Never mind petrol,' and we took the float off him and then spent about an hour driving around Manchester getting drugs.

Bonehead I think we did a pit stop, straight out of Manchester and stocked up on beer and vodka and whatever you do, you know and away we went.

Liam We stop off at some Spar or 7-Eleven, get a load of booze. We get on the ferry, I don't know how we got on the ferry because we are pissed, but we get on this ferry and we are all fucking about drinking.

Jason By the time we got to Harwich there was nothing, they'd drank pretty much everything. Five bottles of spirit, which is kind of going some.

Bonehead And then it just went downhill rapidly.

Phil The Thursday-night ferry to the Dam is obviously full of nuisances. Who goes to Amsterdam on the ferry for the weekend? What kind of clientele?

Noel As fate would have it, West Ham were on a pre-season tour of Holland and there was a load of West Ham fans on this overnight ferry. It's got a casino on it and a fucking nightclub.

Bonehead A disco boat and there were actually people dancing. We were just, Jesus, man, that's not right, you don't dance on a boat, do you? You sit in the bar, yes, or you go to bed if you've got a bed. You don't have a disco on a boat. We really didn't have any money at that point. A bottle of champagne appeared at the table and it was like, 'Where's that come from?' Another bottle appeared and these bottles kept appearing . . . So everyone got drunker and drunker. Liam was going for it and Guigs, weirdly, was going for it. Unlike Guigs, going for it in the bar, but he was up for a bit of fun.

All I remember was it was a bit Benny Hill, the police chasing us around the fucking ferry.

Coyley In my memory, the drink of choice for a few hours before all the excitement started was champagne and Jack Daniel's slammers. So we're having a wonderful little time in the bar and the next thing there is sporadic fighting breaking out all around us. Liam is very excited by the prospect of a lot of chaos going on. He's *very*, *very* excited at this, so he goes and joins in.

Liam Guigs has come back with someone who's collared him saying that we were forging £50 notes. I was like, 'They are real, you daft cunt.' Then a fight broke out. All I remember was it was a bit Benny Hill, the police chasing us around the fucking ferry.

Coyley I keep seeing Liam running, through the windows, along the deck, he's having a great time, he looks like he's in a school playground, chasing leaves.

Phil I know Liam at one point said they ran through a casino and he flung all the roulette shit off the table, as he was running past.

Coyley The next time I see him, he's still running, but he's got policemen running after him.

Bonehead I just remember seeing Guigs getting hurled down the stairs from one deck to another. I was thinking, 'Oh my God, what's going on?'

Coyley I come to the bottom of a stairwell and Liam is on the floor with three or four coppers tying him up. Next to him on the floor is Guigsy. What Liam's done is he's swung for somebody, but completely missed. He's thrown a great punch, but he's not made contact with anybody so he's spun round and ended up on the floor. As soon as he went on the floor, the police have got him as well.

Liam Someone's punched someone, someone's kicked someone, someone's got nicked and then we are handcuffed, me and him.

Bonehead I don't know where Noel went, I think he was being a professional. He must have got his head down, he must have gone to bed or something.

Noel Everybody gets nicked except me. I don't know why I never got nicked. I somehow managed to wangle my way out of it.

Liam Our kid's been fucking reading Shakespeare again, he gets away with it. We get to Amsterdam, straight off the boat and we're nicked. We're sent down to the bottom of the boat into this fucking cell and we had to stay there, little mattress each. We'd been drinking all day so we were pissing in this bucket and it was going everywhere, it was horrible.

Bonehead I got back to me room, me passport was gone, me shoes have gone, everything had gone. So, I reported it and the security are like, 'Yeah, we know your passport has gone, we've taken your passport.'
'Why have you done that?'
'Because you're not going into Amsterdam, you're going to get off this boat and you're going straight home, mate.'
'Why, what have I done?'
'You're with them two, downstairs.'

Jason So, I went round and spoke to the customs or the cops and said, 'Look, I'm kind of missing four people, do you know where they are?'
They went, 'Oh yes, yes, yes, they'll all be getting deported, be getting sent back on the next boat.'
I'm thinking, shit, that's four out of five I've lost, we've not even got to bloody Holland yet. I said, 'Look, I need to give them some train fare at least.'

Welcome to
AMSTERDAM
ETY
S : FOL
DAM
ETY C
S : FOLLOW CAP

FOR IMPROPER USE OF
Exits are located on both sides at the front and
DEPORTED

So he brought them down to this doorway and I gave them £20. 'Okay, say goodbye to your friends.' And this big shutter door came rolling down – I remember seeing everybody bending down looking underneath it getting lower and lower, till you could just see their feet and then they had gone.

Noel I went to see them in these cell things, it was fucking brilliant. I was saying, 'Don't worry about it, I'll call Marcus when I get to Amsterdam. By the way have you got anything left of that float?' Someone had £100, I said, 'Give us that float,' and they said, 'Why, where are you going?'
I said, 'I am going to Amsterdam for the weekend, fuck you lot.'
We went to Amsterdam and turned up at the gig, The Verve played and we got royally shit-faced.

Jason I can remember getting to the hotel and having to ring Marcus and Alan and they're going, 'You fucking what? You what? Are you joking?' Wasn't my fault. What did you want me to do? Fucking slap them on the wrists? 'It's 10 o'clock, get to bed, boys.' A hell of a start that, isn't it? That's a great first day at work that, isn't it? Love it.

Bonehead We just sat in stunned silence. It was just like, 'What the fuck is going on, what are we going to do?' Panic. We were heading back to England, no money, no nothing, no mobile phones. I think Tony had a fiver, so we ate crisps all the way. We didn't even laugh. It was horrible, horrible.

Liam Ended up at Kings Cross with Marcus. There was a load of Americans coming over to see us and he said, 'You've fucking blown it, you've blown it.' He was going, 'You shouldn't be fucking behaving like that.' Who the fuck are you, my fucking dad? This is what happened and we were in the right. They fucking stitched us up. End of. You weren't there. Don't be fucking telling us what it was like and what we should have been doing, we didn't do anything wrong, we were just doing what we were doing. We were trying to just get on with our business.

Bonehead You should have seen the state of us, we looked like you couldn't describe, we were just a mess. I think it was more Marcus panicking about the Japanese record company and the American record company, because they haven't got a sense of humour. None of these people are going to find this funny so he was not happy with us. Everyone's nudging me, 'Bonehead, ask him if could he borrow us an extra £20 so we can get some food.'
I was like, 'You can fuck right off, no, you ask him.' So Liam had to ask him, 'Could you borrow us £20 for some food as well?'

Liam This is called life. It happens, people get into scrapes. We didn't plan it like that. Obviously we would have preferred to get on the boat, have a drink and not be fucking accused of having dodgy £50 notes. Then we would have got there and done the gig and blown everyone to pieces, but it didn't happen like that. It happened like this, so that's life. Them fuckers didn't even turn round and go, 'How are you?' or anything, they were just, 'You ruined the fucking . . .' Fuck off, you knobheads, I've been sat in the fucking bottom of a boat, handcuffed, with piss running around for the however long it takes to get to Amsterdam and back. You've been sat in your fucking BMW.

That's how I'd want my rock-and-roll stars to act. Stick up for themselves, get in a bit of shenanigans and that's it.

Bonehead Marcus was pissed off. McGee, being McGee, thought it was great. He saw that he could say, 'You know what, this is publicity, this is rock and roll.' He was buzzing off it.

Noel This is another reason why I love McGee: I called him and got put through to his office and he said, 'Noel, man, how's it going?'
I said, 'Are you sitting down? I've got some news.'
He went, 'What is it?'
I said, 'Everybody's been arrested on the fucking boat on the way out last night.'
And the only word he said was, 'Brilliant.' I think within twelve hours he had a photographer from the *NME* out there and made a massive deal of it. It was a funny old night.

Liam I know Noel hated it because he wanted to do his thing, but I didn't think it was a bad thing. I thought, for one, we didn't do anything wrong, for two, they were fucking trying it on and three, we stood up for ourselves and that's it. Certainly wasn't going to roll over. That's how I'd want my rock-and-roll stars to act. Stick up for themselves, get in a bit of shenanigans and that's it.

Coyley I think Noel was a bit angry, but a bit excited at the same time. I don't think he quite knew what he should be feeling. In retrospect, I think it should have just been taken for exactly what it was, which is a great little rock-and-roll night out.

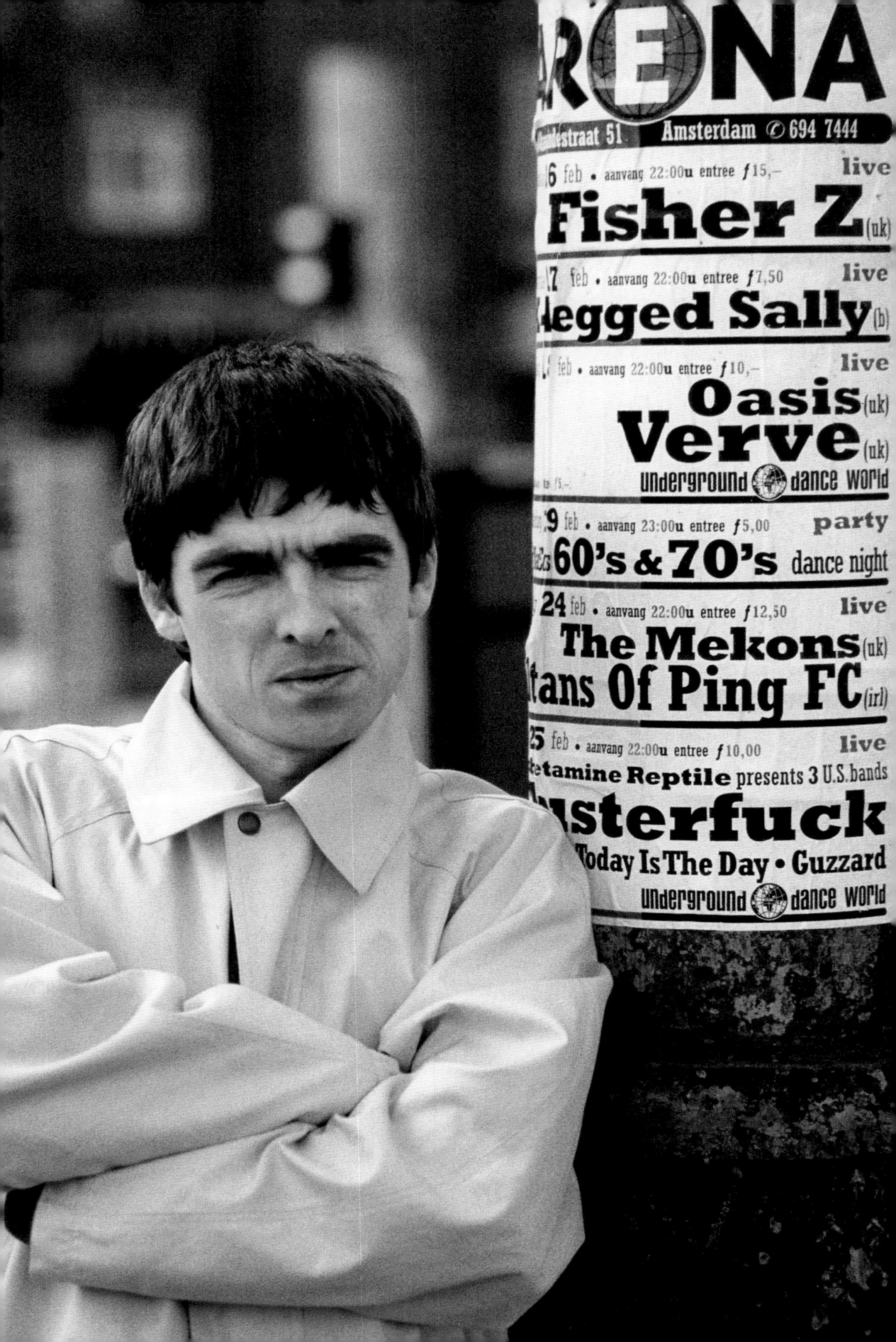
Amsterdam © 694 7444
feb • aanvang 22:00u entree ƒ15,– live
Fisher Z (uk)
feb • aanvang 22:00u entree ƒ7,50 live
Sally (b)
feb • aanvang 22:00u entree ƒ10,– live
Oasis (uk)
Verve (uk)
underground dance world
feb • aanvang 23:00u entree ƒ5,00 party
60's & 70's dance night
24 feb • aanvang 22:00u entree ƒ12,50 live
The Mekons (uk)
Of Ping FC (irl)
feb • aanvang 22:00u entree ƒ10,00 live
Reptile presents 3 U.S. bands
Today Is The Day • Guzzard
underground dance world

Noel Before 'Supersonic' came out we'd done gigs and there were people coming along because they'd read about us in the press, but nobody knew any of the songs. So we were supporting other bands, always playing to this void of an empty dance floor and people stood at the bar. Then on the day that 'Supersonic' came out – bang! The crowd are right there singing those lyrics that you have nonsensically written down at fucking 3 o'clock in the morning about a Rottweiler. We'd had two hard-core fucking years of being in that rehearsal room, rehearsing five nights a week, fucking going nowhere and yet all still believing in it. So by the time we got to be presented to the fucking public . . . they couldn't tell us anything about us we didn't know. People were saying you're great and we were like, of course we are, we're fucking brilliant, we've been great for the last two years.

idn't like . . . all the keness. I see ough all the llshit. I'm a ht cynical stard.

Liam I didn't like TV . . . all the fakeness. I see through all the bullshit. I'm a right cynical bastard. All the, 'Do this rehearsal, look into this camera, can you do it again, can you stop, can you start.' Telling the audience to move here, all this fussing about . . . It was a bit like, 'I don't want to see that side of it.' It was all a bit fucking poncey, 'Do you want to do make-up?'

Noel We never went into a make-up room for years, we were like, make-up? Fucking hell. Beneath us.

Liam When you are on the outside looking in you think it's great, but when you're there it's a bit disheartening. I'd rather just do a load of gigs than do these fucking shitty TV shows where it's all a bit smoke and mirrors. Just get us back to the fucking gigs, man. When we were playing live we were obviously not faking it, so everything just felt good. The minute you had to play the game I found it hard.

Noel *The Word* was quite a big thing. I'd bought this second-hand Super 8 camera because it looked great and I said to Liam, 'When we're doing the musical breaks, pretend you're filming the crowd.' When you see it, he's filming while he is singing and it looks fucking mental.

Liam I can't remember that. I think it just looked good and it was something to do. I don't think I had a tambourine at that time so I thought, have this . . .

Bonehead I remember Marcus saying, 'Look, guys, you are going to have to sign off the dole, man, and Guigs, you've got to leave your job.' Not that we were making money, I think we paid ourselves £100 a week into our bank accounts. But I couldn't get a bank account because I had a County Court judgement against my name, so I was like, 'I can't sign off.'
'Bonehead, sign off, you're in a band, you signed a record deal!'

Noel I remember being in the dole office and, in a glorious moment of fucking symmetry, as I was signing off, Phil Sax, who was the head of A&R at Factory Records, who told us we were too baggy, was signing on. Factory had just collapsed and I was like, 'Ah, how ironic.'

Liam The dole office had been saying, 'Look, you need to get a fucking job, have you looked over there?'
'There's nothing in there that I want to do.'
'Well, do you think I like doing my job?'
'I don't give a fuck about your job, if you don't like doing it, don't fucking do it, but I'm in a band and it's going well.'
Anyway, that time I went in and said, 'Look, I've got a new job and so I'm signing off.'
'What job have you got?'
I said, 'I'm in a fucking band, rock and roll,' and he started laughing. I said, 'Tune in on Thursday, you fucker. Oasis. Number thirty-one.'

Noel We were going to meet this tour manager, a twenty-year-old girl from Queens in New York, and we thought, 'We will fucking make mincemeat of this woman.' But the genius of it was, women were the only fucking people that we weren't insulting at that time, because we were all brought up by our mums. And Maggie was great.

Maggie Mouzakitis Marcus just rang me up out of the blue and said, 'I've got this band, quite small, they're from Manchester. They need somebody to look after them. They've kind of been doing it on their own, but they're getting into a bit of trouble. Would you be interested?' Bonehead was actually managing them on the road as much as he could. He was driving everyone around. He still drove us around, actually, for the first few months of the tour. We didn't have the money to have a driver and he was the only one of the band who had a licence.

Zoe Ball: You're being hyped
as the best band around at the moment.
Is that hype true?
LG: Yeah. The best band about today
on the planet. It's a fact.
THE O-ZONE, BBC1
FIRST TV INTERVIEW
1994

Gents

Noel We are Irish, me and Liam, pretty much. There is no English blood in us, and anybody who knows that will know there is drinking and then there is Irish drinking. Irish drinking can be endless.

Bonehead Yeah, we were known for partying and sitting up and drinking and doing whatever, but you know, we didn't, as a rule, get on stage pissed up, we just didn't.

There is no English blood in us, and anybody who knows that will know there is drinking and then there is Irish drinking.

Liam We'd turn up stone-cold sober and everyone thinks we are fucking off our tits. It was like, 'Excuse me, I've had fucking eight hours' sleep if you don't fucking mind.' There would be people in the crowd shouting, 'You fucking cunt!' just to get your attention. Then there would be people going, 'Do you want a line?' I'm like, 'I'm fucking busy here.' They would be going, 'Do you want me to chop you one out?' 'I'm fucking on stage singing, you fucking maniac.' I would see them physically doing drugs in the gig, getting bounced about, crazy bastards. The fans were a lot worse than us. We weren't fucking raging cokeheads and we weren't Mötley fucking Crüe. It wasn't like, 'Hey man, d'you want a fucking line?' 'Okay, I'm just having me breakfast.' It was never like that. We need to get that cleared up. We were never having a line and going in and doing the gear. When we were working it was always work, work, work. We definitely put the hours in and then we'd get on it, we rarely did anything when we were off our heads. We'd put the work in first and then go mental. At night time the lines would come out and I guess that's when it would all turn to shit. Going on stage pissed was shit, it was like hell, the gigs would be four hours long and if your voice weren't up to scratch it was like, 'This ain't happening again.' You'd learn pretty quick to get your shit together.

Noel There is certainly no drug that I've ever tried that has given me the same fucking 'whoosh' as walking on a stage and playing your songs to these people.

Liam When you did it sober, the gigs were much better and you'd get a pure fucking clean hit. Then you come off and then you'd get wankered.

Noel We were ironing out a lot of drugs on nights in and nights out, and a lot of drink. By today's standards it was insane, even by standards back then it was pretty fucking hard-core. But it was all functioning. Nobody was doing smack and not turning up for shit, let's put it that way. We might have been pissed all the time and McGee might have been pissed all the time, me and Liam and all the people that surrounded it, but shit was getting done. It wasn't tragic in any way.

Noel There was a lot of drugs taken making records. As a rule, me personally, I don't do them during the day anyway. Anything that we ever recorded while we were out of it, we would always end up re-doing it the next day anyway because it was shit. So there's a little tip for you kids. Only when it goes dark; it is a bit more normal when it goes dark, I think. That used to be my mantra anyway: wait till the lights go out, wait till the street lamps come on.

Liam When we started, we were smoking joints on stage but then when it went into arenas it was a bit like, 'These lot have paid quite a bit of money for this, better fucking keep it together.' Obviously there would be days when you'd fall off the rails and shit would happen, but it was to be expected. I thought we kept it together pretty much.

Jason Definitely Noel was more the quiet, silent type and he would sip his drinks whereas Liam and myself and Bonehead were lashing down pints and shots and whatever.

Liam Noel obviously wanted to go to bed and read fucking books on Morrissey, I wanted to go to the bar. That's just the way it is. It wasn't like I was going to the bar to get in the papers. It just makes more sense to me to go to the bar and have a drink, than to go to bed and read a book.

Noel I'm of the opinion that – for want of a better word – partying, and all that shit, that's what time off is for, holidays. When you get to push the button, it is like 'Right, we are getting the show back on the road,' I'm all about the graft then.

Liam He was just as bad as us, without a doubt. There is footage where Bonehead and Guigs and Tony are still playing and we've come off and we've gone back to the dressing room, launching apples and pears and sandwiches and pizzas on their heads. If anyone was taking it fucking seriously it was Bonehead, Guigs and Tony McCarroll. Our kid was right up for the shenanigans as well, but he seems to have forgotten about that.

Jason It just got more chaotic a lot faster than it ever did with most other people.

Leathered at the Riverside

Noel There was a full section of a tour, maybe a full tour, when it was just going off every night, and that's when we had to get security guards. I felt that people were turning up to cause trouble and it was getting me down, if I'm being honest. I remember Liam thinking it was great, but I was saying, 'Hang on a minute, as I fucking recall the rock-and-roll mythology, aren't there supposed to be loads of birds everywhere, never mind fat, fucking dicks in Fred Perry shirts trying to offer us out. Where's the blonde birds with the big tits? Where's that?'

'It's great, it's chaos.'

I am like, 'Fuck chaos, sex is what we need. Women. Fuck fat dudes.'

I felt that people were turning up to cause trouble and it was getting me down, if I'm being honest.

Liam We always had a bit of trouble that followed us around, but I didn't mind it. We had security guards, we're not soft, so fuck it.

Bonehead Anybody will tell you: your reputation follows you. We weren't a bunch of lunatics, you know. Maybe people came to the gigs, trying to get a reaction from us and some gigs did get pretty tense.

Maggie Some of those gigs were quite hectic. I remember me and Phil Smith had to go and hold up the PA stacks at one of them, because they were literally going to fall down on the audience. The audience was bouncing up and down so much that the floor was moving.

Liam People would chuck stuff and it was like, 'I ain't fucking about, we're out of here. I ain't walking on this stage with two eyes and walking off with one because fucking Dermott down the front from Hastings doesn't like our music. Fucking stone missed my eye by about that much, fuck you.' I ain't going blind for rock and roll. A lot of dickheads would stay on but it's not fucking panto.

Coyley Obviously, once you get that reputation, there is always someone who's going to go, 'I'll show him.' If he had never been in the paper having a ruck, that probably wouldn't have happened.

Jason I don't think there was a day went by without something. Right or wrong, you became immune to it. It was like, 'Never mind, everybody got all their fingers? Jolly good, let's get on with it, jump in the van, come on, let's go.'

Liam I should have got a fucking pay rise for them days. It's hard work taking shit off people.

Noel We went on tour with The Verve and this American band called Ace Tone or Acetone, we could never understand what the fuck they were called, but they were great. We were first on the bill playing the Riverside in Newcastle.

Jason We'd parked the van outside and luckily taken the flight cases out. Then I went back out and the van had been nicked. It had me new boots in and a bag of me clothes, I were right pissed off at that.

Maggie It was just like a bad luck show from the start.

Bonehead Newcastle is always a rowdy one. You do a gig up there and they are mad, they are just fucking psychos, man. Place is mad. But great, great gigs. The Riverside gig, I think it was going live on Radio 1.

Noel I don't know what happened but in the middle of 'Bring It On Down' somebody got on stage and fucking smacked me right in the eyeball . . . At that point we didn't have a light show, it was just strobe lights. You know what strobe lights are like, when they flick fast, you don't really know what the fucking hell is going on. Of course there is a fight and loads of people jump in and I hit somebody on the head with my guitar.

Jason Noel leathered this guy with it, proper 'Boing!'

Bonehead Liam's trying to hit this guy with the mic stand, and Noel's trying to hit him with a guitar like, 'Fuck off our stage.'

Noel We kind of did that thing where we decided to fan the flames of the fight instead of getting out of there.

Bonehead The crew were trying to protect the amps because people were throwing bottles of lager. It started getting a bit out of control. We could hear it got really hostile. There was lots of feet banging and hands clapping and a real big murmur in the crowd. The next thing was, 'Pull the plug, the gig's off. It's getting fucking scary out there, there's bottles flying, all sorts of shit. Just get in the minibus and fuck off back to the hotel.'

LG: If you're feeling a bit of a geezer...
Right, whoever threw the fucking bottle, let's have you up here and I'll slap you in front of the crowd.
I'm not here to get fucking things thrown at me. Fucking coconut stall.

Of course there is a fight, loads of people jump in, and I hit somebody on the head with my guitar.

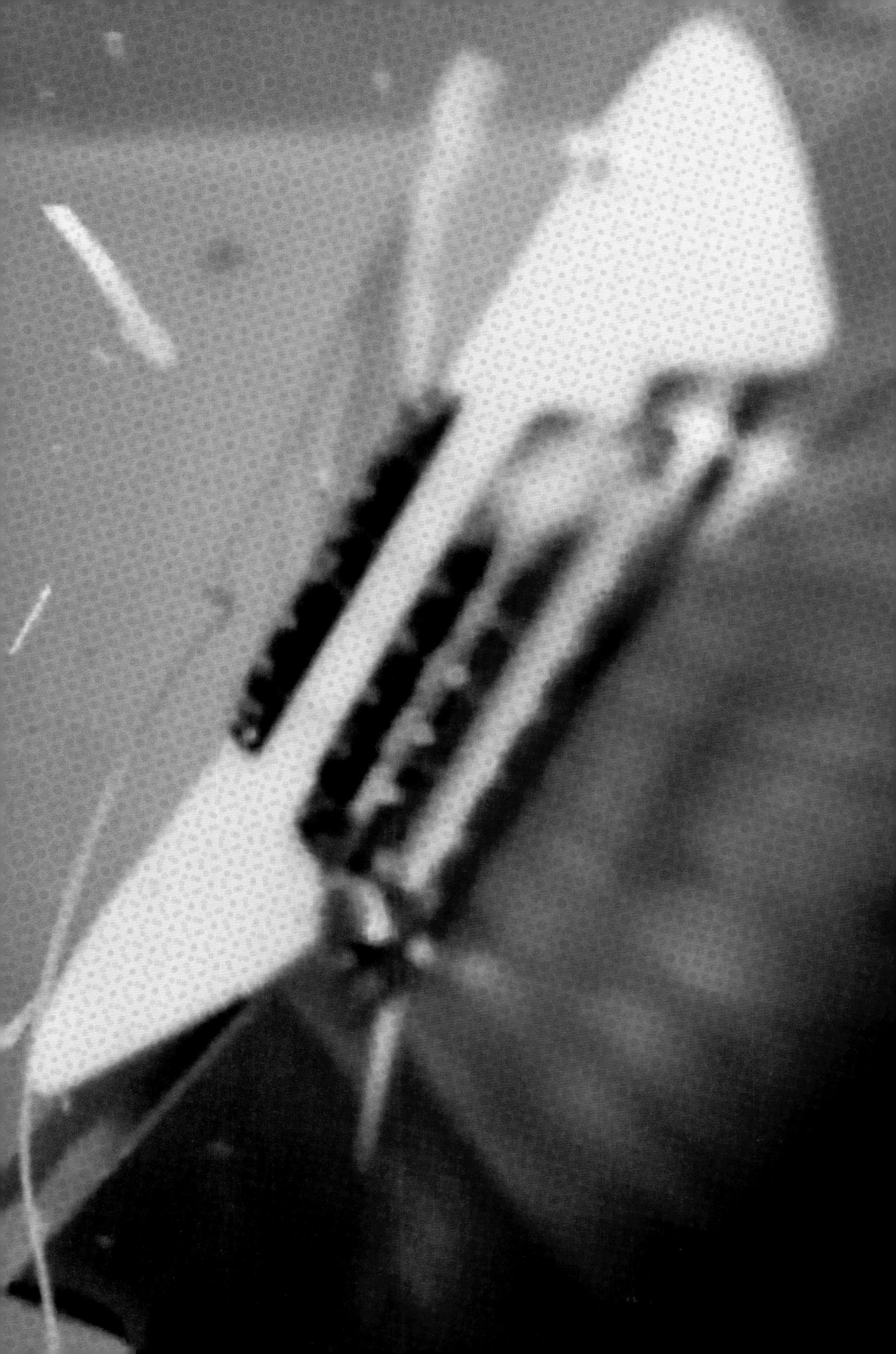

Coyley We get out and as we started getting in the van, there's a few different gangs of people, pointing, 'There they are, there's the fuckers.' We get in the van and we can't get out, there's a car right in front of us and a car right behind us, so we're kind of stuck.

Liam I remember it being fucking moody, they were rocking the fucking van, they mashed the van up to pieces.

Tony The only way the driver could get out was to put it in reverse and shunt the cars out the way, which he did. He didn't give a fuck. Got himself a space and turned and got us back to the hotel.

Noel To leave the venue we had to drive up an alleyway and past the front door. There are people throwing fucking bricks and shit at the van. All very exciting.

Coyley Why they'd want to do something like that, and why that guy got on stage and hit Noel, who would know, but I think the perception is that they are a bunch of thugs from a council estate in Manchester so let's go and fucking have a bit of a do with them, you know.

Noel I ended up with a massive black eye, and had to wear fucking shades for about six weeks, on stage. The other brilliant thing was the guitar had got smashed so Johnny Marr sent me another one, a black Les Paul which he played on *The Queen Is Dead* and I never gave it him back. He claims he borrowed it me, I'm sticking to it to this day that he fucking gave it me. You are not getting it back now, Johnny.

© Jill Furmanovsky

Johnny Marr Yeah, that guitar I bought in early 1986 from John Entwistle and it had been Pete Townshend's guitar, it was a 1960 Les Paul, which buzzed me up. It's on quite a lot of Smiths records that. We never discussed any kind of passing of the baton, I used to tell this story of how we met on a grassy knoll under a full moon, the two of us were there in our shades and I said, 'Here, Noel of Burnage, taketh this Les Paul, ex of The Who and The Smiths and lay down some licks,' and then we drank the blood of a groupie. That was my story. But there is a tradition in rock-and-roll music of giving someone a guitar when you respect them. I did it because I liked him really, I thought he was going to be great.

Noel It's a fucking great guitar and it's got a lot of history, but that just sums up Johnny as a dude really, he's a very fucking generous dude and he is into passing on the light sabres of rock and roll and that kind of thing. I can't believe I just said that.

Jason Johnny sent him a little note, something like: 'This guitar has got a bit more weight to it, you will get a better swing next time.'

I think any band worth their salt is not just about the music.

Liam I enjoyed that night to be fair. He jumps on stage, gives our kid a crack, he got a beating. I know he probably got a bit sore and that, but that geezer got a beating, man. I guess in hindsight it was a bit fucking scary, but I loved it. I'd rather a bit of chaos any fucking time than people just walking off and going, 'That was a fucking splendid gig. Eleven songs and they were all fantastic. All in time, sounded like the bloody record.' Fuck it, man, it's what it's about, innit? It's good looking back. Shitting it at the time though, thinking, 'Fuck, we are going to get fucking battered.' But like you say, you go down fighting, don't you. I like shit like that. No one really got fucking hurt. Our kid's had worse. Have a bit of fucking chaos, come on, it would be boring being great all the time, being perfect.

Noel We were getting the reputation as these bad boys of rock and roll, which was alright, but can we talk about the music first? Because that is what is going to stand the test of time, if you don't have the songs, you don't have anything. But it was the other thing was getting spoke about first.

Liam Well, it wasn't just about the music, was it? Our band was not just about the music and I think any band worth their salt is not just about the music. You need good tunes to accept that kind of behaviour, but if you haven't got that kind of behaviour and you've just got great tunes, then you are pretty boring as far as I'm concerned. It's about both.

The Richard Madeley moment

Noel There was a festival in Sweden and Motörhead, Primal Scream, Verve and Oasis all managed to book into the same hotel. Unbelievable. More unbelievable is that the barman shut the bar at something ridiculous like midnight. We went off the rails that night.

Maggie It started on the way in. It was the first time we'd had a bus. The buses have got these hatches on the upper level and Liam kept on opening up the hatch and coming out. The bus driver was getting really annoyed. He was like, 'Tell him to come in, he can't do that, it's really dangerous!' I was like, 'You fucking tell him that.'

Yeah they were good days. You're young; you've got no fear, have you?

Liam I remember getting on top of the bus a few times, fucking about. I can't see it being that dangerous. I remember jumping off a bus once and breaking my ankle and then hobbling on to do a gig. It was one of them double-deckers. I didn't think it was a double-decker, I thought it was a single one, it took me about half an hour to fucking hit the deck. Yeah, they were good days. You're young; you've got no fear, have you?

Jason Chaos again. It seemed to be just chaos all the time . . . chaos, chaos, chaos.

Liam The hotel was in the middle of nowhere, a little postcard village. They shouldn't have fucking put us there; put us in the city where things are open. They've gone, 'There is no more booze.'
'There is fucking loads behind that bar, fuck off, give it us.' And they didn't.

Noel We basically robbed the hotel bar, told the manager to fuck right off and took all the booze. When that ran out it was like 5 or 6 o'clock in the morning. There was a church across the road from the hotel, I don't know who come up with the idea but it was like, 'The church must have some wine in it.'

Bonehead We were in a church in the middle of Sweden with Primal Scream, fucking weird. Morally is this right? Should we be doing this? You've had a few beers, you're like, 'Fuck, it's rock and roll . . . but should I be doing this shit? You know what, I'm going to go back and say me prayers and ask God for forgiveness, I shouldn't have done that.'

Liam I was still at the bar having an argument and then they come back with some wine, they all had Russian hats on and shit, it was funny, man. We just smashed the gaff up, it was good. I remember some geezer screaming, 'I thought John Bonham was dead.' TVs coming out the window, but fuck it. We were getting the money so we paid the fine. Fuck it.

Noel The next time we're in Sweden, Liam gets nicked for shoplifting.

Liam My moment of madness. That was my Richard Madeley moment. I don't know what the fuck it was. I bought a pair of trainers and on the way out I just thought, 'I'm fucking having them razor blades.' I had them and I put them in my pocket, next minute I got collared. That was it; I don't know why the fuck I did it. You'd think I'd have grabbed more than fucking razor blades, bottle of whisky or something or even robbed the trainers that I'd just forked out £40 for. I think they just give me a fine. I was always getting fined for shit, that's where all my money's gone. Fined here, fined there, just fined. I think they just went, 'Look, what are you doing robbing razor blades, you're a millionaire?'
I just went, 'I don't know, really sorry.' Then they give me a fine and I got back on the bus. Got a bit of grief off the squares and then we carried on.

Noel Just fucking ridiculous. A rock star getting caught for shoplifting.

Jason Fucking razor blades of all things. He could barely grow a fucking beard, what are you nicking razor blades for? Silly fucker.

Noel Liam got arrested and the hotel got destroyed. I wasn't there, I was off cementing international relations with Swedish women, which is what you should be fucking doing. Never mind slinging video recorders around the fucking hotel foyer, that's bollocks. I could never understand hotel rooms being smashed up. That's like *hard work*, you get a sweat on, do you know what I mean? I remember the hotel manager coming into my room and telling me, 'Get out of bed, you're leaving, you're being thrown out.' As he was in the room, fucking screaming and shouting, he was stood against the window. I couldn't understand what was going on until I seen a fucking telephone and a bedside cabinet come flying out of the room above; I was thinking, 'Oh fuck.' I walked past Liam's room and the door was open, I've never seen such destruction. He fucking demolished this room. The Swedish press referred to us as animals.

Jason I think the headline was 'The death of a hotel room' or something like that, and it just showed this mess of a hotel room.

I could never understand hotel rooms being smashed up. That's like *hard work*. You get a sweat on do you know what I mean?

Maggie They're showing this room that's completely trashed, and we're looking at it . . . I was like, that's not even the hotel room. That's not the hotel room! That's another hotel!

Liam Back then you are just thinking, 'Fuck it.' You just get pissed and you look at something and go, 'Bit fucking boring in here, isn't it?' Before you know it you are out on your fucking arse on the street. Lovely. But I guess that's what you do when you're daft and you're young and you feel invincible.

Before you know it you are out on your fucking arse on the street. Lovely.

Maggie It did cause us problems with other hotels down the line though.

Liam We were getting kicked out of shit hotels; we were thinking if we get kicked out of this shit one they might put us in a decent one, for a fucking change.

Jason Bonehead was very minimalist, he used to like taking a little bit of furniture out of his hotel room and making it a little more spacious. There was one time, I think it was in Newport, I knocked on the door and they opened it. I went in and had a look and there is nothing in the bedroom, there is just Bonehead and Tony sat there on the floor. I looked and thought, 'What's wrong with this?' You don't realise there is no furniture at first and then I see the curtains blowing a bit. 'Alright, lads, how you doing, bit of a night, was it?' I can remember leaning out and looking out the window and seeing a bus driving round the debris on the road. Okay, so you've had a good evening. 'You better get the hell out of here quickish.'

Noel We were getting barred from everywhere at one point. We were barred from a full chain of hotels all over England. We were barred from the Columbia in London and we couldn't afford any other hotels.

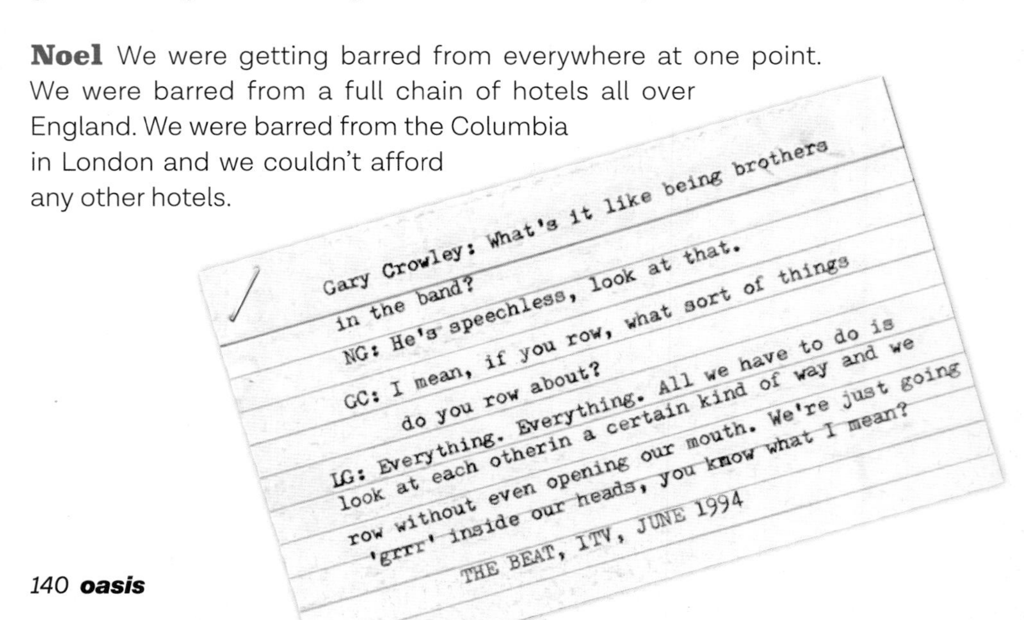
Gary Crowley: What's it like being brothers in the band?
NG: He's speechless, look at that.
GC: I mean, if you row, what sort of things do you row about?
LG: Everything. Everything. All we have to do is look at each otherin a certain kind of way and we row without even opening our mouth. We're just going 'grrr' inside our heads, you know what I mean?
THE BEAT, ITV, JUNE 1994

FOSTER'S
FOSTER'S
DRAUGHT

Maggie I think that might have been orchestrated purposely, that one. Nobody wanted to stay at the Columbia Hotel any more; I think we all kind of had a little hand in that one. That was probably one of the best things they did, actually, banishing us . . .

Noel I wasn't there because I was living in London by this point, but everyone else was staying in hotels in London. We recorded 'Whatever' in Maison Rouge in Fulham or Chelsea. I remember the band turned up, bags and everything. Been barred for life from the Columbia, police were called and all sorts. Somebody fucking smashed up the hotel manager's car.

Liam Even to this day I am fucking banned. Great hotel, man, great hotel. You would get there and there would be about 900 beds in your room. There was never one bed. 'Have I got guests or something?' It was like being in the nick. I can't remember what happened but I guess there were glasses being thrown and shit being thrown out windows and the usual shenanigans.

Jason I think it was Bonehead just doing a little interior decorating again. He was really good at it actually.

I remember Guigs trying to tip a vending machine into the swimming pool. Why? I don't know.

Liam We were never rude to anyone, that is the thing, we were never like, 'Oi, get us a fucking drink!' We just got a bit carried away, having a laugh. Going through the corridor, going for a piss, and you see a nice vase – or a shit one in the Columbia – and you go, 'That would look lovely on Bonehead's head.'

Noel We were fucked at one point, no one would take us anywhere. Trouble seemed to find us at any given point. A lot of it was exaggerated by the press because they were desperate for something. We'd done a gig in Portsmouth and got back to a hotel which had a swimming pool in the bar, or the bar was in the swimming pool, whatever. East 17, who were the One Direction of the day, were also staying there, so there is a lot of screaming schoolgirls. I remember Guigs trying to tip a vending machine into the swimming pool. Why? I don't know. We used to regularly clean bars out in the hotels we were staying in, but tipping stuff into swimming pools, that's fucking slightly mad, now when I think of it. I remember East 17's lot being horrified. I think they checked out. It was a right palaver anyway, a bit of damage, a bit of thievery, a lot of drugs. Yes, Liam copping off with loads of birds, all that kind of shit.

Bonehead Chaos! It followed us, couldn't stop it. Some nights you used to have to sneak out, 'Don't tell Noel.' We'd go out, but something would go off, paparazzi would catch us and we'll have it in rehearsals tomorrow. He will batter us.

Noel At that point it was always going off at gigs and being banned from this and being fucking banned from that, police being involved here and fucking gigs getting cancelled. It was kind of becoming a bit of a pain in the arse, to be honest.

Liam People were still buying the records. You don't buy a scene, do you, you don't buy a scrap, when we put the singles out they were still buying them and going home and getting turned on by them. So the music wasn't getting lost.

Noel The press were loving it. You've got a load of journalists who were vaguely the same age as me, maybe ten years older, they'd never seen this, they'd only read about this before, but it is happening right in front of them.

Liam My main thing was singing and being cool as fuck. I knew my place: my place was singing the songs and I loved that. Second to the singing, was being the ultimate rock-and-roll star. While you are fucking farting about trying to get a guitar sound, I'm going to go to the pub and cause some fucking chaos, or I'd go and have a drink and things would follow me. While they were all trying to be Jimi Hendrix, better guitar players and all that, which is great, I was going, 'That's what you are doing, I'm going to be the face, the geezer who wears the clothes, and I'm going to be the ultimate number-one rock star in this country.' And that is what I was, by doing fuck all and just being cool. I know for a fact when all this ends, at least I'd done what we said on the tin. Bless these bands today, they haven't lived. It ain't big and it ain't brave but it fucking feels good.

Definitely Maybe

Noel The first album is really the purist representation of what we were then, because it was conceived live in a rehearsal room and it was played live in a studio. We'd been playing that for two years before we recorded it. It is perfect that album. It is fucking ten out of ten, perfect. Thinking about it now, I don't think I would change a single note or a single inflection of any lyric or anything.

Noel The cover was fucking haphazardly thrown together on the day. All of Brian Cannon's covers are, then Brian would invent a concept four weeks later and say it's about this or that. It was great to have Brian on board and Brian was one of us, he was the same age as us, wore the same clothes, was into football and all that kind of thing. Brian was one of those guys who came from the Peter Saville school of artwork, which is the artwork is more important than the record. 'The record just sits inside my fucking sleeve.' For a while that was mildly amusing.

Liam Brian Cannon was doing The Verve stuff. We met him and he was a little bit left-field or whatever it is. He just went, 'Let's try it like this.' We never questioned it, we just went, 'Why fucking not?'

Noel I remember the day being long and kind of, 'You lie there; no, you sit there; no, you do that and I'll do that.' We were going to shoot it in somebody's front room, I don't know why. Why is it in Bonehead's front room? I have no idea. I don't know what the concept is of that, 'At home with Bonehead'. All the props came from Phil Smith's bedroom. Bonehead's house, clearly, didn't look exciting enough, didn't have many cool things in it, bar a spider. There is a picture of George Best and of Rodney Marsh, which depict City and United; there's a video still from *The Good, the Bad and the Ugly* – may I point out the worst fucking still, it's not even fucking Clint Eastwood, you would have to watch that film a million times to go, 'Oh yeah, it's that bit' – the globe and the Burt Bacharach picture. I remember saying, 'Get the Burt Bacharach picture,' because the Burt Bacharach influence had already started to take shape in the songwriting. I've always admired his songwriting and particularly that song 'This Guy's in Love With You', which for years – as I'm sat in my flat in Whitworth Street in Manchester, fucking stoned – thought was called 'The Sky's in Love With You'. I thought that was so cosmic; wow, the sky's in love with you, fucking right it is. What? Oh, it's a fella. Okay, still a good tune, but it's a love song. Not psychedelic.

It is perfect that album. It is fucking ten out of ten, perfect.

Liam Obviously he's great and all that, but when they all started listening it was like, 'Fuck's going on here?' But as you dip your toes more into that kind of music you go, 'Ah right, he wrote that, fucking hell, cool, yeah, very smooth.'

Noel The cover is iconic because the album is iconic. If *Definitely Maybe* was the cover of some shit album, it doesn't mean anything. It's given credence because of the music really.

Liam We looked at it and went, 'That will fucking do, can we go to the pub now?' It was just all fucking very odd, but good odd. We always liked the covers.

Noel I can only speak for myself, I felt I knew it would get to number one. I knew all this shit was going to happen. It was just destiny. I don't remember being immensely proud, I remember thinking, 'Right, okay, there is work to be done then.'

Coyley The reaction from our entourage was really quite muted. There wasn't 'Hurrah!' A cheer didn't go up because, well, why shouldn't they be number one.

Liam Job done, we made it, people are into us and we did a great record, we've won the fucking war. We've banished all these cunts out of the way by being us, regardless of all the smashing hotel rooms; that's obviously what people want.

Maggie Those songs, for Noel particularly, had been in his repertoire for a long time, he finally got them on an album and people loved them. That is a huge achievement for five guys back in Manchester who were on the cusp of not making it. You know, something in that period, in 1994, made it possible for them to break through.

Noel It is a funny thing, people are kind of embarrassed about fame and fortune. I've always been of the opinion if you earn it fuck 'em. None of us got given it, we didn't win the pools, we didn't have a trust fund, it was fucking ours. Let's spend it and make some more.

Liam Without a doubt, nothing to be ashamed about, we wanted it. A lot of bands around at that time say they didn't want it – they were just fucking scared, to be honest with you. We were just fucking having it, we were having the lot. We were having fun with it and keeping it real as far as I am concerned.

Johnny Marr *Definitely Maybe* sounded like it had always been there. It's interesting when music or a film does that, just makes sense straight away. So, it has a kind of familiarity about it, but a kind of necessity too. It was just of the moment, without it necessarily relating to any of the bands that were around. Noel's lyrics definitely had a sense of optimism in there, forward thinking and forward motion. However he's expressing himself, he's certainly expressing something that is aspirational. Not necessarily economically aspirational, or socially aspirational, just personally aspirational, you know. Get me out of this feeling, or, I'm going to feel better any minute, but this ain't too bad either.

Without a doubt, nothing to be ashamed about, we wanted it.

Noel *Definitely Maybe* came out and then we didn't see Alan for a couple of years.

Liam One minute he was there, then he weren't there. Obviously it was to do with fucking drugs or something. I don't know, he just wasn't there and it was like, 'Where the fucking hell is McGee? Come on, join the party, what's going down here?' Then he just stopped hanging about and coming out. That's what happens when you play with fire. It is what it is. People, if they are not well they've got to go and get their head sorted. The show must go on, as they say.

Noel I was quite shocked because I was like, 'What? Really? Fucking hell, I didn't know he was that bad.' Goes to show you how much of a fucking haze we were all in. I didn't truly understand what was going on with him. 'Well, fucking hell, so you've had a line, so what?' I've never had a nervous breakdown and I've never been to see a psychiatrist so I don't know, I just couldn't understand it.

Liam I didn't go and see him in hospital. The last thing anyone needs to see when they are in hospital is my fucking head marching through the door.

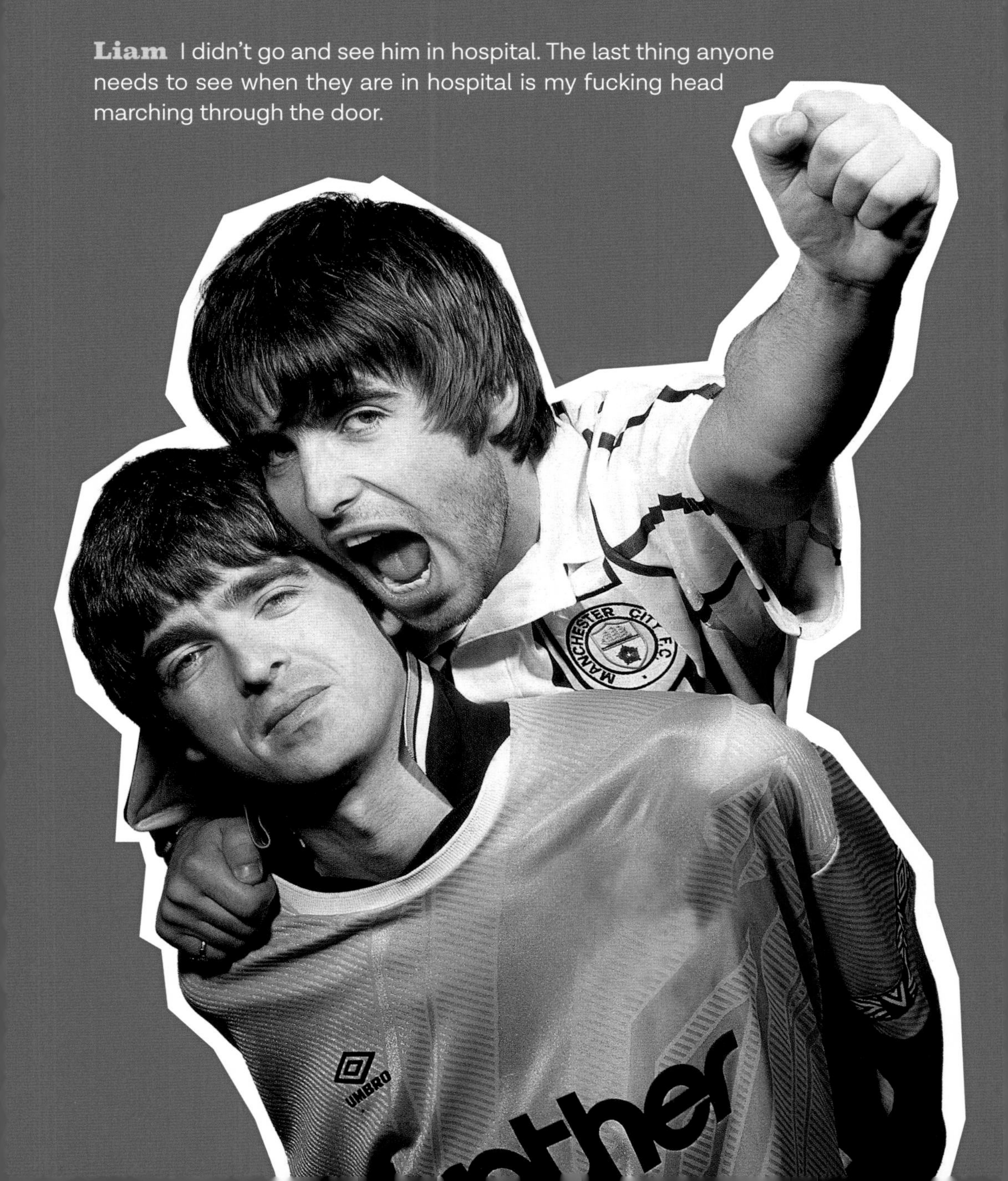

oasis
MEGASTORES
KODAK 5053 TMY
15
KODAK 5

Gunchester

Noel At the start of the nineties Madchester turns into Gunchester. If there's anybody making a few quid there will be someone turning up with a fucking shotgun trying to take it off you. Luckily we don't get involved in all that hairy fucking shit, you know what I mean, but Manchester gigs will always be stressful events.

Liam I remember playing the Haçienda, that was good. I went out for a cig or a beer down the road, come back and one of the bouncers is going, 'What are you doing?'
I was like, 'I'm going in there to do the fucking soundcheck,' and they were getting all fucking lairy like, 'You're not allowed in.'
I was going, 'Come on, I'm in the fucking band.'
That's what it was like back then. But I got in and done the gig.

Noel At the Haçienda, one of our mates was doing the merchandise. It was a fairly small-time operation at that point. He comes back at the end of the night with the take and it's all £50 notes. Surprise, surprise, they are all fake £50 notes. 'I didn't know they were all fake.' What planet are you living on – people are coming to a fucking gig paying with £50? Okay, take one maybe, but not eleven.

Maggie Homecoming gigs are really hard for any band, I think. The crazies came out, the ones that they'd never heard from for years. You know, 'Hey, remember me? We went to school . . .'

Noel It's great when you get on stage, playing in front of your home town and all that, but it can be quite stressful because of the guest list and all that. You've got your family and your connected family and all their mates and everybody trying to get in. It's just a fucking hassle for a week before, trying to get everyone in and then a week after for all the people that you couldn't get in and it's just mither. A fucking pain in the arse playing up there sometimes.

A fucking pain in the arse playing up there sometimes.

Liam I found it hard playing in Manchester. It's easier playing away. Playing at home, I found it a bit uncomfortable. The crowd would be full of people you knew. Your mates that you went to school with jumping up and down at the front, waving at you. Too fucking weird, just weirded me out.

Peggie I thought, 'These young ones, Christ, they're mad.' They were just in there jumping. I thought, 'What the hell are they jumping up and down for?' I couldn't get over it the first time I went.

Liam I didn't like me mam coming to gigs, it done me fucking head in, to be fair, even right up to the big gigs. I would be, 'Don't come.' I'd be stressed out with all the fucking commotion, worried about me mam. Everyone's going fucking nuts and I'd be going, 'Is she alright up there?' Not a good look, is it, when you are trying to rock the joint. I was always on edge; I could never relax with me mam at gigs. Plus, I wanted to get off my fucking box.

Peggie I used to get mesmerised with Liam. I used to look at Liam and think, 'He's got some face, he stands there and just stares the crowd out.'

Fuss in Japan

Bonehead Next thing you know, we were going to Japan. What do you expect? I've never been to Japan. Noel and Coyley had been so they were the old hands. They were like, 'It's mental, Japan, it's really fast and busy and lights and this and that.' So you think, 'Alright, I'm prepared, they told me.' No, nothing ever on this planet was going to prepare me for what we got. It was the other side of the world; I might as well have been doing a gig on Mars.

Noel It was a wild tour, it was fucking great. You felt like a superstar. You couldn't go anywhere.

Jason That was when it actually hit home to them how big they were actually getting. It was very Beatlemania-esque.

You felt like a superstar. You couldn't go anywhere.

Noel We fly into Japan; Guigsy meets a girl on the plane that he ends up marrying, still married to her to this day. That's how weird he is. We were a successful band in England, we were the big noise, but when we got to Tokyo we were met at the airport by fucking thousands of screaming fans. It was like, 'Whoa, fucking hell!' They followed us everywhere, outside hotels, in the fucking foyer, outside your room, in your fucking room, everywhere you went.

Liam I loved going to Japan, it was fucking great, man. They give you presents and stuff and chased you down the road. They speak in a different language, which is perfect when you don't want to really communicate.

Bonehead It was like, 'Oh, Bonehead, Bonehead! Liam! Noel! Guigs! Tony!' and it was just like, fuck off, this is Japan, but they know me name! Yeah, of course they do. How did that happen? How? Tell me, I don't know.

Liam Just hearing that word 'Bonehead' screamed by Japanese people, it's just mental, innit? Bonehead! I loved that.

Bonehead We got to the hotel, go upstairs and every door on the floor that we were on opened and all these heads come out: 'Oasis!' They had pre-booked the fucking hotel. It got me and I loved it, I loved every last minute of it. You'd go out and have a look around the shops and within seconds you were being followed by a lot of fans, and I mean a lot. You turn

I.W.HARPER

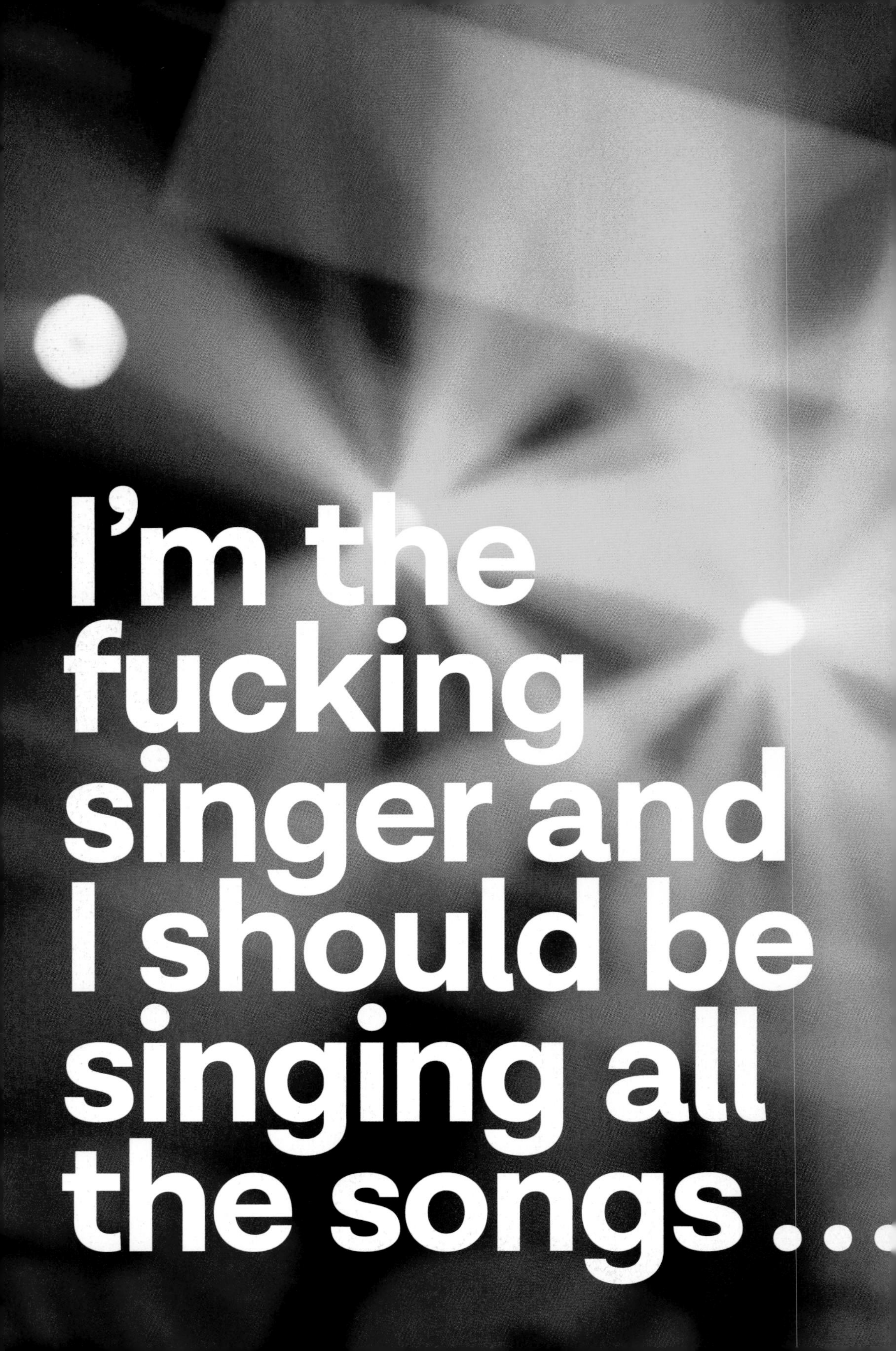

I'm the fucking singer and I should be singing all the songs . . .

滞在日程表

アーティスト名　OASIS　　　予定滞在日程　1994年　9月12日～9月20日

日程	出演先	出演内容	スケジュール	宿泊先
第1日目 9月12日 （月）			来日	六本木プリンスホテル 港区六本木3-2-7 03-3587-1111
第2日目 9月13日 （火）			オフ	同上
第3日目 9月14日 （水）	渋谷クラブクアトロ 渋谷区宇田川町32-13 03-3477-8750	コンサート 前売￥5,500 （550p）	開場18:00／開演19:00 主催：(株)スマッシュ 企画制作：〃／パルコ 後援：エピックソニーレコード	同上
第4日目 9月15日 （木）				同上
第5日目 9月16日 （金）				同上
第6日目 9月17日 （土）			[illegible]	大阪グランドホテル 大阪市北区中之島2-3-18 06-202-1212
第7日目 9月18日 （日）	心斎橋クラブクアトロ 大阪市中央区心斎橋1-9-1 06-281-8181	コンサート 前売￥5,500 （700p）	開場18:00／開演19:00 主催：(株)スマッシュ 企画制作：スマッシュ／パルコ 後援：エピックソニーレコード	同上
第8日目 9月19日 （月）	名古屋クラブクアトロ 名古屋市中区栄3-31-10 052-264-8211	コンサート 前売￥5,500 （550p）	開場18:00／開演19:00 主催：(株)スマッシュ 企画制作：スマッシュ／パルコ 後援：エピックソニーレコード	名古屋クレストンホテル 名古屋市中区栄3-29-1 052-264-8000
第9日目 9月20日 （火）			離日	
第10日目 月　日 （　）				

1994年8月9日

招聘社　株式会社スマッシュ
所在地　東京都港区南麻布3-17-…
TEL　(03) 344…
代表取締役　日　高…

JAPAN TOUR
Sept 1994

Liam I don't think there was any need for that acoustic bit. I like a couple of acoustic songs and that, but going out on the road it was like, 'Come on, let's get the guitars out and fucking have it.' When you've got people stood in front of us, it's a bit of a war, let's fucking blow their heads off. You ain't blowing no one's head off with an acoustic guitar I don't think.

Noel We developed an acoustic bit in the middle of the set in Japan to save Liam's voice. He is resting his voice whilst standing on the side of the stage heckling, smoking and drinking, which is a wonderful remedy for any vocalists out there. Shouting.

Liam There were two reasons for it. I'd say resting the voice would be one. Two, because he fucking wanted to sing songs. I put that one up at the top. He wanted a bit of the fucking action, testing the water for his solo career twenty years later. Maybe my voice was getting a bit tired, I always found it hard singing out in Japan, it was harder, I always had to struggle out there.

Johnny Hopkins There were no drugs on that tour, so the band were completely drug free. That sort of gave it an extra dimension, you know. There was a sense of purity about it. I was going to use the word innocence, but there definitely wasn't any innocence there. It was madness, but at the same time, it was really, really calm.

I like being in a band, I like that gang mentality. I like hanging out with the lads.

Noel The usual: we got barred from the Roppongi Prince Hotel. We are still barred, forever. I was upstairs entertaining – socialising is what I like to say. But I did write 'The Masterplan' that night – I found the lyrics on Roppongi Prince headed notepaper. So while everyone was having a good time I was fucking keeping shop. Like a square.

Liam I like being in a band, I like that gang mentality. I like hanging out with the lads. I'm not one of them that would do a gig and then go, 'I'm off to me room'. I definitely sit down in the bar and have a crack, have a laugh and probably stay up a little bit too late. It's the way it is. Rock and roll to me should have no rules. At some point the rule book comes out and then you are always going to fight against the rules. It is what it is.

I used to come off stage and be floored for ages, it was like a fucking exorcism or whatever the word is.

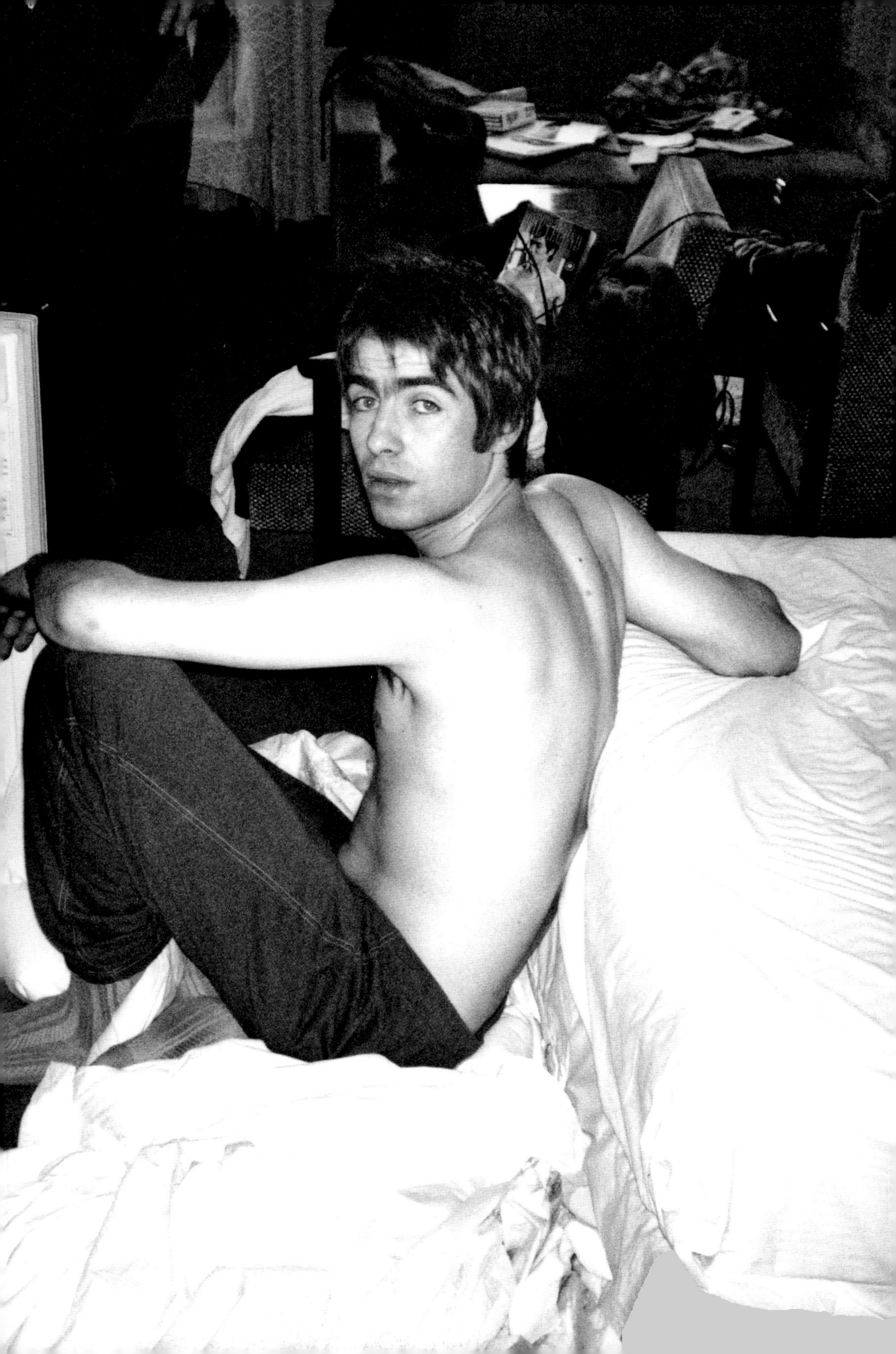

Daniela Soave I was writing for a 'nancy boy's' magazine as far as they were concerned. I knew that I was ten years older than them, I was female, I was a mum and it was all blokes – so I just thought, 'Right, I'm going to stay up later than you, I'm going to drink more than you, I'm going to behave more badly than you,' because I wasn't going to get this story otherwise. The first night in Japan, we'd all been out; it would have been about 3 o'clock in the morning, my head had just hit the pillow, and the phone rang. 'Who's calling me?' So I picked up the phone and it was Liam. 'Get down here now!' So I walked down the corridor back to his room and, honestly, it was like he was shimmering.
He said to me, 'Do you believe in God? Do you think there's a God?'
I looked at him and said, 'How did we get on to this one?'
He said, 'Because. I don't believe there's a God because all that wouldn't have happened.'
And I said, 'What wouldn't have happened?' I didn't know what he was on about. There was all this pent-up, simmering rage.

Liam I think I was getting a bit heavy and a bit deep round them times because all that stuff was happening. There was change going on with the band – this fame and all that stuff. My thing was about me mam going to church; she was going to church every fucking Sunday and then when she got divorced from me dad, she wasn't allowed to take the body of Christ. And I was going, 'That can fuck right off, what's that all about?' She's got to sit in the back and I think that's a bit shit, so I might have been having a bit of a rant about that. You think to yourself, know what, maybe there isn't a fucking God and it's just life and it's just the way it is. I was definitely angry with life, I guess, the shit with my dad and all the other crap, just life in general fucking just pissed me off. I was an angry young man. Singing those songs sort of released it. Released all my shit onto other people but in a good way. It was like everything, all the shit that I had during the day or last week or a month before, just got released, turning it into a good thing. All that fucking angst and all that energy would go through the mixer and come out in a better way. Hopefully people would like it and jump around and get into it. I used to come off stage and be floored for ages, it was like a fucking exorcism or whatever the word is. It was big and powerful and it was great.

I was an angry young man. Singing those songs sort of released it.

Grim up north and all that

Noel My old fella was violent towards me and me mam, that's a fact, but if I could sum him up I would just say he was a shit dad. That's pretty much what it boils down to. A shit dad.

Peggie I have always said that he was jealous of them because they were lads and they were starting to grow up. He thought they should go to bed at 8 o'clock at night; they would be working and coming in with their wages and he'd be, 'Get up to bed.' He was terrible to Paul, he really was. Paul worked with him longer than Noel. Noel got wise to it, Noel couldn't be doing with him. Paul would work from 7 o'clock, because he always worked on the building sites. It could be 11 o'clock before they'd come home. Then Paul would go for his wages on the Friday and he'd say, 'I haven't got them. I haven't got them ready yet.' He'd make Paul beg for his money. Noel wouldn't be bothered, Noel would think, 'I don't want it anyways, I'll get it off me mam.' I would rather give it to them than see them beg. Noel was the one that got it the most because he always thought Noel was out up to no good. I said 'He is only down the road with the rest of the lads his age.' Fifteen, sixteen they were standing on corners. Tommy would always come back and say 'The police are after that lot down the road there.' I said, 'Don't be so stupid.' Then I'd say to Noel, 'What were you doing down Burnage Lane, Noel?' He'd say, 'Doing nothing, I was in somebody's house.' I really think there was something wrong with him. I don't know what it was, whether he had a guilty conscience himself and he was taking it out on them.

Liam He used to knock her about. He never touched me, he used to knock Noel and Paul about a bit, but I never got it. People go, 'You never got a crack,' but you'd seen it. Sometimes you'd want the crack instead of having to witness it. If you're getting booted across the room it has an effect on you. Whether you're getting a kicking or not, you're still watching it. Seeing it and feeling it, they're both shit. They'd be rucking in the morning about something and you'd go to school thinking, 'Fucking hell, this is going to be shit.' You get through school and come home and go, 'I hope the cunt's gone out,' and then you come round the corner and his car would be there and you'd be thinking, 'Fucking hell, here we go.' You get in and then they'd be rucking. So then you'd be going, 'Fucking hell, I hope he goes out with his bird,' or whatever he was doing, because when he'd go out it would be chilled. But then he'd come back a couple of days later, not good. It was shit. It was shit for my mam and it was shit for us and there you go. It was shit, but I guess two fucking doors down it was happening to them as well.

I wanted to
be rich. I was
driven because
I was working
class and poor.

Peggie I remember Noel once saying, 'If you don't get out of here, Mam, and leave him I'm going to kill him.' I thought, 'Oh Jesus, you can't be doing time for the likes of him.'

Liam I remember me mam packing all the gear up and then leaving. My dad had hurt his back at work I think and he was on a mattress downstairs – she left him the mattress. I think my mam's brothers come over, stacked the van up and then off we went to this new house, which was alright. I mean, we had no money, but it was just good me mam not getting a kicking.

My missus thinks my old fella beat the talent into me.

Peggie When Liam was eleven, nearly twelve, I thought, 'This is it; I can't be doing with this any more.'

Paul We only went down the road a couple of miles, he never once came after us and knocked on the door. Wouldn't dare.

Noel To get a new council house when you've already got one, in the eighties, is virtually impossible . . . I remember going with her on endless trips down to the council offices. She finally managed to get one and then she had to do it.

Peggie I left him a knife and a fork and a spoon, and I think I left him too much. Of course it affected them, it definitely affected them. They got very bitter. They never really talk about him.

Liam I just wanted my mam to be alright, that was basically it. I didn't give a shit about me, it was like, 'Look, if that fucker starts again, he's going to get it.' If he continually fucking starts, the older we get, the more he's going to get a proper fucking beating or whatever. I wasn't bothered about us, it was just making sure my mam was happy and safe. Once she's happy, I don't give

a shit about him. It was mainly about getting her away. She thought it was best to get us out of that shit scene. We wanted her to get out of it.

Noel Life didn't change that much apart from fucking idiot Dad wasn't fucking getting on your case all the time about shit. The main thing I liked about at the new house was there was no shouting or slamming of doors. If I am being honest, I don't know many of my mates' dads at that time who were any better. To me everybody's dad was like that. Dads were just twats then because of unemployment and them being men in the seventies, do you know what I mean? Housewives were housewives and kids were kids and the dog got a kick in the arse. Your dad hit your mam, your mam hit the eldest, he hit the middle and eventually the dog got a kick in the bollocks. We never had a dog though, so the goldfish used to get a flick round the gills. Not that we had a goldfish either, we used to do it to next door's goldfish, the fucking orange bastard. It never made it into my songs, any of it, at any point ever. I don't really look at the songs that I wrote then as a reaction to a childhood that really has been overplayed a lot. Why would I want to write about that? I wouldn't want to be playing Knebworth singing a load of songs about domestic violence, trust me. I don't pick the scabs of my childhood at all. We had a shitty upbringing, but it was normal to everybody else of our age, it wasn't like we were locked in a cupboard for thirteen hours a day. It was hard work being working class then and we dug ourselves out of it because we had talent and we chased it and that was it. What drove me on was not the fact that we come from a dysfunctional family, what drove me on was I wanted to be rich. I was driven because I was working class and poor. I wasn't arsed about the fame, I didn't want to go through life being fucking poor. Joining Liam's band was when the light bulb finally came on. It was like, fucking right, this is it. You can't let that kind of thing affect you because then you're carrying that weight all the way through life. I've relieved myself of all that weight a long, long time ago. I know other people's perceptions of it must be vastly different to mine . . . My missus thinks my old fella beat the talent into me. There, what a guy. If I wanted to take revenge on my dad I'd put a fucking baseball bat on his head. I am not that poetic to have done it through the art of song.

Going Val Doonican

Liam He'd always wanted to be the singer, without a doubt.

Noel I always used to think that it's really difficult to be a singer, but now I am a singer, I know that if you are not in the mood, you know what you do? You get in the fucking mood. If you are not going to be in the mood at ten to nine, then fucking get in the mood, sharpish, because there are people out there and they've paid to come and see you. They don't particularly care whether you are fucking brilliant; have the decency to walk on stage and give a little bit of a shit.

Liam I never, ever squirmed out of any fucking gig, man. If anything I done gigs when I shouldn't have done gigs, which fucked up my voice along the way. Yeah, drinking, smoking, staying up all night, I can own that shit, but also doing gigs when you shouldn't be fucking doing gigs, when you are fucking gigged out, when we should have had a bit of time off basically.

Noel All the acts on Creation and some of the acts that inspired them were going to do like an answer to MTV's *Unplugged: Creation Undrugged.* We were doing two or three acoustic songs. Liam developed a phobia of the acoustic guitar.

Liam I'm not into all that acoustic nonsense, fuck off, mate. That was when it all started getting a bit silly, doing acoustic bits. It's like, come on, mate; we're the Pistols here, now we've gone all fucking Val Doonican. No, not for me.

Noel Liam did that thing that Liam always does: there is an offer to do this thing, shall we do it? Yes, fucking great, it will be fucking brilliant. Get to the soundcheck, everything's great. Five minutes before you are due on stage: 'I've got a sore throat, I can't go on.' Which basically in Liam Speak is, 'I'm fucking shitting it so I'm not going on.'

Liam The only time he'd ever do any of them kind of songs would be if my voice was fucked then he'd take over. I'm sure he was slipping shit in my drink so he could do it. I'm sure that fucker was spiking my drinks with fucking gravel or something or some shit.

Noel Singing is hard on the brain, it is hard to get up there and do it. It is even harder if you are staying up till 7 o'clock in the morning drinking and smoking. That's the bottom line, because I fucking said so.

Liam You can't ask a twenty-one-year-old or nineteen-year-old lad to be professional. Can you be more professional? Fuck off. I don't even understand that word, what you on about? It's like trying to take your rawness away.

Noel You can't have it both ways. If you want to do that, great, but something is going to suffer. You can't be on tour and have a great social life. Socialising is for your spare time, it's as simple as that. If a kid is gracious enough to queue up to buy a ticket, you are duty bound to fucking show up. I don't give a fuck if you've got a sore throat, or a sore cat, or a fucking dodgy knee. If they've paid, you play, and if you don't, get a fucking doctor's note.

If anything I done gigs when I shouldn't have done gigs . . .

Liam He was wanting to sing more songs for whatever reason, but you just know where that's going to end, don't you? He'll want to do another one and another one and another one; he was basically after my job. He is within his rights, he writes the songs, but get the fuck out of the band if that's the case. I'm the fucking singer and I should be singing all the songs so if you want to go and do that, then go fucking solo, mate.

Noel Of course, it never dawned on me to be a singer until I actually had to be one. Until he started walking off stage in the middle of gigs and it's, 'Fucking hell, we've got another forty minutes left here, someone better do something.'

Liam He's a great fucking songwriter and always will be, but singing is my gig. If you are going to write the fucking song, at least let me sing it because if you are going to start singing the fucking songs as well as writing them, what am I doing, making the fucking tea? All the coming off and, 'I'm doing this tune,' I was like a fucking yo-yo half the time. On for one song, off for one song, on for one song, off – leave it out, you're taking the piss, mate. Here he comes, here he goes, here he comes . . .

Noel He never rehearsed, he couldn't be arsed, so you didn't really know what anything was going to sound like until you walked out on stage. Which is alright when you are playing the Forum or some fucking pub somewhere in fucking Aldershot. It can be a bit hairy when you are playing a stadium, to 70,000 people, and you think, 'I wonder if he knows the words . . .'

never, ever,
Pearl
rsed getting

could not be

on that stage.

Liam We had a little argument, the fucking amplifier fed back and hurt your fucking ears, the set list got put out wrong, or someone had a fucking bit too much to drink that night. Fuck off. We're in a fucking rock-and-roll band here and everyone's, 'Oh God, he turned up pissed' or, 'He missed the fucking gig.' Fuck off, mate, are you taking the piss? If that's the formula these days, no wonder rock and roll is over.

Noel He was always complaining about monitors, always fucking complaining about the sound on stage. 'Well, why don't you fucking soundcheck like everybody else does?'

Liam The reason I didn't do soundchecks is because I'd be singing in the soundcheck and my voice would go. So I just keep it chilled until the time of the gig. It's fine if you are playing guitar, but even getting up and doing a couple of tunes would fuck my voice up, so I would just wait until 9 o'clock and then hope for the best. I didn't gargle with honey and all that because it wasn't the way I sing. Singing them songs the way I sing them is like being in a boxing match, it's hard work. It takes its wear and tear. I ain't moaning about it, but that's just the way it is.

Singing them songs the way I sing them is like being in a boxing match . . .

Noel I remember Barrowlands well. He didn't have a sore throat; the usual shit, out all night the night before, not bothering to soundcheck. Turns up at the gig, within fifteen minutes he is out of puff. We are halfway through one song, he stops singing, walks off stage. It's Glasgow, they are going fucking mental. The song stops and there is a bit of confusion as to what's happened. You go off stage and say, 'What's happening?' Not only has he left stage, he's left the building, he's fucking gone. So you are there, fifteen minutes into a gig, what are you going to do? You are either going to try and save the night or there is going to be a riot. As luck would have it on this night it was both.

Maggie You don't walk off a Glasgow Barrowlands stage, you know what I mean? That could be quite seriously not good health and safety-wise, or security-wise. You go on there and you finish that gig, because they're quite a rowdy bunch in Glasgow.

Liam If your voice goes it's like the end of the fucking world and there's nothing . . . It's horrible for a singer. It's like a guitarist getting his hands chopped off.

Noel Our security guard came over and said, 'Liam's not coming back. The cars are running so we should leave before there's a riot.' He said it a bit too close to the mic, it went out over the PA thus instigating a fucking riot. Shit started getting thrown everywhere and then we had to fucking leave. Then you go back to the hotel and Liam would show no remorse, he would just be like, 'Fuck 'em.'

Liam I've never walked off because I weren't into it, are you fucking kidding? There is no fucking way. You've got your fans in front of you going mental, the last place I want to be is sat backstage freaking out. I want to be in there getting amongst it. All that nonsense about, 'He didn't give a fuck,' or, 'He walked off,' I never, ever walked off stage because I had a headache or I'd broken a fucking nail or I couldn't be arsed with the gig. Are you kidding? It's because I can't physically fucking sing. A lot of the reason my voice is fucked is because I probably should have walked off. The amount of gigs I stayed on and fucking battled through it because of the shit you get or whatever . . .

Maggie Legally, they have every right not to pay us, so we have to finish the gig. We need to get paid, so get up there and work it. Not finishing the gig was not really something that we could entertain.

Noel I don't mind cancelling a gig before it takes place, if the singer has got a sore throat, I don't mind that. Walking off stage during the second or third song and when there's 15,000, 25,000 people there, that is unforgivable, because how do you get out of that? What happened, nine times out of ten, was I'd take over. And the more I would take over or finish the gig, the more that it became acceptable. So what happened is I really enabled him to behave like that. Nobody wanted to fucking hear me sing lesser versions of 'Live Forever' or 'Supersonic'. Those songs were written for Liam. That kind of thing started to become really regular: Liam just walking off stage, not turning up, turning up and then leaving, or turning up and then sitting in the crowd and basically being a fucking lightweight about it all. Not turning up for stuff, saying you are going to do stuff and then not doing it . . . If you are going to wind me up, that is one fucking way to do it for sure. That went on forever, not turning up at *Jools Holland*, not turning up at *MTV Unplugged*, walking off stage halfway through countless fucking gigs because he'd been on a bender. My whole attitude towards all of this is: that's great, we all love going on benders, I fucking love it, but if I've got shit to do the next day . . . You have got to make a decision, haven't you? I'd rather be on the telly with my band, smashing it, than sitting maudlin in a hotel room at 3 o'clock in the morning, smoking cigs listening to fucking 'Strawberry Fields'. You are not a rock star sitting in a hotel room at 4 o'clock in the morning, listening to fucking 'Bungalow Bill' again, you are just a fucking lad on the piss. When I would be arguing with Liam about shit like this, I would always go back to us in 1991 and say, 'Why were we doing this in the first place? It was to be on MTV.' We used to watch *MTV Unplugged* and think, 'Fucking smash the arse out of that when we do it.' That's what we wanted. What's changed in four years? Why are you now not arsed about going on *Jools Holland*? Why are you now not arsed about fucking turning up for video shoots? When we were starting off we used to laugh at people like Kurt Cobain. He was in the biggest band in the world and was moaning about it all the time, moaning about selling loads of records. We would sit there laughing; 'You fucking knobhead, you cheeky cunt, we are slumming it here at the Boardwalk, damp, shit, fucking awful, playing to six people. I'll take your place any day of the week, give it to me now.' I always used to say, 'We are living someone else's dream.' Think of all the bands who are playing in pubs tonight who are trying to make it, who'd give anything to be on *MTV Unplugged*. Why rehearse five nights a week if you don't want to make it? Why do all those local band nights, to like eleven people, nine of them being your mates and two inquisitive punters, why do all that? In one ear and out the other.

Nobody wanted to fucking hear me sing lesser versions of 'Live Forever' or 'Supersonic'.

Liam I had problems with my throat, just probably caning it too much, just fucking screaming and shouting, I don't know. Oasis was a loud band, having to sing on top of them songs was hard, your voice would pack up and that was it. I never, ever, could not be arsed getting on that stage. I take massive offence at that. There is no way I'm having people going I can't be arsed, that is bollocks. Not having that.

Noel Certainly by the time we get to Knebworth, Liam is not the same frontman as he is eighteen months previous. The more money you get, the more fame, the fucking chicks and the booze and the drugs and all that . . . The entire world is staring at you. I know how I would have felt. You know, you have not even got a guitar to defend yourself with. You are kind of stood there with your balls hanging out singing somebody else's songs. It must be fucking difficult. I am not sure I could do it.

Never been to America,
man, never been. Never
been anywhere, so it
was fucking great.

Shitholes USA

Noel Arriving in New York for the first time, I can look back now and think how it must have been annoying for the rest of the band, because I was like, 'I've fucking seen Times Square, so I'm going to fucking stay in this room and get through this massive bag of psychedelics. See you at the soundcheck.' The Inspirals prepared me because I'd been on stage, even if it was only handing a guitar to somebody, you've kind of been out there in front of a big crowd. And I'd done tours, I'd been on tour buses, I'd been through customs, I knew what it fucking was. American customs fucking blow Liam's mind.

'What is the purpose of your visit?'

'You what?'

'Purpose of your visit.'

'What fucking visit, I'm not visiting, mate, I'm a fucking rock star. I'm here to steal your soul.'

'Okay, you can follow me now, sir, you're coming this way.'

I'm a fucking rock star. I'm here to steal your soul.

Liam Never been to America, man, never been. Never been anywhere, so it was fucking great. It was mega. You could smoke on the planes then and everything. The roads were massive, and the food. I was always, 'Can I get a cheese-and-ham,' then the sandwich come and I'm just sitting there getting rid of twenty layers, launching half of this food into a bin, and then going, 'That's fucking more like it.'

Noel I meet the video director. This guy is talking director bullshit and he said to me, 'I was thinking at the end we should bury the drum kit.' I said, for a laugh, drunk, 'Why don't we bury the drummer?' and he went, 'Great. That's amazing, wow, fucking hell that is amazing.' I remember looking at him thinking, is that how easy this shit is? You just suggest shite randomly? While we're at it, why don't we throw the singer off the Brooklyn Bridge? Fucking hell. Fucking dreadful video. I will tell you what's good about that video, there is a bit where Liam's strapped to a chair, halfway up a wall. Never get him to do that now.

Liam I'd been out all night partying somewhere and I remember getting down to wherever it was on the docks, and there was this chair there. I hadn't had any sleep and I was thinking, 'Bastards are going to put me in that.' It was stupid. I was fucking singing and lip syncing and I could just hear it creaking. I think that sealed the deal for me with videos. Who fucking

screws a chair to a fucking brick wall? That's not cool, is it. The last thing you fucking want is to be stuck on a wall, you want to be on your sofa or in bed.

Jason We did the Wetlands, a great little gig downtown, proper classic New York club, it's great. Once you get the gig out the way, it's head out into New York. It's always a great city is New York, it's got that certain something, a certain energy. I think that night we bumped into the 'rock chicks'.

Noel 'Have you met the "rock chicks"?'
'No. What, are they a band or something?'
They were quite famous for their after-show parties and seemingly every band in the world would end up back at their apartment. One of them, Christine, would end up working for us at one point. She has remained a very dear friend of mine. There were lots of people at these parties, but they weren't wild. Americans are like, 'Fucking hell, a wild party . . . we've got twelve beers and half a joint!' I was like, holy shit, honestly I've had better parties on a fucking Wednesday afternoon.

Liam The beers were always shit until we found a decent Irish bar. Them light beers, how do you get drunk on this shit? There was one in New York, I can't remember what road it was off, but it was just called The Irish Pub and it was amazing. It had the best jukebox, we would go in and stay there till five in the morning. The geezer'd come over and go, 'Right, lads, I've got

Noel There is the legend that Ringo Starr was there and walked out. I was fucking pissed off.

Liam It weren't that bad or I'd have fucking thrown Ringo at him. Here you are, cop for Ringo, you fucking knobhead.

Phil We'd just got hold of the wrong type of drugs and it wasn't good for me personally, that's all I can say. I might plead the fifth on the rest of it.

Liam I don't think it was a triumphant gig, I might have been singing really bad or I might have been just pissed and off me tits. The gig might have been shit but you're allowed them every now and again, aren't you? We've done one bad gig, it's all over. We got the set list wrong, fuck. It just wasn't our night that night, but the next night probably was.

Bonehead It's one of them gigs that you want to just push out of your memory, you know what I mean, fucking ground swallow me up, man.

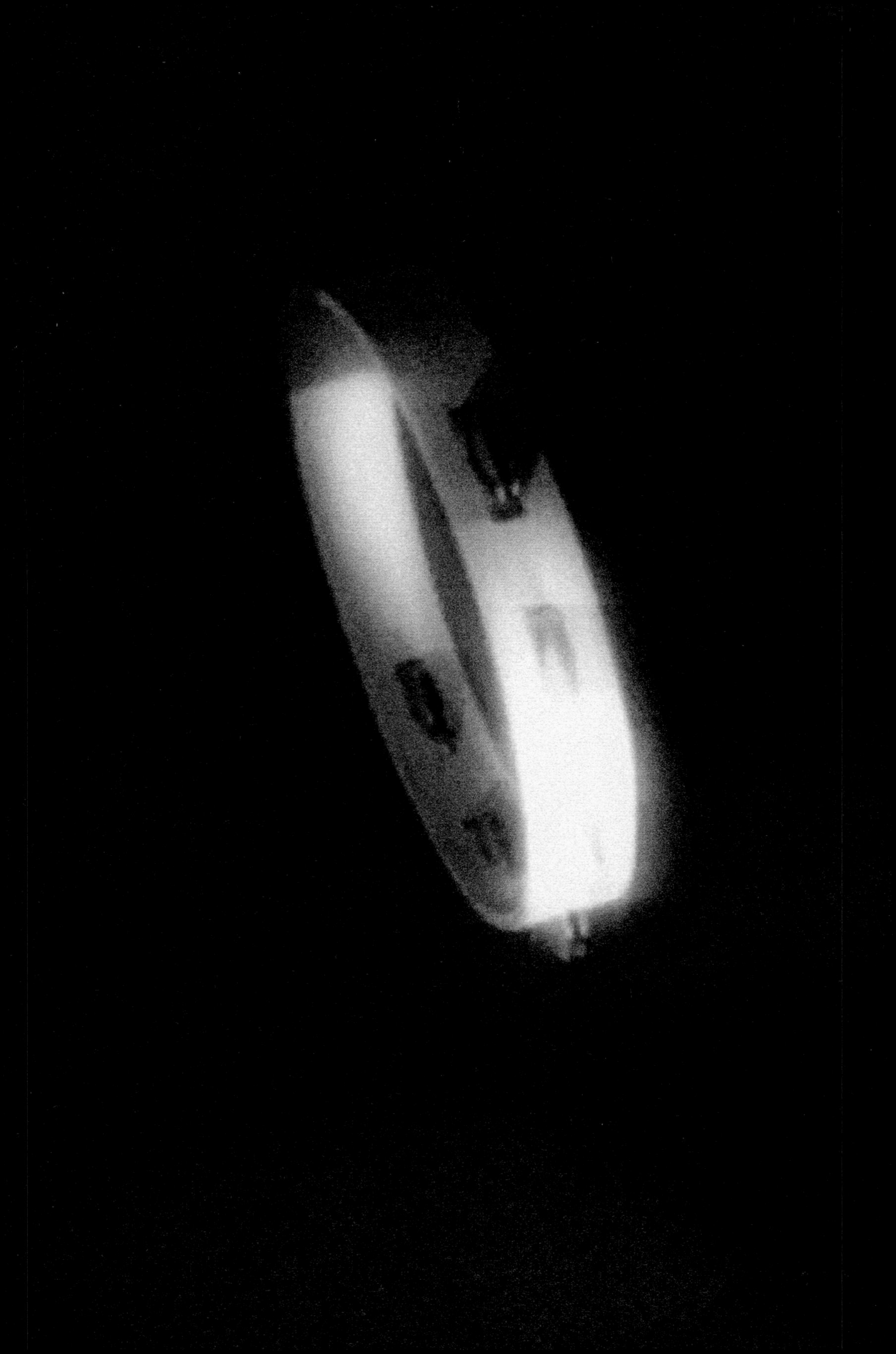

He was lucky I didn't launch a fucking monitor at him or something. Or a fucking drummer.

Noel I felt we could have smashed it in America, I really felt we could have been bigger than anyone that had gone before us because the timing was right, but I also knew that you've got to be able to play when you go there. You won't get anywhere on attitude in the States, you've got to be able to play.

Liam I don't think we were dicking about, I think we were just probably having one of them days. Can't be great all the fucking time. Obviously in Noel's eyes you can, but you just can't. However long we'd been on the road, we had been on the road, I imagine we'd done a lot of gigs and that's just the way it is, everyone just needs to cool out. That's just the way it was, mate. I'm sure they've seen worse than a set list being wrong and a tambourine being thrown at the guitarist. Haven't they had Guns N' Roses and all them other weird fuckers. Jim Morrison played that gig, didn't he?

Coyley I don't know what happened to the band at that point, something changed. The band was never the same. In some respects, that was the night it finished. That was it.

In some respects, that was the night it finished. That was it.

Noel I remember going upstairs to the dressing room and there being a huge argument. Shit got thrown around the dressing room and of course fucking half of the people are on crystal meth so they are not in the best of moods.

Bonehead There was a massive hoo-hah, big screaming match and he said this and he said that. Noel's like, 'You're off your fucking head,' and I was like, 'No, whoa, whoa, no, I'm not, man. No, I'm not.' A fucking raging, serious hangover, do you know what I mean, but no . . .

Noel I might have overreacted a little bit, because it was only one show. In the grand scheme of things it's fuck all.

Liam Maybe it got a bit dramatic, but fuck it, who gives a fuck. We're meant to be a rock-and-roll band, not fucking Boyzone or whatever.

Noel It was a funny old night, but it kind of set the tone for the rest of the nineties because it made me think, actually, you know what, these lot really don't give a fuck.

Liam We were so super real we were never going to keep it together 24/7 twelve months a year. That was never going to happen. Whatever band he was in, it must have been a real fucking shock to him to find out that he was actually in a band full of real people that didn't take themselves too seriously. It must be really fucking hard for him, but a couple of shitty gigs here, there and everywhere didn't freak me out because we'd make up for it along the way.

Noel Why would you put all that at risk? Three years previously, this is the same bunch of people who were rehearsing every night of the week in a dirty fucking rehearsal room in Manchester, going nowhere. But the only reason we were rehearsing every night is to get to this place: L.A. Paid to be there. Fucking hell, it's going to happen and you are going to fuck it up? I was thinking, 'You fucking wankers.'

Liam I didn't worry about it at the time. Water off a duck's back. Tomorrow is another city, another gig. Good night's kip, head down, here we go again.

Noel I don't recall at the time thinking, 'This is it now, the band's finished and I'm going to go fucking solo and be bigger than Cliff Richard.' I don't recall thinking that. I may have been trying to teach somebody a lesson. It failed miserably.

Liam We had a lot of madness around Oasis. We were playing with fire most of the time and that's what it needed. So when shit happened you just had to fucking deal with it because you knew that the good would come and sort it out. I don't know what it is, it's nothing to do with us as individuals, it's the spirit man of Oasis. I truly believe that there was another force out there that was bigger than me and Noel and the band. I know it sounds hippy and shit, but when that fucker had something to say he would say it.

Noel I remember going back to the hotel and seeing Maggie; I think I took the float off her and went to San Francisco.

Maggie Noel was quite upset by it. He came to my hotel room and he said, 'I'm leaving. Can you give me money?'
'What do you want?'
'Well, enough money to go back home.'
I had more money on me, but I said I'd got $700, enough money that he could probably whiz around the States, but not further than that. He said, 'Can I have my passport?' so I gave him the passport. At that point I thought, this isn't really good. He left the room, I remember ringing up Marcus saying he's going to leave, 'No, no, no, he'll be fine, he's going to sleep it off, he'll be fine, he'll be fine.' I don't think so.

Bonehead Noel just got off, you know. Do you blame him? It was one of them, 'Alright, he's got off, he'll be back, he'll be back later, on we go, it's happened before.' No. He didn't come back, he was gone, that was it.

Noel I went to San Francisco from L.A. You know, when I think about it now, that's insane. I left a big bag of charlie in the fucking taxi as well, on the way to the airport. I thought, I better not take that, put it down the back of the seat.

Maggie So, I woke up in the morning and I found a note underneath my hotel-room door. I opened it up, it said, 'I'm leaving.' Bollocks, I knew it. So, I rang up Marcus and told him, so he came over.

Liam I remember some fucking soppy note coming through the door saying, 'How can we go on like brothers when we do this?' It was like, fuck right off. So that got rolled up and then another line of crystal meth went up the old fucking tubular bells.

Noel I don't remember leaving a note. No, that's way too fucking pretentious for me.

Bonehead So, it was like, what do we do? No one knew where he was, he could have been two doors down for all we knew. You know, you have a little bust up, something happens, it goes wrong at a gig, band member storms off, band member comes back, and it's all forgot. So, it was like, alright, we'll just stay put. I really didn't think he'd be gone on a plane, out of it, no way.

Maggie I came up with this idea. We didn't have mobile phones then, so I went down to the hotel and said, 'I'd like to check out Mr Gallagher in room blah, blah, blah,' and they're like, 'Okay.' So I checked him out, I got his bill, and then I traced all the last numbers he rang. The last couple of numbers were U.K. numbers, so he must have been calling his mum or somebody back there. Then I noticed there was one number: 415. I thought, San Francisco? We've just been to San Francisco, and I knew that he had some little fling over there with some girl. So I thought, 'Hmm, he's got to be there.' Me and Marcus were both going, 'He's got to be there.' Marcus is like, 'I know what I'm going to do, I'm going to ring the number, just going to speak to the girl and see what she says, you know, he might not be there.' So, he gets her on the phone, he said to her, 'You know, we're not saying anything, but obviously, you know, legally, if he's missing for another twelve hours, we have to report him as a missing person, you know. We're going to have to get the police involved. We need to find him.' The minute he hung up the phone, he said, 'He's there, I could just tell by the stuff she was saying, he is there.'

Noel I met this girl, I don't know where I met her, it has to have been either at an Inspiral Carpets gig or at an Oasis gig, fuck knows. I must have phoned her and said, 'It's fucked up here, I'm going to come and stay with you for a few days.' I don't know what the fuck I thought I was doing. I was just going to disappear, fake my own disappearance or something. It was quite a traumatic time because you are, at this point, you are the biggest band in England. Thank God you don't have a mobile phone or a computer so you don't realise what the fallout is. I was hanging out there for a couple of days and probably doing too many fucking drugs and blah, blah, blah, not eating and being a bit fucking mad. She was saying, 'You are going to leave and that's it? What are you going to do?'
I guess at that point I'm thinking, 'Actually that is a fucking good point, what am I going to do?' I can't sing, I wasn't a singer then, I'm no frontman. It wasn't like this girl saved me from myself; I just needed a bit of time out, I guess, and a bit of perspective. I wasn't going to burn the house down to prove a point, a few gigs got blown out, what can I say, I had a good time.

I just needed a bit of time out, I guess, and a bit of perspective.

Coyley You meet them people in America, you don't meet them anywhere else in the world, I've met a few like that. They take you under their wing, they're not drawn to normal people, they're spiritual people is what they are, they're not rock-and-roll fans. She was one of them. He didn't come to us, he went to her, and when he come back, he was different. Does she even exist? She probably just turned into a wisp of smoke.

Maggie Marcus's idea was to get Tim Abbott over, because he was quite friendly with them, and get him to go up to San Francisco, find Noel and literally talk him back.

Tim Noel phoned me, about 4 a.m. and I went, 'Alright, Noel, what do you want, you alright?'
'Yeah, I'm alright. What you doing the weekend?'
I said, 'What's it got to do with you, haven't thought that far.'
He said, 'I might be back.'
'What do you mean, you might be back?'
He said, 'Look, can you get Marcus to get me guitars back for me.'
'Hang on, I'll call you back.'
I literally put the phone down and Marcus phoned. He went, 'Boyo, have you heard, Noel's left the band? He's gone AWOL, we don't know where he is.'

'Really,' I said, 'that's funny, he's just fucking phoned me.'

'Has he?'

So, I said, 'Well, get off the line, he's going to call me back.'

So then he calls back and says, 'Oh fuck it, I've had enough of those cunts blah, blah, blah. I'll see you next week. Can you tell Marcus to get me guitars back, and say sorry for everything, sorry but, that's it. I'll see you next week, we'll go out.'

Then Marcus called again, 'Have you heard, have you heard? They've all done meth, they are all off their tits, there's been a massive row, he's took the tour money and he's fucked off.' Right, he didn't say that.

I said, 'Seriously, do you want me over there?'

So, I literally flew out and arrived in L.A. jet-lagged, the day after. Marcus and me sat down, and Bonehead came and joined us and went, 'We fucked it.' Liam was just like, fuck it, another day in the office, what's up with him, he'll be back, won't he?

Bonehead We found out he was in San Francisco and then his next port of call might be home . . . Uh oh, this is serious, you know? We come out of Marcus's room and we all look at each other and go, shit, there's the songwriter gone, there's the chief, I think it's really over, innit? No, we didn't want it to end, why would you, do you know what I mean?

Phil We stayed in L.A. another night, which I presume was just in case he came back. It transpires that he is at this place, but he's not speaking to anybody on the tour: band, crew, Marcus. And if he's not talking to Marcus, then you know it's kind of serious, he's not even going to have a chat. Then we set off on the bus and start driving round America. We just followed the route of the gigs because hotels were booked, where else are we going to go?

Maggie We thought, by the time we get to Texas, hopefully Tim will have done his magic.

Noel Clearly it's not a normal thing to have taken the tour float and a big bag of fucking drugs and run off and left the tour. Clearly that is quite a bizarre thing to do, but I don't remember thinking, 'This is the end' or feeling suicidal or anything like that, at all. That is not the way I am. I kind of sat in this girl's flat, drinking, skinning up whatever, watching fucking American football, I don't know what I'm doing, can't remember. Then a doorbell goes and the girl says, 'There is someone here to see you.'
I was like, 'Wow, fucking hell,' and in walks Tim Abbott.
'Alright, Tim, what you doing here?'

'I've come to find you.'
'How the fucking hell did you find me?'
Then me and Tim went to Las Vegas. I don't remember saying goodbye or thanks. She had a job and all that, I'm sure she was probably glad to see the back of us.

Tim I felt quite happy that he was okay, to be honest. I definitely wasn't that arsed about the agenda of the band or the record label or whatever, as long as he was okay, and he seemed fine. And I felt quite cool, like *The Fugitive*, the one-armed man, you know, you've travelled 7,500-plus miles and nailed your man.

Noel The next gig that hadn't been pulled was in Texas, which gave us a few days. I don't even know why we went to Las Vegas. When I think about it now that is insane. I don't know what I was thinking . . . What a weird place to go while you are in some deep kind of fucking cocaine psychosis. I just remember it being a laugh. I remember we were on the 58-thousandth floor of the Luxor – a nice relaxing few days in a fucking giant jet-black pyramid in the middle of the desert with a boy dressed as a pharaoh bringing you a club sandwich. Mmm. This is going to be fun. I remember going to turn the

tap on in the bathroom and the water coming out like a shower and having a fucking serious conversation with Tim, saying, 'This is fucking amazing, I live in a flat in Manchester which is two fucking floors above a canal and my water fucking comes out of the tap pathetically.' And we had a fucking good hour-long conversation about water pressure. At that point I might have thought, 'You know what, it might be time to get back on tour here.'

Liam I wasn't worried for his state of mind at all, he wasn't the one on crystal meth. He weren't in that bad of a state of mind if he turns round and fucking gets his tour manager to take all the money and book a nice little flight off to Las Vegas. He's got on a flight and hit the hot spots. Not worried about his state of mind at all. I was more worried about the money that he took, without a fucking doubt. Las Vegas, I haven't been there, mate, come back, let us come with you. Fuck the band, the band will be right.

Tim I think he was angry, he was obviously in a fucking weird place and we just caned it for two days. Bought some mescaline of all things, if I remember right.

Noel I was angry; we were better than that. By the time we'd got to Japan we were fucking great. All those years of rehearsals had paid off, we all knew each other, we knew the songs inside out, we knew how to do it. It was like we'd got to America and it had all suddenly gone out the window. Everyone

had forgotten how to do it because, you know, drugs. You have one shot at it in your fucking life. It seemed to me that everyone else was not taking it as seriously, or didn't feel it as seriously, as I did. Maybe they thought they'd made it, I thought there was a lot more work to do. I guess I left a little bit of me on that stage that night and the relationship between me towards everybody else was never the same after that. I think I probably felt after that night, it was more me and them as opposed to us.

I think I probably felt after that night, it was more me and them as opposed to us.

Tim It's about six, seven, the sun was going down and it's getting a bit cold. I made a phone call to Marcus and said, 'Where's the band?'
He said, 'Oh, we're on our way to Austin to do the B sides, is there a happy ending?'
I said, 'Well, I don't know, I'm going to ask him now.'
You know, my whole thing was fucking hell, man, don't walk away now.
I said, 'Look, would you do the B-sides for us? John Lennon would have done the B-sides.'
'Alright, but I'm not fucking speaking to 'em. I'll come back and do the B-sides.'

Owen I didn't know this had happened. Marcus didn't tell me because I was booked to fly out to do these B-sides, so they kept quiet that the band had split up. I walk into the fucking airport in Austin, Texas, off the plane, fucking lovely jubbly, business class, fucking love working for Oasis. There is Tim Abbott and Noel Gallagher, I'm thinking, 'You've come to fucking meet me at the airport, nice one, Noel. A bit much, innit?'
Tim Abbot's like, 'No, we've just arrived from fucking Las Vegas. Didn't you hear?' Noel is like, 'Fucking dickheads.'

Phil I think we'd heard about a couple of days before that he's coming back. It means the band's staying together, rather than going home and signing on the dole again, you know what I mean? So, obviously, by the time he's about to turn up everyone's fucking shitting it because he can put on a stern face, you know, he can unsettle you.

Tim We marched back to the hotel and he walks straight past 'em. But they all smiled, they were like beyond relieved, and it was over to them then, really. I'd done my bit. Brought me man in. And that was it.

Noel I remember the first time we met up after that, there was a band meeting, and me saying, 'All I want to know is what everybody else wants out of it. If it's going to be a load of shit-kickers on the road then let me know, I will fucking go and find something else to do. Or are we going to make it the best that it can be? If we're all here just to fuck about and treat it like a stag do then I'm not interested. I'm in it to be great.' At this point every song I'm writing is amazing and I'm not prepared to fucking piss it up the wall. I don't want to be back on the dole in a year and a half, I don't want to be back in the Boardwalk in two years. We had the chance to make it, like smash it; do you know what I mean? Bigger than all of our heroes, bigger than The Stone Roses could ever even fucking dream about. Bigger than all those bands: The Smiths, New Order, all of them. Bigger than The Jam, everybody.

Coyley He'd just got to the end of the fucking road. That was the end of that era and the new era started when he walked back into that hotel with a skinhead. I always thought, when he came back, you fucking changed, man. Now me, right, I believe in life in outer space and dimensions and different parallel universes and all that. Somebody come and took him. That's what I think, it was that radical. We got a new man back and everything was different after that. We all had to be different, because if you're not with him, you're going home, that was the defining end of the youth of that band.

Liam 'It's going to be like a business and I'm in charge.' Fuck off. What is all that about? I'm all for a bit of fucking discipline, but, 'It's going to be like a business and I'm in fucking charge?' You say shit like that to me and I'm ordering more drinks.

Noel One of the things that I learnt from Inspiral Carpets was if I ever get anywhere, there is no way it is going to be a democracy. They would take forever to decide on anything. That used to drive me mad and I was only a fucking roadie. They used to have band meetings to decide whether to get fucking tea or coffee for the office. I am very good at making decisions. Whether they are right or whether they are wrong, I'll make them on the spot if I have to. I've made some brilliant decisions and I've made some bad ones as well, but at least I've made them. Maybe that's why we got to where we did in double-quick time. Democracy is boring. Democracies are for squares, simple as that.

Jason In the old days it were great; he'd have the best part of a bottle of gin, take his watch off, put it on top of his amp and do the gig. It were great. But then it got a little more . . . As it got a little bigger I guess, everybody changes.

Maggie He probably thought, this is the point where they need to understand that there are some rules that we have to make. We can't just be this chaotic gang from Manchester going around the world. Yes, you can have fun; yes, there's time for this and there's time for that. There's time for the crystal meth and there's time for whatever they were doing, but not before a gig.

> **I guess we all decided to chill out on the drugs and drinking and everyone just fucking get it together a little bit . . .**

Noel My ethos has never changed from day one. We have the band, we work backwards from that. I don't remember ground rules; I'm not the kind of person who would do that. I think the only ground rules ever set down were, 'We are rehearsing five days a week or you are out.' That's it really. Other people might tell you different because I've not got a very fucking great memory about these things.

Liam I guess we all decided to chill out on the drugs and drinking and everyone just fucking get it together a little bit, I guess we did that.

Noel I just wanted to know where their heads were at. I don't recall giving anybody an ultimatum, I don't do ultimatums. I am either in or I am out and that is it. Everybody said the right words and then fucking three weeks later it all blew up again.

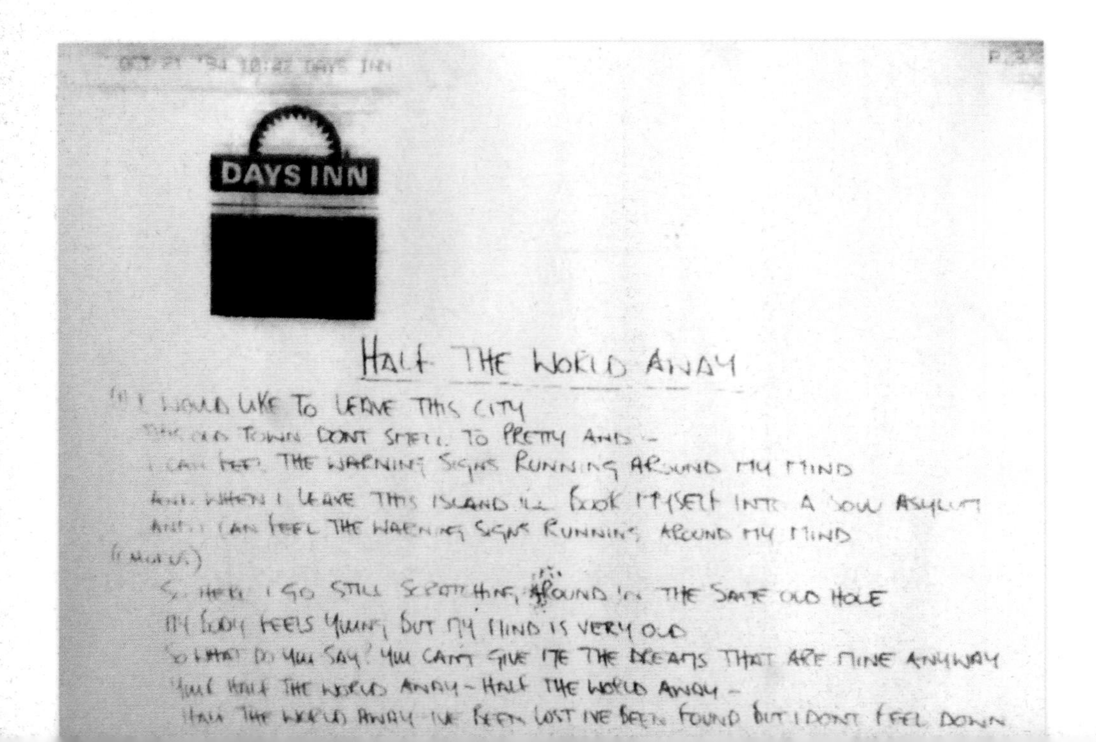
DAYS INN

HALF THE WORLD AWAY

I WOULD LIKE TO LEAVE THIS CITY
TOWN DONT SMELL TO PRETTY AND –
FEEL THE WARNING SIGNS RUNNING AROUND MY MIND
WHEN I LEAVE THIS ISLAND I'LL BOOK MYSELF INTO A SOUL ASYLUM
I CAN FEEL THE WARNING SIGNS RUNNING AROUND MY MIND
(CHORUS)
SO HERE I GO STILL SCRATCHING AROUND IN THE SAME OLD HOLE
MY BODY FEELS YOUNG BUT MY MIND IS VERY OLD
SO WHAT DO YOU SAY? YOU CAN'T GIVE ME THE DREAMS THAT ARE MINE ANYWAY
YOU'RE HALF THE WORLD AWAY – HALF THE WORLD AWAY –
HALF THE WORLD AWAY I'VE BEEN LOST I'VE BEEN FOUND BUT I DONT FEEL DOWN

Lost and found

Phil Being Oasis, they're getting back together by immediately going into the studio and doing some recording.

Noel This was Oasis's fucking cure for everything: 'Let's go in the studio.'

Owen The session was a two-day session in this little fucking shack, brilliant, fucking raining and voodoo dolls and fucking really atmospheric. Noel had his guitar turned up so fucking loud, they're doing '(It's Good) To Be Free' and it's like, 'One fucking take, you cunts . . .'
Shitting themselves, everybody.

Bonehead We're in the studio and it just clicked. I know I, for one, was trying my best and, 'Yes sir, no sir, three bags fucking full sir.' You want me to play that song, yeah, cool, I'll do it, no worries. In, one take, done, yeah. I'm not fucking this one up, ever again, do you know what I mean, no way.

Noel If anything, it was all worth it for that. We did '(It's Good) To Be Free' and 'Talk Tonight'. I'm even thinking 'Half The World Away' was recorded on that session as well.

Liam They were good, they sound dark and all so that was good. Whatever happened was brought out in them tunes, so as far as I'm concerned, job done. We got two great tunes out of that little rendezvous so I'm happy with that.

Owen 'Talk Tonight' was probably the best recording I've ever done with Noel Gallagher, sensational.

Noel I don't have any recollection of actually, physically writing the song. 'Sitting on my own chewing on a bone,' I remember that line came to me on the plane. I don't really recall writing it as such, but it came together very quickly. Funnily enough it was a song that Liam fucking hated and still hates to this day.

Liam 'Talk Tonight' is a killer tune.

Noel 'Talk Tonight' was written about those few days in San Francisco. Out of that kind of traumatic period of, however many days it was, came that great song. So it was meant to be.

It's not about the girl per se, it's about those few days that I spent with that girl. In the song it says, 'Take me to the places you played when you were young,' or something like that. She'd taken me to the playground where she played when she was a kid and she was making sure that I ate because I was doing so many fucking drugs. It's a document of those few days. I can't even remember that girl's name now. When I close my eyes I can't even see her face.

My wife thinks I'm a fucking lump of wood that walks around the house smoking, talking about football. I counter that I am probably the most romantic man she'll ever know because of writing all these songs. As a songwriter, by fucking definition, you are romancing something, even if it is 'Anarchy In The U.K'.

Coyley It shows a little side to him, that you just don't really see, he doesn't show these emotions. In all the time I've known him, he's never shown these things, apart from when you stick him behind the glass with a microphone and then he can let this emotion loose that he's got. The door opens and suddenly you can see Noel, not the Noel that you might perceive him to be, by reading about him, or seeing him on the TV. You see something else, it's like shape-shifting, you kind of see almost a little boy or something, you know, you see something very innocent and very pure. The veil is just lifted away, and you can get a better sense of what somebody's about. They are my favourite moments, it's like a dark room, the door opens and a bit of light comes in and you just see a little bit, and then the door's shut again and he's calling you a twat.

Whatever happened was brought out in them tunes, so as far as I'm concerned, job done.

Noel Buying singles by The Jam and The Smiths is what instilled my obsession with there has to be something as good on the B-side. The Smiths B-sides mean everything to me, The Jam B-sides are fucking unbelievable and it is something I have kept with me to this day, I wouldn't just shove any old shite on a B-side. Unless I had no songs. I had songs I'd recorded, songs that I hadn't recorded, songs I hadn't fully written but it was just a case of finishing them off; in that sense I had a plan of what the first three and a half albums were going to be like. What I didn't know was how great the B sides were going to be because I was writing them on the way. It turns out the plan was flawed because *Be Here Now* should have been all those great fucking B sides, but there you go. At that time when you put out a CD single there were two CDs, what the fuck was all that about? Four B-sides. The thing about those B-sides, although they are tucked away on records and America is still immune to most of them, they made the band what they were in England. They sustained the band for years, the fact that you could just throw in 'Acquiesce' and 'Rocking Chair' and 'Talk Tonight' and 'Half The World Away' . . . all those songs from that period were great. In those days I would write every day because I was so fucking driven and, to my own credit, I recognised that I was in a moment. Everything that I wrote felt great. So I kind of recognised that I was on it. We'd be in hotel bars when we get back after the gig and I'd take me guitar and go up to my room. Liam was kind of sinisterly thinking, 'What's he doing up there?' They are all in the bar fighting with lads, fighting with Geordies till four in the morning and I'd be writing tunes. 'What have you been up to? What you fucking doing up there?'

I guess if Noel was in the bar with me drinking and up all night, the songs would not have got written.

Liam See this is bollocks, I'm not having that. If that's what he was doing then he's a fucking great man and he needs a knighthood, give him one. Fucking put him out of his misery, but when all that madness is going down he is going, 'Oh, you know what, you guys crack on, I'm off to work.' Fuck off, give me a fucking break, who are you trying to kid? I don't think so. Maybe he did write these great songs in the eye of the storm and that, in the madness of parties, but I find that very strange. Having a fucking great time, you're number one, it's all going great, 'Guys, you just crack on there, popping champagne and getting off your box. I must get out of here; I've got a song to write.' Would you? Fuck off. Who are you trying to kid? He just started getting more into the songwriting thing. I guess there are pressures when you've got an album out and it's great and it's doing well. I guess wanting to make a better record will make you go, 'Know what, I'm getting into that.' I was getting into drinking and playing the fucking rock-and-roll star and all that stuff. So me being in my bedroom, that is no place for me. I am mastering my art down at the bar, having it with the fucking people and he's mastering his art up in his bedroom with his guitar. I guess if Noel was in the bar with me drinking and up all night, the songs wouldn't have got written. I get that and it's the big cross that he carries, 'I was in my room doing all the fucking graft.' Well, I was in the bar doing all the graft too, fucking getting mithered to fuck by all the fucking loony fans. While you're sitting trying to get an E-minor together I'm downstairs getting booze poured down my neck and drugs blown up my nose, life's a fucking bitch. It is what it is.

Noel One of the big regrets is not having somebody, when I was writing those B-sides, to say, 'Hang on a minute – to have "The Masterplan" as a B-side, it fucking beggars belief.' Why didn't somebody say that about 'Acquiesce' or all those fucking great songs? I remember being round my house once and having a cassette of 'The Masterplan' and people were just laughing: 'You can't put that out, what are you doing?' I remember getting angry and thinking, 'They asked me to write a song, so I've written one and that's what I've written so it's fucking going on and that's it, that's the end of it.' Little did I know that that kind of thing would very quickly evaporate or I would have kept all those songs. When you are young you've got the power, and you don't realise that you've got it until you don't have it. And then you look back and you think, 'Fucking hell, wow.'

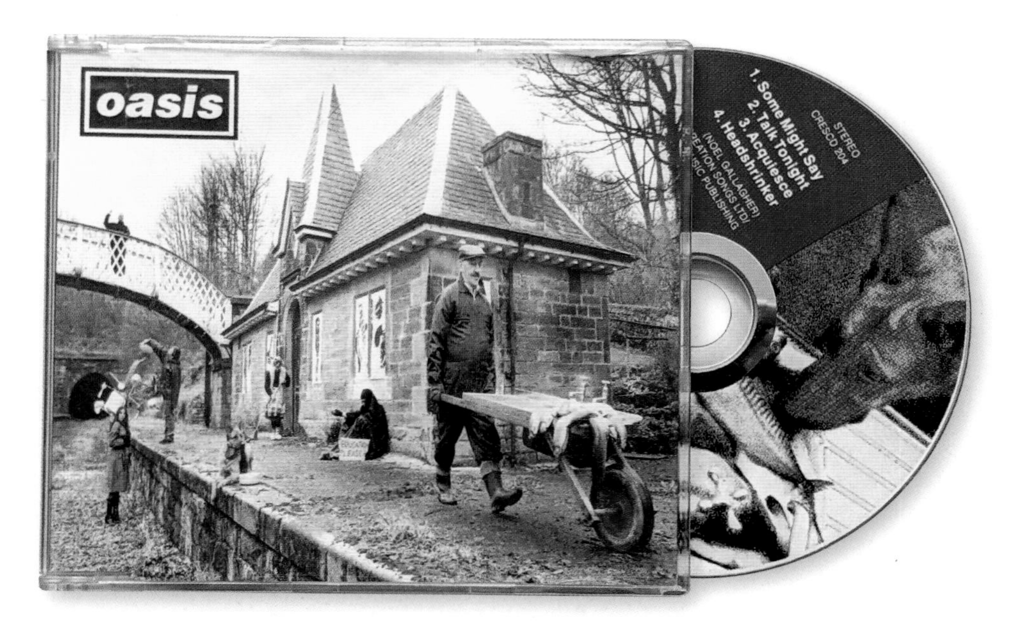

About those videos

Noel The making videos thing had already become boring to us by the fourth single, it was like, 'Have we got to do this again? This is fucking shit.' Somebody came up with the idea to hire the Borderline and get a load of fans in and we'll just do 'Cigarettes & Alcohol' about fifty times and see what happens.

Liam Yeah, that was good, we done it with an Australian guy. His wife worked for some fashion magazine so he brought in loads of models to act like they were our fucking mates and shit. You could see in the video that that's not our crew, it was all a bit ridiculous.

Noel We are all supposed to be getting off with these birds in the toilet. Quite clearly, I'd say a healthy 75 per cent of those girls wouldn't have looked twice at anybody else in the band bar Liam, so it's all a bit fucking ludicrous. Very quickly after that we were spunking hundreds of thousands of pounds on video shoots because he wouldn't turn up. The 'Some Might Say' video for instance . . .

Liam That was a time when I was putting my foot down and going, 'It's a shit idea, we don't have to fucking do it. This is our band, we're fucking in charge now.' We are in charge, we are at the wheel, so if we don't want to do it, we are not fucking doing it. I think we are within our rights. The amount of shit videos we did, they are lucky they got made at all.

Noel You sit in meetings and somebody says, 'This is going to be the idea,' and everybody signs off on it. Then, on the day, while we are all waiting in a motorway service station, he just doesn't turn up. I'm not having that shit. That was a big moment for me. This is going to sound fucking ridiculous; that was the beginning of the end for me. I know the end came a long time after that but I remember thinking, 'If you don't like the fucking idea for the video, say so. Don't have us all sitting here in the fucking rain.'

Liam I wish I'd done it more often to be fair. And I guess Noel reckons that as well because he hates every video that we've ever done, but why didn't he stand up and say something about it?

Noel He came up with the shitty excuse of, 'It's an important song and it needs an important video.'
I was like, 'Have you looked at the fucking words? Are you fucking taking the piss? It mentions sinks full of fishes and dirty dishes, you silly cunt.'

'No, it's an important fucking song and it deserves an important video.' Yeah, whatever, so you had a hangover, did you? We ended up cobbling together some video from the American video for 'Supersonic'.

Liam The one of the stuff in America? Yeah, fucking mega! So we come away with a good video by me not getting up and doing the shit one. Whether it cost forty grand or not, who gives a fuck? That's going to be the first number one, so imagine that, we've got a nice number-one song and we've got a shit video to go with it, I think I done the right thing. He would say that I didn't give a shit, but I couldn't see that. I probably give a shit more than I should have. His thing was, 'What d'you care about videos for? Let the video guy do it.'

I would be going, 'I'm going to be in the fucking video, I want it to look cool.' He would be, 'But you're not a fucking director.'
'I know I'm not a director, but I'm not a fucking actor either, so I think we should get more involved with our videos. You're just leaving it up to these fucking clowns who are just going to do these mad videos and get us in these mad situations and it's going to come across really shit. It's not what we are about. So we should start looking at these scripts that they are giving us before we just go yeah, go and do the video.'

Noel Have you seen how many shit videos we did after? They are fucking all appalling.

Definitely number one

Noel I remember going down to Loco Studios when The Verve were having a weekend off. Owen was doing *Northern Soul* for The Verve . . .

Owen Noel turned up, 'I got a new song. Fucking broke down under the Severn Bridge, in the tunnel. Wrote "Acquiesce".'

Noel 'Acquiesce' . . . it is just so pathetic to me when people say, 'It is clearly about you and Liam.' Oh right, yes, because I forget you were fucking there inside my head when I wrote it, I forget about that so you must be right and I must be wrong. Sorry about that, yes, it must be then. Well, it isn't, in the same way 'Cast No Shadow' is not about Richard Ashcroft. It is dedicated to Richard Ashcroft; it is a whole different thing, isn't it?

Owen So, we record that first, just warm up, one take, boys, here we go. Liam's like fuck it, chorus is a bit high. Noel was, 'I'll fucking sing that then,' you know. Done. Half a day. 'Acquiesce', motherfucker. Then we do 'Some Might Say'.

Noel 'Some Might Say': one of my favourite ever Oasis songs. I remember the demo; I did it on my own, played everything, the drums and all that. I played it on The Verve's equipment because they'd all gone home for the weekend. It's not really a great vocal or anything like that, but the sound of it was great.

Owen Me and Noel listened to it at night. It's a bit fast, because we had the slower version from a year before. But then, the next fucking day, Liam's like, 'That sounds fucking great, I'm going to sing it.'
Midday, cup of tea. Liam destroys it. Fucking hell, we got the wrong backing track, it's too fast. So, it's a really fucked-up recording, but it's beautiful.

Noel There is a strange melancholy to 'Some Might Say' considering I knew it was going to be number one before I even sat and wrote it. It's about the passing of something, a people's anthem about 'one day we'll find a brighter day'. What the fuck, I don't know what all that is about. Looking back on it now, instinctively I may have felt the end of some kind of age of innocence, for us personally, as people about to go and become real rock stars. But when it actually happens to you, you just think, 'Is that it? I don't feel any different.' I don't know what I thought you were supposed to feel; like you were supposed to be sprinkled with some kind of gold dust . . . I just felt like me. I guess it's because I knew in my heart it was going to go to number one anyway. Nobody is really doing the conga, everyone is kind of like, 'This is what we expected.' When you see the footage, me and Marcus just have a quick handshake and that's it really. I see people now get very excited about being number one, I just think, 'You fucking square, it's not that big a deal.' I'm living in a rented flat in Camden, it's shit and the front door doesn't lock properly. It was an awful fucking place. I looked around the shitty flat, I was thinking, 'I've got to get out of this place for a fucking start.' I was thinking, 'It couldn't have been like this for David Bowie, it couldn't have been like this for Marc Bolan or Slade or The Jam.' And you suddenly realise that actually it was like that. You are still the same person, it is just that your record is the most popular that week. It doesn't make you feel any different. I suppose there is the temptation to immediately go and buy a feather boa and a Rolls-Royce, a cane and a top hat and be a fucking dick, do you know what I mean?

You are still the same person, it is just that your record is the most popular that week. It doesn't make you feel any different.

Bonehead Me and Liam went to the pub to celebrate it, we had a couple of beers. I had me daughter with me, she was only about three months old, I remember she was on the table, bouncing on our knees, it was great. I can't explain the feeling. It was like, 'You actually went to number one, fucking hell.'

Noel Creation had never had a number-one single in their entire fucking existence. They were all really very excited and rightly so. They had a party for us and Blur turned up. We were having a great time, everyone is patting each other on the back. There is a downstairs bit and an upstairs bit, and someone came upstairs and said, 'Blur are downstairs, they want to come in.'
We were like, 'Fuck it, whatever.' I didn't see this take place but legend has it Liam went downstairs and got in Damon's face: 'Fucking number one, number fucking one!' As I understand it, Damon had taken that as Liam throwing down the gauntlet.

Liam I can't remember much about that. I think they'd come in to drink our drinks, they're coming to our party, you're going to have a bit of get-you-at-it, end of. It was just a crack, man, it was never malicious. Unless one of them got a bit lippy and then whatever would happen would happen. It was just a joke, just a laugh, man.

Tony

Liam Tony's one of the lads, he was alright, he just wasn't as mad as us I guess. If Noel didn't get on with him then I guess that's their thing. Me and him got on. We weren't fucking best of mates, but we definitely weren't arch enemies, man. He was alright, he was the drummer, he just fucking came, drummed, had a beer, talked about Man United. We all went, 'Nah, nah, nah, not listening.' That was it.

United fan. They are always going to be the first to go.

Noel United fan. They are always going to be the first to go. How Bonehead survived the cull all those years, him being a fucking filthy red, is beyond me because really it should have been a band full of blues. Thinking back on it now, I kind of missed the chance to get rid of the fucker.

Owen I got told that he wound Noel up so much on tour.

Coyley As it went through that American tour, it got really nasty with Tony, but it was unstoppable, and it was very sad, and you'd have to feel for the boy. It manifest itself in verbal abuse and that wasn't very nice but was completely hilarious. Everyone was party to it, I'd hold me hands up and say I give him a terrible time as well, but only for his good and for the good of the band, you know. I did a lot of shouting at him, just trying to get him going, you know.

Bonehead Tony's the person that I shared a room with in the early days, we didn't all have big hotel suites each, we wouldn't have the budget so we would all share. My roommate was Tony, always. We shared rooms in Japan, America and the whole of Europe and England, Scandinavia. I really got on with Tony, I probably got to know him best in the band, I would say. There were a couple of times when Tony would come in the room and he would just pour it out, 'Fucking hell, Bonehead, people are getting at me and people are doing this . . .' Hard for me, you know, to be the guy in the middle. Maybe I'd talk to Noel or talk to Liam, 'Look, go easy on the guy, man, we're in the middle of a tour. If there's a problem, let's deal with the problem when we come off tour.'

Maggie They'd known each other for a while, so it was all a bit of laddish stuff. They were doing it all the time to each other, so I didn't really particularly find a moment where they were bullying him in particular.

Phil The dynamic between them always seemed like he was kind of the odd man out. I was actually his roadie, so I'm not going to join in, I didn't do anything to stop it, however.

Tony Being in that cooped-up environment, I'm not surprised there were relationships not going in the right direction. There were grudges held, it got personal, and it built from there. I was quite a reserved guy. I think people might look at that as some kind of weakness, but it wasn't a weakness, I didn't want to be the leader, I didn't want to be like that. I'll just sit back here and do me drums.

Noel He did start becoming excluded. You know when a pack of feral dogs exclude one puppy, it became like that I think. There was a gradual ramping up of, 'This lad ain't going to be here much longer.' Not because of his personality, I don't give a fuck what people are like.

Tony Noel didn't want to have a relationship with me, I do know that. I tried, but it was uncomfortable. Whether the band then followed suit, stayed on the right side, safe side, of the fence, I don't know.

Owen When we were doing 'Some Might Say', Noel was like, 'Fucking last time he's ever in the fucking studio with me.' I was going to Tony, 'You've got to get your shit together Tony, or it'll be, "Bye, Tony."'

Bonehead They were really struggling with Tony and the drums on 'Some Might Say'. He was too slow, or he weren't in time, or he wasn't doing it right, I don't know. There was a real issue with the drums on it.

Coyley By this time he's history, he is absolutely history, I think they are just looking for the moment where they can just get the scissors out and cut that string and he's gone.

Noel I probably made up my mind, knowing that 'Champagne Supernova', 'Don't Look Back In Anger' and 'Wonderwall', were coming up. He is not playing on them, and if he is playing them, he is fucking not playing them live because I have got bigger ideas for that. The last throw of the dice for us was to get him drumming lessons.

Tony I thought, okay, I'll listen to everybody. I think I was sent to some kind of drum tutor in Wales, a couple of days. Within the first hour he goes, 'I don't really know why you're here,' he said, 'We're doing stuff that I'd planned for tomorrow evening.' I'm going to sit here and say I could have completed that second album; I know I could have carried on. I had an argument with Noel. He said something and I finally went, 'Fuck you, mate,' you know what I mean, fuck off. As he was soundchecking I told him in no uncertain words what I thought of him. I kind of regret that a touch.

Noel We might have had words, it's quite likely, I don't know. Everybody has their own version of events – Liam, Marcus – and somewhere in between everybody's version is mine. You've got to make your own truth out of it, so I wouldn't deny it at all.

Tony Getting the phone call was something I never expected. I think it was lined up that we were going to start rehearsing for the second album the next week, but Marcus phoned and he said, 'Are you sat down? Listen, I've got some bad news for you, the band want you out.' I was shocked, shell-shocked, I was like, oh my God.

Liam Nothing was set in stone, it's no one's God-given right to be in Oasis, if it changes, it changes. They were Noel's songs, so I guess he had more of a say on who is going to sing or fucking drum. He was going, 'Look, the songs are getting better, he's not getting better, I'm not taking him with me.' So what do you do? You don't turn round and go, 'It's either our kid or it's Tony.' I'm sticking with our kid and as bad as that might fucking seem, that's just the way it is, mate. I was gutted for him, but it is what it is, man.

Noel We couldn't have taken him any further. I showed him the drum fill to 'Don't Look Back In Anger' and the look on his face is something I will probably never forget; it looked like I gave him a fucking book in braille and said, 'Read that for us.' It was then I thought, you know what, this is not going to go any further. When most people in the band are a better drummer than your drummer, and you are about to become the biggest band in the world, then clearly someone's got to fucking make that call, do you know what I mean? I'm not one that sits around fucking moaning about a situation. If something is broke, get it fixed. He was going to hold us back, so he had to go. Things like that to me are not a big deal. A drummer? Fucking hell. Worst comes to the worst we will get a drum machine, the song will still be great and it will be in fucking time and it will be cheaper. We were all quite limited, but we fucking tried to make it better. I don't remember him being the kind of guy that was so arrogant that he didn't give a shit. I just feel that that is the limit of his talents, that was it. He was at the fucking ceiling and that was it. He wasn't going to go any further, I knew that as a songwriter, I was going to go further and he wasn't going to come along. I wasn't the best guitarist in the world, but I fucking knew it and I embraced it. Liam wasn't the best and he knew it. Tony wasn't the best and he didn't know it.

Definitely Maybe wouldn't have sounded as good without him drumming on it . . .

Bonehead Tony was a good drummer for the album he played on. I think if Tony didn't drum on *Definitely Maybe* then it probably wouldn't have sounded like it does. Even though he is not the most technical drummer, what he did was absolutely right for that album, and I don't think anyone else could have made it sound like that.

Noel *Definitely Maybe* wouldn't have sounded as good without him drumming on it, he was part of that sound. He was the right man for the job at that time.

Tony Bonehead rang two weeks later, when I wasn't in. Lord knows what sort of phone call that was going to be. But I didn't hear from any of them. Really hurt because it was like one minute I'm surrounded by ten, twelve people that have been your family, if you like, for a couple of years. Then to nobody, is like, fuck. I could never get me head round it and I think about it every day, believe it or not. Still, to this day, it's with me, it's, you know, something missing. But I've just got to accept it.

Noel Tony got kicked out of the band and then obviously it fucking explodes, goes through the roof, so some lawyer contacted him, pissed in his ear, and he thought he was due some mythical figure of £100 million or something laughable. Anyone that ever left Oasis or was asked to leave still gets paid to this day from whatever they played on. We didn't take any of his royalties off him. Anyway, it was going to go to court and he was suing us for lost royalties and all that shit. The amount he was asking for was nonsense. It was clearly nonsense because on the fucking steps of the court he settled for £550,000.

Liam He'd mentioned the confusion over signing the initial record deal and blah, blah. We'd never had a management deal with Marcus, at all. Never signed a piece of paper. We just shook hands down the pub and that was it. So then everybody started to get twitchy, 'Let's sign something.' There were four or five months after that sitting in fucking lawyers' rooms, bored shitless, people clearly speaking the English language, but none of the words making any sense. Forthwith and therefore and blah, blah. Just sitting there and thinking, 'How did it get to this? We signed off five years ago, what the fuck?' Sat in some Gothic castle with a load of lawyers in dickey bows. It's true what they say, that old adage, where there's a hit there's a writ and all that. When shit like that starts happening you do start thinking all the fun's taken out of it. You start a band with the best intentions of just trying to get a big telly and a fit bird and a load of money and the next thing is everybody's fucking hovering round you with briefcases and leather man bags full of shit for you to sign. I guess he felt betrayed, but people come and go in bands and that's the way it is. I don't hold any grudges or anything like that, he does what he's got to do and we do what we've got to do. That's when managers are good. No one really needed to manage the band to do the gigs or get us in a studio, we'd do it ourselves, but when it got to things like that it was 'Over to you, mate. We're out of here.' I guess he's within his rights to sue us. I guess we are in our rights to have a listen to what he's got to say and then take it from there. I certainly didn't feel like going round to his house and filling him in or anything like that. Tony McCarroll was part of my life, we had some good times, magical times. Times that no one will ever have, so it is what it is, that's the way these things go sometimes.

Coyley The band was never the same when he left. As soon as something changes, it's irretrievably gone, some dynamics change forever.

OUVERT

Smedley little mod

Bonehead Tony's gone. It's like, 'Shit, we need another drummer.'

Liam I know Tony weren't the best drummer, but he certainly weren't the worst. Everyone was getting a bit muso going, 'Yeah, but he can do this?' Who gives a fuck? He sits on it, he hits the fucking thing, you are never going to get Keith Moon. A fucking drummer's a drummer, obviously there's great ones and there's shit ones.

It was weird having a cockney in the band and all that, but he was great, he was a brilliant drummer.

Noel I was down the Manor with Paul Weller and got talking to Steve White who says, 'I've got a brother, he plays drums.' So I spoke to him on the phone and said meet me at this cafe in Camden at 3 o'clock, I'll sit outside with a copy of the *Racing Post* and a white carnation, just so you know it's me, and we'll have a chat kind of thing. Now it's a weird way to do business, but there were no mobile phones, no internet, no one had a profile on the web, so it was done the old school way. I remember loads of questionable-looking idiots – aging mods – walking up and I was thinking, 'This better not be him, this better not be him.' Then round the corner comes Alan and he goes, 'Alright, fucking hell.' He sits down and he starts talking about the drums and I'm like, I'm not fucking interested in how you can play, as long as you look the part you'll do. I don't give a fuck who you've been playing with. What we do is not rocket science, you've heard what we do, I can do it and I'm not even a drummer. As he was getting up to leave I said, 'By the way, you're doing *Top of the Pops* on Wednesday.'

Tony I watched it. It was really hard. I've been sacked for apparently not being the best drummer in the world, but our tune's at number one. We've got a fucking number-one album. Why am I sat home, what's wrong? One thing that really hurt me was fucking Bonehead turning to Alan White and they sort of salute. Fucking hell, you don't take long to forget, like. I was pretty affected by that, yeah.

Noel When Alan joined he was a breath of fresh air. It was a great thing to be in a studio with a drummer who, when you were talking to him about a middle eight, backing off on this bit, speeding that bit up, his eyes didn't glass over like he was thinking about a sandwich or something. He was a fucking good lad and he was a proper cockney. It was weird having a cockney in the band and all that, but he was great, he was a brilliant drummer.

Liam Top lad, Alan. Fucking Del Boy cockney dude. I don't give a fuck where you're from. It's not about where you're from, it's where you're fucking at. So it didn't feel strange at all, he was the man for the job at that time. What does he fucking look like? He's cool, a Smedley little mod. When I seen him, I thought, 'As long as he can fucking drum, he's in.'

Bonehead 'Alan White, he's alright'. That would be the thing every day. To the point of boredom, we would keep doing it. 'Alan White, he's alright'. He was alright; liked to drink and a great drummer. An amazing drummer.

It's Thursday night . . .

Noel I cannot overstate the importance of *Top of the Pops* to our generation. In the seventies it seemed so decadent and in the eighties it seemed so glamorous. I remember being gutted when we didn't get on there for 'Supersonic'.

Jason It is a big thing, *Top of the Pops*, because that's what you were brought up on, that's how you discovered your music. If it wasn't for *Top of the Pops* how would you know what band you like or anything like that? It was part of your childhood growing up.

They insisted that you mime, so you could just get pissed all day.

Noel I used to love doing *Top of the Pops*. They insisted that you mime, so you could just get pissed all day. Then all of a sudden everybody started to get up their own arses and say, 'We're playing live.' Well, fuck you, you are an idiot. Still to this day, 'Would you like to play live?'
'Not a fucking chance, mate. No way. If you want me to play live it's going to cost £50 a ticket.'

Maggie It's so unglamorous doing *Top of the Pops*. Back then they used to get us in really early in the morning. Seven in the morning, get the gear

up there, get them up there as well. Then you sit there all day waiting for your call to rehearse. You rehearse it, then you come back and you wait some more. You're just sitting there for eight hours. There's not a lot for them to do other than drink, to be honest.

Noel We had a great time backstage at *Top of the Pops*, a good laugh. I remember once being at *Top of the Pops* and Jimmy Nail was in the charts with 'Crocodile Shoes'. It was a boiling hot day and we were sat outside the BBC bar getting pissed and Jimmy Nail walked past. We all started singing 'Crocodile Shoes', kind of taking the piss. Jimmy Nail turned round, Liam kind of stood up and, I never forget, Jimmy Nail said, 'Aye, he's up and he's stocky.' I don't even know what that fucking means. We ended up going on the piss with him all day, he was top, man. He looked like he would have kicked the fuck out of all of us. Crocodile shoes or no crocodile shoes. I remember when Jon Bon Jovi went solo, his dressing room was opposite us and there was a knock at the door. He said, 'Hey, *Rolling Stone* are writing about you guys.' He gave us the magazine, it was a review of some fucking thing or other, and as he walked out he said, 'When you get to the States you can tell them Jon Bon Jovi showed you your first review.' And as he went into his dressing room, Liam went, 'Who the fuck is Jon Bon Jovi?'

Whitey's second gig

Noel Alan's first real gig was the main stage at Glastonbury. To think of it now, to put him on him like that, fucking hell . . . He's had one warm up, at Bath Pavilions, which was fucking crazy. We are the big noise at this point, so we are headlining the Saturday night, which is the big thing. We came on and the gig didn't work. We started with a new song. That is cardinal rule number one out the fucking window. Why did we start with a new song? And do you know what, it was a new song that didn't have any words, it was an instrumental. When you are at Glastonbury people want the hits and that is one thing I learned from that night; people don't want to hear your new fucking tunes. The worst thing you can say at a festival is, 'Okay, what I'm going to do now, is do a new one.'

'Oh alright, I'll go for a piss then.' We don't want to hear a song that we've never heard of before, fuck that. Play what you are famous for. So we should have done it that night, but we didn't.

I don't remember the gig man, whether it was good or not . . .

Liam I don't remember the gig, man, whether it was good or not, it was what it was. Glastonbury was never that fucking great anyway. I got psoriasis in my hair so I'd been scratching it all day and I remember some geezer coming up to me backstage after it and going, 'Fucking hell, man, it's Liam out of Oasis.' He's going, 'Come here, man, look, look, he's even got fucking cocaine in his hair,' as if I was Ziggy Stardust. It's psoriasis and he's picking it out of my hair, putting it up his nose, wiping it on his gums, and I was sitting there thinking, 'Alright, now I have seen it all now, everyone has lost the fucking plot.' The occasion is great and all that, but that geezer, was the only thing I remember about Glastonbury to be fair.

Noel What I will forever remember Glastonbury for is Robbie Williams. As I recall it, he is hanging round all day and doing his act, the Robbie Williams act. For a while it is funny, but getting into the eighth hour it's, 'Okay . . . wow.' Back then we let everybody stand on the side of the stage. Liam turned round and said, 'Come on, Robbie!' I don't really think he meant for him to come on stage. I just think he meant, 'Come on, hey.' I turn round and Robbie Williams is dancing like MC Hammer to one of my songs on stage at Glastonbury. I remember thinking, 'Oh, that doesn't look good, that does not look good, no.'

Liam Robbie Williams was brought into the camp through Noel's missus back then, Meg Matthews and all that lot.

Noel A friend of ours called Lisa Moorish is friends with Robbie and she came to Glastonbury. We were hanging around with loads of girls at the time and they are all Take That fans, as every fucking female seems to be – it's some kind of by-law that has been passed. She arrives with him and she is one of our mates ergo we are all hanging out together. Then he proceeded to follow us around the world for about a year and everywhere we went there he was. Him and Liam became the best of friends.

Liam It was nothing to do with me. He was there, just acting like a clown. There are all these pictures of him kissing me, that was like photo bombing, him making out as if he was my fucking mate. It was to do with all these fucking hoorahs down here that brought him into the scene, Noel's mates. He was Noel's mate, not mine.

Noel Fuck off, are you being serious? Oh my God, that is truly fucking unbelievable. I am fucking staggered by that. Seriously, those two were like fucking Morecambe and Wise, except not very wise.

Liam If you look at the backstage footage I think you'll see that I'm not there in this crowd. Fucking nothing to do with it. Our kid laughing, 'Yeah, Robbie Williams, yeah!' Where's LG? Disgusted in the corner somewhere, some geezer's fucking pulling my hair to pieces. Selling drugs off the back of my head.

KODAK 5063 TX
0A
1A
2A
6A
7A
8A
24A
25A
26A
30A
31A
32A
Hi res scan MAY 05

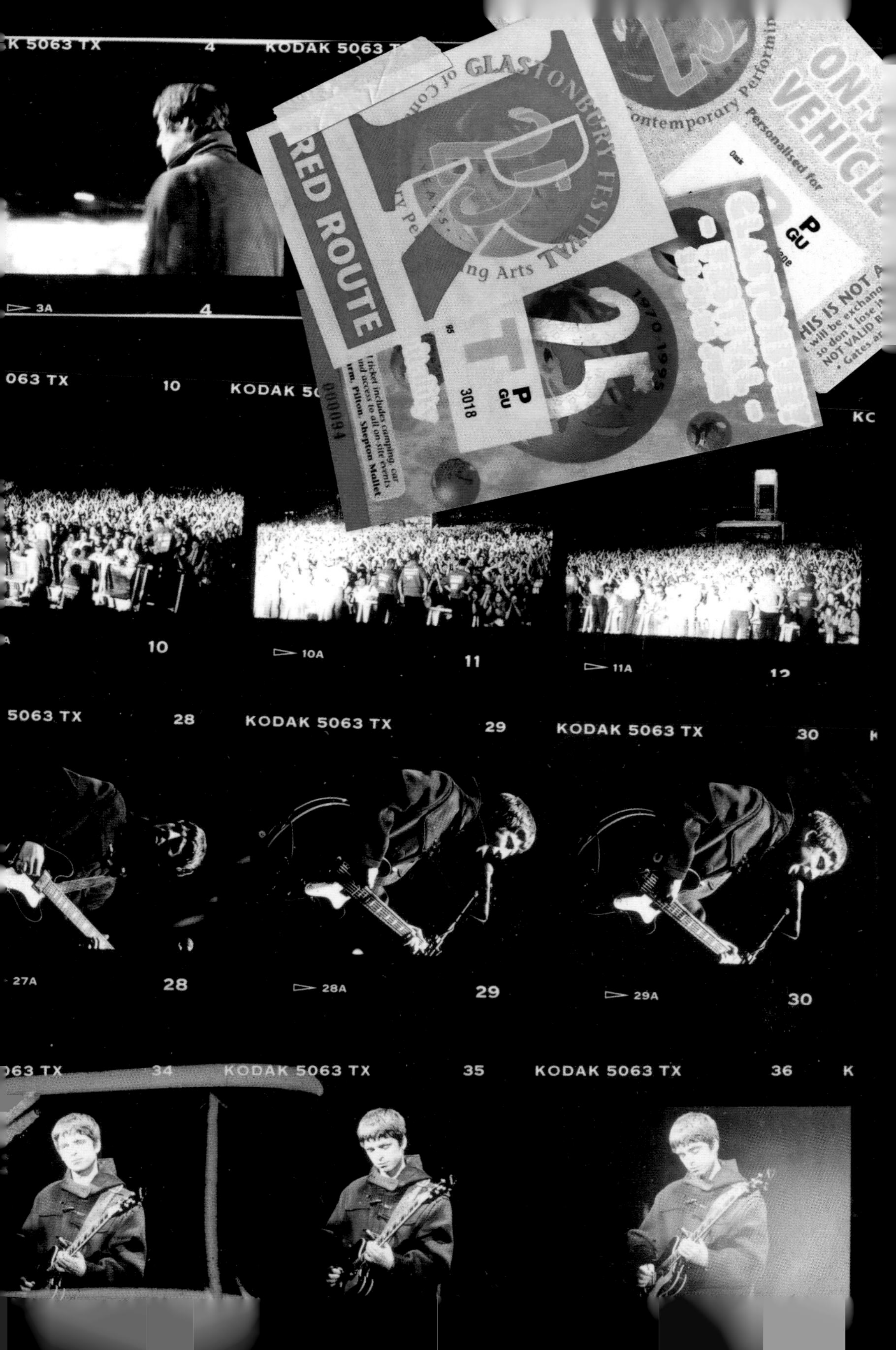
KODAK 5063 TX
RED ROUTE
GLASTONBURY FESTIVAL
ng Arts
Contemporary Performin
Personalised for
ON-S
VEHICLE
1970-1995
25
P
GU
3018
000094
ticket includes camping, car
and access to all on-site events
rm, Pilton, Shepton Mallet
HIS IS NOT A
NOT VALID
Gates a
3A
10A
11A
27A
28A
29A

Rockfield

Noel I don't do things like, 'This is going to be a new direction.' Every time I've tried to do that, to make an artistic move, it's failed miserably because I'm not an artist. I'm just a guy that writes songs; I pick the best ones and then when I'm recording them I take them to the logical conclusion, and then I put them out. Once I put them out I'm going and doing something else. I walk away from artistic decisions, or what's going on in the charts, or he's doing that so I should be doing this. Coming to record *Morning Glory* we are coming straight off the last big, British tour so we are fucking flying at this point. I might have played through them one night on the tour bus on acoustic guitar. I remember one night on a European tour and – most people in bands will know this – you arrive at a hotel at whatever time of day and none of the rooms are ready; you've got to sit on the bus for four hours until some conference for fizzy mineral water has taken place and everyone's left the rooms. I remember Bonehead saying, 'Play us the new tunes then.' I played the band what I had and there was silence and tears.

Is it possible to be walking slowly down the hall, faster than a cannonball? Well, if you do enough drugs it is.

Bonehead He actually sat down on the bus and played them us and we were like, 'Fuck, do it again.' It was incredible to hear them in that form. To hear 'Champagne Supernova' sung like that was pretty raw. Yeah, it got me, it hit me, touched a nerve, pretty emotional.

Noel Looking up and seeing him sniffling I'm thinking, 'Fuck me, it's not that bad. It's pretty good, it could be our *Stairway to Heaven*.' Bonehead was quite an emotional dude, he would always end up crying. Liam was quite an emotional dude, he would always end up fighting. Me and Guigs, as I recall, were quite level-headed, just kind of shrugging our shoulders and, whatever, you know. I am sure the drummer was eternally confused by it all and didn't know what the fuck was going on.

Noel 'Champagne Supernova': I've been asked about it so many times down the years by journalists. Is it possible to be walking slowly down the hall, faster than a cannonball? Well, if you do enough drugs it is. What is a Champagne Supernova? You just have to look out to the crowd and think, who cares what it is about? There is a great sadness in that song, like 'where were you while we were getting high' and yet it has generated the most joyous fucking thing ever. I put it down to magic, I really do put it down to magic and really wish I could tell you what the ingredients were for that magic potion. I haven't got a clue, I haven't got the faintest idea because if I did I'd still be doing it.

Owen They had a rehearsal, because Alan White's joined the band, and Noel is like, 'Come on, producer, you turn up for the rehearsal, check the arrangements.' Turned up and Noel was like, 'Yeah, fucking lovely that is. Good boy, he looks alright, doesn't he?'
It was like, 'Oh, that's alright, yeah, off we go,' and that was rehearsals for the album: half a day.

Noel It was the one record I've ever made that I've never done any demos for. I really didn't know whether any of it was going to work. My one reservation was writing those songs and thinking are we a good enough band to pull it off.

Owen It was May: brilliant time to be recording, sun shining. Rockfield: the posh studio, as well-equipped as anywhere in the world. Kingsley, who set up Rockfield, is a farmer and he built a fucking studio on his farm because he liked music and stuff. Fucking 'Bohemian Rhapsody' in his fucking farm sheds. And the accommodation was fucking magnificent, food was good, it was a good place to work. Out of town a bit, so you didn't get in trouble in town. Just a great recording place, brilliant.

Coyley Noel and Owen were in charge of the recording sessions. I wasn't making the record, so I didn't go. I went and made a record with another band instead. Just like that, gone. There is no discussion because what's the point? The decision is already made. Noel knows what he's doing, he knows what is best for the band and there is no time for sentiment or sentimentality, there is no time for anything. I was broken-hearted, absolutely broken-hearted, yeah.

Noel We went in with the idea to do it the same as *Definitely Maybe,* which was live. On the first day we did two takes of 'Roll With It' and what you hear on the record is the very, very first take. On the master tape you could hear Alan saying, in that cockney accent of his, 'Right, boys, first take, we'll get it first take.' At the mastering I decided we can't have a cockney talking at the beginning of a single and wiped it off. I should have kept it really because it was a poignant moment. The other songs were a bit more intricate. Owen said, 'Well, how Marc Bolan used to do it' – I still don't know to this day whether it's true – 'he would sit and play it with a click track on acoustic guitar and then build it all up from there.'

'Alright, well, let's try that then.' And it's something that I've done from that day forward.

Bonehead Yeah, it was a different way of working than when we did *Definitely Maybe*, but we had a producer who knew what he was doing, a producer who knew us, we knew him. Alan would come in and do his drums, and we'd have this incredible drum track with this incredibly tight bass track Guigs had put in. Then you had Owen being mister enthusiastic, 'Come on!' cheering you on. He'd be like, 'Fucking yes!'

Noel Owen came up with that way of recording for that record and it suited me. It was great because someone can sit in the control room and learn it while you are playing. If you are all in the same room playing a song only you know, you are going to get about fifty takes because somebody fucked up a little bit, fannying about for fucking fourteen hours getting on each other's tits.

Owen *Morning Glory* was incredibly easy – all my sessions with Oasis were incredibly quick and easy – because Noel had the songs and the arrangements and he'd teach the band the songs very quickly, you know. This is the song, boys, here are the chords, we'll have a few run-throughs, then we're going to record it. Two takes.

Noel *Morning Glory* took maybe sixteen days to record.

Liam To be fair, with all the recording side of it, the lads would go down there and just fucking jam and all that, they would be playing guitars all day. I'd just be upstairs watching the box. Or in the pub and that. So I never was really around it, the recording shit.

Owen We were doing a song a day, finished, completely done. That first week of *Morning Glory*, we recorded 'Roll With It' on the first day, 'Hello' on the second day, 'Wonderwall' on the third day, 'Don't Look Back In Anger' on the fourth day. He is like God, isn't he. Then fucking 90 per cent of 'Champagne Supernova' on the Friday. Extraordinary.

Noel We were never the kind of band to sit in there and theorise and try and intellectualise our music. This is the song, I've written it, this is it. I am going to play it how it goes, everybody join in. We weren't artists, we were just a fucking good band who had good tunes and that was it. It didn't take long. Wasn't really fussed about the surroundings. Even to this day the studios don't bother me; it's all about the songs anyway.

Liam Once there was a song finished and there was vocals needed on it and a microphone in front of me, all that other stuff was out. I had this tunnel vision in getting the best vocal and that was it. I was never thinking about, 'Right, let's get the fuck out of here and do this,' even though on some occasions it would be, 'Football's on, let's get it down.' I was pretty good at just fucking having it, let it all out there and just put it together. I never overthink singing that much, it was just one take, two take, three take.

Noel For 99.9 per cent of everybody in the music business this is how it works: get the vocal on as soon as you can. If you've got the bass, the drums and a good vocal you are halfway there. But no, not in Oasis. No, let's record all the fucking music first, all the overdubs and the hand claps and the fucking choirs and the strings and the bells and fucking whistles and then right at the very fucking end, then let's put the vocals on. That became a running theme with Oasis for years. You never really knew what the fuck was happening until the singing was on and the singing was on right at the very, very end.

Liam I preferred singing live because it was one take, you were on your toes more. It was like being in a boxing match or something, whereas in the studios you can cheat a little bit. You can't really cheat when you are at a gig. I preferred that to be honest.

Noel The vibe was pretty good, we were rattling through it there was no fucking about. We were having a good time, the songwriting had taken another step forward. If anybody ever thought I was never going to beat 'Live Forever', we had 'Don't Look Back In Anger', we had 'Morning Glory' and 'Champagne Supernova' and 'Wonderwall' for crying out loud. I remember being into it, everyone else being into it, and then there was the night where it went off.

I never overthink singing that much, it was just one take, two take, three take.

Bonehead We were generally under Owen's orders, 'Fuck off, go to the pub. Get out.' He didn't want us there, you know what I mean, that was Owen's way of working.

Noel That gave Liam a lot of time to fanny about in Monmouth and indulge his greatest hobby which was acting like a fucking buffoon.

Liam That's the problem when you've got these nice villages and nice recording studios, the village, man, it has a pull. You've got to be very strong willed. I always had a pull towards the pub, it was, 'What time we finishing here, shall we go back into town? It's got a bit of a vibe.' I was a sucker for the pub and Bonehead was, Alan was, and all the crew obviously. Guigs wasn't that much and Noel wasn't, so it would be always like, 'He's doing his guitars, he'll be in there all fucking day, won't he, Slash? We could go to the fucking pub.' He would come out and be, 'Right, guys . . . where is everyone?' They are all in the fucking pub. So it would be that kind of thing. I like them kind of vibes but it did cause arguments.

Brian Cannon What a night this was, Liam and I went into Monmouth and we bump into John Robb.

John Robb* I was in the studio next door to Rockfield, Monnow Valley, producing this band called Cable. We'd just finished recording, we went to town for a couple of drinks and there's a bloke with two girls in the seat in front of us. Oh Christ, it's Liam, I go, 'What you doing here?'
He goes, 'We're in Rockfield, we're just doing the album, come down, we're having a party.'
So, we went down there.

Bonehead We knew John Robb, and we invite them all back, drunk. Probably a bad move, because Noel was still working at the studio with Owen, you know, and we bring a bunch of guys we've never met in our lives.

Liam There is a lot of sitting around so what do you do? It's not prison, if I'm not needed today I'm hardly going to sit and fucking watch you play guitar. I'm going to the pub. Obviously you go to the pub, you get wankered, and I probably did bring a few people back one night to have a party; he weren't happy with it and then we smashed the place up and stuff.

Noel He came back with a load of fucking scruffy-looking cunts. They came into the studio, Owen kicked them out and sent them up to the farmhouse.

John Robb Liam goes, 'What do you think of this album?' He's dancing round, 'It's great, isn't it?' And a bloke goes, 'It's crap, it sounds like The Beatles.' The bloke was pissed out of his head, and then he starts prodding Liam with his finger, you can see Liam get rattled. I think, 'I'm just going to leave them to it.' Then there is this massive crashing sound, and the fighting kicks off.

Bonehead It was kicking off outside, big time. I think I hit one of this band that had come down, they were gobbing off, so I punched him.

Liam I liked bringing people in and going, 'Have a listen to this.' Back then it was different. You couldn't do that now because everyone would fucking record it and everyone would take pictures. I shouldn't have brought them

*Musician and award-winning journalist and author.

back, but I thought we were a rock-and-roll band. I thought anything goes
n Oasis. Obviously other people had rules . . . Fuck the rules, man.

Noel I am not generally an obnoxious, belligerent dude. I know people are going to think, 'Fucking hell, I've read your interviews . . .' but generally I'm not. Somebody must have got on my tits. My thing is, when I am working, am fucking working, I don't particularly want people there fucking about. am in the studio, I am in and out. I don't want to be fannying about spending five years making a fucking album.

Bonehead It just exploded into a big, fucking bunch of chaos like you've never seen, man.

Noel Why I had a fight with Liam that night I couldn't fucking tell you, I think maybe one of my guitars got damaged.

It might have bee the bigg fight we ever had

Liam When we had fights we had fights, but we didn't have loads. We had raging arguments, but we weren't fighting. We probably had about five physical fights. We've had hundreds of big, 'Let me at him, hold me back.'

Noel It might have been the biggest fight we ever had.

Liam Probably me not giving a fuck and him trying to write 'Bohemian Rhapsody', and me going, 'Bollocks, let's have it.' That kind of thing. Ying and yang. Him being professional, me being unprofessional. Him turning the band into a square band and me turning it into a fucking oblong band. That kind of shit.

Noel Guigs always used to bring a cricket bat and a set of cricket stumps with him because he is obsessed with cricket. Him and Bonehead playing cricket is a sight to behold I might add. I remember smashing Liam's head in with a cricket bat.

Liam It definitely went off that night, it was bad, man. There were loads of little bits in the corner going 'pfzzz' – electrical bits – loads of broken shi Terrible. The whole studio got smashed to pieces. I think Whitey was locked in a room while I was going to town on the living room. Everything just got fucking blitzed. Noel got Whitey and I think he drove him home, that was it

Noel Me and Whitey left in the middle of the night. As we were leaving the studio Liam appeared out of nowhere and threw a dustbin at the car. I think

it was Alan's car and we'd only known him about two weeks. He was going, 'Fucking hell, mate, what kind of fucking band is this?' He was saying, 'Is that it then? It can't be over, I've only just joined.'
'No, no, don't worry, this is going to happen all the time.'
He was like, 'Fucking hell man, have I joined the fucking Troggs?'

Liam It happened man, but there was no need for the cricket bat round the head.

Owen Next morning, what was that then? Liam's broken his foot, Bonehead's nose gone. I missed it completely, flat out, I was knackered. Strawberry fucking bubble bath from Bodyshop, very nice too, thank you. That was a good week, into bed, zzzz.

Michael Spencer Jones I can remember waking up in the morning and walking into Liam's room, I just couldn't believe the scene of devastation. I wish I had taken a photograph of that room, you know. TV was out the window, still plugged in, double bed was in half. I mean, it was just like, you know, a hand grenade had gone off in the room or something.

Bonehead The studio staff come in and they were cool, like they'd seen it before. They were just like, 'Right, yeah, okay, looks like there's been a bit of a fight, we'll just clean up.'

Michael Spencer Jones They found this local repair man and he came along, and he was saying, 'Oh, we've not seen damage like this since Ozzy Osbourne and Sabbath were recording here, this is nothing.'

Noel There is a cooling-off period. I went to Jersey I believe, I don't know why the fuck I went to Jersey. I think Liam went to Portugal.

Liam I remember going home and I think my foot was fucked and my arm was in a plaster.

Noel Every time there was a scene in Oasis, when we all got back together it was like nothing had ever happened. Nobody would ever mention it. I don't ever recall there being heart-to-hearts. 'Where've you been?'
'Fucking Jersey – where've you been?'
'Been in fucking Portugal, haven't I?'
'How was it?'
'Fucking sunny. How was Jersey?'
'Shite.'
I don't remember a sit-down airing of your grievances, it was always like nothing had ever happened.

Liam Me and him are like telepathic. I felt that we didn't really have to speak to each other half the time because we just knew, you know what I mean? People have that with their brothers and sisters or family and whatever, we didn't really talk that much. I don't think me and our kid ever sat in a fucking room on our own and had these one-on-one questions, nah, fuck, that would be well weird. That would be weird as fuck.

'Hey, Noel, do you believe in God?'

'Hey, Liam, can I just have a quick chat with you there, do you believe in God?' What is it all about? That would be fucking weird, man. No.

We weren't nice, sensitive, middle-class boys that go off sulking if something ain't going right.

Noel We were not brought up in that kind of family. There are five of us in our family: me, two brothers, my mam and dad. We are all close to our mam individually, but not any of the five individuals, none of us, are close to each other. I don't know why that is. That doesn't seem that odd to me, do you know what I mean? Now as you go on through life, involved in other families, you see the Swiss Family Robinson. They are all in each other's pockets, that kind of freaks me out. That's just the way we were brought up.

Bonehead I don't remember it being bad when we got back, I don't remember it being like, someone sulking in that corner. I think it was more of alright, we're back, we're here, we've got a job, let's do it. I don't think there was any bad atmosphere.

Noel We were working-class lads, all of us had Irish parents, so we were little paddy fighters. We didn't give a fuck. We weren't nice, sensitive, middle-class boys that go off sulking if something ain't going right. The kind of band we were, rightly or wrongly, if anybody was ever asking for a smack in the mouth they got one, and then we moved on and that was it.

Liam That's the way I see it, man, that's my mantra: live and let ruck.

Noel The thing blew up that night, and after that we are fine, just normal. Same as things leading up to the night at the Whisky, we are fine. We never let any of it fester. If there was bad blood between anybody it was happening at that particular moment and then when it was gone, it was gone, and we got back in the saddle and did whatever we had to do. We weren't artists. None of us. We didn't give a fuck about that, we were grafters. I never had the feeling

It happened
but there
for the
round

man,
was no need
cricket bat
the head.

that it was over until it was. I never had that feeling. Who was going to leave Liam? What is he going to do? What was I going to do? I didn't have a band, I couldn't sing . . . it was just shit blowing over. I recall buying him a Beatles belt buckle. Maybe I was feeling guilty because I'd fucking broke his arm or something. Maybe I bought it for me and it wouldn't fit.

Liam About £13 they cost, you can get them anywhere. Made up, nice one for that, our kid.

Owen They came back and they had songs to get on with. The sessions slowed down, we did the main tracks in that first week, but then there was a bunch of tunes to do. Like 'Morning Glory' and 'She's Electric'. We were still doing the B-sides, there was a lot of recording to do. We filled up the time.

Noel I live day by day, I don't carry shit with me. Once it's done, it's done. I am not a sulker. When I get back to the studio I forget what went on two weeks ago, I have to finish this record off. We rattled it off pretty quick after that. We weren't the most experimental of dudes anyway. Bonehead had one guitar, I think I had two, we had about four amps and that was it, what are you going to do? Now, when I make records, I am obsessed with FX pedals; I didn't know what they were then. I had a tuner and a digital delay and that's all I had way up until after Knebworth. It's embarrassing.

Liam I knew the songs were a lot richer than *Definitely Maybe*. 'Wonderwall' was the only one that I didn't get at the time, I was, 'I don't know about this man.' It just didn't seem right. I thought it was weird. Then when I heard it I was, 'Ah, now I get it'. I was totally wrong and that's fucking right. I love that tune. Growing up we didn't have much money or anything, there was no carpet or shit like that, no wallpaper and we would just write shit on the walls and have posters up. I always thought that was our 'Wonderwall'. But it's obviously not; apparently it's about some bird. Or maybe not.

Noel In the case of 'Wonderwall' and 'Don't Look Back In Anger' I wanted to sing one of them. Liam hated 'Wonderwall' when he heard it. He actually said, 'I'm not singing it, it's not fucking rock and roll.' I'll sing it then. When it takes shape and everyone is going, 'This is going to be massive,' he decides that he's going to save it from mediocrity and make it what it was. So I said I would sing 'Don't Look Back In Anger'. I'm definitely singing one of them.

Liam I don't really remember the actual day going dum, dum, dum, like a fucking gameshow kind of thing. 'You've got five seconds to fucking pick your song.' I don't remember it like that. I think, if it was, I would have fucking picked 'Don't Look Back In Anger' because, like I said before, at the time 'Wonderwall' was a bit peculiar. I certainly didn't have the choice of singing both; I've never sung 'Don't Look Back In Anger', not even in the shower. I would have fucking nailed it without a doubt. I'd have sang them, easy, but he wanted to sing them, whatever he sung he sung and whatever I sung I sung. I definitely didn't sit in my room crying or ordering up more crystal meth going, 'I want to sing "Talk Tonight".' Did he sing it anyway? Yeah. Done a great job.

Owen Liam was extraordinary, I mean he had this freaky thing where Noel would play the song once, on acoustic guitar, in the control room to Liam, give him the words and then he'd go and he'd fucking sing it. Noel would listen to it to check Liam's got the phrasing for the entire song that he's only just heard the once, melodies, phrasing, the whole fucking caboodle. Noel would go, 'Nice one, Liam.' Noel would fuck off, Liam would just bang another four tracks down. This would be like one in the afternoon, cup of tea, lead vocal done. What the fuck? He's only heard the song once. Five songs, five days, man, fucking amazing.

Liam We'd have a beer, slag our kid off as he'd walked out the door and go, 'Fucking check him out, Slash,' and then we'd get down to proper business. We'd have a laugh, just fucking do it as loud as we could and as raw as fuck. As time went on, some producers were talking this mad language, fucking musos, whereas Owen was, 'Right, take me head off,' and I'd do that. Then that would be it really, we wouldn't sit around and scratch our chins about how great the vocal was, we'd just be raw as fuck.

I would write a song and immediately know Liam was going to sound great singing it.

Noel He was, at that point, great at that, the best in the world. But it's not complicated, it's not jazz fusion. I would write a song and immediately know Liam was going to sound great singing it. I wouldn't have the power for it or the attitude to give it the snarl that was needed, and by the same rule . . .

Liam There was always this thing, 'You can only sing the rock-and-roll songs, your voice ain't up for it.' Leave it out, mate, I'd sing that in my sleep, but he obviously wants to sing it, he wrote it, do what you want. That was definitely the start of Noel picking and choosing the songs he wants to do: 'I'll do this one.' I think, once you open that box . . . It got to the point he was singing half the albums, he is just trying to put me out of a job, the fucker. That's when it started going a bit pear-shaped for me, musically. You've got me in the band; I should be singing all the songs, end of.

Noel One moment that sticks out for me in Rockfield is when we get to do 'Wonderwall'. Outside of the main studio is a wall which is about six foot tall. I don't know who came up with this idea but, 'Why don't we record on the wall?' I was sat on top of a wall with a huge microphone stand facing the acoustic guitar, and headphones going into the studio. I was sat getting the sound right thinking, 'This is fucking amazing, this is like transcendental fucking Beatles shit this.' Then I open my eyes and there's four sheep stood looking at me sideways, chewing grass in the way that they do, 'What the fuck are you doing on that wall?' I vividly remember thinking it was a great idea until I was being given a withering look by a farmyard animal. Actually no, this is shit, can we go back inside now, please, I'm cold?

I want to get in and I want to get it done and I want to get out as quickly as possible.

Owen We were probably a bit young and enthusiastic and dumb and it probably could have been recorded a bit better or posher, but it worked, and we got it down.

Liam I like Owen, he doesn't bullshit, he was just a fucking geezer who knew how to do his thing. To be fair he was winging it just like us. It felt like he was in the band. We'd come back in and he'd go, 'Have a listen, what do you think?'

I'd go, 'Sounds good to me.'

He'd go, 'That's that fucking done.'

If our arms were down we knew that song weren't going on the album or that take was shit. If our arms were in the air, and your drink was in the air, and you'd got a fucking joint in your mouth, you knew we were on it and you go, 'That's done, can we go to the fucking pub now?' It was like that. We'd never sit there and overthink it and go, 'Tell you what, let's just do another fucking five takes or sing it like this.' If it was there, we would be out the door. And that's the way it is, we never judged it, we never thought of going, 'I wonder what the manager will think of it, I wonder what the record company will think of it, I wonder what that kid in fucking Doncaster will think of it?' It was like, fuck them, it was just, 'Does that sound the bollocks to you?'

Noel I want to get in and I want to get it done and I want to get out as quickly as possible. I don't want to sit around for four months fannying about with a bass drum – that is for fucking knobs. Life is too short for that shit. Fucking football is on. I wouldn't argue with Liam's opinion on that.

CHAMPAGNE SUPERNOV

HOW MANY ~~SPA~~ SPECIAL PEOPLE CHANGE
INTO WHAT WE SEE AS STRANGE
WHERE WERE YOU WHILE WE WERE GETTING HIGH?
SLOWLY WALKING DOWN THE HALL?
FASTER THAN A CANNONBALL?
WHERE WERE YOU WHILE WE WERE GETTING HIGH?

OMEDAY YOU WILL FIND ME
AUGHT BENEATH A LANDSLIDE
ITH (MY/A) CHAMPAGNE SUPERNOVA IN THE SKY X 2

AKE ~~[illegible]~~ UP THE DAWN AND ASK IT WHY -
U NEVER DREAM YOU NEVER TRY?
RY AND SPEAK THE TRUTH UNTIL YOU LIE
U WAKE ME + GET ME HIGH
DREAMER LIVES HE NEVER DIES
WIPE THAT TEAR AWAY NOW FROM YOUR EYE

PEAT (A) X 2

PEOPLE BELIEVE THAT THERE GONNA GET AWAY ~~[illegible]~~ FOR THE SUMME
E YOU + I WILL LIVE AND DIE
WORLDS STILL SPINNING ROUND
'T KNOW WHY!

Brothers

Liam Noel: geezer, brother, dude. Miserable bastard. Funny cunt, funny, but miserable . . . Love him dearly, man. Without me he wouldn't be the man he is.

Noel I don't love Liam and I don't hate him. We are just family and families are different; family cannot be summed up in one nifty sentence. What does your mam mean to you? That is too fucking complicated a question. What does your dad mean to you? You can't just say that in an interview. So it's easy to say you love them. But I know what love is. I love my wife. And if that is what love is I certainly don't love Liam, do you know what I mean? I don't hate him either. There is like a bit in the middle that's just . . . you tolerate their shit and vice versa.

Love him dearly, man. Without me he wouldn't be the man he is.

Liam He's a great fucking songwriter. He is my brother first off and that's that.

Noel We used to fight all the time. All the time.

Peggie They are two big egos, the two of them and none of them will back down. Noel is stubborn, always was stubborn. Liam, I know Liam has got a mouth on him.

Jason Brothers is on–off. I've got a brother, we are on–off. You get under each other's skin. You know exactly where to poke that pencil, give it that little nudge and it will just tip it over the edge.

Liam We definitely had little arguments, but I think people are blowing it up a bit too much. These things get blown out of proportion. It's not that bad. The press love it, they love this Mick-and-Keith, Ray-and-Dave kind of thing. It's just the way it is and I didn't mind. If people want to write bullshit, let them write it. I haven't got time for the bullshit.

Maggie When you have that kind of relationship within quite a closed environment like a band, sparks are going to fly somewhere.

Noel I hate the term sibling rivalry because it just sounds ridiculous, but that's effectively what it is. It is a shame because when we first started and we got interest from Creation, Liam would say to me, 'I'll leave everything up

to you because you know what you are fucking doing.' Then it turned on a sixpence when he moved to London. He would question every fucking thing. If there were teabags in the rehearsal room, 'Why is it Yorkshire tea?' Well, because it just fucking is, because it's the best tea in the world, ladies and gentlemen. 'No, I'm not drinking it, get the fucking coffee.' I am exaggerating there, but it was a bit like that.

Liam I think it all stems back to when he bought a stereo, years ago when we shared a room. I've come in one night stoned and, you know when you are going, 'Where the fuck . . . I need to get up and have a piss.' I think I've got up and I couldn't find the light switch so I've pissed all over his fucking new stereo. He's woke up and he's gone, 'What the fuck are you doing?'
'Having a piss.'
I think it basically boils down to that. He's held that grudge. Need to let go of them grudges, brother.

Liam is like a dog

and I am like a cat.

On stage, when he turns to me, and I turn to him, and we just both look at each other, everything just clicks and it just, like, transcends music. And it's only ever me and him that will ever get this . . . And that's what it's about for us.

Noel talking to Steve Lamacq
The Evening Session, Radio 1

Noel Liam was always cooler than me I think. He had a better walk, clothes looked better on him, he was taller and he had a better haircut. And he was funnier. I guess I was more articulate. Liam, clearly, would have liked to have had my talent as a songwriter and there is not a day goes by where I don't wish I could rock a parka like that man. Nobody can rock a parka like Liam. Even in leopard-skin slippers and a parka – nobody gets away with that shit.

Liam He's a great musician and all that and he's got that on me, but believe you me, when I'm stood next to our kid he doesn't get a fucking look in.

Noel I guess the fights are not specifically about the thing that sparked it off really, it is about a power struggle. It is about me being in charge and everybody – from the fucking woman who makes the tea to the guy that signed us and everybody in between – directing everything towards me and Liam being pissed off about it. That is basically what starts it all and right up until the end, that's what it was. A power struggle is what it is. Any psychiatrist will tell you that.

Liam He's a middle kid, isn't he? I guess maybe if you are going to start looking at it in a psychological way, the middle kid's thinking it's all stopping, he's the main man and that and then I fucking pop up and shit on his parade. Psychologically I guess it could come from that. It could come from the fact that I pissed on his sound system one night. It could come from that I'm just better looking or whatever.

Liam, clearly, would have liked to have had my talent as a songwriter and there is not a day goes by where I don't wish I could rock a parka like that man.

Noel There was a period where he was the greatest singer in the world, added to that the greatest frontman, added to that he was a fucking good-looking boy as well, with great interviews and wore great clothes and all that, it was perfect. They were the magic years. There were points where he'd sing 'Rock 'n' Roll Star', 'Live Forever' or 'Morning Glory' and it would be blistering. Nobody sang like he did.

Liam I know Noel thinks I'm brilliant, at the end of the day we didn't need to pat each other on the back and go, 'Oh, you're great.' He gives me the song and I deliver it. I didn't need the pat on the back and neither did he.

Noel Liam is like a dog and I am like a cat. Cats are very independent creatures, they don't give a fuck, right bastards. Dogs: 'Just play with me, play with me, please, please, please throw that fucking ball for me, I need some company.' It is as basic as that. I'm a cat, okay, that is just what I am. I've accepted it, I am a bit of a bastard.

Christine Mary Biller* They are two brothers. Grew up together. One's one way, one is the other. Noel has a lot of buttons. Liam has a lot of fingers. It's that simple really.

Liam I didn't want to hang out with him as much as he didn't want to hang out with me, I had me own fucking mates. None of us are into this kind of sitting around on Christmas Day having dinner with your family. For us it never happened and that's a shame. It's certainly not like we all run around going, 'Come on, Noel, come over, let's go and watch a film, let's go for a walk in the park, show us a picture of your kid and I'll show you a picture of mine.' None of that shit. We weren't like that.

I would much rather not argue with our kid, I'd much rather have good times, go out and have a drink and have a laugh . . .

Noel We've never been that close, never been that way. So it fascinates me that people find it fascinating, but that is not weird to me, that's just the way it is.

Paul Nobody expresses themselves in an Irish family. They don't go, 'I love you, brother,' nobody does that. I think later on in life you have a grudging respect and go, 'He's alright, I suppose.'

Liam I would much rather get on with my brother. Would have much rather got on with him than hate him, but we didn't hate each other for the cameras. We would have disagreements because we'd have disagreements, we are not the same person. He is a bit different, I'm different, and we didn't throw our arms round each other for the cameras, we did it because it was real and that's the way it is. I would much rather not argue with our kid, I'd much rather have good times, go out and have a drink and have a laugh than fucking not see each other. That's shit.

* *'Rock chick' and friend of the band.*

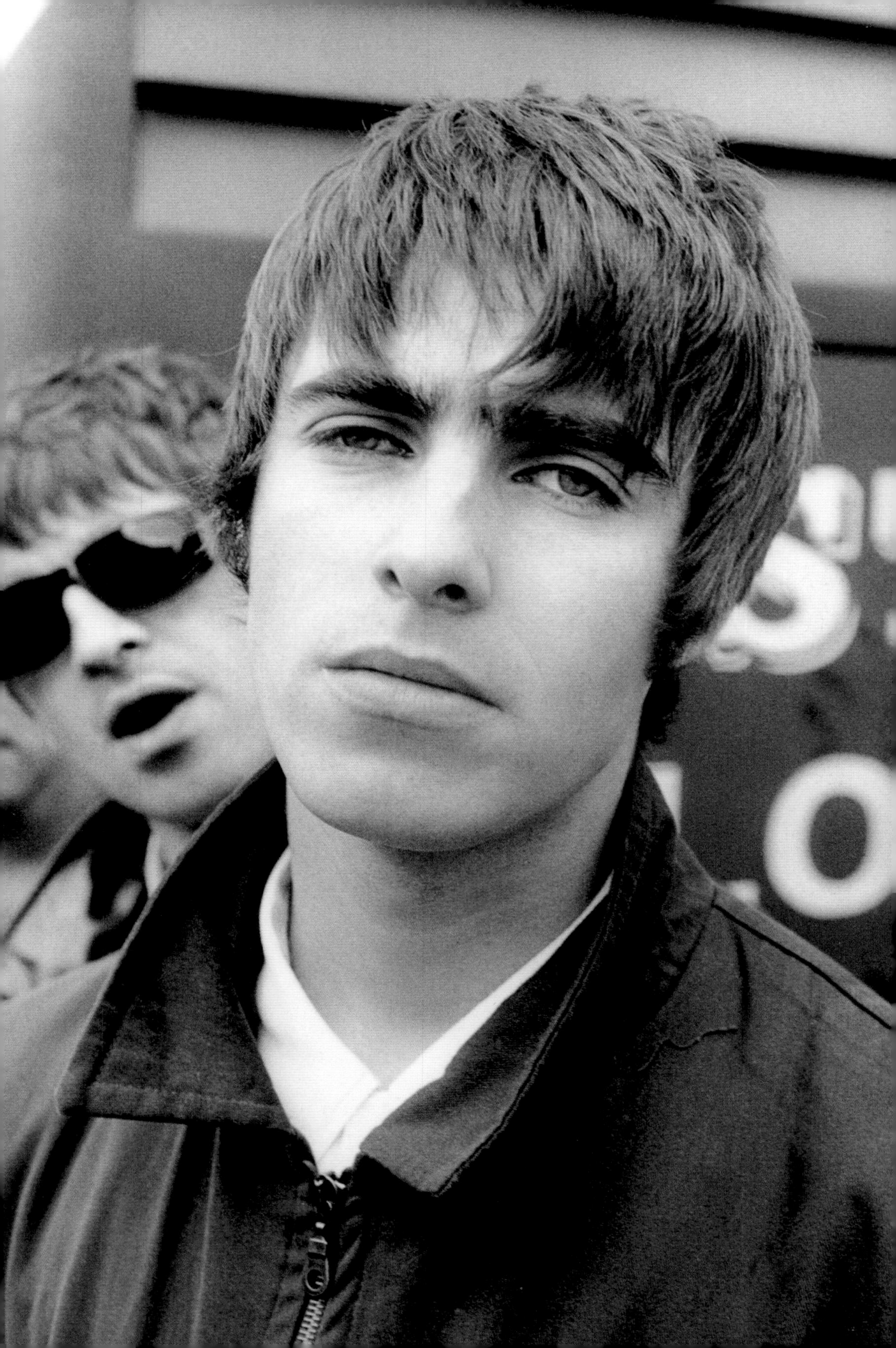

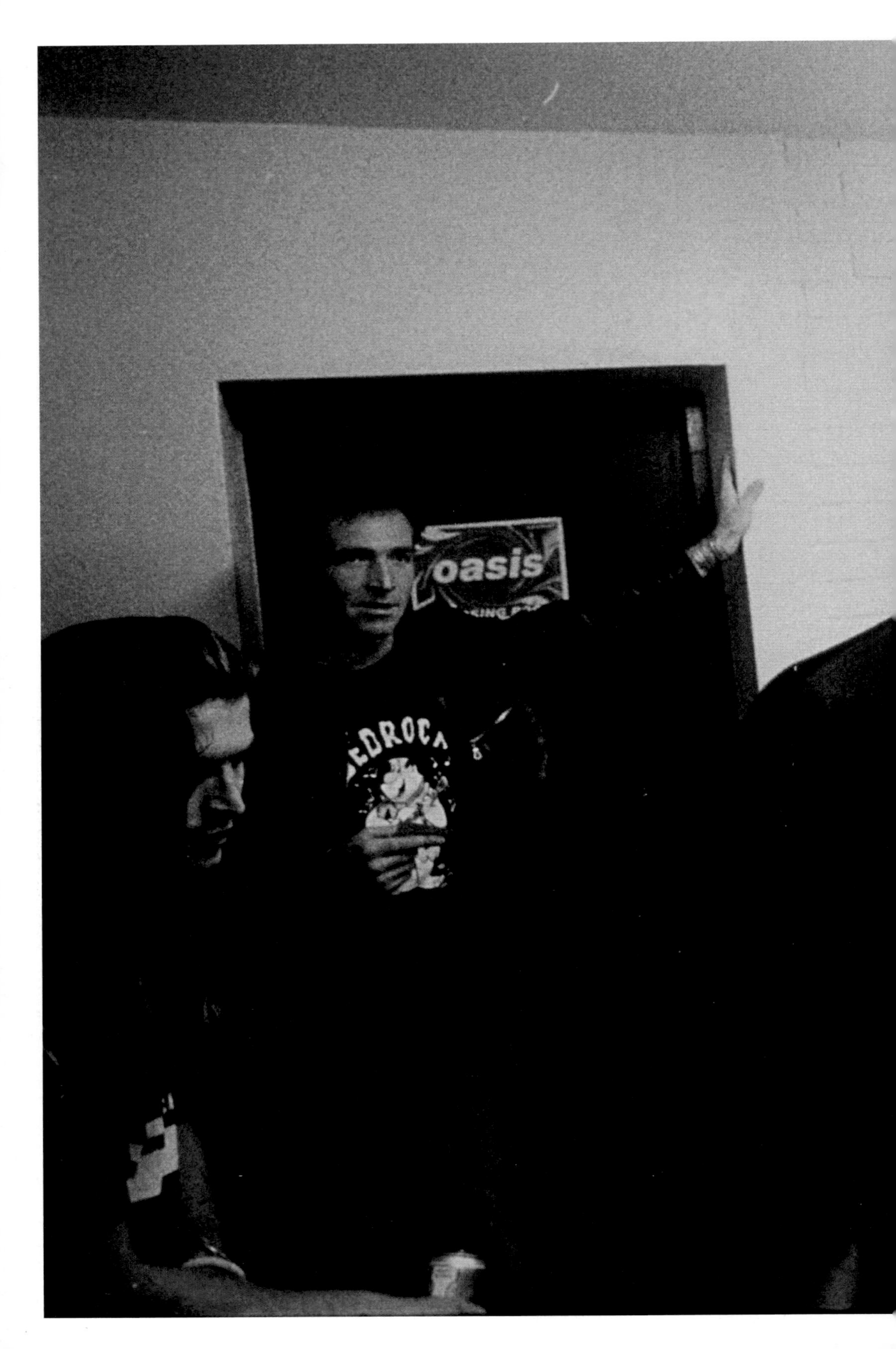
oasis

4A
5
5A
6
21A
22
22A
SAT 22nd APR 95 7:00 PM
SHEFFIELD AREN
SJM Concerts present
OASIS
plus Special Gue
BLOCK 120
DOORS OPEN 6:00PM
£0.00 PLUS £0.0
PROMOTER
SHEFFIELD
ARENA
Ponds
orge
HEFFIELD
ARENA
APRIL 1995
You are invited to join
oasis
AFTER THE GIG, 22 APRIL '95
in the Malcolm Allison
at Sheffield Hallam
11pm till 3am
Pond Street,

Coyley's last stand

Noel The way it works in Oasis is we don't care about where we play or who we play to, just point us at the stage and we do it. We trust the people that work with us, we trust Marcus and our promoters and our lawyers and Alan McGee and all that; that's just shit you don't need to be getting involved in. I remember walking into Sheffield Arena and actually laughing and saying to the promoter, 'When is this going to take place?'
'We are thinking eight weeks from now.'
'You are fucking mad.' It was a huge arena.
For all the bravado in the press and giving it the big one and all that, we always really underestimated what we were worth, which is why bands should never get involved in any decisions like that.

Liam I remember that gig. I remember telling all the crowd to come down from the seats and it went right off, I fucking loved it. I remember I was a bit pissed off because they have bits at the front that are for all the people who have paid more money, I remember it just being a bit quiet down the front and everyone was having it in the stands. And I was thinking, what the fuck is all this about? Everyone's down the front just stood there and everyone's having it at the back. I just asked everyone to come down from the seats and fuck these off from the front and have it a bit. Everyone was just diving out of their seats, coming down the front and then it went off and that was great. That's what them days were about. When we started playing arenas and that the crowd were fucking going mad. I've been to some arena gigs and it's just boring, man, it's like being at Badminton or whatever, it's just boring. Our big gigs didn't lose the fucking chaos. Our big gigs got more chaotic. That's what I felt anyway, I loved it, fucking watching about 20,000 people just fucking all swimming about and going mad. I'd have that any day over 800 or 2,000. The more, madder people in one room the better for me. All the stands were going off, everything was going off, it was just going off and I fucking loved every minute of it. The shittest part about being in a band is not being in the crowd – you want to be part of it. We were up there doing it, and it was all great, and then you would come off and you would be like, 'Fucking hell, them lot looked like they were having it,' and then the crowds were getting bigger and stuff like that. What's going on? At the end of the day you are still just five fucking idiots from Manchester so what were they seeing? Is there something else going on behind us? Part of you just wanted to get in amongst them and see what they were thinking. I don't think we ever captured on record how good we were live really in the early days – we were like a punk band, we were ripping it up, man. No one could fucking touch us. There was all this Britpop going down, everyone was farting about, and we were like a steam train, just ripping your fucking head off. We were like a Rottweiler and then it all kind of got a bit 'Beatley' and a bit more melodic and that.

Jason You certainly need some presence there; he is what fronts the band and no matter who writes the songs or who does the lyrics or whatever, he is the front person. You need somebody with some balls and a bit of character up there, not somebody who is going to whistle a tune in a safari jacket or something. You need somebody who is going to belt it out and get a bit of attitude out.

Liam My mantra was just go on, be your fucking self and all will be good in the world. So I'm pretty good at being me. So it was a piece of piss, it was a doddle, man; all of it was a fucking doddle. If you go on there and you're not being yourself, you're going to get found out, aren't you, and rolled over, so it was a piece of piss, man. Anyone that played with us got a lesson in how to do it, we were a mega, mega band. We weren't frilly or anything like that, we were just like a fucking juggernaut.

Noel Within eighteen months, two years we were selling out football stadiums, but again, with the power of youth and all that, you don't think about it. You don't think about it, you just do it. We didn't really have the foresight; we became an arena band almost immediately, then we became a stadium band almost immediately after that. We jumped ten levels almost immediately, it took U2, fucking hell, twice as long to play stadiums. But no one was really arsed about the show, we were still behaving like we were playing in nightclubs.

Liam It felt beautiful, it felt massive, the sound was better. I could sit here all day, I guess, and play it cool and that, but without a doubt it was fucking amazing playing arenas and that. I didn't feel nervous about it and I give you that straight, it didn't feel like it was too big or anything, it was like, 'Fucking bring it on, can't fucking wait.' But then always thinking on to the next thing, going, 'Right, so we are having a bit of this, what's fucking next after that?' and then we proved it. Knebworth, stadiums, there was always something in the back of my mind or in my soul going, 'This will fucking do for now, can we just get on with it and get to the next thing.'

Noel I don't ever remember being nervous about those shows. I remember being more nervous looking at the itinerary – it just had the word 'helicopter' in it every fucking other page. I was thinking, 'Fucking hell, I'm spending a lot of time in helicopters in the next few months.' That ramps up the odds of dying quite fucking sharpish. Rock stars, baby. We were helicoptered everywhere: helicopter to Slane, helicopter to Loch Lomond, helicopter to the fucking chippy. I was sick of it at the end. Fucking hell they are a pain in the arse. Wearing them fucking headphones . . . helicopters are not great.

Liam Getting off a helicopter the only thing I'm thinking is, 'If that fucking blade messes my hair up, man, I'm walking home.' It was like that, fucking stupid fucking helicopter, fucking messing me hair up. That is seriously where my head would be at.

Noel I remember having a cig on a helicopter on the way to Slane Castle, shouting at the pilot, 'Is it alright if I smoke in here?' I lit up a cig and I don't know why, I asked him if he had an ashtray. He said, 'Open the window.' Flying over these farms in Ireland flicking this cig out of the window, thinking this is fucking bananas.

Liam There were a few people around then that just weren't too clever with the flying business. But getting in a helicopter at your gig? I'm in.

Bonehead We'd done a gig in Belgium the night before, so me being me, I had the world's biggest hangover.

Noel He's quite an emotional dude, Bonehead; honestly it was like trying to get a young calf on to the back of a wagon. 'I am not going on, I am not going on, I'll get a car.'

'It's nine hours in a car.'

'I don't care.'

'Well, we are on in three.'

'Well, I'll run then.'

'Just get on the fucking helicopter.' He wasn't a fearless flier.

Bonehead I'm sure the pilot knew because he just started doing Vietnam fucking stunts, I hated it. Fucking helicopter, stupid invention, what's that about?

Noel I remember we got helicoptered in and congregated at the back of the castle. As we were waiting Johnny Depp appears and introduces himself like we didn't know who he was. And he's going on about 'Live Forever' and *Definitely Maybe*, 'Oh man, those songs.' I was like, wow, fucking hell, holy shit, man. Anyhow, the way Slane is, you look out on to the crowd and it is a hill. Flat down the front, then this big steep hill, and then for some reason the crowd then snakes round this tree and the bars are right at the top. So when we went on, seemingly 80,000 people were at the bar. We came on and plugged in and everybody ran down this hill. It wasn't scary for us, but it was a real sight to see, it was amazing. The gig had to be stopped, there was crushing going on and all that, and it was fucking wild, man, but that's what it was like at our shows.

Peggie It was a great time that, because I had all the family there with me. We had a police escort through Dublin and flying all over, because me sister and her husband had come from a different part. I thought it was great, to me, it was so good, I was so proud of them.

Noel Liam was being a cunt all day, I don't know why. He gets nervous at things like that I think. See that is the difference between me and Liam, I would just embrace the bigness of it all, I love that fucking shit, 80,000 people, 150,000 people, fucking bring it on, I love that. I couldn't understand why he was in a fucking bad mood all day. I do remember there being an altercation afterwards. I am proud to say, or glad to say, I wasn't involved in this particular one. I don't know what had happened, if you are telling me it was him and our Paul then, well, that is nothing new.

Peggie Paul had the sleeve of his jumper all ripped where Liam must have got a hold of him. It wasn't serious, I think it was just something that Paul said to Liam, something was said in jest and that was it. It was forgotten in no time at all.

Liam Cannot remember a fucking thing.

Noel I often wonder what it must be like to have your two brothers be just two brothers and then, on the fucking turn of a sixpence, become these two that are in the press and all that.

Liam Never spoke about it, but he's not that kind of brother to go, 'Look, can I come with you?' He's just like, 'I'm happy for you, go and do your fucking thing.' He was never tagging along like some soppy brother. I guess I'd have killed the pair of them if it was them two that were in a band. There would have been trouble. But he's not like that. He was living in Manchester, so all the shit we were getting up to, being out and about, he'd have people coming and going, 'Your brothers are fucking knobheads.' I guess it was hard, but, like I say, he never come up to me with a sob story, we never even spoke about it. He was always with girls and shit. Don't listen to his stories, sob stories; he had a good time as well.

Noel He has got a unique taste in music, shall we say, and he could talk the fucking back legs off a donkey. I don't really know what that saying means, talk the back legs off a donkey, but he could do it anyway. He very rarely draws breath.

DIAMOND
VISION

Paul Feel left out? Only when they became huge and then certain people didn't put you on the guest list after you'd travelled fucking 200 miles.

Coyley Slane Castle, that's my last gig. I never worked for that band again after that. When I get home, I went to see a doctor with ringing ears. The doctor says to me, 'If I was in your shoes, I would stop doing this because you will go deaf.' So, I left and that was the end of my live life. I'd had enough, as well, I think.

Noel I do recall Mark coming up to me after Slane Castle and saying, 'I don't think I can do it any more.' It was a loud gig. You could feel the fucking volume of it from the stage. Towards the end of the *Definitely Maybe* cycle and starting the beginning of the *Morning Glory* cycle, people were regularly coming into the dressing room going, 'Fucking hell, that was loud.' Now that is alright if it happens once or twice, but it was happening all the time. It means somebody has lost control of the sound out front and they are just turning it up.

It felt like all the people that were with us at the beginning were kind of getting pushed out a bit.

Jason Noel's rig just got louder and louder and louder until it was 120-odd decibels in front of it . . . probably shake your trousers with that.

Coyley My favourite gigs when I was growing up, always left my ears ringing for three days. Three days, so loud, it was bordering on the painful, but you wanted a little bit more, I don't want anybody being able to have a conversation while that band's on.

Noel I think he's said in the past it had got too big for him, which is great, I really admire him for that. I wasn't upset. The thing that upset me the most – not that I was upset crying – was that he was one of my fucking best mates and he wasn't going to be on tour any more. I was gutted that I wouldn't get to hang out with him and listen to his sage advice about various things – which pretty much amounted to a lot of swearing and smoking weed and calling everybody a cunt. If I could have got him a job doing anything else I'd have done it, could have sacked somebody but he didn't want to do monitors. So he retired disgracefully and that was it.

Liam I think he had had enough of the management. They were calling the shots a little bit. People were going, 'It's not a band thing any more, why don't the band tell the fucking management to fuck right off and look after

their crew a little bit better?' I think it was just getting a bit too corporate for him. I don't blame him. One minute you're with the band, and there are five lads, seven lads, ten lads on a bus, you are all having a crack and you're mates. I was never a boss of Coyley and he was never one of my or our employees, we were just all doing it, having the crack and seeing where it went. Then when management come into it, it changes. So one minute you are hanging with your mate and next minute you are in a different hotel or you are on a different bus or you are being told how your day is going to be by some fucking team instead of your mates. I think the fun goes out of it a bit and you kind of think, 'Fuck this shit.' It felt like all the people that were with us at the beginning were kind of getting pushed out a bit. I don't know how and I don't know why, but it felt like it was a bit, 'That fucking Manchester lot that are with them, get rid of them now, get someone else on board.' We should have trapped them, never let them go. Give them a pay rise and just robbed them and kept them there. You'd be like that on the tour bus: 'What do you think you're doing, you cunt? Sit the fuck down, no one gets out of here.' Of course I took it hard. It's a terrible thing that we let them go.

Bonehead I remember it was just like, fucking hell, we got some news, Coyley's going. That's it, he's done. Gutted, oh fuck. It really was like losing a member of the band.

Noel If anybody was a sixth member of that band, it was Mark. We were roadies together with Inspiral Carpets; the first or second batch of demos I ever did were at his house, 'Married With Children' was recorded in his bedroom. He was heavily into Neil Young, Burt Bacharach and all that, and I got a lot of my musical influences from Coyley. I wouldn't have wrote 'Slide Away' or 'Live Forever' if it wasn't for Mark. I wouldn't have wrote 'Half The World Away', I wouldn't have wrote 'Talk Tonight', I wouldn't have wrote the track 'Morning Glory'. He was one of us. The band, Coyley, Phil and Jason, that was it.

If anybody was a sixth member of that band, it was Mark.

Bonehead He wasn't just along for the ride, thinking this could be a good thing, a few quid in the pocket and I'll get to travel around the world with these boys. He wasn't in it for that, that wasn't Coyley's thing. He was there pre *Definitely Maybe*, pre getting signed, he put the time in before any of that happened. Like I said, he believed in what we were doing with a passion, he loved it. He believed it like the sixth member of the band, he really did. I think it meant as much to him as it did to us in that sense. He sacrificed a lot to sit in the Boardwalk every night and develop this sound through Noel's songs. He'd be there with us, he'd record it and he'd help us. He had a great ear for our sound, he really got it, and he was a sound engineer to boot.

Paul To have a sound engineer, that's half the battle. If you've got a sound engineer who knows your sound and can set it on any desk you go to, you're away. But he is a miserable United fan and that has never changed.

Bonehead Even when we had no money, we could bring our own sound engineer with us. Most bands can't afford a sound engineer, so you've got to rely on the guy in house at whatever gig you are at. He's never heard you before, he's just thinking, 'Another twenty minutes of

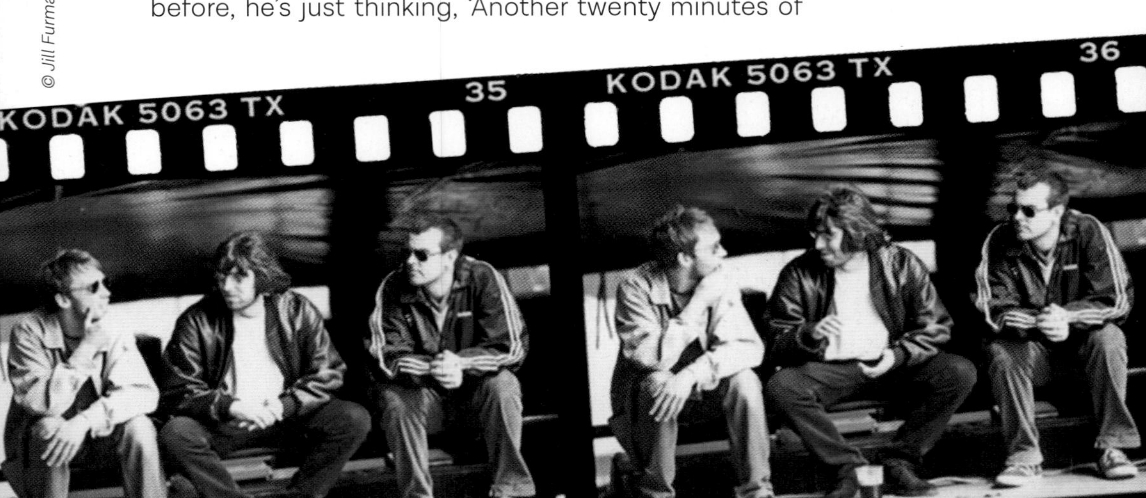

© Jill Furmanovsky

this lot and I'm home for my fish and chips; turn the drums up, turn up the guitar. Bothered if there's a guitar solo there? Fucking arse. Just give me my money and I'm off.' That is your in-house sound engineer. Not all of them, a lot of them do believe in what they do, but we were fortunate we had Mark Coyle – who totally understood us, who knew the songs inside out – doing our sound when we didn't have the money to pay one. That came across. Obviously we were a great band and well-rehearsed, but he was the guy making us sound like us through them speakers for the people who first initially came to see us, which must of helped.

Maggie It was kind of sad to see him go, because he was one of the original guys on the road with us, but he was having problems with his ears, it was quite genuine. This is a problem sound engineers have in general, but I think there was also an element of it becoming a bit too much, it wasn't the vibe that he had originally bought into, you know what I mean?

Coyley I think the best times are when everybody's together and that gang is out on the road, they are the best of times. Doing them little gigs and living in each other's pockets, it's lovely, it's just absolutely fantastic. And then, it does lose something when, you know, the industry grabs hold and all of a sudden, the band are not in your hotel, the band are not on your bus, and in some respects, it's like going to work. Now, I'm not in this business to go to work, I'm in this business to make me feel good. I absolutely love it and live it and the minute I feel like I'm going to work, I'm away. There was a lot of little issues at that time. But you know, the machine goes on and the band goes on, the band had business to do. I am not getting in the way of that.

Noel It made me a little bit sad that he wasn't going to be around because we knew it was going to explode at this point, that it was going to go through the fucking roof. It is a shame that he wasn't around, but there you go. I have more respect for him for bowing out gracefully than waiting to get sacked. I still speak to him regularly. I was a drummer in his band Tail Gunner. We never got to do any gigs, but we did put a record out. It's pretty good actually.

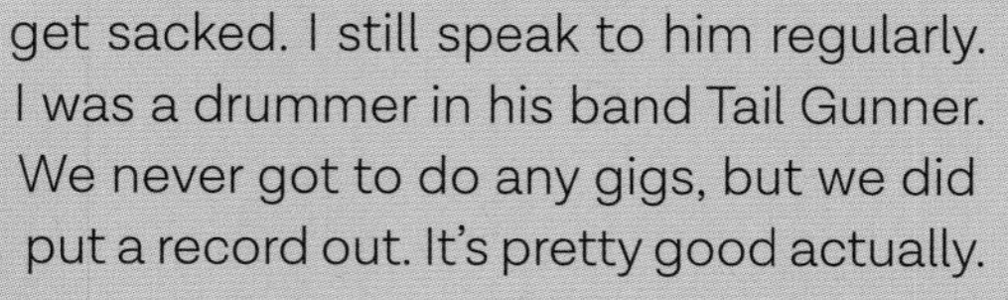

Liam Coyley was cool. Man United fan, deaf as a fucking lamp post man, but I love Coyley.

Coyley Broken-hearted, man, you know, never get over it. Walking away from that. I'm not over it now. But it was the right thing to do.

Say what you say . . .

Noel That whole episode, the 'Country House' thing, it was the best of times and the worst of times.

Liam There was all this Britpop thing bubbling up at that time and we were getting lumped into it. I was like, we are not Britpop, we are not Camden; I always fancied us as a proper fucking rock-and-roll band, a bit more real. They were a bit more flimsy and a bit fucking stupid, a bit silly and, I don't know, there was something comedy about it all. Pulp and all that, they are good bands, but I never felt like we were part of that, I felt like they were just some fucking weird shit going on over there.

Rivalries with other groups and all that, it doesn't really interest me.

Noel I really respected Pulp, still do, I like Pulp and The Verve, but we didn't have a great deal in common with all the bands that were lumped together for the Britpop thing. The northerners we always got on with, everybody south of Watford, quite frankly, you can fuck off as far as we were concerned.

Liam We weren't from London, so people would be, 'Come to this boozer.' Maybe people might have been trying to stir something up, trying to get something going, but it wasn't from us. I remember going to The Good Mixer and just playing pool. It was just another fucking pub to us. I didn't hate them, maybe I fucking did, I just thought we were better than everyone.

Noel Graham Coxon happened to be stood at the bar and we didn't feel like we were giving him shit, we were just being us, do you know what I mean? Kind of taking the piss, but not taking the piss out of him. I didn't feel we were being cuntish towards him. He might have took it the wrong way, but that's not where that rivalry started, that started further down the line. I've always like Graham and he always liked us. I don't know, I think I'm going to blame it on Liam; it's usually his fault.

Bonehead I quite liked them as a band. As people, didn't get on with them all; liked Damon, drummer's alright, Graham is a top guy, bass player is a prick. I like the music, they are alright people, used to bump into them quite regularly in London.

ROLL WITH IT

YOUVE GOTTA ROLL WITH IT
YOUVE TAKE YOUR TIME
OU GOTTA SAY WHAT YOU SAY DONT ANYBODY GET
COS ITS ALL TOO MUCH FOR ME TO TAKE

ONT EVER STAND ASIDE
N'T EVER BE DENIED
U WANNA BE WHO YOU BE IF YOU COMING WITH ME

THINK IVE GOT A FEELING IVE LOST INSIDE
THINK IM GONNA TAKE ME AWAY + HIDE
THINKING THINGS THAT I JUST CAN'T ABIDE

N THE ROAD DOWN WICH YOUR LIFE WILL DRIVE
ND THE KEY THAT LETS YOU SLIP INSIDE
~~S THE NIGHT THAT LIVES BEHIND THE DOOR~~ KISS THE GIRL SHE'S NO
YOU KNOW I THINK I RECOGNIZE YOUR FACE
IVE NEVER SEEN YOU BEFORE

GOTTA ROLL WITH IT
GOTTA TAKE YOUR TIME
OTTA SAY WHAT YOU SAY DONT LET ANYBODY GET IN
S ALL TOO MUCH FOR ME TO TAKE

Liam We were just little fucking shits, but fuck it, why not? It needed shaking up. Everyone was just fucking licking each other's arse and patting each other on the back and it was like, 'You know what, have that fucking tomato on your head,' or whatever it was. It was harmless fun, no one got hurt.

Bonehead Liam's a champion at winding people up, you know, and building it up and getting them going, you know, and he probably, genuinely fucking hated them. Or did he? I don't know, I was never driven by that whole rivalry thing.

Liam It was never malicious, it was just like, you would throw a grape at his head or something when he come in the room, or flick Alex's cigarette out of his mouth or whatever. Just fucking pull Graham's glasses off, it was just a joke. Pinch them up the arse, it was just a laugh.

Noel We weren't competing with anybody and everybody knew it. Rivalries with other groups and all that, it doesn't really interest me. But some lines are just too fucking good not to say them. You can't keep that shit unsaid, do you know what I mean? We were not in competition with Shed Seven or anyone, we were in a different league to that. And even if we actually weren't, I can assure you that everybody in the band absolutely believed that we were in a league with the greats, and at the end of the day that's all that matters. If you believe it, it is. The end.

Looking back now, Blur are alright; when you get out of the bubble they were good. But to me they were a jokey band.

Liam Looking back now, Blur are alright; when you get out of the bubble they were good. But to me they were a jokey band. There was always a bit of a fucking joke going down with it. We were deadly, deadly fucking serious. Maybe a bit too serious. We thought we were on a different planet compared to them. I classed us as classic Led Zeppelin, Pistols, Beatles, Stones, Kinks, a Who kind of band. I never honestly thought of us as that Britpop thing. I was 'Fuck off; you're not putting us in with that lot.'

Noel Why on earth 'Roll With It' is picked as the first single, to this day, blows my fucking mind. And I would have been involved in that decision. Out of all the least favourite of my Oasis songs that would rank pretty fucking high. But anyway, it is chosen. I remember Dick Green coming down and saying, 'We've got a problem with the release date, Blur are planning to release their single on the same day.' So we decide between ourselves we're going to move it back a week. A week later Blur move theirs again. So, alright, somebody wants this face-off in the charts. I am assuming it is Damon because at that point I am not interested, it doesn't mean anything to me. Once they have shifted their single twice I was like, 'You know what, alright then, okay, fucking bring it on.' I quickly ran away with it and got right stuck in.

Bonehead I think it was a clever marketing ploy on the part of Damon, what a clever move, well done Damon. Didn't harm anyone, did it? That hadn't been seen since The Beatles and the Stones, you know, it was that type of thing.

Noel What really fucking annoyed me was the way that the press connived the story that it was us that was moving our single. Call me many things, gobshite, whatever, I am not a liar. And the story was perpetrated that it was us that was looking for the fight, but it wasn't. It was a couple of knobs at the *NME* goading Blur into it as I understand it.

McGee It made us bigger, it's a bit like a boxer taking on a lesser boxer. The minute he put us in the ring, he made us the contenders.

Noel Every interview you are goaded into saying something about Damon and I didn't have enough restraint or self-awareness to think, 'This is a bit unnecessary.' And I am sure the same was happening to Blur. We would get magazine editors asking, 'Will Oasis do the cover?' And we'd say no. 'Well, if you don't do it Blur are going to do it.' The media was playing everybody off against each other. It wasn't manufactured, it was very, very real. Some of the things that were said were totally unnecessary. I was personally as high as a fucking kite most of the time and bang up for a verbal ruck. I remember doing an interview before going on stage and because when I start interviews, I rarely know when to stop, I said, 'I don't mind the guitarist, I've never met the drummer, the other two, I fucking wish they'd get AIDS and die.'

Liam That was a little bit fucking harsh, wasn't it? Mine was more like, I hope they break a fingernail or get a fucking paper cut. I hope they get a boil on their nose the day they were doing the new video, it was always that kind of thing. It was never I hope they fucking die.

Noel It was said with sarcasm and fucking taking the piss. The way she wrote it was quite sinister and then the way it blew up was just the press fucking being the press. She knew it was meant sarcastically, but she chose to write it up in a manner which wasn't. Which is fine, I fucking said it, I accepted all the shit that came with it.

Liam I don't think people should be getting upset about it; it was what it fucking was. I didn't like them; they thought we were loud-mouth cunts, end of. I thought we were better than them, they thought they were better than us, it's just tit for tat nonsense. But I enjoyed it, it was a laugh more than anything, I didn't go home at night and lose sleep over Blur. I'm sure they didn't over us.

Noel Looking back on it now, I think it's one of the greatest episodes ever in British pop music, I think it is amazing. The country came to a standstill for music and wouldn't all of us, as music fans, give anything to be back in those days now? Now music is shit; the charts are flooded with meaningless, media-trained halfwits and the songs are awful, written by songwriting teams in converted garages out in the Home Counties somewhere. Music made by idiots, for idiots.

It was a brilliant and incredible time to be young and have a vested interest in the charts . . .

Liam Thank fuck there were two bands like us around at that time, it would have been very fucking dull, all you would have had was Shed Seven and Echobelly, leave it out.

Noel It was great because it was all about fucking music and all taking place in the charts and on *Top of the Pops*. It polarised everybody: you had to be one or the other, you couldn't be both. In the schools and playgrounds it would be the dude with the side parting, 'Well, actually, I like both groups.' Fuck off, you can be one or the other. I was sat by a hotel pool in Sorrento in Italy on holiday when somebody rang me and said, 'I think it is going to be number two.' I remember feeling slightly disappointed, but not banging-my-fist-on-the-table disappointed. I remember thinking, 'That's a shame.' To be number two in the charts is not a big deal. 'Wonderwall' never got to number one anywhere in the fucking world and that is the song we are most famous for. 'Strawberry Fields' had come out the same week as some tune nobody can remember by Engelbert Humperdinck and only got to number two. It is nothing to be ashamed of. It was a brilliant and incredible time to be young and have a vested interest in the charts, if you were a fan. It's all being played out on *Top of the Pops*, on the news, at award ceremonies and all that. Everybody larger than life, high as fuck; it influenced a whole generation like Kasabian, The Libertines, Razorlight, Arctic Monkeys, The Coral. Yes, we sold a lot of singles off the back of it, but what is important though is *Morning Glory*, the album, has more than stood the test of time. We are all pals now. We straightened it all out with Damon and it's all cool. We seem to have become quite friendly.

PUMA
PUMA

OUT 30.10.95

European Tour Autumn 1995

October	
Mon 2nd	Blackpool Empress Ballroom
Tue 3rd	Stoke Trentham Gardens
Thur 5th	Bournemouth International Ctr
Fri 6th	Gloucester Leisure Ctr
Fri 27th	Stockholm Erikdalshallen
Sun 29th	Copenhagen Grey Hall
Tue 31st	Brussels La Luna

November	
Sat 4th	London Earls Court
Sun 5th	London Earls Court
Tue 7th	Paris Zenith
Wed 8th	Utrecht Music Ctr
Fri 10th	Berlin Huxley's
Sat 11th	Hamburg Grosse Freiheit
Sun 12th	Köln Live Music Hall
Tue 14th	Nantes Trocaderie
Wed 15th	Lille Aeronef
Fri 17th	Leicester Granby Hall
Fri 24th	Edinburgh Ingliston Ctr
Sat 25th	Whitley Bay Ice Rink
Sun 26th	Manchester Nymex Arena

All UK Shows Completely Sold Out

What's the Story

Noel I wasn't expecting the success of *Morning Glory*. I remember Marcus coming down to Rockfield and listening to what we had done in silence and I think Marcus's quote was, 'This is fucking serious now.' I've got to say I didn't think that, I was just writing another load of songs. Marcus was, 'Wow, this is really mature songwriting.' And of course I couldn't be more immature at that point. I remember everyone being knocked out and me being the least impressed by it all. Which is the way it should be I think. I suppose they were blown away because they were expecting another set of 'Rock 'n' Roll Star' and 'Cigarettes & Alcohol' and all that kind of thing. If you do listen to the two albums back to back it is quite a leap going from 'Supersonic', and that kind of thing, to 'Wonderwall', 'Don't Look Back In Anger' and 'Champagne Supernova' barely a year later. It's almost like there was a link that had been missed. We mixed *Morning Glory* at Orinoco Studios in South London. We were on a festival tour and I would fly back on the days off to attend the mixes, and then fly back out to the tour with cassette tapes. I remember Paul Weller seemed to be on the bill everywhere we were; we was sitting in the back of somebody's tour bus with a mix of the track 'Morning Glory'. I think Weller couldn't have thought much of the guitar solo because he said, 'I'll play guitar on it.'
'Oh, alright, cool.'
Somehow we happened to be in London on the same day and I was thinking, he isn't playing on 'Morning Glory', fuck that, but 'Champagne Supernova' could do with a fucking great guitar solo. I remember picking Weller up in a taxi and going to the studio and Owen was lying unconscious on a couch surrounded by tissues. He'd done a bit too much bugle the night before. Weller said to me, 'That is the downside of drugs my friend.' Weller did the mouth organ on 'Swamp Song' and the guitar solo on 'Champagne Supernova' and the little whistling bit that he does.

Owen I mastered it there myself, pure panic, and thinking this fucking doesn't sound as good as *Definitely Maybe*, fuck, it will be alright, it will be alright.

Noel What people don't remember is that album got universally panned by the critics. It is a cliché and truism in the music business that you get ten years to write your first record and ten minutes to write your second, because, if your first record is a success, back the way that the game was then, you want to follow it up immediately. Which is why, I suspect, that on most of the songs on *Morning Glory* there's a verse, then there is a chorus, and that thing is just repeated. There's not a lot of second verses on that record which would lead me to believe that, 'Fuck, we are going in the studio . . . well, that's what I've got, let's just do that.' I often think back to it now when I am labouring over songs, 'Well, nobody mentioned it at the time; nobody mentions that there's only one verse and one chorus in "Wonderwall" bar the odd word.'

Liam I expected people like the *NME* to not get it and not dig it, because half of the shit that was in their paper was fucking ridiculous, so why would they get something that's great? We were on a different page, we were making classic rock-and-roll music that hadn't been made since the Stones or The Beatles, as far as I'm concerned. They just wanted some fast food. I'm glad they didn't get it. As long as the people got it.

Noel They were all expecting another *Definitely Maybe*. I don't really take much notice of reviews, good or bad. I've never read a great review of anything I've ever done. *NME* give it six out of ten. *Definitely Maybe* across the fucking board was a nine-out-of-ten album. That is not what made that record great or what broke that album, it was getting in the van and doing it, the spirit of the group and the chaos that surrounded it, somehow managing to reach people and connect with them. Reviews in *Sounds* or *Melody Maker* or fucking whatever magazine was on the shelves at the time, doesn't mean shit to anyone. I'd re-record the entire fucking thing. I'd do it now if somebody would let me. Because the songs were written and cobbled together on the hoof. We recorded it in sixteen days, we were all pissed, fucking gacked up, fighting, all that kind of thing and we were having a fucking good time. The record is not really about the songs, it is about the meaning people attach to those times. Long may it live, but the record is irrelevant really. It always comes in the top hundred albums of all time, it's always in there. But the best album of all time is not the best album of all time because it's the best album of all time. It is the best album of all time because of the way it made people feel at that time, I think.

The record is not really about the songs, it is about the meaning people attach to those times.

Bonehead You feel a bit invincible, don't you, anything you're going to do is going to turn to gold, and all of a sudden, slap. No, no, no, you can be put down. But then, you get up on stage, after reading some of these

scathing reviews, and you look out and there is all these people bouncing to the ceiling and singing every word and you think, 'It ain't all bad, is it?'

Noel I remember the launch party which was another epic disaster . . . somebody explain to me, whose fucking idea it was to have a launch party in the morning, just because it's got *Morning Glory* in the title. What genius fucking came up with that? I remember getting up and it's 10 o'clock in the morning – what the fuck. My mam was there and a string quartet playing all the tunes, which sounded pretty good as I remember, I don't remember the football, I vaguely remember the gig at the record store, we'd have been fucking wasted by that point surely. That must have been dreadful, one can only imagine what nonsense we came out with that night.

Liam I remember wearing a cool jacket and there was an ice sculpture which I thought was very posh. There was an orchestra, and that's about it really, I think. As soon as a line went up my nose I could have been anywhere, mate.

Noel If your lead single off an album is 'Roll With It' you are not expecting great things, I wouldn't be and I was in the fucking band.

Owen When we were recording it, our main thing was to not fuck it up, and hopefully for it to do as well as *Definitely Maybe*. So, it came out, and there was all the Blur shit and all that, and the bad reviews and all that, but it just kept fucking selling and fucking selling. It turned into the Oasis nation,

MORNING GLORY #2

	11	13	16
NEED A LITTLE TIME TO WAKE UP	✓?	✓	?
" " "	✓	✓	WAKE
" " "	✓!	✓	✓
" " "	✓		?
NEED A LITTLE TIME TO REST YOUR MIND	?	X	X
YOU KNOW YOU SHOULD SO I GUESS THAT	✓	X	?
YOU MIGHT AS WELL			

	11	13	16
WHAT'S THE STORY MORNING GLORY	X	X ✓	✓
WELL		✓	
NEED A LITTLE TIME TO WAKE UP	WAKE		
WELL	?	✓	?
WHAT'S THE STORY MORNING GLORY	✓	✓	✓
WELL	?	✓	?
NEED A LITTLE TIME TO WAKE UP	X	WAKE	
WELL	X	✓	X
WHAT'S THE STORY MORNING GLORY	✓	✓!	
WELL	✓	?	
NEED A LITTLE TIME TO WAKE UP	✓	WAKE ✓	
WELL	X	✓	
WHAT'S THE STORY MORNING GLORY		X ✓	
WELL ?	✓	✓	
	✓	✓	

everyone's buying it, incredible, brilliant. But it wasn't expected, we were just hoping for it not to fuck up.

Noel I remember getting all the phone calls: 'Morning Glory has gone eleven times platinum.' Forty minutes later, 'It's gone fucking fourteen times platinum,' and thinking, 'Alright, we are watching the match, fucking whatever, great, well done, buy a new car, fuck off.' I am not in it for that kind of thing. I make the records and the records are great, and that's all I care about.

Coyley It's a fabulous record, *Morning Glory*, but I think the stamp of that band was *Definitely Maybe*. The two records sound completely different and I always thought that it was so brilliant that they didn't sound the same. I just think that's the greatest, greatest thing.

Noel The week it came out I remember going on stage, plugging in, playing *Morning Glory* and everybody knew all the words immediately. I remember thinking to myself, 'Wow, this is amazing, this really is.' It took ages for people to get *Definitely Maybe*, constant touring for people to know the words; this was instant.

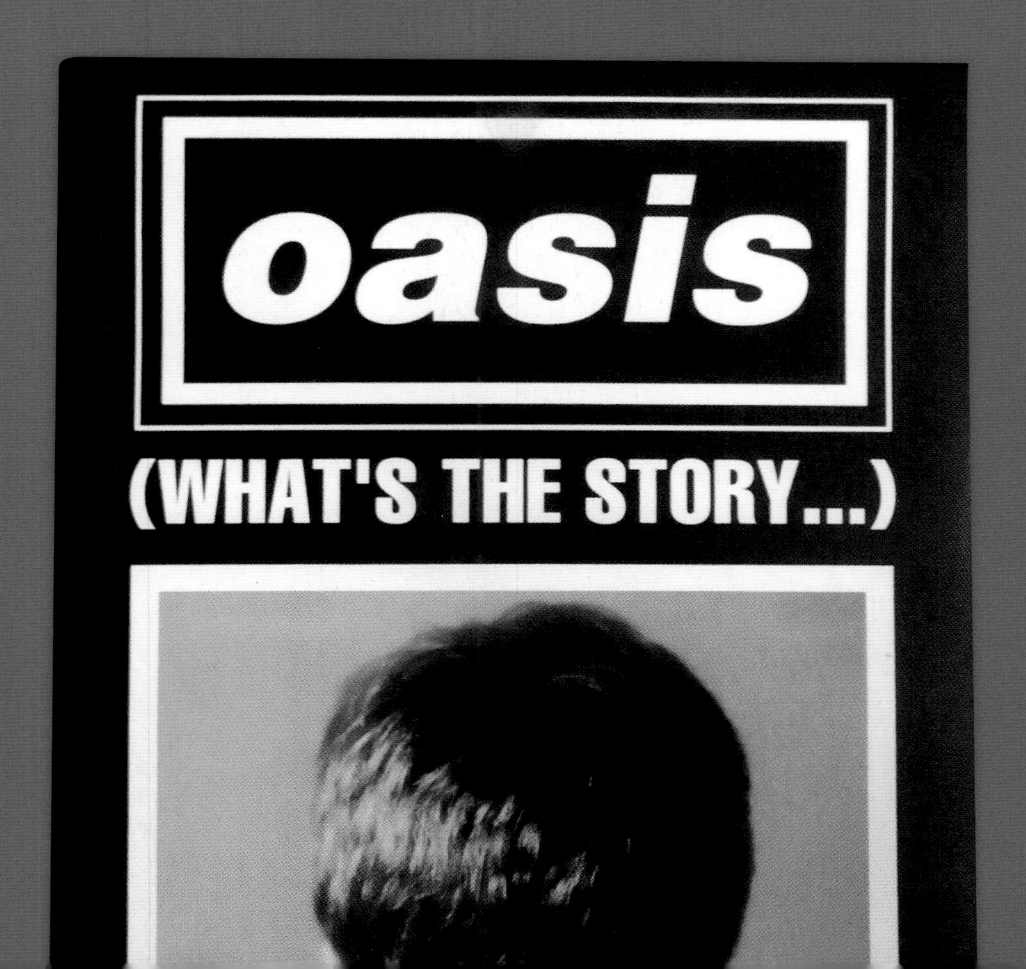

Guigsy

Noel We were rehearsing in Brixton Academy, me and Guigsy got in a taxi back to where we lived. The whole taxi ride he stared out of the window with his back to me. I never thought anything of it; I am a cat, I don't give a fuck. That was just Guigs, probably thinking of the cricket as we are passing the Oval. Go back for the rehearsals the next day, Guigs just doesn't show up. We all wait around; Guigs is half an hour late, then he is an hour late, then it's two hours. Then Marcus arrives and says, 'Guigs wants to leave the band.' I thought, 'Oh fuck, shit, man.'

Bonehead I don't think I noticed it but it'd come to a point where Guigsy just, like, couldn't come on tour. Fuck, what's happened to Guigs? He's had a bit of a fucking breakdown or whatever. And that's when you kick yourself. You're like, shit, we were that blinkered that you couldn't stop and notice that your mate wasn't feeling right or looking right or acting right. There was something wrong with him, could we have grabbed him and sorted it, you know. I don't believe it happened overnight, that sort of thing develops over time, you know, and it builds and it builds and builds until it fucking pops. I think, whatever happened to Guigs, probably was a combination of constant touring and massive pressures to deliver. All eyes were on us; that's not Guigsy's style.

Liam Guigs maybe just didn't like the fame side of it whereas I fucking absolutely fucking loved every minute of it. It was like, bring it on, let's have it. That ain't going to get in my way. I think he had just had enough of it, I think it all got a bit on top. Guigs is a gentle soul, he wasn't lairy or brash, he was a chilled motherfucker. If anyone was going to fucking flip out – with all that thing that was going on, all this fame and all these people and all that stuff – it would have been him. I could have told you that before I was born, he would be the first to flip out. Maybe I'm wrong and I'm being harsh, but I think he just had a bit of a breakdown, kind of. Just got the fear about going on stage and stuff. Maybe smoked too much weed, maybe just needed a breather, but you would have to ask him, I don't think we ever really talked about it.

Noel Guigs is a sensitive soul, any kind of chaos going on, he would shut down. He was suffering from nervous exhaustion, as rock stars always are. My nerves are exhausted or my exhaustion is getting me fucking nervous, I don't know. I've never had it myself, it is a funny thing.

Liam He wasn't like me and Noel, or like Bonehead, we were pretty mental and Guigs was always sort of like the quiet one in the corner, that was his choice. He'd have a couple of drinks and he'd be, 'I'm fucking going to skin up.' So that would be his mood. I think maybe it all got a bit too much for him, or maybe he needed a break – maybe we all needed a fucking break – and he thought, 'Fuck this shit.'

Bonehead Should we have had a bit of time off? Noel's plan of attack was keep it going, got to work, work, work. Fucking hell, we grafted for that. That was our work ethic: play a town, go off and do the next, do the next, then come back to that one we did a few weeks before, do it again before they forget about it, and then go back and do it again. That was our method.

Noel It was in the days of no internet, no social media. Our manager's theory on breaking a band was you have to do the fucking work. We loved it, we weren't a band that couldn't play, we weren't a band that were uncomfortable about playing, we were a band who knew exactly what we were and what we were good at. It's a thing that we kept up right till the end: touring, touring, touring. That's what breaks you. It was gruelling, but you know, if you want to recuperate, that's what time off is for.

Liam I loved touring, but there were certain things where you've got to go, 'Look, this is more important.' We've done that fucking graft, so we need to have a bit of a break. You can go on these tours sometimes and you're not in the right mind and you need a break; your voice fucking goes or your head goes or whatever. A couple of weeks off wouldn't have done anyone any harm, but then I guess it was all about fucking rinsing everything, just getting as much as you can in before it all goes tits up or whatever.

A couple of weeks off wouldn't have done anyone any harm ...

Bonehead Yeah, you get casualties by doing it that way, don't you? Liam's voice is going to suffer, but again, what do you do? Go to bed at 10 o'clock every night? I don't think so, you're not going to stop here, you're going to live it to the full. You only get it once. Have it, and we did.

Liam Being in the bubble was good but you still needed to just lie on a beach, have a holiday with some of the fucking money that you've made. It wasn't necessarily go back to Manchester and sit in a pub with your mates and sniff coke or whatever. It was, 'I just need a fucking minute from all the bullshit, just to recharge the batteries.' You're drinking a lot every night, smoking and all that, and even though it's great, you needed a bit of a break every now and again to then go and hit it hard again. When you don't get them breaks, you end up fucking launching something at someone, having a little ding dong and people are going, 'Oh, they are not taking it serious.'

© Jill Furmanovsky

No, I need a fucking break. A little break on the beach with a pina colada. I would have still got into trouble out there, wouldn't I? You could put me on a remote fucking island and I'd still find a chair to throw at someone.

Noel We were crying out for somebody to just say, 'Know what, this is what you should do,' and give you some kind of Churchillian speech about getting your head down and just fucking go away and do some living and blah, blah, blah. But that never happened. It's ludicrous to think that while I was in the biggest band in the world, I still had nowhere to live. I was still living in a rented fucking house in St John's Wood.

Liam When you are in the fucking bubble you rely on your managers to have a plan. They were just going for the money. We could have quite easily gone, 'I'll have a bit of time off,' go and buy a house, sit in the sun. Instead of sitting back and going, 'You know what, I'm going to let some knobhead that I don't know get behind the steering wheel.' I'd rather fucking steer it over the fucking cliff, because if anyone's going to fuck it up, it's going to be me or our kid or whoever.

Bonehead It can get pretty tiring, it can get pretty hectic for everybody. You were never at home, just constantly on the road, gig, gig, gig. You look at the Oasis 'gigography' and it is pretty full-on. And in between that it wasn't days off: you are doing a video shoot here, you are doing a TV show here, you are doing this, that and the other, here, there . . . or you are in the studio recording.

Liam It was a full-on schedule. I'm very proud that none of us has ended up in rehab or in casualty; we still know what's going on. But all these other cunts are sitting back at home, with their fucking feet up with their families, counting the money, while us monkeys are out on tour losing the fucking plot. Being in the best band – the most important band – of your generation should be a walk in the park. Should be. Things are thrown your way and it's how you deal with stuff like that that makes you great. If any of these bands today or U2 or Coldplay had half of the shit that was thrown our way, fucking hell, they would all be in fucking nobby nuthouses. I'm proud of all the shit that got thrown our way and I'm proud of how we dealt with it. And none of us are in fucking nuthouses, none of us have been to rehab, fucking bang on.

Noel Looking back on it now I never had a day off. If I had a day off on tour I was going to radio stations to do an acoustic session, or I was doing four or five interviews a day. But I don't like days off on tour, they are fucking boring. Once you've been round the world twice they are boring. I've seen Seattle, there is fuck all there. Outside of Tokyo, what are you going to do? Give me work, I much prefer working. It was hard, it was graft, but when you are on the way up, it's exciting because you can feel your thing growing and you don't know what next week is going to bring. Those are the exciting times. You don't know where you are going to be this time next week, you don't know who you are going to meet – fucking John McEnroe one minute and Talking Heads the next.

It was hard, it was graft, but when you are on the way up, it's exciting . . .

Liam Hindsight is fucking great, innit? Oasis was about living in the moment and not making fucking plans and that's what we did. For anyone to be fucking sitting there going, 'If someone had told me this . . . ' then it is contrived, isn't it. We were flying by the edge of our fucking pants and we didn't know what was going on, we didn't know where we were for the next fucking day and it was great, nothing was planned out, and that was the beauty of it.

Just missing my bird

Noel We weren't the kind of band to go, 'Poor Guigs . . .' It was, 'Fuck that cunt, we've got shit to do, man.' Called one of our pals, Scott McCleod, what an amazing episode this is.

Bonehead We went out for Scott McCleod from a little band that we'd played with in the past called the YaYa's. He looked great, he could play bass. What a great job he got, didn't he, you know. Do you fancy being a bass player in Oasis? Fucking yes.

Noel I remember meeting him at the train station, maybe with Liam, a load of fucking paparazzi . . . He got off the train – a lad from Oldham with his bass guitar and the minute he gets off the train it is fucking chaos.
We must have rehearsed and then gone off to the States. We had done a few gigs and on the way to New York to do *David Letterman* I'm sat at the front of the bus and Maggie – fucking God bless that woman, we put her through absolute fucking hell – came up to the front. It was late at night, I was up there fucking smoking, pontificating about how many chimps I'm going to buy when I get back to England and how many rocket ships I'm going to own. She came in half laughing, 'You'll never guess what's just happened . . . Scott's just told me he wants to quit.'

Maggie It's really weird, we'd finished a gig, he said to me, 'Can I have a chat with you?' He said, 'I can't do this any more, I want to go home.'
I said, 'Well, why do you want to go home? You're in Oasis, come on, what's going on?'
'I miss home.'
'You miss home?'

Noel Can't handle it because me and Liam are drinking all the time, arguing all the time, which seven times out of ten escalates into fights about what is the greatest Christmas single of all time. Well, it's fucking 'War Is Over' isn't it?'
'What about Slade?'
'Fuck Slade, Slade are shit.'
'Slade are fucking great . . .'
I think that argument actually did happen at one point. I remember going to the back and he said these words, 'I'm just missing my bird.'

Liam There are people that are cut out for it and there are people that ain't. He might have just gone, 'You know what, this is not for me, this is not what I want.' Either that or his bird must have been fit as fuck, beautiful. Hanging out with the likes of us just didn't float his boat.

Noel You are driving to do *David Letterman* and the bass player that's standing in for the bass player who's got nervous exhaustion is saying, 'I can't handle it any more.' I am laughing. Me, Bonehead and Liam are going, 'We must be the biggest bunch of cunts in the world because our mate is fucking sat at home watching the cricket, this fucking clown here, who's on the dole in England, he doesn't want to be in a band with us. What are we, the biggest three cunts in the world? I think we got slaughtered and high-fived each other.

Bonehead And he went. Tour fever, cabin fever, he couldn't handle it. Why not, I don't know. It wasn't hard. Four gigs.

Noel Marcus was saying, 'Okay, look, this is what we are going to do. We'll just cancel everything,' to a chorus of, 'Whoa, whoa, what the fuck, no fucking way.' That's not how we run shit. If he can't do it then we'll get someone else. Bonehead got up, what's happened? Scott's fucked off home. What? Scott's fucked off home. How are we going to do the show? You are playing bass.

Bonehead We had a run of gigs and the *David Letterman* show which is a massive, massive thing to do, you know, you don't miss out on that. So it was like, 'Bonehead, can you do the bass?' Course I can do the fucking bass. So we went to New York and we did the *Letterman* show with me on bass. It was great, I was comfortable with it and it actually worked.

Liam David Letterman, I've never, ever found him funny. But it's the place where The Beatles done the Ed Sullivan show and all that, so we all go over and stand in a corner where Lennon had and that was great. The best thing about them fucking shows was the mixes; the geezers who used to mix it on American TV made everything sound fucking great. When you do one of them TV shows in England it just sounds like we're in a block party or something. You'll be fucking having it on stage, you go back and listen to it and it will be like two different bands. So I loved doing all them TV shows out in America because you sounded pretty close to how you were live, which was a good thing.

Noel What annoyed me most about it was he borrowed a leather jacket off me, right, which is a Levi's leather jacket, it was a fucking beauty. He's still got it. I think that is what wound me up the most. Wasn't bothered about *David Letterman*, we blagged that, any day of the week; I want that jacket back.

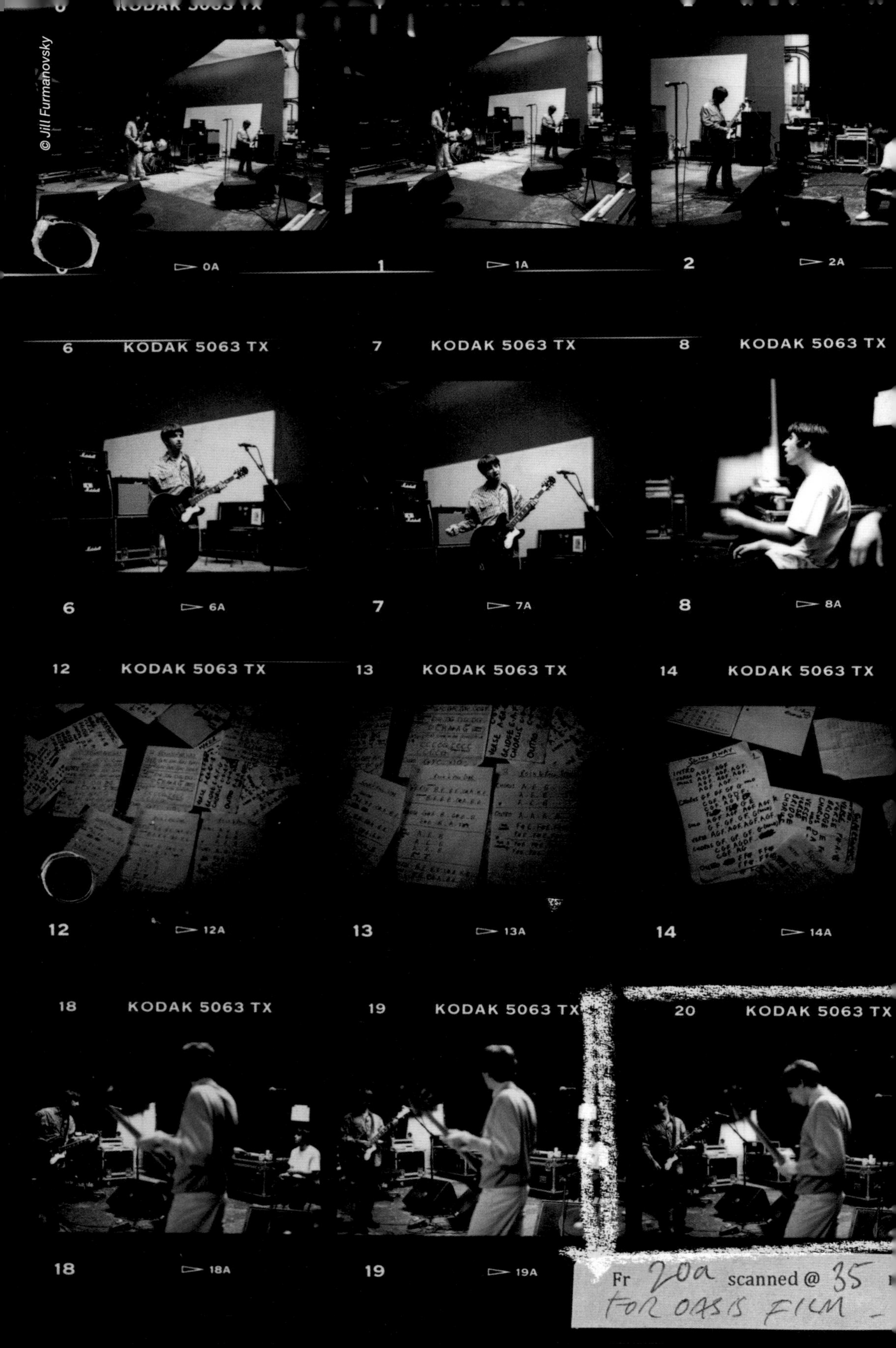

© Jill Furmanovsky
0A
1
1A
2
2A
6 KODAK 5063 TX
7 KODAK 5063 TX
8 KODAK 5063 TX
6
6A
7
7A
8
8A
12 KODAK 5063 TX
13 KODAK 5063 TX
14 KODAK 5063 TX
12
12A
13
13A
14
14A
18 KODAK 5063 TX
19 KODAK 5063 TX
20 KODAK 5063 TX
18
18A
19
19A
Fr 20a scanned @ 35
FOR OASIS FILM

3A
4
4A
5
5A
KODAK 5063 TX
10
KODAK 5063 TX
11
KODAK 5063 TX

Fr 10a scanned @ 35 mb G/S
FOR OASIS FILM - JUNE 16
9A
11A
KODAK 5063 TX
16
KODAK 5063 TX
17
KODAK 5063 TX

15A
Frame 16 scanned @
20MB G/S MAY 14
17A

KODAK 5063 TX
22
KODAK 5063 TX
23
KODAK 5063 TX

21A
22
22A
23
23A

Earls Court

Noel We come back to England, got these gigs coming up, they're sold out, they're going to be monumental. It was all gearing up to be this great thing and we didn't have a bass player. I remember being wherever I was living and the phone going. I picked it up and it was Scott and he said, 'I think I made a mistake.'
I said, 'Fucking too right you have, son.'
'I've thought about it now and I'm up for it.'
'Too late now, see you later.'

Bonehead I seriously thought that everyone was going to turn around and go, 'Bonehead, get the bass on.' Surely not. We can't do Earls Court, two nights as a fucking four-piece.

Noel I think we gave Guigs the first refusal and said, 'Are you going to do this gig or what?' We are doing the gig no matter what. If my fucking mam has got to play bass, I'll teach her to play the bass, that gig is happening. There was a funny request, which the twat in me was thinking I should have accepted: Bruce Foxton offered his services. I was thinking that would be the fucking ultimate to fucking wind Weller up. 'Sorry, man, can't be done.' Peter Hook offered his services. 'Fucking what? Are you taking the piss? No, thanks.'

Bonehead We could have got anyone, we really could. We could have got anyone to just stand in. Choose your bass player, they'll come and do it, surely, we're Oasis.

Noel Guigs was put on medication and he just got back in the saddle and was the same as he had ever been.

Bonehead I remember just getting a message like, 'Guigs is going to do it.' What? Guigs is going to come back and do the gig? It was just like, 'Hallelujah!'

Liam We never fussed about it. It was, 'Have a nice fucking rest, did you, you lazy fucker.' It was kind of that. 'Where have you been?' or 'I didn't even know you'd fucking gone. What, you had a breakdown, when did that happen?' It was shit like that, banter. We never rolled out the carpet or anything. We might have skinned up a few joints for him, but then again we might not have because it might have tipped him back over the edge. It was, 'Hey, you're back, let's get back on with it.'

oasis

EARLS
oasis

COURT
oasis

KODAK 5063 TX
13
KODAK 5063 TX
14
KODAK 5063 TX
12A
13
13A
14
14A
15
DAK 5063 TX
19
KODAK 5063 TX
20
KODAK 5063 TX
19
19A
20
20A
TNA
ODAK 5063 TX
100
DAK 5063 TX
101
DAK 5063 TX
27
yes
24A
25
25A
26
26A
27
ODAK 5063 TX
31
KODAK 5063 TX
32
KODAK 5063 TX
33

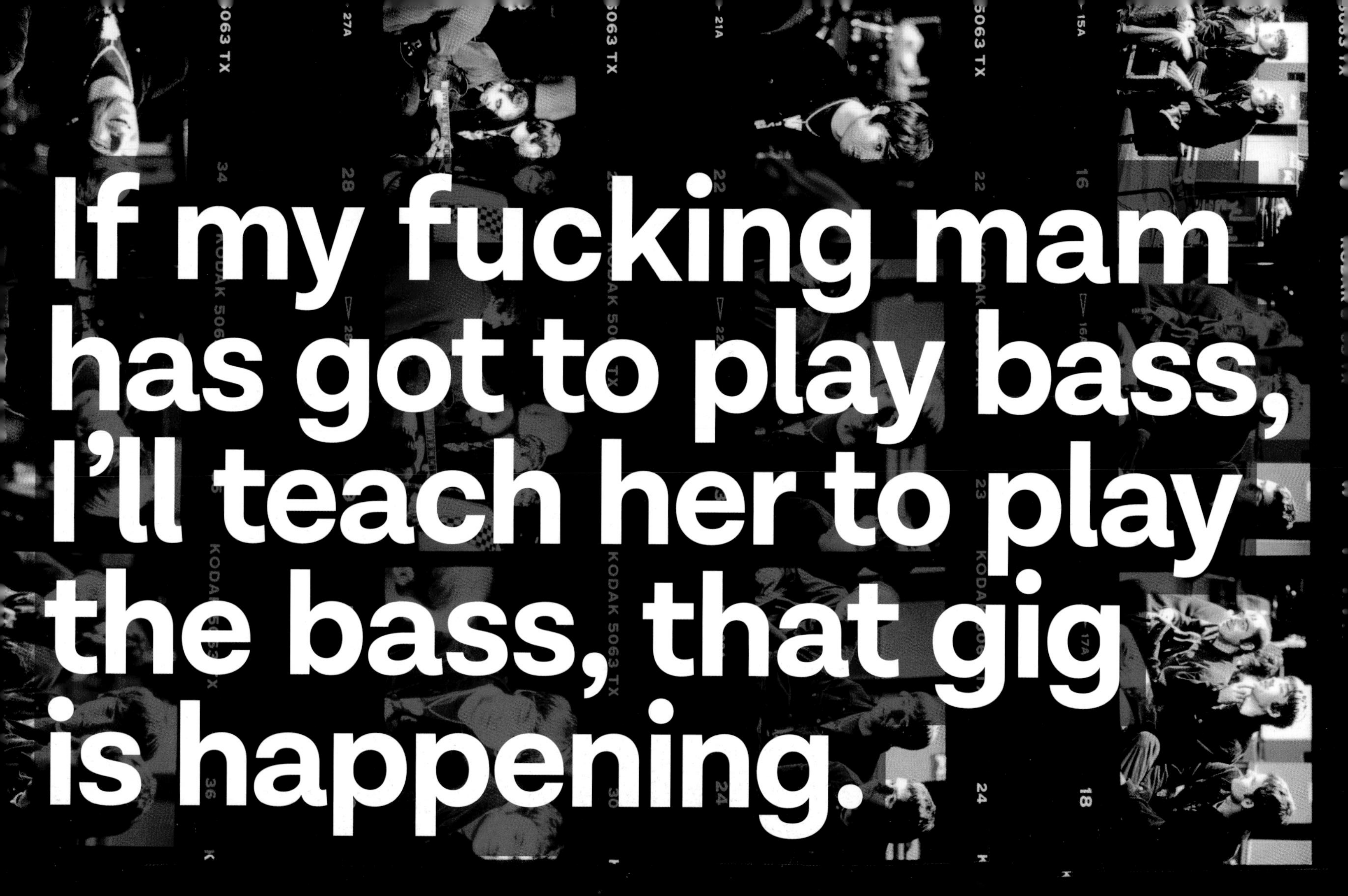
If my fucking mam
has got to play bass,
I'll teach her to play
the bass, that gig
is happening.

Noel I've got to say, me, Liam and Bonehead, to this day, remain pretty straightforward guys. We just weren't that kind of band. I just don't countenance that shit; if you are in a band and you are feeling a little bit down, man, thank God you are not playing in the Boardwalk. It is the survival of the fittest. For my own part in it, most of it is on me. I am writing the fucking songs, everybody defers to me for all the decisions, creatively, fucking everything. I haven't got time to worry about whether the fucking bass player is having a bad day. Just get on with your own shit.

Noel Even if, best-case scenario, Guigs is going to come back, is he going to be up for the biggest gigs of his life?

Bonehead I was really looking forward to it: the brass section were on stage, grand piano, orchestra, Bootleg Beatles supporting us. Real good vibe around the gig. I was just really looking forward to it, buzzing for it, yeah. I used to love, just before the doors open, getting out from backstage and getting into the foyer and watching everybody coming in and filling up. Then I'd get straight back backstage and then, when the lights were down and support bands were on, I used to often get out and get right up to the top, get in a seat or just stand at the back, and watch what was going on, just soak it up.

Tim They hadn't stopped for two years solid, and they got there, it was amazing. One of my greatest memories ever will be walking out Earls Court, first night, with the boys, ready to take on the world. There is a classic one-liner from Liam: I said, 'How you feeling?' I was literally fifty paces from however many thousands of people are out there and he's . . .

Liam 'Here we go, another day, another dollar.'

Bonehead It was good to show a lot of people who might slate you and say, 'Yeah, they are going to burn themselves out, they're going to be over in a year, you won't see them again.' So, it's good for that because it's just like, 'All you lot out there who said that, watch this.'

Maggie It was a big gig. I think it was one of the biggest standing gigs in Earls Court, or something like that. It was like 20,000 people on the floor. It was amazing, but the neighbours didn't like us. We were very loud.

Noel At one point the pogoing started a bit of wave and the ground shook. The residents complained because they thought it was an earthquake. Really it was just 'Cigarettes & Alcohol' kicking in. Those gigs were unbelievable. Unbeknown to us, Kasabian are there, two young lads, Tom and Serge, only fourteen and fifteen. Their mum and dad has picked them up outside after

oasis

the gig. Obviously we have since become friends and they said that's what kind of sparked it for them, what kicked it off.

Maggie We had loads of complaints from neighbours, but you know who the neighbours are. They're all Chelsea, you know? It was all very 'my candelabra moved across my table while I was eating dinner.'

Bonehead There ain't many bands that can go out and do Earls Court the way we did it and get that reaction from a crowd. It was pretty dull on stage, you know. Thank God we had a good light show going down, because we weren't doing much, there was no choreography involved, no nothing. Liam was prowling about, he was cool, but they just go berserk. There was a big element of danger. I'm sure people were like, 'What's going to happen next?' Even I did on stage. You never knew what was going to happen, it could be all hunky-dory backstage before you go on, and then it just goes off. That danger combined with the music and just, I think, the presence, the attitude of us, you know. The crowd just reacted.

Noel I remember them being great gigs. I probably said something outrageous in the press about one of the greatest gigs of all time. I am not sure at the time you really appreciate it, they are just more big gigs. I've got to say I don't remember Bono being there, I don't remember Madonna being there, I don't remember Elton John being there. I only recently read that all those people were there. I kind of remember some of the gig because I've seen bits of it on the telly. I don't remember it because we were fucking hammered, I suppose, for a couple of days afterwards. I think we played 'Hey Jude' over the PA at the end, we all came out of the dressing room and the entire crowd had stayed behind and were all singing it. So there was something in the air celebrating something or other.

Liam There were days when you were in the zone where you could just stand perfectly fucking still while there is all this chaos going on around you. All these kids are just leaping about and the sound's pumping, roaring, and I would just be fucking stood completely fucking still, like a boxer, thinking this is the best feeling in the world. Pure control, not feeling the need to join the madness or bust a move, just absolute fucking still, but yet all this craziness is going on around. When you had them moments, where it was pure still while all this shit was going on, that was like nirvana, the highest peak, man, it was the bollocks.

I would just
be stood
completely
fucking still,
like a boxer . . .

On the list

Noel Famous people had started turning up. There was a show in New York when John McEnroe traipses into the dressing room followed by Duran Duran. As they walked in, I would always walk out. Not because I don't like John McEnroe – I fucking love John McEnroe – or Duran Duran, but I just don't have anything to say to them. So I would kind of pass them on to Liam and then do one, sharpish.

Liam I remember John McEnroe coming, he was fucking cool. Dave Gahan was around at the time, he was pretty scary, but a top guy.

Jason Where you are getting actors and models and pop stars and rock stars and people like that wanting to come along to see the show, it changes the dynamics of it. You are hanging round a different set of people and different ideas.

I guess you get a sense of people who you know you have got a great deal in common with, and they are usually songwriters.

Noel We are in L.A. and we are out with Kate Moss and Johnny Depp. I think it was Johnny who said, 'You should play at the club tomorrow night, a secret show.' Now if Liam hadn't been in the car that gig would never have happened, but he was, 'Yeah, yeah, let's do it.' I can't be fucking arsed, it's the size of a shoebox, who gives a fuck, a load of rich kids, 'Yeah, yeah, man, fucking hell, Oasis.' Honestly, he was like a Jack Russell with a chicken leg on Christmas Day. So we had to go and do it. I don't like doing secret gigs, I'm not getting paid, fuck that. I'm not doing it for a laugh, I'm doing it to be mega fucking wealthy. Has it gone down in the annals of being a legendary gig? No, it's forgotten. There were about sixty people there. A waste of a fucking night out. On my night off an' all. I just want to go out and get hammered and go back to the hotel and swan around in an expensive robe and drink fucking red wine and smoke a cigar.

Liam I think we did a gig somewhere in L.A. and then I think we done a second gig that night at the Viper Room. I don't think we could get all our

gear across so we just used what was there, which was pretty shit. But we done a little mad, raucous gig; it was good.

Noel I liked Evan Dando, still do to this day, he's a good lad, heart of gold, it's like having a huge hound dog on tour with you, a lovely guy. I love Evan, I fucking do love him and I love his tunes and he is a fucking good guy, man, but he was a bit crazy for a bit.

Liam He was well off his box and I loved every minute of him. He was a dude. I don't know what we were doing with him, what he was doing there or why. He was mad and he was a lovely guy. We were just getting off our tits and doing drugs and getting wasted, man. I remember he would always break up TVs and fucking phones and put them in his bag. And then we would be going through airports and they would be, 'You can't go through, this is like a bomb!' We would be saying, 'What you doing?' He was a bit weird, man. I can't remember much about it but we had some good times. I've not seen him fucking since. He sort of just come in, caused a bit of chaos, beautiful chaos, and fucking left.

Noel I guess you get a sense of people who you know you have got a great deal in common with, and they are usually songwriters. But I don't consider

those people celebrities; Bono's not a celebrity, neither is Johnny Marr or Weller, they are fucking the same as me, do you know what I mean?

Peggie I just love meeting them. 'Oh, you're awful small, you're not like what you look on the television.' I said to Noel, 'How is that Bono?'
He says, 'Oh, he's fine, Mam.'
I said, 'He's a small little fella, isn't he?'

Noel I do remember one gig in Manchester, being in the dressing room afterwards and going to get a drink, going for the bottle of whatever it was, and George Michael getting there before me. I was kind of looking, 'Fucking hell, there's George Michael in our dressing room. How fucking mad's that?' Mental.

Liam I remember George Michael used to come to some of our gigs. I remember going into a bar and he was sat there with fucking Bananarama. He started popping up at a few gigs, he was into us. I liked George Michael, he was cool. Bananarama cool, still cool. People just want to come and have a nosey, don't they? They didn't get the red carpet rolled out for them. We'd never really get into arse-licking them. I've been to gigs to see big bands and the red carpet is rolled out and it's all a lot of fucking, 'Oh great,' slapping each other on the back. We didn't give a fuck who was coming to

© Jill Furmanovsky

see us, how many films you've been in or how many records you've sold. It was, 'Oh look, there's such and such over there,' and that would be it. I was never fazed by it because I was never impressed by it. Only Paul McCartney. We were in L.A. and he was mega and I was proper freaked out by it, but he was great. I met Ringo. Anyone else I was never star-struck by. Certainly wasn't star-struck by Bono and I mean that, and that's not a diss on him. I was more struck by John McEnroe, to be fair, he was cool. But I'm not really fazed by these people that come to see us.

Noel If John Lennon walked in here now I'd shake his hand and all that. It was his music that matters to me, you have got to understand they're just like you, but more talented versions of you. Paul McCartney is a fucking Scouser with a bass. There's fucking millions of them, but he is more talented than all of us, you know what I mean? Paul McCartney doesn't have the answers, Paul Weller doesn't have the answers, Peter Hook certainly doesn't have the answers, Johnny Marr ain't got them, George Harrison never had them, Iggy Pop, Lou Reed. They are just like you, but a different version of you.

People just want to come and have a nosey, don't they?

Making headlines

Noel They were crazy times. There was a period of about six months where it was everything that I said or Liam did got on the news.

Liam You give a lot of your privacy away when you are a band like that. I think what you get back from it is well worth it: you get to travel the world, you make shitloads of money, you make great fucking music, you touch people, you change people's lives, you make their lives more important and they make your lives that little bit more important.

Noel There is a saying within the music business, 'I only release records to promote my latest interviews,' which is true, because really I'd much rather be doing an interview, to be honest. Writing songs is difficult; talking shit is easy. In fact that is going to be the title of my audiobook: *Talking Shit is Easy*. In that period, it seemed that everything that I said, ended up on the front page of the newspapers. People started to think it was an act. I remember somebody saying in an interview somewhere: 'the master of manipulating the press' and I was thinking, 'I am just answering stupid questions with stupid answers.' Interviews are fucking nonsensical most of the time. Until it gets to be really fucking important, like, 'What do you think about Rwanda?'
'Does she sing with the Happy Mondays?'
'What?'
Whereas at the beginning it's like, 'Where do you get your flares from?' So I was well prepared for the silly questions, which, still to this day, do get a silly answer. When I am doing an interview with a journalist, to me that is as important as the records. I am there to be a gobshite and cause as much fucking trouble as possible. That was my thing, and to polarise opinion, so when people put a microphone in front of me, I would have said, 'The Beatles, who the fuck are they? Fucking better than them and the Sex Pistols; we're the Sex Beatles, fuck them all.' That is what I would have said. I would dread picking up the *NME* because I would think, 'I know this is going to look bad in print.' But it's usually all said with a smile on my face, just taking the piss really. I was always 'blasting' people: 'Gallagher Blasts Royal Family'. Or 'ranting': 'Rocker's Amazing Rant About Gerbils and Bourbon Biscuits'. I get bored in interviews very quick.

Maggie When they got to the stratospheric era of '95, '96 they were in the press all the time. And they were still quite young, really, to be thrust in that kind of spotlight without having any sort of knowledge of how to deal with it.

Noel We never started an interview by saying, 'By the way, you do know that we use cocaine?' Somebody asks you, 'Do you take drugs?' and you say yes. I never boasted about it at all, but I won't tell a lie in interview, I try not to anyway. I'm not saying, and I wasn't saying at the time, 'This is a fantastic lifestyle choice and everybody should fucking get involved in this shit.' Clearly I wasn't saying that.

Liam Drugs were just done because there was fuck all to do. That's what people do, innit? Before you go up your own arse, you take whatever's there, don't you? It wasn't a big deal. Started off with, I guess, weed, partial to a fucking can of gas every now and again just to blow your head. I was doing mushrooms at the age of fifteen so a little bit of acid every now and again was like nothing. My mind had already been fucking tilted to one side, so a little bit more won't hurt . . . a bit of coke, a bit of speed and bit of everything, man.

Writing songs is difficult; talking shit is easy.

Noel You grow up on a council estate in Manchester, you are surrounded by it. I used to bunk off school to glue-sniff. Then speed and weed, or draw as we used to call it, and magic mushrooms and acid and fucking cocaine, ecstasy. They were all just there. But it wasn't like it was our whole fucking reason for being. Nobody was on heroin, nobody fucking died. As far as I can remember, the only thing that died was my creativity.

Liam People would be going, 'You can't be fucking saying that.' Well, I ain't going to lie about it and then get caught out doing drugs. Yeah, I do drugs; write about it and fucking move on. That's the way it is. I know loads of people that are in the public eye that are secret little sniffers and that. The minute it comes out on them, they are fucked. Do you do drugs? Yes, we do and I will do them off your fucking head in a minute if you fucking carry on fucking asking silly questions. Yes, we do; move on.

Jason Wow. Rock-and-roll band taking drugs and drinking, are you joking? Good God, people will be reading in libraries next.

Noel The *News of the World* once ran a story that I had a £4,000-a-week cocaine habit, so we worked out that if it is so much a gram, it meant that it was something like sixteen grams a day you were doing. Now, twenty-four hours in a day . . . you work it out. I managed to keep a very successful band going, all over the fucking world! It's not nice for your mam to read because the next time you see her she is going to think you are going to look like a fucking big, fat, plastic bag full of piss or something.

Peggie I used to go to the shop and see them on the front of the paper and think, 'Oh God, is there anybody looking at me buying the paper?' Then you would go out and somebody would say, 'I've seen your boys in the paper,' and you think, 'Oh God.' There's always someone out there to say, 'That Liam is a right so and so, who does Noel think he is?'
I'd read it and then I'd think, 'Is it true?' Then I'd say to Liam, 'Is that true, Liam, what's in the paper?'
'No, no, no, Mam.'
You think, 'Well, is he telling me lies?' You think to yourself, 'Something must have happened for it to be on the paper.'

Paul I suppose it wasn't nice for her. It would have been nice to know what was going on, but all you get out of Liam is, 'No, Mam, don't believe what you read in the press,' and all you get out of Noel is, 'Wasn't me.'

Liam I certainly wasn't hamming it up because we were in a fucking band. I was doing more drugs before I was in the fucking band. You don't join a band to do drugs, if you do that you're a fucking sad cunt. I've never hyped anything. That was my life, man, that's how we behaved. If I'd seen it from any other band then I'd go, 'Yeah, they're hyping it,' but we had a laugh. Play a great gig and then you get fucking drunk and then shit would happen, that's life. I was getting up to shit like that before I was in a band, that's what my life was.

I can't give you a story about my drug hell. I can't because I didn't have one.

Noel Whatever I do is not an act, if I am being asked a question I give an honest answer, apart from when I'm lying through my teeth, which is a good 75 per cent of the time, I have to say. Or making shit up. I might bend the truth, or spin a good yarn for a laugh, but I'm not a liar. If there is one thing I have always been, it's true to myself. That's the way we were and if people didn't like it, fuck 'em, I'm not asking anybody to like it, I'm not asking anybody to like me. I'm just asking them to buy my records and, more importantly, buy the merchandise because that is where the real money is. Liam has that thing where he is the nicest guy in the room, then he's had a drink and he will pick a fight with a microphone, for no reason. There isn't a person in the Oasis organisation that's not been on the end of some of those moments, but that's just the way it was.

Liam I didn't need pushing. If anyone had said, 'Can you go out and drink tonight and go and start a ruckus?' they'd have got the sack because it would have been like, 'You're not right for me.' We didn't need pushing in that direction, that was there whether we wanted it or not. There's always some fucking crackpot around the corner, just waiting to get inside my head, or maybe I'm the crackpot around their corner, waiting to get inside their head. I remember going to the pub some days at 11 o'clock, by half twelve people would be dancing on the tables and snorting cocaine. Come in for a quiet pint and read the paper, now all hell's broke loose. One thirty in the afternoon, everyone's just fucking having it. Good times.

Noel I'm not a tragic character and I can handle drink and drugs, I don't have a problem with any of that, I can go toe to toe with anyone. I'm not a tragic fucking Pete Doherty character; I got most things in my life under control. I can't give you a story about my drug hell. I can't because I didn't have one. It was fucking great up until the point where I just said, 'That's it, no more.' But I can't sit here and tell you I had a shit time because I didn't, it was a great time.

Liam I was too off me tits to care. Obviously it pissed you off when there is people trying to trip you up . . . I kind of loved all of it, that is what I signed up for. They are outside me door because I am fucking interesting, not because I'm boring, so let's have it.

Noel I've always been able to get up in the morning, look myself in the mirror, and the same thing at night. I'm fucking cool about it, I really don't give a fuck. It means nothing to me what people write about me or say about me or feel or believe.

Liam I'm the frontman, I'm the face of the band, so I guess they are going to go for it. Fuck it, let them have it. It would do your head in at times when they were writing shit about you and fucking waiting outside your house, but it happened and it come and it went. I guess when you put yourself out there for it you are going to get shit.

Noel I don't know what I was going to see him for, but I remember having to fight through a scrum of photographers. I thought my house was bad – there was quite a few outside mine – but there was tons of them outside his house. I thought, 'Fuck that, I couldn't live like that.' Poor fucker.

Liam The only thing I didn't like was people with cameras putting it in your face. If you are going to take a picture, take it from over there; don't be putting it in my face. That ain't being a moaning fucking famous person or whatever, it was just 'You are in my fucking personal space here, knobhead. Get out of it.' If there were people with cameras in my face and I'm a bit drunk, my thing is, 'You are in my space, you are going to get a fucking clout.' They invented a long lens for a reason, to stand over there and take a fucking

picture, not to put a long lens up my nose. I enjoyed that as well, even though it cost me quite a few quid and that, but I enjoyed a photographer having a dig every now and again.

Noel I'm going into these interviews thinking, 'This is it, I'm a fucking rock star, this is it.' And I'm sat opposite a journalist and within thirty seconds I'm thinking, 'You're a fucking knobhead,' asking me the same questions as a guy asked me two months ago. I get bored quickly and think, 'Alright, I'm going to jazz this up a little bit now and fucking threaten to kill Prince Andrew or something, see if that gets interesting for fifteen minutes.' The journalists were either trying to be on side with you by, 'Yeah, man, fucking hell, it's so rock and roll, dude, got any coke?' Or they were trying to get you to slag Liam off, or Damon, all the time – I did enjoy slagging Liam and Damon off, I have got to say. It wasn't till much later on in the game that I became at peace with it and thought, 'Okay, a silly question deserves a silly answer so fucking here goes.' One of the most notorious quotes was 'taking drugs is like having a cup of tea'. And it was, and it is to young people. That fucking story went all over the world and it was crazy for a bit. It was at the *NME* Awards, and I remember saying I'll only do interviews in the toilets. All the cubicles are full of people doing drugs so I said, 'Drugs are just like having a cup of tea, everyone's on fucking drugs; all these people here are taking drugs, I bet there are people in the Houses of Parliament on drugs.' I didn't think anything of it. I went home and I drove up to Birmingham with Paul Weller, we were recording a Ronnie Lane EP for some thing or other. I didn't have a mobile phone, so the next afternoon somebody is flicking through Ceefax to look for football scores and I see my name pop up on it. I said, 'Whoa! Go back . . . what the fucking hell is that?' And there was this story. So I drive back to London and I got the driver to drive past my house. Honest to God it was insane. There were TV presenters with the lights on, fixing themselves to go live to the fucking 6 o'clock news. About six weeks before, Brian Harvey had been kicked out of East 17 for saying something about ecstasy. I'm watching the news in Tim Abbott's office and Michael Howard, the then Home Secretary, is on the TV giving an interview about what I'd said. Fucking hell, this is insane. Michael Howard came out with, 'This is a disgrace. I hope Oasis see sense and send him the same way as Brian Harvey.' Fucking hell; we laughed out loud and went out on the piss.

Liam It was a headline and bring it on; I'd rather that than, 'Oasis went to bed at 9 o'clock with a cup of cocoa and were up at six in the morning doing yoga.' It had all been done before and it will all get done again. Listen, if anyone in this band turns round to you and tells you that they were not happy with these kind of headlines, they are fucking liars. We were going, 'Yeah, cool. Nice.' It's better than having some fucking Tory MP, or whatever, on there doing coke. I'd rather us be on the front than any other cunt. I loved it.

Hypocrisy has never been in short supply in England, particularly in the media and in politics.

Noel The thing that was funny was the hypocrisy of the British press who would write these things, this outrage. Well, I've taken drugs with you. I have been in toilets in clubs in London with you, you fucking liar. And although not everybody in the House of Commons is on drugs, clearly, as the years have gone by we've come to know some people have been and are. It is the hypocrisy of politicians who are up to far worse than some rock star trying to live his fucking dreams. Hypocrisy has never been in short supply in England, particularly in the media and in politics.

Liam They loved it as much as us, because there was nothing else about. It was boring and it ain't big and it ain't fucking clever, but it was right for the times. They might have been exaggerating, but there was definitely some truth in it I guess.

Partridge

Alan

ge in

ack

IVE

VALLACE

Steve
ving chat
hell Alan
opping as
y Ferrino.
chphrase
Knowing
l be back
hat.
o created
ater Paul
k on the
spring.
said: "It
ut Alan's
-screen."
unned by
after the
Records
into the
nd-up.
only No.
string of
th tickets

g to patch
ife with
Cole, mum
girl. She
over his

s new

lling

IVE

VALLACE

Michael
first guest

DRUGS ARE JUST LIKE A CUP OF TEA

POT SHOT: Noel Gallagher attacked 'hypocrite' MPs

Noel It did make life quite difficult though, getting around was difficult, couldn't go out of the fucking house because you got followed everywhere.

Liam There are some days where you go, 'Fucking hell, I can't go down the fucking shop, me private life is fucking all over the papers,' but that's just the way it is. I got used to it very quick and moaning about it ain't going to fucking change it. There is two ways to go: I stop being in a band and go be a hermit and then that's my dream over, or I fucking shut the fuck up, get down to the pub, have a pint, chill the fuck out and just fucking get on with it. No one's stopping me from doing what I am meant to be doing, and that was being in the fucking greatest rock-and-roll band of my era. A couple of little shitty stories here, a couple of little fucking digs by the press, fucking water off a duck's back, give a shit, bring it on.

My band is better than every single tabloid newspaper put together.

Noel The *News of the World* and all the Sunday papers and all the fucking tabloids were out to get a story because surrounding the band was chaos. Stories were regular fucking occurrences and here was a band who were bigger than anybody else who didn't give a fuck. The drugs and fucking women and all that shit and the chaos and all that. I was always of the opinion, this shit doesn't matter, we're better than this. My songs are bigger than your newspaper. My band is better than every single tabloid newspaper put together. It doesn't matter to me. We were better than that.

Liam Whatever they were saying, it was pretty much true. I was a bit lairy, I was out taking drugs . . . it's all there to see, it's pretty much true, but I wouldn't say I'm a cunt. They were going, he's this and he's that and it's like, no, I'm a good person. But all the rest of it is true, without a doubt. But I didn't care what they wrote, I'd never sit there and go, 'I'm fucking suing them, they've said this about me.'

Noel Wherever we went the press followed. So I got the big house in Primrose Hill, turned it into a nightclub, then never went out for two years. When you have got responsibilities only to yourself, fucking live life to the full, man. Those parties at Supernova Heights went on forever, forever and ever. There were quiet times, but they were only to clean the place up. Everybody was round at my house all the time. I can't name names, but we all had a great time. It brings a smile to my face when I think about it, and outside the house was never less than fifteen, twenty kids, the press, all the time.

Liam I went round a few times to lend some sugar and shit, but never got the invite. I might have wrote on his wall every now and again on the way back from the shops. Liam is God. Liam is cool – kiss kiss.

Noel They weren't 'parties'. I don't remember there being any loud music. It was where everybody that I knew hung out. I remember once there was a random scruffy guy sat at the kitchen table. 'Excuse me, mate, do I know you?' And he said, 'No.'
'Well, what are you doing here?'
'I delivered pizzas here about six hours ago.'
'Well, what are you doing?'
'Everyone said it was alright to stay.'
'Take your fucking crash helmet with you and your moped and get out.'

© Jill Furmanovsky

Phil Obviously, they took guitar music as far as it's been in Britain since The Beatles. I know Led Zep were doing that, but Led Zep were never that visible, you know. They were the most visible band since The Beatles and the Stones, because everybody knew who they were. They weren't just the biggest band, they were the biggest people in Britain at that time, Noel and Liam, without a doubt. Everybody had an opinion on them. Everyone in the street is treating you different, your world is different.

When people ask you for your autograph for the first time it is a fucking bizarre thing.

Noel When people ask you for your autograph for the first time it is a fucking bizarre thing.

Liam If people stop and ask you for an autograph, it takes you two seconds to do it. You can make or break someone's day, and I'm not about breaking people's day. I never, ever go out of my way to make people feel

bad. I hope I didn't anyway. I rarely have a shitty day, nine out of ten I was buzzing. I was always up and still am. I've had a lot of shit go down and that and so have other people, but I always try and keep a positive outlook, it will all be right tomorrow.

Noel I kind of envied the rest of the band for being able to take a step back. I think for me and Liam it was slightly different, we were in it, immersed in it, and probably still are. So it must have been nice to take a step back and just go, 'Holy fuck, how did it get like this?'

Bonehead We were kings, yeah, we really were. We were just living that dream and that's exactly what we wanted, that's why you join a band.

Liam I did exactly what it said on the tin: rock-and-roll star, that's me, boom. I just absolutely fucking love it, and loved it, and still do love it. I wake up with a smile on my face. I'm always out to have fun, definitely.

I did exactly what it said on the tin: rock-and-roll star, that's me, boom.

Noel He's into the chaos for sure; he causes a lot of it.

Jason Chaos just came their way constantly, it was quite an interesting band like that. If it wasn't one thing it was another and it just kept on coming and coming and the band were getting bigger and bigger and more and more chaos ensued. When you are involved in it, it doesn't faze you. You just think, 'Oh well, another incident, another issue, another something that's happened. Surely nothing else can go wrong. Oh it can.'

Liam As much as it was getting big, I still was trying to keep an element of fucking rawness to it. They wanted to drive the rawness out of it, hence the rule book getting invented, and shit like that.

Maggie Anything that starts becoming stratospheric has to become a business, unfortunately. That's the harsh reality of life. Things that do stop being fun, which is quite difficult when you have one of the main band members thinking of it as a rock-and-roll thing, it's quite hard to say that the rock-and-roll image doesn't really exist. It is a business, at the end of the day. There's a part of you that has to accept that.

Awards ceremonies

Liam Always had a good time at awards ceremonies, was never there for the whole 'Aren't we great' business, we were there for a piss-up and a night out sitting round a table with your mates. Getting drunk, hurling abuse at a load of people that are not as good as you. Amazing.

Noel The first year at the Brits was great. We were there, Blur were there, Pulp, all the new wave of bands and there was a sense that at some point, some woman is going to come in with a clipboard and say, 'Hang on a minute, you lot are not supposed to be here.' The Brits were something; it wasn't even that popular then, it was just something you'd seen on the telly that fucking Take That were on and dippy boy bands. I think Suede were on it once and it was like, 'Oh my God, Suede are on the Brits. Wow.' Then, for want of a better word, indie music started to become mainstream. My management said, 'You have to be there by such a time so the car's coming to pick you up.' I remember it was just a minicab. I get to Alexandra Palace before the fucking doors open. I get out of this taxi, streets are lined with all these people, I don't even think the photographers have turned up yet, and I go up to the door and they wouldn't let us in. I had to sit in the kitchen with the fucking staff making the dinner for an hour and a half to wait for the show to fucking start and then come out and go up the fucking red carpet thing. Sat next to me was Annie Lennox. She was saying 'I really like your work,' and I was like, 'I haven't got a job, what you talking about, work?' We were doing fucking gack under the table and all sorts, it was insane . . . Annie Lennox was fucking disgusted.

We were there for a piss-up and a night out sitting round a table with your mates.

We got presented an award by Ray Davies and I am sad to say that I didn't fully appreciate that at the time. I mean, I love their music so much. If it wasn't for Ray Davies there wouldn't be Britpop, he's one of my fucking idols. He's written some of the best songs that this country's ever produced. Then the next year by Pete Townshend. Pete Townshend, one of my other idols; if there wasn't Townshend and The Who there wouldn't have been punk rock. I remember Damon getting Best Band and saying, 'We should be sharing this with Oasis.'

NME
voted by

Liam I remember running over to their table – I was doing a Kanye before he was doing it – I remember going to Graham, 'Give us that, that's ours, that,' and he's going, 'Fuck off, man.' I was going, 'You know it's ours though,' just winding him up, just taking the piss basically. Yeah, I had a good night, we drank lots and lots of stuff, it was great.

Owen The Brits in 1996, that was a night. Off our tits. Fucking Guigs had some really strong fucking skunk. It was so big that they had to give them everything, didn't they, but they didn't give us best producer, you know, Brian fucking Eno won it. Me and Noel were like, 'What the fuck?' He gave me their Oasis Best Album award, 'You have that, Owen.' Next day I had Creation on the phone going, 'Can we have the award back?' I was like, 'No, he fucking gave it to me.' Fuck off. We were on the table next to Tony fucking Blair. Tony, come here. Cherie, give us a fucking cuddle, love. Oh, what a night, Jesus Christ.

Liam I didn't even know who Tony Blair was. What's his record, like? Is he a jazz musician or something? Who the fuck's Tony Blair? Never heard of him. I'm not into politics in records, full stop. Give a shit about John Lennon's political side, Bono's and all that? I am not into it, man. We make music to have a good time, all that fucking shit is a personal thing. I think it shouldn't be involved with music. I've always thought, 'Once you start opening that box you are going to end up like a self-righteous fucking idiot,' and my words

have come true. Once you go down that route of politics, man, and all that bollocks, you ain't coming back, mate . . . If there's anyone who can bullshit you more than yourself, that's a politician. You sit round a table with them, they will fucking get inside your head, mate, and before you know it you will be one of them. You know what, you can fucking enjoy your canapés and your champagne, I'm off to Wetherspoons to speak to the real people. We must have really got on people's tits. The likes of Michael Hutchence must have been just like, 'Oh, for fuck's sake, these cunts are going to be here.' But that's the way it is, man. We earned our right to be there. As much as they were thinking that, we were thinking, 'Are these cunts still fucking here? Have we not got rid of these lot yet? Who let these lot in?' We would always flip it back to them.

Noel In the case of Michael Hutchence, the week before he'd said something about Oasis and I used to live for shit like that. I've never really picked a fight with anyone but I used to live for the day people say, 'Fuck, have you heard what what's-his-name said about you?' I'd just be like, 'Oh, he has no idea what's going to happen to him now.' Before we went on stage I met him backstage and he kind of apologised, 'I was only having a laugh,' and I was saying, 'Mate, we'll fucking see who's having a laugh in about fifteen minutes.' Some lines are just too good not to say.

Liam God bless his soul. He did nearly get brained because he was being a fucking cheeky cunt. I think it was something about Paula Yates, God bless her soul, I think someone had said she was getting a bit frisky and that. I was pissed and he said something like, 'I'll fucking come down on you like a fucking can of Fosters,' something like that. 'Well, fucking let's have it,' and I picked the fire extinguisher up, but didn't hit him with it. That was the end of that, but he was having a bit of me and I was having a bit back.

Noel I remember at the Brits once we were just sat round the table talking to our pals. Somebody came and said we had got to leave. The award ceremony had finished, everyone had gone home and we were just sat round these tables like the fucking guys from the Monty Python sketch, just talking shit. It was like 3 o'clock in the morning or something, we'd been there all night.

Liam We were a bit arrogant, I guess we were a bit cocky and that, probably had too much to drink. Every time we did an awards ceremony we definitely didn't just sit there, I don't think I ever ate any bit of food at these awards ceremonies, I think it was just proper fucking sessions. We just drank too much and we would hurl abuse at people, it was just fucking banter, man. Come on, everyone needs to fucking lighten up. Maybe we went a bit overboard every now and again, but where I'm from we just give each other shit on a daily basis and it's just a laugh. People from up north can take it a bit more, I guess some people couldn't take it. It was never malicious, we were just having a fucking laugh.

Noel I don't have any of my awards, I give them away when I get them. If you are sat besides me at a table at an awards ceremony and you ask for it, I'd give it to you. I think I kept one Ivor Novello and that's because my missus put it in her handbag. That was it. I've given the rest away, they don't mean anything. I don't have any of my gold discs. It doesn't make you any better, any greater a person or better a songwriter, or the record any more popular. It's what you're doing on stage that matters.

Liam You come to my house, there is not one award up, there is not one fucking disc up of our great achievements. They're all under the fucking stairs, just sitting there. I've never put them up because that's not what I'm about. I've been round some people's houses and it's like there are all the fucking awards up. What is this, the fucking eighties or the seventies? Mine are just under the staircase. So award ceremonies are good just for the simple fact – to get wasted and hurl some abuse at some poor shit indie band.

We treated being the biggest band in the world with the contempt it deserved and deserves.

Noel We treated being the biggest band in the world with the contempt it deserved and deserves. There was a lot of, 'I can't be arsed doing that.' Do we want to do the Super Bowl? That's a ball ache, no, takes two weeks. So we never played the game in any way, we couldn't be bothered. Too busy having a good time. We won five MTV awards one year and we forgot to go to the ceremony. We just went to the pub, and we were kind of, 'Aren't we supposed to be doing something?'

'And the winner is Oasis . . .' and there are five empty seats. We were in the fucking Warrington pub in Maida Vale getting pissed. We never got a single fucking play on MTV after that, they hated us. MTV fucking hated us, the *NME* fucking hated us, Epic, our record label, loathed us, but the people loved us and that's what carried us through to this day.

Music Club B
Sponsored
Britannia
BRIT
Britannia
BRIT
BRIT
BRIT
BRIT
BRIT

Take that look from off your face

Noel *Morning Glory* took on a life of its own after 'Wonderwall' and 'Don't Look Back In Anger' came out consecutively as singles, then it just blew up all over the world. I wrote 'Don't Look Back In Anger' and I didn't think anything of that song when I wrote it. I thought, 'Yeah, that will be good, I can hear it live, it's got a great chorus and I think that will be great for the next record.' As I recall it, we were doing this gig in this strip club in Paris. I can't remember the gig, but I remember that there'd been a fight – there seems to have always been fights, fucking mental. I decided I wasn't going to go out that night. I took my guitar and went back to my hotel room. If I'd have known that night what I know now about 'Don't Look Back In Anger', I would never have finished it. The lyrics would never have been right. If you were to think that night, this song is going to outlive me, you'd think I'd better come up with a better first line than 'Slip inside the eye of your mind', whatever that is supposed to mean. I sat down and I wrote it and it took me one night, not even all night. It took me a couple of hours maybe to write that song, give or take a few words – Liam came up with the Sally bit – and I didn't think anything of it. Then you put it out and it takes on a life of its own.

Liam I don't remember it being in Paris, I can remember this distinctly, it was in America. I was sat at the back of this hall and I heard him playing this fucking tune and I was going, 'What's he singing there?' And in my head, this is my version, I've gone, 'He shouldn't sing that there, he should sing "So Sally can wait . . ."' Maybe I was stoned or maybe it was just a dream. But it wasn't about 'I wrote that bit' or 'I want credit for that' give a shit. Then next minute that comes out. It was a top tune.

Noel When I am frustrated with the songwriting process, or I feel like I'm not cutting-edge enough, I always go back to that night and think, 'No, just write it, let the people decide.' That song is played at weddings, funerals, football matches, concerts . . . It's an extraordinary song and it's extraordinary because the people made it extraordinary. I say to people, 'I only wrote it, my part in it was that minimal.' The way that I write songs is just to write it, fuck it, just let it go. Then record it and when you are recording it, make it the best that it can be. Then let other people decide if it's great or shit, if it's their favourite song or it's their worst song. Years later I was at an Ian Brown gig and a girl came up to me and she said, 'Can I ask you a question? Is Sally in "Don't Look Back In Anger" Sally Cinnamon?' And I thought, 'Fucking bastard; why didn't I think of that?'

DONT LOOK BACK IN ANGER . .

SLIP INSIDE
THE EYE OF YOUR MIND
DONT YOU KNOW YOU MIGHT FIND
A BETTER PLACE TO PLAY
YOU SAID YOU'D ONCE NEVER BEEN
ALL THE THINGS THAT YOU'VE SEEN
ARE GONNA FADE AWAY

START THE REVOLUTION FROM YOUR BED
THEY SAY THE BRAINS YOU HAVE WENT TO YOUR HEAD
STEP OUTSIDE THE SUMMERTIMES IN BLOOM
STAND UP BESIDE THE FIREPLACE
TAKE THAT LOOK FROM OFF YOUR FACE
COS YOU AIN'T EVER GONNA BURN MY HEART OUT

SO SALLY CAN WAIT, SHE KNOWS ITS TOO LATE, AS WERE WA
MY SOUL SLIDES AWAY, BUT DON'T LOOK BACK IN ANGER I H

(2) TAKE ME TO THE PLACE WHERE YOU GO
WHERE NO-BODY KNOWS IF ... T OR DAY
PLEASE DON'T PUT YOUR
OF A ROCK N ROLL BAND
WHO'LL THROW IT ALL AW

REPEAT BRIDE + CHO

Tous les jeudis de 21 à 2 h : un concert, une soirée, un DJ (Grebo), de la musique proposés par Les Inrockuptibles. Pertinent, non ?

16 juin Oasis ★
23 juin Compulsion
30 juin Swell
7 juillet Shed Seven

★ Les Inrocks à l'Erotika ★

L'Erotika ★ 62, boulevard de Clichy ★ Paris XVIIIe ★ Métro Blanche

Les Inrockuptibles

The Point

Noel I remember those gigs at The Point being, like, wildly emotional with the crowd and all that. I remember it being breathtaking, the energy and the vibe in the room, it was unbelievable. You could feel it, and it was almost like we were not worthy of that.

Peggie I remember in The Point them all shouting, 'Peggie! Peggie! I was up in the balcony with all these fans down in the thing screaming, 'Peggie! Peggie!' Let me move away. I was so embarrassed.

Bonehead My family, when we got six weeks off school in summer, we'd spend that whole six weeks without fail in Ireland, as did Noel and Liam.

Liam Beautiful, it always just seemed really sunny, and we used to get there and go up to my uncle's farm, fuck about in the hay. He lived a bit outside of town so we would walk into town and shit and just fart about. Robbing cigs and stuff and just messing about really. Me old fella would never come with us so it was always, like, pleasant times, me mam would have a good time with her family and that, so that was good.

The energy and the vibe in the room, it was unbelievable.

Peggie They used to love going over to Ireland, they'd run wild over there, out in the fields and doing all sorts of things. That was their summer holiday, never went nowhere else.

Noel There is rage in Oasis music and let me explain that to you. If I say to people there's rage in the music, people might think about screaming and shouting, but you can rage joy. When the Irish are sad they are the saddest people in the world, when they are happy they are the happiest people in the world. When they drink they are the most drunk people in the world, there is one rule for the Irish and different ones for everybody else. Oasis could never have existed, been as big, been as important, been as flawed, been as loved and loathed, if we weren't all predominantly Irish.

Paolo Hewitt* It's that immigrant thing. It's that sense of identity. Ireland saw them as their own and everybody was in such a good mood. So we got back to the hotel, there were loads of us in the bar.

* *Journalist and author of* Getting High: The Adventures Of Oasis *and* Forever The People.

Paul That hotel was on Grafton Street and it's pretty open; anyone could walk in off the street because Ireland didn't really have security around rock bands in them days. My dad, we'd seen him skulking around and I thought, 'What the fuck is this going on?'

Liam All I remember is someone turned round and said, 'Your old fella's over there with a journalist trying to get summat goin'.' I was about to kill him. I just thought it was a shitty thing. I'm there with me mam's family and all that. He was from Dublin so they just turned up, trying to get some fucking vibe, just to get a story. It's the cheapest shot, cheapest bullshit. I thought, 'You are going to get it, you cunt.' If he thought he was brave enough to come up and have a bit of a fucking square-off, then he's fucking foolish. In front of the press, with your son? It's just pathetic, what a sad cunt. See you later, mate.

All I care about is the music. In the end none of this will matter, when it's all said what will remain is the songs.

Maggie It's really quite a tasteless thing to do, isn't it? Obviously they were setting this up and they were banking on a confrontation that they wanted to film and get in the papers.

Noel The thing with me old fella turning up is, no one had actually seen him. I never actually saw him. I was coming out of the bogs and somebody says, 'Don't go out there because the *News of the World* are out there and we think your dad's with them.'

I was like, 'Okay, cool, whatever.' I don't want to cause a scene. I'm kind of long since over whatever was going on with my old fella, that doesn't bother me in the slightest. I don't know what our Paul was doing. Me mam had clearly gone to bed. I did think Liam was a little bit upset.

Liam There was a scene and I was trying to take the dignified approach. Fuck him, he doesn't mean anything to me any more. We were better than that. I was fucking steadfastly not going to get involved.

Paolo Hewitt Noel was saying to Liam, 'Do not react.' He's at one end of the bar and Noel and Liam are at the other. Liam's going 'I'm going to fucking kill him'. And that's when they locked horns, 'Don't fucking react, you're not going to react.' To give Noel his due credit, he contained Liam, because Liam would have gone for him, and Noel protected his brother. Got him out of there because the press were obviously looking for the big fight.

Liam Without a doubt, I was ready to fucking kill him. Without a doubt. But then the only people who would have got that would have been the *News of the World*. I would have killed him if I'd got near him. I got calmed down by our kid I think, but, no, I didn't dwell on it, just moved on to the next fucking heap of shit.

Noel You don't really know what the papers have said to your dad. They might have spun him some yarn about, 'I know Noel and Liam; I know they'd really like to meet you.' You don't know. Tabloid journalists – the ones who are out there to set people up, secret filming, phone tapping and all that kind of shit – are, as we've since found out, fucking scumbags. I can only say the only way to deal with it is rise above it.

Liam Round about that time there was a story every couple of fucking months. Poor old fucking dad's trying to get in touch with his famous fucking sons, but they are cunts because they don't want nothing to do with him. All that bollocks.

Paul That is not cool, turning up with the *News of the World*. I was in that hotel and he still never mentioned me. It doesn't bother me, but if you are going to have a reconciliation, we are three, not two. Or is there only two that makes money?

Peggie Oh, that was awful that. Really, I couldn't believe that happened, because it really spoiled their night. He knew how Liam would get up on his high horse, because Liam was just waiting for something like that. I think Noel was more nervous because Noel doesn't like anything like that. He had no intentions of making up with them. That was his last thought, but of course he got paid from the *News of the World*. And then, of course, it was all over the papers the next day. He's always said it was me and the three kids that ruined his life. But I would just like him to know that it wasn't us that ruined his life. He ruined his own life.

Noel You know, it was kind of long since over, whatever was going on with my old fella. All I care about is the music. In the end none of this will matter, when it's all said and done, what will remain is the songs.

Home fixture

Noel We knew that the two records had captured something, probably we didn't even know what it was. I would probably hazard a guess we were as amazed as anybody else really, but not overwhelmed. We were never that blown away by it. When Maine Road sold out – fucking hell, it sold out in twenty minutes, a football stadium – I don't ever recall going, 'What? Wow! Fucking get my mam on the phone!' I just wanted to say, 'When is it? April? Right, well, give us a fucking shout in January.'

It was our first stadium gig and it was kind of a celebration of an era . . .

Liam Maine Road was a great gig, man. Just playing there – getting on the pitch, going backstage in the ground and all that – was great. I used to go and see City there all the time. I first got into City through a teacher who was at St Bernard's. I think, it was Mr Walsh. He was a City fan and he'd get tickets to give to kids and that. He'd get ten tickets and he'd bring along whoever had not been a cunt that week. I don't know how I got one. Anyway we'd go and sit there behind the goals and just have it and that. It was great, man.

Bonehead Maine Road and the streets surrounding it really are a stone's throw from where I was born. I was a United fan as a child, but all my friends were City fans and if City played at home, everybody walked down to the ground to stand in the Kippax and cheer on City, you know. I was really familiar with the ground, inside, outside and surrounding it, so it meant that much to me to do it, and to do two nights.

Noel It was our first stadium gig and it was kind of a celebration of an era and a generation. We still felt like a cult band, we still felt that we weren't really in the mainstream at that point. So selling out two nights at a stadium was cool as fuck.

Liam I remember turning up at the gig, I can't remember what I was wearing, but I think there was an old training kit in the corner in one of the dressing rooms so I thought, 'I'll fucking have a bit of that.' Whacked it on over me gear. 'Yeah, man, we're in. Be playing for them next.'

MANCHESTER CITY
FOOTBALL
CLUB

© Jill Furmanovsky
KOD•AK TX 6043
UMBRO
PRESSPART
Lloyds Metal Group plc
makro

Noel The fact that I'm a Man City supporter, and have been all my life, to be offered to play a gig like that was laughable, to do two nights was even more funny. To become a stadium band . . . stadium bands were like U2 and Guns N' Roses. Stadium bands have stadium music and use stadium tactics. We didn't have any of that. We walked into a football stadium playing the same game that we'd played in the Boardwalk two years before. I didn't know what stadium tactics were until Johnny Marr pointed them out to me once. He said, 'How have you managed to do that without resorting to stadium tactics?'
I was like, 'What are these tactics?'
And he said, 'All the people over there, sing . . .'
I was like, 'Oh that, yeah, can't be arsed with that.'

We walked into a football stadium playing the same game that we'd played in the Boardwalk two years before.

Bonehead We walked out on the stage and it was like, 'Fuck, oh my God.' I could see everything I knew about that place, I could see those houses through the gaps, I could see the floodlights. You know when people say, 'It's a homecoming gig.' I was just getting emotional because that really was a homecoming gig, for all of us.

Noel When I walked out on stage at Maine Road I wasn't detached from it, I fucking loved it. It never paralysed me: the size of this shit or the amount of press or the record sales or the attention that we were getting; that was all great, I loved all that. It never consumed me in any way whatsoever. I felt like a) I fucking deserved it and b) I could get more. I'll have more of that, I don't give a fuck.

Peggie That was the best day, people would be coming to our door and I'd think, 'I can't believe this is happening.' It was great.

Noel You had to queue up and get your guest-list tickets, and people were giving blag names and all that shit. Someone was saying the United players were outside and getting shit off loads of City fans and we were like, 'Good, fuck them off.'

Liam I remember United players wanting tickets, it was like, 'Fuck 'em, always beating us fucking ten nil.'
I loved them gigs, they were great. There was always lots of fighting and shit down the front. I know in this day and age people go, 'Yeah, but I got

piss thrown over me one night at an Oasis gig.' You're lucky you didn't get your fucking nose broke. I guess if lads want to scrap each other, let them fucking scrap each other. If they start hitting birds and shit like that, then that's out of order, but if people want to get a bit argy down the front, I fucking loved it.

Noel The chaos surrounding those gigs was fucking insane. Maine Road is right in the middle of Moss Side, a great place for a football ground, not a great place to do a gig. It's a tasty area and it was a bit of a scene outside. Helicopters all over. Even friends of ours from Manchester were coming in saying, 'Man, it's fucking rough out there.' We didn't see any of it going on, but you read it in the newspapers and all that. I remember playing on stage and looking down towards the north stand and there are people on the roof who couldn't get in.

The chaos surrounding those gigs was fucking insane.

Maggie Outside was like Beirut. You stepped outside and there was people hanging off fire engines, trying to get into the venue. We had a fire alarm go off in one of the stands so obviously the fire engines were coming in and they looked at it as an opportunity, to jump on and try to get in. At the far end people had gotten onto the roof over the terrace. I don't know how they got up there, but they found some ladder from some neighbour. There was like fifteen of them dancing away. The police come, they take the ladder away. It was like, 'Oi, you're supposed to get them down. They might hurt themselves!' It was absolute chaos.

Noel The songs were made for the stadium environment. They sounded good in my flat, then they sounded good in the Boardwalk, then they sounded good at the Apollo, sounded better at GMEX. Then they sounded better at Maine Road, and better at Knebworth.

Liam The small gigs were great and all that because they were a bit punky, but our sound was too big for it, I think. We were better off in arenas and stadiums.

Noel I can understand if a dude is stood on stage, singing an intimate song about how his dad shit on his cornflakes when he was a child and he's never recovered from it, I can understand if the meaning is getting lost when there are 50,000 people. 'Cigarettes & Alcohol', let me tell you, you

can't fucking play that to enough people. Seventy thousand people didn't seem enough, 125,000 . . . You was thinking there should be more people here to listen to this. The second night at Maine Road, sitting in a box on my own watching the stadium empty with the floodlights on I'm thinking, 'Where does it go from here?' Slowly watching the stadium emptying, trying to appreciate it, but then, almost immediately, bang, you are into a party. Then you can be anything from seventeen hours on the piss to four days, you know what I mean, so by the time you have pulled your head out of your arse there is something else happening. All of those gigs at that point, were a level up. They just became this massive celebration of the level that we were at. I look back on them all with great pride and fondness and I can only say from my own part, I had a fucking great time.

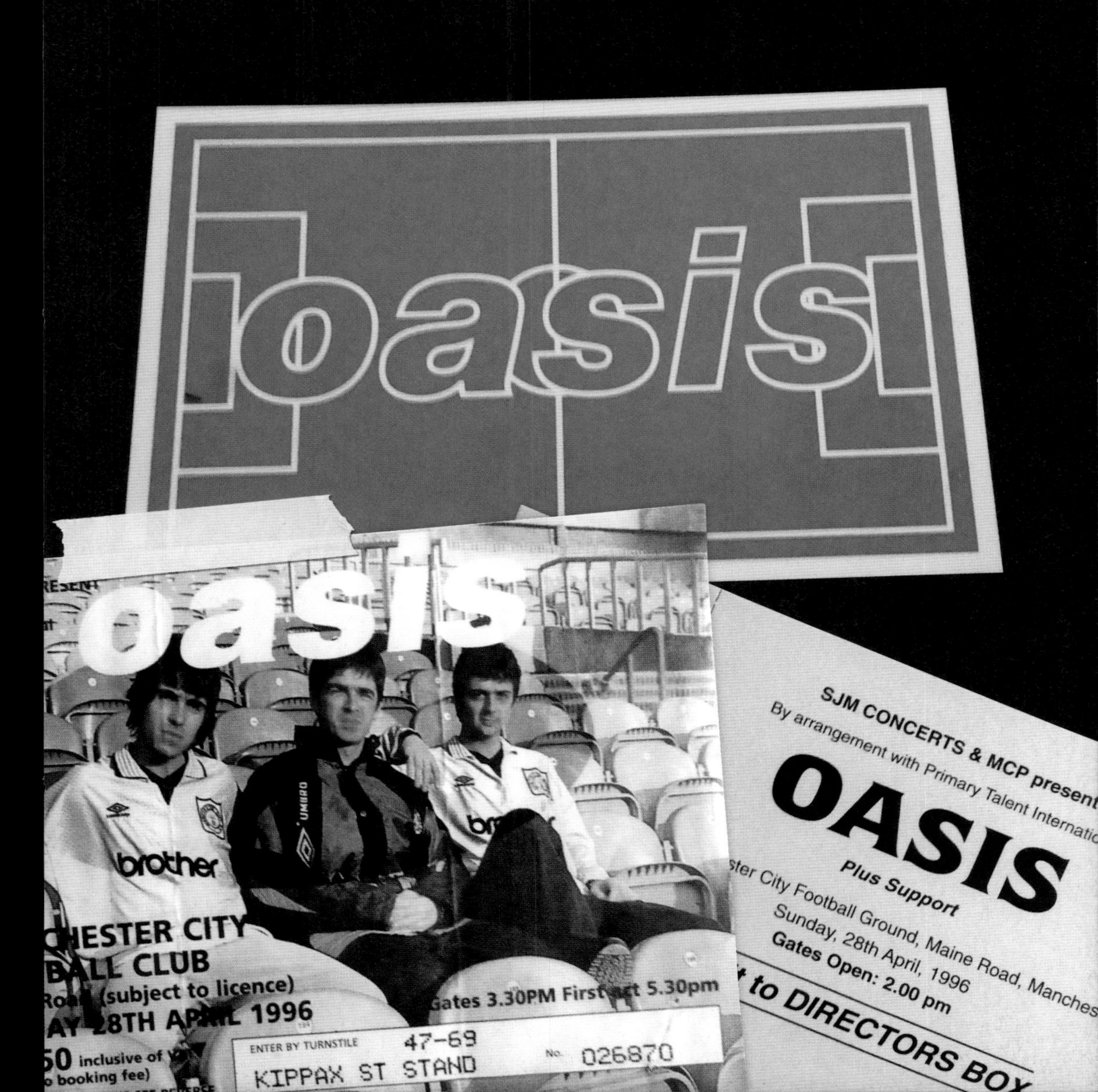

© Jill Furmanovsky
KODAK 5063 TX
FOR OASIS FILM - JUNE 16
Fr 13 scanned @ 35 mb G/S

The songs were made for the stadium environment.

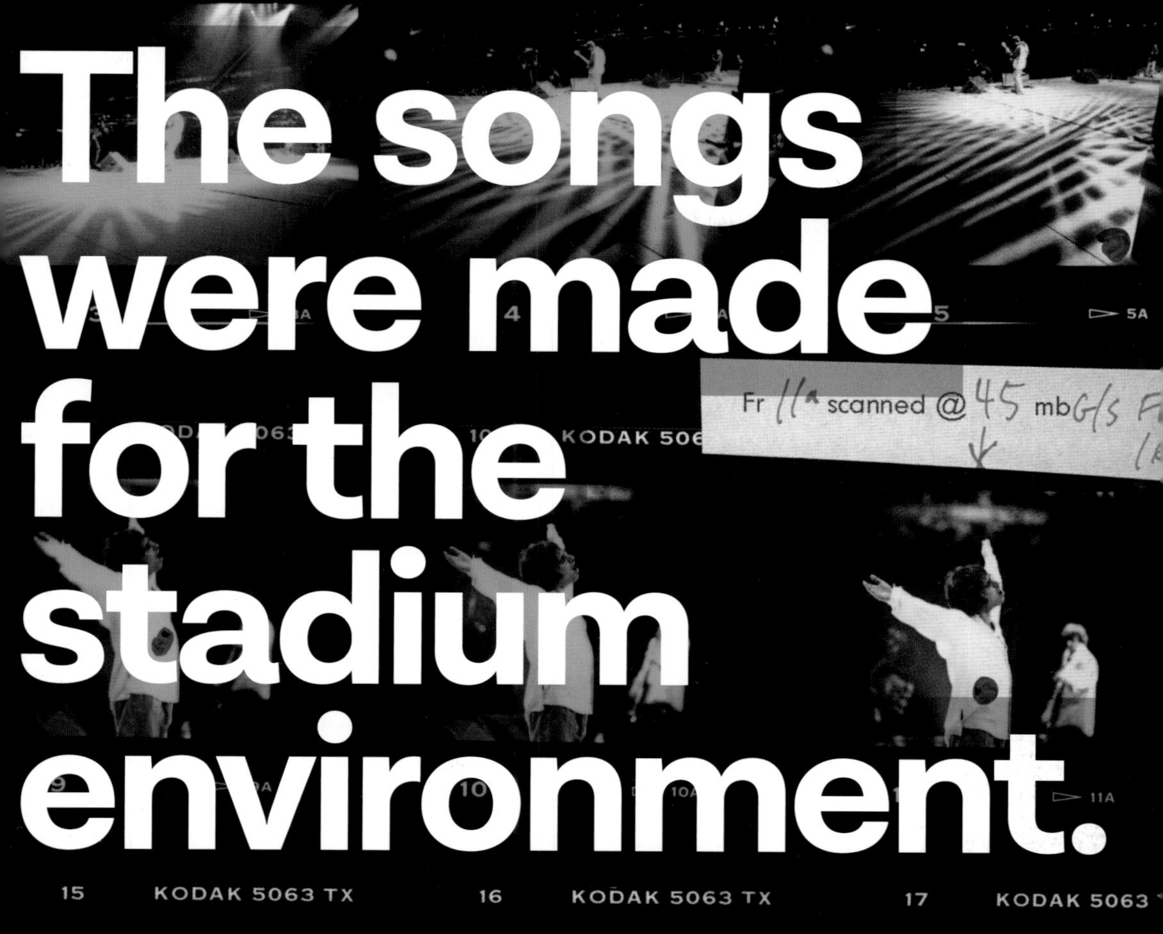

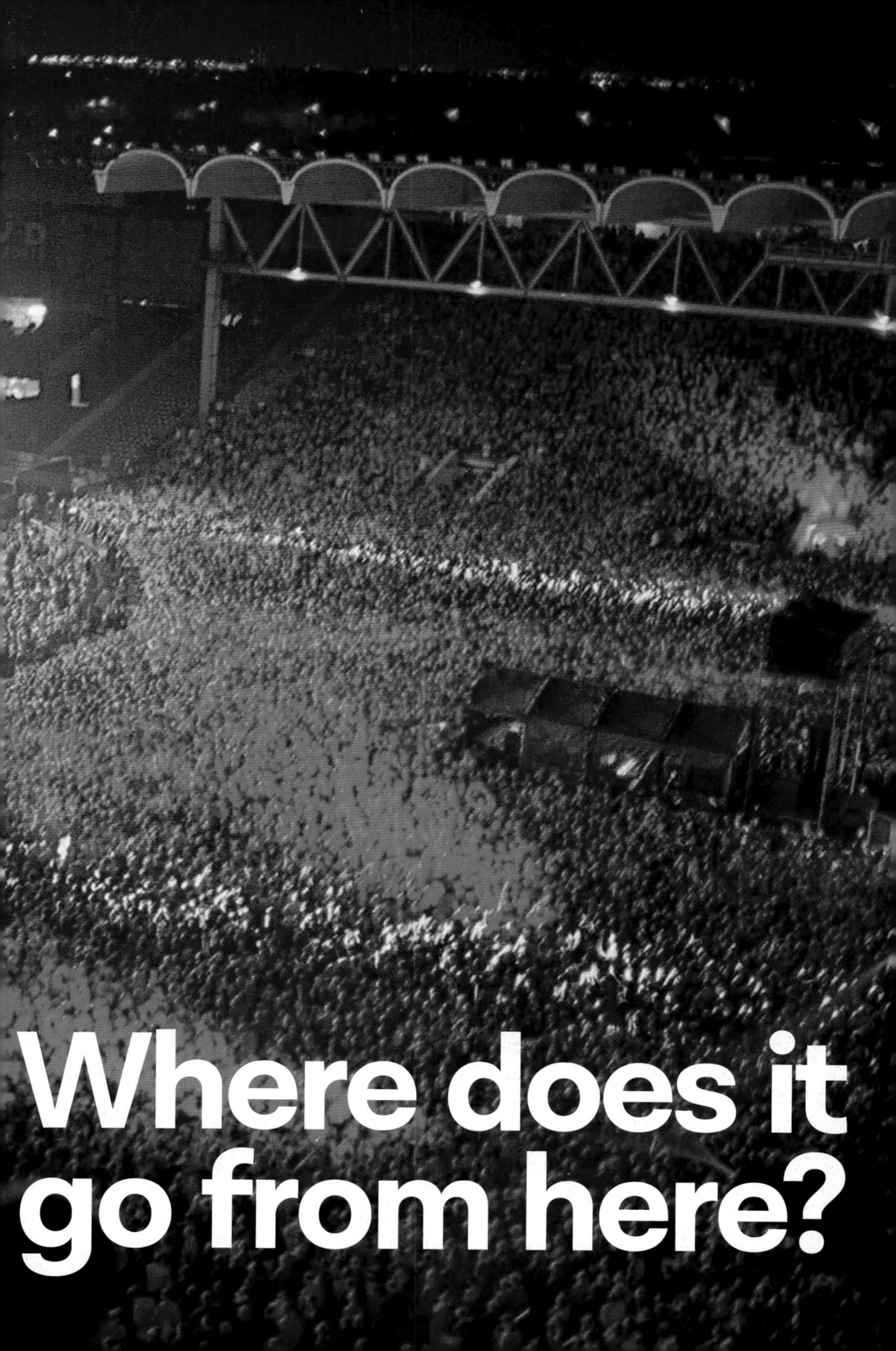

Where does it go from here?

Knebworth

Noel Somebody suggested that the next thing we should do should be these things at Knebworth and we all just went, 'Yeah, whatever.'

Owen I remember Marcus Russell saying to me, 'Do we book these big gigs while they can? Everyone's going, it's fucking brilliant but it's like a fucking truck going down a hill – it's burning and the wheels are fucking falling off – so I don't know when it's going to end, Owen.'

Noel The band never talked strategy. Ever. I don't remember sitting in a room with anyone mapping out the future. I'd heard of the name 'Knebworth'. It didn't really have much significance to me until somebody said it was Batman's house in the film. I thought, 'Oh, that gaff, nice.' If I'm being taken to Knebworth and asked what I thought, the deal's already been done. They are not taking me there for me to go, 'No, I'm not interested.' Somebody's already made their mind up. But somebody's only got to whisper in my ear, 'These are the biggest gigs of all time . . .' Yes, that's me, you've just sold it to me. Thank you very much. 'You'll be bigger than Led Zeppelin and Pink Floyd . . .' Let's do it. They had been booked. This is what we are going to do. We all know what we are doing, the songs are great, I'm great, you're great, the rider's great. Let's fucking have it. As long as it doesn't rain it should be good.

Maggie Everyone was taking a big leap of faith that they were going to sell these shows out. The first one, I think, went on sale and it sold out in, like, minutes, or something ridiculous like that.

Noel They would be talking figures and this, that and the other; I was worried about where the aftershow was going to be. I remember walking round what became the site having the same feeling that I did when they took me to Sheffield Arena and Maine Road, just laughing going, 'This is ludicrous,' and then, 'Are you sure we are going to sell this out?'

Bonehead We'd reached a level where I was really confident. Two nights, alright, we can do it.

Liam I can't believe that we only did two nights. Whose fucking great idea was that? We should still be playing there right now.

Noel The only time I was aware of what it would mean was when the two nights sold out immediately. I think the idea was put to us that we should put another two nights on sale and I think for the first time ever, we kind of backed away. I don't know, maybe we had got the sense that it had become too big. Then you find out afterwards that we could have done seven nights and really we should have done. It was like a year where we went from the Manchester Academy to Knebworth, there were three big jumps: Sheffield Arena to Maine Road to Knebworth. But the shift from the rehearsal room in the Boardwalk to the stage in the Boardwalk with eleven people was massive. Then the shift from the eleven people in the Boardwalk to it being 700 people was huge. They are all massive jumps and none more important than the other, I might add. But the jump from the university circuit to the arenas was huge. I'd only ever seen proper big bands in arenas. Then going from that to stadiums – fucking football stadiums – what? What's endearing to me when I see the footage of the stadiums, is there is no production in it at all, it's just a load of lads on stage playing guitar.

I can't believe that we only did two nights. Whose fucking great idea was that? We should still be playing there right now.

Liam Same performance, but better gear, you know what I mean? And more people. But believe you me, every gig, was treated the same as far as I'm concerned.

Noel Now people play stadiums and it's fucking bells and whistles, lasers up the arsehole and all sorts of shit going on, backing singers and fucking all the rest of it. You see other bands playing those big gigs and it is like fucking Cape Canaveral, it is like NASA. If you do that gig now, fuck me, there would be like a million dollars spent on the screen. We had swirly cardboard cut-out things, they weren't even in colour, they were black and white. There was an atrocious lack of foresight about how it looked. It wasn't a statement at all, we just couldn't be arsed. It's difficult enough to come up with a chorus, far less a meeting about how it's going to look. I am not sure whether it would have worked if there'd been the obligatory second stage down the end of a catwalk. We were still behaving like a bunch of shit-kickers, but I suppose that is part of the charm. Our mantra I guess was it's all about the songs. I don't remember, at the time, feeling dwarfed by it and to everyone's credit I don't remember anybody shitting it that much. They might have been shitting it in private, but I guess we were so confident in the songs that we were doing, we were just rolling into it. My whole theory is, don't worry about those people out there, they'll have a great time whatever. The stage remains the same size, don't look at

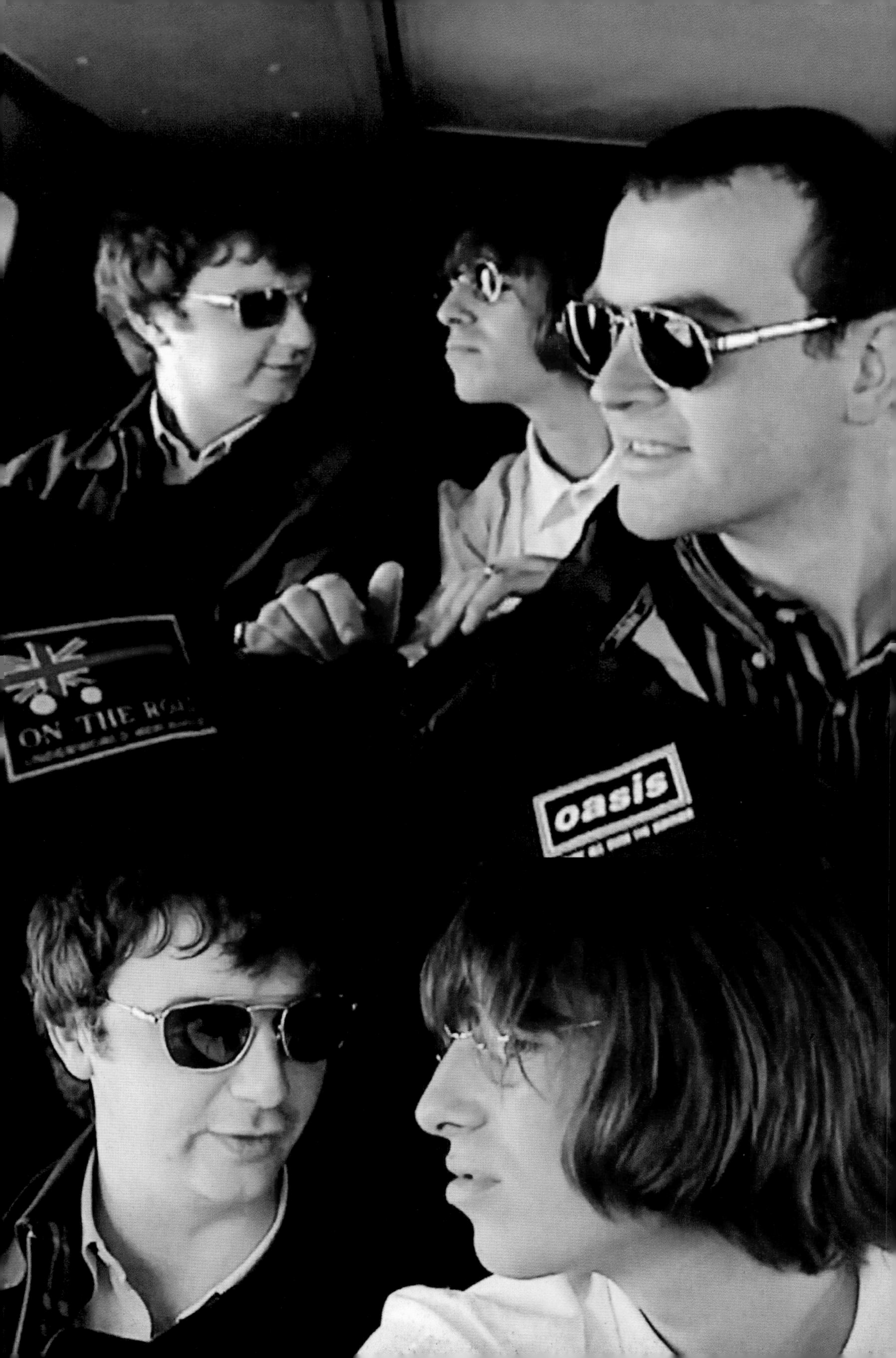
oasis

Same performance, but better gear, you know what I mean? And more people.

them, just stand there with your head down, don't worry about it. You know the songs inside out, just play them. Maybe buy a new shirt, that's it.

Liam I was never sat backstage fucking biting my nails, nervous, wondering how it was going to go. It was, 'Is everybody in, right, can we fucking get on with it?' and that was it, and still is today. The small gigs I find a bit odd when people are right in your face, I find them more nerve-wracking. The big gigs were a piece of piss. Never had stage fright. Never, ever went, 'Can't do this,' not once. Was never overawed by it. Obviously you have a bit of nervous energy, but we were never going, 'Fucking hell, I don't think I can do this gig. Fuck me, shit. What am I going to do. Fucking hell, I'm going to lose the plot. Fucking head's gone shit.' It was just like, fucking bring it on.

Noel We're going to fly into Knebworth in a helicopter and I said to Coyley and Phil, 'You're getting on the helicopter,' because they were our two oldest mates. 'You should be there to see it with us. This thing that we're all part of.'

Phil We just got in the helicopter with them and flew to Knebworth. It is the only way to get to a gig really.

Noel When we were up at the highest in the helicopter it didn't look that impressive, it looked so small. As we were getting near to the ground it's like, 'Fuck me, it's massive.' There is no fairground, there is not a lot of other shit going on, there is no second stage, it's effectively a load of fucking Oasis fans in a field getting pissed.

Bonehead That was a good helicopter, I was a dab hand by then. Knebworth, how we getting in? Helicopter. It's like phff, cool, I can do them now. When you're flying across the countryside and you can see it in the distance, you're like, shit, is that us? It was a great, great feeling though: helicopter coming down, you can see loads of people, everyone knew it was the band arriving. It was just brilliant, best thing ever.

Liam Being on stage and looking out at them people, it's a fucking head fuck. I guess being in that crowd and watching that band, that would have been fucking amazing, I wish I'd have gone to that gig.

Noel When the people are with you and they've taken those songs into their hearts, it's easy. There is no point in getting nervous, you might as well just be one of them and enjoy the gig.

Phil In the end they're just doing a gig. They just bowled on stage and did a gig, wherever it was, whether it's a hundred people or, you know, 150,000, they didn't seem to bother any more than in the early days.

Noel 'This is history.' By this point we know that nearly three million people have applied for a ticket. It was historic. It is very unlike me to get swept away in the moment, I'm usually a bit reserved about things like that. I think I might have had a drink before I went on, so I could well have got swept up in the moment of it all.

Phil Probably felt like history in my brain, you know, it was my history. I always believed, I never doubted that that band would do something.

When the people are with you and they've taken those songs into their hearts, it's easy.

Liam It was biblical, man, it felt fucking biblical. All the rest of it is a load of bollocks really, sitting about all day, doing interviews, talking . . . It felt fucking great. It just felt fucking real. The rest of it, you take it or leave it, but that hour and a half, whatever it was, was just like, 'This is the shit, man, this is really fucking proper.'

Noel Now how does John Squire end up on stage at Knebworth? I don't know, he certainly wouldn't have suggested it. I think it would have been either me or Liam and he agreed to do it. It was a bit odd because he never hung out with us. He kind of turned up, hung out on his own tour bus then did the tune, never seen him again. I thought that was a bit weird – I thought he might be, I don't know, one of the lads for the weekend, but there you go.

Liam John Squire, best guitarist in the world today, and was back then. We were all, 'Fucking hell, man; he's up for playing with us.' We were made up. We were fucking stoked, man. That was good, he was cool. John's very quiet and you can tell there is a lunatic inside dying to get out, but he's very reserved and chilled. We'd always have to play it cool around him because he is not into fussing about or jumping up and down . . . he is a bit of a chilled guy. We'd always try and hold it down around him, but top man.

Noel It is something you can't define. You can't bottle it and fuck knows how you make that happen again, you just don't try. You are just happy that it happened to you once in your life and I guess if you were in the crowd, you were happy that you were there, in your twenties while the band was there in their twenties, and that's it.

Liam I can't even describe it. I would say that gig was more about the fans than us. The fans: I hate that fucking word. More about the people than the band.

Bonehead It went above and beyond and further than anything I ever imagined, so that was the pinnacle for me.

Noel Even though it's huge, it's a low-key thing. We are still the same band that was playing in the Boardwalk, we are wearing virtually the same clothes. All that's happened is it's caught fire and all these people have got on board. There were no sing-alongs, no dropping down of the chorus from 'Wonderwall' and the singer going, 'Everybody over there . . .' None of that shit. In, out, put the kettle on, there's a rider to get through here. That was the ethos.

Liam After the first night I got a bit carried away. I woke up the next day and there were people knocking at my door going, 'Come on, you've got to go and do a soundcheck.'
I was like, 'What the . . . who, what, where, what fucking soundcheck?' Forgot that we had to do it all again. I remember not having a change of clothes, but I had this big woolly jumper, so I said, 'I'll have to wear that,' because I didn't want to wear the same jacket twice.

Noel Every night after the first night is a breeze because you know that it's steadily going to get better. Whatever sense of anticipation or excitement, and whatever level of nerves you might feel. The second night, honestly, it was like fucking playing in my front room, I was that relaxed about it. How it used to work with Oasis was the second night would always be the first night to go on sale, so you'd go out and you'd do the first night and that would be amazing and then you'd think, 'Fucking hell, all the diehard fans are coming tomorrow, all the nutcases are coming tomorrow.'

Two days after Knebworth I was no bigger a rock star than I was two days before it.

Liam People have come up and gone, 'I was at Knebworth.'
'Alright, cool, me too.'
We never really sat and thought about it. Looking at it, it's insane, it's massive.

Noel Well, yeah, you'd have to be a dick to be in the midst of that and not think that this is significant. Even if you wouldn't be pompous enough to think this is a significant moment in popular culture, you would certainly think it's a significant moment in my life. I knew it was significant. But it didn't change my life in any way. Two days after Knebworth I was no bigger a rock star than I was two days before it. I wasn't perceived any differently, the band weren't perceived any differently. I remember it being immediately slagged off in the fucking papers. Some tabloid journalist was going on about the price of the food or some shit like that, which is a very British thing to do. 'Wasn't it great?'
'Fucking hell, what about the queue for the toilets, though?' Whatever.

Liam It was amazing, it was beautiful, it felt like it was fucking justice, it was us and the world. It was up there with any kind of fucking religion, to get them people in that crowd without fucking religion is something. It's a big, powerful thing. Deep, man, deep but beautiful. I just wanted more of it. I've heard Bonehead bang on a million times, 'We should have split up, blah,

blah, blah.' What for? Do what? Go and be a fucking car mechanic? Just have a bit of time off, go and do it all again next year. All them people that were there, or that couldn't get tickets, let's go and fucking play for them next year or whatever.

Noel We should have at least gone away and given the impression that we'd split up. We should have gone away and done a bit of living. A fucking shitload of money was about to fall into our laps: all the royalties were starting to come in from the first two records, the tours, Knebworth and Maine Road, and that kind of thing. We should have gone off and done that thing that rock stars are supposed to do.

Looking back on it now, it did feel like the end of something as opposed to the beginning of something.

Bonehead That was my life. That band was my life. It went above and beyond anything I ever imagined. That was the pinnacle for me. It was like, where the fuck do you go from here? My attitude then was more, give me more, give me more, give me more, I want more of this, keep going. Now, looking back, I honestly think we should have just went, thank you, every one of yous, for getting us here. A final bow, on that stage. We were Oasis and goodnight, you know, and walked off.

Noel Looking back on it now, it did feel like the end of something as opposed to the beginning of something. I had a sense, even at Knebworth, that it was never going to happen again. You couldn't have envisaged it getting any bigger than that. The only thing to be bigger than that would be to do more nights. So it couldn't get any bigger than that, it could just get more repetitive: four nights next year, seven nights the year after . . . It was right on the cusp of the analogue age going into the digital age. Walking out on that stage that afternoon there were no mobile phones. It was the pre-digital age, it was the pre-talent-show-reality-TV age; things meant

more. It was a great time to be alive, never mind a great time to be in Oasis. We were about to enter a celebrity-driven culture. I've always thought that it was the last great gathering of the people of the old age before the birth of the Internet and phones with cameras and video phones and iPads and all that shit. I do say it quite often, and I said it then, people think I'm a fucking lunatic, I thought we were going to be the last great rock band. What I meant by that was we did it in the traditional way. It's no coincidence that things like that don't happen any more I don't think. I think that the Internet and the digital world has taken a lot of the magic out of rock and roll. The biggest music phenomenon was a band that came from a council estate. I just think in the times in which we live, it would be unrepeatable. We should be worried about that, because where's it going to be twenty years from now?

I loved every minute of it. It meant the be-all and end-all, man.

Liam I just wanted it all there and then, do you know what I mean? I just wanted it to happen in one big fuck-off explosion of madness. I loved every minute of it. It meant the be-all and end-all, man. Life or death. I'd do it all again in a heartbeat. It's easy to say that in hindsight we should have gone, 'Right, let's knock it on the head.' And do what? Go back to your fucking house and count your fucking money? It doesn't work like that. Bollocks to that. When you score a hat trick in the first twenty minutes against United, you don't turn round to your fucking bench and go, 'Here you are, let me out of here now, I've done my fucking bit.'

Noel I wish I was that cool, back then, to have taken a step back and gone, 'Let's just all go our separate ways for a bit.' Imagine doing those gigs and then just splitting up. We should have disappeared into a puff of smoke. It would have been fucking unbelievable. If it would happen to me again, knowing now what I should have known then, we would have just gone off stage and said something cryptic and then disappeared for five years. Then slowly just made a great album or wrote better songs as opposed to just carrying on, carrying on and carrying on. But, you know, it was my idea to keep going, I keep on fishing for it because I am an addict. I'm addicted to the thing. That's what shit-kickers do, they ride it until the wheels come off then they are desperately running around for the Sellotape, trying to put it back together again.

Liam Just because you can't get any bigger or any higher, doesn't mean to say you can't keep doing it, do you know what I mean? You shouldn't just stop because you kissed the sky. Give it a fucking love bite.

© Jill Furmanovsky

You shouldn't just stop because you kissed the sky. Give it a fucking love bite.

No regrets

Liam People have asked me before, are you surprised at how big Oasis got? No, I thought we would be a lot bigger, that's where my head was at.

Noel I'd often think, 'What was it all about?' It happened in such a short space of time, two and a half years from signing off to walking out on that stage. When we got to Knebworth, we'd only just become rock stars; two and a half years previous to that we were a bunch of shit-kickers. You might think it took a lot of savvy business people saying, 'We are going to do this and we are going to do that,' but there was no smart business moves or anything like that. What happened is just magic. When you say it now, it is so ridiculously simple: you had a band who had the tunes, who were grafters, who came from nothing and wanted it all. Were willing to put themselves out there and inspired a generation of people to believe in them. We were the last. We were the greatest. Nothing anybody does could be as big as Oasis.

It is about the way that we made people feel, and people will never forget that.

Liam Just pure self-belief, man. Self-belief, not arrogance, self-belief. Arrogance is someone who turns round and goes, 'Oh yeah, we're going to be the best band in the world,' knowing deep down that the tune you're writing ain't even going to get out the studio. It was pure, pure fucking self-belief and that was it. It would never get repeated. Not because we were greater or better than anyone else, but just because we didn't actually give a flying fuck. We've definitely got a table upstairs with the big boys, whether they like it or not.

Noel I guess if you were to rationalise everything that we ever did, you couldn't say that anybody that was ever in Oasis, me included, was the best in the world at anything. As a lyricist, I wouldn't even come in the top twenty; guitarist, wouldn't even come in the top twenty. Liam as a singer, not the best. Bonehead as a rhythm guitarist, wasn't even the best guitarist in Oasis, never mind the fucking world. Guigs, no chance. But when it all came together, in a venue or in a field, it did something that was great. We made people feel something that was indefinable. I thought we were the greatest and I still do and that's because of all the elements. Because of Liam and me and the songs and the words and the fans and whoever else was in the band at the time. We made a thing and it was great. I don't consider myself to be that great. I wrote the songs, but to me that is a very small part, that's just kind of like giving birth to a child. This thing goes on and becomes fucking huge because of the people. I do believe that certain songs have written themselves, like 'Slide Away' and 'Don't Look Back In Anger' and 'Live Forever' and 'Wonderwall' and 'Champagne Supernova'. I do believe that if I hadn't had written those songs that night, somebody else would have written them somewhere else. Neil Young calls it sitting beside the rabbit hole, and when the rabbit pops up you are quick enough to grab it by the ears. Keith Richards says they fall out of the sky. I liken it to going fishing, if you are not fishing you are not catching anything, and if I hadn't have been writing and hadn't picked up my guitar on that particular day, then that song would have sailed on down the fucking stream and somebody else would have written it.

Liam We were properly, properly into it and we weren't fucking lazy; we grafted like fuck and we weren't cunts. We took the piss out of people and we had banter with people and obviously we got shit back if anyone was funny enough to take us on, it was just a crack. Yeah, we took drugs, we fucking slapped a photographer, all fucking true, but that is it.

Noel It is not about the songs and it is not about the fucking clothes and the attitude and the headlines and the scandal. It is about the way that we made people feel, and people will never forget that. Don't ask me to intellectualise it because I don't fucking know. I was only on the stage and in the studio, but people will never, ever, ever forget the way that you made them feel.

Liam We had a laugh, man. Anyone who says we didn't have a laugh obviously weren't there. It was twenty-four hours of the day fucking fun.

Noel If it wasn't a proper fucking laugh – regardless of all the getting kicked off aeroplanes and thrown out of hotels at 4 o'clock in the fucking morning – we wouldn't have stayed together for eighteen years or whatever it was. That in itself tells you something.

Liam It was everything and more than I asked for and I wouldn't have changed a fucking thing. I enjoyed the chaos, I enjoyed the hard graft, all the riches, I enjoyed all the fucking nonsense.

Noel There was chaos fucking surrounding the band and right in the middle it was fucking worse. Whatever the chaos going on, screaming kids and this, that and the other, in the dressing room it was worse. There was shit getting smashed up and people fucking pissed falling over and it was great. We didn't have that calm in the eye of the storm, in the eye of the storm was a worse fucking storm, it was pissing down.

Liam The wheels were never on the fucking bus. We were on the bus, but the fucking wheels were rarely on it.

Noel In the end, when it is all said and done, what will remain is the songs and you can have an opinion on them for as long as you like. None of that stuff about hotel rooms is greater than the impact that 'Live Forever' had on a generation. None of that stuff about drugs and not turning up for gigs and what was said in the press will ever be as great as 70,000 people singing 'Don't Look Back In Anger'.

Liam We just knew that we were here to take this band as far as we could without turning into knobheads and licking arse and being what all these other fucking clowns had been before us. We had to keep it super, super fucking real, for us first of all, and for the real people that are out there. The real people that needed something, instead of some fucking soft-soap bullshit band. We were a Mike Tyson band; just come in, fucking blazing, knocking everyone out. We were never going to do ten rounds and it was never going to get to points, we were just going to come in and fucking have it. No regrets, no fucking regrets. Even the fuck-ups were great. When we fucked up, we fucked up fucking great and we did it with a laugh. Even the shit bits were great. It was just mega.

Noel Oasis's great strength was the relationship between me and Liam. It is also what drove the band into the ground in the end.

Peggie Sometimes I do look back on it and I wish it had gone in a different way, but it hasn't. It's caused a lot of problems with them two. But it's their lives and you have to just let them get on with it.

Liam I get where she's coming from, because obviously me and our kid ain't got a relationship any more, but who is to say if we had been a pair of fishmongers we'd not have still slapped each other with a bit of trout every now and again. I know people in all walks of life that don't speak to their brothers for a long, long time and they are not in rock-and-roll bands. At the end of the day I weigh it up and I go, 'Did the good times outweigh the bad times? Fucking 100 per cent!' I'm happy with that. As far as I'm concerned I'm still fucking up. I come in with nothing so if I leave with nothing – even if I leave with 10p in my back pocket – I'm still up, man. It's all good, man. I didn't join Oasis to fucking make lots of money, we did it to make music and whatever comes with that. We were never driven by money, I certainly weren't anyway. It was more about just having a good time and getting in a bubble and playing your music and having a fucking laugh. The money side of things, I don't know how much money we made, I don't know how much money we lost, didn't care. I've had big houses and I've had small houses; I've had big bar bills; I've had this and I've had that and it doesn't really matter. I've had good times, I'd rather that than a big bank account, sitting there going like, 'Yeah, I was in a pretend rock-and-roll band and I got up to fuck all really. I've got 90 million quid in the bank.'

Noel People make the mistake of thinking that the people on the stage are defining something; what if no one turns up? The people who are making it what it is are the people in the crowd. So we can all sit here and suck each other's ball bags about fucking 2.6 million people applying for tickets, but, you know, what's great about that is the 2.6 million people. Not anything that we did. It means shit without the people. The people were with us and I felt that, and I felt quite invincible in feeling that. Really, the people are as important as the songs and the chaos and the nonsense and the interviews and the fucking drugs and the fucking this, that and the other. What should never be forgotten is what the fans did, or how special they made it. The songs are not great because of anything that I did, I only wrote them. They are made special night after night after night, decade after decade, because of the people. When I sing 'Champagne Supernova' on acoustic guitar in a soundcheck and there are eleven people there, it's good, it's fucking good, but it's not extraordinary. When you are singing it five hours later to 12,000 people in Glasgow, *that's* extraordinary. Not because of anything I'm doing, I'm playing it the same, it's the love and the vibe and the passion and whatever it is, the rage and the joy that's coming from the crowd, if anything that is what Oasis was. There's a chemistry between the band and the audience. There's something magnetic drawing the two to each other.

I'd do it all again in a fucking heartbeat, the exact same way.

Liam We meant every fucking last fucking breath of it. Everything. Every fuck that got fucked, every fucking swear word that come out of my mouth, every fucking table that got chucked, every fucking drink or whatever that went up our nose or in our mouth, I meant every fucking minute of it and I wouldn't change anything. I'd do it all again in a fucking heartbeat, the exact same way. It was a miracle that we even got there. It was everything and more than I asked for.
Yeah, we felt untouchable, man.
Supersonic, even.

13